Cave and Karst Systems of the World

Series editor

James W. LaMoreaux, Tuscaloosa, USA

More information about this series at http://www.springer.com/series/11987

Gheorghe M. L. Ponta
Bogdan P. Onac
Editors

Cave and Karst Systems of Romania

Springer

Editors
Gheorghe M. L. Ponta
Groundwater Assessment Program
Geological Survey of Alabama
Tuscaloosa, AL
USA

Bogdan P. Onac
School of Geosciences
University of South Florida
Tampa, FL
USA

and

"Emil Racoviță" Institute of Speleology, Romanian Academy
Cluj-Napoca
Romania

ISSN 2364-4591 ISSN 2364-4605 (electronic)
Cave and Karst Systems of the World
ISBN 978-3-030-08088-4 ISBN 978-3-319-90747-5 (eBook)
https://doi.org/10.1007/978-3-319-90747-5

Printed on acid-free paper

This Springer imprint is published by the registered company Springer International Publishing AG part of Springer Nature
The registered company address is: Gewerbestrasse 11, 6330 Cham, Switzerland

To all cavers and scientists who have investigated the endlessly fascinating karst and cave world of Romania

Foreword

Limestones make up 4400 km^2 (1.8%) of Romania's surface, which apparently is not much but they still host all known types of karst landscape and more than 12,000 caves showing a variety of genetic and morphological features. There are active caves traversed by tumultuous rivers, caves at higher levels with quiet chambers and large detrital deposits, and walls that witness the various phases of riverbed deepening. At higher elevations are hydrologically dry cavities in which the water percolating from surface has generated spectacular speleothems that create the magnificent underground landscape that makes the caves famous.

Caves are developed in the three mountain chains, namely the East and South Carpathians, Apuseni Mountains, as well as close to the Black Sea in the southeastern part of the country, known as Dobrogea. With respect to their geological age, some of the limestones are Paleozoic, but the vast majority belong to the Mesozoic, most notably the Jurassic and Upper Cretaceous. They are part of the Alpine fold belts (Upper Cretaceous) and few are constituent of the Neozoic formations (especially Miocene).

In almost all limestones, natural cavities exist across the altitude scale from the sea level to over 1500 m in elevation. Caves are generally easily accessible, which explains why since ancient times they have been known by the inhabitants of these places, as evidenced by some documents of great value to show that the Romanian caving can boast a venerable past. Here are some examples:

Prehistoric/Neolithic artifacts, such as tools, pottery, and even skeleton remains, were found in many caves. Extraordinary are, for example, foot imprints discovered in hardened moonmilk or the invaluable cave paintings, dating back to more than 30,000 years, depicting animals (mammoth, horses, reindeers) and humans (possibly shamans). Another example of early use of caves was passed on through toponymy. For example, Peştera Zmeilor (Dragon Cave), which is famous for its skeletal remains of cave bears, was considered by locals as the resting place of fairy beings from the past. A valuable testimony is that of Pythagoras, who mentions that the Dacians (Romanian's ancestors) were immortal, since their King Zalmoxis who was apparently considered dead after a battle had in fact retreated into a cave and emerged after 10 years alive.

The first permanent dwellings in caves dates back to the times when monks started to use them as refuge, places of worship, or even as monasteries to defend the Christian faith against the plague of the Ottoman persecution. These are many, but just to name a few, the cave of Sfântu Grigore (St. Gregory) Decapolitul next to the Bistriţa Monastery and the little church that was erected under the portal of the Ialomiţei Cave (Bucegi Mountains), the foundation of which is lost in the darkness of time. Important is also the Sfântu Andrei (St. Andrew) Cave near Ion Corvin Village (Dobrogea), which was home of Apostle Andrew the Savior, who in the first century brought Christianity to the territory of the present-day Romania.

First scientific information about Romanian caves dates back to the eighteenth century and refers to geographical, geological, and archeological findings. The authors were generally speaking scientists, but it took some time before the information about caves gradually begins to penetrate into collective consciousness. This led to visits and exploration tours organized by curious adventurers, and thrill seekers, which were the forefathers of modern cave exploration that took off after World War II.

At the beginning of the twentieth century, a new scientific discipline of biology was born, namely the biospeleology. The founding father was the Romanian Scientist Emil G. Racoviță, who dedicated his life to the study of organisms living in caves and which, because of the permanent darkness, the high humidity, and the precarious food, have developed special anatomical and physiological abilities. He studied the fauna of over 1000 caves in France, Spain, Corsica, Algeria, Slovenia, and Romania, presenting them to the world. Recognized for his scientific merits, he has been appointed Professor at the University of Cluj (Romania), where on April 27, 1920, he founded the world's first speleological institute, meant to serve as a global center of biospeleology research with collaborators around the world. This fact brought Romania's speleology and caves into a privileged position. This book responds to the growing interest in cave and karst studies.

Simultaneously with the development of the speleology as a science, the more explorative and amateur caving movement took off. The many cavers organized themselves into various clubs and associations and begun a vast activity of exploration and mapping, thus boosting the speleological patrimony of the country. If in the 1930s there were roughly 500 known caves in Romania, nowadays there are more than 12,000! It is the merit of three generations of cavers, professionals (research institutes and universities), and amateurs (tourists, climbers) alike, who were extremely active after 1950.

In conclusion, I would like to say that I am happy that my name is present alongside the most well-known, knowledgeable, and performers of Romanian speleology.

Bucharest, Romania — Marcian Bleahu
Honorary Member of the Academy of Romanian Scientists

Preface

Romania has many unique karst landscapes and a great variety of caves as documented in this book, but is not a true karst country. The karst-prone rocks cover about 5500 km^2 representing only $\sim$2.3% of the total surface of the country and occur as scattered islands mainly along the Carpathians. After an introductory section that presents the general geological, hydrogeological, and karst settings of Romania, the book follows the geographic structure of the Systematic Catalogue of the Romanian Caves published by Goran in 1982. The presentations begin with caves located in the East Carpathians, continue along the South Carpathians, and then cover the Apuseni Mountains. The caves of the Someş Plateau and Dobrogea are described before the final section, which includes a number of specific topics, such as evaporite karst, ice caves, bat fauna, cave biology, show caves.

In Romania, the scientific interest in caves and karst dates back to 1776 when J. Fridvaldszky included in his famous treatise "*Mineralogia Magni Principatus Transylvaniae*," a detailed description of sulfur, alum, and calcite cave deposits. By the turn of the nineteenth century, professional geographers and geologists were conducting extensive field campaigns and described or mentioned in their publications karst and caves from Bihor Mountains, Vârghiş Gorges, and Banat. In 1907, the Romanian Biologist Emil G. Racoviţă published the *Essay on biospeological problems,* widely considered the founding book of biospeleology. Soon after (1920), he established the world's first Speleological Institute in Cluj-Napoca, which truly catalyzed the systematic research of the Romanian caves. Between 1920 and 1956, sustained field campaigns conducted by E. G. Racoviţă and his collaborators successfully described over 250 caves, the majority of them being documented in two volumes of the monograph *Enumération des grottes visitées* (Enumeration of visited caves).

Shortly after 1956, when the Institute of Speleology was reorganized carrying this time Emil Racoviţă's name ("Emil Racoviţă" Institute of Speleology (ERIS)), with offices in Bucharest and Cluj-Napoca, the recreational caving movement began in several major cities throughout Romania: Cluj-Napoca, Reşiţa, Bucureşti, Arad, etc. If in 1945 about 500 caves were known, the 1965 map of the Romanian karst regions published by ERIS included 984 caves. Their number increased exponentially to over 6800 by 1982, reaching about 12,300 in 2017. This significant increase was the result of the fact that in 1976, the amateur speleology movement receives the status of a sport discipline. As such, the Central Committee of Sport Speleology was formed as a subsidiary of the Romanian Federation for Tourism and Climbing, and an annual competition (Speosport) began between the caving clubs, which acquired points (and ranking) for each new cave discovered or meters of galleries added to known caves or shafts. After the totalitarianism regime collapsed in 1989, the cave exploration activities decreased significantly as Romania opened its borders and cavers found more attractive opportunities to travel and explore karst areas outside the country. However, a handful of gifted amateurs continue to work along with researchers in their mutual interests in caves, and this book has been prepared by both.

The reader must consider this book no more than an introduction to the cave and karst of Romania. It was written with the intention of capturing representative examples of caves, which host remarkable fauna, minerals, fossils, artifacts, human traces, or large perennial ice deposits. If none of these, then the caves and shafts were included simply because of their length or depth, passage morphology, volume of void, or peculiar speleogenesis. It is our hope this book will be a useful source of information and a stimulating introduction to the world of karst and caves, and entice readers to delve further into this captivating topic. The cooperation of the many authors (more than 70) has made the editor's tasks feasible. Spencer Coca, Joe Kearns, Oana Moldovan, Afina Lupulescu, John Mylroie, Arthur Palmer, and Robert Scharping are thanked for language corrections. This book would not have been possible without the support of the Springer International Publishing.

Tuscaloosa, USA — Gheorghe M. L. Ponta
Tampa/Cluj-Napoca, USA/Romania — Bogdan P. Onac

Contents

The Founder of Biospeleology and World's First Speleological Institute 1
Gheorghe Racoviţă

"Emil Racovită" Institute of Speleology: World's First Research Unit Dedicated to Karst and Cave Studies 5
Ioan Povară

Geology of Romania 9
Ioan Balintoni

Karst and Caves of Romania: A Brief Overview 21
Bogdan P. Onac and Cristian Goran

The Systematic Catalog of the Romanian Caves 37
Cristian Goran

Karst Hydrogeology 41
Gheorghe M. L. Ponta

Rodna Mountains: Izvorul Tăusoarelor Cave (Pestera de la Izvorul Tăusoarelor) 47
Bogdan P. Onac, Virgil Drăguşin, Felix Papiu, and Crin-Triandafil Theodorescu

Persani Mountains: Karst of Vârghis Gorge 57
Daniel Veres, Marian Cosac, George Murătoreanu, and Ulrich Hambach

Leaota Mountains: Rătei Cave 65
Maria-Lidia Nuţu-Dragomir and Gigi-Paul Dragomir

Piatra Craiului Mountains: Grind Pit (Avenul de Sub Coltii Grindului) 73
Spencer F. Coca

Caves of the Făgăras Mountains 79
Virgil Drăguşin, Ionuţ Mirea, Augustin Nae, and Maria-Laura Tîrlă

Căpătânii/Parâng Mountains: Polovragi Cave–Oltetului Gorge Karst Area 83
Gheorghe M. L. Ponta, Gheorghe V. Aldica, and Radu Dumitru

Karst of Sebes Mountains 93
Gheorghe M. L. Ponta

Sebesului Mountains: Răchiteaua–Sipot Cave System 105
Raul-Bogdan Tomuş and Radu-Ciprian Breban

Sureanu Mountains: Valea Stânii–Ponorici–Cioclovina cu Apă Karst System 113
Raul-Bogdan Tomuş, Radu-C. Breban, and Bogdan P. Onac

Retezat Mountains: Jiul de Vest–Cernisoara Basins 131
Ioan Povară and Gheorghe M. L. Ponta

Southern Carpathians: Pârgavului Cave 143
Doru Bădescu, Horaţiu Roman, and Iulia Bădescu

Mehedinti Mountains: Cioaca cu Brebenei and Closani Caves 149
Ioan Povară, Virgil Drăguşin, and Ionuţ Mirea

Mehedinti Mountains: Martel and Lazului Caves 157
Gheorghe M. L. Ponta, Gheorghe V. Aldica, and Tiberiu Tulucan

Mehedinti Mountains: Isverna Cave 165
Mihai Baciu, Gabor Mogyorosi, Virgil Drăguşin, Zsolt Gyurka, and Szabolcs Storozynski

Mehedinti Mountains: The Cave from Mohilii Creek (Ascunsă Cave) 171
Virgil Drăguşin, Marius Vlaicu, and Emilian Isverceanu

Mehedinti Plateau: The Zăton-Bulba Karst System 175
Ioan Povară and Cristian Lascu

Mehedinti Plateau: Epuran-Topolnita Karst System 183
Cristian Goran and Ioan Povară

Caves of Cerna Valley 203
Gheorghe M. L. Ponta

Cernei Mountains: Caves Conveying Geothermal Fluids at Băile Herculane 213
Ioan Povară, Horia Mitrofan, Bogdan P. Onac, Constantin Marin, Eugen Niţu, Daniela Ioniţă, Alin Tudorache, and Mădălina Vişan

Aninei Mountains: Captare Cave (Pestera de la Captare) 227
Petr Barák, Tomáš Svoboda, and Gheorghe M. L. Ponta

Aninei Mountains: The Ponor-Plopa Cave System 233
Silviu Constantin

Aninei Mountains: Buhui Cave 243
Eugen Creţu

Domanului Mountains: Comarnic Cave (Pestera de la Cantonul Comarnic, Pestera din Ogasul Ponicovei) 249
Ioan Povară

Domanului Mountains: Exploratorii 85 Cave (Pestera Exploratorii ′85) 257
Günther Karban

Domanului Mountains: Poiana Gropii Pit (Avenul din Poiana Gropii) 265
Günther Karban

Metaliferi Mountains: Zidită Cave 271
Ferenc L. Forray

Bihor Mountains: Dârninii Cave (Pestera din Peretele Dârninii) 277
Felix Papiu and Bogdan P. Onac

Bihor Mountains: Zgurăsti–Poarta Lui Ionele Cave System 283
Christian Ciubotărescu

Bihor Mountains: Coiba Mică–Coiba Mare–Tăuz Cave System 293
Christian Ciubotărescu and Bogdan P. Onac

Bihor Mountains 303
Gheorghe M. L. Ponta

Bihor Mountains: Drăcoaia Cave 319
Jozef Psotka and Vladimír Papáč

Bihor Mountains: Vărăsoaia Cave System (V5) 327
József Zih, Katalin Zih-Perényi, and Ludovic Mátyási

Bihor Mountains: Valea Rea Cave 335
Katalin Zih-Perényi, József Zih, Paul-Erik Damm, Szabolcs Szűcs,
and Bogdan P. Onac

Bihor Mountains: Humpleu–Poienita Cave System 347
Philipp Häuselmann

Codru Moma Mountains: Vascău Karst Plateau 357
Gheorghe M. L. Ponta, Mihai Besesek, and Valentin Alexandru Radu

Pădurea Craiului Mountains: Meziad Cave 367
Ioana N. Meleg, Marius Robu, Daniela R. Borda, Ludovic Mátyási, Călin Ghemiş,
and Viorel T. Lascu

Pădurea Craiului Mountains 375
Gheorghe M. L. Ponta

Pădurea Craiului Mountains: Stanul Foncii Shaft
(Avenul din Stanul Foncii) 393
Felix Papiu

Pădurea Craiului Mountains: Izbândis Cave System 399
Ferenc L. Forray

Pădurea Craiului Mountains: Vântului Cave (Wind Cave) 405
Bogdan P. Onac

Somes Plateau: Măgurici Cave (Pestera de la Măgurici de la Răstoci) 419
Bogdan P. Onac and Tudor Tămaş

Dobrogea: Movile Cave 429
Serban M. Sarbu, Cristian Lascu, and Traian Brad

Dobrogea: Limanu Cave 437
Virgil Drăguşin, Daniela Dimofte, and Ionuţ Mirea

Evaporite Karst of Romania 443
Maria-Laura Tîrlă

Eastern Subcarpathians Bend: Salt Karst: Meledic Plateau and
Slănic Prahova 451
Gheorghe M. L. Ponta

Ice Caves in Romania 455
Aurel Perşoiu and Bogdan P. Onac

Cave Minerals of Romania 467
Bogdan P. Onac

Caves Discovered by Mining Activities and Mined Caves 475
Bogdan P. Onac

Cave Biology 485
Oana Teodora Moldovan and Ruxandra Năstase-Bucur

Bat Fauna in Caves of Romania 493
Ioan Coroiu

The Archaeology of Caves in Romania 501
Mircea Anghelinu and Adina Boroneanț

Show Caves of Romania 519
Ioana N. Meleg, Marius Robu, Daniela R. Borda, Călin Ghemiș, Ludovic Mátyási, and Viorel T. Lascu

Cave Protection in Romania 537
Oana Teodora Moldovan

Annex: Cave Survey 543
Gheorghe M. L. Ponta

Contributors

Gheorghe V. Aldica National Institute of Materials Physics, Măgurele, Romania

Mircea Anghelinu Faculty of Humanities, "Valahia" University of Târgoviște, Târgoviște, Romania

Mihai Baciu The Group for Underwater and Cave Exploration (GESS), Bucharest, Romania

Ioan Balintoni Department of Geology, Babeș-Bolyai University, Cluj-Napoca, Romania

Petr Barák Czech Speleological Society (Devon Caving Club), Lhota Rapotina, Czech Republic

Mihai Besesek Speowest Speleological Association, Arad, Romania

Daniela R. Borda "Emil Racoviță" Institute of Speleology, Romanian Academy, Cluj-Napoca, Romania

Adina Boroneanț "Vasile Pârvan" Institute of Archaeology of the Romanian Academy, Bucharest, Romania

Traian Brad "Emil Racoviță" Institute of Speleology, Romanian Academy, Cluj-Napoca, Romania

Radu-C. Breban Proteus SpeleoClub, Hunedoara, Romania

Radu-Ciprian Breban Proteus SpeleoClub, Deva, Hunedoara, Romania

Doru Bădescu Laboratory of Geology and Paleontology, Faculty of Geology and Geophysics, University of Bucharest, Bucharest, Romania

Iulia Bădescu Département d'Anthropologie, Université de Montréal, Pavillon Lionel-Groulx, Montréal, QC, Canada

Christian Ciubotărescu Speleological Association for Environmental Protection and Karst "Sfinx" – Gârda, jud. Alba, Romania

Spencer F. Coca Spenco Rockoil Projects, Zărnești, Brașov, Romania

Silviu Constantin "Emil Racoviță" Institute of Speleology, Romanian Academy, Bucharest, Romania

Ioan Coroiu Department of Taxonomy and Ecology, Babeș-Bolyai University, Cluj-Napoca, Romania

Marian Cosac "Valahia" University of Târgoviște, Târgoviște, Romania

Eugen Crețu Terassond, Bad Bibra, Germany

Daniela Dimofte "Emil Racoviță" Institute of Speleology, Romanian Academy, Bucharest, Romania; Faculty of Geology and Geophysics, University of Bucharest, Bucharest, Romania

Gigi-Paul Dragomir Total E&P, Stavanger, Norway

Virgil Drăguşin "Emil Racoviţă" Institute of Speleology, Romanian Academy, Bucharest, Romania

Radu Dumitru Asociatia Speologică "Focul Viu", Frumoasă, Bucureşti, Romania

Ferenc L. Forray Department of Geology, Babeş-Bolyai University, Cluj-Napoca, Romania

Călin Ghemiş Muzeul Ţării Crişurilor, Dacia Blvd, Oradea, Romania

Cristian Goran "Emil Racoviţă" Institute of Speleology, Romanian Academy, Bucharest, Romania

Zsolt Gyurka Hungarian Speleological Society, Isaszeg, Hungary

Ulrich Hambach BayCEER & Chair of Geomorphology, University of Bayreuth, Bayreuth, Germany

Philipp Häuselmann Swiss Speleological Society, Bäriswil, Switzerland

Daniela Ioniţă Department of Analytical Chemistry, Faculty of Chemistry, University of Bucharest, Bucharest, Romania

Emilian Isverceanu Underwater and Cave Exploration Group (GESS), Bucharest, Romania

Günther Karban Munich, Germany

Cristian Lascu "Emil Racoviţă" Institute of Speleology, Romanian Academy, Bucharest, Romania

Viorel T. Lascu Centrul pentru Arii Protejate şi Dezvoltare Durabilă Bihor, Oradea, Romania

Constantin Marin "Emil Racoviţă" Institute of Speleology, Romanian Academy, Bucharest, Romania

Ioana N. Meleg "Emil Racoviţă" Institute of Speleology, Romanian Academy, Bucureşti, Romania

Ionuţ Mirea "Emil Racoviţă" Institute of Speleology, Romanian Academy, Bucharest, Romania

Horia Mitrofan "Sabba Ştefănescu" Institute of Geodynamics of the Romanian Academy, Bucharest, Romania

Gabor Mogyorosi Hungarian Speleological Society, Budapest, Hungary

Oana Teodora Moldovan "Emil Racoviţă" Institute of Speleology, Romanian Academy, Cluj-Napoca, Romania

George Murătoreanu "Valahia" University of Târgovişte, Târgovişte, Romania

Ludovic Mátyási Geo Prospect, Ştei, Romania

Augustin Nae "Emil Racoviţă" Institute of Speleology, Romanian Academy, Bucharest, Romania

Eugen Niţu "Emil Racoviţă" Institute of Speleology of the Romanian Academy, Bucharest, Romania

Maria-Lidia Nuţu-Dragomir Institute of Geodynamics of Romanian Academy, Bucharest, Romania

Ruxandra Năstase-Bucur "Emil Racoviţă" Institute of Speleology, Romanian Academy, Cluj-Napoca, Romania

Bogdan P. Onac Karst Research Group, School of Geosciences, University of South Florida, Tampa, FL, USA; "Emil Racoviţă" Institute of Speleology, Romanian Academy, Cluj-Napoca, Romania

Vladimír Papáč Speleoclub Drienka, Rimavská Sobota, Slovakia

Felix Papiu "Omega"/CSER Speleological Club, Cluj-Napoca, Romania

Aurel Perşoiu "Emil Racoviţă" Institute of Speleology, Romanian Academy, Cluj-Napoca, Romania

Gheorghe M. L. Ponta Geological Survey of Alabama, Tuscaloosa, AL, USA

Ioan Povară "Emil Racoviţă" Institute of Speleology of the Romanian Academy, Bucharest, Romania

Jozef Psotka Slovak Museum of Nature Protection and Speleology, Liptovský Mikuláš, Slovakia

Gheorghe Racoviţă "Emil Racoviţă" Institute of Speleology, Cluj-Napoca, Romania

Valentin Alexandru Radu Speowest Speleological Association, Vladimirescu, Romania

Marius Robu "Emil Racoviţă" Institute of Speleology, Romanian Academy, Bucureşti, Romania

Horaţiu Roman Faculty of Geology and Geophysics, University of Bucharest, Bucharest, Romania

Serban M. Sarbu "Emil Racoviţă" Institute of Speleology, Romanian Academy, Bucureşti, Romania; Department of Biological Sciences, California State University, Chico, USA

Szabolcs Storozynski Hungarian Speleological Society, Budapest, Hungary

Tomáš Svoboda Czech Speleological Society (Devon Caving Club), Lhota Rapotina, Czech Republic

Szabolcs Szűcs Evosoft Hungary Ltd., Miskolc, Hungary

Crin-Triandafil Theodorescu Luşca, Bistriţa-Năsăud, Romania

Tudor Tămaş Department of Geology, Babeş-Bolyai University, Kogălniceanu, Romania

Raul-Bogdan Tomuş Proteus SpeleoClub, Deva, Hunedoara, Romania

Alin Tudorache "Emil Racoviţă" Institute of Speleology of the Romanian Academy, Bucharest, Romania

Tiberiu Tulucan Arad, Romania

Maria-Laura Tîrlă Faculty of Geography, University of Bucharest, Bucharest, Romania; Research Institute of the University of Bucharest (ICUB), Bucharest, Romania

Daniel Veres "Emil Racoviţă" Institute of Speleology, Romanian Academy, Cluj-Napoca, Romania

Mădălina Vişan "Sabba Ştefănescu" Institute of Geodynamics of the Romanian Academy, Bucharest, Romania

Marius Vlaicu "Emil Racoviţă" Institute of Speleology, Romanian Academy, Bucharest, Romania

Katalin Zih-Perényi Institute of Chemistry, Eötvös Loránd University, Budapest, Hungary

József Zih Ponor Ltd., Budapest, Hungary

The Founder of Biospeleology and World's First Speleological Institute

Gheorghe Racoviţă

Keywords
Cave • Biospeleology • Research • Cluj-Napoca Romania

Racoviţă, Emil Gheorghe (born November 15, 1868, in Iaşi, deceased November 19, 1947, in Cluj) is a world-renowned Romanian biologist (Fig. 1), the founder of biospeleology as a scientific discipline and of the world's first speleological institute.

Emil, the son of Eufrosina and Gheorghieş Racoviţă, attended the primary school in Iaşi where he had I. Creangă as a teacher, and then, he continued at the "Institutele Unite" High School in the same city, being the student of P. Poni, A. D. Xenopol, and G. Cobălcescu. In 1887, he went to Paris to attend the Law School, but he also attended the lectures given by L. P. Manouvrier at the School of Anthropology, and then those at the Sorbonne School of Sciences, where he had as teachers, among others, the renowned zoologists H. de Lacaze-Duthiers and G. Pruvot. In 1889, he obtained his degree in Law, and in 1891, in Natural Sciences; in 1896, he was very successful in defending his doctoral thesis entitled *Le lobe céphalique et l'encéphale des Annélides polychètes (Anatomy, Morphology, Histology).*

Since 1900, he has been working at the Arago Laboratory of Banyuls-sur-Mer (France) as Vice-Director and Lecturer at the Faculty of Sciences. On February 1, 1920, he has been appointed Professor of Biology at the Faculty of Sciences in Cluj (Romania), and in April 26, the Director of the Institute of Speleology. In the same year, he was elected Full Member of the Romanian Academy. In June 13, 1926, he gave his well-known reception speech *Speleology—a new science of the ancient subterranean mysteries*. For three consecutive years (1926–1929), he was elected

Fig. 1 Emil G. Racoviţă in 1921 (photograph from the archive of G. Racoviţă)

President of the Romanian Academy. Between 1922 and 1926, he was also a Senator of the University of Cluj and became its Rector during 1929–1930.

Emil Racoviţă started his naturalist activity during the years of his studies by conducting marine biology research at the Roscoff and Banyuls-sur-Mer oceanic stations. The work of this first period was part of his doctoral thesis and established him as a recognized biologist worldwide. As a result, he was chosen as a naturalist of the Belgian Antarctic Expedition, initiated and led by Marine Lieutenant A. de Gerlache de Gomery. Deployed on board the Belgica vessel between August 16, 1897, and November 18, 1899, this expedition completed Racoviţă's scientific formation in the same way the famous Beagle expedition helped C. Darwin crystallize his evolutionary theory. It was the first scientific expedition to be undertaken in the Antarctic, and the first that spend the winter at the extreme latitude of 71° 31′. The substantial contribution of the Romanian naturalist was reflected in the observations and studies carried out on marine birds and mammals, particularly on whales, as well as in the collection of a huge scientific material, from which more than 60 studies were published by various specialists.

Returned to France, Emil Racoviţă continued his marine biology research at Banyuls. On July 16, 1904, during a cruise on the Balearic Archipelago, he visited the famous Cueva del Drach (Dragon Cave) on Mallorca Island (Spain), where he discovered his first cave animal, an aquatic

For queries and clarification, contact:
Bogdan P. Onac,
Email: bonac@usf.edu

G. Racoviţă (Deceased) (✉)
Cluj-Napoca, Romania

G. M. L. Ponta and B. P. Onac (eds.), *Cave and Karst Systems of Romania*,
Cave and Karst Systems of the World, https://doi.org/10.1007/978-3-319-90747-5_1

crustacean isopod that he will describe a year later under the name of *Typhlocirolana moraguesi*. Recognizing the importance of cave animals in deciphering the evolution processes, he started to dedicate to researches on the subterranean domain, laying the basis of a new scientific discipline—**biospeleology**. He formulated with remarkable clairvoyance its main aims and directions in the memorable *Essai sur les problèmes biospéologiques* (1907). With this work, he opened the series on subterranean biology published by zoologists around the world under the common title of "Biospeologica." Accompanied by his collaborator, R. Jeannel, he explored over 1000 caves in France, Spain, Corsica, Algeria, Slovenia, and Romania, collecting a significant faunistic material. Along with R. Jeannel, he initiated the first cadaster of the investigated caves, published in seven series under the title *Enumération des grottes visitées* (1907–1929).

In 1920, returning to his homeland at the request of the Transylvanian Dirigent Council, he founded the Institute of Speleology in Cluj, world's first scientific research institute dedicated to the study of the subterranean domain. The results of the researches were published in the journal *Lucrările Institutului de Speologie*, one of the most important speleological publications at that time. At the University of Cluj, he had a sustained and extensive activity in organizing the Society of Sciences (1920) who was the president until his death, and published the Bulletin of the Society (since 1921). He also initiated and led the first Congress of Naturalists in Romania (Cluj, April 18–21, 1928) and had an essential contribution in the elaboration of the first nature conservation law (July 4, 1930) and the establishment of the Committee on Natural Monuments (1931), who was the member until 1947. He was concerned in improving the organization and functioning of the University library by coordinating the publication of a *Catalog of scientific and medical journals* (1926). He was active in the Transylvanian Association for Romanian Literature and Culture (ASTRA) founded in Sibiu in 1861 and in which was active until 1946. E. Racoviţă collaborated with the organization of the Ethnographic Museum in Cluj (1928) and the foundation of the first tourism society in Transylvania, the "Frăţia Munteană" (1922).

Emil Racoviţă's scientific work is marked by many valuable contributions to the progress of biology, new and bold ideas, some of which have preserved their validity to this day. He raised the systematics from the level of a descriptive discipline to the rank of applied phylogeny, considering that species can only be studied and understood in the context of their historical development, geographical distribution, and their ecological relations. He has set precise criteria for describing new species, by militating for the adoption of unique rules. He formulated an original definition of the species as "... *an isolated colony of consanguineous,*" thus summarizing the most important characteristics of this fundamental biological unit. He vigorously promoted the need for studies on homogeneous groups of animals, the so-called phyletic series, the only ones able to bring to light the true natural history of various taxonomic categories. Applying consistently such principles in his research on isopod crustaceans, he provided through numerous published papers, true models of scientific analysis. He brought to the world attention the complex problems of the subterranean domain, establishing its extension, the origin of the cave animals, and the ways of colonizing the cave habitats. He outlined the phases of biospeleological research development, anticipating from the beginning the transition from descriptive research to experimental research, carried out in laboratory setup inside caves. He formulated a series of basic principles, which have led to the formulation of the evolutionary neo-Lamarckist theory. He proposed replacing the classic notion of "species" with the more comprehensive "species line" that includes the entire historical sequence of forms that shaped the present species. He emphasized the particular role that geographic, ecological, or physiological isolation has in diversifying species by preserving and amplifying some small initial variations. He specified the evolutionary significance of the adaptation process, for which he proposed the more appropriate term "accommodating" and which he defined as the tendency of the bodies to achieve a perfect fit with their living environment. He introduced the notion of "seclusion" to designate the direct action of environmental factors (as opposed to the compensating reactions of adaptation), emphasizing on the decisive role played by evolution in the formation and improvement of the artificial environment. He also made important contributions in establishing nature conservation rules and environmental principles on which the establishment and management of nature reserves should be based.

Through his entire activity and undeniable value of his work, Emil Racoviţă is among the most important Romanian scientists and is part of the great scientific and cultural personalities of the humanity. His merits are confirmed by the many titles and distinctions he was granted: Member (1893) and Honorary President (1925) of the Zoological Society of France; Knight of the Leopold Order of Belgium; Member of the Romanian Geographic Society (1899); Member of the Geography Society of Paris; Correspondent Member of the Society of Physicians and Naturalists of Iaşi (1900); Knight of the French National Order Legion of Honor (1902); Member of the Entomological Society of

France (1906); Correspondent Member of the Zoological Society and of the Royal Geographical Society of London (1910); Correspondent Member of the Society of Natural Sciences of Barcelona (1922); Honoris Causa of the University of Lyon (1923); Member of the Paris Biology Society (1925); Grand Officer of the Star of Romania Order (1928); Member of the Romanian Society of Geology (1930) and its President (1934); Member of the Spanish Society of Natural History (1930); President of the Society of Sciences of Romania (1932); Founding Member of the Paris Biogeography Society (1935); Correspondent Member of the Paris Medical Academy (1944); and Member of the Zoological Society of London (1947).

"Emil Racoviță" Institute of Speleology: World's First Research Unit Dedicated to Karst and Cave Studies

Ioan Povară

Keywords
Emil Racoviţă • Cave • Karst • Biospeleology
Cluj-Napoca • Romania

Introduction

The "Emil Racoviţă" Institute of Speleology (ERIS), ***first of its kind in the world***, performing research activity for more than 97 years, owns its existence to the initiative of the great scientist ***Emil G. Racoviţă (1868–1947), a biologist, polar explorer, and founder of biospeleology***. The establishment of the ERIS became official when the Law no. 19.11 from April 26, 1920, was published in "*The Official Monitor of Romania*" (in Romanian: "*Monitorul Oficial*") no. 86 from July 20, 1920. The institute was founded as an independent research unit within the Faculty of Sciences at the University of Cluj, and E. G. Racoviţă served as Director and Manager between 1920 and 1947. The research trends initiated by E. G. Racoviţă were continued until 1949 by R. Jeannel and P. A. Chappuis, two close collaborators of the Romanian scientist. In 1951, within the Geological Committee of the People's Republic of Romania, two research groups were assembled in Bucharest and Cluj, respectively. Their mission was to prospect and identify phosphate deposits, an opportunity that led to the exploration of a large number of caves, which were simultaneously surveyed and investigated for their fauna.

On **June 21, 1956**, the Institute of Speleology was reorganized, under the leadership of Professor C. Motaş, within the Ministry of Education, and after a period in which it was successively subordinated to the Romanian Academy and the Ministry of Education, it moved under the aegis of the Romanian Academy, as a national research unit within the Biological Sciences Section (Government's Decision 656/June 5, 1990). Currently, the "Emil Racoviţă" Institute of Speleology is structured in four departments in Bucharest (Biospeleology and Karst Edaphobiology; Geospeleology and Paleontology; Karstology, Karst Inventory, and Protection; Hydrogeochemistry) and one in Cluj-Napoca that includes researchers in both geo- and biospeleology fields.

"Emil Racoviţă" Institute of Speleology conducts high-quality theoretical and applied research on karst and cave issues. Its research activities combine the local and regional perspective with a global application. The main *fields of cave research* include:

- Taxonomic, phylogenetic, biogeographic, and ecological studies of subterranean environments, aiming to highlight their diversity and complexity;
- Mineralogical and geochemical studies;
- Quaternary climate oscillation reconstructions based on speleothems, ice, and guano deposits;
- Studies of the fossil mammal fauna;
- Study of the karst system complexity and distribution, and regional karst research;
- Karst environment protection/conservation and its vulnerability to pollution: The Hydrogeochemistry Laboratory (ISO/CEI 17025:2005 accredited) is able to analyze groundwater and surface water samples.

During the institute almost 100 years of activity, its researchers have published more than 2000 papers in Romanian and international scientific journals, more than 300 books at well-known national and international publishing houses, out of which 40 received various awards. More than 150 national and/or international research projects have been undertaken, and 62 scientific meetings have been organized.

I. Povară (✉)
"Emil Racoviţă" Institute of Speleology of the Romanian Academy, Calea 13 Septembrie 13, 050711 Bucharest, Romania
e-mail: ipov.iser@gmail.com

G. M. L. Ponta and B. P. Onac (eds.), *Cave and Karst Systems of Romania*,
Cave and Karst Systems of the World, https://doi.org/10.1007/978-3-319-90747-5_2

Books and reference papers published since 1956

- the monographs of the *Fundata, Gura Dobrogei*, and *Limanu* caves, as well as of the caves from the *Vârghiş Gorges*;
- Recherches sur les grottes du Banat et d'Olténie (published by CNRS in Paris);
- Peşteri din România (*Caves of Romania*) (1961);
- *Initiation à la biologie et l'écologie souterraines* (Jean-Pierre Delarge, Paris);
- *Stygofauna Mundi: A faunisitic, distributional, and ecological synthesis of the world fauna inhabiting subterranean waters*;
- Speologie—Ghid practic (*Speleology. A practical guide*);
- Peşteri din România (*Caves of Romania*) (1976);
- Modelling of the chemical speciation in natural aqueous systems;
- Peşteri din România. Ghid turistic (*Caves of Romania. Tourist guide*);
- Carstul Munţilor Pădurea Craiului (*The karst of the Pădurea Craiului Mountains*);
- Peşteri scufundate (*Submerged caves*);
- Chiroptere din România (*Chiroptera of Romania*);
- Geologia regiunilor carstice (*The geology of karst terrains*);
- Biodiversitatea în mediile subterane din România (*Biodiversity of the subterranean environments of Romania*);
- Masivul Piatra Craiului (*Piatra Craiului Massif*);
- Scărişoara Glacier Cave. Monographic study (*Romanian Academy Award 2004*);
- Coleoptera din peşteri. Adaptări, texonomie, ecologie, feromoni, comportament, protecţie (*Coleoptera in caves. Adaptations, taxonomy, ecology, pheromones, behavior, protection*);
- Fauna României, Insecta, Coleoptera (*Romania's Fauna, "Insecta", Coleoptera*) (vol. X, fasc. 6);
- Encyclopaedia Biospeologica (*3 volumes*);
- The first ecological reconstruction of underground environment from Romania—Cioclovina Uscată Cave;
- Diversitatea lumii vii (*The diversity of the living world*);
- Life and death at Peştera cu Oase. A setting for Modern Human emergence in Europe (Oxford University Press);
- Valea Cernei. Morfologie, hidrologie, ape termale (*Cerna Valley. Morphology, hydrology, thermomineral waters*);
- Reconstituiri paleoclimatice pe baza mamiferelor mici din depozite carstice. Studiu de caz—Dobrogea Centrala (*Paleoclimate reconstructions based on small mammals from karst deposits. Case study: Central Dobrogea*);
- Fauna României, Chiroptera (*Romania's Fauna, Chiroptera*);
- The paleontology of the cave bear bone assemblage from Urşilor Cave of Chişcău—Osteometry, palaeoichnology, taphonomy, and stable isotopes (*Romanian Academy Award 2015*);
- Ice caves (Elsevier).

The biospeleological expeditions with results published by the Romanian Academy include: *Bulgaria 1983*; *Cuba* (4 volumes entitled "*Résultats des expéditions biospéologiques cubano-roumaines à Cuba*"); speleological expedition to *Venezuela*; and the expedition to *Israel*, for collecting cave and soil fauna (*Soil fauna of Israel*).

Over the last 15 years, the scientists from the ERIS have conducted important research activities that translate in 482 papers and 49 books published; 470 international scientific meetings attended; 22 national and international scientific meetings organized; 62 contracts and 14 Romanian Academy and Romanian Science Foundation (UEFISCDI) awards received. The research activities are supported by laboratories equipped with state-of-the-art facilities, operated by specialists from all ERIS's departments.

The most important research grants during these recent years were focused on:

- monitoring the conservation status of caves and bats species or the human impact on show caves;
- karst and related terrains vulnerability in Romania;
- climate archives in karst—an integrated approach to the study and modeling of abrupt climate oscillations;
- chemical reactions within the cave environment: mineralogical, geochemical, and geochronological aspects;
- paleoclimate reconstructions based on interdisciplinary studies of ice deposits from caves of Romania;
- stable isotope signature in cave guano as archive of past environments;
- identification of hypogene caves on Cerna Valley (Romania) based on stable isotope analysis;
- remote connections during the climate changes from Western and Eastern Europe, based on contemporaneous speleothem climate archives, dated to the last interglacial;
- thermal system resilience to anthropogenic and natural disturbances;
- the development of the "*Natura 2000*" network;
- the diversity and the metabolic activity of the ice caves microbiome, as a response to climate changes and to the anthropogenic activity.

The "Emil Racoviţă" Institute of Speleology of the Romanian Academy is currently collaborating with more than 15 Romanian institutions (e.g., Institute of Biology,

"Grigore Antipa" National Museum of Natural History, National Institute for Research and Development in Tourism, University of Bucharest, "Babeş-Bolyai" University in Cluj-Napoca, Academy of Agricultural and Forestry Sciences "Gheorghe Ionescu-Şişeşti") and more than 30 foreign partners (University of Melbourne, University of Miami, University of Bergen, University of Bremen, University of South Florida, Chinese Institute of Zoology, Royal Belgian Institute of Natural Sciences, Department of Evolutionary Biology and Environmental Studies at the University of Zürich, Institute of Soil Biology from České Budějovice, Polish Geological Institute, Stuttgart State Museum of Natural History, Natural History Museum from Verona, etc.).

Being committed to improving the knowledge and the protection of the subterranean realm, the "Emil Racoviţă" Institute of Speleology has organized yearly the following scientific and applied field courses to educate young cavers: *National School of Biospeleology, National School of Chiropterology, National Stage of Subterranean Surveying and Mapping, National Stage of Karstology and Geospeleology, and National Field Karstology Studies*. At the same time, ERIS provides environmental consulting assistance to apply its own research findings to improve cave and karst resource management.

Since its establishment in 1920 as the world's first cave research entity, the "Emil Racoviţă" Institute of Speleology has been a leading international center for karst and cave research. The mission of ERIS is to design, conduct, educate, and disseminate research findings worldwide. ERIS's mission and vision for the future is to continue its dynamic national, regional, and international leadership in cave and karst research to help policymakers, educators, and the general public to properly manage karst waters, karst landscapes, and their biospeleological and physical resources.

Geology of Romania

Ioan Balintoni

Abstract

Located at the southwestern end of the Trans-European Suture Zone (TESZ), the territory of Romania includes several major Alpine terranes of East European (Moldavian and Scythian platforms) or of West European affinity (Foreapulian, Getic, Euxinic terranes and the North Dobrogean Orogen). The pre-Alpine terranes from the basement of the Alpine terranes of West European affinity have a peri-Gondwanan provenance of Avalonian type (late-Neoproterozoic peri-Amazonian), Cadomian type (late-Neoproterozoic peri-North-African), or Carpathian type (essentially Ordovician, peri-North-African). Carpathian-type terranes were described in the Apuseni Mountains (Someş, Biharia, Baia de Arieş), in East Carpathians (Bretila, Tulgheş, Negrişoara, Rebra), and in the Getic Domain of South Carpathians (Cumpăna upper unit of the Sebeş-Lotru Terrane, Leaota, Bughea, Caraş, Padeş, Făgăraş). Cadomian type includes the Lotru lower unit of the Sebeş-Lotru Terrane. Drăgşan and Lainici-Păiuş terranes from the Danubian Domain of South Carpathians, the Histria and Altîn Tepe from Central Dobrogea, the east Moesia from South Dobrogea and Boclugea, Megina, Orliga, Uzum Bair from the basement of the North Dobrogean Orogen are all Avalonian type. The carbonate rocks are important in the basement of Baia de Arieş, Negrişoara, Rebra, Făgăraş, and Lainici-Păiuş terranes. The Mesozoic carbonate rocks anterior to the Alpine terranes amalgamation cover significant surfaces in the Carpathians and Dobrogea, where Romania's main karst regions occur. Karst-like features are also known on evaporites of Cenozoic age occurring in the Transylvanian Basin and Carpathian foreland.

Keywords

Carpathian peri-Gondwanan Terranes • Euxinin Terranes European Craton • Scythian Platform • North Dobrogean Orogen

Introduction

For more than a century, Romanian geologists have focused their research in collecting detailed information (geology, geophysics, tectonics, etc.) needed in developing geological maps and models. To our knowledge, only two Romanian geology overviews are written in English (Burchfiel and Bleahu 1976; Săndulescu 1994), and all the others are exclusively in Romanian (Oncescu 1957; Mutihac and Ionesi 1974; Săndulescu 1984), thus hidden behind a language barrier. Several new publications dealing with local and regional studies contribute substantially to developing a deep understanding of the Romanian geologic history. Many of these studies are cited within this chapter, which provides a very detailed presentation of the crystalline basement units (spatial distribution, orogenic epochs, ages, etc.) on which rest the sedimentary cover, including the karst-forming rocks.

The territory of Romania is located at the south-eastern end of the Trans-European Suture Zone (TESZ) (e.g., Pharaoh 1999; Fig. 1) and includes geotectonic units associated with both the East European Craton (EEC) and Western Europe, the latter one of peri-Gondwanan provenance. With respect to the Alpine structure of Romania, several major terranes (geological units) with pre-Alpine lithospheric basement can be identified (Fig. 2). Using the terminology proposed by Balintoni (1997), these are:

I. Balintoni (✉)
Department of Geology, Babeş-Bolyai University, Kogălniceanu 1, 400084 Cluj-Napoca, Romania
e-mail: ioan.balintoni@ubbcluj.ro

G. M. L. Ponta and B. P. Onac (eds.), *Cave and Karst Systems of Romania*,
Cave and Karst Systems of the World, https://doi.org/10.1007/978-3-319-90747-5_3

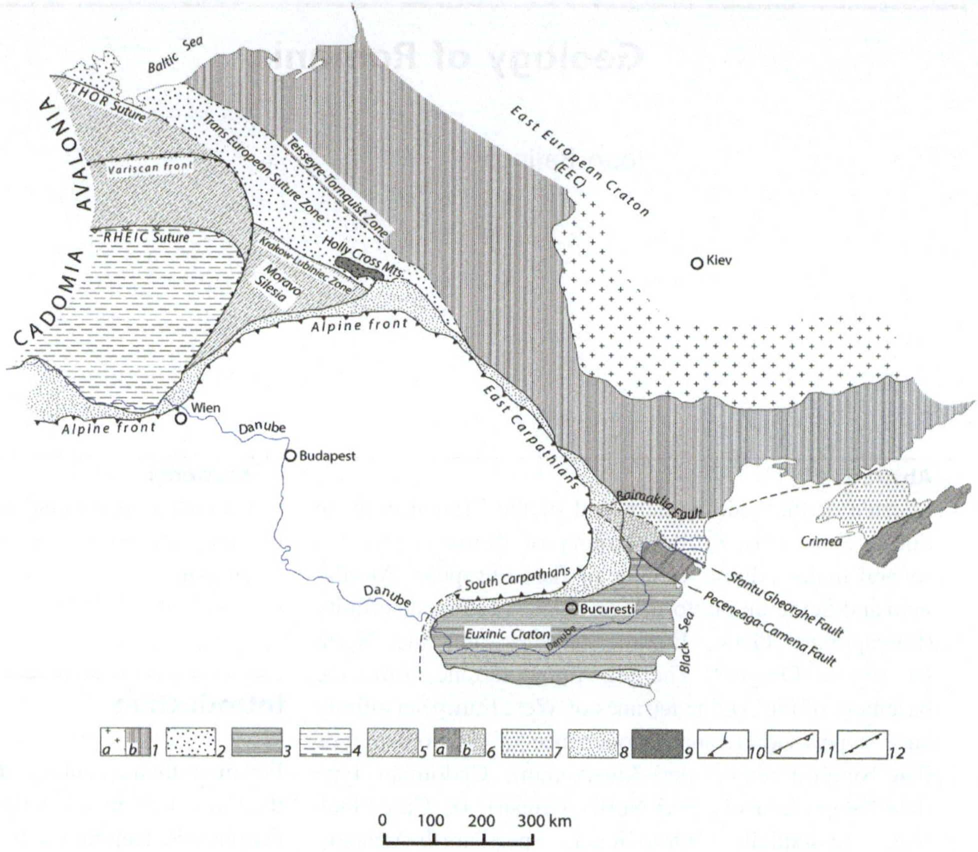

Fig. 1 Sketch map of the TESZ and adjacent areas. Compiled from Săndulescu (1984), Balintoni (1997), Pharaoh (1999). *1* East European Craton (a Ukrainian Shield, b platform cover), *2* Trans-European Suture Zone, *3* Euxinic Craton, *4* Cadomia, *5* Avalonia, *6* North Dobrogean–Crimean Orogen (a outcrop, b covered), *7* Scythian Platform, *8* Carpathian Foreland Basin, *9* Holy Cross Mountains, *10* Alpine Front, *11* Rheic Suture, *12* Variscan Front

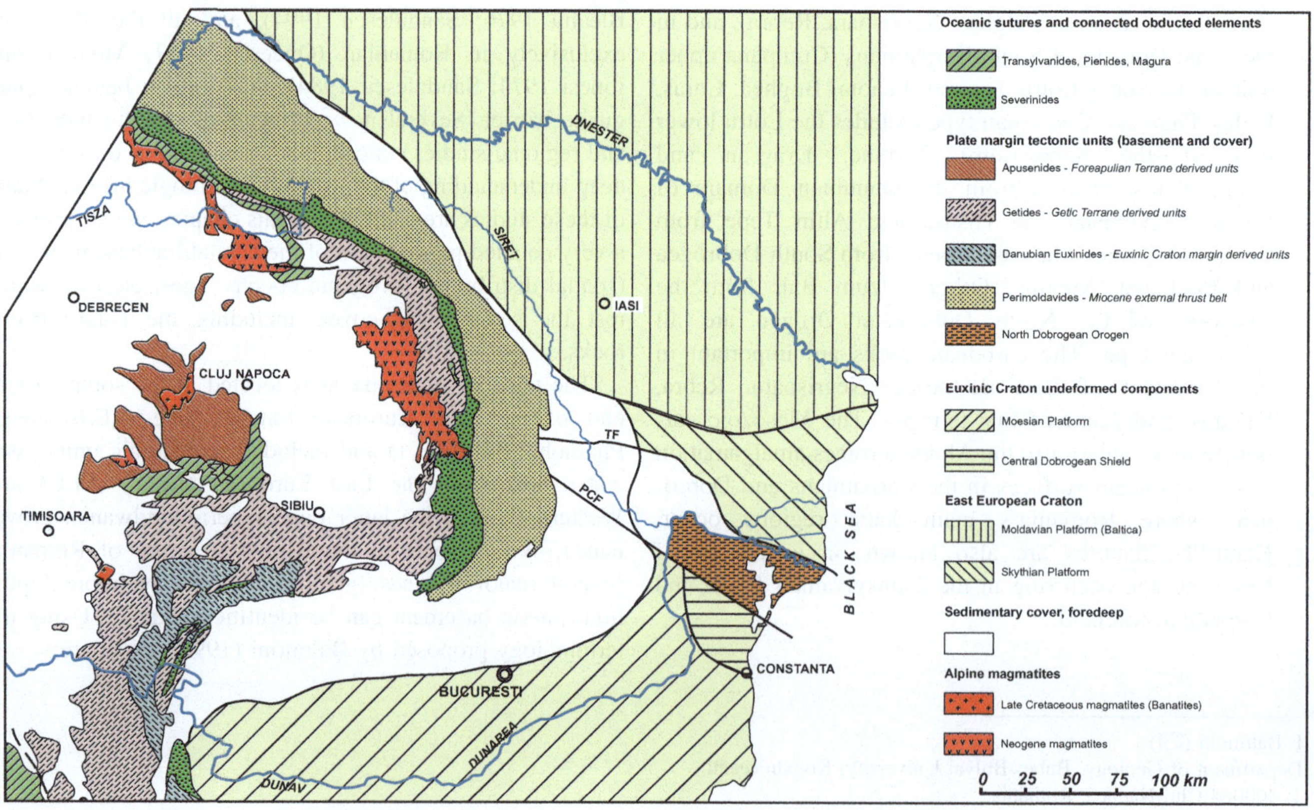

Fig. 2 General Alpine structure of the Romanian territory compiled from Săndulescu (1984) and Balintoni (1997)

(1) Intensely deformed terranes:
  (a) the Getic Terrane with the East Carpathians and the Getic Domain of South Carpathians as major components
  (b) the Foreapulian Terrane with the Apuseni Mountains
  (c) the North Dobrogean Orogen.
(2) Marginally deformed terranes: the Euxinic Terrane with its Danubian Domain of South Carpathians.
(3) No visible deformed terranes: the EEC and its south-western margin, the Scythian Platform.

Except the EEC, the basements of the other terranes have a peri-Gondwanan (peri-Amazonian or peri-North-African provenance (Nance et al. 2008). The European peri-Gondwanan terranes record post-Grenvillian thermo-tectonic events associated with the Pan-African (Avalonian-Cadomian), Caledonian, North-African Ordovician, and Variscan orogenies, or they were generated during these orogenies (e.g., Nance et al. 2008; Balintoni and Balica 2013).

In Alpine terms, the Moldavian Platform (the western margin of EEC) represents the East Carpathians foreland and the Euxinic Terrane is the foreland of the South Carpathians. The North Dobrogean Orogen thrusting over the south-western margin of EEC, the Scythian Platform forms its foreland. Solving the history of the Alpine terranes is an intricate task that includes: the identification of the pre-Alpine terranes from the basement of the Alpine ones and understanding their pre-Alpine history and the recognition of the Alpine sediments anterior and posterior to the Alpine terranes amalgamation. In addition, the Alpine terranes history also incorporates an extensive magmatism, well known in the western part of Romania (the banatites) and in the western volcanic chain of the East Carpathians.

Boundaries of the Alpine Terranes

Săndulescu (1984, 1994) proposed a largely accepted model of the Alpine continental and oceanic units on the Romania territory and around it. According to this author, between the Foreapulian and Getic terranes, the main Tethyan Suture Zone remnants are visible. Also, in the front of the Getic Terrane are known the remnants of the Severin rift or ocean. Between the Severin rift remnants and Peceneaga–Camena Fault is situated the Euxinic Terrane, and the North Dobrogean Orogen overthrusts the Scythian Platform between Peceneaga–Camena and Sf. Gheorghe faults.

Schmid et al. (2008) use for the Forealpine Terrane the name of Tisza, for the Getic Terrane the name of Dacia, and Moesia for the Euxinic Terrane. The Tisza and Dacia terranes separated each other and against the surrounding terranes by branches of the Alpine Tethys. The Alpine Tethys branches between the Tisza and Alcapa terranes and Tisza and Dacia terranes are connected by great faults with the Meliata–Maliac Ocean and farther with the Neotethys. The terminal part of Alpine Tethys between Dacia Terrane and the eastern continental blocks is called the Ceahlău–Severin Ocean. As a conclusion we remark that overall the two models are similar, recognizing on the territory of Romania the same number of continental blocks and oceanic branches.

The Basement of the Alpine Terranes

The Foreapulian Terrane

Balintoni et al. (2010a) separated in the basement of the Foreapulian Terrane from the Apuseni Mountains the following pre-Alpine terranes: Someş, Baia de Arieş, and Biharia. The zircon LA-ICP-MS U/Pb ages on orthogneiss and metagranitoid samples from the basement of these terranes yielded early Paleozoic protolith ages (Table 1). Except a late Cambrian age yielded by the Biharia Terrane metagranites, the other ages are early to middle Ordovician. These results suggest an Ordovician orogeny for the geotectonic context in which the above terranes have been generated (Balintoni and Balica 2013). The Someş Terrane rocks (dominantly orthogneisses and paragneisses) suggest a tectonic setting of active plate margin, whereas the Biharia Terrane rocks, essentially metabasites and metagranites, indicate a rift environment. The Baia de Arieş lithology that includes large limestone lenses is interpreted as passive plate margin deposits. The metamorphic mineral associations in the Someş, Biharia, and Baia de Arieş terranes rocks suggest two metamorphic events: First of them is probably of Ordovician age, while the latter one, with an intricate thermal path, is Variscan.

The Getic Terrane

The Getic Terrane in East Carpathians

Balintoni and Balica (2013) found in the basement of the Crystalline-Mesozoic Zone the following pre-Alpine terranes: Bretila, Tulgheş, Negrişoara, and Rebra. Likewise Apuseni Mountains, the zircon LA-ICP-MS U/Pb ages on ortho-rocks sampled from these terranes yielded Ordovician ages (Table 2). Thus, the East Carpathians pre-Alpine terranes also originated during an Ordovician orogenic event. The same authors consider the Bretila Terrane a continental margin magmatic arc (rock assemblage of metagranite, orthogneiss, paragneiss, and amphibolite); the Tulgheş Terrane should expose the mixed infill of a back-arc basin (basinal rock assemblage with thick black quartzite lenses and associates manganese ore bodies, followed by recurrent metarhyolite layers with Kuroko-type base metal ore bodies); the Rebra and Negrişoara terranes are interpreted to constitute the inboard continental margin characterized by metaterrigenous layers, thick carbonate rock layers containing Mississippi Valley-type syngenetic mineralizations, and occasional distal metavolcanic layers. The metamorphic history of the East Carpathians terranes is similar with that of the Apuseni Mountains: a pre-Variscan metamorphic event and a convoluted Variscan thermo-tectonic history. The Bretila Terrane in the East Carpathians can be correlated with the Someş Terrane in Apuseni Mountains, whereas Negrişoara and Rebra terranes with Baia de Arieş. The Tulgheş Terrane in East Carpathians suggests an evolved stage from a rift to a back-arc basin.

The Getic Terrane in South Carpathians (Getic Domain)

Balintoni et al. (2010b, 2014a) described in the Getic Domain of South Carpathians the following pre-Alpine terranes: Sebeş–Lotru, Leaota, Bughea, Caraş, Padeş, and Făgăraş. The U/Pb zircon ages of the igneous protoliths from these terranes are listed in Table 3. The Sebeş-Lotru Terrane has a composite structure with a lower unit (Lotru) and an upper unit (Cumpăna) of late Neoproterozoic and Ordovician age, respectively. The other five terranes are Ordovician in age similar to the ones from the Apuseni Mountains and East Carpathians.

The Sebeş–Lotru, Leaota, and Caraş terranes originated in a magmatic arc tectonic setting (the quartzite and carbonate rocks are quite scarce, mica schist subordinate, paragneiss locally important, and rock assemblage is

Table 1 U/Pb zircon ages of the igneous protoliths from the peri-Gondwanan, Carpathian-type terranes in the basement of the Apuseni Mountains

Terrane	Metamorphic unit	Sample codes and types	Best age (Ma)
Someş	Someş	165, 166 Iara orthogneiss	459.8 ± 2.7
Baia de Aries	Baia de Arieş	171 Pociovaliştea augen gneiss	470.8 ± 5.0
		186, 268-9 Lupşa metaporphyroid	467.8 ± 3.8
		178 Mihoeşti metagranite	467.8 ± 4.7
		181-Muncelu metagranite	467.1 ± 3.9
Biharia	Biharia	170, 290, 291 Biharia metagranite	495.0 ± 2.1
		327, 328, 328b Biharia metabasite	477.8 ± 3.2

Table 2 U/Pb zircon ages of the igneous protoliths from the peri-Gondwanan, Carpathian-type terranes in the basement of the East Carpathians

Terrane	Metamorphic unit	Sample codes and types	Best age (Ma)
Bretila	Bretila	255-Anieş augen gneiss	464.0 ± 3.0
		257-Hăghimaş granitoid	469.2 ± 6.5
Tulgheş	Tulgheş	10-476-Zugreni metarhyolite	462.6 ± 3.1
Negrişoara	Negrişoara	10-475-Pietrosu porphyroid	461.1 ± 5.2
Rebra	Rebra	256-Nichitaş orthogneiss	447.9 ± 2.8

Table 3 U/Pb zircon ages of the igneous protoliths from the peri-Gondwanan Cadomian-type (Lotru metamorphic unit) and Carpathian-type (the other metamorphic units) terranes in the basement of the Getic Domain (South Carpathians)

Terrane	Metamorphic unit	Sample codes and types	Best age (Ma)
Sebeş–Lotru	Cumpăna	271-Căpâlna orthogneiss	458.9 ± 3.5
		283-Latoriţa orthogneiss	466.0 ± 4.2
	Lotru	279-Godeanu migmatic dyke	549.3 ± 3.9
		275-Frumoasa metagranite	587.5 ± 3.8
Leaota	Lereşti	221-Clăbucet orthogneiss	479.0 ± 5.2
		224-Lalu metagranite	475.0 ± 4.6
Caraş	Caraş	220-Naidăş metagranite	481.7 ± 3.2
	Bocşiţa-Drimoxa	215-Brădeşti metagranite	479.1 ± 2.7
Padeş	Padeş	239-Veţel orthogneiss	450.5 ± 2.9
Bughea subduction complex		223-Albeşti metagranite	467.8 ± 5.9

dominated by orthogneiss, granitoid, metabasite, and metaultrabasite). The Padeş Terrane origin is in a back-arc environment as that of the Tulgheş Terrane from East Carpathians. The Făgăraş Terrane, typical for passive continental margin environment, comprises metaterrigenous rocks (mica schist, paragneiss, quartzite), sometimes amphibolite, and especially thick layers of carbonate rocks that host Cu–Pb–Zn and magnetite mineralizations. The Bughea Terrane, located in the eastern part of the South Carpathians, marks a structural boundary of Variscan age between the Sebeş-Lotru Terrane in lower position and Leaota Terrane in upper position. It consists of a predominantly semipelitic to mafic melange and the associated Albeşti Granite of Ordovician age.

Except the Lotru unit of the Sebeş-Lotru Terrane of end-Neoproterozoic Cadomian-type, all the other ones record an initial pre-Variscan Ordovician metamorphic history, followed by a Variscan complex evolution, including an eclogite event for the Sebeş-Lotru and Bughea terranes. In accordance with their petrology, the Getic Domain pre-Alpine terranes can be correlated with those of East Carpathians and Apuseni Mountains.

The Carpathian terranes of Cambrian–Ordovician age and North-African provenance were named by Balintoni and Balica (2013) and Balintoni et al. (2014a) "Carpathian-type" terranes.

The Euxinic Terrane

As a continental lithospheric block, the Euxinic Terrane exemplifies a pre-Alpine craton, built up by the Moesian Platform, the Central Dobrogean Shield, and the Danubian Domain of South Carpathians, this one strongly deformed during the Variscan and Alpine orogenies (Balintoni 1997).

The Euxinic Terrane in South Carpathians (Danubian Domain)

Its basement consists of two pre-Alpine terranes: Drăgşan and Lainici-Păiuş (Liegeois et al. 1996; Balintoni et al. 2011a). The Drăgşan Terrane is a composite one, comprising a lower orthogneiss assemblage (Făgeţel), a middle metabasic–ultrabasic assemblage (Straja) and an upper mica gneiss unit (Dobrota) (Berza and Seghedi 1975). An orthogneiss sample from the Făgeţel assemblage yielded an U/Pb age of 808.6 ± 1.9 Ma (Balintoni et al. 2011a). For the Straja assemblage, Balica et al. (2014a, b) published four U/Pb ages of 501 ± 1.3, 516.7 ± 3.3, 578 ± 1.5, and 621.1 ± 4.5 Ma. According to Liegeois et al. (1996), the Straja assemblage of the Drăgşan Terrane has an intra-oceanic origin (an island arc). In this case, the Făgeţel assemblage probably represents a fragment of the continental margin to which the Straja assemblage docked; the Dobrota assemblage suggests a post-docking sedimentary cover.

The Lainici-Păiuş Terrane basement was divided into a lower "Carbonate-Graphitic" Formation consisting of marble, graphite mica gneiss, amphibolite, and calc-silicate gneiss, and an upper "Quartzitic and Biotite-Gneiss" Formation with minor marble, graphite mica gneiss, amphibolite and calc-silicate gneiss (Berza 1978). In accordance with these rock assemblages, the Lainici-Păiuş Terrane basement suggests an initial passive continental margin origin. However it became an active one later, how is documented by the Tismana, Suşiţa, Novaci, Olteţ, and Arsasca granitoid plutons with ages between 601.0 ± 2.2 and 587.4 ± 1.3 Ma (Table 4) intruding the Lainici-Păiuş Terrane basement (Balintoni and Balica 2012). The deposition age of metasedimentary rocks are older than ca. 600 Ma (the age of Tismana pluton) and younger of ~622 Ma, the age of the youngest detrital zircon grain found in a metaquartzitic sample (Balintoni et al. 2011a).

The initial metamorphic history of the Danubian Domain terranes is associated to the late Neoproterozoic Avalonian–Cadomian orogenic belt. The Drăgşan Terrane shows a medium grade initial metamorphic event, and the Lainici-Păiuş Terrane basement contains low-pressure and high-temperature metamorphic parageneses (Berza and Seghedi 1983). During the Variscan orogeny when the Danubian Domain was an upper plate, several granitoid plutons intruded its basement (Table 5, from Balintoni et al. 2014a). According to Balintoni et al. (2011a), the Danubian Domain terranes have an Avalonian provenance.

Table 4 U/Pb ages of the Cadomian granitoid plutons in the basement of the Lainici-Păiuş Avalonian terrane, the Danubian Domain (South Carpathians)

Terrane	Pluton name	Best age (Ma)
Lainici-Păiuş	Tismana	600.5 ± 4.4
	Suşiţa	588.1 ± 3.1
	Novaci	591.5 ± 4.1
	Olteţ	587.3 ± 2.6
	Arsasca	595.8 ± 7.2

Table 5 U/Pb ages of the Variscan granitoid plutons in the basement of the Drăgşan and Lainici-Păiuş Avalonian terranes, the Danubian Domain (South Carpathians)

Pluton name	Best age (Ma)
Retezat	309.7 ± 5.1
Buta	303.7 ± 2.4
Parâng Latoriţa	285.7 ± 1.8
Parâng Jieţ	297.7 ± 3.4
Culmea Cernei	286.8 ± 4.2
Frumosu	303.4 ± 2.9
Furcatura	316.4 ± 2.9

The Euxinic Terrane in Central Dobrogea (Central Dobrogean Shield)

In central Dobrogea, the basement of the Euxinic Craton crops out and is bounded by the Peceneaga–Camena and the Capidava–Ovidiu faults in north and south, respectively. The Central Dobrogean Shield consists of two tectonic units: the lower Altîn Tepe Group (Krautner et al. 1988a, b) made up of metasediments metamorphosed up to the amphibolite facies, and upper, very low-grade Histria Formation turbidites (Seghedi and Oaie 1995). Balintoni and Balica (2016) consider the two rock assemblages as terranes. The Histria Terrane was interpreted to be Ediacaran based on the medusoid *Nemiana simplex* Palij imprint (Oaie 1992). This age was confirmed by subsequent work of Zelazniewicz et al. (2009) and Balintoni et al. (2011b), both papers reporting detrital zircon ages between 633 and 579 Ma. Regarding the age of the Altîn Tepe Terrane basement, the youngest detrital zircon ages peak of 512 Ma reported by Balintoni and Balica (2016) suggests a maximum late Cambrian deposition age for it. The Histria Terrane does not show a Phanerozoic metamorphic history, and its rock assemblages suggest passive continental margin deposits. The initial metamorphism of the Altîn Tepe Terrane basement was probably Ordovician, connected to the Caledonian history of the Euxinic Craton margin.

The Euxinic Terrane in South Dobrogea (East Moesia)

In South Dobrogea, the basement of the Moesian Platform is covered by sediments and it was intersected only by boreholes. According to Seghedi et al. (2005), the pre-Jurassic Palazu thrust fault brings over the Histria Formation three discordantly superposed sequences of metamorphic rocks with different histories; from down to up these are Ovidiu Group (granite gneiss cut by pegmatite veins), the Palazu Group containing a banded iron formation (BIF), and the Cocoşu Group (volcano-sedimentary association). Balintoni et al. (2014b) reported for the Ovidiu Group granite gneiss two U/Pb Archean ages of 2781 ± 43 and 2895 ± 62 Ma. These are the first Archean ages obtained on the Romanian territory. The Moesian Platform basement was also penetrated by boreholes in the Romanian Plain (metamorphic and magmatic rocks), but no protolith ages were generated until now. Considering all the above information on the Euxinic Craton basement, a rather complex structure and history are evident; this is what characterizes the old continental fragments.

The Tişoviţa Terrane

The Tişoviţa Terrane basement is a mafic–ultramafic rock assemblage described as an ophiolite, and until recently it was considered to indicate the suture between the Drăgşan and Lainici-Păiuş terranes (e.g., Krautner 1996–1997). However, new isotopic data on rocks from its basement and the correspondent Deli Jovan and Zaglavak massifs, south of Danube (Zakariadze et al. 2006; Plissart et al. 2012; Balica et al. 2014b) restricted its age at middle Devonian. Concomitantly, Negulescu et al. (2014) established the Variscan age of the medium- to high-grade metamorphism affecting the Poiana Mraconia tectonic unit basement, which is thrusted over the Tişoviţa Terrane. Because such metamorphic event characterizes the Sebeş-Lotru Terrane basement (Medaris et al. 2003), subducted under the Danubian Domain basement, very probably the Tişoviţa Terrane is a remnant of the Rheic Variscan suture between the Getic Terrane basement and the Danubian margin of the Euxinic Terrane basement.

The North Dobrogean Orogen

The basement of the North Dobrogean Orogen consists of four pre-Alpine terranes, known as Boclugea, Megina, Orliga, and Tulcea (e.g., Seghedi 2012). Because “Tulcea” is also the name of a Cimmerian tectonic unit, we propose the name of “Uzum Bair” Terrane for the metamorphic basement of Tulcea tectonic unit, a place where it crops out.

Boclugea Terrane basement consists of quartzites and phyllites, an assemblage of biotite grade metasediments (Seghedi 2012). The Hamcearca and Chetros granite bodies around 600 Ma intruding the Boclugea Terrane basement (Balintoni et al. 2013) attest its Neoproterozoic age.

Megina Terrane basement is dominated by amphibolites and orthogneisses (Seghedi 2012). Two orthogneiss samples yielded U/Pb ages around 510 Ma (Balintoni et al. 2013); thus, the Megina Terrane basement is younger than the one of the Boclugea Terrane. Its initial metamorphism is probably early Ordovician, because it is covered by Paleozoic sediments beginning with the Priopcea quartzites, apparently of middle to late Ordovician age (Balintoni et al. 2013).

Orliga Terrane basement consists of micaceous quartzites with subordinate paragneiss, metabasic rock, and marble, the metamorphic degree reaching the sillimanite zone (Seghedi 2012). An Orliga metaquartzite yielded a group of U/Pb ages on detrital zircon around 510 Ma (Balintoni et al. 2010c). This is the maximum possible age of its rock assemblages. The initial metamorphism of Orliga basement happened during the Variscan thermo-tectonic events (Balintoni et al. 2010c, 2013).

The age of the *Tulcea Terrane* basement is not known, but the Agighiol Granite of ca. 300 Ma (unpublished data) proves it was involved in the Variscan orogeny. The pre-Alpine history of the North Dobrogean Orogen basement components is complicate and poor understood until present. The above-presented data attest Avalonian–Cadomian (Boclugea), Caledonian (Megina), and Variscan (Orliga) metamorphic events. During the Variscan orogeny, the Boclugea, Megina, and Uzum Bair terranes were altogether in the position of upper plate and the Orliga Terrane in that of lower plate. Detrital zircon data (Balintoni and Balica 2016) attest a peri-Amazonian provenance for all the North Dobrogean Orogen pre-Alpine terranes.

The East European Craton

In Romania, the East European Craton components are presented by the *Scythian* and *Moldavian* platforms. The first one is located southward of the Moldavian Platform, and it is separated from this unit by the Baimaklia Fault (Seghedi 2012). At surface, the Sf. Gheorghe Fault constitutes its boundary with the North Dobrogean Orogen (Seghedi 2012). Beneath the North Dobrogean Orogen it is continuing until the Peceneaga–Camena Fault (Săndulescu 1994). The basement of the Scythian Platform in this region consists of magmatic rocks, granite, diorite, and gabbro that yielded K/Ar ages of 790 and 640–620 Ma (Neaga and Moroz 1987). The magmatic basement is unconformably overlain by undeformed Vendian deposits. According to this description, no similarities exist between the basements of the Scythian Platform and the North Dobrogean Orogen components.

The *Moldavian Platform* represents the western part of the East European Craton. It is dipping by normal faults, beneath the East Carpathians thrusting front; thus its margin is not known. The basement of the Moldavian Platform was intercepted by drillings. A description of its basement is provided by Ionesi (1994): plagioclase paragneiss with almandine and sillimanite, or biotite and hornblende; orthogneiss with oligoclase and microcline, sometimes with ocular structure; pink granite with muscovite and biotite. Balintoni et al. (2014c) sampled two drill cores and dated the orthogneisses by zircon U/Pb LA ICP MS method. One sample yielded a Concordia age of 2071 Ma and the other a Concordia upper intercept of 2072 Ma. These ages are bracketed by the interval indicated by Kuznetsov et al. (2010) for the undifferentiated complexes surrounding the Archean nuclei of Sarmatia.

The Pre-Permian Covers of the Pre-Alpine Terranes Basement

The Foreapulian Terrane Components

In the Apuseni Mountains is known a pre-Permian cover on the Biharia Terrane basement, called the Păiuşeni Lithogroup (Balintoni 1997). It is very low-grade metamorphosed and consists of metaconglomerates and metapsamites, subordinate metapelites, alternating with metamagmatites that vary from basic to acid rocks. Their age is post-Ordovician and can be associated with a Variscan active plate margin.

The Getic Terrane Components in East Carpathians

In the East Carpathians, there is a pre-Permian cover on the Bretila Terrane basement (Krautner and Krautner 1970). Lithologically, it consists of basic metavolcanics and metasediments that include conglomerates and phyllitic rocks, white and black quartzites, and limestones. Balintoni (1997) called this sequence the Rodna Lithogroup. As in the Apuseni Mountains, the Rodna Lithogroup is the cover of a Variscan active plate margin.

The Getic Terrane Components in South Carpathians

In the South Carpathians, Balintoni (1997) describes the pre-Permian Banat Lithogroup, supposedly in genetic relationship with the Sebeş-Lotru Terrane. The Banat

Lithogroup shows two contrasting rock assemblages: one consisting of acid and basic metavolcanics, including metaultrabasites, whereas the other one includes (meta)sedimentary rocks. At least partially the first assemblage can be in a Variscan tectonic connection with the basement of the Sebeş-Lotru Terrane.

The Euxinic Terrane Components

The Danubian Domain in South Carpathians

A synthesis of the interesting fossiliferous Ordovician–Carboniferous formations transgressive on the Drăgşan and Lainici-Păiuş terranes basement was given by Berza and Iancu (1994). Balintoni (1997) described them as the Timiş and Jiu lithogroups, both transgressive on the Drăgşan and Lainici-Păiuş terranes, respectively. The Timiş Lithogroup includes basic volcanics that are missing in the Jiu Lithogroup. In general lines, the Timiş Lithogroup suggests some characteristics of a rift sedimentation, while the Jiu Lithogroup consists dominantly of platform detrital rocks, including limestones.

The Moesian Platform

According to Ionesi (1994), the first sedimentary cycle of the Moesian Platform begins in Cambrian and ends in late Carboniferous. The sea deposits (sandstone, clay, shale, limestone, dolomite) totalize 6500 m in thickness. It is worth noting the sedimentary gap between Ordovician and middle Silurian, the presence of some Silurian basic tuffs, and a quite significant sea retreat during the late Carboniferous–early Permian.

The North Dobrogean Orogen Components

Within the North Dobrogean Orogen are known Paleozoic fossiliferous deposits resting on the Megina, Boclugea, and Uzum Bair terranes basement (see description bellow).

The Megina Terrane Paleozoic

The Megina Terrane unmetamorphosed Paleozoic begins with the middle to late Ordovician Priopcea quartzites (Balintoni et al. 2013), followed by the Silurian Cerna Formation consisting of dark gray limestones, shales, and black argillites, and the lower Devonian Bujoare Formation, formed by a succession of white and gray limestones, quartzitic sandstones, and black slates (Seghedi 2012).

The Common Deposits for the Megina and Boclugea Terranes

These deposits, described under the name of Carapelit Formation, have a late Paleozoic age and are contemporaneous with the Variscan tectonics and magmatism. It consists of lower, gray alluvial deposits, followed by continental red beds and an upper volcano-sedimentary succession (Seghedi 2012). On the Boclugea Terrane are not known pre-Carapelit deposits.

The Uzum Bair Terrane Paleozoic

According to Seghedi (2012), the Uzum Bair basement includes the following Paleozoic formations: Dealul Horia, Rediu, and Beştepe. Dealul Horia Formation consists of sandstones dominates proximal turbidites, a succession of fine-grained, greenish sandstones and siltstones. The deposits are ascribed to the Ordovician based on their geometric position below the Silurian Rediu Formation, which comprises a lower, siliceous member and an upper member of black or gray slates. The Beştepe Devonian Formation consists of a lower member of siliceous rocks (cherts, siliceous shales with scarce, thin pelagic limestone interbeds) and an upper member of distal turbidites.

The Scythian Platform

In accordance with data discussed by Seghedi (2012), over the Vendian–lower Cambrian assemblages follow a succession dominated by siltstones and mudstones with interbedded sandstones and limestones in the upper part (Ordovician–middle Silurian). The lower Devonian includes fine-grained clastics, and the acid tuffs are conformably interbedded with the sandstones. The middle Devonian–lower Carboniferous incorporates evaporate-bearing dark carbonate successions, often bituminous. The upper Devonian is missing, and the lower Carboniferous consists of massive limestones and dolomites.

The Moldavian Platform

Ionesi (1994) mentions a late Vendian–Devonian megacycle on the Moldavian Platform. During this time interval, several depositional gaps exist, especially between late Cambrian and late Ordovician, and also during the early Silurian. The Silurian–Devonian deposits are mainly carbonates and at their upper erosional part developed a karst landscape, concealed today by younger sediments.

Pre-Alpine Terranes and Pre-Alpine Orogenies

The origin and evolution of the pre-Alpine terranes are connected to the great orogenies associated to the Plate Tectonics. We will discuss shortly such events recorded by their pre-Alpine terranes.

The Foreapulian Terrane

The three pre-Alpine terranes from the basement of the Foreapulian Terrane originated in the North-African Ordovician orogen, somewhere in the northeastern African part of the Gondwana (Balintoni and Balica 2013; Balintoni et al. 2014a). The Someş, Biharia, and Baia de Arieş terranes migrated toward Laurussia during the late Devonian–Carboniferous periods and docked to Laurussia during the Variscan orogeny. The Someş and Biharia terranes were the upper plate and Baia de Arieş Terrane suffered continental underplating.

The Getic Terrane in East Carpathians

All four pre-Alpine terranes from this region have a similar pre-Alpine history with those from the Apuseni Mountains. During the Variscan orogeny, the Bretila Terrane was the upper plate, and the Tulgheş, Negrişoara, and Rebra terranes suffered underplating. The Variscan nappes were preserved very well in the basement of East Carpathians (Balintoni 1997).

The Getic Terrane in South Carpathians

In South Carpathians, the lower Lotru Unit of the Sebeş-Lotru Terrane shows an Avalonian–Cadomian origin and a North-African provenance. The upper Cumpăna Unit, Ordovician in age, has its origin in the North-African Ordovician orogen, and the superposition between the two units probably happened during the Variscan orogeny. The other pre-Alpine terranes from the Getic Domain originated as parts of the North-African Ordovician orogen and migrated toward Laurussia during the Variscan orogeny. The Sebeş-Lotru Terrane suffered continental subduction under the Danubian Domain components and eclogitic grade metamorphism (Medaris et al. 2003). Leaota Terrane (and probably Caraş Terrane too) suffered underplating beneath the Sebeş-Lotru Terrane back margin. The docking history of the Făgăraş and Padeş terranes is not yet well understood.

The Danubian Domain in South Carpathians

The Drăgşan and Lainici-Păiuş pre-Alpine terranes represent Avalonian entities of Pan-African age. The Drăgşan Terrane has an intra-oceanic origin and the Lainici-Păiuş Terrane originated along a passive continental margin that became active later. They probably docked gently to the Euxinic Craton margin during the Caledonian orogeny and were involved as an upper plate in the Variscan orogeny. Many granite bodies intruded the Danubian basement during that time.

The Moesian Platform

In the East Moesia, the Ovidiu Group originated during the late Archean global orogenic events. The Palazu Group was probably metamorphosed within the 1777–1620 Ma interval according to data presented by Giuşcă et al. (1967). These ages correspond to the Rio Negro–Juruena orogenic province in Amazonia (Cordani and Teixeira 2007). The Cocoşu Group shows a late Neoproterozoic Pan-African initial history, as the Histria Formation (Krautner et al. 1988a, b). Regarding the Paleozoic sedimentary cover, the gap between Ordovician and the middle Silurian, as well as the presence of the basic tuffs in Silurian, could be associated with the Danubian terranes docking. The sea retreat during the late Carboniferous–early Permian records the Variscan events all around Euxinic margin.

The Central Dobrogean Shield

The Histria Terrane does not record post-Pan-African thermo-tectonic events, and its overthrusting on the Altîn Tepe Terrane is probably Variscan. The Palazu thrust, certainly pre-Jurassic in age, can be also pre-Ordovician, because westward of Danube it is covered by Paleozoic formations beginning with the Ordovician (Krautner et al. 1988a, b). Therefore, the Palazu thrust established the final Pan-African relationship between the South Dobrogea basement and the Histria Terrane.

The Scythian Platform

The pre-Vendian magmatic rocks from the Scythian Platform basement suggest a Neoproterozoic orogenic history. The lower Devonian acid tuffs probably indicate Caledonian events in relationship to the Avalonian terranes docking. The Scythian Platform was also involved in the Variscan orogenic events, as documented by the Permian magmatic events (Seghedi 2012).

The Moldavian Platform

The basement of the Moldavian Platform records orogenic events around 2100 Ma and during the Mesoproterozoic, well known within the EEC basement (Kuznetsov et al. 2010).

Alpine Sediments Anterior to the Alpine Terranes Amalgamation

The Variscan orogeny was followed by a period of orogenic collapse and crustal extension when the Alpine terranes have been individualized, with oceanic tracts between them.

The Foreapulian Terrane

The Permian deposits are characterized by basic and acid volcanics in alternation with detrital sediments. During Triassic, the sedimentation turns suddenly from detrital (at the beginning), to one essentially limy during the rest of the period. The carbonate sequence can reach in some regions 1500 m in thickness (Ianovici et al. 1976). The Jurassic begins with detrital sediments, but again, in its greatest part, is carbonatic, with different intercalations of pelitic rocks. In some areas of Romania (e.g., Pădurea Craiului Mountains), a very well-developed late Jurassic paleokarst landscape is known (Bleahu 1972, 1989). The depressions (dolines, shafts, etc.) are filled with early Cretaceous bauxites (Pop and Mîrza 1977). The Neocomian–Turonian formations (pre-Gosau) that can reach a cumulate thickness of 3000 m are also dominantly carbonaceous. Toward the upper part only, the detrital intercalations become more and more significant.

The Getic Terrane in East Carpathians

On the Getic Terrane in East Carpathians, the sedimentation begins in Triassic, with detrital deposits (early part) and mainly limestone afterward. The Jurassic is characterized by impure carbonate deposits, along with clay and jasper toward the final part. During the Neocomian, the arenites and flysch deposits (Lunca Formation) dominate.

The Getic Terrane in South Carpathians

In Banat area, a sandstone–conglomerate formation with coal of late Carboniferous–early Permian age was documented. The Triassic, essentially carbonates, appears only locally. The lower Jurassic is equally terrigenous and limy. The rest of Jurassic and the lower Cretaceous are dominated by limestones.

The Danubian Domain in South Carpathians

The upper Carboniferous–Permian deposits have a detrital and volcano-sedimentary aspect. The Triassic sediments are missing, whereas the Jurassic begins with Gresten-type facies, followed by dominant carbonate deposits that continue also during the Cretaceous.

The Moesian Platform

The Permo-Triassic cycle is predominantly detrital with volcanic intercalations. The Jurassic–Cretaceous cycle begins with detrital sediments in the lower Jurassic, followed by carbonate deposits during the rest of the interval.

The Central Dobrogean Shield is covered in places by carbonate Jurassic deposits. In the North Dobrogean Orogen, some of the Cimmeric tectonic units preserve pre-tectonic sediments as for example the Tulcea Nappe. The sedimentation begins with a detrital lower Triassic, followed by a carbonate middle Triassic and again a detrital upper Triassic that continues until the Oxfordian.

The Scythian Platform hosts a volcano-sedimentary formation of Permian–lower Triassic. This is followed by a limy middle Triassic, a detrital upper Triassic, and a carbonate Jurassic with a detrital sequence in its middle part.

The Moldavian Platform has no pre-Valanginian Permo-Mesozoic deposits.

The Alpine Terranes Amalgamation

The Apuseni Mountains, the Carpathian chain, and the North Dobrogean Orogen are characterized by nappe structures. These nappes were generated during the Alpine terranes amalgamation, including the oceanic tracts between them. The nappe ages are late Jurassic to early Cretaceous in the North Dobrogean Orogen and middle Cretaceous to Tertiary in the Apuseni Mountains and the Carpathian chain. For the Apuseni Mountains and Carpathian chain, Săndulescu (1984) proposed a nappe classification based either on their generation time (Cretaceous or Tertiary) and position against the foreland (marginal to internal) or on their provenance place. In this scheme, the Cretaceous nappes are Dacides and the Tertiary ones Moldavides.

The Dacides are divided in marginal, external, median, and internal. After the provenance place, there are Transilvanides and Pienides. Therefore, the Foreapulian margin nappes are Internal Dacides; the sheared Getic margin represents the Median Dacides; the Severin Ocean provided the External Dacides; the Euxinic Craton margin in South Carpathians furnished the Marginal Dacides; the Tertiary nappes in the East Carpathians are the Moldavides; from the Main Tethys originated the Transilvanides and Pienides.

In contrast, Balintoni (1997) suggested a nappe classification according only to their provenance: Internal Dacides = Apusenides; Median Dacides = Getides; External Dacides = Severinides; Marginal Dacides = Danubian Euxinides; Moldavides = Perimoldavides; to these, the Transilvanides and Pienides are also included. In the North Dobrogean Orogen were described the Neocimmeric Măcin, (Consul), Niculițel, and Tulcea nappes.

The Transilvanides, Pienides, and Severinides

The Transilvanides and Pienides in their greatest part are remnants of the oceanic tract between the Foreapulian and Getic terranes. The Severin Ocean separated the Getic Terrane from the foreland terranes. The *Transilvanides* are known in the Apuseni Mountains, where overthrust the Apusenides, and in the East Carpathians, where they sit over the eastern Getides. They consist of Mesozoic magmatic rocks including ophiolites, flysch formations, and thick sequences of carbonate rocks. The *Pienides* outcrop northward of the North Transilvanian Fault (Săndulescu 1984) shows both Cretaceous and Tertiary deformations, respectively; flysch and large carbonate blocks (klippen) dominate their lithology. The *Severinides* spread beneath the Getides from the Maramureş County until the Banat area and consist especially of volcanics and flysch formations.

The Alpine Post-amalgamation Sediments

According to Săndulescu (1984), such sequences include the post-tectogenetic covers, the foreland basins and the intra Carpathian basins. The rising of the Alpine chains restricted the sedimentation areas to basins between the mountainous branches, in the front of Carpathian chain, or within the orogen components. The post-tectogenetic covers followed the nappe generation; they were again deformed when the tectogenesis resumed. The foreland basin formed in the front of the Carpathian chain on the overthrust foreland margin. Between the Apuseni Mountains and the Carpathian chain developed the Transylvanian and the Pannonian basins, with several gulfs on the western side of the Apuseni Mountains. Extensive salt deposits of Badenian age in the Transylvanian Basin and of Burdigalian and Badenian age in foreland basin are known. Most of these salt bodies are characterized by the particular diapir tectonics. In the Transylvanian Depression, beneath the Transylvanian basin deposits, a Paleogene–Oligocene post-tectogenetic cover with Eocene and Oligocene limestones has been documented.

The Alpine Magmatism

In the Banat area and in the Apuseni Mountains (W Romania), an intense late Cretaceous volcanic activity (intrusive and extrusive) is known under the name of "banatitic" magmatism. However, its geotectonic context is still controversial. Along the East Carpathians, in their western part, a Tertiary volcanic chain connected to the subductional history of the Severin Ocean lithosphere was emplaced. A Tertiary magmatism also affected the Apuseni Mountains' basement. It has an extensional genesis associated to the clockwise rotation of the Foreapulian Terrane around the western spur of the Euxinic Craton.

References

Balica C, Balintoni I, Berza T, Hann HP (2014a) New data on the origin of the Drăgşan Group, South Carpathians, Romania. In: Bucur II, Săsăran E, Har N (eds) "Ion Popescu Voiteşti" Ann Scientific Session Abs vol, Cluj University Press, pp 4–5

Balica C, Balintoni I, Berza T (2014b) On the age of the Carpathian-Balkan pre-Alpine ophiolites in SW Romania, NE Serbia and NW Bulgaria. In: Proceedings of XX CBGA Congress, Bull Shk Gjeol 1/2014 Special Issue, Tirana, Albania (24–26 September 2014), pp 196–197

Balintoni I (1997) Geotectonica terenurilor metamorfice din Romania. Ed Carpatica, Cluj-Napoca, p 178

Balintoni I, Balica C (2012) Avalonian, Ganderian and East Cadomian terranes in South Carpathians, Romania, and Pan-African events recorded in their basement. Min Petrol 107:709–725

Balintoni I, Balica C (2013) Carpathian peri-Gondwanan terranes in the East Carpathians (Romania): a testimony of an Ordovician, North-African orogeny. Gondwana Res 23:1053–1070

Balintoni I, Balica C (2016) Peri-Amazonian provenance of the Euxinic Craton components in Dobrogea and of the North Dobrogean Orogen components (Romania): a detrital zircon study. Precambrian Res 278:34–51

Balintoni I, Balica C, Ducea M, Chen F, Hann HP, Sabliovschi V (2009) Late Cambrian-Early Ordovician Gondwanan terranes in the Romanian Carpathians: a zircon U/Pb provenance study. Gondwana Res 16:119–133

Balintoni I, Balica C, Ducea MN, Hann HP, Sabliovschi V (2010a) The anatomy of a Gondwanan terrane: the Neoproterozoic-Ordovician basement of the pre-Alpine Sebeş-Lotru composite terrane (South Carpathians, Romania). Gondwana Res 17:561–572

Balintoni I, Balica C, Ducea MN, Zaharia L, Chen F, Cliveţi M, Hann HP, Qiuli L, Ghergari L (2010b) Late Cambrian-Ordovician northeastern Gondwanan terranes in the basement of the Apuseni Mountains, Romania. J Geol Soc London 167:1131–1145

Balintoni I, Balica C, Seghedi A, Ducea MN (2010c) Avalonian and Cadomian terranes in North Dobrogea, Romania. Precambrian Res 182:217–229

Balintoni I, Balica C, Ducea M, Stremţan C (2011a) Peri-Amazonian, Avalonian-type and Ganderian-type terranes in the South Carpathians, Romania: the Danubian domain basement. Gondwana Res 19:945–957

Balintoni I, Balica C, Seghedi A, Ducea MN (2011b) Peri-Amazonian provenance of the Central Dobrogea terrane (Romania) attested by U/Pb detrital zircon age patterns. Geol Carpat 62:299–307

Balintoni I, Balica C, Ducea MN, Hann HP (2014a) Peri-Gondwanan terranes in the Romanian Carpathians: a review of their spatial distribution, origin, provenance, and evolution. Geoscience Front 5:395–411

Balintoni I, Balica C, Munteanu M (2014b) The age of the East European Craton (EEC) basement in the Moldavian Platform (Romania). In: Bucur II, Săsăran E, Har N (eds) "Ion Popescu Voiteşti" Ann Sci Session Abs Vol, Cluj University Press, pp 8–9

Balintoni I, Balica C, Seghedi A, Munteanu M (2014c) The Ovidiu Group in south Dobrogea, the first proved Archean suite in the Romanian Basement. In: Bucur II, Săsăran E, Har N (eds) Ion Popescu Voiteşti, Ann Sci Session Abs Vol, Cluj University Press, pp 6–7

Balintoni I, Balica C, Seghedi A (2013) New data regarding the pre-Alpine Geology of North Dobrogea (in Romanian). In: Bucur II, Săsăran E, Har N (eds) "Ion Popescu Voiteşti" Ann Sci Session Abs Vol, Cluj University Press, pp 4–7

Berza T (1978) Mineralogic and petrographic study of the Tismana granitoid massif (in Romanian). DS Inst Geol Geofiz 53:5–176

Berza T, Iancu V (1994) Variscan events in the basement of the Danubian nappes (South Carpathians). Rom J Tect Reg Geol 75 (2):93–104

Berza T, Seghedi A (1975) On the presence of kyanite in the amphibolite complex of the Drăgşan series from the Motru Basin (in Romanian). DS Inst Geol Geofiz LXI:11–20

Berza T, Seghedi A (1983) The crystalline basement of the Danubian units in the Central South Carpathians: constitution and metamorphic history. An Inst Geol Geofiz 61:15–22

Bleahu MD (1972) Karst of Rumania. In: Herak M, Stringfield VT (eds) Karst. Important karst regions of the Northern Hemisphere, Elsevier, Amsterdam, pp 341–353

Bleahu M (1989) Paleokarst of Romania. In: Bosak P, Ford DC, Glazek J, Horacek I (eds) Paleokarst. A systematic and regional review, Elsevier, Amsterdam, pp 231–251

Burchfiel BC, Bleahu M (1976) Geology of Romania. GSA Spec Paper 158, p 83

Cordani UG, Teixeira W (2007) Proterozoic accretionary belts in the Amazonian Craton. Geol Soc Amer Mem 200:297–320; Hatcher RDJ, Carlson MP, McBride JH, Martinez Catalan JR (eds) 4-D Framework of Continental Crust

Giuşcă D, Ianovici V, Mânzatu S, Soroiu M, Lemne M, Tănăsescu A, Ioncică M (1967) On the absolute age of the crystalline formations from the Carpathian Orogen foreland (in Romanian). St Cerc Geol Geofiz Geogr 12:287–297

Ianovici V, Borcoş M, Bleahu M, Patrulius D, Lupu M, Dimitrescu R, Savu H (1976) Geology of the Apuseni Mountains. Ed Acad RSR, Bucuresti (in Romanian)

Ionesi L (1994) Geology of the platform units and the North Dobrogean Orogen. Ed Tehnică, Bucureşti (in Romanian)

Kräutner HG (1988) Precambrian in the Carpathians: east Carpathians. In: Zoubek V, Cogné J, Kozhoukharov D, Kräutner HG (eds) Precambrian in younger fold belts: European Variscides, the Carpathians and Balkans. Wiley, pp 625–638

Kräutner HG (1996–1997) Alpine and pre-Alpine terranes in the Romanian Carpathians and Apuseni Mountains. In: Papanikolaou D (ed) Terrane Mmaps and terrane descriptions. IGCP Project no. 276. Ann Geol des Pays Helleniques, Athens, pp 331–400

Kräutner HG, Berza T, Dimitrescu R (1988) Precambrian in the Carpathians: south Carpathians. In: Zoubec V, Cogne J, Kozhoukharov D. Krautner HG (eds) Precambrian in younger fold belts: European Variscides, the Carpathians and Balkans. Wiley, pp 639–660

Kräutner HG, Krautner F (1970) Crystalline formations from the northern slope of the Rodna Massif. DS Com Stat Geol, Inst Geol LV(1):173–196 (in Romanian)

Kräutner HG, Mureşan M, Seghedi A (1988) Precambrian of Dobrogea. In: Zoubec V, Cogne J, Kozhoukharov D, Kräutner HG (eds) Precambrian in younger fold belts: European Variscides, the Carpathians and Balkans. Wiley, pp 361–379

Kuznetsov NB, Natapov LM, Belousova EA, O'Reilley SY, Griffin WL (2010) Geochronological, geochemical and isotopic study of detrital zircon suites from late Neoproterozoic clastic strata along the NE margin of the East European Craton: implications for plate tectonic models. Gondwana Res 7:583–601

Liegeois JP, Berza T, Tatu M, Duchesne JC (1996) The Neoproterozoic Pan-African basement from the Alpine Lower Danubian nappe system (South Carpathians, Romania). Precambrian Res 80:281–301

Medaris G, Ducea M, Ghent E, Iancu V (2003) Conditions and timing of high-pressure Variscan metamorphism in the South Carpathians, Romania. Lithos 70:141–161

Mutihac V, Ionesi L (1974) Geologia României. Ed Tehnică, Bucureşti

Nance RD, Murphy JB, Strachan RA, Keppie JD, Gutierez Alonso G, Fernández-Suárez J, Quesada C, Linnemann U, D'Lemos RS, Pisarevsky SA (2008) Neoproterozoic-early Palaeozoic tectonostratigraphy and palaeogeography of the peri-Gondwanan terranes: Amazonian v. West African connections. In: Ennih N, Liegeois JP (eds) The boundaries of the West African craton. Geol Soc London, pp 345–383

Neaga VI, Moroz VF (1987) Die Jungpalaozoischen Rotsedimente im Sudteil des Gebietes zwischen Dnestr und Prut. Z Angew Geol 33 (9):238–242

Negulescu E, Săbău G, Ducea MN, Massonne HJ (2014) Sm-Nd garnet and U-Th-Pb monazite ages in the basement units of the South Carpathians: new data and a consistency check. In: Proceedings of XX CBGA Congress, Bull Shk Gjeol 1/2014 Special Issue Tirana, Albania (24–26 September 2014), pp 212–215

Oaie G (1992) Traces of organic activity in the Greenschist Series of central Dobrogea Romania. St Cerc Geol Acad Rom 37:77–81

Oncescu N (1957) Geologia Republicii Populare Române. Ed Tehnică, Bucureşti

Pharaoh TC (1999) Palaeozoic terranes and their lithospheric boundaries within the Trans-European Suture Zone (TESZ): a review. Tectonophysics 314:17–41

Plissart G, Diot H, Monnier C, Mărunţiu M, Berger J (2012) Relationship between a syntectonic granitic intrusion and a shear zone in the Southern Carpathian-Balkan area (Almaj Mountains, Romania): implications for late Variscan kinematics and Cherbelezu granitoid emplacement. J Struc Geol 39:83–102

Pop G, Mîrza I (1977) Les paléokarst Éocrétacé des Monts Pădurea Craiului et sa signification dans la mécanisme de genèse des bauxites. Bull Sci Géol Strasbourg 30:51–58

Săndulescu M (1984) Geotectonics of Romania. Ed Tehnică, Bucureşti (in Romanian)

Săndulescu M (1994) Overview on Romanian Geology. Rom J Tect Reg Geol 75(2):3–16

Schmid SM, Bernoulli D, Fügenschuh B, Maţenco L, Schefer S, Schuster R, Tischler M, Ustaszewski K (2008) The Alpine-Carpathian-Dinaridic orogenic system: correlation and evolution of tectonic units. Swiss J Geos 101:139–183

Seghedi A (2012) Palaeozoic formations from Dobrogea and pe-Dobrogea-an overview. Turk J Earth Sci 21:669–721

Seghedi A, Oaie G (1995) Palaeozoic evolution of North Dobrogea. In: Săndulescu M, Oaie G, Grădinaru E, Rădan S (eds) Field guidebook, central and north Dobrogea, IGCP Project no. 369 "Comparative evolution of Peri-Tethyan Rift Basins", Mamaia, p 75

Seghedi A, Berza T, Iancu V, Mărunţiu M, Oaie G (2005) Neoproterozoic terranes in the Moesian basement and in the Alpine Danubian nappes of the South Carpathians. Geol Belgica 8:4–19

Zakariadze G, Bayanova TB, Karpenko SF (2006) Age, composition and origin of the paleo-oceanic gabbros of Deli Jovan massif (NE Serbia). Ofioliti 31:223–224

Zelazniewicz A, Bula Z, Fanning CM, Seghedi A, Zaba J (2009) More evidence on Neoproterozoic terranes in Southern Poland and southeastern Romania. Geol Quart 53:93–124

Karst and Caves of Romania: A Brief Overview

Bogdan P. Onac and Cristian Goran

Abstract

Romania is not regarded as a country with an overabundance of karst resources, although it is home of many unique karst landscapes and caves as documented in the chapters of this book. Karst-prone rocks cover at least 5500 km^2 (Fig. 1) representing ~2.3% of the total area of the country (238.397 km^2). Worth noting is that the low percentage is mainly because extensive areas in which karst forming rocks exist, are covered by younger, Pliocene, and Quaternary age deposits.

Keywords

Karst • Non-carbonate karst • Caves • Speleogenesis Romania

Introduction

Romania is not regarded as a country with an overabundance of karst resources, although it is home of many unique karst landscapes and caves as documented in the chapters of this book. Karst-prone rocks cover at least 5500 km^2 (Fig. 1) representing ~2.3% of the total area of the country (238,397 km^2). Because some authors considered only the soluble rocks (limestone, salt, gypsum) while others also included metamorphosed and non-carbonate lithologies, the percentage above varies from one publication to another, and will likely change in the future as more research is in progress (Bleahu and Rusu 1964; Orghidan et al. 1965; Senco 1968; Bleahu 1972; Ponta 1998; Onac and Constantin 2004; Bălteanu et al. 2012). Worth noting is that the low percentage is mainly because extensive areas in which karst forming rocks exist are covered by younger, Pliocene, and Quaternary age deposits (Bleahu 1972).

This chapter briefly reviews the general climatic conditions of Romania, highlighting their influences upon karst processes. It is then followed by a presentation of carbonate and non-carbonate lithologies based on which the distribution and types of karst are exemplified. This chapter also introduces the reader to the major speleogenetic processes and cave types documented in Romania.

Environmental Conditions

Climate controls the dynamics of karst processes in many ways. First, it is the amount of precipitation that matters in karst and cave development because the higher the rainfall amount is, the greater the dissolution of karst rocks will be. Second, the temperature modulates the availability of water that effectively contributes to the capacity for dissolution via processes such as evaporation and freezing. Both precipitation and temperature play a major role in the type of vegetation cover, which in turn is responsible for the amount of biogenic carbon dioxide produced in soil. The presence of this compound is crucial in speleogenesis and also responsible for modeling buried karst rocks. Climatically, warm and wet conditions would promote higher biological activity and hence, more suitable for karst development.

Located in east-central Europe, Romania is characterized by a temperate and continental climate. However, the presence of the Carpathian range, which acts as a barrier for the atmospheric flow, greatly modulates the precipitation and temperature regime in different parts of the country as shown in Fig. 2 (Sandu et al. 2008; Dumitrescu and Bîrsan 2015).

B. P. Onac (✉)
Karst Research Group, School of Geosciences, University of South Florida, 4202 E. Fowler Ave., NES 107, Tampa, FL 33620, USA
e-mail: bonac@usf.edu

B. P. Onac
"Emil Racoviță" Institute of Speleology, Romanian Academy, Clinicilor 5, 400006 Cluj-Napoca, Romania

C. Goran
"Emil Racoviță" Institute of Speleology, Romanian Academy, Calea 13 Septembrie 11, Sector 5, 050711 Bucharest, Romania
e-mail: cristian.goran@gmail.com

G. M. L. Ponta and B. P. Onac (eds.), *Cave and Karst Systems of Romania*,
Cave and Karst Systems of the World, https://doi.org/10.1007/978-3-319-90747-5_4

Fig. 1 Karst areas of Romania. The distribution of limestone areas is modified from Orghidan et al. (1965) and Senco (1968). The occurrences of karst on evaporate rocks are used with permission from Tîrlă (2018)

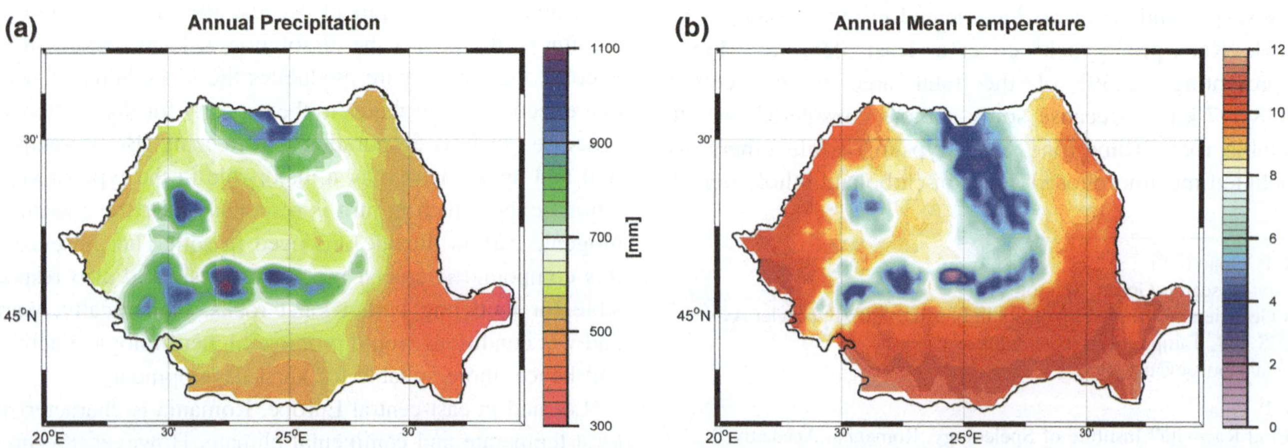

Fig. 2 Mean annual precipitations and temperature in Romania. Data from Dumitrescu and Bîrsan (2015), available at https://doi.pangaea.de/10.1594/PANGAEA.833627

For instance, the Atlantic air masses influence only the western and central (inner Carpathian basin) parts where rainfall is higher, whereas across the Carpathians (east), climate is more continental. The southern part is under Mediterranean airflow influences, thus more maritime climate conditions prevail (mild winters).

Climatically, the major karst areas of Romania are identified as Dfa (wet, warm continental), Dfb (wet, temperate continental), and Dfc (cool continental) on the Köppen climate classification map (Fig. 3). The natural vegetation of the karst regions varies according to elevation (less with latitude) differences. The low land areas are covered by deciduous forest composed predominantly of oak or/and mesophytic (beech, basswood, hazel, maple, etc.) association. Fir forests develop on the limestone escarpments in the SW karst of Romania.

From the perspective of this book, illustrating the distribution of the mean annual temperatures and precipitations across Romania (Fig. 2), as well as the Köppen climate subtypes (Fig. 3), should be enough for readers to understand the relationships between climate conditions and the intensity of karst processes.

Geologic Settings

The previous chapter (see Balintoni 2018) abundantly illustrated the long and complex geologic history of Romanian territory, but since the karst-prone rocks received limited attention, is now the time to address this topic. Karst and caves are widespread in all major geomorphological units of Romania and develop on a variety of lithologies (Goran 1983; Onac and Constantin 2004; Ponta and Ursu 2017), which are further divided into carbonate and non-carbonate rocks, and discussed below along with their tectonic settings.

Carbonate Rocks

From a tectonic point of view, the carbonate units occur in the following geologic settings (Balintoni personal communication): (1) pre-Alpine terranes basement, largely represented within the terranes of passive continental margin origin [e.g., Rebra and Negrişoara in the East Carpathians (Rodna Mountains, Ţibău area, Bistriţei Mountains and south of the

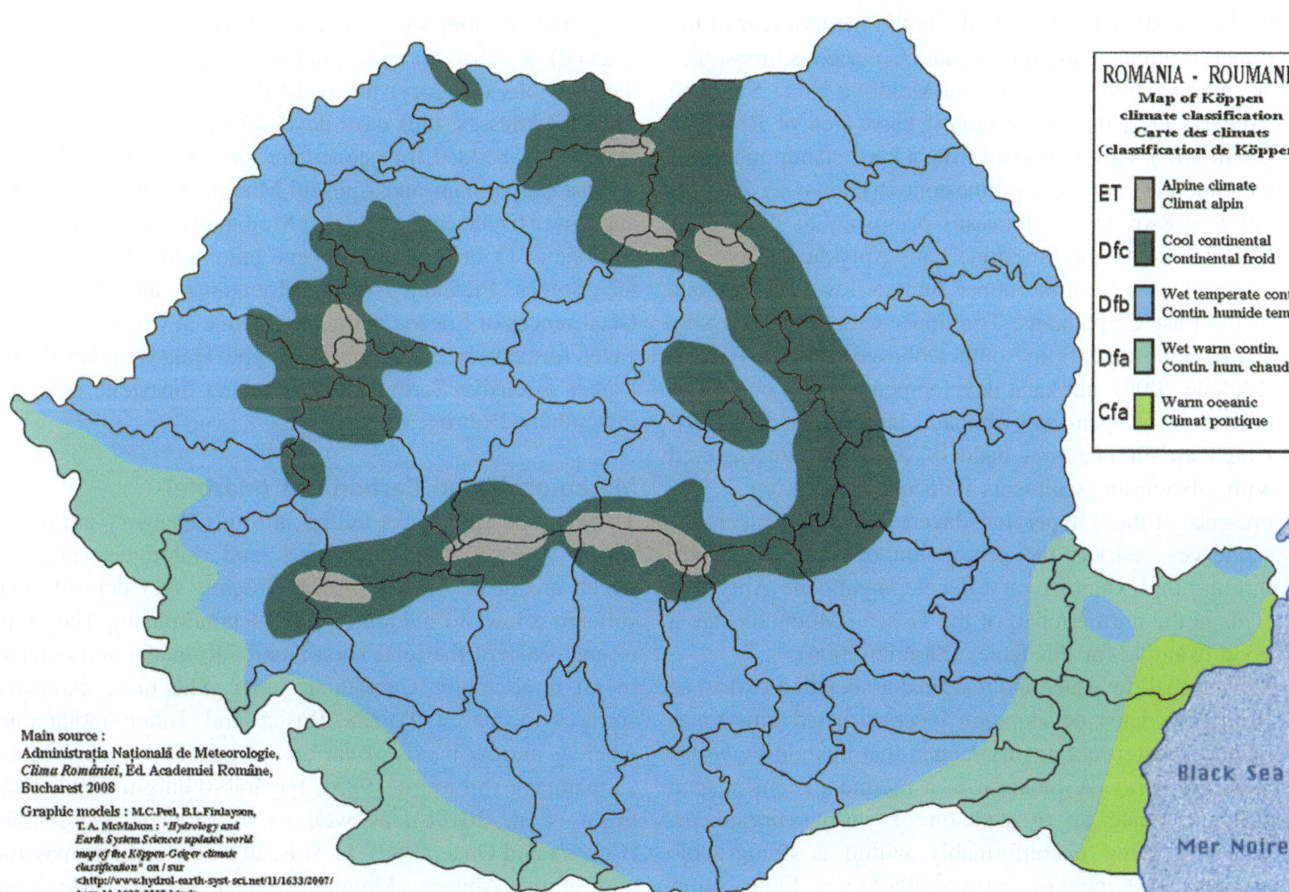

Fig. 3 Köppen-based climatic classification of Romanian territory. Available from https://commons.wikimedia.org/wiki/File:Romania_map_of_Köppen_climate_classification.png, under CC BY-SA 4.0 license

Gheorgheni town area)]; Făgăraş Terrane in South Carpathians (Făgăraş Mountains and especially Poiana Ruscă Mountains); Baia de Arieş Terrane in the Apuseni Mountains and Preluca Massif; Lainici-Păiuş Terrane in the South Carpathians Danubian Domain), (2) pre-Permian cover of the pre-Alpine terranes (Rodna and Păiuşeni lithogroups in Rodna and Apuseni Mountains, respectively), (3) Alpine sediments anterior to the Alpine terranes amalgamation (comprises all the Mesozoic pre-Albian carbonate rocks outcropping in the Carpathians and the pre-Cretaceous ones in the North Dobrogean Orogen), and (4) Alpine sediments posterior to the Alpine terranes amalgamation (limestones in the Transylvanian Basin) (Fig. 1).

The vast majority of the carbonate rocks are incorporated in Mesozoic- and Cenozoic-age sequences, half of which belong to various Jurassic and lower Cretaceous lithostratigraphic units. Genetically, these rocks were laid down in two settings: (1) open sea and (2) shallow/marginal marine assemblages.

1. The open-sea carbonates include: *pelagic limestones*, which are exclusively of Mesozoic age and outcrop over relatively large areas in the Apuseni Mountains, where they form well-layered sequences surpassing 100 m in thickness (Bleahu 1972, 1974). In the western part of the Banat Mountains, the Jurassic and Cretaceous limestones are the backbone of the Reşiţa-Moldova Nouă Syncline (Bucur 1997), the single largest karst area of Romania (>600 km^2). Less important from a karst standpoint is the red, nodular, and marly limestones (known as Ammonitico Rosso; Upper Jurassic) occurring in the Almăj Mountains and the limestone with Aptychus (a type of marine fossil) from the upper Jurassic-lower Cretaceous of the East Carpathians. The upper Cretaceous massive chalky limestone from south Dobrogea is also pelagic (Melinte 2006), but karst development on it is poor. The other type of open-sea carbonates includes *mix deposits*, which are in fact thin bedded carbonates interlayered with siliciclastic sediments (siltstone, shales, etc.). The presence of these impervious layers within the calcareous sequences restricts the percolation of water to depth, limiting the extent of karst development. The Aptychus Beds in the northern part of the Trascău Mountains are a good example for this facies (Săsăran 2006).
2. Marine shallow water and marginal associations. Most of Romania's karst develops on bioconstructed limestones (shallow water reef environments) that include *bioherms* and *reef cores*. A bioherm is a mound or lens-shaped structure made up of skeleton of various organisms, which is lying unconformably within a stratigraphic sequence that includes various lithologies. Such examples are the middle Kimmeridgian limestones in Dobrogea, some limestones in Gosau facies within the Apuseni Mountains, and a small number of reef cores in the carbonate massifs of Pădurea Craiului and Dâmbovicioara Pass (Bleahu 1972). It is important to remember that the reef limestones are primarily built up of lime mud resulted from destruction of bioherms. Depending on a number of environmental parameters acting at the time of deposition (e.g., influx of siliciclastic sediments, sea-level fluctuations, and evaporation) and during diagenesis, a large diversity of limestone types were formed. It is beyond the scope of this chapter to discuss each of them. Rather, we focus on the type of bedding (i.e., *massive limestones* or *dolomites* and *bedded limestones*), since this characteristic of carbonate rocks is crucial for speleogenesis.

The first category includes sequences of carbonate rocks well in excess of 100 m in thickness that are common for the middle to upper Triassic-age deposits in the East Carpathians and Apuseni Mountains, as well as upper Jurassic and lower Cretaceous limestones with Nerinea throughout the Carpathians and Dobrogea and with Pachiodontes and corals across the Carpathians, respectively.

The *bedded limestones* are very common and typically deposited on both sides of the reef as either micritic (fine grained) or detrital (coarse grained) limestones. Compare to the previous category, the bedded limestones never reach great thicknesses, thus cave development is restricted. Karst features on bedded limestones are known from the Triassic of East Carpathians and Apuseni Mountains, middle Jurassic and lower Cretaceous (Apuseni Mountains), middle to upper Jurassic of Dobrogea, and Eocene nummulitic limestones in the Someş Plateau, Rodna Mountains, and Dobrogea. Occurrences of Eocene detrital limestone in which small-size caves formed were documented in the Transylvanian Basin. Rather extensive Sarmatian-age bedded limestone outcrops in southern Dobrogea.

Metamorphosed Carbonates (Marble)

The karst-bearing rocks falling into this category are represented by crystalline limestones and dolomites (marble), which are mainly pre-Variscan Orogeny reef deposits that suffered a low- to medium-grade metamorphism. They represent ~8% of the total karst area of Romania and outcrop in all three major Carpathian units. The most extensive surfaces occur in Poiana Ruscă and Bihor mountains, whereas narrow bands of various lengths also occur in East Carpathians (Silvestru 1985), Făgăraş (Giurgiu 1990; Drăguşin et al. 2018b), as well as in Muntele Mare-Gilău (Cocean and Onac 1989). In Trascău and Metaliferi massifs, part of the Apuseni Mountains, there are also Mesozoic low-grade metamorphosed carbonate rocks (Balintoni personal communication).

Since marble has a very low porosity and accordingly a negligible permeability, water can only percolate along fractures and joints. As a result, the karst landscape is in general limited to small-scale solutional fissure-controlled features (linear karren), poorly developed dolines, and very short, but sometimes interesting caves.

Non-carbonate Rocks

Evaporates (Salt and Gypsum)

Because Romania sits in the temperate climate belt, the spatial occurrence of evaporite karst is limited (Fig. 1). Massive, thick layers of salt rocks occur in the Miocene formations from the Transylvanian Basin (Sovata-Praid, Turda), Getic Depression (Ocnele Mari), and in the Carpathian Band Zone (Meledic Plateau, Slănic Prahova) (Senco 1968; Giurgiu et al. 1980; Povară et al. 1982; Mărunțeanu and Ioane 2010). When salt deposits are exposed to surface, the dissolution proceeds rapidly generating various types of karrens, dolines, natural bridges, or even valleys. Except for the Cave 6S from Mânzălești (Meledic Plateau), now the longest in Europe (3234 m; Giurgiu 2010), all the others are modest in size and extent. Due to the high solubility of salt, both surface and underground karst features are ephemeral and ever-changing.

Since at many locations other rocks cover the salt layers, subsurface dissolution causes the formation of collapse and suffusion dolines. This process is exacerbated by anthropogenic activities that are responsible in some salt mining areas for massive collapses, in which large salt lakes formed (e.g., Bride Lake/Lacul Miresei at Slănic Prahova; Dordea et al. 2013).

Gypsum deposits, while relatively frequent in many sedimentary sequences spanning the time period between middle Paleozoic and Pleistocene (Patrulius 1976), were deposited as thin layers interbedded with sandstones, marls, and salt. Taking into consideration these lithostratigraphic characteristics and the high solubility of gypsum should come as no surprise that the karst on this type of rock (compared to salt) has a much-limited distribution in Romania (Senco 1968; Bleahu 1972; Ponta 1986; Onac and Istvan 1994). In fact, outcrops of Eocene and lower Miocene gypsum (sometimes anhydrite too) are known from a handful of locations, such as: Meseș Mountains, Cheia (Cluj County), Valea Seacă (Hunedoara County), Vrancea, and Buzău Subcarpathians (Ponta 1986), and in the Getic Depression at Nucșoara (Viehmann and Mac 1966; Bulgăreanu 1997; Băicoană and Breban 2000). At all these locations, the karst landscape consists of a variety of karrens, dolines (some as large as 100 m in diameter and 20 m deep), and small-size caves. For more information on evaporite karst, readers should refer to the chapter by Tîrlă (this book).

Siliciclastic

Most siliciclastic deposits are composed of quartz grains of various sizes and clay minerals cemented together usually by quartz as well. Except for sandstones or conglomerates in which the cementing agent is a carbonate, nearly all other siliciclastic rocks are very poorly soluble. However, when time is long enough, dissolution processes acting upon conglomerates, sandstones, siltstones, or even loess may produce interesting karst-like features. Examples of this type of pseudokarst are given by Stănescu (1963), Senco (1968), Giurgiu (1992), and Mârza et al. (1995) from Mesozoic deposits in East Carpathians (e.g., Rarău, Ceahlău, Ciucaș, and Bucegi) and from various locations across the Romanian Plain. Caves in upper Oligocene age sandstones were described by Domșa (1988) from the Someș Plateau.

Shales and clays are not dissolving per se, but when interbedded between more resistant rocks, they can be eroded away, leaving behind shelter caves like those described by Băicoană and Breban (2000) at Râpa Roșie (Alba County).

Igneous and Metamorphic

Although there are abundant outcrops of igneous and metamorphic rocks in Romania, the development of karst-like features was only reported from some restricted areas (Fig. 1) in which volcaniclastic deposits, tuffs, andesites, gneisses, crystalline schists, and skarns are present (Senco 1968; Tulucan 1986; Băicoană and Breban 2000; Tămaș et al. 2000; Onac 2002). Probably the most interesting pseudokarst on volcanic rocks was the one documented by Naum and Butnaru (1967) from some very highly weathered pyroclastites in the Călimani Mountains. The study investigated karrens, dolines, and short caves, leading the authors to introduce the term *volcano-karst*. Later, Balintoni (1970) shown that these voids resulted from the removal of sulfur and secondary silica by infiltration waters under physicochemical conditions identical to those existing in the oxidation zones of the complex sulfate ore deposits. Unfortunately, the mining of sulfur in large-scale quarry operations led to the destruction of all cavities from this area.

Peculiar type of cavities were described by Moréh (1991, 2009) from the volcaniclastic deposits outcropping along the Mureș Gorge in Călimani and Gurghiului mountains. Most of them display a tube morphology (up to 25 m in length) with oval cross section (0.4–1.2 m in diameter) and are in fact tree mold caves. The vast majorities are horizontal to subhorizontal, but a few short pits (upright tree molds) were also discovered.

The surface karst-like features on all igneous and metamorphic rocks are poorly developed. Except for some skarn-hosted caves, which could be rather spacious and large (hundreds of meters), all the other has modest sizes. Furthermore, speleothems are rare, but certain caves associated with skarn and polymetallic ore deposits display exotic mineral assemblages (Mârza and Silvestru 1988; Onac 2002).

Types of Karst

The karst of Romania is predominantly carbonate (over 98%) and is distributed around and within two major geographic units: the East and South Carpathians, Apuseni Mountains, and Dobrogea. The two-karst provinces are very different due to their geological evolution, spatial development, overall relief, and conditions of karstification. In both, the karst rocks are distributed around geomorphological structures (island blocks) made of crystalline and igneous rocks of Precambrian and Paleozoic age. In the Carpathians, these mountain blocks are dispersed, elevated, and tectonically fragmented, some of them occurring as narrow limestone stripes (also known as bars) and crystalline dolomites, whereas in Dobrogea they form a unitary and strongly peneplaned massif (Măcin Mountains).

On the edge of the pre-Mesozoic mountain crystalline blocks, carbonate platforms developed during the Triassic, Jurassic, Cretaceous, and Cenozoic sedimentary cycles, each important tectonogenetic phase being preceded by a karstification stage. The karst formed on Mesozoic carbonates was periodically fragmented, relocated, or buried under siliciclastic deposits (the Carpathians, North Dobrogea), whereas the one formed during Cenozoic cycles was largely suspended due to the uplift of the erosion platforms (e.g., Transylvanian Depression, South Dobrogea). A proof for the existence of the karstification cycles is the Upper Jurassic paleokarst filled with bauxite in the Pădurea Craiului and Şureanu mountains (Papiu et al. 1971; Pop and Mârza 1977), the petroleum structures and hydrothermal aquifers existing in the buried karst of the Getic and Pannonian depressions, and the entire submerged littoral karst of the Black Sea Platform in Dobrogea.

The Carpathian karst includes the following subunits: (i) the mountainous chain of the East and South Carpathians, and the Apuseni Mountains, in which karst developed on thick (hundreds of meters) crystalline (marble) and Mesozoic carbonate rocks, folded and strongly tectonized. In addition to these, Paleogene basinal limestone forming monoclines occurs on the southern part of the Rodnei Mountains and small areas along the Timiş-Cerna Corridor, the Huedin-Păniceni Depression and the western part of the Apuseni Mountains are covered by Miocene limestones; (ii) Transylvanian intra-mountainous depression, with Cenozoic limestone rocks, which may contain interbedded siliciclastic layers; (iii) the Subcarpathians with isolated Mio-Pliocene evaporites (salt and gypsum); (iv) the South Carpathians foreland depressions (Getic and Pannonian), in which the Mesozoic carbonate platform is buried up to 2000 m. The distribution and general appearance of the karst structures resulted from the orogenic fragmentation of the mountain chain (pre-Sarmatian) and because of the uplift and erosion that happened in the Carpathians during Quaternary.

Following morphogenesis, the Mesozoic carbonate platforms preserved at surface or near the edge of the crystalline mountainous blocks remained more unitary, generating plateaus. When the margin of the mountains or of the intra-mountains tectonic corridors were fragmented by longitudinal and transversal faults, limestone bars (rooted/embedded or suspended) formed. Higher up in the Carpathians as well as along the border of graben-type depressions, the differential weathering, and the tectonic fragmentation generated small isolated massifs made of crystalline and sedimentary limestone and dolomites; similar feature may have formed on Mesozoic olistoliths (Bleahu and Rusu 1964; Bleahu 1971, 1972; Bleahu et al. 1976). In all this morphological diversity, organized in relation to the direction and position of the mountain axis, one must note the difference between the distribution and size of the karst structures in the three Carpathian branches: in the East Carpathians, the unit with the most nappe structures and overthrusts, the karst is scattered and occurs predominantly on isolated massifs (Hăghimaş is the only plateau); in the South Carpathians, the mountains with the highest altitudes (over 2000 m), the isolated massifs are rare and the karst appears in the form of limestone stripes and longitudinally elongated plateaus located on the outer mountain ridge or on the edge of the tectonic corridors (Piatra Craiului, Căpăţânei, Vâlcan, Mehedinţi, and Cerna); the Banat and Apuseni Mountains are generally of lower elevation (highest peak is 1849 m) and less fragmented, thus, extensive limestone surfaces occur, predominantly as calcareous plateaus. In addition, isolated massifs (Metaliferi, Trascău) and narrow limestone stripes were formed along tectonized ridges (Dognecea, Codru Moma, Trascău, Muntele Mare). Another characteristic of the Carpathian karst is that its diversity is in direct relationship with the age and continuity of the soluble rocks, and the total mountain relief (Fig. 4).

The three morphotectonic types (plateau, stripe, and isolated massif) regarded as genetic karst landform classes became greatly diversified following the morphohydrographic (fluvial erosion) and karst evolution they experience since the Late Pliocene. Each of these types, depending on the size of the carbonate unit (horizontal and vertical), altitudinal location, and the relationship with the valleys or the surrounding landscape (Fig. 5), is further subdivided into three genetic-evolutionary types of karst massifs (Goran 1983; Bălteanu et al. 2012).

Suspended unitary karst plateaus (Figs. 4, 6a and 9b) are extensive, well-karstified surfaces displaying relatively flat

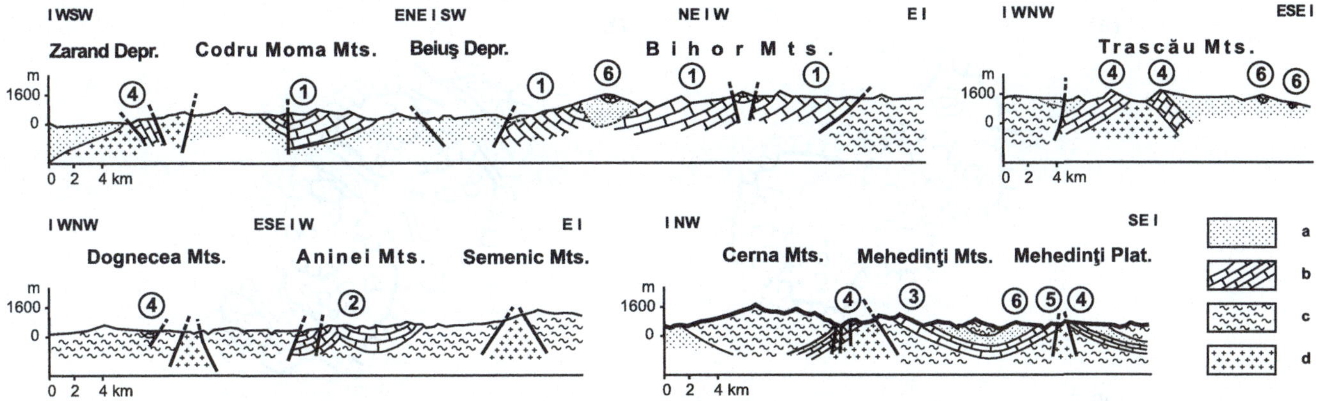

Fig. 4 Relationship between mountain areas and the distribution of karst types in the Apuseni Mountains, Banat Mountains, and west of the South Carpathians (modified after Goran 1983). *1* Suspended unitary plateaus; *2* dissected high plateaus; *3* unitary limestone bars; *4–5* leveled limestone bars; *6* isolated massifs; *a* non-carbonate sedimentary rocks; *b* carbonate rocks; *c* crystalline schists; *d* igneous rocks

Fig. 5 Distribution of the types of karst areas in the Romania (geological limits modified after Bleahu and Rusu 1964). *1* Unitary suspended karst plateaus; *2* dissected high karst plateaus; *3* subsided and partly covered karst plateaus; *4* unitary limestone bars; *5* fragmented limestone bars; *6* leveled limestone bars; *7* isolated limestone massifs

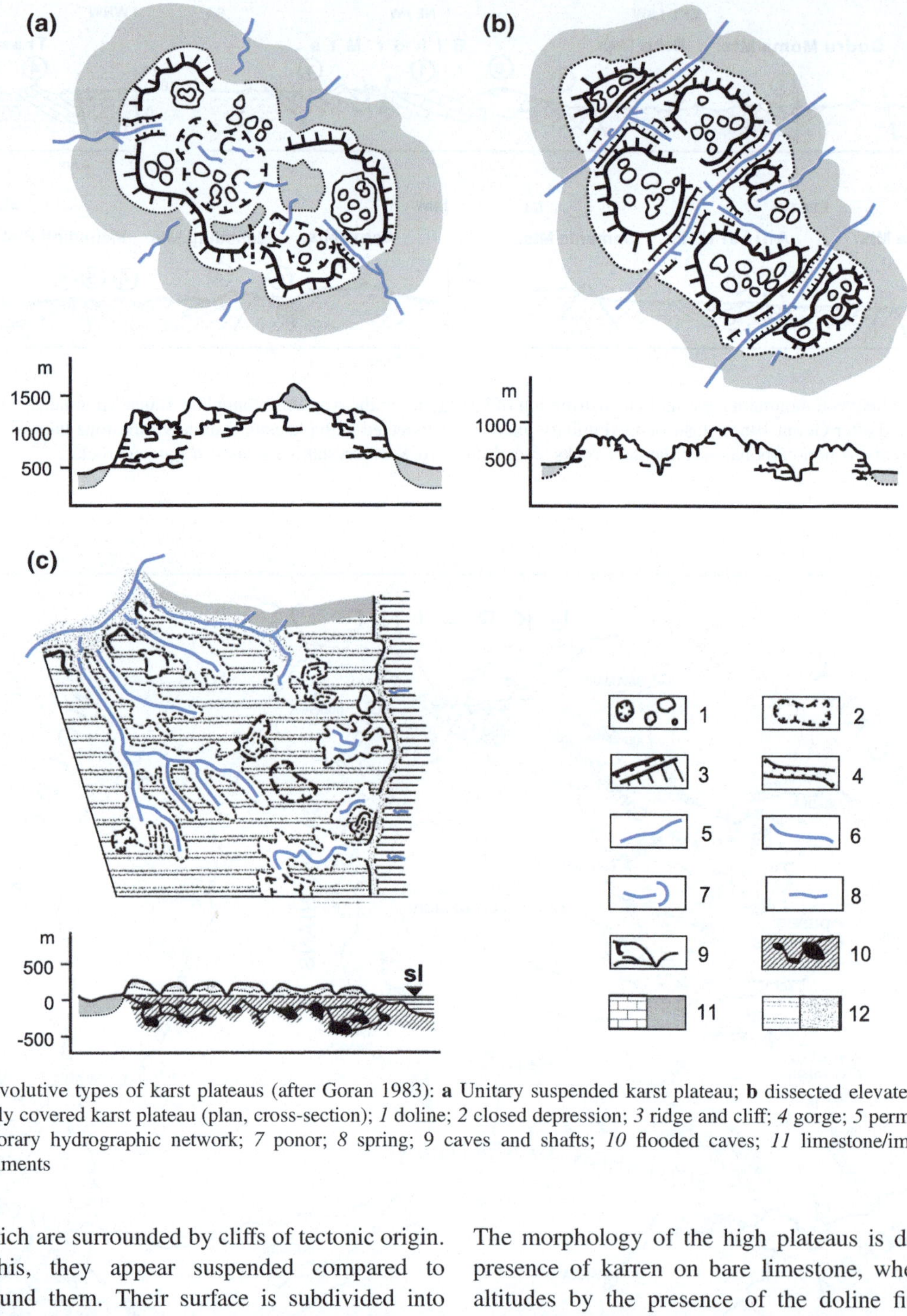

Fig. 6 Genetic-evolutive types of karst plateaus (after Goran 1983): **a** Unitary suspended karst plateau; **b** dissected elevated plateau; **c** unitary subsided and partly covered karst plateau (plan, cross-section); *1* doline; *2* closed depression; *3* ridge and cliff; *4* gorge; *5* permanent hydrographic network; *6* temporary hydrographic network; *7* ponor; *8* spring; 9 caves and shafts; *10* flooded caves; *11* limestone/impervious rocks; *12* loess/alluvial sediments

landscapes, which are surrounded by cliffs of tectonic origin. Because of this, they appear suspended compared to depression around them. Their surface is subdivided into endorheic basins formed due to the disorganization of superficial drainage and from the development of large karst forms (blind valleys, uvalas, closed depressions). They have autogenous evolution and clearly defined karst systems, with large shafts and caves whose drainage is radial toward springs and resurgence-type caves located at the foothill of the plateau. This type of karst is found in Retezatul Mic (1800–2000 m), Bihor (1000–1300 m), Poieni Plateau (1000–1200 m), Mehedinţi Mountains (1000–1200 m), Pădurea Craiului (800–900 m), and Vaşcău (800–900 m). The morphology of the high plateaus is dominated by the presence of karren on bare limestone, whereas at medium altitudes by the presence of the doline fields covered by superficial deposits and forest vegetation.

Fragmented high plateaus (dissected by rivers; Figs. 4 and 6b) appear in the central and western part of the Banat Mountains (Locvei and Aninei), where the limestone forms the Reşiţa-Moldova Nouă synclinorium, which is fragmented by a permanent hydrographic network, transversally crossed by the Danube, Nera, Miniş, Caraş, and Bârzava rivers. Plateaus located at altitudes between 200 and 300 m, in the south of the region, and 600–800 m in the north are delimited by cliffs and deep gorges. On the corrugated surfaces of the plateaus, the presence of dolines and blind valleys is ubiquitous, fact that allowed the development of drainage systems, which discharge through numerous caves situated in the valley at the bottom of the gorges. The same type of plateau, but covering smaller surfaces, made up of less soluble rocks (tightly bedded, softer, impure, and ~ 100 m in thicknesses), develops on the Paleogene limestones of the Someş Plateau (Purcăreţ-Boiu Mare) (Onac et al. 2010).

Buried karst plateaus (Fig. 6c) are located outside the mountain range (Romanian Plain, West Plain and Hills) and formed in the subsiding zone of the pericarpatic foreland, where detrital sedimentation fossilized the Mesozoic carbonate platforms. On the paleosurface of these limestones, a karst interface with closed depression and valleys was highlighted by means of drilling and geophysical methods.

Unitary limestone bars/stripes (Fig. 7a) are axially elevated tectonic structures, delimited by high cliffs, and which are bypassed by the hydrographic network coming from higher mountainous areas at their extremities. Depending on altitude and their Quaternary evolution, the unitary stripes have either calcareous ridge appearance (Piatra Craiului, Buila-Vânturariţa, Scăriţa-Belioara) or suspended, narrow, and elongated plateau (Hăghimaş, the southern part of the Mehedinţi and Cerna mountains). The ridges, which have been modeled during the Pleistocene under periglacial conditions, display an obvious tectonostructural landscape (asymmetric slopes), with few karst features (dry valleys, linear karren, and shafts developed along deep fractures) and widespread scree deposits along the flanks, whereas the plateaus have uneven surfaces with a succession of prominent mounds, separated by doline fields and alignments of suspended valleys. In the ridges, there are vertical shafts developed along fault lines (Grind Pit in Piatra Craiului) or gravitational fractures connected to slope retreat (Buila-Vânturariţa and Scăriţa-Belioara), while in the slopes of the neighboring valleys there are multi-level caves and high discharge springs.

Fragmented limestone bars/stripes (Figs. 7b and 9a) form when limestone bands are dissected by transversal valleys coming from the high mountain area and incised deep canyons (Trascăului Mountains, 1200 m, Balşa-Ardeu-Cibul in Metaliferi Mountains, 800–1000 m, Vârtoape-Piatra Cloşani in Mehedinţi Mountains, 1000–1400 m, and Băniţa-Roşia-Taia from Şureanu Mountains, 1000 m). The interfluves of the bars are elevated, have steep slopes and rolling or flat surfaces, where dry valleys, dolines, and karren occur abundantly. The caves are multi-level, depending on the stages of deepening of the valleys where the drainage resurgences. The underground networks can be simple lateral discharges of the interfluves (Râmeţi, Intregalde, Cernişoara, Motru Mare, Sohodul de Runcu), underground meanders of the valleys (the caves of Lazului, Martel, Muierii, Polovragi), or stream piracy of neighboring valleys (Jiul de Vest-Cerna Spring, Cerna Olteţului-Tărâia-Olteţ). It is worth noting that because of the width

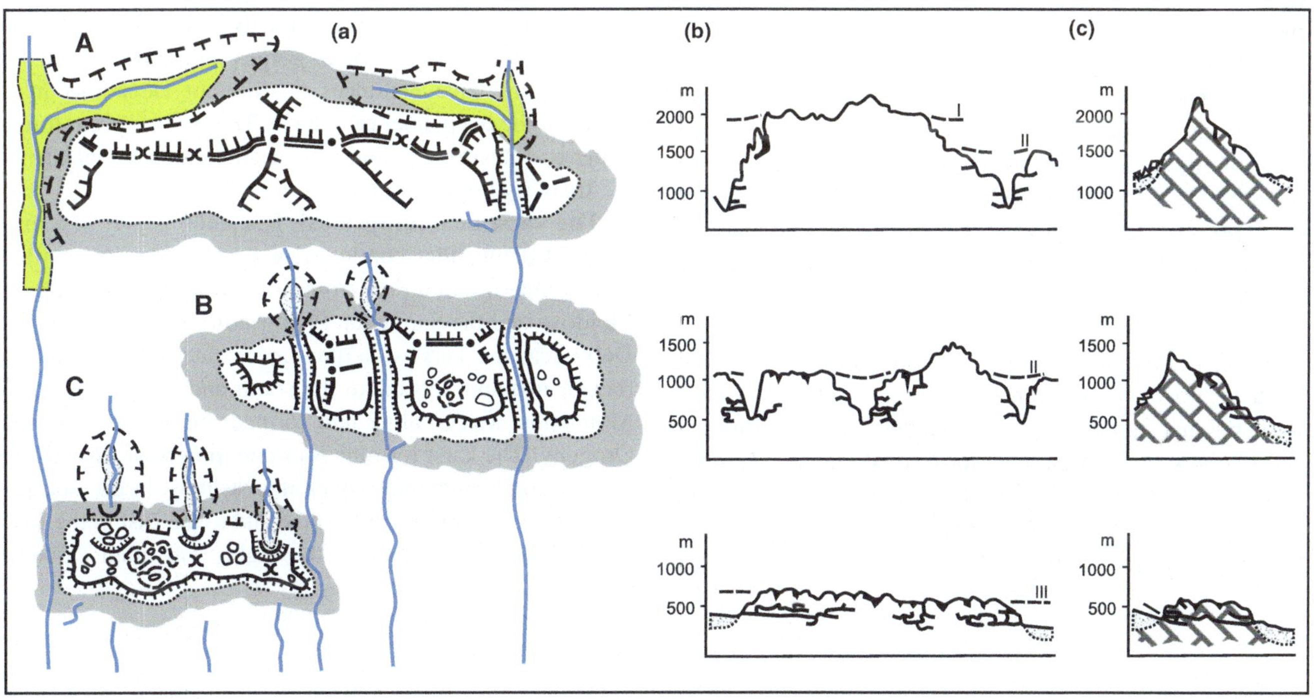

Fig. 7 Genetic-evolutive types of limestone bars (after Goran 1983): *A* Unitary limestone bars; *B* fragmented limestone bars; *C* leveled limestone bars; **a** plan; **b** longitudinal section; **c** cross-section; I, II, and III—the Carpathians peneplains (for legend see Fig. 6)

of the limestone stripes in the western part of the Şureanu Mountains and the southern of the Vâlcan Mountains, they function as bars, but morphologically (interfluves and karst landforms) better resemble the fragmented plateaus.

Leveled limestone bars/stripes (Fig. 7c) are the most karstified of all limestone bar types (Mehedinţi Plateau, Moneasa-Dumbrăvita). Their landscape has low altitudes and is flattened (resembles a karst plain), preserving now suspended remains (saddles, dry valleys, antithetic steps) of transversal paleohydrographic networks, and karren fields, groups and alignments of small dolines, which mark the route of underground drainage. The streams flowing from impervious rocks disappears underground at the contact with limestone in the upstream flank of the bars, through deep ponors and evolve in closed fluviokarstic depressions (Zăton, Ponorel, Ponorăţ, Fundătura Ponor; Fig. 9c). These alluviated depressions periodically flooded and with the appearance of polje, form regressively, from the ponor of the blind valleys toward the impervious basin of the captured valleys. The karst drainage is made through ponors and cave entrances, the network that crosses the limestone bar has large multi-level galleries. The drainage directions can be transversal on tectonic alignments (Moneasa, Ponorici-Cioclovina), or when the bars have a complex karst organization (Jiu-Cerna, Zăton-Bulba, Epuran-Topolniţa systems), it is longitudinal and the springs are located at one end of the bar (Goran 2001).

Isolated ridge massifs (Fig. 8a) appear in most mountainous units, on interfluves and are either reefs or olistoliths (Maramureş, Rarău, Giumalău, Bucegi, Metaliferi, Trascău mountains) or residual witnesses detached by differential erosion from the impermeable rocks around (Făgăraş, Poiana Ruscă). They have limited extension, rocky surface, and because they occur on the summit, are completely isolated from the hydrographic network of the region. The isolated massifs may appear as ridges (Strunga-Bătrâna-Guţanu, Oslea) or hillocks with cylindrical (e.g., Piatra Singuratică), conical (Lespezi, Târnovul), or pyramidal (Pietrele Doamnei) morphologies. Their karstification is weak, with few karren and rarely caves.

Isolated slope massifs (Fig. 8b) are located in an intermediate position between the ridges and the valleys, implying that streams flowing down the slope sometimes cut short gorges or disappear underground (Postăvaru, Trascău). Under these conditions, karstification is more active and small-size caves develop on upstream or downstream side of the isolated limestone massif.

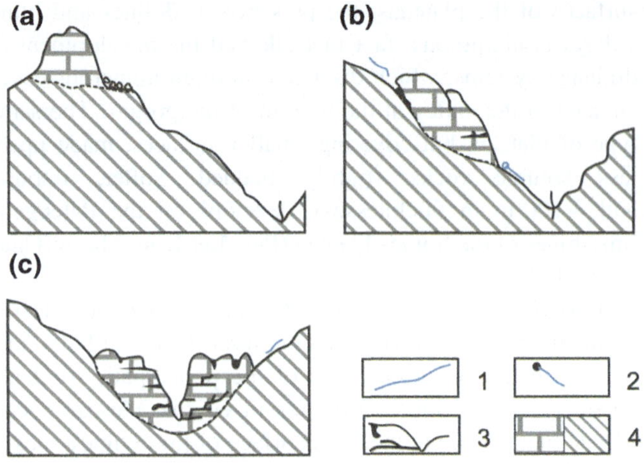

Fig. 8 Genetic-evolutive types of isolated limestone massifs (after Goran 1983): **a** isolated ridge massif; **b** isolated slope massif; **c** isolated valley massif; *1* hydrographic network; *2* spring; *3* caves and shafts; *4* limestone/impermeable rocks

Isolated valley massifs (Fig. 8c) are small-size lithologic units (olistoliths, reefs), fragmented by tectonics and erosion, buried, and then exposed in the axis of valleys, which cross them through short sections of gorges (Lupşa, Ialomiţa, Aiud, Feneş). The upper part of these isolated massifs may have been leveled by erosion (e.g., terraces) allowing the formation of karren and small dolines (Bucegi, Leaota). The limestone massifs outcropping along valleys can be highly karstified, thus hosting well-organized aquifers and large caves through which the hydrographic network is drained underground (Bolii Cave, Natural Bridge from Grohot). In the vertical walls of some gorges (Vârghiş, Aiud), the existence of cave entrances at different elevations marks stages in the deepening of valleys. These caves were formed either by transverse drainage or as subterranean meanders of the main hydrographic path.

The karst of Dobrogea (SE Romania) is part of a much wider structure that continues in the neighboring northern and southern countries (fossil reefs in the Republic of Moldova and the Sarmatian Platform in the NE Bulgaria). In Dobrogea, the karst structures are located at low altitudes (between 0 and 250 m) to the south and east of the Măcin Mountains, a crystalline unit formed during the Hercynian Orogeny. The karst forming rocks are progressively younger toward south and they cover increasingly larger surfaces, reaching westward to the Danube and to the east to the Black Sea, where a large part of the highly karstified carbonate platforms was transgressively cover by recent sedimentary deposits. Three other factors are important for the karst evolution in Dobrogea: (i) Miocene-Pliocene epeirogeny, which fragmented the landscape in large blocks that moved vertically one from another; (ii) the Quaternary Black Sea-level oscillations (10–20 m) influenced the organization of the karst by modifying the local hydrologic base level; (iii) part of the landscape is covered by thick sequences (up to 40–60 m) of loess interbedded with paleosoils that fossilized most of the large karst forms; (iv) the current semiarid climate (precipitation 400–450 mm/year) prevents water from reaching the karst rocks and is responsible for the gradual disappearance of the permanent hydrographic network. The types of karst closely follow Dobrogea's geological evolution and are very different in its three major tectonostructural units.

North Dobrogea karst develops on limestones and dolomitic limestones of Triassic, Jurassic, and Cretaceous ages, in a series of faulted, heavily leveled anticlines and synclines in the Tulcei Hills (NE part) and in a syncline with bedded Jurassic and Cretaceous limestones within the Babadag Plateau (in the S). Due to the intense pediplanation of the region and the loess cover, the karst landscape appears in the form of isolated masses, with the appearance of inselbergs or small ridges aligned along the main tectonic directions. The most important calcareous areas are in the form of cliffs, in the right bank of the Danube at Somova-Câşia, Mahmudia, Murighiol-Dunavăţ, or on the shore of Razelm Lake (Babadag-Enisala-Jurilovca). The very few caves existing in this region are small and appear on the slopes of the Taiţa and Slava valleys. It is worth mentioning the existence, on small limestone outcrops in the agricultural land south of Mahmudia, of kamenitzas with diameters of 80–100 cm.

Central Dobrogea karst is formed on Jurassic reefal limestones, arranged in alignments of synclines. Casimcea Syncline is the most important karst area of the region, which forms a 35 km long leveled bar, between Mireasa, Cheii (Fig. 9d), and Taşaul Lake. Some of the Casimcea Valley waters (fed from impervious areas) drain through the limestone bar and discharge through a series of springs on the shores of the Taşaul Lake. Near the lake, it was discovered a flooded cave with large calcite speleothems (Taşaul Lake Cave), formed when the Black Sea level was lower.

The bar has a flattened surface on which small dolines are present. Except for is northern part where a steep slope (30–50 m) occurs, the entire structure lies at the same elevation with the surrounding plateau. It is fragmented by a few small gorges traversed dry valleys (Mireasa, Cheii, and Visterna), in the slopes of which open a number of caves. Among these, the Bats Cave from Gura Dobrogei (over 500 m) and La Adam Cave, are especially important for the bat colonies, Quaternary mammal fossils, and the archaeological remains (Dumitrescu et al. 1962–1963; Bleahu et al. 1976).

Spectacular from a karst landscape point of view is the bioconstructed (bioherm) structures occurring in the Jurassic limestone. These outcrop in the Cheii Gorge and in the right bank of Casimcea Valley, and when affected by erosion, the soft sediment (loess) was removed leaving behind an interesting landscape dominated by cylindrical towers (reef structures) separated by chimneys (Fig. 9e).

South Dobrogea karst is situated in a platform region stretching between the Danube and the Black Sea (Fig. 6). In this area, a quasi-continuous, 500–600-m thick carbonate sequence consisting of three superposed and fossilized karst units of Jurassic, Cretaceous, Eocene, and Sarmatian age exist. The dominant type of karst is the semi-buried plateau (Fig. 6c), in the structure of which, the most karstified unit, the Jurassic limestone is buried at the bottom of the stack and discharges its aquifers through numerous springs located along the coast (Zamfirescu et al. 2010).

The largest surface of the karst is made up of the Sarmatian limestone, which occupies the eastern half of the plateau and forms a sizable endorheic area that includes large closed depressions, partially filled with loess (Negru Vodă, Topraisar, Cobadin; Ilie 1969) and dry valleys, which have a typical canyon appearance such as Mangaliei and Canaraua Fetii. Ponors exist in the thalweg of these valleys filled with siliciclastic sediments, whereas in their slopes some small-sized caves and karst springs with high flow occur. Canaraua Fetii is a wide canyon-type valley, with flat stream bed and steep slopes, which has an almost continuous cliff/overhang that is perforated by small caves and shelters. The most important caves in the region are Limanu (3.5 km) and Movile, both covered in dedicated chapters later in this book (Drăguşin et al. 2018a; Sarbu et al. 2018).

Types of Caves and Speleogenesis

Information on ~12,300 caves has been archived at the "Emil Racoviţă" Speleological Institute in Bucharest (Goran 2018), but only 6816 were included in a printed catalog of caves of Romania (Goran 1982, 1989). Using data from various publications, 8128 caves (maps and various information) are compiled in an online database at www.speologie.org. The greatest concentration of caves occurs in the karst areas of the South Carpathians and Apuseni Mountains (Fig. 1). Vântului (Wind) Cave în Pădurea Craiului Mountains is the longest (>50 km; Szilágyi Palkó et al. 2007), whereas Vărăşoaia cave system (V5) in the Bihor Mountains is 653 m deep (Zih et al. 2018).

From a genetic point of view, the following four major cave types can be distinguished in Romania: dissolution, erosion, shelter, and fissure (Bleahu 1982). The emphasis on

Fig. 9 Illustrative examples for various types of karst: **a** Turda Gorges are carved in a fragmented limestone bar of the Trascău Mountains (photograph courtesy of M.-L. Tîrlă), **b** Damiş, a suspended karst plateau in Pădurea Craiului Mountains (photograph by B. P. Onac), **c** Zăton fluviokarst depression in Mehedinţi Plateau (photograph by C. Goran), **d** Cheii Gorges are cut in a low-elevation leveled reefal limestone bar in central Dobrogea, where bioherm structures (**e**) are well-exposed along the road (photographs by B.P. Onac)

this book is on dissolution caves, but before dwelling into more details on this group, the other types are briefly discussed as well. Both erosion and shelter caves are mainly developed in non-carbonate lithologies (siliciclastic and volcanic) and are rather short and unspectacular. The first type forms by suffusion and mechanical abrasion caused by moving water along fissures, whereas the latter results when weathering selectively removes sediments and weak rocks (clay, loess) that are overlain by more resistant ones. Caves belonging to these two groups were described by Băicoană and Breban (2000) from various locations in Hunedoara and Alba counties, by Popescu (1988) in loess deposits from Dobrogea, and by Exner (2002) in the outskirts of Suceava City.

Fissure (or crevice) caves are known in Romania in all rock types. They are commonly found in steep mountainsides where due to gravitational sliding of cliffs, cracks or network of cracks are widened. Suggestive examples of caves and shafts of this origin were reported by Onac (1987) from Runcului Gorge (Muntele Mare-Gilău Mountains) where they reach up to 432 m in length and 48 m in depth.

The dissolution caves are by far the most common in Romania. They form when acidic water passes through pores and fissures enlarging them. Taking into the consideration the source of chemical aggressiveness, the caves of this type are assigned to two broad categories: *epigenic* and *hypogenic*. The origin of epigenic caves is related to meteoric water that gains its solutional capacity from surface or near-surface sources (e.g., CO_2 produced in soil), whereas the hypogenic caves owe their genesis to aggressive solutions of whose origin is below the surface.

The patterns of epigenic caves depend on the type of recharge and structural characteristics of the bedrock and can be of many types (Palmer 2007). In this review, we choose to discuss and provide only the most significant examples. The vast majority of caves have *branchwork* passages (single or multiple), which are the result of recharge through point sources (ponors, dolines, etc.) and flow along fractures or/and bedding planes in gentle or steep dipping carbonate units. The following caves and shafts are good examples of this type: Bulz, Vântului, Vadul Crişului, Osoi, Avenul din Şesuri, Coiba Mare, Huda lui Păpară, Avenul din Poiana Gropii, Comarnic, Bulba, Martel, etc. Some of these caves have their passages developed on two or more levels. Vântului Cave, for instance, is a textbook example of water table cave; the underground stream kept pace with the entrenchment of the Crişul Repede Valley, producing distinct tiers.

When recharge is dominated by sinking streams and storm waters, the highly aggressive water is injected under pressure (phreatic conditions) in all available spaces surrounding the main passage causing them to enlarge rather rapidly. This type of widening produces irregular single-passage caves with *anastomotic mazes*. The largest is Peştera din Pârâul Hodobanei (Vălenaş et al. 1982; Damm and Mitrofan 2005), in which anastomotic passages are superposed on the basic levels of cave development. There are good examples of such passages in Meziad, Topolniţa, and Muierii caves, which are/were fed by large sinking streams that often caused massive flooding events.

Caves displaying *network* mazes (passages intersect in a grid-like pattern) form either when the soluble rocks are highly fractured (Izvorul Tăuşoarelor, median section of Topolniţa, parts of Şugău) or when tectonic and diffuse recharge are working hand in hand. Limanu Cave (see chapter by Drăguşin et al. 2018a) is a good candidate for this category, although recently it has been proposed a hypogenic origin (coastal mixing; Onac and Drăguşin 2017).

Hypogenic caves have several possible origins, but however, all share some common particularities. Among these we mention the typical network and spongework cave patterns, which are unrelated to surface recharge, absence of any solutional features due to rapid water flow (e.g., scallops), presence of cupolas but no vadose morphologies (e.g., shafts, canyons), and occurrence of exotic minerals. Considering these, Onac and Drăguşin (2017) classified the hypogenic caves of Romania in three groups: (i) sulfuric acid caves related to rising of hydrogen sulfide-rich thermal waters in the lower part of the Cerna Valley (SW Romania) and in the vicinity of Mangalia on the Black Sea coast, (ii) cavities created in various rocks due to dissolution caused by circulation of hot metasomatic and hydrothermal ore fluids, and (iii) caves formed in the seacoast mixing zone. Within each group there are several well-documented caves, some representing outstanding, world-class examples of hypogenic speleogenesis (e.g., Movile, Diana, and Valea Rea caves). For further details, see the chapter by Onac and Drăguşin (2017) and the references therein.

Acknowledgements Dr. M. Ionita is thanked for kindly providing Fig. 2 included in this chapter. We are grateful to Dr. M.-L. Tîrlă who shared with us the distribution map of evaporates within Romania.

References

Băicoană M, Breban RC (2000) Pseudocarstul. Fenomene endo şi exo. Clubul Speol Proteus Hunedoara, p 45

Balintoni I (1970) Relaţia dintre vulcano-carst şi epigeneza în zăcămîntul de sulf din Munţii Călimani. DS Inst Geol LV I(2):5–11

Balintoni I (2018) Geology of Romania. In: Ponta GML, Onac BP (eds) Caves and karst systems of Romania. Springer International Publishing, Cham, pp 9–20

Bălteanu D, Jurchescu M, Surdeanu V, Ioniţă I, Goran C, Urdea P, Rădoane M, Rădoane N, Sima M (2012) Recent landform evolution in the Romanian Carpathians and Pericarpathian regions. In: Lóczy D, Stankoviansky M, Kotarba A (eds) Recent landform evolution. The Carpatho-Balkan-Dinaric region. Springer, Dordrecht, pp 249–286

Bleahu M (1971) Les surfaces d'applanissement karstique des Carpates Roumaines et leur evolution. Studia geomorphol Carpatho-Balcanica, pp 41–48

Bleahu MD (1972) Karst of Rumania. In: Herak M, Stringfield VT (eds) Important karst regions of the Northern Hemisphere. Elsevier, Amsterdam, pp 341–353

Bleahu MD (1974) Morfologia carstică. I. Condiţionarea geologică şi geografică a procesului de carstificare. Ed Ştiinţifică, Bucureşti, p 590

Bleahu M (1982) Relieful carstic. Ed Albatros, Bucureşti, p 296

Bleahu M, Rusu T (1964) The karst of Rumania. A brief outlook. Rev Roum Géol Géophys, Géogr, série Géogr 8:193–202

Bleahu M, Decu V, Negrea Ş, Pleşa C, Povară I, Viehmann I (1976) Peşteri din România. Ed Ştiinţifică şi Enciclopedică, Bucureşti, p 415

Bucur II (1997) Formaţiuni mezozoice din zona Reşiţa-Moldova Nouă. Ed Presa Univ Clujeană, Cluj-Napoca

Bulgăreanu VCA (1997) Solution lakes on gypsum evaporites: lake Invârtita (Nucşoara), Romania. Int J Salt Lake Res 6:55–66

Cocean P, Onac BP (1989) Le relief développé sur les calcaires cristallines de la partie sud-est du massif de Gilău-Muntele Mare. Studia UBB Geol-Geogr XXXIV(1):11–20

Damm PE, Mitrofan H (2005) Peştera din Pârâul Hodobanei. Speomond 9:15–19

Domşa M (1988) Câteva consideraţii preliminare asupra unor cavităţi în gresie din zona Surduc-Salaj (Podişul Someş). Peştera 2:43–53

Dordea D, Nedelcu M, Gorie J, Necşoiu I (2013) The salt mountain. GeoExPro 10(6):80–84

Drăguşin V, Dimofte D, Mirea I (2018a) Limanu Cave. In: Ponta GML, Onac BP (eds) Caves and karst systems of Romania. Springer International Publishing, Cham, pp 437–441

Drăguşin V, Mirea I, Nae A, Tîrlă L (2018b) Caves of the Făgăraş Mountains. In: Ponta GML, Onac BP (eds) Caves and karst systems of Romania. Springer International Publishing, Cham, pp 79–82

Dumitrescu A, Bîrsan VM (2015) ROCADA: a gridded daily climatic dataset over Romania (1961–2013) for nine meteorological variables. Nat Hazards 78(2):1045–1063

Dumitrescu M, Samson P, Terzea E, Rădulescu C, Ghica M (1962–1963) Peştera "La Adam", staţiune pleistocenă. Lucr Inst Speol "Emil Racoviţă" I–II:229–284

Exner M (2002) Pseudocarst în Lunca Sucevei. Ecocarst 3:46

Giurgiu I (1990) Peşterile descoperite la peste 2000 m altitudine, completare esenţială a imaginii carstului din România. Bul Club Speol "Emil Racoviţă" 13:62–75

Giurgiu I (1992) Pseudocarst in Romania. Cerc Speol 1:85–88

Giurgiu I (2010) Peştera (the cave) 6S de la Mânzăleşti (subcarpaţii Vrancei, România), one of the largest salt cave in the world. Nat României 21:233–234

Giurgiu I, Silvăşanu G, Vădeanu T (1980) Observaţii asupra carstului pe sare din zona Meledic (Munţii Buzău). Peştera 6S de la Mânzăleşti. Comparaţie între formele carstice dezvoltate în sare şi calcare. Bul Club Speol "Emil Racoviţă" 7:55–66

Goran C (1982) Catalogul sistematic al peşterilor din România. Ed CNEFS, Bucureşti, p 496

Goran C (1983) Les types de relief karstique de Roumanie. Trav Inst Spéol "Émile Racovitza" 22:91–102

Goran C (1989) Le spéléogramme de Roumanie. Misc Speol Rom 1:335–346

Goran C (2001) Coordonatele structurale şi distribuţia formelor în configuraţiile fluviocarstice. Rev Geomorfol 3:25–32

Goran C (2018) The systematic catalogue of the Romanian caves. In: Ponta GML, Onac BP (eds) Caves and karst systems of Romania. Springer International Publishing, Cham, pp 37–39

Ilie I (1969) Tipurile morfogenetice ale carstului dobrogean. Studiul geografiei asupra Dobrogei. Soc Ştiinţe Geogr RSR, Bucureşti, pp 59–64

Mârza I, Silvestru E (1988) First mention of the hydrothermal karst phenomenon in Romania, associated to the Neogene metasomatic sulphide ore deposits from Rodna Veche. Stud UBB Geol 33:77–81

Mârza I, Viehmann I, Stăncioiu R (1995) Stalactites argilo-sableuses sousaériénes. Trav Inst Spéol "Émile Racovitza", XXXIV:191–195

Mărunţeanu C, Ioane D (2010) Salt karst in Mânzăleşti—Romania. In: Evelpidou N, de Figueiredo T, Mauro F, Tecim V, Vassilopoulos A (eds) Natural heritage from East to West. Springer, Heidelberg, pp 105–109

Melinte MC (2006) Cretaceous-cenozoic paleobiogeography of the southern Romanian Black Sea onshore and offshore areas. Geo Eco Marina 12:79–90

Moréh K (1991) Cavităţi din Defileul Mureşului între Topliţa şi Deda. Bul Speol (Galerii) 13:78–84

Moréh K (2009) Peşteri de mulaj din Defileul Mureşului dintre Topliţa şi Deda. Speomond 14:40–43

Naum T, Butnaru E (1967) Le volcano-karst des Călimani (Carpates Roumaines). Ann Spéléol XXII:727–755

Onac B (1987) Grottes de traction gravitationnelle dans les Gorges du Runc (S.E. de Muntele Mare-Carpates Occidentales). Studia UBB Geol Geogr 32(2):87–92

Onac BP (2002) Caves formed within upper Cretaceous skarns at Băiţa, Bihor County, Romania: Mineral deposition and speleogenesis. Can Mineral 40:1693–1703

Onac BP, Constantin S (2004) Europe, Balkans and Carpathians. In: Gunn J (ed) Encyclopedia of caves and karst science. Fitzroy Dearborn, New York, pp 328–331

Onac BP, Drăguşin V (2017) Hypogene caves of Romania. In: Klimchouk A, Palmer AN, De Waele J, Auler AS, Audra P (eds) Hypogene karst regions and caves of the world. Springer International Publishing, Cham, pp 257–265

Onac BP, Istvan D (1994) Le karst sur la roche Miocene du Nord-Ouest de la Roumanie. Trav Inst Spéol "Émile Racovitza" XXXIII:173–178

Onac BP, Vereş D, Tămaş T, Montana Baia-Mare CS, Todoran V (2010) The hydrogeology of Cuciulat-Purcăreţ-Boiu Mare karst area, northern part of the Someş Plateau, Romania. In: Orăşeanu I, Iurkiewicz A (eds) Karst hydrogeology of Romania. Belvedere, Oradea, pp 329–341

Orghidan T, Puşcariu V, Bleahu M, Decu V, Rusu T, Bunescu A (1965) Harta regiunilor carstice din România. Lucr Inst Speol "Emil Racoviţă" IV:75–104

Palmer AN (2007) Cave geology. Cave Books, Dayton, p 455

Papiu VC, Mînzatu S, Iosof V, Udrescu C, Giuşcă R (1971) Constitution chimico-minéralogique de la formation bauxitifère du Bassin de Hateg. CR Inst Géol Géophys 57(2):77–123

Patrulius D (1976) Introduction to the geology of carbonate rocks and evaporites from Romania. In: Patrulius D (ed) Carbonate rocks and evaporites, Series No. 15 Guidebook, Int Colloq Carbonate Rocks and Evaporites, pp 5–12

Ponta G (1986) The evaporite karst from Romania. Le Grotte d'Italia XII(4):407–415

Ponta G (1998) Karst of Romania. In: Yuan D, Liu Z (eds) Global karst correlation IGCP 299. Science Press, Beijing, pp 193–207

Ponta G, Ursu V (2017) The highlights of Romanian karst. In: Moore K, White S (eds) Proceedings of 17th International Congress of Speleology, Sydney, 2, pp 78–81

Pop G, Mârza I (1977) La paléocarst éocrétace des Monts Pădurea Craiului et sa signification dans le mécanisme de génèse des bauxites. Bull Soc Géol Bull 30(1):51–58

Popescu G (1988) Observaţii asupra pseudocarstului în loess. Cerc Speol Ploieşti, pp 25–31

Povară I, Cosma C, Lascu C, Bulgăreanu VAC (1982) Un cas particulier de karst dans les repos de sel (Slănic-Prahova, Roumanie). Trav Inst Spéol "Émile Racovitza" 21:87–93

Sandu I, Pescaru VI, Poiană I (eds) (2008) The climate of Romania. Ed Acad Române, Bucureşti

Sarbu S, Lascu C, Brad T (2018) Movile Cave. In: Ponta GML, Onac BP (eds) Caves and Karst Systems of Romania. Springer International Publishing, Cham, pp 429–436

Săsăran E (2006) Calcarele Jurasicului superior-Cretacicului inferior din Munţii Trascău. Ed Presa Univ Clujeană, Cluj-Napoca

Senco V (1968) La carte du karst et du clasto-karst de Roumanie. Rev Roum Géol Géophys Géogr Ser Géogr 12(1–2):35–41

Silvestru E (1985) Aspects of karstification in the crystalline limestones on the southern slope of Rodnei mountains. Theor Appl Karstol 2:87–95

Stănescu I (1963) Contribuţii la cunoaşterea reliefului din masivul Ceahlău. An Şt Univ AI Cuza 9:193–194

Szilágyi Palkó P, Deák Z, Dascăl T (2007) Peştera Vântului 50 ani. Delroti SRL, Cluj-Napoca

Tămaş T, Munteanu G, Feier N, Zaharia L, Lascu I (2000). Pseudokarst in volcanoclastic rocks: the caves from Ciceu Corabia. Măgura Hill, NW Romania. In: Onac BP, Tămaş T (eds) Karst studies and problems—2000 and Beyond, Cluj-Napoca, pp 135–139

Tîrlă M-L (2018) Evaportie karst of Romania. In: Ponta GML, Onac BP (eds) Caves and karst systems of Romania. Springer International Publishing, Cham, pp 443–450

Tulucan T (1986) Clasificarea genetică a fenomenelor endovulcanocarstice din România. Aspecte ale repartiţiei acestora în lanţul munţilor Carpaţi. Bul Speol 10:121–135

Vălenaş L, Halasi G, Czako L (1982) La morphologie et la hydrologie des conduits submergés du Bassin de la Vallée Gîrda (Monts Bihor). Nymphaea X:183–194

Viehmann I, Mac I (1966) Observatii asupra carstului din ghipsurile Munţilor Meseş si cheilor Turzii. Lucr Inst Speol “Emil Racoviţă” 5:67–74

Zamfirescu F, Danchiv A, Popa I, Popa R, Rudolph-Lund K (2010) Southern Dobrogea. In: Orăşeanu I, Iurkiewicz A (eds) Karst hydrogeology of Romania. Ed Belvedere, Oradea, pp 351–360

Zih J, Perényi K, Mátyási L (2018) Vărăşoaia cave system. In: Ponta GML, Onac BP (eds) Caves and karst systems of Romania. Springer International Publishing, Cham, pp 327–334

The Systematic Catalog of the Romanian Caves

Cristian Goran

Romania is a country with an old speleological research and caving tradition, despite the fact that the surface of the karst-prone rocks is very small (2.3%; Onac and Goran 2018). Interest and information about caves date back to the second half of the eighteenth century when mineralogical observations (Fridvaldszky 1767) and the topographic survey and description for military purposes of the Veterani Cave (Anonymous 1789) in the Danube Gorge were published. Over the next century, the interest of the Austro-Hungarian geographers for the caves in Transylvania resulted in the publication of one of the world's first speleological inventory by Bielz (1884). The book enumerates and describes 73 caves, to which 20 more were added in a subsequent publication by the same author in 1886.

The actual cave research in Romania began after the establishment of the world's first Speleological Institute in Cluj (1920), following sustained field campaigns conducted by E. G. Racoviță and his collaborators. In two volumes of the monograph *Enumération des grottes visitées* (Enumeration of visited caves), the location, description, environmental conditions, and in many cases, cartoons, maps, or photographs from over 250 caves are published (Jeannel and Racovitza 1929; Chappuis and Jeannel 1951). After 1950, following the example of Racovita, two teams of professional speleologists from Bucharest and Cluj started a systematic research of the caves in the main karst regions of Romania. At the same time, other biologists, geologists, or geographers published the results of their speleological research, which made it possible to produce a first inventory of the Romanian caves (Orghidan et al. 1965). The paper contains a list of 984 caves and shafts for which information was available from other published studies or about which the authors had direct information. The caves are classified/ordered by morphological units (mountains, plateaus, etc.) and located on a map that shows the distribution of carbonate rocks in Romania (scale 1:500.000).

After 1960, and especially in the 60–70s, an important contribution to creating a national cave catalog and gathering knowledge of the speleological heritage had the emergence of the caving clubs (amateur cavers). These formed in 1975 the Sports Speleology Commission (SSC) within the Romanian Federation of Tourism-Climbing, which began re-exploring known caves and at the same time started to investigate new karst areas. As a result of their activity, the Romanian Cave Catalog containing an inventory of 2000 caves was published by Bleahu and Povară (1976). The novelty of the catalog consisted in the introduction of a decimal classification of the caves, each karst zone or hydrographic basin having its own numerical code, followed by the cave order number (assigned chronologically).

The SSC activity and the appearance of the cave catalog made it necessary to establish a system of evidence of speleological discoveries and research. This task was assigned to the author of the present chapter, who in 1978 organized a national caving archive, the so-called The Cadastre of Romanian Caves (CRC) at the "Emil Racoviță" Institute of Speleology (ERIS) in Bucharest. Each person who would discover, extend, map, or remap a cave had to send a form known as *Announcement and confirmation form* to CRC along with cartographic or photographic evidences (location, cave map, entrance photograph, etc.). The filing author/club would then receive from CRC the cave code confirmation number of the discovery, which authenticates the copyright over the cave. Through the cadastre, the SSC also regulates the methodology of cave inventory, the exploration priorities, verifies cave data and the correctness of the cartographic representations and, in exceptional cases, organizes actions to approve the caves' parameters (length, depth, etc.). Between 1977 and 1989, the SSC organized for all its affiliated craving clubs an annually speleology competition (Speo-Sport), during which all caving activities (from discovery and exploration to research and protection) performed over a year-period were awarded. The contest was

C. Goran (✉)
"Emil Racoviță" Institute of Speleology, Romanian Academy, Calea 13 Septembrie 11, 050711 Bucharest 5, Romania
e-mail: cristian.goran@gmail.com

G. M. L. Ponta and B. P. Onac (eds.), *Cave and Karst Systems of Romania*,
Cave and Karst Systems of the World, https://doi.org/10.1007/978-3-319-90747-5_5

an extraordinary incentive, with over 1000 caves discovered and several hundred others re-investigated almost each year. The avalanche of discoveries and a series of large campaigning meant to systematic investigate various karst areas in Romania required the publication of a new catalog, which included more than 6800 caves (Goran 1982).

The publication in 1982 of the new, more explicit and organized catalog allowed a large number of cave discoveries and speleological research to be carried out by intensive exploration, so that the number of caves recorded at CRC reached 11,500 by 1990, a year with major political and social changes in Romania. With the change of the Romanian political regime (1989), the interest in cave exploration has fallen dramatically. Some of the members of the caving clubs have either focused on other speleological activities or on their own professional career. After a period of investigations, in 1996 the Romanian Federation of Speleology was founded as an autonomous structure proposing a technical, logistic, and communicative reorganization of the activity of amateur speleologists. Announcing the discoveries to CRC has not been considered a necessity, the caving clubs became legal entities with their own strategies and means of capitalizing on the activity, among which the publication of regional catalogs with the investigated caves.

Under the new conditions, the CRC has turned into the Romanian Karst Cadastre (RKC) and become a compartment of ERIS, while remaining the legal depository of speleological and karstological information nationwide. The number of caves recorded in the RKC is about 12,300, but because many are missing GPS locations and some of the old landmarks have deteriorated, when a new catalog is being developed, probably a series of doubled caves will be identified.

The Systematic Catalog of the Romanian Caves produced by Goran (1982) contains 6816 entries (all over 5 m in length or depth) documented by the end of the 1981 caving year. Each cave is provided with its location and identification data. It was developed based on the information published or transmitted by the caving clubs to CRC and is systematized in four major sections: the decimal catalog, the alphabetical catalog, the bibliography of the caves, and the speleogram of Romania. The Decimal Classification System of Caves (Bleahu and Povară 1976) was revised and completed, establishing an equivalence between the geological (initial) and geographic classification of the geomorphological units. The caves were ordered from upstream to downstream in each basin or karst region, and those developed in non-carbonate rocks were highlighted by a lithological specification and presented separately. In the book, the cave characterization follows the order of the catalog numbers, in the form of separate tables with column for position #, situation of marking/identification in the field, official name, synonyms, location, number of entrances, altitude (absolute or relative), mapping status, hydrological type, length, vertical range (positive and/or negative), source and date of the information. The listing also includes caves mentioned in various publications, which in the meantime have been destroyed or never found again.

The Decimal Classification System of Caves (Bleahu and Povară 1976; Goran 1982, 2002) covers the entire area of the country in which karst-prone rocks outcrop. Based on this classification, the territory is divided into geographic or karst units of up to the fourth order, using numbers as follows: The first digit indicates a large geographic province; the second, a mountain group or other large subunit; the third figure denotes the landscape unit; and the fourth, identifies the hydrographic basin or karst sector. For example, Scărişoara Ice Cave's code is 3448/8, which translates as follows: 3 = Apuseni Mountains, 4 = Bihor Mountains, 4 = closed basins and high plateaus, 8 = Ocoale Close Basin, 8 = number of the cave (order) in the basin. For caves in non-carbonate rocks, a lithological specification is added in front of the catalog number, separated by a dash (01—salt, 02—gypsum, 03—conglomerate, 04—sandstone, 05—volcanic agglomerate, 06—magmatic rock; crystalline shale, 08—loess, 09—sand or clay).

Romania's speleogram is a statistical analysis of the speleological inventory performed on the basis of the cave catalog information (Goran 1982, 1989). Presented in a table-type format, it summarized for all categories of geomorphological units and types of rocks, the total, average, maximum, minimum, or percentage values for the main parameters and characteristics of the caves (density, altitude, number of entrances, length, vertical range, etc.). As an example, the 6816 known caves in 1982 in Romania had the density of the underground network in karst areas of 136.8 m/km^2, the average development of 97.1 m, the average vertical range of 9.3 m, and 103 caves with a cumulative length of 4822.6 m were formed in non-carbonate rocks.

Because the number of caves almost doubled over the past 35 years, Romania's current speleological inventory is significantly different from the published catalog of Goran (1982). The reason for this is the exploratory speleology that lately focused on the systematic investigation of some karst areas and on extending certain cave networks using modern techniques. Even if for some karst units there are speleological inventories or recent monographs, the idea of publishing a new, more systematic and complex catalog of caves in Romania should rank high among the priorities of the "Emil Racoviță" Institute of Speleology.

References

Anonymous (1789) Geschichte der Veteranischen Höhle: Mit einem Plan. Ein Beytrag zur Geschichte der Kriege zwischen Oestrreich und der Pforte. Frankfurt/Leipzig, p 33

Bielz A (1884) Beitrag zur Höhlenkunde Siebenbürgens als Vorarbeit einer Anthologie dieses Landes. Jahrb des Siebenbürg Karpathen-vereins IV:1–66

Bielz A (1886) Zweiter Nachtrag zu meinen Beiträgen zur Höhlen-kunde Siebenbürgens. Jahrb des Siebenbürg Karpathenvereins VI:171–178

Bleahu M, Povară I (1976) Catalogul peşterilor din România. Ed CNEFS, Bucureşti, p 53

Chappuis PA, Jeannel R (1951) Enumération des grottes visitées, 1927–1949 (8e série). Arch Zool Exp Gén 88:81–230

Fridvaldszky J (1767) Mineralogia Magni Principatus Transylvaniae. Claudiopoli, Typis Academicis Societatis Jesu, p 206

Goran C (1982) Catalogul sistematic al peşterilor din România. Ed CNEFS, Bucureşti, p 496

Goran C (1989) La spéléogramme de la Roumanie. Misc Speol Rom 1:335–346

Goran C (2002) Cadastrarea carstului. Ecocarst 3:5–18

Jeannel R, Racovitza EG (1929) Enumération des grottes visitées, 1918–1927 (7e série). Arch Zool Exp Gén 68(2):293–608

Onac BP, Goran C (2018) Karst and caves of Romania: a brief overview. In: Ponta GML, Onac PB (eds) Caves and karst systems of Romania, Springer International Publishing, Cham, pp 21–35

Orghidan T, Puşcariu V, Bleahu M, Decu V, Rusu T, Bunescu A (1965) The map of Romania's karst regions. Lucr Inst Speol "Émil Racoviță" IV:75–104

Useful Websites

http://asztrorege.blogspot.ro/2010/07/pestera-lui-veterani.html
http://www.zobodat.at/pdf/Hoehle_040_0011-0015.pdf
www.speologie.org

Karst Hydrogeology

Gheorghe M. L. Ponta

Abstract

The availability of groundwater in karst varies widely due to complex structure, geology, and degree of karstification. The karst aquifer system has an extensive network of interconnected joints, fractures, dissolution/solution cavities, within the carbonate group, that affect their capacity to store and transmit water from the land surface to springs. At the boundary between noncalcareous and karst formations, streams are sinking underground, generating caves.

Keywords

Karst hydrogeology • Spring • Ponor • Hydrogeological legend

Introduction

The availability of groundwater in karst varies widely due to complex structure, geology, and degree of karstification. The karst aquifer system has an extensive network of interconnected joints, fractures, dissolution/solution cavities, within the carbonate group, that affect their capacity to store and transmit water from the land surface to springs. At the boundary between noncalcareous and karst formations, streams are sinking underground, generating caves.

In limestone terrain, the rainwater percolates downwards under the combined influence of capillarity and gravity through fissures, fractures, joints, and along bedding planes through the *epikarst* and the *vadose zone* to reach the aquifer system. The deeper zone in the aquifer system, where all the openings are filled with water, is known as *phreatic zone*. The caves located in this zone are totally submerged. This aquifer system is a reliable source for water supply.

G. M. L. Ponta (✉)
Geological Survey of Alabama, 420 Hackberry Lane, Tuscaloosa, AL 35401, USA
e-mail: gponta@yahoo.com

The location where a surface stream disappears underground in a karst region is known as ponor (sinking stream). Sinking streams may be classified according to their flow (partial or total loss), distance (single point or diffuse water loss along a streambed), perennial or temporary running penetrable (cave) or impenetrable inlet (Fig. 1).

The spring flow from fractured and cavernous limestone is significantly more than from other materials as a seep of water springing from hillside, but both are called springs. Many of the springs are structurally controlled, occurring where fault planes intersect the surface. Springs may be classified or grouped in several ways according to their (1) mean flow or discharge, (2) geologic setting (rock structure), (3) mean temperature, and (4) chemistry of their water (LaMoreaux and Tanner 2001). A modified table after Meinzer (1923) and Springer et al. (2008), with a new column showing the relationship between the springs' symbols used on Fig. 1, and classification of springs based on magnitude of mean flow is shown in Fig. 2.

A Brief History of Karst Hydrogeology Research in Romania

Although Romania has only about 4602 km^2 of calcareous rocks (Senco 1968), they are spread all across the country, with springs used for public and domestic water supply for a very long time. The presence of numerous ponors (sinking stream) gave the name of several villages like Ponoare (Mehedinți), Ohaba Ponor (Sebeş Mountains), and Ponoare Valley (Vaşcău Plateau). In the karst literature, the word "ponor" is considered to have a Slavic origin, but the Romanian root should also be taken into account for future references. Mihuția performed the first investigation of the Boiu Spring in 1901 using coal dust as a tracer, which proved the connection between Câmpeneasca Cave and Boiu (Mihuția 1904). This was the first recorded tracer test performed in Romania.

G. M. L. Ponta and B. P. Onac (eds.), *Cave and Karst Systems of Romania*,
Cave and Karst Systems of the World, https://doi.org/10.1007/978-3-319-90747-5_6

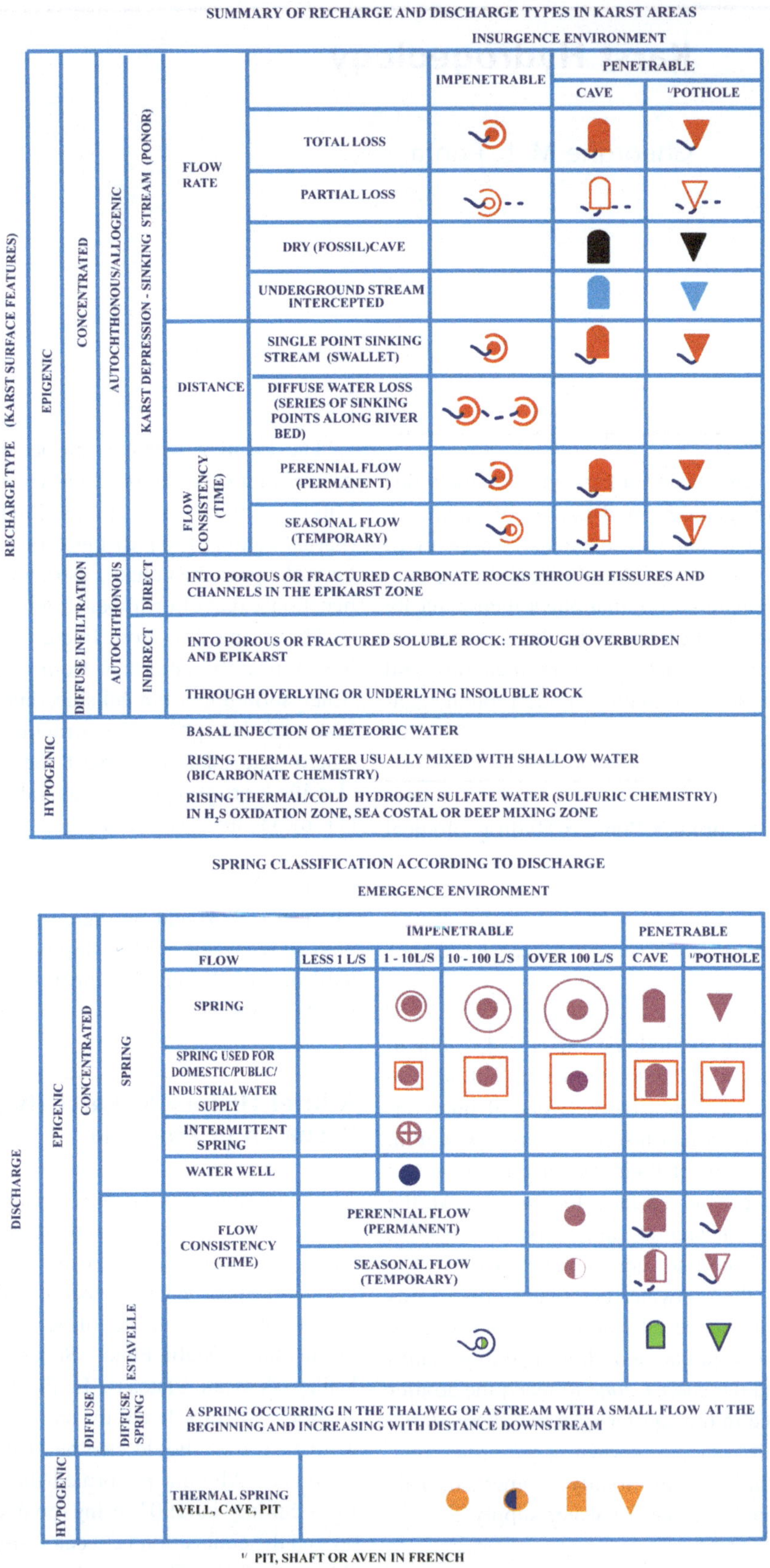

Fig. 1 Summary of recharges and discharge types in karst areas

Meinzer, 1923					Symbol on Figs. 3 and 6	Springer et al., 2008				
Magnitude	Flow (ft³/s, gal/min, pint/min)		Flow (L/s, mL/s)		Flow (L/s)	Magnitude	Flow (gal/min)		Flow (mL/s, L/s, m³/s)	
	From	To	From	To			From	To	From	To
1st magnitude	> 100 ft³/s		>2,831 L/s		>100	Eight	>44,880 gal/min		>10.0 m³/s	
2nd magnitude	10 ft³/s	100 ft³/s	283 L/s	2,831 L/s	>100	Seventh	4,488 gal/min	44,880 gal/min	1.0 m³/s	10.0 m³/s
3rd magnitude	1 ft³/s	10 ft³/s	28 L/s	283 L/s	>100	Sixth	448.8 gal/min	4,488 gal/min	100 L/s	1.0 m³/s
4th magnitude	100 US gal/min	1 ft³/s (448 US gal/min)	6.3 L/s	28 L/s	>10	Fifth	100 gal/min	448.8 gal/min	10 L/s	100 L/s
5th magnitude	10 gal/min	100 gal/min	0.63 L/s	6.3 L/s	>1	Fourth	10 gal/min	100 gal/min	1.0 L/s	10 L/s
6th magnitude	1 gal/min	10 gal/min	63 mL/s	630 mL/s	<1	Third	1 gal/min	10 gal/min	0.1 L/s	1.0 L/s
7th magnitude	2 pint/min	1 gal/min	16 mL/s	63 mL/s	<1	Second	0.12 gal/min	1 gal/min	10 mL/s	100 mL/s
8th magnitude	less than 1 pint/min	less than 1 pint/min	0.00	<8 mL/s	<1	First	0	0.12 gal/min	0	10 m L/s
0 magnitude	no flow (sites of past/historic flow)	no flow (sites of past/historic flow)	no flow	no flow	no flow	unmeasurable	no discernable flow to measure			

Fig. 2 Classification of springs, based on magnitude of mean flow (modified after Meinzer 1923 and Springer et al. 2008)

In the second half of the twentieth century, Orghidan et al. (1965) and Senco (1968) with the "Emil Racoviță" Institute of Speleology (ISER) and the Institute of Geography, respectively, published the first karst maps of Romania. About the same time, M. Şerban, D. Coman, I. Viehmann of the ISER Cluj Napoca; and I. Povară, T. Constantinescu, G. Diaconu, C. Goran, and C. Lascu of the ISER Bucureşti conducted numerous dye studies in Apuseni Mountains and Southern Carpathians, respectively. V. Trufaş worked extensively in Sebeş Mountains, whereas M. Pascu in Cerna Valley.

In the 1970s and 1980s, G. Simion, I. Lazu, and I. Orăşeanu, along with N. Terteleac, I Iurkiewicz, M. Mitrofan, G. Ponta, R. Strusiewicz, and E. Strusiewicz from the Geological and Geophysical Prospecting Company (formally IGPSMS), Bucureşti, began a systematic karst inventory and dye/tracer studies in the Western part of the Country (Cerna—Jiul de Vest area and Apuseni Mountains, respectively). Because in the late 1970s and early 1980s there was a shortage of hard currency (US dollars) in Romania, the researchers above teamed up with the Institute of Physiques and Nuclear Engineering Bucureşti (E. Gaşpar) and conducted numerous tracer studies with radioactive (tritium) and activable isotopes (e.g., In-EDTA). During the same period, S. Bulgăr, P. Niță, and P. Miță with the Meteorological and Hydrogeological Institute from Bucureşti conducted studies in the Cerna Valley.

After 1990, the new generation of hydrogeologists, D. Slăvoaca, R. Slăvoaca, R. Bandrabur, G. Bandrabur, and G. Dragomir, under the supervision of I. Orăşeanu, performed new dye studies and repeated some of the old ones, this time with fluorescein and rhodamine. An extensive history of Romanian Karst hydrogeology can be found in Orăşeanu (2010).

Karst Hydrogeological Maps

In the second part of the last century, the Geological Institute of Romania developed a geological (scale 1:50,000) and a hydrogeological (1:100,000) mapping programs. In the 1980s, the hydrogeological maps (1:100,000) were completed in most of the lowlands/plains of Romania, and a new hydrogeological program in the Carpathian Mountains was ready to begin. In the mid-1980s, the Commission on Hydrogeological Maps within the International Association of Hydrogeologist (IAH), in cooperation with International Association of Hydrological Sciences (IAHS) and UNESCO, prepared a revised edition of the International Standard Legend for Hydrogeological Maps (ISLHM), published as a UNESCO technical paper in 1983. The modified version of the legend was used to initiate the Romanian Hydrogeological Mapping program for karst terrains at scale 1:50,000. Since the program started, one map was printed (Vaşcău) and another four are in the draft form (Zece Hotare, Poiana Horea, Pietroasa, Pui).

In 1995, Struckmeier and Margat published *Hydrogeological Maps, a Guide and a Standard Legend*. The regional aquifers are shown in blue (intergranular aquifers) and green (fissured aquifers), and brown is used for rocks with local groundwater resources (light brown) or little or no usable groundwater (dark brown). The karst aquifers are included in the section *Fissured Aquifers, Including Karst Aquifers* (Ponta 2003). Because a large volume of data related to groundwater/surface water in karst areas was available, they are presented on the karst hydrogeological maps of this book (Fig. 3), using a range of pink to differentiate aquifers in karst terrains. The traditional range of green was used to represent the fissured nonkarst rocks (Ponta 2003). This legend is very similar to the one used by Institute of Geology and Geophysics in Bucharest (Romania) for their karst hydrogeological maps at 1:50,000 scale.

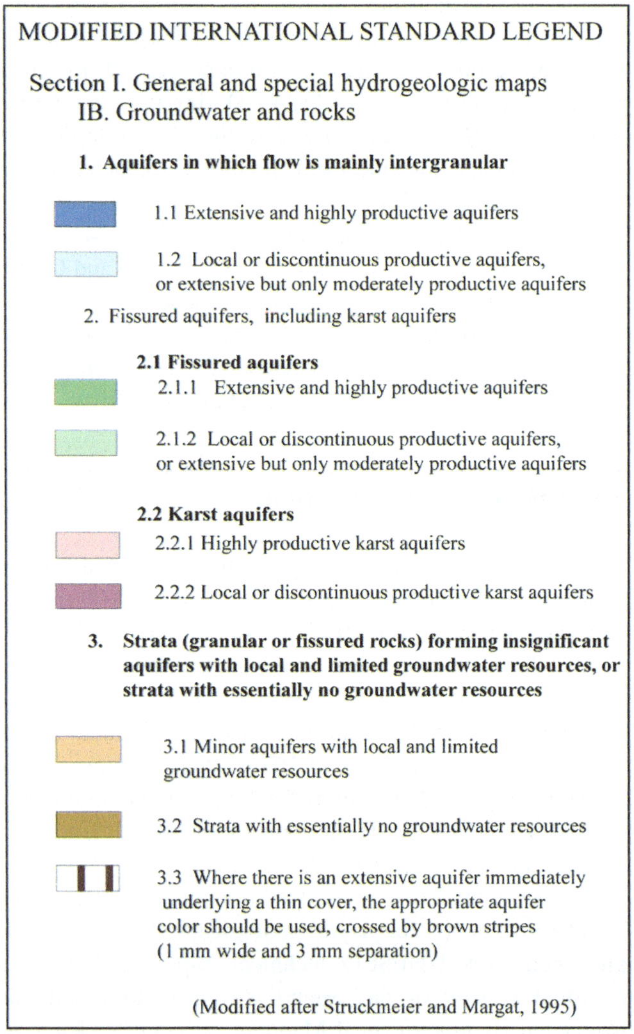

Fig. 3 Modified International Standard legend for hydrogeological maps, after Struckmeier and Margat (1995)

Based on the discharge rates of springs and wells, constancy or type of rocks (more or less karstified), presence of caves, tectonic/structural control, and karst surface features, the karst aquifers can be subdivided into two subcategories as: *Highly productive karst aquifers* and *Local or discontinuous productive karst aquifers* (Ponta 2003).

Method of Investigation Tracer and Dye Studies

Worldwide, to define the recharge area of a spring, meaning to trace the path between sinking points and resurgences, is performed with different types of fluorescent dyes (fluorescein, rhodamine), lycopodium spores, or chemicals such as table salt. Up to the mid-1970s, the majority of the dye studies in Romania were performed with fluorescein. In the mid-1970s and 1980s, E. Gaşpar set up the methodology for utilizing radioactive tracers as ^{82}Br (Bromine), ^{131}I (Iodine), and In-EDTA activable tracers.

In-EDTA

Here, we provide a few details on In-EDTA, a frequently used tracer in hydrogeology. Indium under the complex EDTA form can be detected at concentrations as low as 10^{-12} g/L, and the necessary quantity for a tracer study is very small. In addition, In-EDTA is a very safe and environmentally friendly. The tracer is extracted from the collected water samples through precipitation process with bismuth hydroxide and is captured on a nuclear membrane (Gaşpar 1994). After drying at the room temperature, the precipitate is removed from the filter and then warm-encapsulated in polyethylene sheet. The capsules containing the precipitate in the form of Indium hydroxide powder are activated through neutron irradiation with the help of a pneumatic tube in the nuclear reactor, along with the reference sample, and the intensity of the 417 keV radiation is measured (Gaşpar 1994).

Dye Studies

A dye trace study has to be planned based on understanding of the hydrogeologic setup of project site, surrounding area, and karst inventory. At the spring, well, or in the creek, the charcoal bags have to be attached to a small boulder and positioned at locations that are protected from sunlight (Fig. 4). The personnel collecting the charcoal bags have to use new disposable latex gloves at each location to avoid cross-contamination. Also, it is recommended that the person launching the dye should not to be part of the monitoring

Fig. 4 Charcoal bag used at dye studies

team to avoid accidental contamination of the charcoal bags or water samples. The charcoal bags should be introduced in labeled plastic bags upon retrieval and immediately stored in a cooler with ice. Chain-of-Custody procedures have to be employed. A new charcoal bag will be placed at the same location in the stream or spring. Locations for the deployment of charcoal bags for the execution of the dye tracer test have to be provided in a spreadsheet (date and time recorded) and shown on a map.

Background (Preinjection) Monitoring

Detergents, bathroom cleaners, pigments for inks and dyes, antifreeze, industrial wastes, naturally occurring mineral fluorescence, and residual dye from the previous studies may be all sources of fluorescence in water. Background fluorescein (the dye which will be used) at the injection site and spring and well will be monitored by installing charcoal bags at locations prior to initiation of the dye tracer test.

The charcoal bag will be removed for evaluation prior to injection of the dye and replaced with a new one based on a pre-established monitoring program. The dye tracer test has to be designed to maintain environmental and human safety during the performance of a tracer test. Using too little tracer can lead to no detection or an ambiguous result. Using too much dye will result in a long duration of a green-colored water flow, which will be a concern for the authorities and public, mainly if the source is used as water supply. Always individuals and authorities have to be informed prior to initiation of a tracer test. There are numerous methods to estimate the quantity of the dye, with one frequently used being based on the Worthington and Smart empirical formula (Goldscheider and Drew 2007):

$$M = 1.9 \times 10^{-5}(LQC)^{0.95}$$

where M is the required trace quantity, i.e., mass (kg), L is distance in (km), Q is discharge (L/s), and C is the target peak concentration (μg/L).

Some researchers believe that a dye study based solely on charcoal bags should be considered only a qualitative study. For a relevant quantitative investigation is recommended that the charcoal bags to be back up by water samples. For a dye study that unequivocally documents the connection between two points, the results should be at least three times higher than the values of the background monitoring.

References

Gaşpar E (1994) Trasori în sisteme acvifere. Ed Acad Române, Bucureşti, p 234

Goldscheider N, Drew D (2007) Methods in karst hydrogeology. Taylor & Francis, p 264

LaMoreaux PE, Tanner JT (eds) (2001) Springs and bottled waters of the world. Springer International Publishing, Cham, p 315

Mihuţia S (1904) A vaskohi meszko-fensik hidrographiai viszonyai. Foldr kozlemenyek XXXII:1–31

Meinzer OE (1923) Outline of ground water hydrology, with definitions. U.S. Geological Survey Water Supply Paper 494, Government Printing Office, Washington, DC, p 71

Orăşeanu I (2010) Short history of the hydrogeological investigations of the karst in Romania. In: Orăşeanu I, Iurkiewicz A (eds) Karst hydrogeology of Romania. Ed Belvedere, Oradea, pp 19–26

Orghidan T, Puşcariu V, Bleahu M, Decu V, Rusu T, Bunescu A (1965) Harta regiunilor carstice din România. Lucr Inst Speol "Emil Racoviţă" 4:75–104

Ponta G (2003) Proposed modifications of the international standard legend for large-scale hydrogeological maps in carbonate rocks. International conference on karst hydrogeology and ecosystems. Western Kentucky University, Bowling Green, Kentucky

Senco V (1968) La carte du karst et du clasto-karst de Roumanie. Rev Roum Géol Géophys géogr, ser Géogr 12(1):35–41

Springer AE, Stevens LE, Anderson DE, Parnell RA, Kreamer DK, Levin AL, Flora SP (2008) A comprehensive springs classification system: integrating geomorphic, hydrogeochemical, and ecological criteria. In: Stevens LE, Meretsky VJ (eds) Aridland springs in North America: ecology and conservation. University of Arizona Press, Tucson, AZ, pp 49–75

Struckmeier WF, Margat J (1995) Hydrogeological maps: a guide and a standard legend. Int Assoc Hydrogeologists, 17:177

Rodna Mountains: Izvorul Tăusoarelor Cave (Pestera de la Izvorul Tăusoarelor)

Bogdan P. Onac, Virgil Drăguşin, Felix Papiu, and Crin-Triandafil Theodorescu

Abstract

Izvorul Tăuşoarelor Cave, which held for many years the record for the deepest cave in Romania, is renowned for its steeply descending galleries, tectonic-controlled speleogenesis, and a rich assemblage of sulfate minerals (gypsum, mirabilite, arcanite, bassanite, epsomite, konyaite, leonite, and syngenite) that form a variety of speleothems. The enigmatic spherical concretions, known as "Tăuşoare balls," formed simultaneously with the limestone bedrock during its early diagenesis. The existence of hundreds of bones of cave bear (*Ursus spelaeus*) and brown bear (*Ursus arctos*) suggests the two species cohabited during the Quaternary in this part of the country. The cave represents the hibernation site for four bat species.

Keywords

Minerals • Sulfates • Speleogenesis • Tectonics
Limestone concretions • Paleontology

B. P. Onac (✉)
Karst Research Group, School of Geosciences, University of South Florida, 4202 E. Fowler Ave., NES 107, Tampa, FL 33620, USA
e-mail: bonac@usf.edu

B. P. Onac
"Emil Racoviţă" Institute of Speleology, Romanian Academy, Clinicilor 5, 400006 Cluj-Napoca, Romania

V. Drăguşin
"Emil Racoviţă" Institute of Speleology, Frumoasă 31, 010986 Bucharest, Romania
e-mail: virgil.dragusin@iser.ro

F. Papiu
"Omega"/CSER Speleological Club, N. Iorga 1A, Cluj-Napoca, Romania
e-mail: papiufelix@yahoo.com

C.-T. Theodorescu
Luşca, 5, Năsăud, CP 425200 Bistriţa-Năsăud, Romania
e-mail: triandafil96@yahoo.com

Geographic, Geologic, and Hydrologic Settings

The cave entrance is situated at an altitude of 942 m in the southwestern part of the Rodna Mountains, on the eastern side of Bârlea Massif, in the left (east) bank of the Izvorul Tăuşoarelor Valley (Fig. 1). The easiest way to reach the cave is following the National Road DN 17D from Năsăud to Rebrişoara, from where a forest, unpaved road along Gersa Valley ($\sim$28 km) passes within 100 m from the cave entrance.

Izvorul Tăuşoarelor Cave (hereafter Tăuşoare) develops in an N-S-oriented narrow stripe of pure to sandy upper Eocene (Priabonian)–lower Oligocene (Rupelian) limestone interbedded with thin layers of black bituminous shales (Munteanu 2004) (Fig. 2). Both limestone and shale contain disseminated fine-grained pyrite in framboidal form. This 60 m thick karst-forming unit (strata dip $\sim$65° toward NW; Silvestru and Viehmann 1982) overlies a bed of conglomerates of middle Eocene age, which in turn lays transgressively and discordantly on the crystalline basement of Rodnei Mountains (Kräutner et al. 1989). Sandstones and a highly weathered rhyolite body outcrop in the vicinity of the cave, providing the bulk of the allogenic sediments transported underground. These are an important source of Na, K, and Mg, which once released into solutions combine with another ion and contribute to the genesis of various cave minerals. The contact between the sedimentary and crystalline rocks is along a fault line-oriented N-S (Silvestru and Viehmann 1982).

A series of tectonic and microtectonic studies in and around Tăuşoare Cave revealed that the main directions of fractures in the region are W-E, N-S, and N 35°E. The first coincides with two major faults that delimit Rodna Mountains to the north and south. The second is parallel to the one separating the sedimentary and metamorphic units, whereas the third one, although mostly affecting the crystalline rocks, was traced in the limestone as well (Silvestru and Viehmann

G. M. L. Ponta and B. P. Onac (eds.), *Cave and Karst Systems of Romania*,
Cave and Karst Systems of the World, https://doi.org/10.1007/978-3-319-90747-5_7

Fig. 1 Location of Izvorul Tăușoarelor Cave in the Rodna Mountains

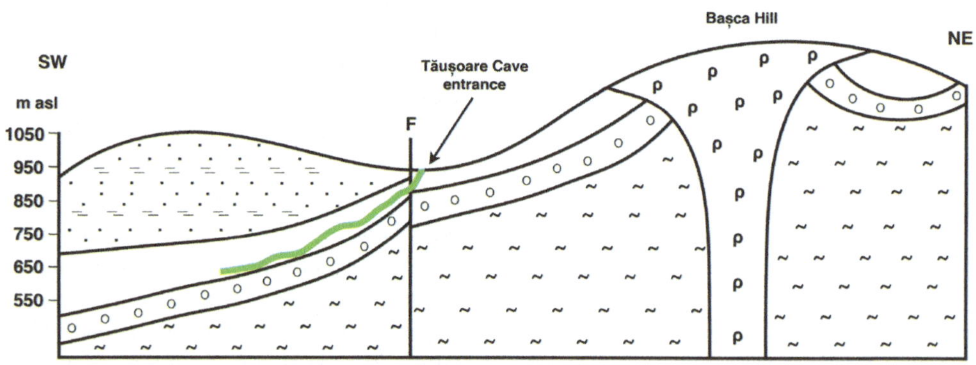

Fig. 2 Geological cross section in the surroundings of the Izvorul Tăușoarelor Cave (geology after Kräutner et al. 1989)

1982). Furthermore, Silvestru (1984) depicted three types of fissures in the cave generated by rupture strain, shear, or tension, all apparently playing a significant role in speleogenesis. Along the same line, Fabian (1984) concluded that the strike of the tectonic effort in the Tăușoare area is N 76°W, being an important factor in the development of the cave.

From a hydrogeologic point of view, the cave is part of the Tăușoare-Zalion hydrokarstic system of the Telcișor catchment area (Iurkiewicz 2010). The water of the Izvorul Tăușoarelor Valley sinks ~300 m upstream from the cave entrance, reappears in various underground locations, and flows along different cave passages to ultimately resurface in the Izvorul Rece (Cold Spring) situated at ~5.7 km from the terminus point of Tăuşoare Cave (Viehmann et al. 1964), at an altitude of 550 m. The hydrologic connection was reported by the above authors after successfully conducting two tracer tests on December 1, 1957, and June 13, 1961, respectively.

History of Exploration

Izvorul Tăuşoarelor Cave was discovered in 1955 by Leon Bârte, a teacher from Parva village, who contacted I. Viehmann then a teacher himself in Năsăud, but whom a year later joined the "Emil Racoviţă" Speleological Institute (ERIS) in Cluj. Between 1956 and 1957, a team led by I. Viehmann, L. Bârte, and others from ERIS (M. Şerban, T. Rusu) explored the cave between the cave entrance and the Dining Room (Sala de Mese), discovering the first part of the Headway Gallery (Galeria de Înaintare) and the Z Gallery (also known as the "700-Step" Gallery (Galeria de 700 paşi; Bleahu et al. 1976). In 1957, wood ladders were placed along the main route to facilitate the access down to the Dining Room. The exploration and mapping of the cave continued between 1956 and 1971, when a number of new rooms and passages were discovered. Among these are: Balls' Room (Sala Bilelor), Sugar Cubes' Chamber (Sala Cuburilor de Zahăr), the Amphitheater, Gypsum (Galeria Gipsului), Steps (Galeria Săritorilor), Kilometer's (Galeria Kilometrului), Cosmuţa, Climbers' (Galeria Alpiniştilor), and Dry (Galeria Uscată) galleries (Fig. 3). By the end of 1971, the cave reached 5.1 km in length and became the deepest in Romania as the Old Ponor (Sorbul Vechi) was 350 m below the cave entrance (Viehmann 1973). Izvorul Tăuşoarelor Cave has been declared a scientific reserve and placed under the protection of the Romanian Natural Monuments Commission in 1965.

In July 1971, during a Belgian-Romanian expedition led by I. Viehmann, L. Vălenaş (then a Geography student) discovered a short passage (now known as the Vălenaş Gallery) at the end of which he entered into a whole new part of the cave, mainly a canyon-type (Belgians' Gallery) with an underground stream whose water disappears after ~250 m between some boulders in the New Ponor (Sorbul Nou). When mapped the deepest point of this passage (−356 m) was found to be below the Old Ponor, thus becoming the deepest point in the cave. Upstream from the junction with the Vălenaş Gallery, the team explored 600 m of new passages. Consequently, the discoveries made by Vălenaş added 1075 m to the total length of the cave (Viehmann 1973) (Fig. 3).

Beginning with 1974, the Emil Racoviţă Caving Club from Cluj took over the exploration activities, with M. Domşa, C. Popa, D. Moldovan, R. Patalita, E. Silvestru, V. Bocîrnea, O. Moldovan, M. Foia, R. Mihăilaş, S. Obrejan cumulating the most hours spent in exploration and mapping (Popa 1988).

Fig. 3 Map of the Izvorul Tăuşoarelor Cave

Their effort resulted in a considerable extension of the Gypsum Gallery and the discovery of three new systems [Sasca, Mezei, Scholars (Elevilor)], and Girls (Fetelor) Gallery, along with some small side passages (Popa 1988; Papiu 2007). In 1981, M. Domşa and C. Popa initiated and led the efforts of completely remapping the cave. On this occasion, the Mountain's Room (Sala Muntelui) is surveyed along with other less important passages near the cave entrance. The length of the cave reaches 13 km, and the deepest point was found to be at −347.5 m. Between then and now, the only significant discovery was made by D. Nicoară and M. Domşa who found in 1985 the Leon Bârte Room, part of the "New System" (Domşa 1990). In its easternmost part (Cave Bear Room), they uncovered bones of *Ursus spelaeus* and *Ursus arctos* and reached the highest elevation of the cave (+80 m).

In order to develop a management plan for the cave, the Local Council of the Bistriţa-Năsăud City requested a new, more detailed map. The process was completed in 2013, and to the surprise of many, the new measurements indicate the cave's total length to be only 8650 m and the deepest point is in the Old Ponor (−329 m), whereas the highest elevation of +80 m is in the northernmost part of the Cave Bear Room (Fig. 3).

Cave Description

Immediately after passing the entrance gate, a metal ladder eases the descent of a 12 m pit that gives access to a passage with its floor covered by breakdown blocks. This continues for 50 m, after which it becomes a steeply descending, large and lofty fracture passage called the Headway Gallery. Along it flows a stream that receives two tributaries from NE galleries and another one that drains down the Parallel Gallery (Fig. 3). After climbing down the 4th ladder, the Headway Gallery continues as an active stream passage with waterfalls and abundant sediment deposits all the way to the Old Ponor. From the 4th ladder, a metal bridge over the stream on the Parallel Gallery take us at the entrance in the Z Gallery, a descending dry passage that ultimately leads to the Dining Room, which is followed by the Balls' Room.

During the early exploration of the Tăuşoare Cave, 13 enigmatic spherical concretions (8–30 cm in diameter and each weighing 2–4 kg) were found on the floor of the Balls' Room (Fig. 4). Originally they were interpreted as grinders, i.e., limestone cobbles rounded as they swirled around by eddies at the bottom of potholes (Şerban et al. 1961). This explanation failed when similar concretions were found embedded in the limestone wall at various locations in the cave (e.g., Mezei System; Fig. 4). A new hypothesis was put forward by Fabian and Viehmann (1979), who suggested the concretions formed by subaerial weathering of limestone within some so-called karst pockets. The most recent study addressing the origin of the "Tăuşoare balls" used microfacies, chemical, and mineralogical analyses on both limestone and concretions (Munteanu 2004). The author concluded that these nodules formed during diagenesis, at the same time when carbonate sediments were turned into limestone.

Upstream (east) from the Dining Room begins the ascent into the Kilometer's Gallery, which ends in the Mountain's Room. Characteristic to both these passages is the huge amount of limestone breakdowns accumulated on their floor and the presence of the second permanent water stream entering the cave in multiple locations in the upper part of the Mountain's Room. South from the Dining Room is the entrance in the Gypsum Gallery, named after its beautiful gypsum flowers and crusts covering a section of it (Fig. 5).

Fig. 4 Carbonate nodules ("Tăuşoare balls") detached (**a**) and embedded (**b**) in the limestone wall (photograph by C. Theodorescu)

Fig. 5 Gypsum flowers (photograph by C. Theodorescu)

In this gallery, we come across the third underground river that flows toward the Headway Gallery. Following it upstream, after just 50 m the passage becomes rather narrow and the water emerges from a narrow fracture at the base of the cave wall. The gallery continues with a climb up on a fossil passage (from place to place the river appears but then disappears again) along which some chimneys and large limestone blocks need to be negotiated. Next, climbing a 3-m waterfall and squeezing through the ironically named Fats' Alley (Aleea Graşilor), we enter the Leon Bârte Room, part of the "New System." In this sector, the cave has several large, ascending passages that all join in the Cave Bear Room, from where a NE trending gallery reaches the highest elevation in the cave (+80 m). The third stream flowing through these galleries originates from this part of the cave.

From the Dining Room, a passage that opens in the SE wall near the ceiling represents the access point into the Mezei labyrinth system. Near the west end of the Dining Room, in its northern wall, a gallery gives way into the Sasca System that ultimately connects with the Headway Gallery. Downstream on this gallery, following any of the fossil passages opening in its left wall, we can return into the Dining Room passing through the Sugar Cubes' and Balls' Room. However, the easiest access is along the Climbers' Gallery. Continuing from the confluence of waters coming from Climbers' and Headway galleries, on the right side is the Girls Gallery and a system of small passages partially filled with sediments. From here, the cave continues down to −329 m, where three of the four underground rivers disappear in the Old Ponor. At present, this is the deepest point of the cave. Back in 1971, the New Ponor was explored to a depth of −356 m. However, during a large flash flood that occurred in 2005, the end part of the cave was filled with sediments; thus, the last survey indicates this point is only −302 m below the cave entrance.

Moving downstream from the Girls Gallery, the stream passage can be bypassed via the Dry Gallery that ends in the Amphitheater Room, a large circular chamber filled with very fine sand. From its proximity, the Vălenaş Gallery connects to the Belgians' Gallery along which the fourth river flows. Seventy meters downstream from the intersection point of the two galleries the New Ponor is reached. If followed upstream, the Belgians' Gallery is steeply ascending, presents several dozens of small waterfalls, and has a labyrinth of small corrosion and collapse galleries that form the Scholars System. The new map shows that between this system and the Headway Gallery there are only 3 m, but these are completely filled with breakdowns and sediments.

Speleogenesis

The hypothesis that tectonics was crucial to the development of Tăuşoare Cave was first advanced by Viehmann and Şerban (1962–1963). They noticed that most passages (dry or active) are very high and narrow, typical for water flowing along major fractures in the bedrock (Fig. 6). Authors also documented the presence of significant amounts of sediments (from large cobbles to fine-grained sand) along these galleries, clearly indicating vigorous flow and catastrophic floods at times. At various locations throughout the cave, the water also exploited bedding planes, widening some sections (Fig. 7) or creating flat-ceiling, broader and much lower galleries (Viehmann et al. 1964).

The relationship between tectonics and speleogenesis in the karst of the Rodna Mountains and specifically in Tăuşoare Cave was further substantiated by using tectonograms and detailed cave morphology observations (Silvestru and Viehmann 1982; Silvestru 1984). The studies revealed that most cave passages are oriented E-W and N-S, mirroring the major fault line directions in the region. The galleries bearing N 40° W represent tension fissures that were exploited by underground streams. The depth of the cave (−329 m) is explained by step faulting and a high limestone beds dip (20–22°; Silvestru 1984). Without discounting any of the above hypotheses, Onac and Cocean (1996) suggested Tăuşoare is a drawdown vadose cave formed by a sinking stream in an interfluvial setting.

Considering the most recent speleogenetic theories (Audra and Palmer 2013) and taking into account cave passages morphology and the abundance of sediment infill, we consider Tăuşoare a typical epiphreatic (floodwater) cave. Its underground stream is fed by the sinking water of the Izvorul Tăuşoarelor Valley (ca 300 m upstream from the cave entrance) that fluctuates greatly in discharge. Surface morphological evidences hint to several fossil ponors along the valley, including the present cave entrance (Viehmann et al. 1964). Runoff from adjacent insoluble rocks is the main source of recharge, which explains the chemical aggressiveness of water. Judging on the size of the cobbles and boulders present in various parts of the cave, severe floods were likely common in the past, especially during snowmelt

Fig. 6 The Headway Gallery has a typical morphology for a passage developed along a fracture (photograph by C. Theodorescu)

events. Dissolution caused by largely allogenic water and abrasion by water-borne sediments greatly enlarged the network of fractures/cracks and bedding planes (Figs. 6 and 7). Before the hydrologic breakthrough was achieved (the water from Izvorul Tăuşoarelor Valley started to discharge through the Izvorul Rece Spring), the enlargement of passages proceeds at slower pace. This is because the amount of water draining through the hydrokarst system was limited. It is mostly during this period in the cave evolution that abnormally high discharge (during floods) caused water level to rise behind natural constrictions (e.g., collapse, sediment infill) along the stream passages. At these times, water was injected into adjacent fractures and fissures or along bedding planes, enlarging them and creating some of the network mazes. Evidences of periodic floods are the thick rhythmic sediment sequences (Baciu 1990) that partially or completely fill many side passages (Fig. 8). The existence of bypass galleries (e.g., Scholars and Sasca systems, and two others along the Gypsum Gallery), which represent looping tubes formed above the normal water table under hydraulic pressure, attests short intervals of phreatic conditions triggered by floodwater fluctuations in the epiphreatic zone.

Cave Climate and Radon Concentration

There is a limited amount of information regarding the physical parameters (temperature and relative humidity) characterizing the underground climate of the Izvorul Tăuşoarelor Cave, which were mostly collected between 1956 and 1961 at 12 stations (Viehmann et al. 1964). It appears from this dataset that the mean annual temperature

Fig. 7 Fracture- and bedding plane-controlled passage in the Kilometer's Gallery (photograph by C. Theodorescu)

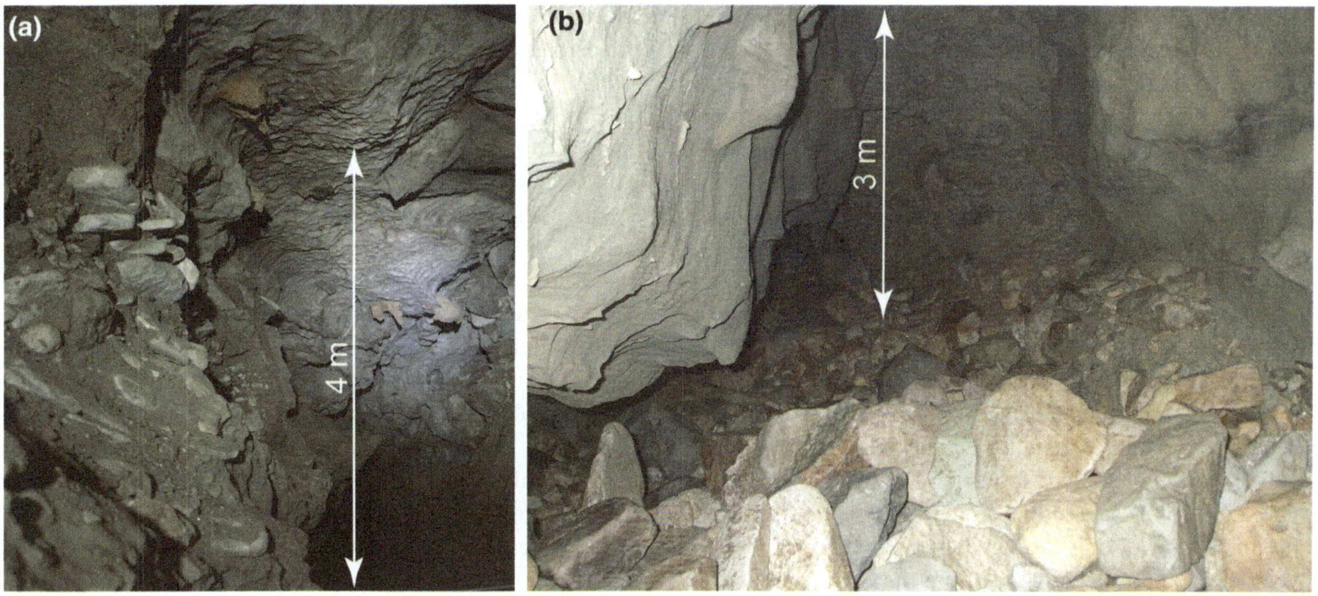

Fig. 8 Alluvial sediments in Gypsum (**a**) and Kilometer's (**b**) galleries (photograph by C. Theodorescu)

in the cave is 7.25 °C, varying between 6.5 and 8 °C. Occasional measurements undertaken by Viehmann and Onac (unpubl. data) between 1982 and the late 1990s show that while the warmest temperature never exceeded 8 °C, the lowest reached 5.8 °C. This difference was explained by changes in the size of the cave opening (which was enlarged) and the fact that at various times, the cave was completely sealed by a metal door. A Tinytag PLUS 2 temperature logger placed in the Mezei System between March and July 2015 recorded an average temperature of 6.77 °C, with a standard deviation of only 0.02 °C. The relative humidity varies between 85% (close to the entrance) to 100% in the deeper parts of the cave (e.g., Dining Room, the Amphitheater, Balls' Room).

It is worth noting a strange coincidence observed twenty years after the cave was discovered, that is, the occurrence of sulfate speleothems (see below) right after a major breakthrough that opened a new cave passage at the lower end of the Dining Room. A strong current of air has been felt ever since in this part of the cave.

Due to its complex configuration and topography, the ventilation regime is not yet fully understood. During winter, air moves throughout the cave from the entrance, down along the Z Gallery, into the Dining and Balls' rooms and up the Gypsum Gallery toward Cave Bears's and Mountain's rooms. In the summer, however, the aerodynamic exchanges with the exterior cease almost completely. The airflow is weaker this time of the year because the difference between the outside and inside air density is small. This pattern of air circulation was confirmed by radon concentration measurements at eight stations between the entrance (0 m) and Balls' Room (−190 m). The study showed higher variability of the radon values during the winter, but all remained below 1500 Bq m^{-3} (Cucoş Dinu et al. 2016). Instead, the summer values (except the one near the entrance) are constant and very high (between 2500 and 3000 Bq m^{-3}; Fig. 9), indicating a poorly ventilated environment that allows radon to build up. The study concluded that in order to avoid serious health issues, any professional activities during summer should not exceed 7–9 h/month (Cucoş Dinu et al. 2016).

Speleothems and Minerals

Ever since it was discovered, the cave became famous for its gypsum flowers (anthodites), crystals, and crusts abundantly covering the walls and ceiling of the Gypsum Gallery

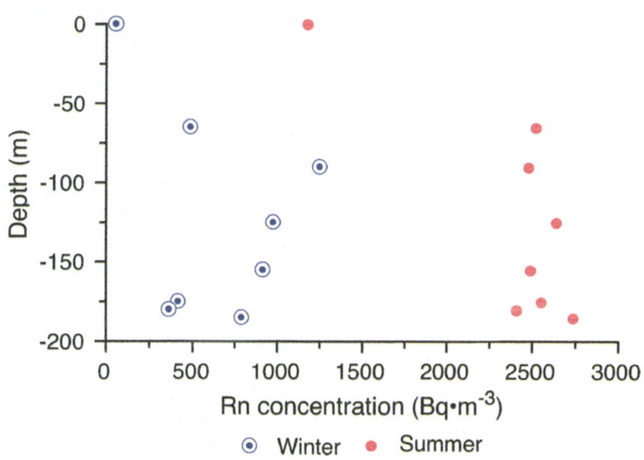

Fig. 9 Radon concentrations in eight locations throughout Izvorul Tăuşoarelor Cave

(Viehmann and Şerban 1962–1963; Viehmann et al. 1964). A detailed mineralogical investigation carried out on gypsum samples from this location revealed the presence of bassanite ($CaSO_4 \cdot 0.5H_2O$), a mineral that has never been documented from a cave environment prior to its discovery (Jude 1972). Another study aiming to describe the gypsum speleothems and provide a glimpse on their genesis was published by Onac (1987). All these studies consider oxidation of pyrite present in limestone (and black bituminous shale) and in the overlying sandstone as the main source for the sulfate ion. The process implies pyrite being oxidizes to sulfuric acid, which then reacts with limestone-producing calcium sulfate.

A sample of white, delicate needle-like crystals collected from the Dining Room allowed the identification of mirabilite, $Na_2SO_4 \cdot 10H_2O$ (Moțiu et al. 1977). The authors also reported the presence of thenardite (Na_2SO_4), but apparently this mineral was only a dehydration product of mirabilite when the sample was removed from the cave. The discovery of mirabilite triggered additional mineralogical research (Domşa 1988; Silvestru 1990), mainly because it was noted that the white efflorescences covering small areas (up to 1.5 m^2; Fig. 10a) of the cave floor appear and disappear from time to time. These new studies added two new minerals (arcanite, K_2SO_4 and epsomite, $Mg_2SO_4 \cdot 7H_2O$) and explained their genesis in different ways. All three studies agree on the source of the sulfate (resulted from the oxidation of pyrite present within the bituminous limestones), but have completely different views on the origin and transport of Na, K, and Mg ions.

During a sampling trip, one of the authors (BPO) noted that the appearance of the sulfates previously described from the Dining Room is different. When they occur on areas of the cave floor that is covered by sandy-clay sediments, white cotton-like and delicate acicular crystals form, whereas efflorescences are typically found on fragments of banded sandstone or on loose material resulting from their weathering (Fig. 10b). The collected samples were analyzed by means of X-ray diffraction, scanning electron microscopy, and microprobe. The results indicated that apart from mirabilite and epsomite, three new and very rare sulfates also occur. These are syngenite [$K_2Ca(SO_4)_2 \cdot H_2O$], leonite [$K_2Mg(SO_4)_2 \cdot 4H_2O$], and konyaite [$Na_2Mg(SO_4)_2 \cdot 5H_2O$], the last two never cited from a cave environment before (Onac et al. 2001). The origin of these minerals is strictly related to the reaction between the slightly acidic waters (see above) and Ca, Na, K, and Mg anions provided by limestone, sandstone (partly or completely weathered) and clay sediments. Flooding events that often occur in the Dining Room could also explain the genesis of sulfates. The water coming from the Gypsum Gallery is enriched in sulfates, thus impregnates the sediments where it reacts with various metal cations. This scenario could explain the changes in the abundance of these speleothems at different time periods.

Carbonate speleothems are represented by stalactites, stalagmites, columns, pools, flowstones, eccentrics, and a very unique, thick shelfstone projecting out from the cave wall like a table. After this speleothem, the hall was named the Dining Room. With very few exceptions (some aragonite stalagmites), all these speleothems are composed of calcite and are mainly found in the Balls' Room, Dining Room, and along the Gypsum Gallery (Viehmann 1975).

Geochronology

Uranium–thorium dating was performed on a series of stalagmites from the cave, at Centro Nacional de Investigación sobre la Evolución Humana (Spain) and at the University of

Fig. 10 **a** Efflorescences of mirabilite and leonite carpeting cave floor and blocks in the Dining Room, **b** close-up of the sulfate deposit (photograph by C. Theodorescu)

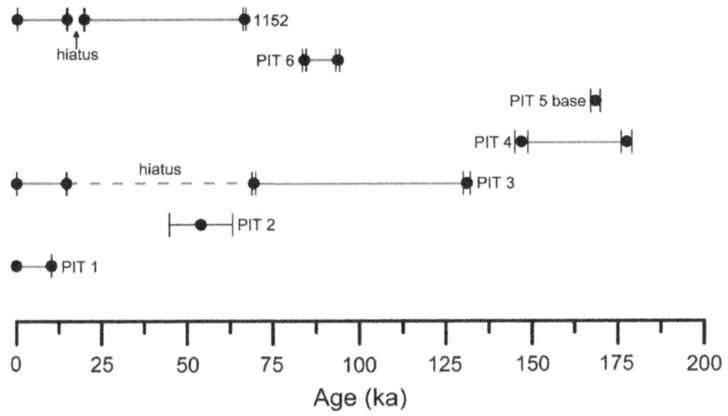

Fig. 11 Age and growth intervals for seven stalagmites dated by means of U-series method

Melbourne (Australia), by D. Hoffmann and J. Hellstrom, respectively. The preliminary data offer an insight into the depositional periods of calcite as a reflection of climate variability, as well as into the age of the cave (Fig. 11). The oldest age obtained for the base of stalagmite PIT 4 from the Mezei System (177.5 ± 1.5 ka) indicates that this particular part of the cave was already formed at that moment. This stalagmite stopped growing at ~ 147 ± 2 ka, implying its growth happened entirely during the glacial conditions of the Marine Isotope Stage 6 (MIS 6), or throughout a warmer period that occurred shortly before the MIS 6 glacial maximum.

The analysis of stalagmites, PIT 3 and 1152, shows that calcite deposition took place almost uninterrupted between the Last Interglacial and the present. Stalagmite PIT 3 started forming at ~131 ± 1 ka but had a long growth hiatus between 69.5 ± 0.5 and 14.7 ± 0.1 ka. On the other hand, stalagmite 1152 formed between 66.8 ± 0.4 and 0.5 ± 0.03 ka, with a short hiatus during the Last Glacial Maximum, between roughly 20 and 15 ka.

The lower part of two other stalagmites, PIT 2 and PIT 5, have been dated to 54 ± 9 and 168.5 ± 1.3 ka, respectively, indicating that they could be used in replication experiments for MIS 6 and MIS 3. The deposition of speleothems during extended cold periods when soil organic activity is reduced might be explained by the presence of disseminated pyrite within the limestone and overlaying rocks that promotes a dissolution mechanism involving sulfuric acid, as described at Corchia Cave in Italy (Regattieri et al. 2014).

Microbiology

The microbiological and enzymological studies performed by Manolache et al. (1991) indicated that the number of aerobic heterotrophic bacteria (alive and dead) and the mean value of the enzymatic indicator are low compared to other caves because of the absence of clay, which would favor the growth of bacteria and accumulation of enzymes. The authors suggest that the lack of clay and other floor sediments could be an explanation for this situation. A small number (1.85×10^6 to 2.67×10^6 per gram) of ammonifying, denitrifying, sulfate-reducing, and iron-reducing bacteria, as well as *Azotobacter,* were documented in some of the samples. Over 90% of them are cocci and less than 5.6% are rods.

Cave Fauna

Compared to its size, the cave is poor in terms of fauna. Except for an endemic troglobiont diplopod (*Romanosoma bârtei* Ceuca, 1967), a detailed study conducted by Nitzu et al. (2008) revealed the following groups: Arachnida (*Porrhomma* sp., *Taranucnus bihari*), Collembola (*Deuteraphorura silvaria, Protaphorura armata, Plutomurus unidentatus, Desoria violacea*), Insecta (*Trechus latus, Duvailus(Duvaliopsis) pilosellus, Quedius mesomelinus*), and other undetermined species belonging to Gastropoda, Acari, Opilions, Trichopteran, and Diptera. Four species of bats (*Myotis myotis, M. blythii, Rhinolophus hipposiderus, R. ferrumequinum*) are hibernating in Izvorul Tăuşoarelor Cave (Chiş 2010; Coroiu et al. 2014).

Paleontology

In the Cave Bear Room (Sala Ursului de Cavernă) it was discovered an important bone deposit of *Ursus spelaeus* and *U. arctos,* some of which are in anatomic connection. Remarkable is the fact that the bones of these two species of bears were found together in various parts of the room and at different levels of the alluvial stratum. These may indicate that the two species cohabited the cave for some time, which is an interesting finding not only for Romania, but for Europe too. The investigated bones have relatively large dimensions, thus invalidating the assumption advanced by Jurcsák et al. (1981) that the size of the cave bears could be inversely proportional to the cave's altitude (Domşa and Popa 1988).

Cave Conservancy

Izvorul Tăuşoarelor is an A class protected cave (scientific reserve), meaning access is only granted for scientific or exploration purposes. This kind of activities, however, requires special permissions from the Romanian Speleological Heritage Commission and the Bistriţa-Năsăud Museum Complex, the custodian of this cave. Any individuals or groups holding a valid authorization will be guided by the cave custodian. Beginning with 2007, the cave was included in the European Ecological Network Natura 2000 (indicative ROSC10193) as a special area of conservation (Gavriloaie et al. 2016).

Acknowledgements Drs. D. Hoffmann and J. Hellstrom are thanked for kindly providing the U-series ages reported in this chapter.

References

Audra P, Palmer AN (2013) The vertical dimension of karst: controls of vertical cave pattern. In: Schröder JF (ed) Treatise on geomorphology, vol 6, Academic Press, San Diego, pp 186–206

Baciu N (1990) Studiul granulometric şi mineralogic al depozitelor aluvionare din Peştera Izvorul Tăuşoarelor (Mţii Rodnei). Peştera 3:111–124

Bleahu M, Decu V, Negrea Ş, Pleşa C. Povară I, Viehmann I (1976) Peşteri din România. Ed Ştiinţifică şi Enciclopedică, Bucureşti, p 415

Chiş VT (2010) Bat species (Chiroptera) identified in the Rodna Mountains - Rodna Mountains National Park and adjacent areas (Eastern Carpathians, Romania). Transylv Rev Syst Ecol Res 9:1–14

Coroiu I, Culisir A, Kluppel R, Theodorescu C, David A, Stermin A (2014) Seasonal dynamics of Mouse-Eared bats (*Myotis myotis/blythii*) in Northern Romania. Int Zool Cong "G.Antipa" Museum, Bucharest. Book of Abstracts, p 62

Cucoş Dinu A, Călugăr MI, Burghele BD, Dumitru OA, Constantin C, Onac BP (2016) Radon levels in Romanian caves: an occupational exposure survey. Environ Geochem Health. https://doi.org/10.1007/s10653-016-9878-1

Domşa M (1988) Geneza mirabilitului din Peştera Tăuşoare, Munţii Rodnei. Peştera 2:190–204

Domşa M (1990) Noi date asupra geologiei Peşterii de la Izvorul Tăuşoarelor (Munţii Rodnei, România) notă preliminară. Peştera 3:70–73

Domşa M, Popa C (1988) Noi resturi de urside discoperite în peşterile României. Peştera 2:156–183

Fabian C (1984) The karst phenomena study from the theory of systems point of view. Theor Appl Karstol 1:29–34

Fabian C, Viehmann I (1979) Morphogenese des billes calcaires dans la caverne de Tăuşoare (Monts Rodna, Roumanie). Trav Inst Spéol "Émile Racovitza" 18:209–214

Gavriloaie C, Rusu C, Petrescu-Mag IV (2016) The present day status of a remarkable protected area: Tăuşoare Cave, Romania. ELBA Bioflux 8(2):56–63

Iurkiewicz A (2010) Rodna Mountains. In: Orăşeanu I, Iurkiewicz A (eds) Karst hydrogeology of Romania. Ed Belvedere, Oradea, pp 29–32

Jude L (1972) Les anthodites des Monts de Rodna. Trav Inst Spéol "Émile Racovitza" 11:343–347

Jurcsák T, Plopiş R, Ignat D, Şerban M, Popa E (1981) Date privind fauna fosilă a Peşterii Urşilor (M. Bihor). Nymphaea 8–9:161–257

Kräutner HG, Kräutner FL, Szász L, Seghedi I (1989) Geologic map of Romania, Rebra Sheet (20c). Scale 1:50.000, Inst Geol Geofiz, Bucureşti

Manolache E, Drăgan-Bularda M, Kiss Ş (1991) Microbiological and enzymological researches in the Tăuşoare Cave and in the Great Cave of the Firea Valley. Studia UBB Biologia 36(1):9–51

Moţiu A, Viehmann I, Strusiewicz R (1977) Découverte des nouveaux mineraux dans la grotte Tăuşoare (Monts Rodna). Trav Inst Spéol "Émile Racovitza" 16:211–216

Munteanu C-M (2004) Studiul complex al depozitelor calcaroase Priabonian-Rupeliene din regiunea Peşterii Izvorul Tăuşoarelor (Munţii Rodnei), cu consideraţiuni supra genezei nodulilor carbonatici. Unpubl BSc Thesis, Universitatea Babeş-Bolyai Cluj-Napoca, p 117

Nitzu E, Popa I, Nae A, Iuşan C (2008) Faunal research on the invertebrates (Coleoptera, Ortophtera, Collembolan and Araneae) in the Rodnei Mountains Biosphere Reserve. Trav Inst Spéol "Émile Racovitza" XLVII:3–52

Onac B (1987) Unele aspecte privind modul de apariţie şi originea gipsului din Peştera Tăuşoare şi Peştera Vântului. Bul Speol 11:33–37

Onac BP, Cocean P (1996) Une vue globale sur le karst roumain. Kras i Speleol 8(XVII):105–112

Onac BP, White WB, Viehmann I (2001) Leonite $K_2Mg[SO_4]_2$ $4H_2O$, konyaite $Na_2Mg[SO_4]_2$ $5H_2O$, and syngenite $K_2Ca[SO_4]_2$ H_2O from Tăuşoare Cave (Rodnei Mts., Romania). Min Mag 65(1):1–7

Papiu F (2007) Remember Tăuşoare. Speomond 12:24–27

Popa C (1988) Peştera de la izvorul Tăuşoarelor. Peştera 2:60–70

Regattieri E, Zanchetta G, Drysdale RN, Isola I, Hellstrom JC, Dallai L (2014) Lateglacial to Holocene trace element record (Ba, Mg, Sr) from Corchia Cave (Apuan Alps, central Italy): paleoenvironmental implications. J Quat Sci 29:381–392. https://doi.org/10.1002/jqs.2712

Silvestru E (1984) The relationship between tectonics and karstification in the cave from Izvorul Tăuşoareler (Rodna Mountains). Theor Appl Karstol 1:35–42

Silvestru E (1990) On the genesis and evolution of mirabilite in The Cave of Izvorul Tăuşoarelor (Romania). Trav Inst Spéol "Émile Racovitza" 29:79–83

Silvestru E, Viehmann I (1982) Étude de microtectonique comparée dans le Karst des Monts Rodna (Roumanie). Trav Inst Spéol "Émile Racovitza" 21:63–67

Şerban M, Viehmann I, Coman D (1961) Peşteri din România. Ed Meridiane, Bucureşti (available in Romanian, French, German, and Russian)

Viehmann I (1973) Résultats de le expedition belgo-roumain de la grotte de Tăuşoare (Monts Rodna, 1971). Proc 6th Int Congr Speleol, Oloumouc (CSSR) VIII:229–232

Viehmann I (1975) Quelques considération sur le aragonite et les cristallisation excentriques des grottes. Trav Inst Spéol "Émile Racovitza" 14:141–147

Viehmann I, Şerban M (1962–1963) Notă preliminară aspra Peşterii Izvorul Tăuşoarelor. Lucr Inst Speol "Emil Racoviţă" 1–2:179–207

Viehmann I, Rusu T, Şerban M (1964) Complexul carstic Tăuşoare-Zalion (Munţii Rodnei). Lucr Inst Speol "Emil Racoviţă" 3:21–48

Persani Mountains: Karst of Vârghis Gorge

Daniel Veres, Marian Cosac, George Murătoreanu,
and Ulrich Hambach

Abstract

Vârghiş karst, with its limited spatial development but well-preserved features, is a peculiarity in the karst inventory of Romania. It shows several well-marked karstification levels, and a relatively high number of caves, the majority harboring thick clastic deposits. However, its most important asset is that it hosts numerous traces of past human occupation. For example, recent investigations of rock shelter Abri 122 produced one of the most significant Middle Paleolithic lithic assemblages for Romania, covering the time interval from Marine Isotope Stage (MIS) 5/7 to MIS 3. The occurrence of a volcanic ash layer within Bear's (Ursului) Cave originating from the Ciomadul volcanic complex (East Carpathians) and dated to ~43/50 ka highlights the potential of Vârghiş karst in preserving such isochronous marker horizons, calling for further research.

Keywords

Karst • Caves • Middle Paleolithic • Vârghiş
Romania

D. Veres (✉)
"Emil Racoviţă" Institute of Speleology, Romanian Academy, Clinicilor 5, 400006 Cluj-Napoca, Romania
e-mail: daniel.veres@ubbcluj.ro

M. Cosac · G. Murătoreanu
"Valahia" University of Târgovişte, Târgovişte, Romania
e-mail: cosac_marian@yahoo.com

G. Murătoreanu
e-mail: muratoreanug@yahoo.com

U. Hambach
BayCEER & Chair of Geomorphology, University of Bayreuth, Bayreuth, Germany
e-mail: ulrich.hambach@uni-bayreuth.de

Introduction

Karst in general and especially caverns and rock shelters are well known for their suitability in conserving crucial evidence of past human occupation, from the cradle of mankind to more recent periods. Since a range of discoveries in Carpathian caves provided a significant number of modern human remains, interdisciplinary research directed toward augmenting the archaeological data and understanding past environmental conditions has been carried out in the region quite intensively. However, the cave-based Middle Paleolithic (MP) in the Romanian Carpathians is still based almost exclusively on collections without chronological control, whereas the limited radiometric data available lack well-defined archaeological contexts. In this respect, the importance of the Vârghiş karst area in Transylvania (Orghidan and Dumitrescu 1963; Dénes 2003), although of limited spatial extent, arises from its high potential in providing a compelling record of past human occupation, from MP to very recently (Cosac et al. 2018; Veres et al. 2018). For example, interdisciplinary research in the archaeological profile Abri 122 has produced the most important MP lithic assemblage in the Carpathian region to date. Overall, the samples recovered from this site depict a distinctive MP occurrence among the analogous datasets in Romania. Its main characteristics reside not only in the fact that it is the first MP industry recovered from an East Carpathian karst area, but also in the intriguing technological inventories and typological features. It exhibits several technological options including discoid, centripetal Levallois, and Kombewa (Cosac et al. 2018), whereas the toolkit consists mainly of quartzite, lydite, opal, and volcanic rocks, all locally available materials. Furthermore, the toolkit displays few denticulated, notched, and truncated items, alongside many sidescrapers of different types and bifacial implements (leafpoints, knives) in various stages of manufacture/rejuvenation. To date, such particular mix of typological features remains unprecedented in the MP lithic assemblages from the Carpathian area, thus calling for further research.

G. M. L. Ponta and B. P. Onac (eds.), *Cave and Karst Systems of Romania*,
Cave and Karst Systems of the World, https://doi.org/10.1007/978-3-319-90747-5_8

Geographic, Geologic, and Hydrologic Settings

The Vârghiş karst area (670 m a.s.l.) is part of Basin 1200 in the Romanian karst inventory (Goran 1982) being positioned at the northern edge of the Perşani Mountains (East Carpathians; Fig. 1). It is the most important karst area in this geotectonic unit that comprises mostly Mesozoic wild-flysch, andesitic and pyroclastic, and locally, Quaternary basaltic rocks (Bleahu et al. 1976) (Fig. 2a). Here, the Vârghiş River, a tributary of Olt River, crosses a 3.5 km long band of Triassic and Jurassic limestones (Fig. 2a, b), which host on 56 km^2 a well-developed exokarst and around 124 caves (Orghidan and Dumitrescu 1963; Dénes 2003). Although most caves are small, many of them contain thick successions of clastic sediment infilling with evidence of past human occupation (Orghidan and Dumitrescu 1963; Dénes 2003). The Vârghiş River has deepened in the limestone bed for about 350 m in several successive stages (Fig. 2c), with the main periods of karst formation highlighted by the presence of four fossil geomorphological levels of caves and conduits (Orghidan and Dumitrescu 1963). There are also active (or temporarily active) conduits extending beneath the valley floor and through which most of the Vârghiş River water flows (such as Cave No. 1200/96, Activul Galben—1200/102, Cave No. 1200/111, Active Cave No. 1200/96, Levis Cave—1200/117, Răsăritul Apei Cave—1200/45) (Dénes 2003).

History of Exploration

The first reports of the Vârghiş karst are linked to the description of Peştera Mare de la Mereşti Cave (The Great Cave from Mereşti, from hereon PPM) by J. Fridvaldszky in 1767, the first cave described in Transylvania, and among the first ones in Europe. This was followed by more detailed reports by J. Benko (1774), J. E. von Fichtel (1780), and J. Kleinkauf (1793). In 1836, I. Fekete also published a detailed description of PPM, whereas in 1868, B. Orbán reported a comprehensive investigation of the area. In 1876, A. Hoch also describes elements of macrophages in PPM

Fig. 1 Location of karst area in the Perşani Mountains

(Păunescu 2001; Dénes 2003, 2005). If the interest for the Vârghiş karst manifested at such an early stage in regional speleological research, the archaeological potential was first evidenced by F. Podek in the early nineteenth century. In 1935, J. Teutsch, a pioneer of Transylvanian Paleolithic research, performed the first documented archeological excavations, followed by M. Mottl in 1941–1942 (Jungbert 1979; Păunescu 2001; Dénes 2005). With reference to current standards, these early studies drew attention to the general archaeological potential, but the caves investigated by F. Podek cannot be identified, and the reports of M. Mottl did not include excavation plans or descriptions of the archaeological and faunal material (Murătoreanu et al. 2015). An ample description of the most famous caves and

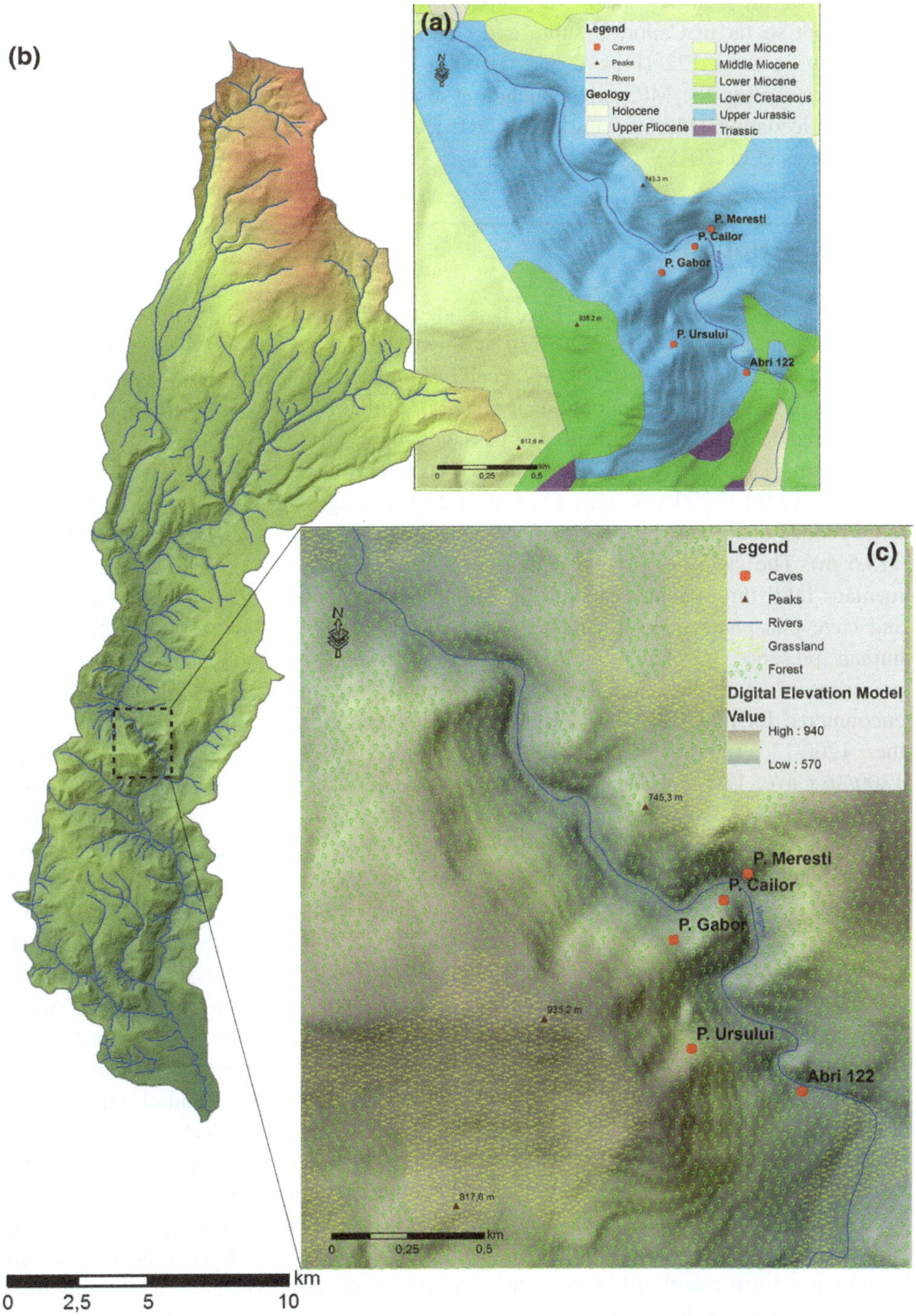

Fig. 2 Geologic (**a**) and geomorphologic (**b**) maps showing the location of the Vârghiş Gorge within the Vârghiş Valley, with the caves in which archaeological investigations were conducted in the period 2014–2017 (**c**)

rock shelters throughout the Vârghiş karst was performed by Orghidan and Dumitrescu (1963), a comprehensive work that remains to the current day the most important interdisciplinary study of this area. In 1969, L. Roşu carried out excavations in Cailor and Tătarilor caves, but the research was published by Păunescu (2001). Between 1971 and 2005, speleologist I. Dénes re-evaluated the known caves and identified new ones (Dénes 2003, 2005). In 2014, a re-evaluation of the archaeological and karst research potential of the Vârghiş karst was initiated with studies carried out so far in Cailor, Ursului, and Gabor caves and especially in Abri 122 (Figs. 2 and 3), with exceptional results concerning the MP period, including also the direct radiometric dating (Murătoreanu et al. 2015; Cosac et al. 2018; Veres et al. 2018).

Karst Description

Four main stages of speleogenesis have been identified in the Vârghiş karst at 5, 20, 40, and 70–120 m above the valley floor, and some of these erosional levels can be connected to remnants of river terraces on a regional scale (Orghidan and Dumitrescu 1963). The length of cave passages reported in the Vârghiş karst totals 7502 m (Dénes 2003). The best known is PMM (1200/14), with 1527 m of cave passages, large rooms, and a vertical development of 50 m (–34.5 m; +15.5 m). The cave has several entrances, and thick sedimentary infill throughout its length, including thick guano and cave phosphate deposits and numerous traces of past human presence (Orghidan and Dumitrescu 1963). The presence of multiple entrances is a common feature encountered for many caves in Vârghiş karst; for example, the 1200/13 Tătarilor Cave, 1200/18 Ursului Cave, 1200/36 Cave, 1200/8 Cave all exhibit multiple entrances, whereas other caves are accompanied by rock shelters, Abri 1200/122, 1200/3 Cave, and Ursului Cave 1200/36 (Orghidan and Dumitrescu 1963; Dénes 2005). The rock shelter identified as Abri 122 (Figs. 3 and 4) is located on the right side of the valley, in the lower third of the slope, at 625 m a. s.l, and roughly 30 m above the Vârghiş riverbed (Dénes 2003). It consists of a sheltered limestone platform, apparently continued with an in-filled small cave room. The rock shelter with a surface of approximately 30 m^2 has been the subject of intense archeological research during the last decades (Cosac et al. 2018 and references therein). Ursului Cave (Figs. 3 and 4), also recently re-investigated, is located on the higher slopes, in a precipitous terrain today. It consists mainly of a large cave room, with three entrances, and hosts thick clastic deposits (Orghidan and Dumitrescu 1963), including the first report of tephra layer in an archaeological context in the Carpathian karst (Veres et al. 2018).

Cave Sediments, Speleothems, and Radiometric Dating

The vast majority of caves within Vârghiş karst preserve thick sedimentary deposits, and many contain documented traces of human occupation from the Middle Paleolithic up to the Medieval Period, including numerous human osseous remains, of unknown ages (Orghidan and Dumitrescu 1963). From our limited observation of the sedimentary suite within Ursului and Gabor caves, the infill consists mainly of a lower bed of poorly sorted small alluvial pebbles of wildflysch, quartzite, and volcanic rocks, poor in animal remains, capped by a thick horizon of limestone scree and a fine (presumably aeolian) loamy matrix, usually rich in bone remains and containing also lithics. Toward the cave entrances, a thick horizon rich in humus matter and plant litter, with abundant faunal remains and ceramics, was identified in most caves previously investigated (Orghidan and Dumitrescu 1963; Bleahu et al. 1976). Although direct radiometric dating is limited to the luminescence ages reported in Veres et al. (2018) from Abri 122, it has been suggested based on detailed paleontological surveys of several caves that most cave infill dates back to the Middle Pleistocene (Orghidan and Dumitrescu 1963; Dénes 2003).

The caves within the Vârghiş karst are rather poor in speleothems, and where they existed, have largely been destroyed (i.e., PPM; Peştera Calului; Cave No. 1; Dénes 2005). Speleothems are still present in Ursului Cave, Cave 1200/36, Lublinit Cave 1200/9, Formation Cave, and the Cave 1200/74 (Orghidan and Dumitrescu 1963; Dénes 1995, 2005). During the archeological survey of Ursului Cave (Figs. 3 and 4), a macroscopic lens of volcanic ash was discovered in the stratigraphic profile. This important stratigraphic marker horizon has been linked based on glass shard chemical data to an eruption around $\sim$43/50 ka (Veres et al. 2018) from Ciomadul Volcano, located ca 30 km to the east. Providing that this or more volcanic ash layers are traced within other cave sequences from Vârghiş karst, they might allow for a better integration of records, on a wider regional scale. However, multi-method chronological investigations of the sedimentary suite were performed so far at Abri 122 and Ursului Cave only, with ongoing research in Gabor Cave (Figs. 3 and 4). The archaeological inventory indicates that the technological assemblage of lithics found within the two main levels of human occupation at Abri 122 pertains to the MP industries (Cosac et al. 2018). Establishing, however, a reliable chronology for the archeological sequence at Abri 122 proved rather difficult. On the one hand, the radiocarbon dating of animal bones and charcoal was complicated by limited amounts of preserved collagen/carbon within the dated bone remains, as well as the age of the bone or charcoal samples at or beyond the

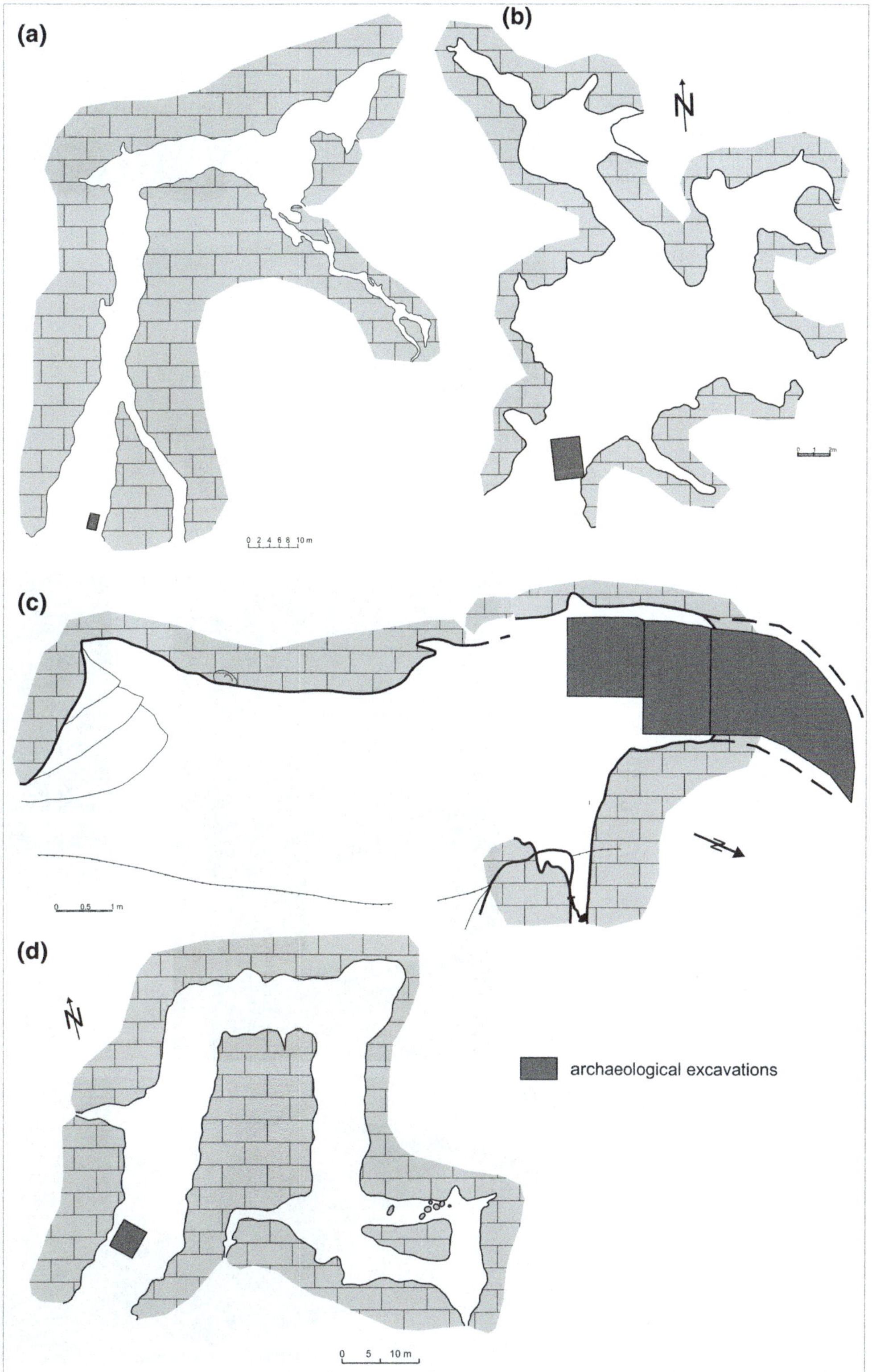

Fig. 3 Maps of the archeologically surveyed caves between 2014 and 2017: **a** Calului Cave; **b** Ursului Cave; **c** Abri 122; **d** Gabor Cave

upper limit of the method. Nonetheless, based on these results (Veres et al. 2018), as well as the reconstructed fossil micro-mammal assemblages (Cosac et al. 2018), the upper span of the MP occupation level at Abri 122 reaches into MIS 3. Optically (OSL) and infrared stimulated luminescence (IRSL) dating of silt-sized grains indicate ages of >100 ka for the lowermost lithic-rich horizon, whereas the radiocarbon dating of animal bones from the same layer

Fig. 4 Excavation details of some of the caves surveyed for their archaeological content between 2014 and 2017: **a** Calului Cave; **b** Gabor Cave; **c** Ursului Cave; **d** Abri 122 (the copyright of all photographs in this figure belongs to the authors)

returned infinite ages. Following these data, it appears that the MP assemblage documented at Abri 122 is currently one of the most important such inventories in the Carpathian region (Cosac et al. 2018) and can be assigned in a broad age range from MIS 5/7 to MIS 3.

Cave Minerals

In a rather basic mineralogical study conducted by Diaconu et al. (2006–2007), they employed X-ray diffraction analyses on a limited number of samples collected from Cave nr.

9 (also known as the Cave with Lublinit). They identified mainly calcite and hydroxylapatite, alongside detrital quartz and clay minerals. The presence of calcite is directly related to precipitation of calcium carbonate within the cave environment, whereas the hydroxylapatite resulted from the reaction between calcium and phosphate ion leached from guano and osseous remains found within the cave.

Paleontology

The preservation of a very rich and diverse fossil bone inventory in the Vârghiş caves was already reported by Orghidan and Dumitrescu (1963). More recently, archaeological research in Abri 122 identified osseous remains of *Bos/Bison, Ursus spelaeus, Capra,* and *Canis lupus*, in direct association with the MP lithics. Remains of several micro-mammal species have also been recorded, including *Microtus arvalis, Microtus gregalis, Lagurus lagurus, Arvicola terrestris, Cricetulus migratorius, Cricetus cricetus, Sorex araneus,* and *Microtus nivalis*, as well snakes *Natrix sp., Colubrinae indet.* (cf. *Zamenis sp.*), amphibians *Rana sp.* (*Rana temporaria, Pelophylax sp., Hyla sp.*), fishes *Osteichthyes indet.*, and birds *Aves indet.* (Cosac et al. 2018).

Cave Conservancy

The complex natural reserve of Vârghiş Gorge, established in 2000, covers an area of 10.07 km^2 (Dénes 2003). Its custodians are Asociaţia Speo—Turistică şi de Protecţia Naturii "Lumea Pierdută" ("Lost World") (http://www.vargyasszoros.org/) Baraolt, alongside Ocolul Silvic Privat Baraolt, and Asociaţia Carpaterra Braşov. In addition, any cave research requires a permit from the Romanian Speleological Heritage.

Conclusions

Previous investigations corroborated by recent surveys highlight the importance of karst in the Vârghiş region as an insufficiently explored archive of past human presence in the Carpathian area. The significant number of small caves and rock shelters compensates the limited development of large caverns within this small karst area. Most of the caves preserve thick sedimentary suites and a very rich fossil faunal inventory; it is expected that ongoing research within Abri 122 in special, and the Vârghiş karst in general will further increase the number of caves harboring lithic industries, allowing for a better understanding of this particularly important site within the Carpathian karst and Romanian Paleolithic research.

Acknowledgements The authors wish to thank Mrs. I. Dénes and the staff of Baraolt "Lost World" Association for the unconditional support offered during the archaeological research. DV acknowledges the support from project PN-II-ID-PCE-2012-4-0530 "Millennial-scale geochemical records of anthropogenic impact and natural climate change in the Romania Carpathians."

References

Bleahu M, Decu V, Negrea Ş, Pleşa C, Povară I, Viehmann I (1976) Peşteri din Romania. Ed Ştiinţifică şi Enciclopedică, Bucureşti, p 415
Cosac M, Murătoreanu G, Veres D, Niţă L, Schmidt C, Hambach U, Radu A, Cuculici R, Buzea DL, Mărgărit M, Dumitraşcu V, Vasile Ş, Petculescu A (2018) Multi-proxy archaeological investigations of a Middle Palaeolithic occupation context in Eastern Transylvania, Romania. Quat Int 485:115–130 https://doi.org/10.1016/j.quaint.2017.09.014
Dénes I (1995) "Gyökérsztalagmitok" a Vargyas-szoros 74-es számú barlangjában. Acta Hargitensia 11:44–46
Dénes I (2003) Rezervaţia naturală complexă Cheile Vârghişului. Speomond 7–8:8–11
Dénes I (2005) Peşterile din Secuime. Speomond 9–10:33–38
Diaconu G, Dumitraş D-G, Marincea Ş (2006–2007) Mineralogical analyses in two caves from the Perşani Mountains. Trav Inst Spéol "Émile Racovitza" XLV–XLVI:113–129
Goran C (1982) Catalogul sistematic al peşterilor din Romania. Ed CNEFS, Bucureşti, p 496
Jungbert B (1979) Repertoriul localităţilor cu descoperiri paleolitice din Transilvania (II). Acta Musei Napocensis 16:407–414
Murătoreanu G, Cuculici R, Cosac M, Radu A, Niţă L, Veres D, Buzea DL (2015) Potenţialul arheologic al carstului din Cheiele Vârghişului—Munţii Perşani. Etape preliminare în realizarea unui Sistem Informaţional Geografic. Arheovest III 2:591–604
Orghidan N, Dumitrescu M (1963) Studiu monografic al complexului carstic din defileul Vârghişului. Lucr Inst Speol "Emil Racoviţă" 1–2:69–178
Păunescu A (2001) Paleoliticul şi mezoliticul din spaţiul transilvan. Ed Agir, Bucureşti
Veres D, Cosac M, Schmidt C, Murătoreanu G, Hambach U, Hubay K, Wulf S, Karátson D (2018) New chronological constraints for Middle Paleolithic (MIS 6/5-3) cave sequences in Eastern Transylvania, Romania. Quat Int 485:103–114 https://doi.org/10.1016/j.quaint.2017.07.015

Leaota Mountains: Rătei Cave

Maria-Lidia Nuțu-Dragomir and Gigi-Paul Dragomir

Abstract

Rătei Cave with its 7224 m length is located in the Leaota Mountains (South Carpathians). It develops in Middle and Upper Jurassic limestones and is formed by water enlarging post-Albian fractures. A small dam constructed near the cave entrance collects the water into a reservoir from where it is distributed for public supply. The water is calcium carbonate type, with a discharge between 40 and 115 L/s. The Rătei Cave stands out by erosional and dissolution morphology and a few speleothems.

Keywords

Drinking water source • Cave morphology
Speleothems

Introduction

Rătei Cave entrance is located on the northern side of the Rătei Creek, 32 m above its thalweg and 200 m downstream from the Rătei Gorges in Leaota Mountains (Fig. 1). The Rătei's Creek large catchment area comprises crystalline rocks, with waters which sinks underground at the contact with Middle and Upper Jurassic limestones and along post-Albian fractures/joints existing in the Rătei Gorges, generating the cave with the same name.

Geologic Settings

The crystalline basement of the area belongs to Lereşti Formation (Leaota Series), which consist of rocks metamorphosed in the greenschists facies (355–357 ± 4 My; Axente et al. 2005) that are overlain by Jurassic sandstones, limestones, jaspers and Cretaceous deposits represented by Raciu Breccia (Aptian), Bucegi Conglomerates (Albian) and alternations of sandstones and polymictic conglomerates (Fig. 2). The Jurassic carbonaceous deposits in which the cave is developed consist of limestones with bivalve and brachiopods (Bajocian), overlain by oolitic limestones (Bathonian), massive limestones (Oxfordian) and cherty limestones (Kimmeridgian) (Fig. 2).

Hydrologic Setting

The discharge of Rătei's underground stream varies seasonally; over the 2012–2016 period, the measurements recorded a minimum of 41 L/s (October 2012) and a maximum of 111 L/s (April 2013). The hydrological regime of Rătei Cave is controlled by a few distinct underground streams that eventually confluence about 50 m downstream of Sump 3 (Fig. 3).

The main stream enters the cave via Sump 5 at the upstream end of the Waterfalls Passage and sinks at its downstream end to emerge in Sump 3. According to Dragomir (2002), this stream represents 2/3 of the total discharge of the cave outlet. A left side tributary is encountered between Sumps 3 and 4 (shown as +3.5 m on the Access Passage on Fig. 3). It has a small flow rate and constant temperature. Another important stream appears in Titans Chamber, flowing also through the lower section of the Breakdown Chamber, and sinks at the bottom of the northern wall of the Great Chamber to re-emerge, according to Povară et al. (1973), at the Sump 4. A small stream emerges at the end of the Gravels Passage, flows through the

M.-L. Nuțu-Dragomir
Institute of Geodynamics of Romanian Academy, J.L. Calderon 19–21, Bucharest, Romania

G.-P. Dragomir (✉)
Total E&P, Stavanger, Norway
e-mail: gigidragomir@yahoo.com

G. M. L. Ponta and B. P. Onac (eds.), *Cave and Karst Systems of Romania*,
Cave and Karst Systems of the World, https://doi.org/10.1007/978-3-319-90747-5_9

Fig. 1 Location of Rătei Cave in Leaota Mountains

Meanders Passage to disappear in the Water Shaft and to re-emerge also in Sump 4.

Although Slăvoacă et al. (2010) suggest that 70% of the supply of the aquifer associated to Rătei underground stream is provided via sinking points developed along Rătei Creek and its tributaries, the latest field observations show that the swallet situated upstream in Rătei Gorges is inactive for several years now, even during rainy periods. Also, on the left side tributaries of Rătei Creek, no sinking points were observed in streambeds, at the contact between conglomerates and limestones. Thus, we may assume that the cave streams actually derive from diffuse seepage.

All streams combine their flows 120 m downstream of Sump 4, and record a discharge at the cave entrance ranging from 41 to 111 L/s. Variations in the water temperatures were noticed between 2012 and 2016, ranging from 3.9 °C (March 2016) up to 9.6 °C (September 2015). The amplitude of the temperature variations for all streams is very low, between 0.3 and 2.0 °C, which is inconsistent with a supply preferentially from swallets, but rather from diffuse infiltration. According to Slăvoacă et al. (2010), the general chemical type of the water is calcium carbonate, with 603 mg/L total dissolved solids (TDS) content.

History of Exploration

The first mention of Rătei Cave dates back to the nineteenth century, time since its resurgence began to be used for public water supply. Patrulius (1969) conducted geologic investigations in the area and suggested, based on his observations made in the lower-level passages that apparently the cave has modest dimensions compared to the extent of the karst area. Povară et al. (1973) made an extensive study of the cave focusing on morphology, hydrology, sedimentology and proposed a genetic model for the cave.

Since 1970, the "Focul Viu" Caving Club explored the cave and surrounding area. They managed to enter the main network and discovered two major dry passages. The explorations continued in the early 1980s, when along with "Hades" Caving Club, new discoveries were made and the cave was resurveyed.

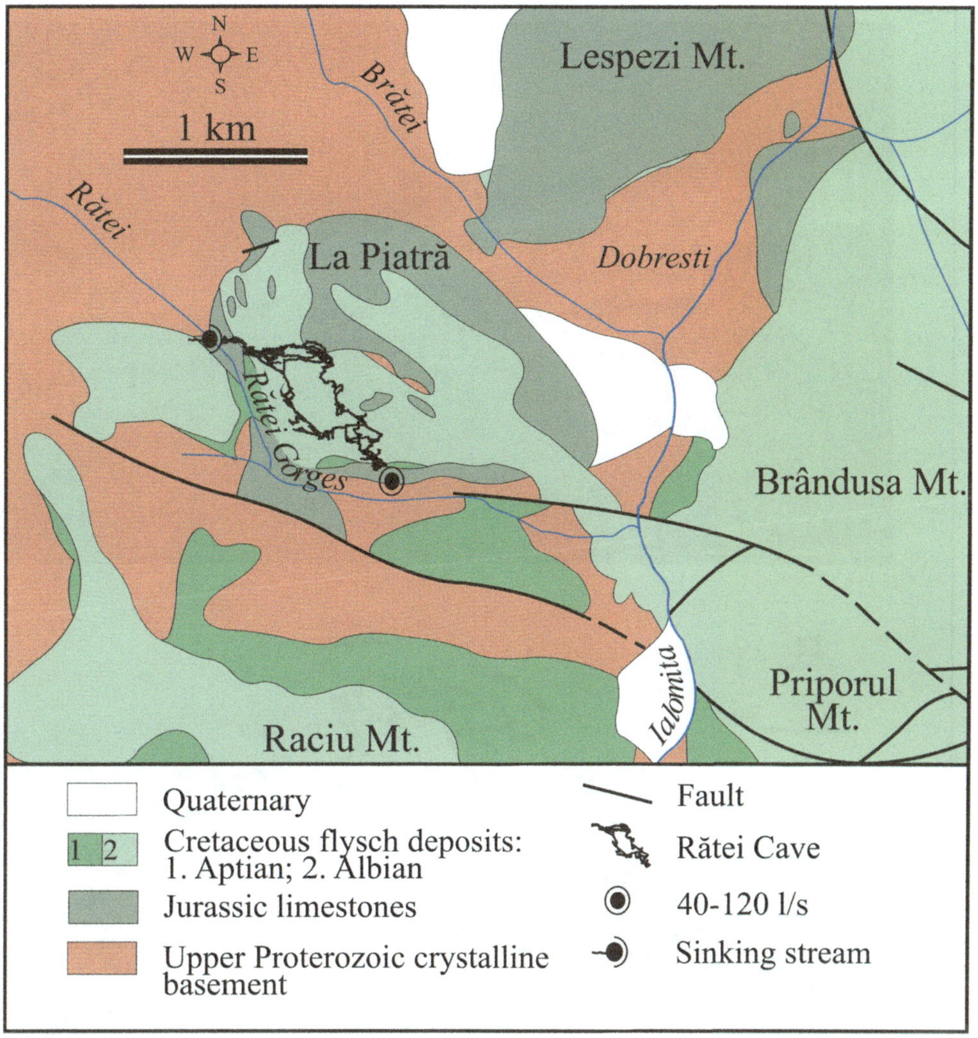

Fig. 2 Geological map of west Leaota Mountains (modified after Patrulius 1969)

Cave Description

Ratei Cave is an anastomotic cave type (Palmer 2007) with a total length of 7224 m (Fig. 3). A dam built at the entrance supplies drinking water for the Târgovişte City located 50 km away. Near the entrance (~40 m), due to the water rise, the stream passage ends in two sumps (3 and 4). These can be by-passed through a narrow loop (150 m long) that reaches the underground stream in the Access Passage (Fig. 3), then ending in the Waiting Chamber. Erosional blades and scallops are visible on the walls. The ceiling reveals the contact between limestones and breccia along a fault line (Fig. 4a, b).

Forty meters from the Waiting Chamber is the Dry Confluence, which represents the starting point for a large loop. Towards southeast, the Rimstone Dams Passage (Fig. 5a) begins and extends above the cave entrance area (Fig. 6, profile B-B′). The main loop began with the Straight Passage and continues with Waterfalls Passage that ends in a narrow sump (Sump 5). Some 60 m before this sump, a wet area occurs with abundant dripping originating from a large flowstone named Medusa. Between the flowstone and the opposite wall, the entrance in the New Passage (Fig. 5b) is located; this is exclusively carved in the Raciu Breccia up to the Little Gate. Between this point and the Breakdown Chamber, occasionally, the walls are coated with twinned gypsum crystals.

The Breakdown Chamber is a relatively large, descending room with a stream flowing in its lower part. Continuing towards west, after a short meander is the entrance in the Titans Chamber, which is the largest in the cave. It is characterized by chaotic breakdowns and is where the highest elevation (+69 m) is reached in Ratei Cave. This stream is recharged by water losses along Rătei Creek, appears in Titans Chamber and sinks at the bottom of the northern wall of the Great Chamber. Some 200 m before the Breakdown Chamber is the entrance into the Great Labyrinth (with corrosion blades protruding from the ceiling), a steeply descend gallery towards the Great Chamber and the Clay

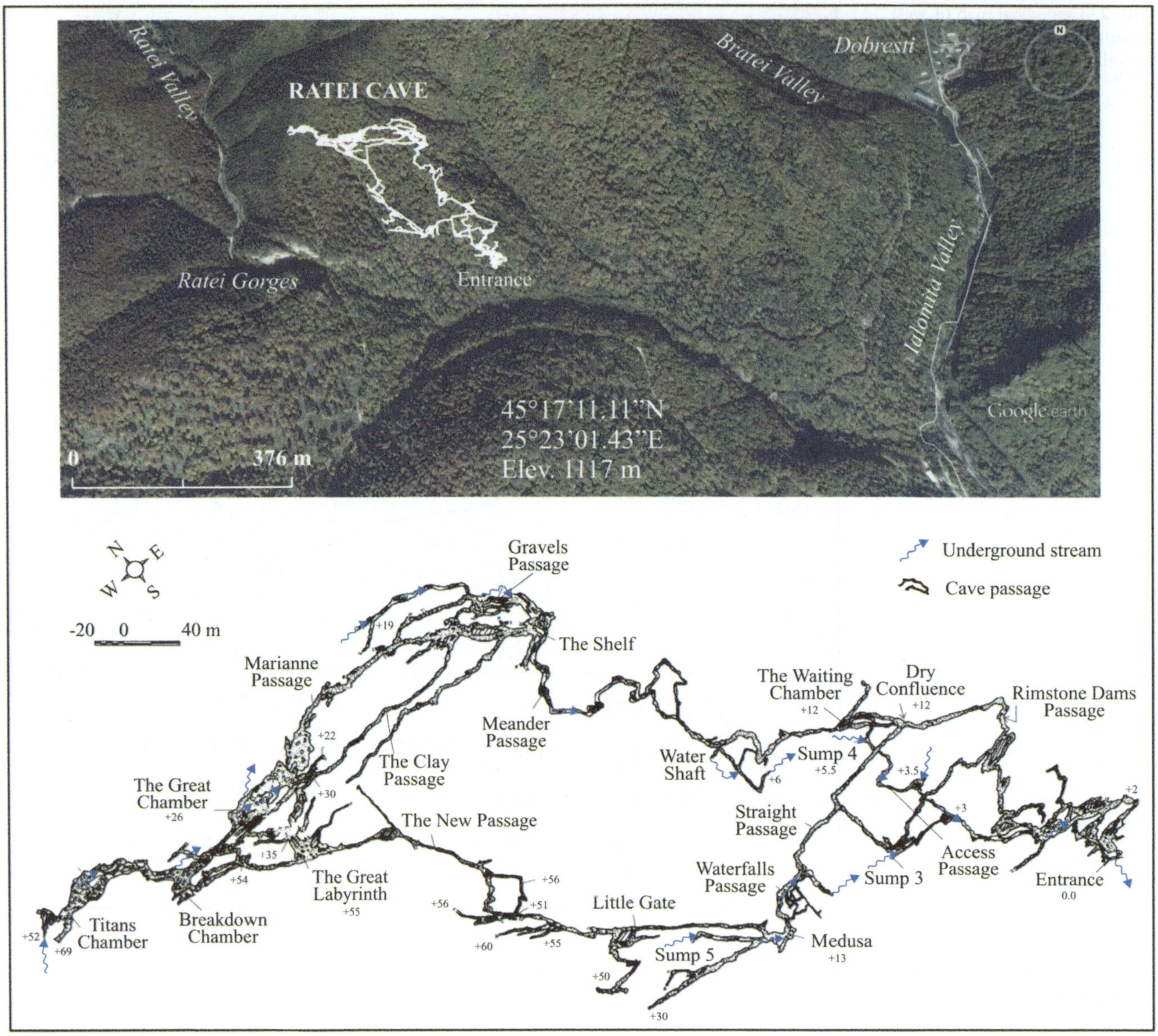

Fig. 3 Detailed plan of Rătei Cave (modified after Dragomir 2002). In inset, the cave plan viewed onto Google Earth map

Passage. In the Great Chamber, during high flow, the stream splits into two branches: the largest water amount flows northward, whereas the remaining runs eastward. During drought, the entire stream flows northwards, reappearing at the end of the Gravels Passage, then flows down into the Meanders Passage, disappear again in the Water Shaft and re-emerge in Sump 4.

East of the Great Chamber, Marianne and Meanders passages begin. The first one is a phreatic tube of modest heights (~0.8 m), hosting on its floor several 1–2-m-diameter water pools with calcite rafts and accumulation of *Ursus spelaeus* and other mammals bones. The Meanders Passage is by far the most spectacular in the cave with meanders and erosional ledges. The loop closes back in the Waiting Chamber.

Speleogenesis Remarks

The initiation and subsequent evolution of Rătei Cave was shaped by the Paleogene tectonic and the Quaternary climate. The process leading to the formation of Rătei Cave was the enlargement of ante- and post-Albian fractures system during Tertiary times under the action of groundwater. This system extends mainly along three directions: N-S, E-W and NW-SE. The same fractures system has been

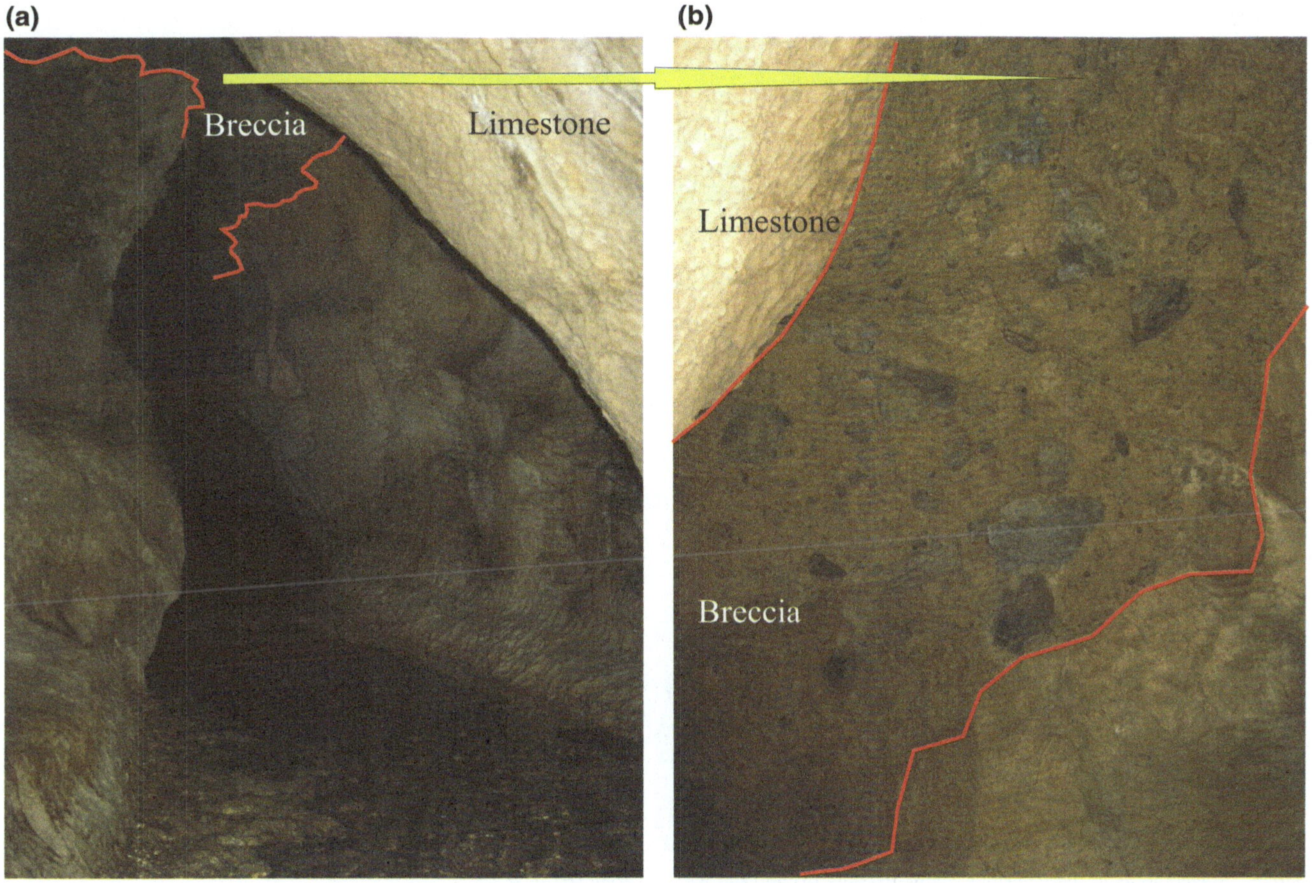

Fig. 4 Lithological contact between limestone and breccia seen in the on Access Passage: **a** general view, **b** detail from the passage ceiling

identified both within the Jurassic limestones and within the Cretaceous Bucegi Conglomerates.

According to Velcea (1961), in Bucegi Mountains there were recognized two glacial phases, Riss and Würm, and one interglacial phase Riss-Würm. Since these climatic conditions during the Quaternary times were identified also in Ialomiţa Cave (located on Ialomiţa River at 1530 m altitude), we can assume that the same conditions influenced the speleogenesis of Rătei Cave as well.

Patrulius (1969) first suggested that the passages of Rătei Cave are developed on several levels, but only later, Povară et al. (1973) have identified three distinct levels. The upper one is situated at 66–50 m above the entrance, and it is represented by the New Passage, the Titans Chamber and the upper part of the Breakdown Chambers. The median level, located at 35–15 m above the entrance, includes all the other large chambers at the end of the cave, the Straight Passage, the Rimstone Dams Passage, the Clay Passage and the Marianne Passage. At last, the lowest one is situated at only 5 m above the entrance and comprises the Meanders Passage, the Waterfalls Passage and the Entrance Passage.

As suggested by Povară et al. (1973), the median level could be syngenetic with a glacial phase, whereas the lower level corresponds to a post-glacial phase. Regarding the upper level, the lack of evidences makes impossible to affirm if this level belongs to a syn- or periglacial phase. The morphology of the passages is the result of tectonics and underground stream discharges variations. The cave passages were formed usually along fractures, by vertical entrenching or lateral widening. The vertical entrenching is associated to interglacial and interstadial periods with high water discharges, and the lateral widening related to glacial periods with low water discharges, respectively.

It is assumed that initially, the present day stream from Titans Chamber flowed somewhere in the upper part of the Breakdown Chamber and the New Passage. Afterwards, through a series of consecutive sinks (Fig. 6, profile A-A′), via the Great Labyrinth, the Breakdown Chamber and the lower part of Titans Chamber, the stream was directed through different galleries towards the Meander Passage. The latter one has been permanently traversed by a stream, the older streambed levels being preserved as a vertical succession of protruding ledges that extends over 10–20 m elevation range. From the Meander Passage, the stream flowed outwards through the Rimstone Dams Passage (Fig. 6, profile B-B′).

Fig. 5 Speleothems in the Rimstone Dams Passage (a) and the New Passage (b)

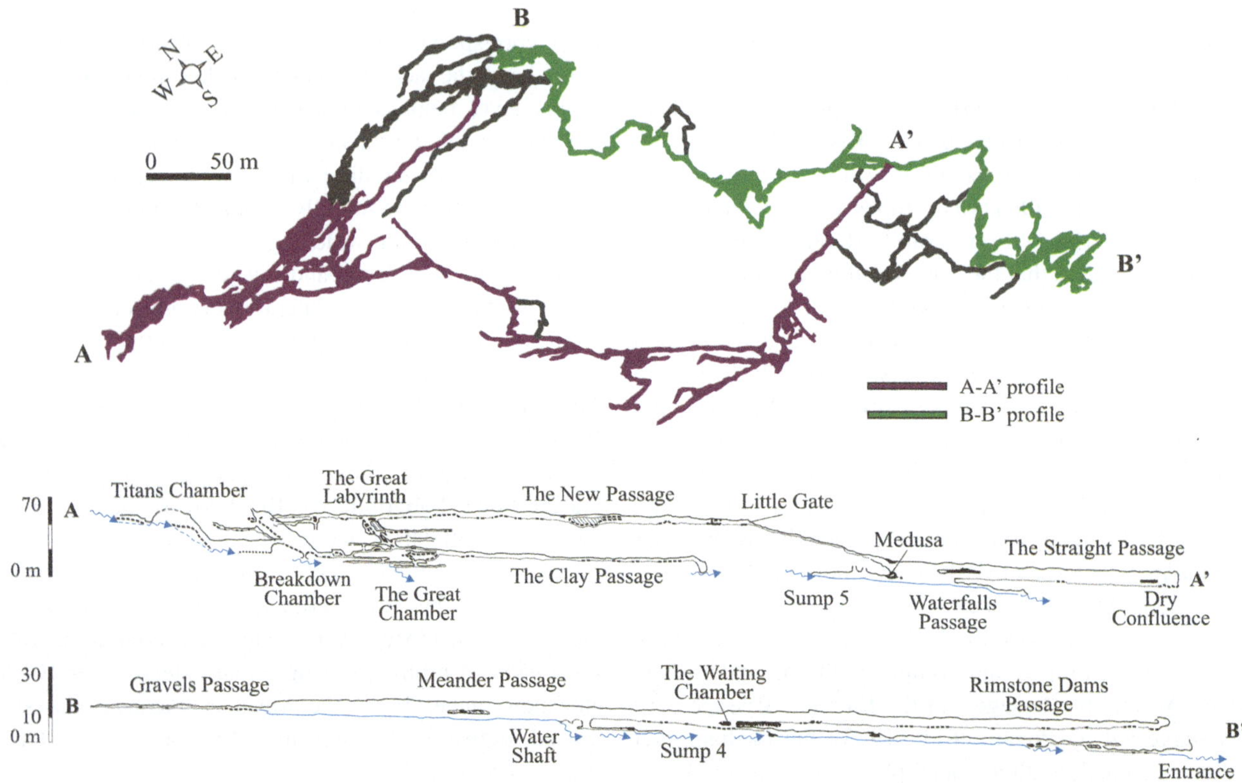

Fig. 6 Plan of Rătei Cave showing the location of the two longitudinal profiles (modified after Dragomir 2002)

Later in the cave history, the stream that enters now via Sump 5 seems to have flown at a higher level than the current one, entering into the New Passage upstream of the Little Gate, to continue its flow down in a phreatic regime through a narrow inclined passage, resembling the hydropower pressure tunnels, at the lower part of the New Passage towards the Straight Passage, and finally merging the other stream coming from the Meander Passage, in the Dry Confluence area (Fig. 6, profile A-A′).

Speleothems and Cave Minerals

A variety of common calcite speleothems (stalactites, stalagmites, columns, flowstones, pools, calcite rafts, etc.) exist in different parts of the cave. A few aragonite stalactites and gypsum crusts were identified in the upper level by Diaconu (1983). The presence on the cave walls of small pockets filled with brownish deposits of some unspecified iron oxides and hydroxides suggests the existence of pyrite in the limestone. If so, oxidation of pyrite would produce sulfuric acid, which after reacting with limestone precipitates gypsum.

Cave Bats

During 2012–2016, in Rătei Cave were recognized up to 80–100 singular specimens of bats belonging to *Pipistrellus pipistrellus* Schreber (Fig. 7a) and *Myotis myotis* Borkhausen (Fig. 7b) species, mainly on the active passages, but also in dry galleries like Rimstone Dams and Meander passages. For the first time in March 2016 a hibernation colony was observed (Fig. 7c) in the Titans Chambers, composed of 15 specimens of *Myotis myotis*. No maternity colonies were observed.

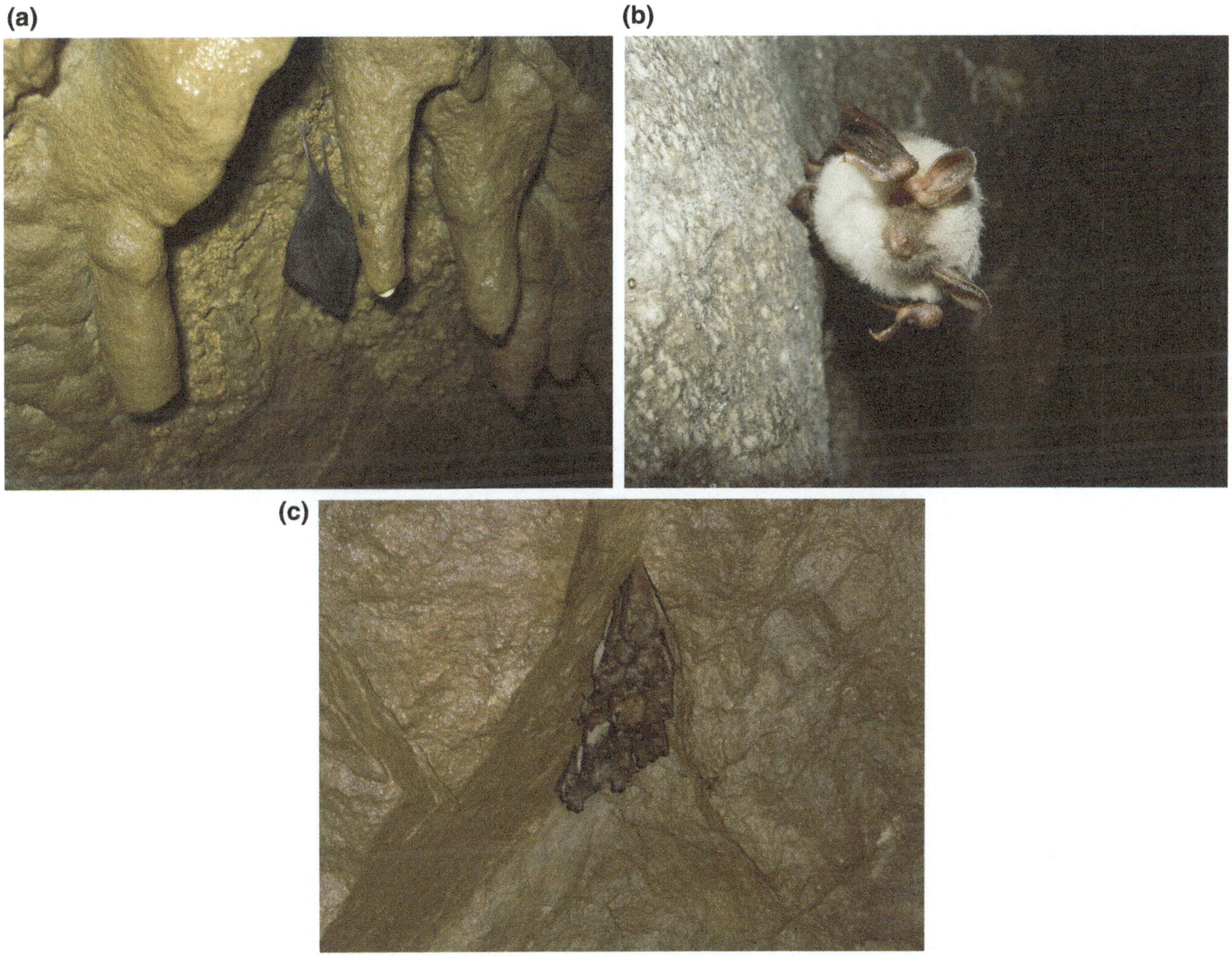

Fig. 7 Species of bats encountered in Ratei Cave: **a** *Pipistrellus pipistrellus*, **b** *Myotis myotis*, **c** Hibernation colony of *Myotis myotis* located in the Titans Chambers

Conclusions

The Rătei Cave appearance is the result of streams flowing along preferential pathways, controlled by Paleogene tectonic and repeated Quaternary climate changes. The few underground streams that control the hydrological regime of the cave display seasonal variations of the water discharges. The Rătei Cave is mostly known for its spectacular erosional morphology and less for the abundance of speleothems.

Acknowledgements We acknowledge the merits of Dr. Cristian Panaiotu, who has been the true soul of "Focul Viu" Caving Club during the entire period when the cave survey was performed, and whose efforts resulted in all the graphic representations related with the Rătei Cave. Authors are also grateful to Dr. H. Mitrofan for his insights on this manuscript.

References

Alexe A, Maluski H, Iancu V (2005) $^{40}Ar/^{39}Ar$ geochronology in the eastern part of the South Carpathians: evidence for pre-alpine tectonothermal history. Rev Roum Geol 48–49:3–27

Diaconu G (1983) Sur la paragenèse minoralogique gypse-aragonite rencontrée dans quelques grottes de Roumanie. Trav Inst Spéol "Émile Racovitza" 22:81–90

Dragomir GP (2002) Peştera Rătei. Ecocarst 3:28–33

Micalevich-Velcea V (1961) Masivul Bucegi. Studiu geomorfologic. Ed Acad RSR, Bucureşti, pp 150

Palmer AN (2007) Cave geology. Cave Books, Dayton, p 454

Patrulius D (1969) Geologia Masivului Bucegi şi a Culoarului Dâmbovicioara. Ed Acad RSR, Bucureşti, p 321

Povară I, Lascu C, Goran C, Herdlicka W, Jibotean M (1973) Observations préliminaires dans la grotte de Rătei (M-ts. Bucegi, Carpathes Meridionales de Roumanie). Livre Cinquantenaire Inst Spéol "Émile Racovitza", Ed Acad RSR, Bucureşti, pp 633–642

Slăvoacă D, Slăvoacă R, Istrate A (2010) Bucegi Massif. In: Orăşeanu I, Iurkiewicz A (eds) Karst hydrogeology of Romania. Ed Belvedere, Oradea, pp 49–53

Piatra Craiului Mountains: Grind Pit (Avenul de Sub Coltii Grindului)

Spencer F. Coca

Abstract

Discovered in 1985 at an altitude of 2020 m above present sea level, the "Avenul de sub Colții Grindului" Cave (hereafter Grind Pit) represents Romania's first Alpine-type cave included on the list of world's deepest caves. With a depth of 540 m, the Grind Pit was for decades the deepest cave in Romania and has only recently been surpassed by the Vărășoaia cave system in the Apuseni Mountains. The cave develops in the Upper Tithonian limestones belonging to the Grind Formation. The highly tectonized, bedded limestone dips 75° to 85° toward east, and therefore, pit's verticality is high. The origin of the pit is related to a massive tectonic breccia zone, formed during the uplift of the Carpathians, which is traversed by a fault that facilitated karstification. The stream flowing through the cave is recharging multiple karstic springs (medium flow: 400 l/s) located near the town of Zărnești, four kilometers away. The exploration activities are concentrated at −540 m in an effort to surpass the current terminus of the cave choked by a large boulder. The expected total vertical potential of the Grind Pit is 1270 m.

Keywords

Alpine karst • Pit • Tectonics

Introduction

Even though existence of a well-developed endokarst in the Piatra Craiului Mountains was for many years unexpected, systematic exploration over the past decades proved the contrary. Discovered in 1985, the Grind Pit is the main access point into the (hypothetical) largest Alpine-type karst system in Romania. The Grind Pit is located in the South Carpathians, on the eastern flank of the Piatra Craiului Mountains (Fig. 1).

S. F. Coca (✉)
Spenco Rockoil Projects, 54A Piatra Craiului, 505800 Zărnești, Brașov, Romania
e-mail: scoca@spenco.ro

Geographic and Geologic Settings

From a geomorphological point of view, Piatra Craiului Mountains represent a 20 km ridge oriented NE-SW, separated from the surrounding mountain units by the Bran–Rucăr corridor (east), the Dâmbovița and Tămaș basins in the west, the Brașov Basin (north), and the Dâmbovicioara Basin to the south. The hydrologic drainage of the area is controlled by the Bârsa River and its main tributary, Râul Mare in the north, and Dâmbovița and Dâmbovicioara rivers in the south.

Due to the strike and dip of the limestone bedding planes, different landforms developed on the eastern and western flanks of the Piatra Craiului ridge. The eastern flank is characterized by a relatively gentle slope, whereas on the western side, the landscape is rugged and steep. Unique in the Romanian Carpathians is the high density of residual rock morphologies (spires, towers and needles, arches) (Constantinescu 2009).

The Grind Pit is located on the eastern flank of the Piatra Craiului ridge at an altitude of 2020 m above present sea level, in a steep gully, south of the Piscul Baciului Peak (Figs. 2 and 3). The sedimentary record of the Piatra Craiului Mountains displays a mixed siliciclastic–carbonate stratigraphy of Jurassic and Cretaceous ages (Popescu 1966; Bucur 1978). The carbonate sequence hosting the Grind Pit belongs to the Grind Formation, and it is characterized by a 1050 m thick section of Kimmeridgian–Tithonian limestones deposited in a shallow marine basin (Coca 1998). From a facies perspective, the strata of the Grind Formation are characterized by cherty limestones deposited as calciruditic debris flows at the base, followed by massive beds of tidalites and biohermic back-reef sediments (Coca 1998; Bucur et al. 2009; Mircescu et al. 2014). Discordant Neocomian deposits, including marls and glauconitic limestones (Hauterivian), marly limestones with patch reefs (Barremian), and

G. M. L. Ponta and B. P. Onac (eds.), *Cave and Karst Systems of Romania*,
Cave and Karst Systems of the World, https://doi.org/10.1007/978-3-319-90747-5_10

Fig. 1 Location of Grind Pit in the karst of Piatra Craiului Mountains

polymictic conglomerates (Aptian) overlay the Jurassic sequences. The Mesozoic deposits were uplifted and subaerially exposed during the main phase of the Alpine Orogeny at the end of Cretaceous. Karstification occurred soon after the erosion of the clastic Cretaceous cover. Associated with this uplift, a dense network of intersecting faults/joints developed; these represented the main dissolution paths in the carbonate deposits of the Piatra Craiului Mountain's ridge.

History of Exploration

Although first speleological explorations in the Piatra Craiului Mountains began in the second part of the nineteenth century (Bielz 1884), it was the merit of A. Prox to investigate all known caves and to summarize his findings in two scientific papers (Prox 1968, 1984). He was also the first to hypothesize a great endokarstic system and paleokarst in the Piatra Craiului Mountains. One of Prox's and other explorer's mistake was to direct their attention to the deep pits developed in the overlying Neocomian conglomerates, like Funduri (−142 m) and Vlăduşca (−64 m), and to neglect the high altitude, more difficult to access limestone areas. In contrast to these karst explorations, Constantinescu (1980, 1984) was the first who attempted speleological studies in the area. Many cavers concentrated their efforts on the high altitude part of the ridge, especially W. Gutt, A. Zakarias, and I. Dobrescu; the last one discovered the entrance of the Grind Pit in 1985. Popescu (1988) reached a depth of −165 m, where a detrital plug stopped further cave exploration. Many digging sessions led in 1994 to the continuation of the cave down to the actual depth of −540 m, where a large-scale boulder choke was encountered (Coca 1997, 2000). To date, "Focul Viu" and "Avenul Braşov" caving clubs continues the exploration of the cave.

Cave Description

The speleological parameters of this cave are relevant for understanding the characteristics of the karstification processes. The cave has uncommon dimensions for this region and consists of an array of shafts interconnected by small horizontal passages (Fig. 4). Except for the uppermost part,

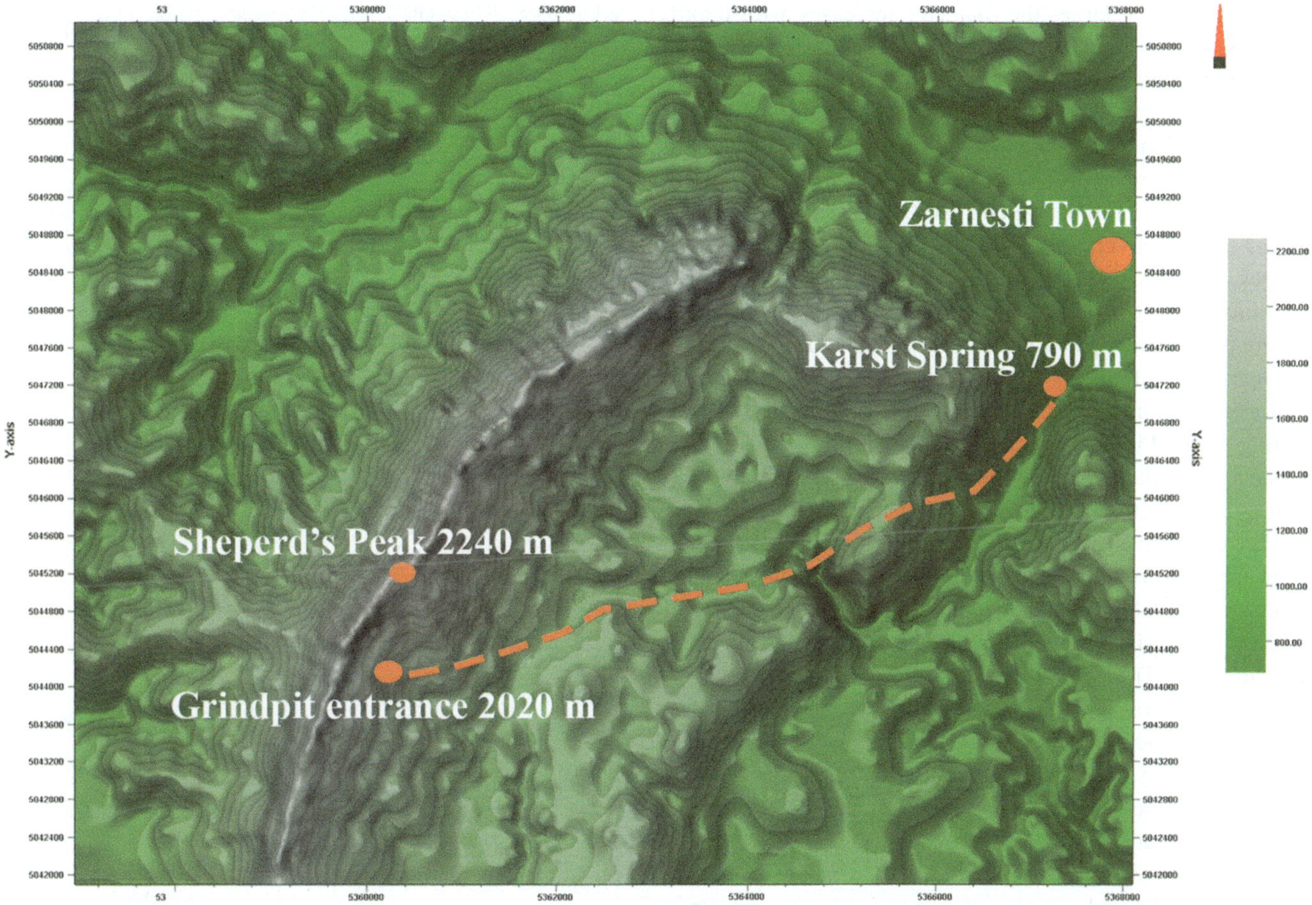

Fig. 2 Topographic map of the Piatra Craiului Mountains showing the location of the Grind Pit entrance at 2020 m altitude on the eastern flank. The dashed line represents the hypothetical drainage path toward the karst springs near the town of Zărneşti

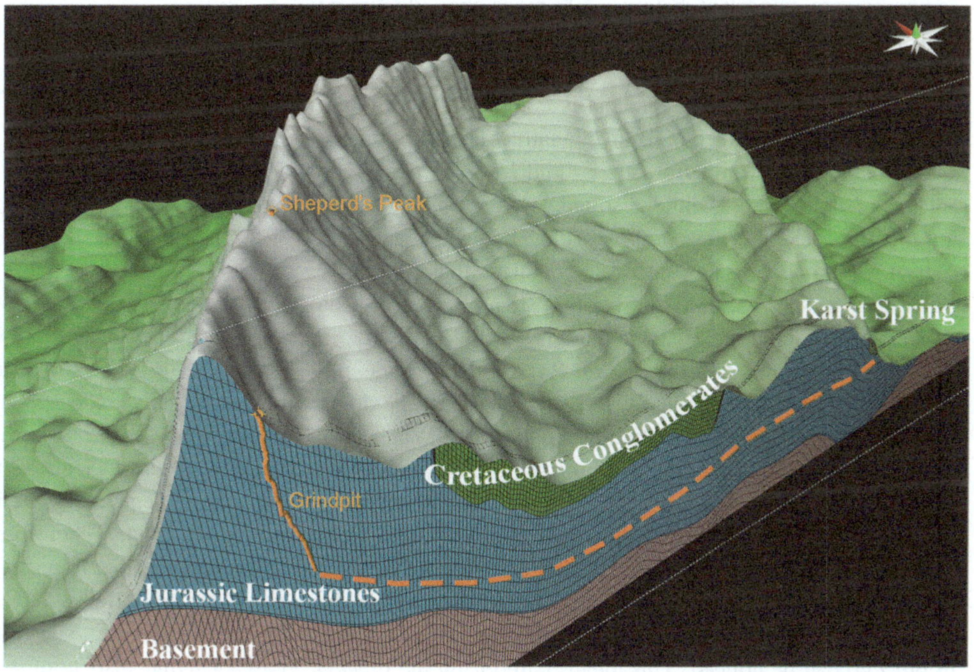

Fig. 3 Cross section through a 3-D geomodel of the Piatra Craiului Mountains showing the spatial distribution of Jurassic limestones and the mapped path of the Grind Pit in solid line

every pit is very large; the diameter increases with depth, reaching a maximum of 30 m. The shape varies from elliptic to circular, and their depths range between 14 and 69 m. Noticeable is the abrupt morphology change following the very narrow passage (only 19 cm wide!) at −288 m depth. In the upper part (above −288 m), passages are more vertical due to their development along fault planes, whereas in the lower part, subvertical passages are guided by bedding planes. There is also a change in the orientation of these two sections. Above −288 m, the cave has a SW orientation, whereas the lower part trends toward NE, and the direction changes several times within few meters (zigzag pattern), following the intersection plane of two major faults. Comparing Grind Pit's general orientation with the surface joint pattern, remarkable similarities can be observed, all indicating the structural control in the cave development.

The entrance is also located on a major fault trending ENE, noticeable from the top of the ridge. Abundant corrosive features are visible throughout the cave providing evidence of highly aggressive waters as a major factor in the pit's development. Thin, blade-like echinoliths of different sizes are present on almost all walls. A close-up investigation reveals a selective dissolution of the micritic parts and sparite-cemented cracks forming a small-scale box work pattern. Corrosive features can be seen above the mentioned detrital plugs, where temporary stagnant waters widened the bases of the pits. Speleothems are rare. These include small stalactites, stalagmites, and widespread calcite curtains in the upper part of the pit. Clastic deposits cover the whole range from gravels to clays, the latter ones only found below the detrital deposit plugs, clearly originating from their weathering. These clastic sediments are important below −400 m as they document high water-flow stands. The mentioned detrital plugs (boulder chokes) are tectonic breccia lenses, developed between strata during orogenic uplift. The breccia is composed of different size lime clasts embedded in a clayey matrix. The components are autochthonous (only from upper and lower strata of the cave passage) and neither sorted nor rounded. This matrix-supported breccia comprises diagenetically cemented portions, partly encrusted by calcite, and areas that are poorly consolidated and washed out by the stream. The breccia appears at a depth of −130 m and is visible (varying in thickness) down to −540 m where it forms a 10 m wide and 30 m high cone that completely seals the cave passage down below, making human penetration impossible. Water-flow disappears between the clasts, washing out the clayey matrix. Even after heavy rainfalls and spring snow melting periods, there was no significant water standing above the plug, indicating unrestrained drainage. In the upper parts of the cave where breccia patches are still hanging in cracks and on the walls, preferential corrosion (higher in limestones and lower in breccia) can be noticed. This means (supposing the breccia filled in earlier times the space of the actual cave) that enlargement of the cave was primarily a result of corrosive processes and only subordinately, mechanically washout of breccia. The water-flow in the pit is permanent and well organized. The water appears at a depth of −70 m and collects on the way other streams at −100, −150, and −285 m. Because the surface above the cave consists of bare limestone (no soil or vegetal cover), infiltration rate is very high, along numerous joints. Therefore, sudden rainfalls or snow melts have catastrophic impacts, endangering explorers' life. The water temperature is 4 °C and freezes in winter down to a depth of −50 m. All vertical shafts below −120 m form cascades that also contribute to corrosion due to sprinkled water vapor acting on the shaft's walls.

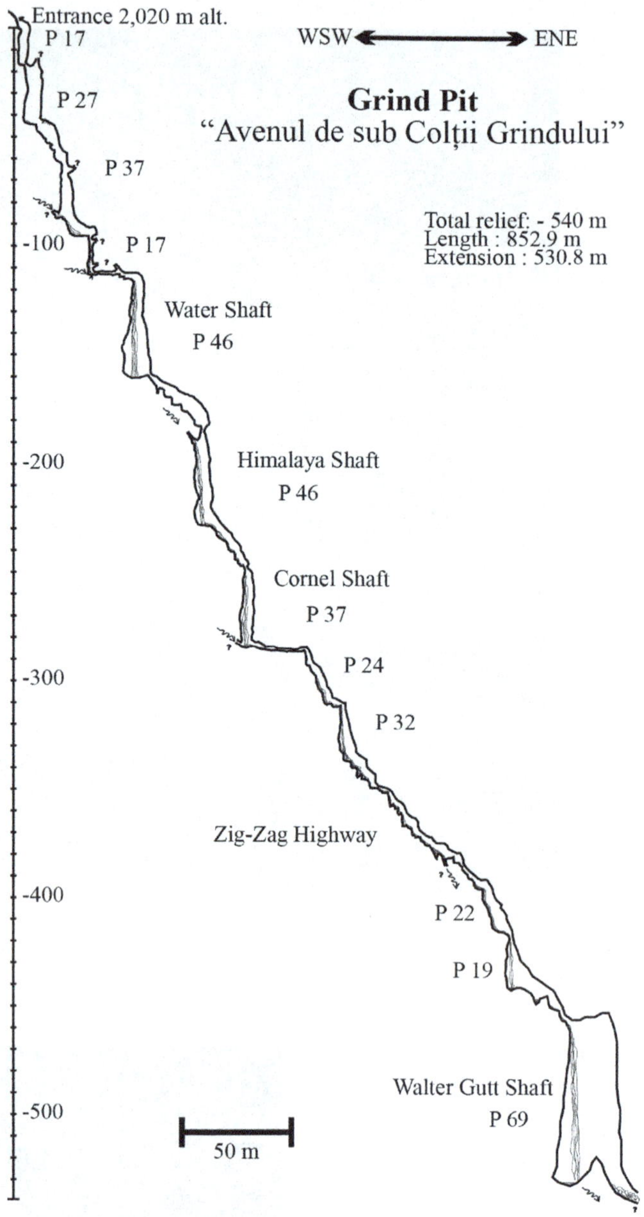

Fig. 4 Projected profile of the Grind Pit showing the succession of shafts that lead to the boulder choke at −540 m in depth

Speleogenesis

Speleogenetic work is in progress, and the main questions to be answered are: (1) How such a large cavity formed in steep (almost vertical) strata where water drainage is fast and therefore not enough time for large-scale dissolution was available, and (2) where are the other, similar caves that must exist in the Piatra Craiului Mountains in order to explain the hydrologic drainage of a 22.9 km^2 limestone catchment area? (Constantinescu 2009). Certainly the water drained through the Grind Pit represents only a small fraction of the water emerging through the karst springs located near the town of Zărneşti in the northern part of the ridge (Orăşeanu 2010). Our investigations showed that the response of the Zărneşti karst springs to heavy rainfall ranges between 16 and 21 days.

The genesis of the cave might be associated with the presence of tectonic breccia at the intersection between faults/joints and bedding planes, which facilitates the dissolution process. However, still a substantial amount of water is needed to explain the large-scale karstification visible at shallow depths in the Grind Pit. In this respect, geomorphological research pointed out the possible role some small-size glaciers formed during Pleistocene in the Piatra Craiului Mountains may have had; the melting waters of these local glaciers could have been the source for very aggressive cave-forming stream (Constantinescu 2009).

Conclusions

The Grind Pit is the first and only large-scale cavity discovered so far in the Piatra Craiului Mountains. The tectonic and structural control on the Grind Pit development is obvious since the cave passages follow the intersection plane of two major faults, which also facilitated the aquifer's recharge along it. Also, the highly porous tectonic breccia accelerated the penetration of water and subsequently a fast karstification. The general inclination of the cave passages is similar to the direction in which the beds are dipping, a fact that clearly suggests a mixed structural-stratigraphic control in the cave development. The cave continues below the breccia plug at the actual terminus of the pit at −540 m below surface. The underground stream is well organized and likely connected to the karst springs located near the town of Zărneşti. Exploration is still going on with the hope to enter Prox's postulated major karst system. Presuming that other pits with similar hydrologic function (water collectors) exist, and considering the total discharge of the aquifer (2200 l/s), we expect many other large and/or deep caves will be discovered in the Piatra Craiului range.

Acknowledgements I would like to thank D. Sigheartău and R. Moga for the many unforgettable days spent exploring and surveying the Grind Pit. This paper is dedicated to W. Gutt, who spent his entire life searching for the large Piatra Craiului cave system and who managed to pass this onto us.

References

Bielz EA (1884) Beitrag zur Höhlenkunde Siebenbürgens, Jahrbuch des SKV IV:24–25

Bucur II (1978) Microfaciesurile calcarelor albe din partea nordică a masivului Piatra Craiului. Consideraţii biostratigrafice. DS Şed Inst Geol Geofiz LXIV:89–105

Bucur II, Săsăran E, Iacob R, Ichim C, Turi V (2009) Upper Jurassic shallow-water carbonate deposits from some Carpathian areas: new micropaleontological results. In: Popa M (ed) Proceedings of the 8th symposium of IGCP 506, Marine and non-marine Jurassic: global correlation and major global events, Bucharest, Romania, 1:13–14

Coca S (1997) Avenul de sub Colţii Grindului—deepest cave in Romania. Proc 12th Int Congr Speleol, La Chaux-de-Fond 4:87–90

Coca S (1998) Stratigraphy and sedimentology of the Piatra Craiului Group (Jurassic), Romania: formation of the Dacian passive continental margin. Unpublished M.Sc. Thesis, University Salzburg

Coca S (2000) Avenul de sub Colţii Grindului. Ecocarst 1:21–22

Constantinescu T (1980) Le karst de Piatra Craiului. Particularités de la genèse et de l'évolution du karst. Trav Inst Spéol "Émile Racovitza" 19:203–217

Constantinescu T (1984) Carte de la circulation des eaux souterraines dans le massif de Piatra Craiului. Trav Inst Spéol "Émile Racovitza" 23:75–79

Constantinescu T (2009) Masivul Piatra Craiului. Studiu geomorfologic. Ed Univ Bucureşti, p 164

Mircescu CV, Bucur II, Săsăran E (2014) Dasycladalean algae from Upper Jurassic-Lower Cretaceous limestones of Piatra Craiului Massif (South Carpathians, Romania) and their relationship to paleoenvironment. Studia UBB Geologia 59(1):5–27

Orăşeanu I (2010) Dâmbovicioara Passage. In: Orăşeanu I, Iurkiewicz A (eds) Karst hydrogeology of Romania. Ed Belvedere, Oradea, pp 55–68

Popescu G (1988) Avenul de sub vîrful Grind. Cerc Speol Inst Petrol Gaze Ploieşti, pp 103–109

Popescu I (1966) Contribuţii la cunoaşterea ştratigrafiei şi structurii geologice a masivului Piatra Craiului. DS Şed Inst Geol Geofiz LII (2):157–176

Prox A (1968) Der Grosse Königstein. Siebenbürgisches Archiv, Dritte Folge Bd 7:1–68

Prox A (1984) Über die Genese der Königstein-Schächte, Naturwis Forsch Siebenbürgen II. Böhlau Verlag, Köln, pp 337–354

Useful Websites

http://avenulgrind.blogspot.ro/
http://www.avenul.ro
http://www.foculviu.ro

Caves of the Făgăras Mountains

Virgil Drăguşin, Ionuţ Mirea, Augustin Nae, and Maria-Laura Tîrlă

Abstract

The Făgăraş Mountains consisting mostly of impervious metamorphic rocks are an unlikely location for karst development. Yet, due to the occurrence of a series of marble stripes interbedded within the metamorphic rocks, several caves developed in two areas in the central part of the mountains: Piscu Negru (at ~1200 m) and Muşeteica-Râiosu (between ~2100 and ~2400 m). Average air temperature is around 6.5 °C inside the Piscu Negru caves and 3.3 °C in Muşeteica. Cave No. 1 from the Piscu Negru Mine is the longest cavity in the massif (804 m), whereas M3-R2 Cave contains the speleothems from the highest altitude in Romania. Cave fauna is poor, and some bat species were found at the highest elevations of their habitat. The caves preserve reliable speleothem records for future studies on past climate variability and interactions between glaciations and karst.

Keywords

Marble • Alpine karst • Speleothems • Bats
Făgăraş Mountains • Romania

Introduction

The Făgăraş Mountains, made mostly of impervious metamorphic rocks, are an unlikely location to find karst landscape. Yet, due to the existence of a series of crystalline limestone stripes interbedded within the metamorphic rocks, several caves were discovered by the Emil Racoviţă Caving Club. Some of these caves have a high scientific importance as they host speleothems that could offer information on past environmental variability or on the uplift rates of the Carpathian Mountains. In addition, some of them are shelters for hibernating bat colonies. These caves were described in detail by Giurgiu (2006) and Giurgiu et al. (2006), thus this chapter focuses on recent advances in the scientific knowledge of these caves and their surroundings.

Geographic, Geologic, and Hydrologic Settings

The geologic history of the Făgăraş Mountains goes as back as Middle-to-Late Ordovician, as documented by the crustal construction time of the Romanian Carpathians basement, which is consistent with data from the Alpine and Variscan Europe (Iancu et al. 1998; Balintoni et al. 2014). Among many evidences which support this statement, the U-Pb dating of a paragneiss from the Cumpăna metamorphic unit shows the age of 503–498 Ma (Balintoni et al. 2014). The poly-metamorphic terranes were involved during the Cretaceous in the Alpine tectonic evolution of the South Carpathians (Iancu et al. 2005).

Our interest focuses on a lithotectonic assemblage dominated by large crystalline limestone stripes (lenses) and amphibolite, hosted by micaceous metapelites (Dimofte 1962; Pană and Erdmer 1994). The crystalline soluble rocks in the Făgăraş Mountains outcrop only on their central and western parts, with the largest area occurring in the Buda-Muşeteica area (Dessila-Codarcea et al. 1968). They range in thickness from 500 to 1200 m (Kräutner 1980), generally striking on a W-E direction and dipping south, with a maximum dip in the ridge crest area (Giuşcă et al. 1977).

The caves appear almost everywhere in the central and western parts of the massif, where the crystalline limestone outcrops (Fig. 1). Muşeteica area hosts caves at the highest

V. Drăguşin (✉) · I. Mirea · A. Nae
"Emil Racoviţă" Institute of Speleology, Romanian Academy, Frumoasă 31, 010986 Bucharest, Romania
e-mail: virgil.dragusin@iser.ro

M.-L. Tîrlă
Faculty of Geography, Department of Regional Geography and Environment, University of Bucharest, 010041 Bucharest, Romania

M.-L. Tîrlă
Research Institute of University of Bucharest (ICUB), Mihail Kogălniceanu 36-46, 050107 Bucharest 5, Romania

G. M. L. Ponta and B. P. Onac (eds.), *Cave and Karst Systems of Romania*,
Cave and Karst Systems of the World, https://doi.org/10.1007/978-3-319-90747-5_11

elevation in Romania, M3-R2 Cave being at 2430 m above sea level (Giurgiu 1990; Tîrlă et al. 2016). The caves from Piscu Negru (Fig. 2) were discovered during mining exploration, Cave No. 1 (804 m) being the longest in the massif (Giurgiu 2006). During spring rainfalls and snow melt in April 2016, the mine entrance collapsed and the access to the caves was completely obstructed.

Cave Morphologies and Genesis

The marble caves in the Făgăraş Mountains have narrow, steep passages (~35–45°), developed on planar structures (fault, foliation planes, or lithologic contacts). Speleothems are rare in the caves from Piscu Negru area where sharply edged walls and pressure tubes with pools and waterfalls plunging over fault scarps stand as the main features (Fig. 2b). In contrast, the caves in Muşeteica are subfossil remnants of some formerly larger karst systems, now reduced in size and very dispersed due to slope retreat. The morphology and survey of M3-R2 cave were recently re-assessed (Tîrlă et al. 2016; Fig. 2a).

Speleothems and U-Th Dating

Scientists of Emil Racoviţă Institute of Speleology are presently studying speleothems from the Muşeteica area in order to reconstruct past climate variability and to infer mountain uplift rates on longer timescale (over 1 Ma). A small stalagmite from the M3-R2 Cave was dated using the U-Th method to have grown between 124.8 ± 1.7 and 123.5 ± 1.7 thousand years ago, respectively (Drăguşin 2013). A flowstone from the same cave appears to be as old as 233,000 ± 35,000 years (Hellstrom pers comm). This last figure gives a minimum age estimate of the cave.

Microclimate of Muşeteica and Piscu Negru Caves

Results of a 24-h study conducted in September 1993 in the M3-R2 Cave and Cave No. 1 from the Piscu Negru Mine have outlined the differences between their microclimatic patterns (Bogdan et al. 1995). The atmosphere of M3-R2 Cave is strongly influenced by the outside, high-mountain

Fig. 1 Location of the marble caves in the Făgăraş Mountains

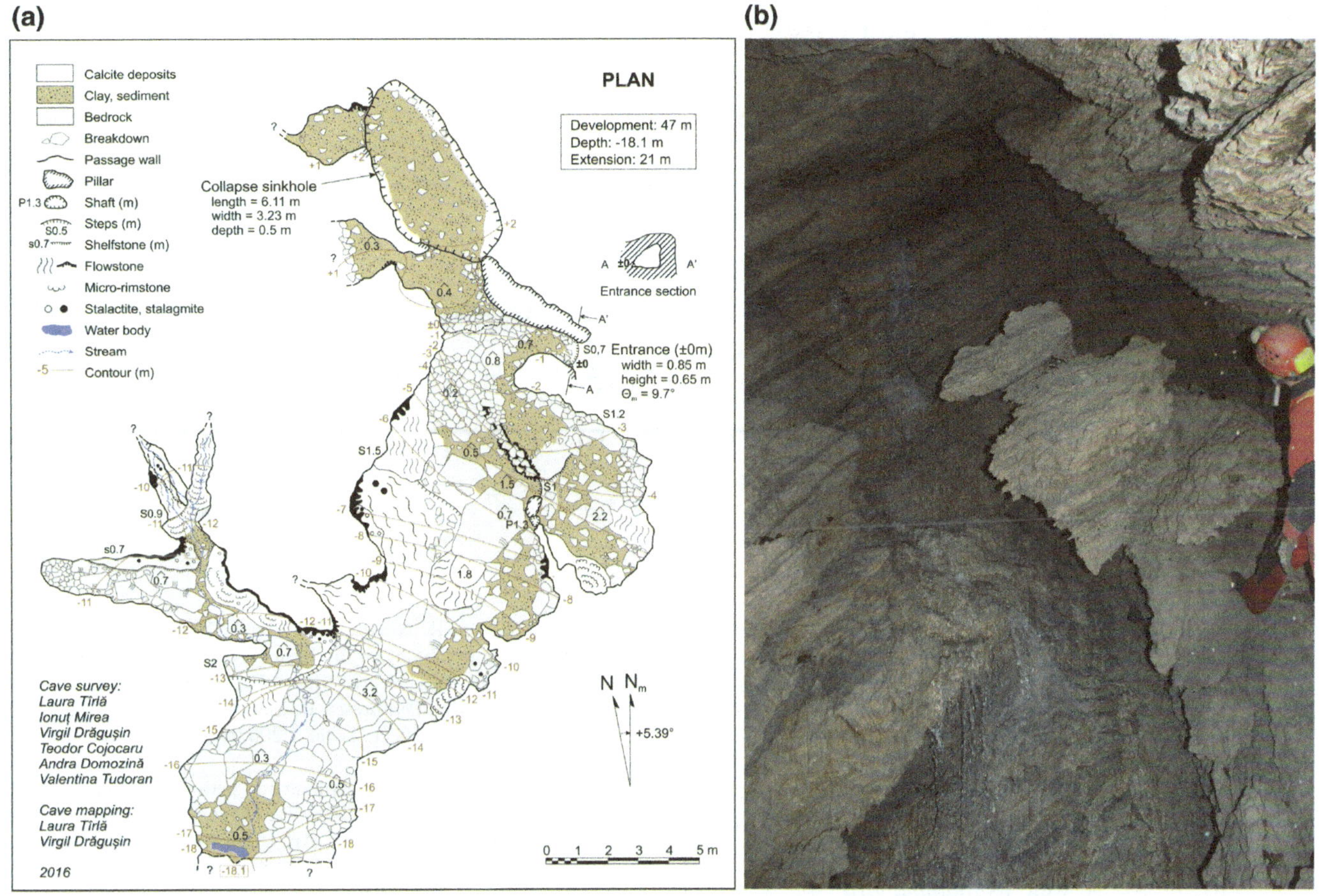

Fig. 2 **a** Map of M3-R2 Cave (used with permission from Tîrlă et al. 2016), **b** wall morphology and waterfall in Cave No. 1 from the Piscu Negru Mine (photograph by L. Tîrlă)

climate, and is characterized by a perturbation meroclimate in the first 8–10 m from the entrance, and a transition meroclimate in the rest. Contrastingly, Cave No. 1 from the Piscu Negru Mine has a stability meroclimate due to its remoteness (350 m) from the mine entrance.

Since 2014, the caves in Mușeteica and Piscu Negru are part of a monitoring program developed by the "Emil Racoviță" Institute of Speleology. Temperature, relative humidity, and CO_2 concentrations are recorded in the M3-R2 Cave and Cave No. 1 from the Piscu Negru Mine.

In the M3-R2 Cave (~50 m long, 20 m deep; Tîrlă et al. 2016), temperature has a seasonal variability between ~2–3 °C in early summer and 4–5 °C in autumn. Minimum temperature is recorded in June–July when there is still snow cover and the entrance is likely obstructed by snow. As snow cover melts and cave ventilation recommences, the temperature rises until September with large fluctuations imposed by outside temperature variability (Fig. 3). Being steeply inclined, the cave is heavily ventilated during the snow-free season, with CO_2 concentrations almost equal to those measured outside (200–300 ppm). Inside M1 Cave from the Mușeteica Valley, we recorded temperature values between August and September 2017. The values rise gradually from 2.8 to 2.9 °C but then decrease abruptly to 2.8 °C in late September, probably shifting to a winter mode.

Cave No. 1 from the Piscu Negru Mine shows a smoother seasonal pattern of variability on the active passage, with the highest temperature (T) recorded in September 2014 (8.2 °C) and the lowest in April 2015 (5.4 °C), reaching again 7.4 °C in late September 2015 (Fig. 3). The shift in T values in April 2015 seen in Fig. 2 is a result of the temperature logger being moved from a location that was close to the underground river in a position which is better sheltered from air circulation and river influences. An air current is always present in the cave, and it could explain the relatively low CO_2 concentration, between 2930 ppm in August 2014 and 610 ppm in April 2015. The high summer values could reflect an important input from soil organic activity, while the low values in spring might reflect a depletion of the soil CO_2 reservoir over the winter.

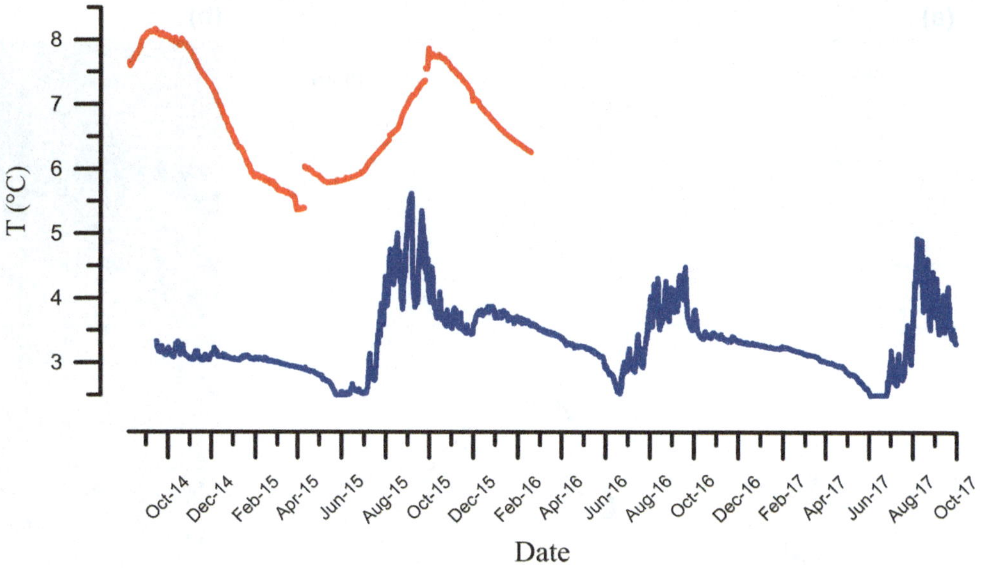

Fig. 3 Temperature profiles for Cave No. 1 from the Piscu Negru Mine (red) and M3-R2 Cave (blue) (Color figure online)

Cave Biology

The mine and the caves at Piscu Negru represent a hibernaculum for a bat colony composed of *Myotis myotis* (Borkhausen 1797), *Rhinolophus ferrumequinum* (Schreber 1774), and *R. hipposideros* (Bechstein 1800). Lepidopteran *Scoliopteryx libatrix* (Linnaeus 1758) and *Triphosa dubitata* (Linnaeus 1758) as well as *Salamandra* sp. were also observed in the mine galleries. The severe environmental conditions in the Piscu Negru and Muşeteica caves have determined some microorganism species, otherwise ubiquist, to develop particular adaptations leading to new, extremely resistant strains: *Sarcina flava*, *S. lutea*, *Bacillus subtilis*, *Staphylococcus aureus*, and *S. albus* (Bogdan et al. 1995). In the M3-R2 Cave, the collembolan *Paronychiurus* sp. (Giurgiu 2006), and the spider *Tenuiphantes cristatus* (Menge 1866) were identified.

Acknowledgements This work was financially supported by the FREem project (IFA-CEA C4-08/2014) and ICUB grant for young researchers (13055/2017). We are grateful to J. Hellstrom from the University of Melbourne for providing U-Th age for the flowstone in the M3-R2 Cave.

References

Balintoni I, Balica C, Ducea MN, Hann H-P (2014) Peri-Gondwanan terranes in the Romanian Carpathians: a review of their spatial distribution, origin, provenance, and evolution. Geosci Front 5 (3):395–411

Bogdan M, Macrea R, Negrea M, Sârbu I (1995) Observaţii asupra climatului şi microbionţilor din Peştera M3-R2 din Muşeteica şi din Peştera 1 de la Piscu Negru (Munţii Făgăraş). Cerc Speol 3:21–26

Dessila-Codarcea MD, Dimitrescu R, Stancu I (1968) Geological map of Romania, Sibiu Sheet. Scale 1:200,000 (L-35-XIX). Inst Geol Geofiz, Bucureşti

Dimofte C (1962) Observaţii geologice şi petrografice asupra cristalinului din versantul sudic al Munţilor Făgăraş, între Valea Oltului şi Vârful Surul (in Romanian). An Univ Bucureşti, Ser Geol XI (32):17–41

Drăguşin V (2013) Late Pleistocene climate variability recorded in stalagmites from Romania. Unpublished Ph.D. Thesis, Babeş-Bolyai University, Cluj-Napoca, Romania

Giurgiu I (1990) Peşterile descoperite la peste 2000 m altitudine, completare esenţială a imaginii carstului din România. Bul Club Speol “Emil Racoviţă” Bucureşti 13:62–75

Giurgiu I, Vlădulescu M, Silvăşanu G (2006) Peşterile de la Piscu Negru. Invitaţie în Carpaţi 59:41–52

Giurgiu I (2006) Peşteri din Muşeteica. Invitaţie în Carpaţi 67:28–57

Giuşcă D, Anastasiu N, Popescu GC, Şeclăman M (1977) Observaţii asupra şisturilor cristaline din zona centrală a Masivului Făgăraş (Cumpăna—V. Cârţişoara). An Univ Bucureşti, Ser Geol XXVI:3–17

Iancu V, Mărunţiu M, Johan V, Ledru P (1998) High-grade metamorphic rocks in the pre-Alpine nappe stack of the Getic-Supragetic basement (Median Dacides, South Carpathians, Romania). Miner Petrol 63:173–198

Iancu V, Berza T, Seghedi A, Gheucă I, Hann H-P (2005) Alpine polyphase tectono-metamorphic evolution of the South Carpathians: a new overview. Tectonophysics 410(1–4):337–365

Kräutner HG (1980) Lithostratigraphic correlation of Precambrian in the Romanian Carpathians. An Com Geol Geofiz LVII:229–296

Pană D, Erdmer P (1994) Alpine crustal shear zones and pre-Alpine basement terranes in the Romanian Carpathians and Apuseni Mountains. Geology 22:807–810

Tîrlă L, Drăguşin V, Mirea I, Cojocaru T (2016) Speleomorphology of M3-R2—the highest cave in the SE Carpathians. Rev Geomorfol 18:54–62

Căpăţânii/Parâng Mountains: Polovragi Cave–Oltetului Gorge Karst Area

Gheorghe M. L. Ponta, Gheorghe V. Aldica, and Radu Dumitru

Abstract

The Polovragi Cave–Olteţului Gorge karst area is about 3 km^2 in size and is traverse by the Olteţ River through a spectacular narrow gorge, which represents the natural boundary between Parâng Mountains to the west and Căpăţânii Mountains in the east. These mountains are part of the Southern Carpathian. Focul Viu Caving Club surveyed the Polovragi Cave between 1975 and 1985. In parallel with the survey activities in the main cave developed on the first level, several smaller caves were identified on both sides of the Olteţului Gorge. These caves are located on the second and third levels, which are interrelated with different peneplanation events that occurred in the Carpathians. In 2000, Focul Viu began working in the cave again, replacing gates, removing trash, digging to open new passages, and performing underwater explorations. By the end of 2011, new passages were found in the Hope Chamber and Costin Gallery, extending the total length of the cave to 10,793 and 92 m in vertical range. The Polovragi Cave along with two other smaller caves is presented in this chapter. The Polovragi Cave is the third longest cave in the Southern Carpathians and the eleventh in Romania. One of the smaller caves hosts an important bat colony, and the other one, with 800 m of passages, hosts *Ursus spelaeus* bones and deposits of saltpeter. The Bones Cave (Peştera cu Oase) located on the western side of the gorge (opposite side of Polovragi Cave) at a higher elevation is an old meander of the Olteţ River.

Keywords

Cave • Gorge • Dye studies • Fractures

Introduction

The Polovragi Cave–Olteţului Gorge karst area (3 km^2) is traversed by the River Olteţ through a narrow gorge, which forms the natural boundary between Parâng Mountains to the west and Căpăţânii Mountains in the east, both parts of the Southern Carpathians (Fig. 1). A limestone ridge of Jurassic age is crossed by Olteţ River, generating steep gorges, 5–10 m wide at the base and 300 m high (Fig. 2). Twenty-five meters above the bottom, the gorge opens up to 40 m, where several cave entrances, including the one of Polovragi Cave, may be found. A second caves' level at 75–100 m above the bottom of the gorge, with 24 caves on the western side of Olteţ River, including The Bats Cave (Peştera Liliecilor), is located (Ponta and Aldica 1986). The Bones Cave is located on a third cave level, at 150 m above the thalweg of the river. On the eastern side of Olteţ River, 23 caves were surveyed. These caves levels are interrelated with the peneplanation of the Carpathians.

A few hundred meters into the limestone, part of the waters of the Olteţ River sinks underground through an impenetrable ponor, generating the Polovragi Cave. The cave formed on the east side of the river, by successive captures of the Olteţ River along E–W fractures (Ponta and Aldica 2009). Olteţ River, with an estimated discharge of 500–600 L/s, represents the mainstream flow, which carries over the surface and groundwater of an area where crystalline rocks and Mesozoic sedimentary deposits cover a granitic intrusion.

G. M. L. Ponta (✉)
Geological Survey of Alabama, 420 Hackberry Lane, Tuscaloosa, AL 35401, USA
e-mail: gponta@yahoo.com

G. V. Aldica
National Institute of Materials Physics, Atomistilor 105 bis, 077125 Măgurele, Romania
e-mail: aldica2000@yahoo.com

R. Dumitru
Asociatia Speologică “Focul Viu”, Frumoasă 31, Bucureşti 1, Romania
e-mail: carst_webmaster@yahoo.com

G. M. L. Ponta and B. P. Onac (eds.), *Cave and Karst Systems of Romania*,
Cave and Karst Systems of the World, https://doi.org/10.1007/978-3-319-90747-5_12

Fig. 1 Location of Polovragi Cave–Olteţului Gorges karst area within Romania

Geographic, Geologic, and Hydrologic Settings

The Polovragi Cave–Olteţului Gorge karst area is developed in a small island of limestones of the Cerna–Jiu sedimentary zone of the Danubian Autochthonous (Săndulescu 1984), which consists of mid-grade metamorphic basement, often intruded by granitoide and molasses deposits of Permian to Carboniferous age, transgressively covered by Mesozoic sediments. The limestone island is located on the eastern end of the Danubian Autochthonous, at the interface with the Getic Domain Nappe.

Cerna–Jiu sedimentary zone is characterized by the development of an extensive carbonate platform, which was formed starting from Middle Jurassic almost to the end of Lower Cretaceous. It is bounded by Cerna Valley in the west and continues on the southern slopes of the Vâlcan Mountains, extending east of the Jiu Valley as small islands of limestones reaching Olteţului Valley, in the vicinity of Polovragi village. Cerna–Jiu area functioned mostly as a marginal sedimentary area, where reef facies developed especially in Jurassic and Early Cretaceous, ending with the accumulation of wildflysch-type deposits.

The N–S-oriented hydrographic network in the area is dominated by the Olteţ River, collecting the surface and underground waters in the area, including the spring of Polovragi Cave. During geologic times, the changes in the base level of the Olteţ River controlled the development of the four known levels in the cave. The geology of the area is complex; a relatively small area of sedimentary rocks, including limestone disposed in a sequence of grabens, horsts, or monoclinal structures, overlies a granitic and crystalline basement. The crystalline rocks are part of the Olteţ Thrust and have been penetrated by Şuşiţa granitoides and are overlain by Mesozoic deposits, 100–500 m thick, depending on their position in the micrograben.

Based on belemnites findings (Nedelcu 1978), the calcareous sandstone deposits that outcrop at the south entrance in the Olteţ Gorges on both sides of the river are Middle Jurassic in age. These deposits are overlain by up to 400 m of Tithonian (Upper Jurassic) limestones. Microscopic

Fig. 2 Olteț Gorges (from Ponta and Aldica 2009, with permission)

analysis showed that this unit is formed from several limestone layers, 50–80 m thick, separated by calcarenites. The limestones are strongly fissured and fractured, with breccia fragments 0.5–2 mm in size identified along the fracture planes (Nedelcu 1978). The joints and fractures are part of two distinct systems: one oriented N–S and the other one NW–SE. Cretaceous flysch represented by black calcareous marls and green clays with coal beds overlies the Jurassic limestones (Ponta and Aldica 2009).

The geological map of Trifulescu (unpublished) completed with G. Ponta's observations (hydrogeologic mapping/karst inventory of the area; Fig. 3) shows that the limestones of Polovragi Cave–Oltețului Gorge karst area are separated from the limestone outcrops of Cerna–Oltețului Valley (Fig. 2) and Dry Valley (Valea Seacă), a possible continuity being under the crystalline formations of the Getic Domain, as shown on the geological map 1:200,000 (Codarcea et al. 1967).

In 1970, the Institute of Meteorology and Hydrology from Bucharest conducted a dye study in the water losses that occur in the Cerna–Oltețului Valley and demonstrated their connection with the Polovragi Spring (Niță and Bulgăr 1971). More recently, Bandrabur and Bandrabur (2010) conducted two dye studies in Cerna–Oltețului Valley (4050 m away) and Dry Valley, 2200 m away. A cross section shows the dye traveled under crystalline formations of the Getic Domain through limestone deposits toward the Polovragi Spring at an average velocity of 57.86 or 20 m/h, respectively. Although the geological map 1:200,000 shows continuous limestone strips under the crystalline formations, a connection between those two karst areas at these velocities is very unlikely.

For comparison, a dye study conducted in Thailand in 2012 (Ponta et al. 2013) at Tham Lumphini Suan Hin Spring, a through cave with only 200 m of submerged/impenetrable passages versus 4,050 m (the length between Cerna-Oltețului Valley – Polovragi Spring), a flow velocity of 35 m/hour was recorded, 1.5 times slower than the dye study mentioned above. Based on data provided by Bandrabur and Bandrabur (2010), the excess flow of the Polovragi Spring of 234 and

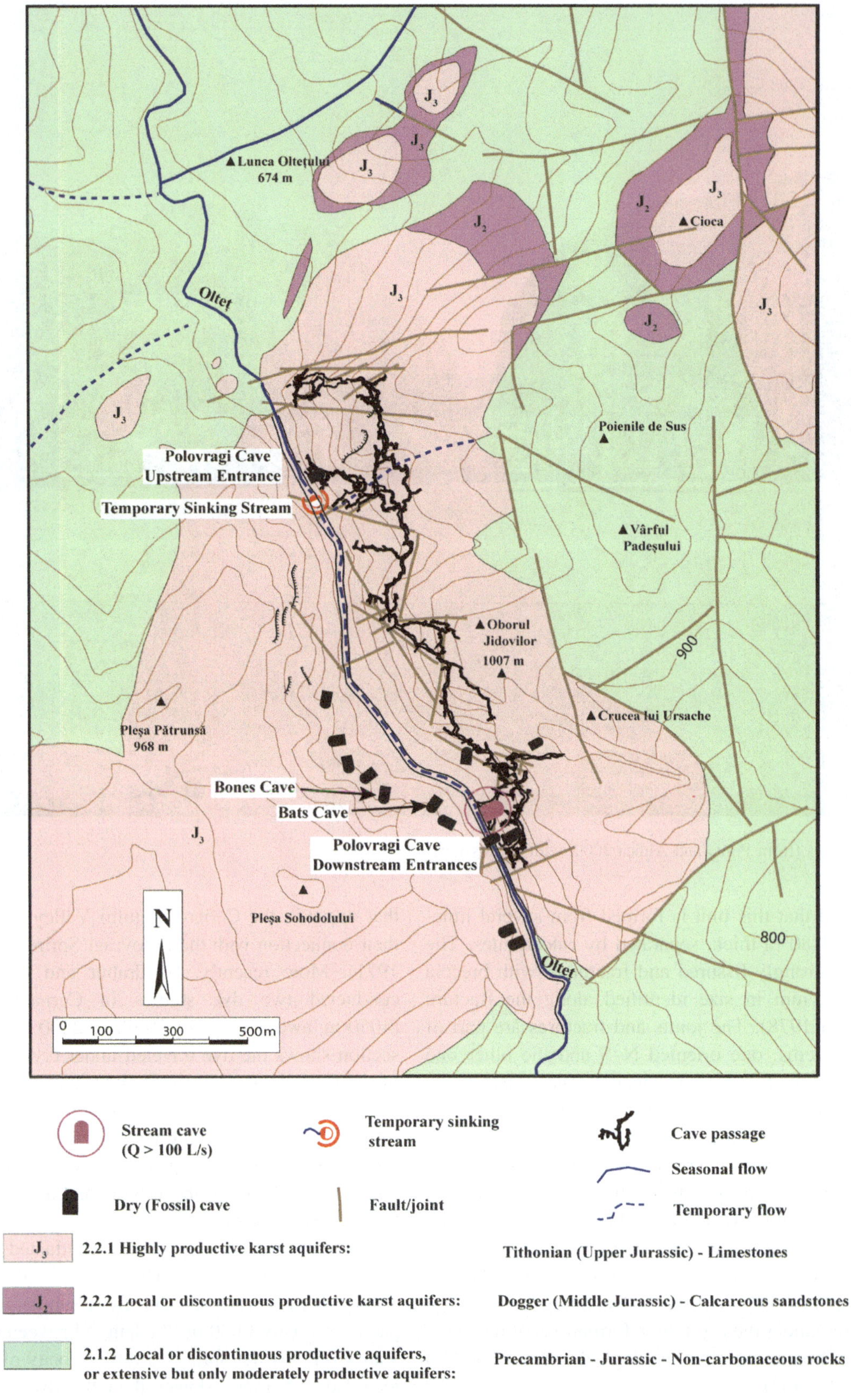

Fig. 3 Karst hydrogeologic map of Polovragi Cave-Olteţului Gorges karst area (geology modified after Trifulescu unpublished)

114 L/s, respectively, was assumed to come from Cerna-Olteţului and Dry valleys. Further detailed studies of the geology of the area, an updated karst inventory, and new dye studies are recommended.

Also, Niţă and Bulgăr (1971) assume a connection between Galbena River in the west with the springs on the western side of the Olteţ River mentioned by Bandrabur and Bandrabur (2010), and the hypothetical flow path is perpendicular (E–W) on the general N–S trend of the hydrographic network in the area. The recharge region of the limestones outcropping in the western side of the gorges is sufficient to explain the springs' flow located opposite to the Polovragi Cave. It is also possible that some of the waters sinking along the Olteţ River are partially recharging the springs located on the western side of the gorges, through conduits or unknown cave passages as mentioned above. The losses along the Olteţ River may also recharge the springs located on the western side of the gorges, assumption suggested by Bandrabur and Bandrabur (2010). The Bones Cave located on the western side of the Oltet Gorge at a higher elevation (Fig. 3) represents an old meander of the Olteţ River and could be a confirmation of this hypothesis.

History of Exploration

The main entrance of the cave was known for a long time as cave of Pahomie from Polovragi. Joanes made a brief description of the cave in 1868, followed by A. Vlahuţă in "Romania Pitorească" (1901). R. Jeannel and E. Racoviţă reported for the first time the location of the cave in 1929. In 1951, Chappuis and Winkler published a description of the cave (Bleahu et al. 1976). Iancu et al. (1961) published a cave-related paper that includes a map of 961 m of passages. In 1974, Focul Viu Caving Club from Bucharest began an extensive study of the area, surveying the cave, and conducting geologic, geomorphologic, and tectonic observations.

In 1975 and 1976, Niphargus Praha and Liliacul Arad speleo clubs joined the exploration and mapping of the cave (Goran 1982). A new description of the cave was included in the 1976 edition of the book *Peşteri din Romania* (Caves of Romania; Bleahu et al. 1976). In 1980, F. Păroiu made a cave-diving attempt in the upstream sump of the Active Chamber/Sala Activului (Sump 6), followed in 1982 by Ş. Sârbu, M. Oancea, and I. Puşpurică who explored Sump 6 (50 m long and −6 m deep) and Sump 7 to a depth of −22 m (Puşpurică and Sârbu 1985). Also, the five underwater passages located downstream of the Active Chamber were traversed, making the connection with the cave's resurgence (Peştera de la Resurgenţă; Goran 1982) from Olteţului Gorge. Brief descriptions of the Olteţ River and Polovragi Cave are presented by Goran et al. (2006).

In 2000, Focul Viu Caving Club began working in the cave again, replacing gates, and removing trash, combined with digging and underwater explorations. By the end of 2011, new passages were found in Hope (Speranţele) Chamber and Costin Gallery, the cave becoming 10,793-m-long and 91-m-vertical range (Dumitru pers. comm.).

Caves' Descriptions

Polovragi Cave

The Polovragi Cave is located on the east side of the Olteţ River and has six entrances, three dry and three active. The upstream entrance (dry; 2 × 3.5 m) is at the northern end of the gorge, at about 15 m relative altitude (645 m asl), and initially functioned as a sinking stream (Aldica and Ponta 1983). At the bottom of the gorge, in the same area, the Olteţ River is partially sinking underground through a ponor (630 m asl), generating Polovragi Cave's stream. In the upper section of the gorge, at the contact between limestones and non-calcareous rocks, a diffuse infiltration occurs along joints and fissures. The rest of the three entrances are 1.4 km downstream at the Olteţ River thalweg (6 × 1, 4 × 1, and 4 × 1 m; 605 m asl). Through these entrances, during heavy rains or melting snow, karst springs fed by the cave stream (one permanently and two temporally) emerge to the surface. The main entrance of the Polovragi Cave (7 × 11 m; 630 m asl) is located ~25 m above the springs. Next to it is a smaller one (1 × 0.4 m), both being dry (Fig. 4).

Polovragi Cave with 10,793 m of passages and 91-m vertical range (−62 m, +29 m) has a branching index of 6.81 and an extension of 1520 m. It is a cave with a stream at water table (Palmer 2007). For description purposes, the cave was divided as follows: the Access Passage (Zona de Acces), the Upstream Section (Zona Amonte), and the Downstream Section (Zona Aval) (Fig. 4). The Access Section, with 1224 m of passages, is between the upstream entrance and the Wonder Chamber (Sala Minunilor). The passages are in general small, formed along E–W- and NE–SW-oriented fractures, and developed on three different levels, most of them being formed by surface sinking streams. The Access Passage, with an average height of 1 m, has frequent changes in direction. This area was for a long period of time completely submerged (under the phreatic conditions). This hypothesis is sustained by the cave morphology (horizontal ceiling, narrow passages), negative corrosion forms (ceiling pendants), and large argillaceous deposits. In two areas were the passages narrows, a strong airflow exists, and a group of anemolites formed. In most of this section, those three cave levels are distinct, but in some areas, they are interconnected, forming one large borehole

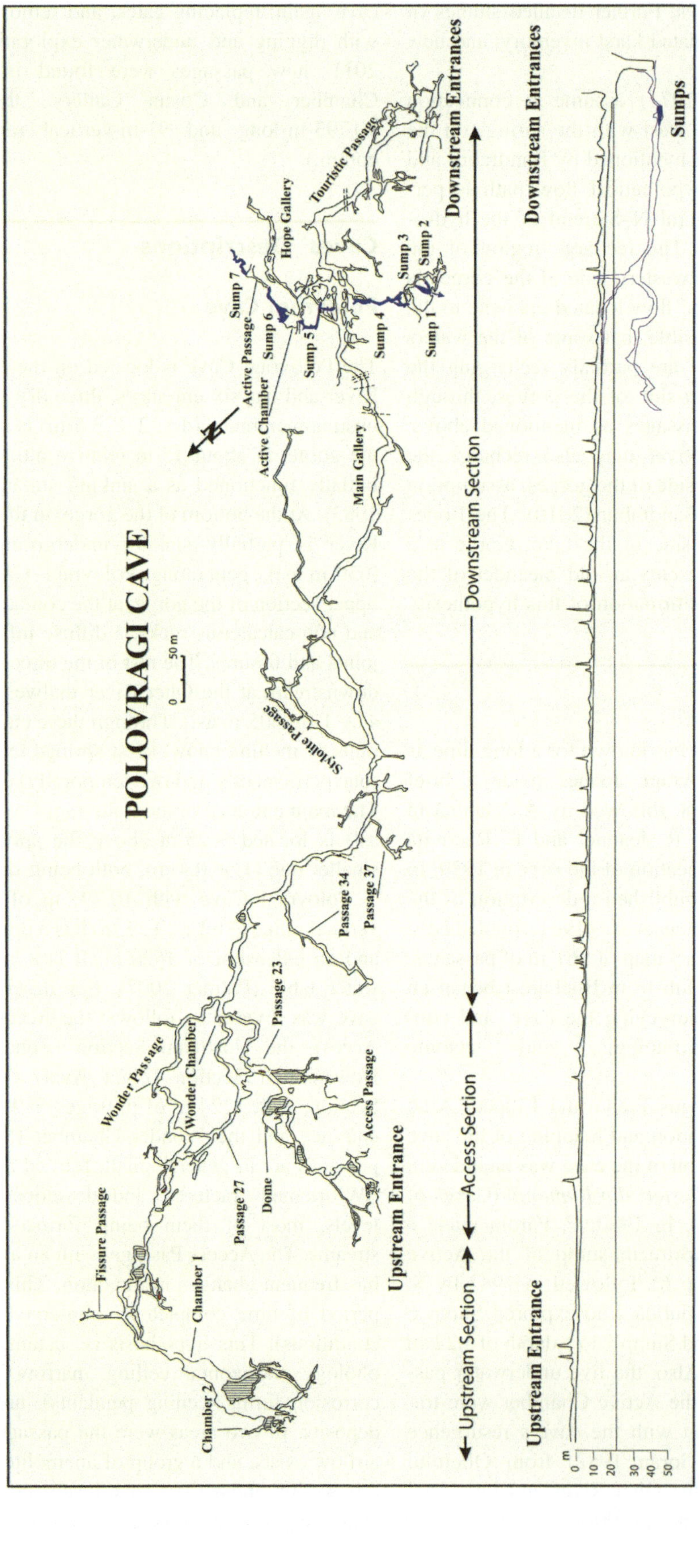

Fig. 4 Simplified map of Polovragi Cave (surveyed by Focul Viu Caving Club)

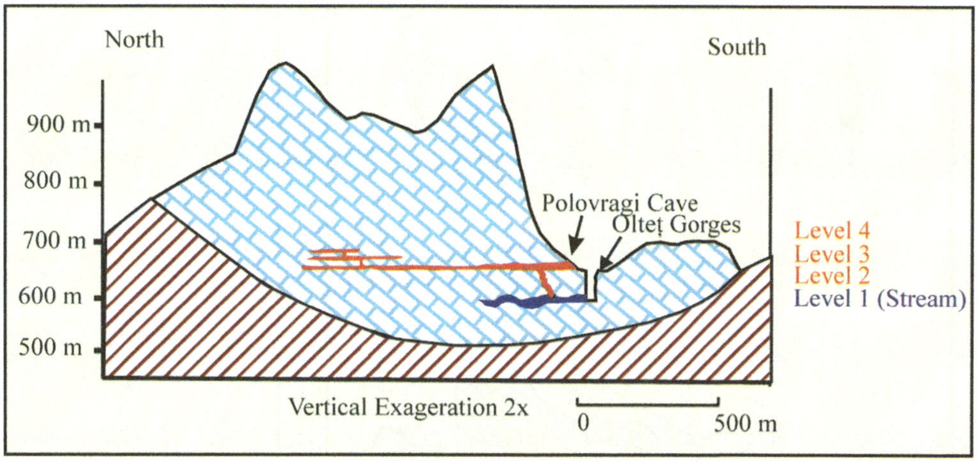

Fig. 5 Stages of passages development in the Polovragi Cave (modified after Ponta and Aldica 2009, with permission)

well decorated with stalagmites up to 4 m high and 2 m in diameter (e.g., the Dome/Domul).

Upstream of the Dome, the Passages 27, 23, and the Wonder Passage, with a total length of 2880 m, form the Upstream Section (Fig. 4). Passage 27 has two upper level galleries with chimneys recharged by sinking streams located on the side of the mountain. The main borehole in this section originated near the limestone/non-calcareous rock interface and is developed on three levels, 13 m apart. Their morphology varies from bare passages to well decorated sections, with calcite deposits occupying most of the passage. At the intersection of the fractures system mentioned earlier, the three levels merge in large rooms as Chambers I and II (Sălile I and II).

The Downstream Section of the cave with 5067 m is the longest one. The side passages mapped at west of the main gallery were formed by partially sinking streams displayed along Olteţ River, and those close to the main entrance represent former outlet conduits. Downstream of the Wonder Chamber, the main passage is oriented N–S, is 3–4 m high, and has sand and clay on its floor. Beyond Passage 37, the main gallery narrows (1 × 1 m), and the floor is covered with thick argillaceous deposits, gravel, boulders, up to the intersection with Stylolite Passage (Galeria Stilolitelor). Past the junction with this gallery, the main gallery becomes wider and taller (borehole type) up to the downstream entrance of the cave. In this area, a narrow passage descends to the lower level of the cave (Active Passage). The waters are coming in through Sump 6 (12 m in depth and 54 m in length, followed by a second 30-m-long sump (Sump 7), the end of which was not yet reached. Between this sump and the sinking point on the Olteţ River, the path of the stream is unknown. It is possible to be formed by flooded conduits, penetrable only for divers. The subterranean stream crosses the Active Chamber (Sala Activului), sinking in a series of five downstream sumps, the water coming out in the Olteţ Gorges. The fractures oriented NE–SW are controlling the development of the cave in this section. As shown in Fig. 3, the cave passages generally follow the fractures, faults, and joints identified at the land surface by geologic mapping. The cave map identified some of those features during the survey and illustrates the important role of the tectonics in the genesis of the cave. Polovragi is a multi-level cave, with three fossil levels and one stream level formed at the elevation of Olteţ River. The second level (Fig. 5) has the largest passages being active for the longest time.

Bats Cave (Peştera Liliecilor)

The cave was discovered and surveyed in 1975. It is situated on the western slopes of the Olteţ Gorges at higher relative altitude (about 75 m above the thalweg of Oltet River), along a second erosional level, corresponding to an earlier stage of karstification, which is interrelated with the peneplanation of the Carpathians. The cave has a medium size chamber, which ends with an ascending side passage, where the highest elevation of the cave is reached. A 0.5-m-thick guano deposit, generated by the bats colony in the cave, covers the floor. It is one of the largest bat colonies observed in a small cave, hundreds of bats flying all over during the exploration and survey of the cave (Fig. 6).

Saltpeter Cave (Peştera cu Şalitru/Peştera cu Oase/Peştera nr. 4 din Cheile Olteţului)

The cave is located in the western side of the Olteţ River and Polovragi Cave, at about 200 m upstream from its main/touristic entrance. The Saltpeter Cave, an old meander of the Olteţ River, is a large borehole with five entrances grouped in two areas. At the main entrance is a large block, which can be observed from the road along Olteţ River. The cave has a large passage, which narrows toward the

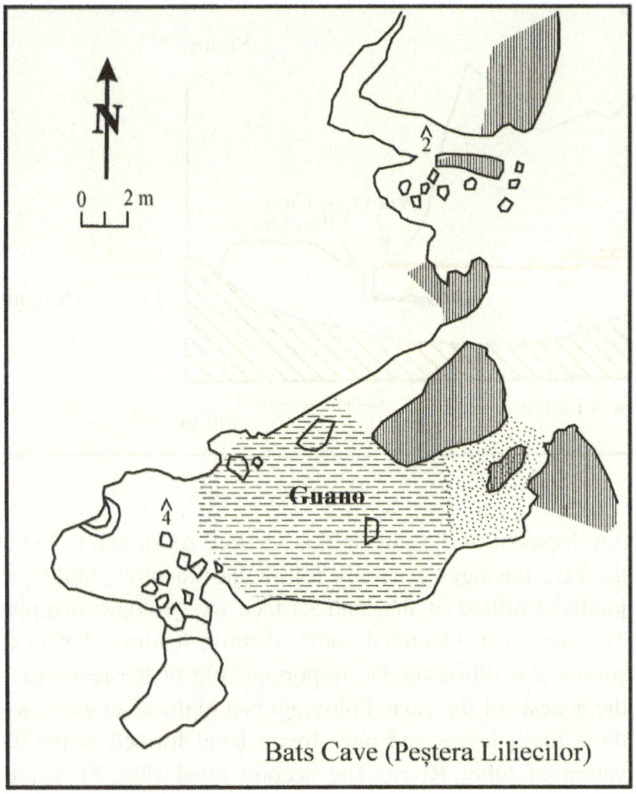

Fig. 6 Map of the Bats Cave (surveyed by Focul Viu Caving Club)

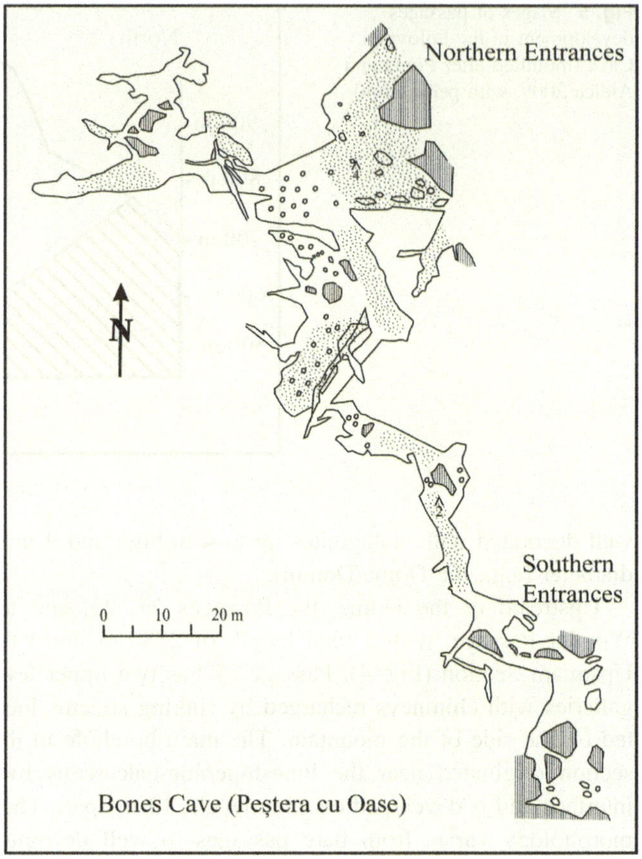

Fig. 7 Map of the Bones Cave (surveyed by Focul Viu Caving Club)

downstream entrances. In the cave was found a deposit of *Ursus spelaeus* bones, from where the name the Bones Cave comes from. The locals named it the Saltpeter Cave because of the saltpeter deposits mined from the cave to manufacture gunpowder (Fig. 7).

Cave Mineralogy

The mineralogical analyses performed by Diaconu et al. (2008) in Polovragi Cave identified the presence of three mineral species: hydroxylapatite, brushite, and taranakite, all in relation to the presence of bat guano accumulated over the limestone bedrock.

Cave Fauna

In Polovragi Cave were identified three bat colonies. Two of them are for hibernation and one for maternity. The following bat species appear in all colonies: *Myotis myotis*, *Myotis blythii*, *Myotis emarginatus*, *Myotis bechsteinii*, *Rhinolophus hipposideros*, and *Rhinolophus ferrumequinum*. The first colony (380 individuals) observed in December 2005 was located in the touristic section of the cave; this had the highest number ever recorded, and the dominant specie was *R. ferrumequinum*. The second colony (*M. myotis/M. blythii*) of 155 bats was identified in the upper section of the cave in the vicinity of a large rimstone pool, whereas the third colony was observed in Bats Gallery (Galeria Liliecilor) (Chachula 2009).

Anthropology

Although known since the nineteenth century, the archeological investigations in Polovragi Cave begun only between 1965 and 1967; they revealed a variety of artifacts as early as the Dacian Kingdom, the Copper Age, and the Neolithic Period. Some of the archeological material is probably related to Polovragi Fortress that has been built around 150 BP (Boroneanţ 2000).

Cave Conservancy

About 800 m of passages within Polovragi Cave are open for tourists. The features that characterize this section of the cave include various speleothems and large-sized halls. For more details, readers are invited to inspect the *Show Cave* chapter authored by Meleg et al. in this book. The Polovragi Cave–Olteţului Gorge karst area is part of the Protected Natural Area 2.440 located in the Natura 2000 ROSCI0128-"Nordul Gorjului de Est," site managed by the S.C. Butterflyeffect S.R.L. The touristic reservation is administrated by the "Alexandru Ştefulescu" Museum of Gorj County and the speleological reservation by Focul Viu Caving Club. Overall, Polovragi is a class B protected cave with A and C sectors, respectively. The Gallery 27, the Wonder Gallery, and the Stream Gallery are all class A sectors, whereas the show cave section belongs to class C. Both upstream and downstream entrances are gated.

Acknowledgements We are thankful to Focul Viu Caving Club Bucharest, Liliacul Arad, Nyphargus Praha, and GESS Bucuresti and all our friends and colleagues for their support during the exploration and research activities at these caves. We are also grateful to G. Diaconu (Senior Researcher with the "Emil Racoviţă" Institute of Speleology), the scientific coordinator of Focul Viu Caving Club between 1976–1980, for his advices and recommendations for fieldwork activities.

References

Aldica G, Ponta G (1983) Observaţii morfologice în Peştera Polovragi. Stud Cerc Speol 3:25–28

Bleahu M, Decu V, Negrea Ş, Pleşa C, Povară I, Viehmann I (1976) Peşteri din România. Ed Ştiinţifică şi Enciclopedică, Bucureşti, p 415

Bandrabur G, Bandrabur R (2010) Parâng and Căpăţânii Mountains. In: Orăşeanu I, Iurkiewicz A (eds) Karst hydrogeology of Romania. Ed Belvedere, Oradea, pp 69–75

Boroneanţ V (2000) Arheologia peşterilor şi minelor din România. Ed cIMeC, Bucureşti, p 370

Chachula O (2009) Monitorizarea coloniilor de lilieci in perioada hibernarii din pestera Polovragi. In: Dumitru R (ed) Focul Viu 40 de ani de speologie. RAO, Bucureşti, pp 139–141

Chappuis PA, Jeannel R (1951) Enumération des grottes visitées, 1927–1949 (8e série). Arch Zool exp gén 88:81–230

Codarcea A, Răileanu G, Bercia I, Marinescu F, Mutihac I, Pavelescu M, Stancu I (1967) Geological map of Romania, Târgu Jiu Sheet (L-34-XXX). Scale 1:200,000, Inst Geol, Bucureşti

Diaconu G, Dumitraş D, Marincea Ş (2008) Mineralogical analyses in Peştera Polovragi (Olteţului Gorges) and Peştera Muierii (Galbenului Gorges), Gorj County. Trav Inst Speol "Émile Racovitza" XLVII:89–105

Goran C, Constantin S, Horoi V (2006) Karst areas in the Southern Carpathians between Cerna and Olt rivers. In: Constantin S (ed) Field Trips in the karst of Romania. Selected routes from Southern Carpathians, Bucharest, pp 76–77

Goran C (1982) Catalogul sistematic al peşterilor din Romania. Ed CNEFS, Bucuresti, p 496

Iancu S, Lupu S, Ilie I (1961) Contribuţii la studiul geomorfologic al Peşterii Polovragi. Anal Univ "C. I. Parhon" Ser şt nat geol-geogr 27

Jeannel R, Racoviţă EG (1929) Énumération des grottes visitées, 1918–1927 (7e série). Arch Zool exp gén 68(2):293–608

Nedelcu V (1978) Prospecţiuni şi studii geologice pentru minereuri auro-argentifere şi sulfuri polimetalice în formaţiuni cristalofiliene în Munţii Căpătinii. Arh Inst Geol Geofiz, Bucureşti

Niţă I, Bulgăr A (1971) Contributii la studiul scurgerii apei pe râurile influenţate de carst in zona hidrografică dintre Cerna si Gilort. Hidrotehnica 16:9

Palmer AN (2007) Cave geology. Cave Books, Dayton, p 454

Ponta G, Aldica G (1986) Peşteri din Cheile Olteţului. Bul Speol 10:245–269

Ponta G, Aldica G (2009) Geology and tectonics of Polovragi Cave—Romania. In: 15th International Congress of Speleol, Kerville, Texas, 3:1638–1642

Ponta G, Memon B, LaMoreaux J, Julawong J, Wongsawat S (2013) Karst landforms in the Saraburi Group limestones, Thailand. In: Land L, Doctor DH, Stephenson BJ (eds) Proceedings of 13th Multidiscussion Conference of Sinkholes and the Engineering Environmental Impacts of Karst, Carlsbad, New Mexico, pp 299–309

Puşpurică I, Sârbu Ş (1985) Consideraţii asupra morfologiei zonelor submerse din Peştera Polovragi. Stix 2:20–24

Săndulescu M (1984) Geotectonica României. Ed Tehnică, Bucureşti, p 336

Vlahuţă A (1901) Romînia Pitorească. Ed. Socecu, Bucureşti, p 280

Further Reading

Aldica G (2009) Focul Viu la Polovragi. In: Dumitru R (ed) Focul Viu 40 de ani de, speologie edn. RAO, Bucureşti, pp 58–66

Bleahu M (1972) Karst of Rumania. In: Herak M, Springfield VT (eds) Important karst regions of the Northern Hemisphere. Elsevier, Amsterdam, pp 341–353

Cocean P (1995) Peşterile României. Potenţial turistic. Ed Dacia, Cluj-Napoca, p 264

Orghidan T, Negrea Ş, Racoviţă G, Lascu C (1984) Peşteri din România. Ghid turistic. Ed Sport-Turism, Bucureşti, p 504

Useful Websites

www.speologie.org
www.karstgeology.com

Karst of Sebes Mountains

Gheorghe M. L. Ponta

Abstract

Şura Mare Cave with its enormous entrance and 11,694 m of passages is the tenth longest cave in Romania and the fourth one with respect to its vertical range (+425 m). The karst system penetrated vertically and horizontally the entire sequence of limestone deposits of the Ponorici Cioclovina-Ohaba Ponor Plateau. Şura Mare Cave is known for its long underground stream, spectacular domes, and large well-decorated chambers. Dye studies performed in the early eighties prove the connection with Dosul Lăcşorului and Fundătura Hobenilor shafts, making this karst system one of the most important in Romania.

Keywords

Karst • Caves • Tracer studies

Introduction

Şureanu Mountains are located in the western part of the Southern Carpathians (Fig. 1) being divided by Sebeş River in Cibin Mountains (East) and Sebeş Mountains (West). The karst is present in the western part of the Sebeş Mountains, where the limestones deposits are divided by Strei River in Ponorici Cioclovina-Ohaba Ponor Plateau to the northwest with Şura Mare, Cocolbea, Ponorici-Cioclovina caves and Bojiţa–Tecuri Plateau to the southwest with Şipot karst system, which include in its watershed the Răchiţeaua Shaft, Valea Clenjii Cave, and Ponorici II swallet. Numerous tracer studies were conducted in both plateaus to define the recharge areas of the main springs. For a complete account for the history of tracer studies in the area, the readers are encouraged to examine Trufaş (1986, 1978), Orăşeanu et al. (1991) and Bandrabur and Bandrabur (2010).

Geographic, Geologic, and Hydrologic Settings

The Sebeş Mountains karst area is developed in limestones belonging to the sedimentary zone of the Getic Domain Nappe (Mutihac and Ionesi 1974; Săndulescu 1984), which was completed as a structural unit of the Southern Carpathians at the end of Cretaceous, during the Laramic orogenesis. In the present structure, the Getic Domain occupies a larger area than the Danubian Autochthonous, forming mostly the mountainous edifice of the Southern Carpathians.

The Getic Domain in the Cibin and Sebeş Mountains consists of mesometamorphic crystalline schists, which are part of the Sebeş-Lotru Terrane and are represented by gneiss, paragneiss, amphibolites, and mica schists (Stilla 1972).

The Getic sedimentary deposits were removed by erosion in many places. Where the Getic Domain functioned as trench, the sedimentary deposits are still present on extended areas. The largest karstifiable outcrop in Romania known as the Reşiţa-Moldova Nouă sedimentary zone, located in the western part of the Southern Carpathians (Banat), is the classic place where the Getic sedimentary cover was developed. It was also preserved in Dognecea, Şipot, Vânturariţa, and Lotru areas. Smaller outcrops of sedimentary deposits are found in the marginal areas of the post-tectonic depressions, which in the Paleogene–Neogene (earlier knows as Tertiary) functioned as intra-mountainous depressions, as Lotru, Petroşani, and eastern part of the Haţeg Depression.

In the Sebeş Mountains is characterized by the development of Permian, Jurassic, and Cretaceous deposits (Savu et al. 1968; Mutihac and Ionesi 1974). Permian deposits outcrop on smaller areas in the vicinity of Cioclovina village and are represented by conglomerates and red sandstones up to 50 m in thickness. The Jurassic is present only on the

G. M. L. Ponta (✉)
Geological Survey of Alabama, 420 Hackberry Lane, Tuscaloosa, AL 35401, USA
e-mail: gponta@yahoo.com

G. M. L. Ponta and B. P. Onac (eds.), *Cave and Karst Systems of Romania*,
Cave and Karst Systems of the World, https://doi.org/10.1007/978-3-319-90747-5_13

Fig. 1 Location of Sebeş Mountains karst area within Romania

eastern margin, overlaying the crystalline rocks and the Permian deposits. The Lower Jurassic deposits are 100–125 m thick and consist of sandstones and conglomerates with argillaceous layers developed in Gresten facies; they are present in a smaller area in Cioclovina village (Pop et al. 1985).

Middle to Upper Jurassic (Aalenian-Lower Oxfordian) is a predominantly carbonate sequence (calcareous sandstones, bioclastic limestones, marly limestones, and biomicritic rocks) of about 70 m thick, unconformably disposed over the older sedimentary formations and the metamorphic basement of the area. Upper Oxfordian—Tithonian age limestones, about 100–150 m thick, appear on two distinct facies: (a) basin-like, represented mainly by micrite and biomicritic, and (b) a reefal one, made off red micritic limestones probably corresponding to the Tithonian, which is 50–70 m thick. The lower part of these limestones is locally dolomitized (Pop et al. 1985).

The limestones of Upper Jurassic–Lower Cretaceous age are an important sequence of about 300 m thick which overlain unconformably the metamorphic basement and older sedimentary formations. These limestones cover large surfaces in the area and present a wide variety of carbonate textures bearing biomicritic, biopelmicritites, pelmicrites, and their sparitic correspondents. They are discontinuously overlain by the Upper Aptian—Albian bauxite complex. These aluminum-reach (±iron) sediments filled the paleokarst features (dolines, potholes, etc.) forming bauxite ore deposits having lens-like shapes of variable sizes, from insignificant appearances to 800 × 500 m, and over 20 m thick (Papiu et al. 1971).

Alluvial, proluvial, scree accumulations, and calcareous tufa of Quaternary age are common, mainly along valleys and at the base of the cliffs surrounding the plateaus.

Groundwater in Karstified Fissured Rocks

Highly productive karst aquifers occur in the Mesozoic limestones, where groundwater movement occurs along well-developed secondary porosity (fractures and joints). These interconnected features serve as conduits leading water from the sinking streams to the springs. The total thickness of the karstic aquifers situated in Mesozoic limestones is about 500 m (Fig. 2). The crystalline formation of Precambrian age and Permian-Lower to Middle Jurassic non-carbonaceous rocks constitutes the impermeable basement for the karst aquifers in the area (Ponta and Terteleac 1989).

The Strei Valley incised the limestone deposits as deep as 400–500 m, generating two distinct plateaus: (1) Ponorâci Cioclovina–Ohaba Ponor on the right side (NW) and (2) Bojiţa–Tecuri, on the left side (SW). The edges of the plateaus are marked by vertical cliffs, in which fossil and active cave entrances are located.

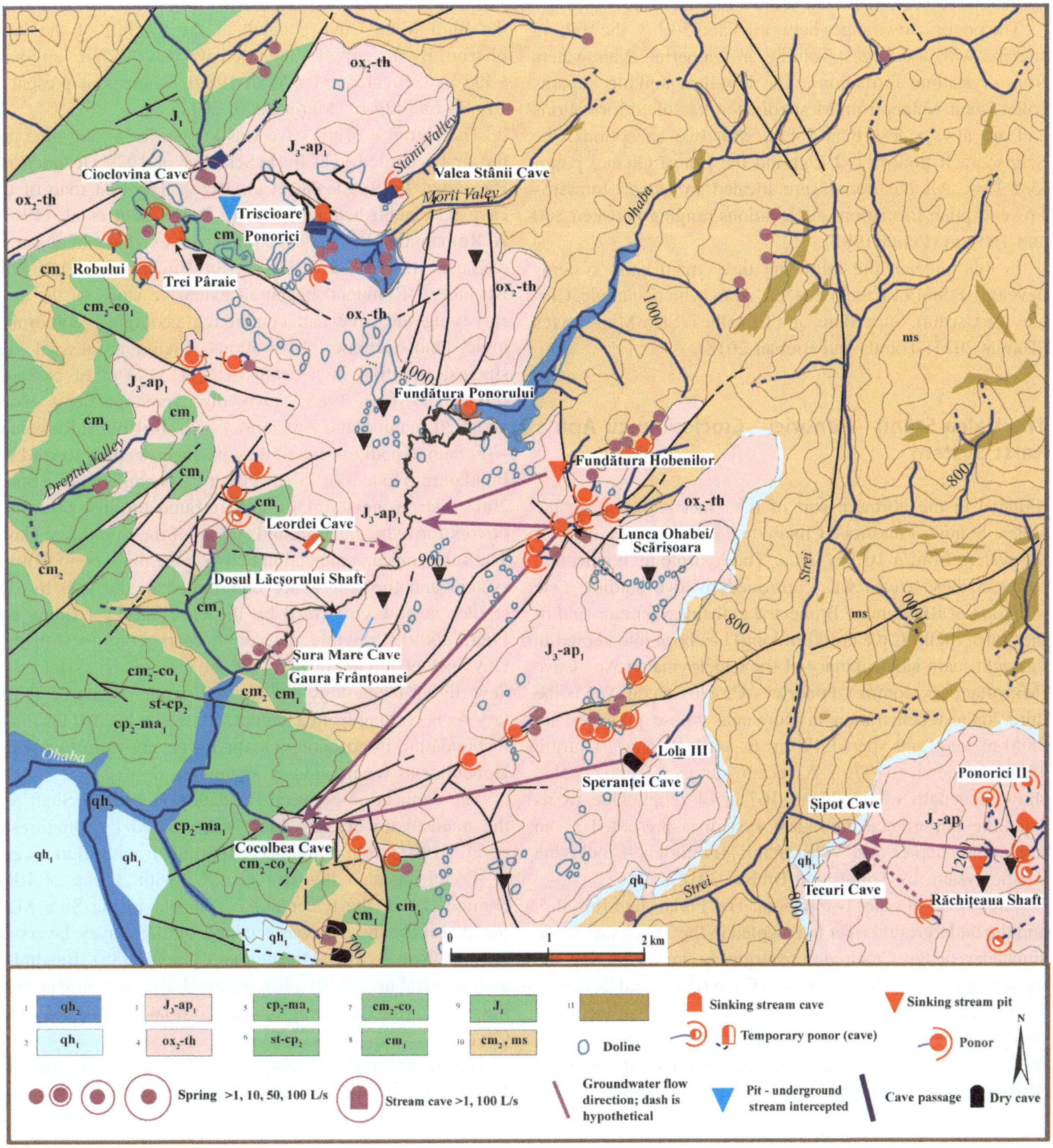

Fig. 2 Karst hydrogeologic map of Sebeş Mountains karst area (geology modified after Pop et al. 1985). Key for the legend is available in the karst hydrogeology chapter

The Ponorici Cioclovina—Ohaba Ponor Karst Plateau

The groundwater in this area is hosted in Aalenian-Lower Oxfordian, Upper Oxfordian–Tithonic, and Berriasian–Lower Aptian deposits, the last one showing the greatest extent, covering 45 km^2. In this area, 21 ponors and 14 karst springs have been identified (Ponta 1991, 1998).

The ponors flow ranges between 5 and 50 L/s; the sinking streams with the largest volume are gathering their waters from crystalline formations. The low-flow sinking streams collect their waters from Mesozoic non-karstified formations and are always less than 1 L/s, with two exceptions: the Scărişoara II ponor (1–2 L/s) and the Lunca Ohabei ponor (3–5 L/s). All the ponors are located in Upper Jurassic–Lower Aptian formations at elevations ranging between 800 and 1000 m (Ponta 1991).

In this plateau, the following three main karst systems develop: Ponorici-Cioclovina cu Apă, which include Cioclovina Uscată), Şura Mare, and Cocolbea (Şura Mică) caves (Tomuş 2011; Tomuş and Breban 2018).

The Valea Stânii—Ponorici—Cioclovina cu Apă Karst System

Situated in the northern part of the limestone plateau, the Valea Stânii—Ponorici—Cioclovina karst system collects the waters from a 10 km^2 area. The hydrologic connection between these caves was confirmed by dye studies. One kilogram of Rhodamine B injected at the downstream end of the Valea Stânii Cave (1400 m long) has been intercepted in the right main side tributary of the Cioclovina cu Apă Cave (Mitrofan pers. comm.). Ponorici—Cioclovina cu Apă is the third longest penetration in Romania, whose passages are 7890 m long and spread 170 m in vertical range (Giurgiu 1976; Iurkiewicz 1976). A recent survey (Tomuş 2011) shows a length of 4272 and 167 m of relief. Due to its scientific importance, this karst system is presented as an individual chapter (see Valea Stânii—Ponorici—Cioclovina karstic system by Tomuş and Breban 2018).

The waters of other two ponors, Trei Pâraie, and Robului, and the underground river intercepted in the Triscioare Shaft, situated near the Cioclovina resurgence, are recharging the mainstream of Cioclovina cu Apă Cave (Ponta and Terteleac 1989; Ponta 1998). The minimum average combined discharge of the swallets, which recharge the system, is about 30 L/s, and represents 20% of the multi-yearly average discharge rate of the Cioclovina Spring (125 L/s).

Sura Mare Cave System

The Şura Mare Cave system that includes Dosul Lăcşorului and Fundătura Hobenilor shafts is located in the Ponorici Cioclovina—Ohaba Ponor Plateau. Several perennial rivers flowing on noncalcareous rocks are sinking underground at the crystalline rock/limestone boundary through ponors, generating several cave systems. The most important one is the Şura Mare Cave situated at the foothill of the karst plateau. By the end of 1989, the cave was 6797 m long with +405 m vertical range (Ponta 1989), whereas at present is 11,694 m/+425 m (Marin 2000, 2012).

The entrance of the cave was known for a long time and the first recorded visit by I. Schädler in 1929 (Jeannel and Racovitza 1929; Bleahu et al. 1976). The first map of the cave was published in 1967 by Dumitrescu et al. (1967). A Romanian—British expedition surveyed about 2 km of passages in 1969. In 1976, Focul Viu Caving Club from Bucharest (Romania) began an extensive study of the area, surveying the cave and conducting geologic, geomorphologic, and tectonic observations, including several dye studies, resulting in a detail map of 3143 m of passages (Ponta 1978). In 1985, newly discovered passages ended about 20 m under the Fundătura Ponor. In 1998, a Romanian—French expedition found new passages leading toward the Fundătura Hobenilor Shaft/Ponor (Chachula and Bondar 2003). The waters sinking underground through these two ponors form the underground stream of the Şura Mare Cave. The subterranean connection between the main ponors of the system and the resurgence determined in the past by dye studies is now confirmed by the cave survey (Aldica and Ponta 2009). Presently, B. Tomuş (pers. comm.) with Proteus SpeleoClub are resurveying the entire karst system with new digital technology and found that the stream of Tributary A_6 is coming from a temporary swallet located upstream of Fundătura Ponor along Ohaba River, not from Fundătura Hobenilor as we previously assumed.

The connection between Fundătura Hobenilor Shaft and the mainstream of Şura Mare was proved by Dumitrescu et al. (1967), by a tracing experiment with fluorescein. G. Ponta repeated this test in 1987 with 3.5 kg of Rhodamine B, but the tracer did not appear in the Şura Mare outlet. Likely the dye was retained by the clayey layers on the way to the cave (Mitrofan and Ponta 1985). Bandrabur and Bandrabur (2010) also repeated the experiment with fluorescein, the dye being found 126 h later in the Cocolbea Cave.

About 300 m downstream from the entrance in the Şura Mare Cave, on the right side there are three karst springs

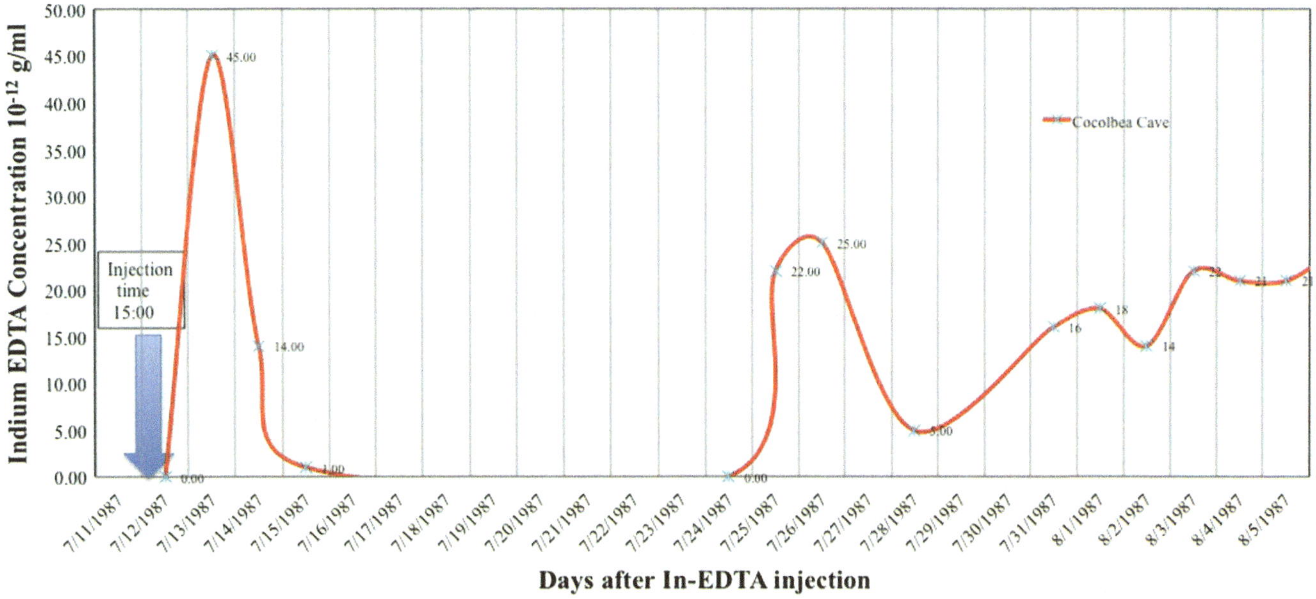

Fig. 3 Tracer (In-EDTA) breakthrough time curve Lola III Ponor—Cocolbea Spring

spread over 15 m, having a cumulated discharge of 13–15 L/s. The average temperature of these springs is +16 °C, whereas that of the water of the main river is between +5 and 14 °C. The average discharge of the stream flowing through Şura Mare Cave is 150 L/s, whereas the total average flow into the ponors is 53 L/s, which represents 35%. Very close to the Şura Mare entrance is Gaura Frânţoanei Cave/Spring (1200 m long) with an average yield of 2 L/s.

On July 12, 1987, 50 g of In-EDTA was injected in Lola III ponor (Fig. 3) to define the watershed between Cocolbea and Şura Mare river caves. Because the tracer was detected in both springs, mainly in Şura Mare, the experiment was repeated in June 1988 with 1 kg of fluorescein. The water samples analyzed by M. Constantin ("Emil Racoviţă" Institute of Speleology, Bucharest) confirmed the connection between Lola III ponor and Cocolbea Cave (Ponta and Marin 1989), the tracer being found after 10 days from the injection moment.

At the same time, 25 g of In-EDTA was injected in Lunca Ohabei ponor (Fig. 4). The tracer appeared in both springs, with higher concentration in Şura Mare Cave. The rainfalls from those days are reflected in the general form of the graph. The appearance of the tracer in the Cocolbea Spring is determined by a main tectonic fault, and the lag between the two appearances is due to the fact that vadose flow toward Cocolbea Spring is discontinuous. It is very possible that during low flow, some of the ponors recharge one spring whereas at high flow a different one. More detailed karst inventory and tracer studies are recommended in this fractured area to better define the recharge zone of these springs.

Sura Mare Cave Description

A large catchment area is feeding a spectacular underground river that carves Şura Mare Cave in Upper Jurassic–Lower Cretaceous pink limestones. It hosts spectacular domes and rimstones (gours) left behind by dripping water and at about 11.7 km in length, it has one of the longest underground rivers in Romania (Fig. 5).

The cave entrance is a 37 × 13 m portal, located at the upper end of a 400 m long gorge, which began in the vicinity of Ohaba Ponor village (Dumitrescu et al. 1967). In the entrance chamber, a sand beach is formed along the left sidewall, which ends upstream with two large boulders. The depths of the water in this area are 1.2–1.5 m. The entrance chamber is known as the Bats Chamber (Sala Liliecilor) and was named by M. Dumitrescu in 1967, due to the presence of hibernation bat colonies during winter (Murariu et al. 2007). The right side of the chamber is a large cone of sediments covered by clay, atop of which Neolithic artifacts and domestic animal skeletons were identified.

Beyond this chamber, the main passage narrows to 4–6 m, with boulders on the floor, which at one point dammed the stream, resulting in a 15 m long lake with a depth of 1.5 m. The height of the passage is 30–40 m, but occasionally, the limestones blocks stuck between the walls created a false ceiling at only 7 m above the floor. About 500 m into the cave, the number of underwater boulders diminishes, and the first 50 m long lake (Big Lake/Dorna Mare) appears, with an average depth of 2 m. Upstream of this lake the passage narrows to 2 m, having a canyon

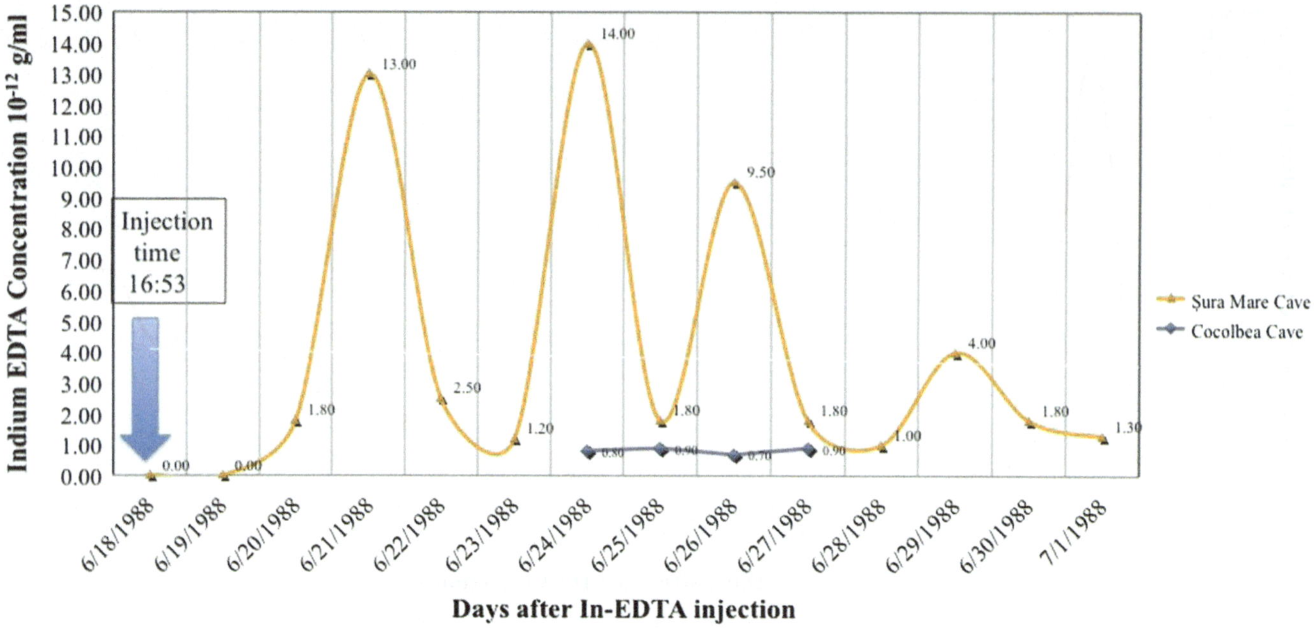

Fig. 4 Tracer (In-EDTA) breakthrough time curve Lunca Ohabei/Scărişoara Ponor—Cocolbea Spring (modified after Ponta 1991, with permission)

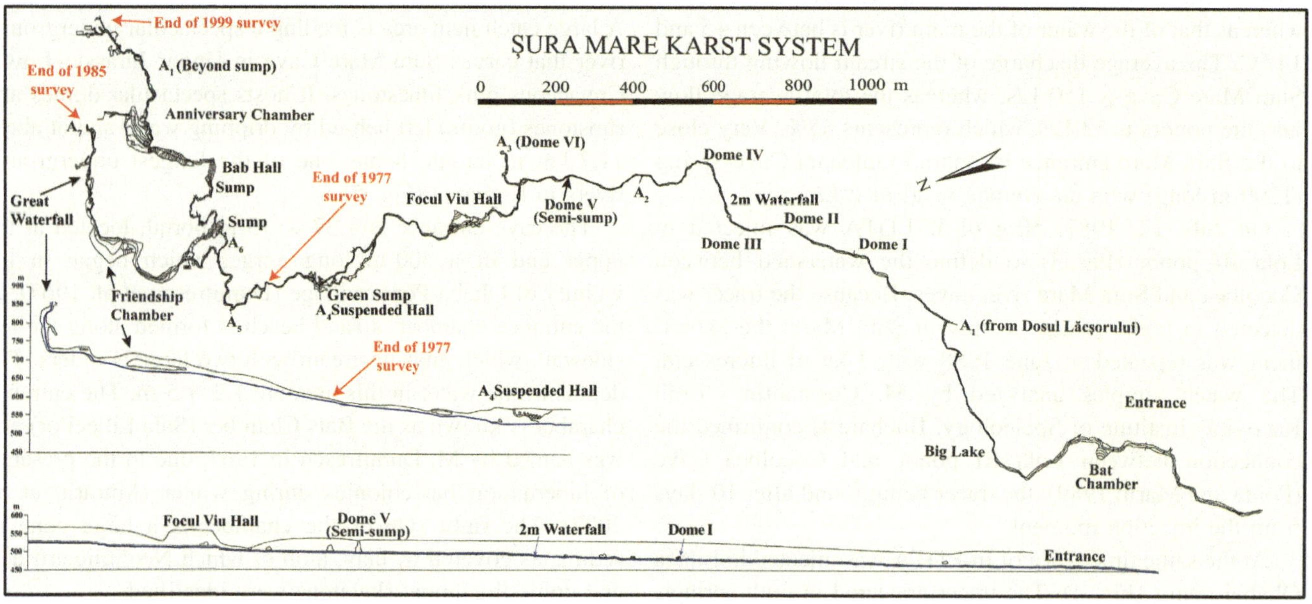

Fig. 5 Şura Mare karst system (*Map digitized by* G. Lazăr and T. Marin; *Survey team* G. Ponta, G. Halasi, R. Muller, O. Vasiliu, V. Barbu, T. Popescu, T. Tulucan, C. Lascu, S. Sarbu, A. Solomon, D. Borodan, G. Drăghici, C. Panaiotu, G. Ionescu, I. Puspurica, T. Marin, V. Bondar, G. Lazar, F. Papiu, D. Bondar, V. Băltăreţu, B. Tomuş, V. Bogdan)

aspect, and the first small (+0.5 m) waterfalls appear and occasionally flowstones form on the walls.

At 931 m from the entrance, the first left side tributary (A_1) appears with a yield of 1–2 L/s. The stream can be followed upstream for only 7 m where ends in a rockfall. A dye study conducted with 1.5 kg of Rhodamine B by T. Negoiţă on April 30, 1984, in the Dosul Lăcşorului Shaft, which in straight line is just 200 m away, was intercepted in this tributary (Mitrofan and Ponta 1985). At this confluence, the direction of the main passage changes 90° to NNE. The average width continues to be 1–2 m, occasionally larger, where domes occupy the entire passage. At 1050 m from the entrance is the first 10 m long dome (Dome I), with a gallery above, unexplored at that time.

For the next 1100 m, three additional domes are encountered. Only the last one (Dome V) located at 2150 from the entrance (Kalitzky pers. comm. 1984; Ponta 1989) can be passed through a free dive, which at the slightest flood becomes a sump. Between domes III and IV, a 2 m waterfall is located, with a terrace present in the left wall. Upstream of the waterfall, the second stream known as Tributary 2 (A_2) enters the cave at the thalweg level with a flow of 1 L/s. During rainy season this stream has a vauclusian aspect. Tributary 2 was reached for the fist time by I. Giurgiu in 1976. It is possible that this stream originates in the Leordei Cave, but a hydrologic link has not yet been tested.

After a few small waterfalls, a new dome is situated transversally to the main gallery, with the waters of Tributary 3 (A_3) coming from the ceiling (a 23.5 m high waterfall), and has a total length of 238 m and +63 m vertical range) of passages developed on two levels. The lower level (active) ends in an impenetrable sump. The upper level is 120 m long and is situated at +20 m above the active level of the A_3. In this section, the height of the main passage is 60 m.

Beyond this point, the main passage widens, and boulders are more frequent along the stream, which form a 1.7 m high waterfall at the downstream entrance in the Focul Viu Hall (earlier known as Mendip Hall). This hall is the largest by far in the cave, with the river flowing along its west/right wall, and a 35 m high rise formed by boulders cemented with calcite on the east/left. On the top, a spectacular dome with rimstone pools at the base and numerous speleothems are present (Fig. 6). Worth noting is the discovery of a small stalagmite, which has in its upper part a bat covered with calcite (Fig. 7). In the vicinity is another dome that acts as a natural bridge above the stream, reaching the western wall of the main passage.

Just few hundreds of meters upstream of Focul Viu Chamber is the Suspended Hall (Sala Suspendată), which has its floor covered with numerous boulders that create several levels. The lower one, situated immediately above the mainstream is formed by a deep rimstone pool, which is recharged by the waters of Tributary 4 (A_4), which ends in the Green Sump (Sifonul Verde), named by the English team the "Static Sump" At the same level, but on the opposite side of the main passage, a well-decorated gallery with a large rimstone pool was mapped. The main passage between this point and the sump has canyon morphology and hosts several small waterfalls occur. The sump (final sump of the British expedition) was passed through a fossil passage located 9 m above the active stream, but a few years later, G. Halasi passed the sump with scuba gear.

On the other side of the sump the morphology of the gallery changes, the frequency of waterfalls increases up to Tributary 5 (A_5), which merge with the mainstream above a right side terrace. The water is coming out from a narrow (0.5 m high) gallery, which becomes impenetrable after just 5 m; the discharge is about 1 L/s.

The main passage that continues upstream (E) is steep and has numerous waterfalls, the highest one being 2 m. The direction of the passage changes toward NW at the confluence with Tributary 6 (A_6) (5 L/s), which ends after 10 m in a sump and a calcitic flowstone. This sump was passed in 1998 by a French-Romanian expedition. Upstream and downstream of this confluence, in the main passage's wall, entrances to side galleries can be observed. One of this fossil meander located 4–5 m above the stream was map. After crossing two short sumps along the stream of A_6, the cave opens shortly in the Bab Hall, which continues with the largest room of the cave, the Anniversary Chamber (Sala Aniversării; dedicated to the 30th anniversary of Focul Viu Caving Club). The floor of this room is covered by numerous boulders, and well-decorated areas are present. The upper end continues with a medium size gallery, A_1 (Beyond Sump), which ends in a vertical chimney. With over 2 km of new passages and reaching +425 m above the cave entrance, this was the most remarkable discovery performed by T. Marin, D. Bondar, G. Lazăr, and their French friends (Marin 2000).

From the confluence with A_6, the cave continues 60 m toward NW, then after another 60 m return to the previous NNE direction. In this section, numerous waterfalls are present, the highest one being 6 m. Upstream, the gallery widens suddenly and the floor is covered with boulders, giving the impression of a leveled, almost horizontal, surface. In the right side opens the Friendship Chamber (Sala Prieteniei), which is 202 m long and has a +28 m vertical range. The entire floor is covered with cascading rimstone pools and a large dome having an unexplored chimney above.

Upstream of the lower entrance in the Friendship Chamber, the gallery is steeper and has canyon morphology, ending in two 7 m high waterfalls. The mainstream (A_7) discharge is 50 L/s and is coming from the 100 m high Great Waterfall/Cascada Mare (Petrescu 2009). A secondary stream with 10 L/s appears in the cave from the right. A few meters before the 7 m waterfall is another large room with a false floor formed by boulders stuck between the galleries walls. This room elevation is the same with the upper end of the Friendship Chamber. Over the past 20 years, several new expeditions explored and mapped new passages, extending the cave's length to 11,694 m and +425 vertical range (Marin 2000, 2012).

Fig. 6 Dome in Focul Viu Hall (photograph courtesy of C. Lascu)

Fig. 7 Bat bones covered with calcite (photograph courtesy of C. Lascu)

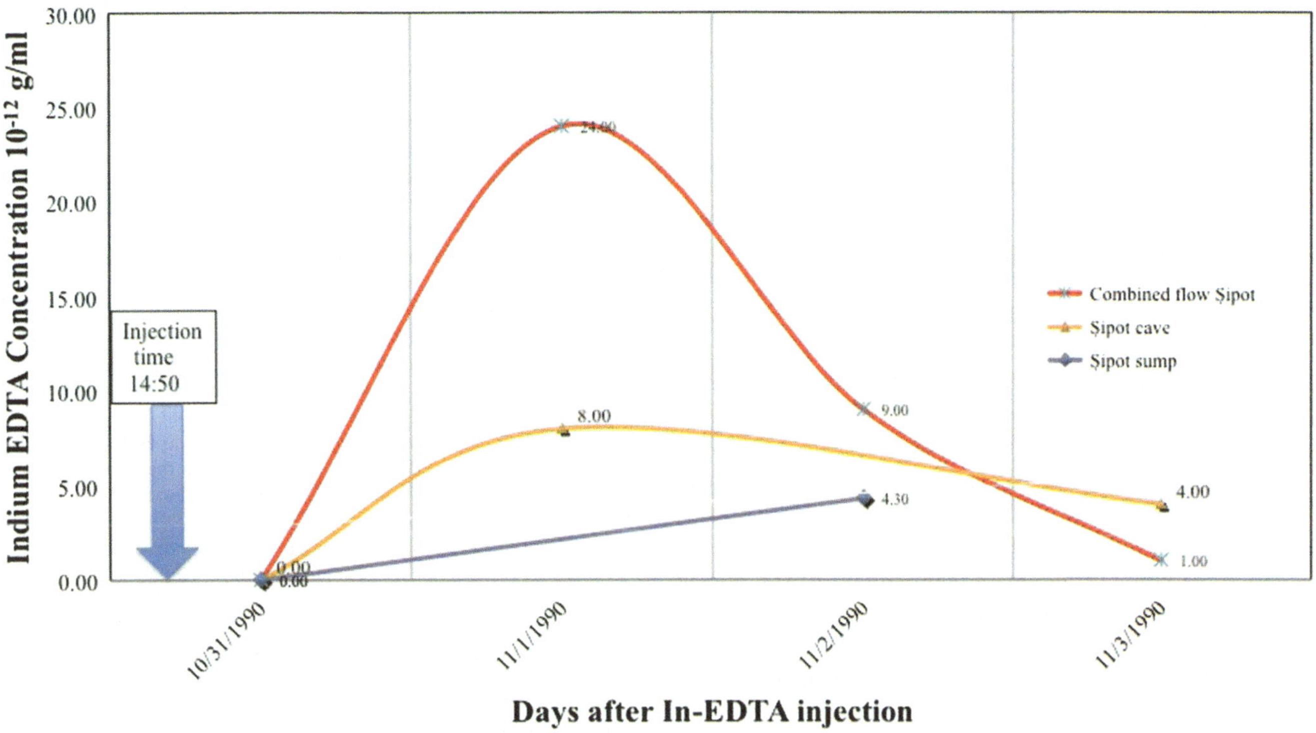

Fig. 8 Tracer (In-EDTA) breakthrough time curve Ponorici II Ponor—Sipot springs (modified after Ponta 1991, with permission)

The Cocolbea (Sura Mică) Karst System

The Cocolbea (Şura Mică) Cave is the main resurgence of the southern half of the Cioclovina–Ohaba Ponor Karst Plateau. Situated at the intersection of two major fractures, the water emerges from a 125 m long cavity. The multi-year average discharge of the cave is 50 L/s and is theoretically recharged by 11 ponors located along or in the close proximity of two faults. The ponors are concentrated in two points: Lunca Ohabei and Lola Valley. The cumulated discharge of these ponors is 10 L/s, which only represents 20% of the resurgence yield.

The fact that all these ponors belong to the Cocolbea resurgence is hypothetical. Only three experiments with tracers have been carried out, one with fluorescein and two with In-EDTA. The fluorescein and one of the In-EDTA experiments were conducted in the Lola doline whereas the second in the Lunca Ohabei ponor corresponding to the location named Scărişoara. Details about these tracers' experiments are presented above in the Şura Mare karst system chapter.

The Bojita–Tecuri Karst Plateau

The karst aquifers in this area lie in Upper Jurassic–Lower Aptian carbonate deposits. The plateau is surrounded by vertical cliffs at the foothill of which there are two major springs, Şipot and Izvoreni (not on the map as they are located south of the area shown on Fig. 2). Their waters come from 9 ponors concentrated in the northern half of the limestone area between 1092 and 1221 m in elevation. The resurgences are situated between 817 and 350 m altitude, suggesting a 400-m-thick karst aquifer. The water divide between Şipot and Izvoreni resurgences is only partially known. The geological and cave surveying allow us to suppose that the majority of waters on the plateau are drained toward Şipot.

A gauging station situated on the Crivadia River recorded a mean annual runoff of 253 L/s, corresponding to an annual precipitation of 418 mm. The response of a rainfall is rapidly observed on the hydrograph (Ponta 1991). Izvoreni and the two outlets of the Sipot springs, and their cumulated discharge were monitored.

In 1986, Orăşeanu and Venţel performed tracer tests in the area, showing the connection between three sinking streams on the plateau with the Sipot springs; the results were published five years later (Orăşeanu et al. 1991). On October 31, 1990, the author of this chapter injected 20 g of In-EDTA in Ponorici II ponor (located 2 km away from the Şipot springs) and the tracer was found in Sipot springs after 48 h, proving the underground connection between Ponorâci II and Şipot springs (Fig. 8). The graph presents two curves of different amplitude, influenced by the rain event that occurred during the test. Even both Şipot outlets present sump at the entrance, the main underground stream appears to flow under vadose conditions.

Izvoreni resurgence drains an important area of limestones located south of the study area, which was less studied, and thus far the watershed/recharge area of the spring has not yet been defined.

Cave Conservancy

Given the exceptional value and uniqueness, Şura Mare Cave is included in the A class of protection (scientific reserve), meaning access is completely prohibited, except for exploration/survey and scientific purposes. Permits from the Romanian Speleological Heritage (Comisia Patrimoniului Speologic) and the Administration of the Grădiştea Muncelului-Cioclovina Natural Park are needed in order to access this cave.

Acknowledgements We are thankful to Focul Viu Bucureşti Caving Club who surveyed Şura Mare, one of the most stunning cave in Romania, with support of cavers from Cristal Oradea, GESS Bucuresti, "Emil Racoviţă" Institute of Speleology in Bucharest, and Vertikum Budapest. For the wellbeing of the most spectacular underground stream in Romania, Şura Mare Cave, it is our hope that future explorations will end before making the connection with one of the sinking streams, thus preventing Şura Mare to become a through cave.

References

Aldica G, Ponta G (2009) Peştera Şura Mare - Jurnalul turelor, bilanţul realizărilor. In: Dumitru R (ed) Focul Viu 40 de ani de speologie. Ed RAO, Bucureşti, pp 99–104

Bandrabur G, Bandrabur R (2010) Sebeş Mountains. In: Orăşeanu I, Iurkiewicz A (eds) Karst hydrogeology of Romania. Belvedere, Oradea, pp 77–88

Bleahu M, Decu V, Negrea Ş, Pleşa C, Povară I, Viehmann I (1976) Peşteri din România. Ed Ştiinţifică şi Enciclopedică, Bucureşti, p 415

Chachula O, Bondar V (2003) Şura Mare - Present şi perspective. Nota 1. Ecocarst 4:19–22

Dumitrescu M, Orghidan N, Orghidan T, Tanasachi J, Puşcariu V, Avram Ş (1967) Contribuţii la studiul peşterilor din judeţul Hunedoara. Lucr Inst Speol "Emil Racoviţă" VI:9–88

Giurgiu V (1976) Peşteri din Romania mai lungi de 2 km. Caietul Cerc Speol Emil Racoviţă 4:9–10

Jeannel R, Racovitza EG (1929) Énumération des grottes visitées, 1918-1927 (7e série). Arch Zool exp gén 68(2):293–608

Iurkiewicz A (1976) 4 zile în Ponorâci-Cioclovina cu Apă. Caietul Cerc Speol Emil Racoviţă 4:238–252

Marin T (2000) 130 ore de bivuac pentru cea mai spectaculoasă explorare a anului 1998–Şura Mare. Speomond 5:11–15

Marin T (2012) Explorări în Peştera Şura Mare. Speomond 16:2–5

Mitrofan H, Ponta G (1985) Preliminary note on the karstic system of Şura Mare (Sebeş Mountains). Theor Appl Karstol 2:79–87

Murariu D, Decu V, Gheorghiu V, Nistor V (2007) An impressive aggregation of bats hibernating in the Southern Carpathians (Şura Mare Cave—Romania). Studia Chiropter 5:21–27

Mutihac V, Ionesi L (1974) Geologia României. Ed Tehnică, Bucureşti, p 646

Orăşeanu I, Venţel R, Gaşpar E (1991) Tracers investigations in Poiana Tecuri area (Şureanu Mountains, Romania). Theor Appl Karstol 4:211–212

Papiu VC, Mânzatu S, Iosof V, Udrescu C, Giuşcă R (1971) Alcătuirea chimico-mineralogică a formaţiunilor bauxitifere din bazinul Haţegului. CR Inst Geol 57(2):77–123

Petrescu D (2009) Şura Mare cascadă finală. In: Dumitru R (ed) Focul Viu 40 de ani de speologie. Ed RAO, Bucureşti, pp 105–108

Ponta G (1978) Peştera Şura Mare. Aragonit (Bul Cerc Speol Amatori "Liliacul" Arad) 1:50–59

Ponta G (1989) Pestera Şura Mare (M-ţii Sebeş). Bul Speol 12:75–100

Ponta G (1991) New tracing experiences in the Sebeş Mountains—Rumania. In: Proceedings of international conference on environmental changes in karst areas. Universita di Padova, pp 195–204

Ponta G (1998) Karst of Romania. In: Yuan D, Liu Z (eds) Global Karst correlation IGCP 299. Science Press, Beijing, pp 193–207

Ponta G, Marin C (1989) Karstic aquifers in Sebeş Mountains. In: Proceedings of 10th international congress on speleol, Budapest, pp 463–465

Ponta G, Terteleac N (1989) Hydrogeological map, Sheet Pui. Scale 1:50.000. Inst Geol Geofiz, Bucureşti

Pop G, Hârtopanu I, Hârtopanu P, Szasz L, Popescu G, Bandrabur T, Stelea I, Moisescu V (1985) Geologic map of Romania, Pui Sheet (106a; L-34-95-A). Scale 1:50.000, Inst Geol Rom, Bucureşti

Savu H, Pavelescu M, Stancu I, Lupu D (1968) Geologic map of Romania, Orăştie Sheet (26). Scale 1:200.000, L-34-XXIV. Inst Geol Rom, Bucureşti

Săndulescu M (1984) Geotectonica României. Ed Tehnică, Bucureşti

Stilla A (1972) Zona Pui şi bazinul Haţeg - unităţi structurale distinctive în Carpaţii Meridionali. DS Inst Geol LVII I(5):173–176

Tomuş RB (2011) Monografia carstului din Şureanu. Hunedoara, Club Speleol Proteus, p 244

Tomuş RB, Breban RC (2018) Răchiţeaua-Şipot cave system. In: Ponta GML, Onac BP (eds) Karst and cave systems of Romania. Springer International Publishing, Cham, pp 105–111

Trufaş V (1978) Apele subterane din Munţii Sebeş. Univ Bucureşti, St Geogr, pp 193–208

Trufaş V (1986) Şureanu. Ghid turistic. Ed Sport-Turism, Bucureşti, p 177

Further Reading

Bleahu M (1972) Karst of Rumania. In: Herak M, Springfield VT (eds) Important karst regions of the Northern Hemisphere. Elsevier, Amsterdam, pp 341–353

Buduran SE, Sopincean CA, Sopincean MS, Isfan MR, Stan CO, Gherman LL (2013) Peştera Ponorâci-Cioclovina cu Apă. Ed Karina, Deva, p 78

Cocean P (1995) Peşterile României. Potenţial turistic. Ed Dacia, Cluj-Napoca, p 264

Goran C (1982) Catalogul sistematic al peşterilor din România. Ed CNEFS, Bucureşti, p 496

Orghidan T, Negrea Ş, Racoviţă G, Lascu C (1984) Peşteri din Romania. Ed Sport-Turism, Bucureşti, p 504

Sopincean CA, Buduran SE, Sopincean MS, Chachula OM, Stan CO, Gherman LL, Isfan MR (2015) Peştera Şura Mare. Ed Belvedere, Oradea, p 78

Useful Websites

www.speologie.org
www.karstgeology.com

Sebesului Mountains: Răchiteaua–Sipot Cave System

Raul-Bogdan Tomuş and Radu-Ciprian Breban

Abstract

Răchiţeaua–Şipot cave system is one of the most complex and interesting underground feature in the Şureanu Mountains karst, and future speleological explorations will likely prove its scientific and hydrologic significance. It is composed of Răchiţeaua Shaft, a 2430 m long and 232 m deep stream cave and the Şipot Sump Cave, a 1270 m long river passage with six challenging sumps. Răchiţeaua Shaft is impressive through its vadose shafts, corrosion, and erosion features occurring especially along the main horizontal gallery, and also for the abundant speleothems (flowstones and rimestone pools) that mainly decorate the upstream tributary gallery.

Keywords

Karst plateau • Vadose shaft • Sump • Spring

Introduction

The Răchiţeaua–Şipot area of the Sebeş Mountains (also known as Şureanu) belongs administratively to the commune of Baru Mare (Hunedoara County), being one of the most important karst area (basin 2066 in Goran 1982) in Romania (Fig. 1). This is a karst plateau, developed between 800 and 1318 m elevation, within which occurs the three resurgences from Şipot and a number of active valleys (Poiana, Ponorâci, Tăul fără fund, Teiul Lung, and Valea Clenjii). There are 55 known cavities in the 2066 karst basin, among which the most important are: Tecuri, Răchiţeaua, Valea Clenjii, and Şipot Sump (Fig. 2).

Geographic, Geologic, and Hydrologic Settings

The karst area of the Şureanu Mountains is situated in the S and SE area of the massif, occupying a total surface (including the Vârtoape Plateau) of 134 km^2 (see Fig. 2 in Chapter 13). Its western border is the Cioclovina–Fizeşti alignment, to the south is the Strei Valley and then Jiul de Est River, whereas toward northeast is an imaginary line connecting the Poiana Omului, Paltinului Peak, Jigorel Peak, and Bolii Cave. East of the Bolii Cave, the karst area is limited to only one limestone stripe of about 1 km wide that extends all the way to the village of Răscoala. The Şureanu karst develops between 490 and 1326 m in the Ohaba-Ponor and Poiana-Tecuri area, respectively, but the average elevation is less than 1000 m. The maximum size of the karst region is 25 km in the Cioclovina—Piatra Roşie section (Trufaş 1986).

The geological formations that make up this area belong to the Getic Domain Nappe, the Pui geosyncline sedimentary zone and the Haţeg intra-mountain depression (Savu et al. 1968; Stilla 1972). Crystalline rocks are developed in the western, northern, and eastern parts of the area. The sediments in this region belong to the following sedimentary cycles: Jurassic-Lower Cretaceous, Albian s.l., and Upper Cretaceous. Coarse sandstones and quartz conglomerates of Lower Jurassic age appear at Teiul Lung. The sparitic limestone belonging to Middle Jurassic outcrop near Teiul Lung, west of Pleşa Peak, Piatra Brâncusi, and Purcarului Peak. The Upper Jurassic rocks consist of carbonate sequences represented by two facies, a deep one with fine-layered limestones and a reefal one deposited in a shallow-water basin (Pop et al. 1985). The Upper Jurassic–Lower Cretaceous reef limestone is the most common karst-forming unit in the area.

The region was uplifted as a result of the short-lived Austrian Orogeny that happened during Albian. All previously deposited carbonate rocks were subjected to intense

R.-B. Tomuş (✉) · R.-C. Breban
Proteus SpeleoClub, Deva, Hunedoara, Romania
e-mail: bogdantomus@yahoo.com

R.-C. Breban
e-mail: radu_breban@yahoo.com

G. M. L. Ponta and B. P. Onac (eds.), *Cave and Karst Systems of Romania*,
Cave and Karst Systems of the World, https://doi.org/10.1007/978-3-319-90747-5_14

Fig. 1 Map of the Romanian karst areas showing the location of the Răchiţeaua-Şipot cave system in the Şureanu Mountains

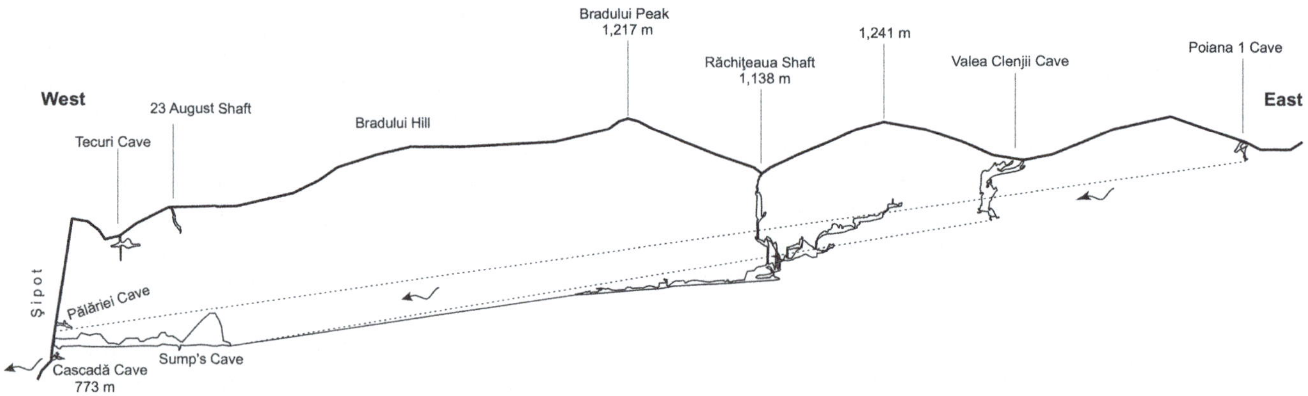

Fig. 2 Cross-section between Poiana and Şipot Sump caves illustrating the underground drainage and locations of the major cavities in the area

weathering, leading to strong karstification processes. In paleokarst features such as dolines and other closed depressions, red clays, detrital rocks, and bauxites are accumulated (Papiu et al. 1971). Today, most of these lenses are either mined or almost completely eroded.

The Upper Cretaceous sedimentary cycles represent formations accumulated over a long period of orogenic restlessness. It is precisely because of the genetic conditions that almost each of the Upper Cretaceous stratigraphic entities is disposed discord and transgressive over the units underneath

(Stilla 1985). Marine and torrential Paleogene to Neogene deposits of siliciclastic sediments occur in the southern part around the village of Merişor. Alluvial, proluvial, scree accumulations, and calcareous tufa of Quaternary age are common, mainly along valleys.

From a tectonic point of view, there are a series of transversal reverse faults (NE–SW) and also a major N–S one that bounds the Bojiţa-Tecuri Plateau to the west (Stilla 1985).

The hydrological network of the area consists of several watercourses (Ponorici, Valea Clenjii, Stânca Trăsnită, Poiana, Tăul fără fund, Răchiţeaua, Teiul Lung) that all originate on the crystalline area. Diffuse losses occur shortly after they come in the contact with limestones, and especially in well-developed ponors, many of them representing actual cave entrances. Based on tracer experiments carried out over time by Trufaş (1978), Orăşeanu et al. (1991), and Bandrabur and Bandrabur (2010), it was established that at least under normal flow conditions, all these losses are organized underground in two distinct hydrokarst systems:

(1) Şipot 2 Cave (40 L/s), which drains the waters that sunk in the Teiul Lung Cave, the ponors from the Poiană, Stâna Trăsnită, and Tăul fără Fund, and
(2) Şipot Sump Cave and Waterfall Cave (total discharge of 200 L/s) that collect their waters from losses in the Valea Clenjii Cave, Gârla Vacii Ponor (S of Clenjii Valley), Ponorâci Ponor (Ponorâci II), the Bojiţa Ponorâci (Ponorâci I), and Răchiţeaua Shaft (Fig. 2).

At high flows, it appears that the two systems communicate with each other.

The first dye studies with fluorescein in the region were conducted in 1969 by V. Trufaş in the Poiana and Ponorâci ponors, resurfacing in one of the three Şipot springs and in the Şipot Sump Cave, respectively (Trufaş 1978). The tests conducted in 1981 by the "Emil Racoviţă" Caving Club from Bucharest with C. Marin ("Emil Racoviţă" Institute of Speleology, Bucharest) concluded that there is a connection between Stânca Trăsnită and Valea Clenjii Cave. During the explorations made by members of the Proteus Hunedoara Caving Club in the Răchiţeaua Shaft, they surveyed a left (east) tributary to the main gallery, reaching a point that is 130 m away and 53.5 m below the terminus of the 8-m-long Bojiţa Cave.

By comparing discharges and the orientation of galleries, it is plausible that the water sinking in the Ponorâci Ponor is also drained through the Răchiţeaua Shaft. In addition to these two large hydrokarst systems, the existence of a perched aquifer is inferred from the water fluctuations of a lake in the Tecuri Cave, which rise immediately after periods with abundant rainfall or snow melting.

History of Exploration

Răchiţeaua Shaft was known since 1970 as a lake (4 × 2 m) filling a ponor. When its water were drained underground on November 27, 1977, members of the "Emil Racoviţă" Caving Club from Bucharest (CSER) succeeded to enter the shaft to a depth of 18 m. The presence of some unstable collapsed blocks stopped the advance and when the team returned in April 1978, the shaft was filled with alluvial sediments (Vlădulescu and Giurgiu 1994). In August 1980, CSER Bucharest begins an unclogging action, but fails to re-open the cave.

Between November 1982 and April 1983, several ample digging activities took place at which cavers from various caving clubs participated (CSER Bucharest, Piatra Roşie Petroşani, Hidrocarst Vulcan, Politehnica Cluj-Napoca, "Emilian Cristea" Alba Iulia, Proteus Hunedoara), but because successive floods filled the cave entrance once again, the activities were stopped. Beginning with 1984, the exploration activities were led by the Piatra Roşie Petroşani Caving Club, which initiated a mechanized process using air compressor, pneumatic hammer, winch, and electric power. Cepromin Cluj-Napoca, CSER Bucharest, Emilian Cristea Alba Iulia, Proteus Hunedoara, and Zarand Brad were all involved in this "mining" operations aiming to secure the entrance section of the shaft from collapsing and filling with alluvial sediments. Their hard work was plentifully rewarded as it led to new and significant discoveries. In 1986, at the National Speleological Convention (Speosport), Cepromin Cluj-Napoca and Piatra Roşie Petroşani displayed the first map of the shaft, which at that time was 1491 m long and had a vertical range of 287.5 m.

In November 1987, Piatra Roşie Petroşani and Zarand Brad caving clubs secured the entrance shaft with a metal pipe ending the digging operations. Between 1987 and 1993, CSER Bucharest explored and resurveyed the shaft (D = 1312 m, d = 219 m). Since 1997 the explorations are continued by cavers of Proteus Hunedoara Club, which in the period 1999–2001, climbed several chimneys and discovered over 1000 m of new galleries. They also resurveyed the entire cave (2000 and 2001) to the present length and vertical range of 2430 and 243 m, respectively (Fig. 3) (Tomuş 2011).

Sipot Sump Cave The cave entrance was discovered in 1969 by V. Trufaş from the Faculty of Geography at the University of Bucharest. In August 1971, a team of the CSER Bucharest led by V. Trufaş, explored the cave to a sump situated ~10 m from the entrance (Fig. 4). A collaboration established in 1983 between CSER Bucharest and the Underwater and Cave Exploration Group/GESS (divers M. Codescu, N. Grigore, V. Lascu, and G. Silvăşanu) led to

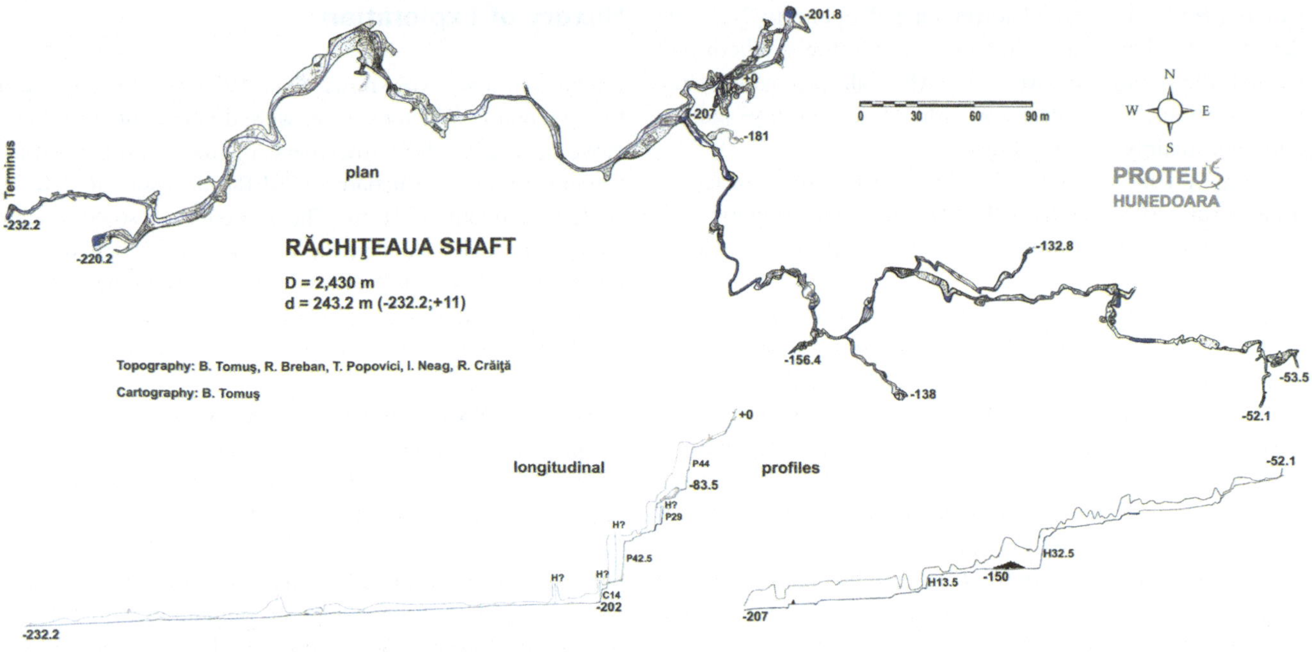

Fig. 3 Plan and longitudinal profiles of Răchiţeaua Shaft

Fig. 4 Map and profile of the Şipot Sump Cave

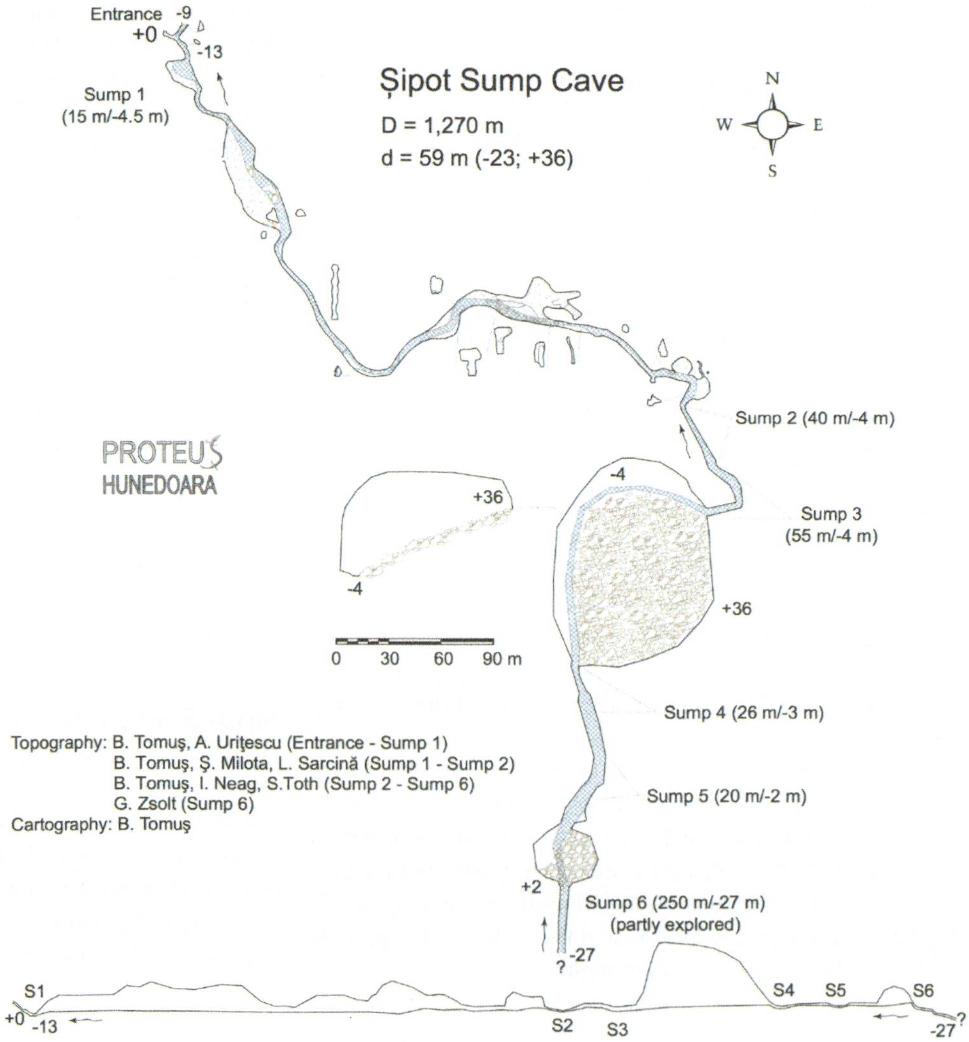

a series of diving activities, which ultimately allowed to successfully cross the first sump (15 m in length and 5 m in depth) and discover 808 m long gallery before reaching the second sump (Giurgiu and Lascu 1985). Between January and September 1984, Sump 2 (20 m, −6 m) and Sump 3 (20 m, −11 m) were passed and the exploration of the fourth sump begun, the cave reaching a development of more than 1000 m. In October 1986, a larger team returned to the cave, with S. Sarbu, R. Lovin, and M. Codescu crossing sumps 1–3, and the first two also passing Sump 4. S. Sarbu then dived Sump 5 and found a 20 m room at the end of which he plunged into Sump 6, but had to abandon and return after 25 m because he reached the end of his guideline.

In November 2003, a team consisting of Ş. Milota, L. Sarcină, and B. Tomuş passed Sump 1 and completely resurveyed the gallery between S1 and S2. Five years later (July 2008) the team that included B. Tomuş, I. Neag, and S. Toth crossed Sump 1 (15 m, −4.5 m), S2 (40 m, −4 m), S3 (55 m, −4 m), S4 (26 m, −3 m), and S5 (20 m, −2 m), remapping the cave beyond S2 (Fig. 4). During that expedition, B. Tomuş plunged Sump 6 for 105 m and to a depth of −17 m, from where it continues to descend to at least −22 m. The reason why the lengths and depths of sumps S2–S5 differ from one dive to another is because the alluvial sediment transported by the underground river continuously modifies their morphology.

In October 2012, the National *Speosub* team, made of 13 cave divers and an impressive support crew, managed to advance in Sump 6 to a distance of 250 m and a maximum depth of 27 m (diver: G. Zsolt). At this time, the real chances of further exploration in Sump 6 are very low due to technical and logistical difficulties. Therefore, the main objective is to intercept a fossil level of the cave.

Description of the Caves

Răchiţeaua Shaft is one of the best-developed ponors in Romania. The entrance is located at the base of a limestone cliff and access is through a 15 m deep metallic tube that vertically penetrates the alluvial sediments accumulated in the entrance area. At the end of the pipe begins a narrow, 5m high, heavily descending and meandering gallery through which, after 50 m, the first pit, a drop of 44 m (P44; Fig. 3) is reached. From here follows a succession of large vadose shafts (29, 42.5, 14 m), with grayish-white, well-corroded walls (Fig. 5), at the bottom of which (−202 m below the

Fig. 5 Vadose shaft (P42.5 m) in Răchiţeaua Cave (photograph by B. Tomuş)

entrance) begins a horizontal gallery (Venţel 1987). In this area, there are two parallel chimneys (heights of over 40 m) through which most of the water enters the cave, fact that makes the sequence of shafts to actually represent a tributary of the main drainage.

Continuing along the stream passage, the gallery is large (10 m high and 2–15 m wide; Fig. 6) and has typical canyon morphology, displaying many corrosion and erosion features such as ceiling channels (Fig. 7), septa, notches, and meanders. After 100 m from the base of the 14 m waterfall (−207 m), an important left tributary is intercepted. This upstream section represents half of the known development of the cave. The water flowing along this gallery appears to be supersaturated with respect to calcium carbonate as ropes rigged over some of the waterfalls in the cave for 2–3 years were found covered by a thin (1 mm) calcite crust.

Downstream from the confluence, in the median section, the gallery is richly decorated with speleothems (Fig. 8). Because of the reduced hydraulic gradient, along this gallery, sand and gravels deposit occur in the riverbed and form side terraces. Finally, the water disappears through the sediment fill in the Terminus point of the gallery, at −232 m below the entrance.

Şipot Sump Cave is located at the highest elevation among the three Şipot resurgences and has two small openings. Between the entrance and Sump 1 there are numerous limestone breakdown blocks. After S1, the main gallery has comparable sizes with those in Şura Mare and Cioclovina cu Apă, and furthermore, the cave houses one of the largest rooms (volume wise), discovered and surveyed so far in the Şureanu karst (100 m diameter and 60 m high). The large variety of corrosion and erosion morphologies, the presence of some paleokarst features and speleothems, as well as the great potential for new discoveries, makes Şipot Sump one of the top caves in Şureanu Mountains.

The cave known to date is composed of a single, stream gallery with average/constant heights of at least 20 m. There are several large rooms formed at the intersections of fractures, in which collapsed blocks of limestone form massive piles, sometimes covered by sand or speleothems. In

Fig. 6 Canyon passage in the horizontal section of the Răchiţeaua Shaft (photograph by B. Tomuş)

Fig. 7 Large ceiling channel in Răchiţeaua Shaft (photograph by B. Tomuş)

Fig. 8 Massive flowstones in the middle section of the horizontal gallery in Răchiţeaua Shaft (photograph by B. Tomuş)

addition to the difficulties posed by passing all 5 sumps, the existence of Sump 6, which is the longest and deepest, will make underwater explorations very difficult in the future.

Cave Climatology

The investigated area, except for the Şipot springs, is situated between 1000 and 1300 m in altitude, where the annual average temperature is 5 °C; for this reason, snow remains on ground from October until April. The general configuration of the cavities in the region favors the formation of ice deep underground (Răchiţeaua Shaft, Valea Clenjii Cave), preserving it sometimes even in the summer (Tăul fără Fund Cave). Temperatures measured in the Răchiţeaua Shaft on May 9, 1998 was 8 °C at the entrance, 6 °C at the base P29, 8 °C at the confluence with the left tributary, and 9.2 °C in the Terminus point (Fig. 3).

Cave Conservancy

Both caves presented in this chapter are class B of protection. This implies that any visits (e.g., exploration, research, or adventure caving) require permits issued by both the Romanian Speleological Heritage (Comisia Patrimoniului Speologic) and the Administration of the Grădiştea Muncelului-Cioclovina Natural Park.

References

Bandrabur G, Bandrabur R (2010) Sebeş Mountains. In: Orăşeanu I, Iurkiewicz A (eds) Karst hydrogeology of Romania. Ed Belvedere, Oradea, pp 77–88

Giurgiu I, Lascu V (1985) Peştera sifonului de la Şipot (Munţii Sebeş). Styx 2:25–31

Goran C (1982) Catalogul sistematic al peşterilor din România. Ed CNEFS, Bucureşti, p 496

Orăşeanu I, Venţel R, Gaşpar E (1991) Tracers investigations in Poiana Tecuri area (Şureanu Mountains, Romania). Theor Appl Karstol 4:211–212

Papiu VC, Mânzatu S, Iosof V, Udrescu C, Giusca R (1971) Alcătuirea chimico-mineralogică a formaţiunilor bauxitifere din bazinul Haţegului. CR Inst Geol 57(2):77–123

Pop G, Hârtopanu I, Hârtopanu P, Szasz L, Popescu G, Bandrabur T, Stelea I, Moisescu V (1985) Geologic map of Romania, Pui Sheet (106a). Scale 1:50.000, L-34-95-A, Inst Geol Rom, Bucureşti

Savu H, Pavelescu M, Stancu I, Lupu D (1968) Geologic map of Romania, Orăştie Sheet (26). Scale 1:200.000, L-34-XXIV, Inst Geol Rom, Bucureşti

Stilla A (1972) Zona Pui şi bazinul Haţeg - unităţi structurale distinctive în Carpaţii Meridionali. DS Inst Geol LVII I(5):173–176

Stilla A (1985) Géologie de la région de Haţeg - Cioclovina - Pui - Bãniţa (Carpathes Méridionales). An Inst Geol 66:91–179

Tomuş RB (2011) Monografia carstului din Şureanu. Hunedoara, Club Speleol Proteus, p 244

Trufaş V (1978) Apele subterane din Munţii Sebeş. Univ Bucureşti, St Geogr, pp 193–208

Trufaş V (1986) Şureanu. Ghid turistic. Ed Sport-Turism, Bucureşti, p 177

Venţel R (1987) Avenul Răchiţeaua (munţii Sebeş). Bul Speo 11:111–114

Vlădulescu M, Giurgiu I (1994) Ponorul Răchiţeaua (munţii Sebeş). Cerc Speol 2:22–23

Useful Websites

http://www.speosub.ro/?p=1187; Information on the National *Speosub* team

https://www.speologie.org; search for 2066 Basin

Sureanu Mountains: Valea Stânii–Ponorici–Cioclovina cu Apă Karst System

Raul-Bogdan Tomuş, Radu-C. Breban, and Bogdan P. Onac

Abstract

The discovery of a *Homo sapiens* skull and a rich assemblage of rare minerals in Cioclovina Uscată Cave, along with a significant fossil ossuary in Cioclovina 2 Cave, the major hydrologic connection between Ponorici and Cioclovina cu Apă, complemented by the diversity and beauty of speleothems in Valea Stânii Cave, are just a few reasons why the Valea Stânii–Ponorici–Cioclovina cu Apă karst system represents a landmark for the Romanian karst. The scientific value of this karst region is topped by the numerous Dacian artifacts found at surface.

Keywords

Ponor • Spring • Cave minerals • Speleothems
Homo spaiens • Archeology

Introduction

The investigated area is located on the eastern part of the Şureanu Mountains (also known as Sebeş Mts.) at the springs of the Luncanilor Valley, on the territory of Boşorod commune, Hunedoara County (Fig. 1). The Cioclovina zone (karst basin 2063 in the Romanian Cave Catalogue of Goran 1982) is a suspended karst plateau, which develops between 490 m in the Ohaba-Ponor area and 1326 m in the Poiana-Tecuri area, but the mean altitude is less than 1000 m. Out of 71 known cavities, only three exceed 1500 m in length: Ponorici–Cioclovina cu Apă/Water, Cioclovina Uscată/Dry, and Valea Stânii (Tomuş 1999). These are also the ones having a vertical range of more than 100 m. The karst system distinguishes itself through some important caves from an archaeological point of view (Cioclovina cu Apă, Cioclovina Uscată, Cioclovina 2, Cioabe/Shred), mineralogic (Cioclovina Uscată), paleontological (Cioclovina Uscată, Cioclovina 2, C.S.U. Shaft), speleothems/esthetic (Valea Stânii, Ponorici–Cioclovina cu Apă, Cioclovina Uscată, Mielului/Lamb).

Geographic, Geologic, and Hydrologic Settings

Şureanu Mountains are part of the South Carpathians and are bounded to the north by the Mureş Valley, Cindrel Mountains to the east, the lower part of Streiului Valley on the west, and Haţeg and Petroşani basins to the south. The karst area of the Şureanu Mountains is situated in the S and SW of the massif and covers an area of 134 km^2, surface that also includes the Vârtoape Plateau. It is bordered on the west by the Cioclovina–Fizeşti alignment, in the south by the Streiului Valley and then by the Jiul de Est River, whereas to the northeast by the Poiana Omului–Paltinului Peak–Jigorel Peak–Bolii Cave. From the Bolii Cave, it continues as a narrow (1 km wide), N-S-oriented limestone stripe almost to the village of Răscoala. The maximum area of the karst is over 9 km in the right slope of Petros and 25 km in the direction of Cioclovina–Piatra Roşie.

From a geological point of view, the Cioclovina region consists of crystalline rocks belonging to the Getic Domain Nappe and sedimentary formations of the Haţeg Basin (Stilla 1972; Săndulescu 1984). The crystalline schists are represented by rocks with a high degree of metamorphism, such

R.-B. Tomuş (✉) · R.-C. Breban
Proteus SpeleoClub, Deva, Jud, Hunedoara, Romania
e-mail: bogdantomus@yahoo.com

R.-C. Breban
e-mail: radu_breban@yahoo.com

B. P. Onac
Karst Research Group, School of Geosciences, University of South Florida, 4202 E. Fowler Ave., NES 107, Tampa, FL 33620, USA
e-mail: bonac@usf.edu

B. P. Onac
"Emil Racoviţă" Institute of Speleology, Romanian Academy, Clinicilor 5, 400006 Cluj-Napoca, Romania

G. M. L. Ponta and B. P. Onac (eds.), *Cave and Karst Systems of Romania*,
Cave and Karst Systems of the World, https://doi.org/10.1007/978-3-319-90747-5_15

Fig. 1 Location of the Valea Stânii–Ponorici–Cioclovina cu Apă karst system in the Şureanu Mountains

as quartzites, quartzite schists, garnet mica schists, muscovite mica schists, and granite gneisses (Balintoni 1997). Sedimentary formations are grouped within four depositional cycles (Stilla 1985). In the Lower Permian cycle, the Sebeş crystalline schists are transgressively and discordantly overlain by a red-violet detrital formation with coarse conglomerates at the base and poorly cemented micaceous sandstones in the upper part. The sedimentary units ascribed to the Jurassic–Lower Cretaceous cycle rest unconformably over the Permian rocks changing from detrital (base) to carbonaceous (top). Beginning with Dogger (Middle Jurassic), the lithological units rest in concordant contact with the Lower Jurassic formations made of grayish-yellowish, well-layered and fossil-rich sandstones. During the Upper Jurassic, sedimentation continues with spathic calcareous sandstone, over which a pale pink limestone unit with nodular chert was deposited. Massive, fossil-rich reef limestones accumulated throughout Neocomian (Lower Cretaceous) and continued in the Barremian–Lower Aptian (Lower to Middle Cretaceous) period, making up the bulk of the carbonate deposits in the area described by Savu et al. (1968) and Stilla (1985). For details on the geology of the region, readers are directed to Fig. 2 in the chapter on the Karst of Sebeş Mountains.

After the deposition of the last term of the Jurassic–Lower Cretaceous stratigraphic sequence, followed a relatively long period of uplift, during which a well-developed karst landscape formed on the limestone bedrock. In the negative forms (dolines, uvalas, shafts, etc.), aluminum-rich detrital sediments accumulated during Upper Aptian and Albian generating bauxite deposits (Papiu et al. 1971).

From a tectonic point of view, the Cioclovina area is an integral part of the Haţeg Basin, whose plicative (folds) and disjunctive (rupture) elements are also found here (Stilla 1985). The Ţâfla–Piatra Roşie Syncline and the Cioclovina Anticline belong to the first category. The disjunctive elements include the South-North marginal fault Ciopeia–Băieşti–Rotunda–Cioclovina-Piatra Roşie, marking the line along which the crystalline basement is thrusted over the sedimentary rocks, as well as the Cioclovina Fault with a NE–SW direction revealed between Rotundei and Feţii peaks (Savu et al. 1968).

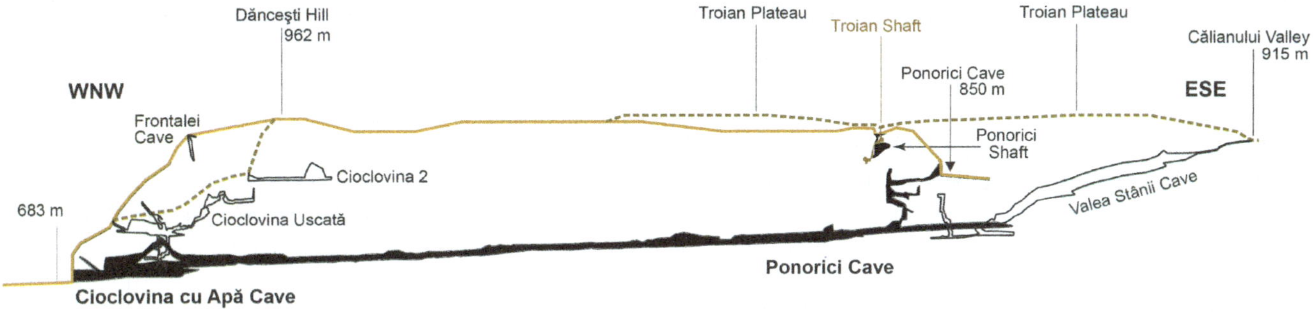

Fig. 2 Profile in the investigated area showing the location and relationship between various caves that are part of the Valea Stânii–Ponorici–Cioclovina cu Apă karst system

From a hydrogeological point of view, the main resurgence of the 2063 Karst Basin is the watercourse emerging from the Cioclovina cu Apă Cave, which forms the Luncani Valley. This river originates mainly from the confluence of the Ponorici and Călian streams that flow over crystalline rocks before disappearing underground through a number of ponors at the contact with the limestone bedrock. The water of the Ponorici Stream is lost in the ponors on the left side of the valley or in a doline formed in front of the cave. The amount and location of the losses depend on the volume of water and the degree of ponors infilling. The water of Călian Valley is lost in two ponors (on both sides of the valley) just before the entrance into the Valea Stânii Cave. The groundwater flow directions have been established by two fluorescein tracer tests (Bandrabur and Bandrabur 2010), which indicated that the water of the Călianului Valley flow across Valea Stânii Cave, then through the Ponorici–Cioclovina cu Apă, where confluences with the underground flow of the Ponorici Stream to then resurface through the latter cave (Fig. 2).

Besides these main courses, there are other streams with low flow. On the Triscioare Plateau, there are several ponors (Trepâraie, Cerbului, Robului) that drain the waters toward the stream flowing through Cioclovina Uscată Cave and probably discharge in the Văratec Valley (Tomuş 2011). On the Troian Plateau (about 3.3 km^2), there are no perennial creeks; however, the existence of a series of dolines valleys in the SE part of the plateau makes us believe that the water collected by these dolines appears in the Cioclovina cu Apă Cave, through its left tributary located in the middle section of the cave.

The dolines in the area W and N of the plateau are probably responsible for the springs occurring at the limestone-crystalline contact in the western, north-western, northern extremities of the plateau, as well as a small part of the water in the Călian Valley (Tomuş 2011).

History of Exploration

The Ponorici–Cioclovina karst area gained scientific interest at the turn of the nineteenth to the twentieth century, when the first papers describing the presence in the Cioclovina Uscată Cave of both osteological deposits and traces of human habitation were published. Chronologically, the following dates are of significant importance in the scientific investigations of the caves from the Cioclovina karst area.

1878—First paper in which G. Téglás makes a reference to the Cioclovina Uscată Cave;
1880–1915—Several geological and karst publications mention the Cioclovina Uscată and Cioclovina cu Apă caves (Bielz 1884; Halaváts 1898; Roska 1912a, b);
1897—In a paper describing the Ponorici Cave, L. Abafi Aigner included maps of the chambers located near the cave entrance;
1917—Z. Schréter publishes the first description of the Cioclovina Uscată Cave and provided estimations regarding the volume of the phosphate deposit;
1919—G. Götzinger provides a detailed study of the phosphate accumulation;
1932—Schadler describes *ardealite*, a new mineral from the Cioclovina Uscată phosphate deposit;
1932—I. Gherman publishes the first karst monograph of the investigated region;
1942—Discovery and description of a *Homo sapiens fossilis* skull from Cioclovina Uscată Cave by Simionescu (1942) and Rainer and Simionescu (1942);
1953—Dumitrescu et al. (1955) uncovered in the Cioclovina cu Apă Cave a collection of artifacts assigned to Hallstatt Culture (early Iron Age);
1955—Cioclovina Uscată Cave, and a ~40 ha surrounding perimeter was declared a scientific reserve under the patronage of the Romanian Academy;

1959—the SALEM team of the "Emil Racoviță" Institute of Speleology makes the junction between the Ponorici and Cioclovina cu Apă caves (Negrea 1974);
1973—"Emil Racoviță" Caving Club from Bucharest explored and surveyed Ponorici–Cioclovina cu Apă Cave;
1983—M. Băicoană and members of the Proteus Speleo-Club Hunedoara discovered Valea Stânii Cave;
1984—Proteus SpeleoClub Hunedoara discovered the stream passage in the Cioclovina Uscată Cave;
1985—Proteus SpeleoClub Hunedoara and "Emil Racoviță" Caving Club (Bucharest) conducted a tracer test, that confirmed the connection between Valea Stânii and Cioclovina cu Apă caves;
1985—Proteus SpeleoClub Hunedoara discovers Cioclovina 2 Cave;
1986—CS Focul Viu Bucureşti resurvey the main passages of the Ponorici–Cioclovina cu Apă Cave (scale 1:200) with a suspended mining compass (BCRA Grade V) by a team lead by G. Aldica and I. Aurel. The map and the cross section were presented in the same year at the 1986 Speosport meeting;
1997—Proteus SpeleoClub and CS Focul Viu remapped the Valea Stânii Cave;
1998—Members of the Proteus SpeleoClub Hunedoara find a 113-cm-triangular calcite monocrystal in Cioclovina Uscată Cave;
1999—Asociația Sporturilor Montane Hunedoara (ASMH)/ Hunedoara Mountain Sports Association discovered ASMH Cave from where the stream in Cioclovina Uscată Cave originates;
2003—Berlinite and hydroxylellestadite, two high-temperature phosphate/silicate minerals formed during guano combustion were described from Cioclovina Uscată Cave (Onac and White 2003; Onac et al. 2006; Onac and Effenberger 2007);
2005—Members of the Proteus SpeleoClub Hunedoara crossed a sump in the Cioclovina cu Apă Cave and found a room in which the underground streams from Cioclovina Uscată and Cioclovina cu Apă confluence.

Description of the Caves

The Valea Stânii–Ponorici–Cioclovina (Fig. 2) is an important karst system of the Şureanu Mountains due to the number and complexity of the caves, their morphology, and speleothem/minerals diversity.

Valea Stânii Cave (also known as Valea Călianului Cave) was discovered in 1983 by M. Băicoană (Proteus Speleo-Club Hunedoara) and is one of the most beautiful caves in the Şureanu Mountains. It develops in two types of limestones, the transition between them being easily observed due to changes in the morphology (cross section) of the Gallery. The first part of the cavity, dominated by collapses and large spaces, develops on the contact between the green shales (crystalline basement) and pink-white, reddish limestone of Upper Oxfordian–Tithonian age containing nodular cherts. In the second part, the Main Gallery develops in reefal limestone (Barremian–Aptian inferior), and the morphology of the passages is dominated by meanders and dissolution levels. Apparently, in the Gallery beyond the second sump (Fig. 3), the cherty limestone is once again present (Milota pers. comm.).

From a hydrogeological point of view, two distinct underground streams flow through the cave. The first is represented by the water of the Călianului Valley, which enters the cave, but then disappears in a ponor located immediately after the Great Chimney (Marele Horn/H30; Fig. 3). At high flow rates, the ponor cannot drain the entire amount of water, so it continues its course until the confluence with the second stream, which comes from the waters lost in dolines of the Troian Plateau. After the two underground streams confluence, the main watercourse continues for another 300 m (straight line) on a Gallery that is above the upstream end of the Ponorici–Cioclovina cu Apă Cave (Fig. 2). Moving upstream on the second tributary, Milota and Săsăreanu passed Sump 1, a 7 m long dive, then continued for $\sim$236 m until Sump 2. In August 1998, Ş. Milota dived this 13 m long and 5 m deep sump and found that the cave continues along a large but challenging (from an exploration point of view) Gallery; thus, it has not been mapped yet.

Two chimneys were explored in the cave (Fig. 3): the Great Chimney (H30) has great potential to continue, but the exploration at its upper part is hampered by the presence of large unstable collapsed blocks. The second chimney is 23 m (H23), but beyond this, only a small extension was found.

Ponorici Shaft (Avenul Ponorici) represents the upper level of the Ponorici–Cioclovina cu Apă cave system, having certainly a direct, yet undiscovered communication with the Ponorici Cave (Fig. 2). The large entrance leads through a narrow passage into a small well-decorated room. From this point, slipping between large speleothems/limestone blocks and after passing another narrow strait, the passage reaches the ceiling (bell-type) of a chamber. At the base of the 26 m-vertical drop, there is a 45 × 12 m room having an uneven floor covered with massive limestone breakdowns.

Ponorici–Cioclovina cu Apă Cave is one of the longest hydrological breakthroughs in the Romanian karst, with a straight line distance between the two entrances of 1320 m and a total length of its galleries of about 4300 m (Figs. 2

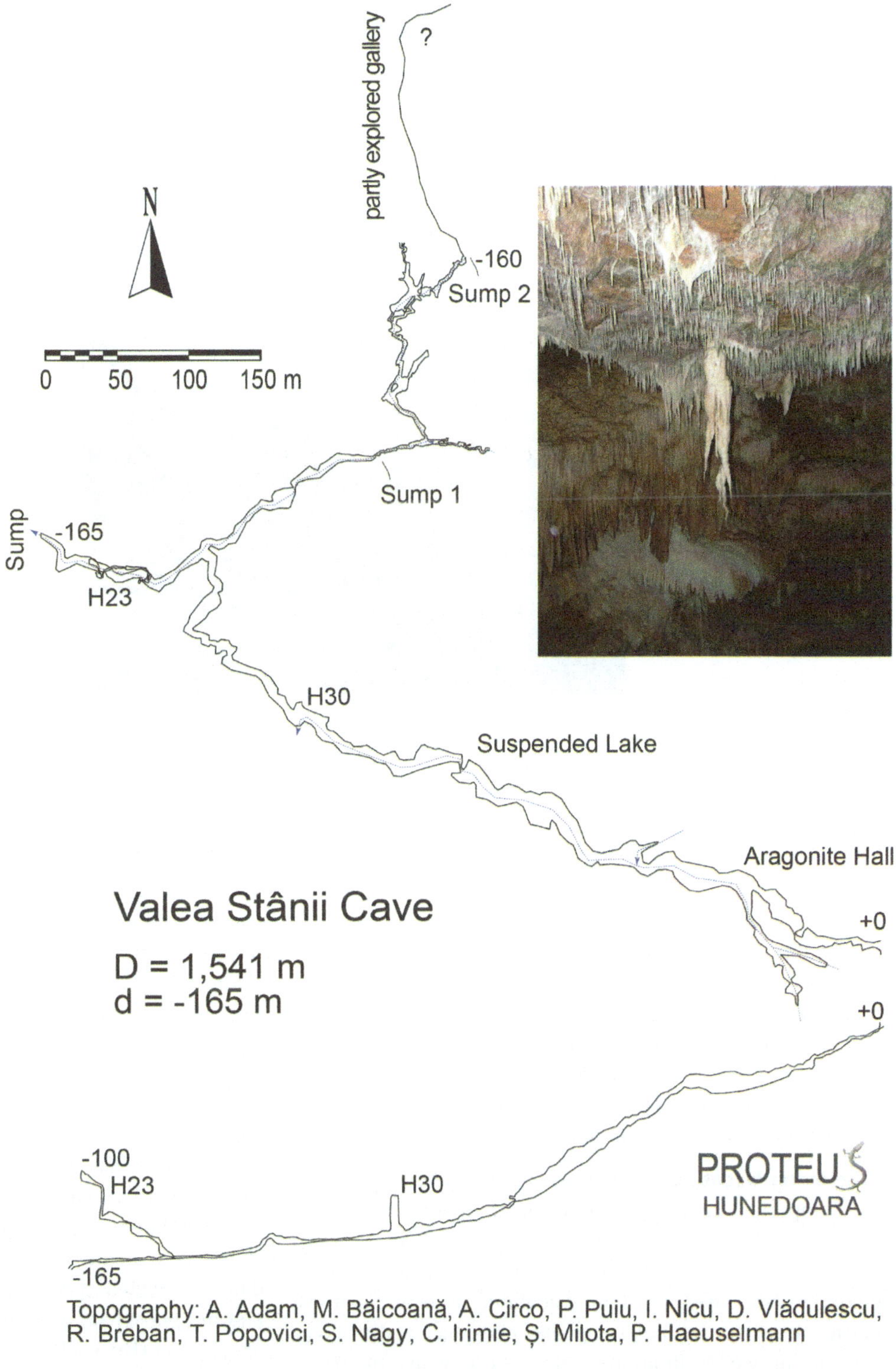

Fig. 3 Plan and profile of the Valea Stânii Cave with an image in inset showing a highly decorated passage within the cave (photograph courtesy of D Herlea)

and 4). The high esthetic value of the cave is given by the galleries morphology and speleothems, as well as by archeological and paleontological remains.

From the Ponorici entrance, the cave begins with a large, horizontal dry/fossil Gallery followed by the Ponorici Hall (Sala Ponorici) from where a 23 m-vertical drop ends into a descending Gallery, which leads to the Great Confluence (−117 m). Once the mainstream passage of the cave has been intercepted, a succession of waterfalls having different drops (14, 12, 2 m, etc.; Fig. 5a) needs to be descended (Mitrofan 1988). Continuing downstream, the size of the Gallery remains about the same (morphology changes thought; Fig. 5b–d) until leaving the cave through the

Fig. 4 Ponorici–Cioclovina cu Apă Cave. **a** Plan and profile. **b** Speleothems in the Canopy Gallery (photograph by R.-B Tomuş)

Cioclovina cu Apă entrance. The Canopies Gallery (Galeria Baldachinelor) is beautifully decorated by a variety of speleothems (Fig. 4 inset).

The cave features two right side galleries, a short fossil passage (Candle Gallery/Galeria Lumânărilor) and the Gallery of the Great Rimstone Pools and Shelfstone (Galeria Marilor Gururi şi Planşee), which drains the waters from the Valea Stânii Cave and Ponorici ponors.

Cioclovina Uscată Cave is from a scientific viewpoint the most important cave in the Şureanu Mountains, because it hosts one of the oldest early modern human skulls from Romania (29,000 ± 700 ^{14}C yrs BP), 29 minerals (some unique to this cave), has four Paleolithic horizons, and last but not least, for its massive guano phosphate sedimentary deposit, in which thousands of fossil remains of *Ursus spelaeus* were identified. At present, due to the explorations conducted by cavers of the Proteus SpeleoClub Hunedoara, the cave has a total length of 1406 m and 140-m-vertical range (Fig. 6).

Morphologically, the cave can be divided into two distinct parts: a mixed phreatic and vadose sector between the natural entrance all the way to the terminal part of the Bivouac Room (Sala Bivuacului) and a vadose one consisting of the two ascending passages (Section "A" and "B"; Fig. 6) branching out from the Main Gallery (Galeria Principală), as well as the Gallery after the Chimney (Galeria de după Horn) and the upstream end of the cave (Section "98" in Fig. 6). The Main Gallery of the cave is largely devoid of speleothems, which were vandalized by tourists (in the nineteenth to early twentieth century) or destroyed during mining of the guano phosphate. However, this part of the cave has large rooms (Fig. 7a), beautiful corrosion features (tubes of 5 m in diameter, ceiling and wall notches, pendants), as well as a peculiar "paleokarst" formation (Fig. 7b), and the phosphate deposit containing a rich paleontological association (Roska 1912b; Simionescu 1942; Rădulescu and Samson 1959), and unique minerals (Marincea et al. 2002; Onac et al. 2002, 2006, 2007).

From the Bivouac Room through a narrow passage and after descending, a 10 m shaft, the 60 m long underground river (Fig. 8) is intercepted; its upstream and downstream sides end in sumps. The only continuation is along an ascending canyon-type Gallery that ends shortly at the foot

Fig. 5 Common features and morphologies in the Ponorici–Cioclovina cu Apă Cave. **a** The second shaft in Ponorici Cave. **b** Cioclovina Hall (also known as the Treasure's Hall), **c–d** Canopies Gallery (all photographs by R.-B Tomuş)

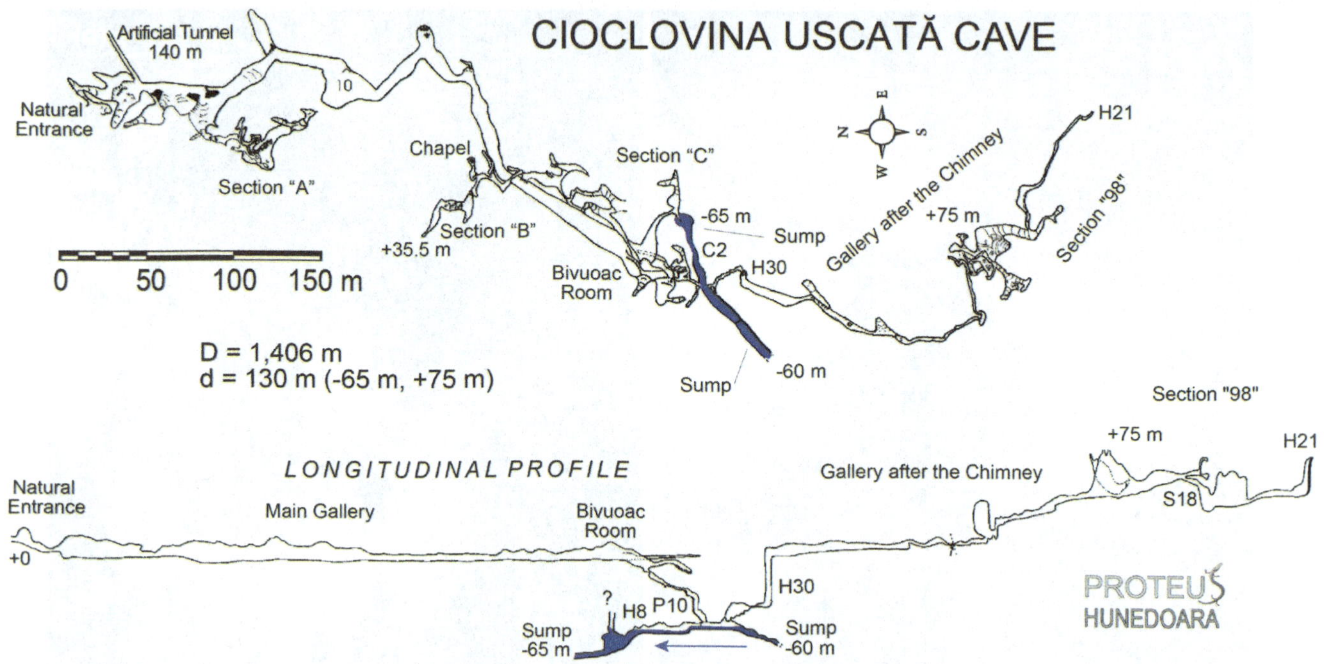

Fig. 6 Map and profile of the Cioclovina Uscată Cave

of a 30 m chimney, which lead into the Gallery after the Chimney and Sector "98" (Tomuş 2011). Due to difficulties associated with this climb, the second part of the cavity is in a perfect state of preservation. In the Gallery after the Chimney, a large diversity of ordinary and rarer (eccentrics, helictites, heligmites, shields, and a 113 cm tall monocrystalline stalagmite) speleothems can be observed.

Worth mentioning is that much of the fossil passage between the cave's natural entrance and the Bivouac Room was almost entirely filled by an impressive phosphate-rich deposit (Fig. 9) believed to have formed due to the presence of large accumulations of bat guano and cave bear remains. Between 1912 and 1941, more than 30,000 m^3 of guano sediments were mined and used as fertilizer (Breban et al. 2003). The mining operations were ample and included an artificial tunnel to facilitate the access the guano phosphate deposit, a railway and wagons inside the cave, an inclined cableway from the cave entrance to the Luncanilor Valley, and even a milling plant to grind the cave bear bones.

The Evolution of the Valea Stânii–Cioclovina cu Apă Karst System

The Triscioare–Ponorici–Fundătura Ponorului area had a special geomorphological evolution, which included several stages until it reached the current configuration with the three main valleys: Călian, Ponorici, and Ponorului, and their associated ponors. The incision of the hydrographic network in the Ponorici area happened in four stages that are outlined below (Fig. 10). In the *first phase*, the Ponorici Valley seems to have been connected with those of Ponorului and Călianului, forming a larger river that drained into the NE part of the Haţeg Depression, between the Arsului and Robului hills (Fig. 10a). Toward the end of this first stage or the beginning of the second one, the formation of the Cioclovina Uscată Cave was initiated, fact that explains the origin of the underground allogenic sediments, transported by the waters from the Poiana Omului area.

The *second stage* (Fig. 10b) begins with the underground capture of the Ponorului Valley in the Fundătura–Balaj area. Originally, the river formed by Ponorici and Călian was drained toward the present Luncanilor (Morii) Valley through the south of the Dănceştilor Hill, but then shifted on the northern slope, along the current doline valley that extends all the way to the Cioclovina saddle.

In the *third stage*, the Călianului Valley begun to evolve independent from Ponorici stream, thus the three major catchments in the area become fully functional (Fig. 10c). This stage marks the onset of the processes that culminated with the genesis of Şura Mare, Ponorici, and Stânii Valley caves. No detailed studies are available with respect to the speleogenesis of Valea Stânii Cave, but Mitrofan (1979) offered a comprehensive account on the role of tectonics in the development of the Ponorici–Cioclovina cu Apă Cave. Accordingly, it is assumed that decompression forces driven by directional tension triggered the opening of fractures along which the galleries evolved. Furthermore, the confluences and diffluences are rigorously caused by the presence of decompression fissures converging on the synclinal

Fig. 7 Cioclovina Uscată Cave. **a** The Bivouac Room with its historical inscriptions on the ceiling left behind by miners and tourists (photograph by R.-B. Tomuş). **b** The paleokarst exposure along the Main Gallery (photograph by B. P Onac)

Fig. 8 Image of the stream intercepted in the lower part of the Cioclovina Uscată Cave (photograph by R.-B Tomuş)

Fig. 9 Depiction of the guano phosphate deposit near the Natural Entrance (**a**; photograph by B. P Onac) and in the Bivouac Room (**b**; photograph by R.-C Breban)

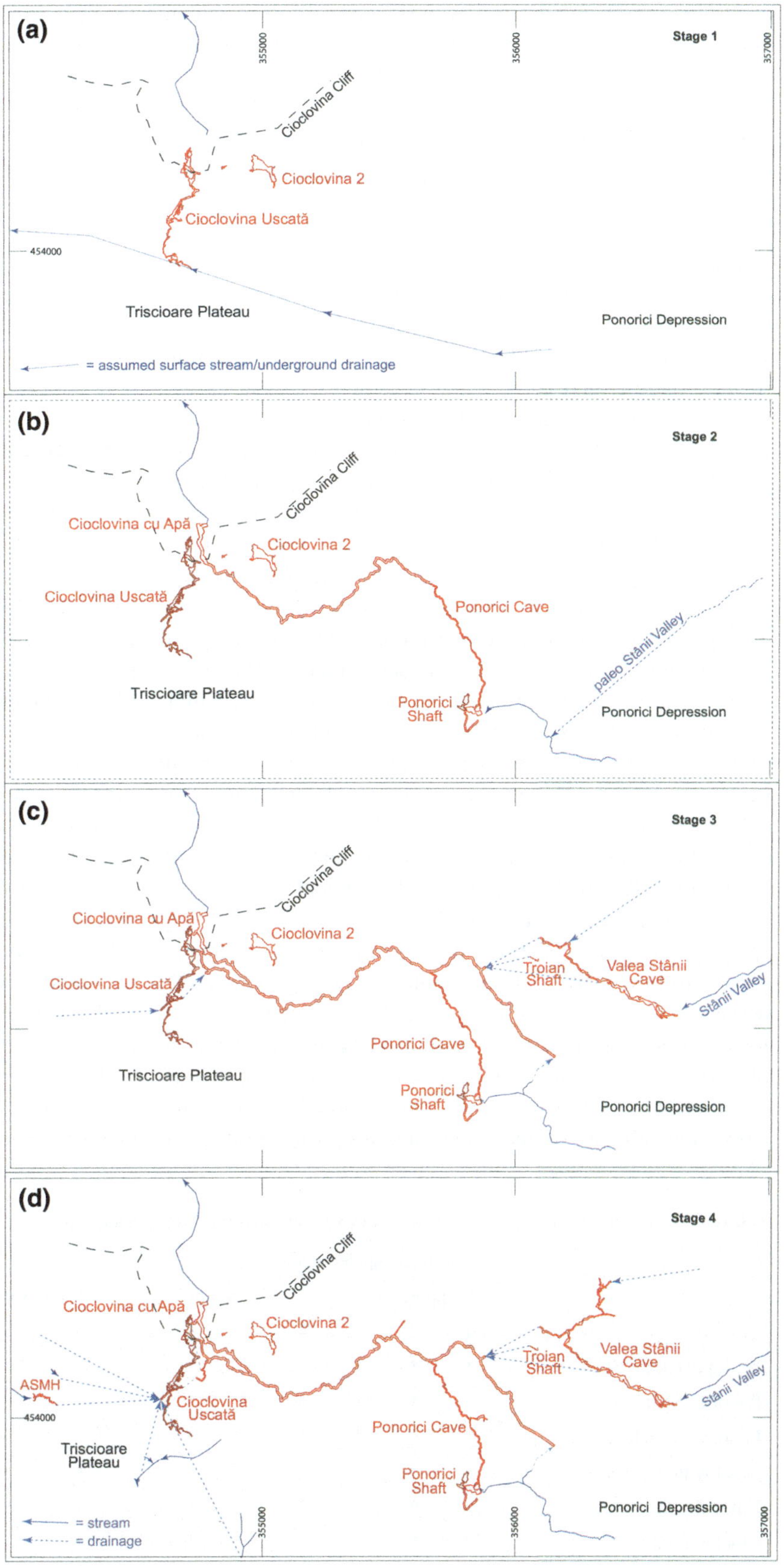

Fig. 10 **a–d** Stages in the evolution of the Valea Stânii–Ponorici–Cioclovina cu Apă karst system (see text for details)

axial plane and being created by the tangential forces, which appeared at the curvature of the periclinal endings of the geological structures (e.g., the confluence of the Ponorici Gallery with the Gallery of the Great Rimstone Pools and Shelfstone, the diffluence at the end of the Canopy Gallery, etc.). Another interesting observation recorded by Mitrofan (1979) is that waters, which enter the cave via epikarst never succeed in forming long galleries, even if flowing through tectonically heavy decompressed areas.

At present (the *4th stage*), the three valleys have reached the impervious bedrock and cave passages that formerly were characterized by vadose flow (e.g., the entrance section of the Ponoroci and parts of the Cioclovina Uscată caves) became fossil/dry galleries. The waters lost on the Troian Plateau follow different underground routes to ultimately join the stream in the Gallery of the Great Rimstone Pools and Shelfstone of Ponorici Cave. The streams sinking on the Triscioare Plateau eventually end up in the Ponorici–Cioclovina cu Apă Cave after flowing a short distance through the lower part of the Cioclovina Uscată Cave (Fig. 10d). As a result of the evolution through these stages, the water of the Ponorici Valley is drained underground to Ponorici–Cioclovina cu Apă Cave and from there to the Luncanilor Valley. Given this fact, both Ponorici and Călianului valleys remained suspended onto a surface generated by an erosion cycle previous to the present one, to which Luncanilor Valley belongs.

For the section located between the natural entrance and the Bivouac Room (Main Gallery) of the Cioclovina Uscată Cave, Häuselmann et al. (2010) suggested the following speleogenetic stages: the first one corresponds to a phreatic phase, followed by vadose conditions during which the deposition of the first sediments and the precipitation of the large flowstones occurred. The third stage is once again a phreatic one, with the complete removal of the previous accumulated sediments and the corrosion of the paleo-speleothems. As the water table dropped, the flow of the cave stream entered a vadose regime. Most of the current sediments were deposited during this time, but a gradual increase of the water level in the cave was inferred toward the end of this phase as a result of the natural entrance being clogged by sediments. Then, follows a period when new galleries formed in the Section "B" with the subsequent transportation of a part of the Main Gallery's sediment through them. The clogging of the link between the Main Gallery and the Gallery after the Chimney, the formation of the 30-m shaft (H30 in Fig. 6), and the return of the Main Gallery to vadose conditions represent the current state in the evolution of the cavity.

Cave Sediments and Speleothems

The sediment in Cioclovina Uscată Cave is mostly sandy-silt and its thickness drops from ~ 14 m near the entrance to about 7 m in the Bivouac Room (Häuselmann et al. 2010). The total thickness of the sediment accumulation reaches 21 m. In cave observations and sedimentological analyses pointed out the existence of four main lithological complexes, which from bottom to top include: (1) medium-size pebbles in a sandy matrix directly overlying the limestone bedrock (0.5–2 m thick), (2) the lower sandy complex; medium-size sterile sand and silt reaching up to 3 m in thickness, (3) the upper sandy complex comprises the richest phosphate sediment (up to 8 m), and (4) the sandy–silty complex (8 m), which hosts numerous vertebrate remains and limestone breakdowns. All these sequences developed in a typical cave channel lithofacies (Bosch and White 2004) and hold key information that allowed reconstructing the speleogenetic phases for the section between Bivouac Room and cave entrance (Häuselmann et al. 2010).

Except for the major shafts in the investigated area, which are moderately decorated, in all caves discussed in this chapter, speleothems are diverse and abundant (see Table 1).

Table 1 Types of speleothems identified in various caves of the investigated area

Cave	Carbonate speleothems														Others	
	Dripping				Seeping			Capillary			Pool					
	1	2	3	4	5	6	7	8	9	10	11	12	13	14	15	16
de dupa Troian Shaft	*	*	*		*	*			*							
Ponorici Shaft	*	*	*	*	*	*	*									
Ponorici–Cioclovina cu Apă	*	*	*	*	*	*	*	*	*		*	*		*		
Cioclovina Uscată	*	*	*	*	*	*	*	*	*	*	*	*	*	*	*	*
Valea Stânii	*	*	*	*	*	*	*	*	*		*	*	*	*		
Cioclovina 2	*	*	*	*	*	*	*	*	*		*					*

1 Stalactites, *2* Stalagmites, *3* Columns, *4* Cave pearls, *5* Canopy/Veils, *6* Flowstones, *7* Rimstone pools, *8* Capillary, *9* Eccentrics, *10* Shields, *11* Calcite rafts, *12* Shelfstones, *13* Spars, *14* Pool pearls, *15* Gypsum crystals and crusts, *16* Phosphate earthy masses and crusts

Stands out the triangular monocrystalline stalagmite (113 cm in height) from Cioclovina Uscată and the massive columns, canopies, domes, veils, and stalactites from Valea Stânii and Ponorici–Cioclovina cu Apă caves.

Mineralogy

The presence of the infilling material attracted attention because of its richness in vertebrate bones (mainly *U. spelaeus*) and the presence of phosphate nodules/impregnations. Chemical analyses performed at the beginning of the twentieth century on these sediments indicated high contents of phosphate, turning the cave into a valuable source of fertilizers (Horváth 1916; Schréter 1917; Horusitzky 1918; Gotzinger 1919). Depending on depth and location of the guano phosphate sediment along the Main Gallery, these studies estimated concentrations of P_2O_5 between 1.5 and 30%. This elevated interest on Cioclovina Uscată's guano phosphate deposit has had both beneficial and detrimental effects on the cave. On the positive side, the mining of sediments led to significant mineralogical, paleontological, and archeological discoveries. To highlight just a few of these, it is worth noting that a new mineral species, namely ardealite (Table 2) was first described from this cave by Schadler (1932). It followed significant paleontological, archeological, and anthropological findings, which are presented in more details later in this chapter. The downside part of all mining operations is that they heavily altered the cave morphology and sediment stratigraphy, destroyed

Table 2 Minerals identified in Cioclovina Uscată Cave

Mineral	Chemical formula (according to IMA-CNMNC[a])	Class
Calcite	$CaCO_3$	Carbonates
Aragonite	$CaCO_3$	
Burbankite	$(Na,Ca)_3(Sr,Ba,Ce)_3(CO_3)_5$	
Gypsum	$CaSO_4 \cdot 2H_2O$	Sulfates
Kröhnkite	$Na_2Cu(SO_4)_2 \cdot 2H_2O$	
Bassanite	$CaSO_4 \cdot 0.5H_2O$	
Hydroxylellestadite	$Ca_{10}(SiO_4)_3(SiO_4)_3(OH,Cl,F)_2$	
Goethite	$FeO(OH)$	Oxides and hydroxides
Romanèchite	$(Ba,H_2O)_2(Mn^{4+},Mn^{3+})_5O_{10}$	
Hematite	Fe_2O_3	
Birnessite	$(Na,Ca,K)_{0.6}(Mn^{4+},Mn^{3+})_2O_4 \cdot 1.5H_2O$	
Todorokite	$(Na,Ca,K,Ba,Sr)_{1-x}(Mn,Mg,Al)_6O_{12} \cdot 3\text{-}4H_2O$	
Atacamite	$Cu_2Cl(OH)_3$	Halides
Ardealite	$Ca_2(PO_3OH)(SO_4) \cdot 4H_2O$	Phosphates
Berlinite	$AlPO_4$	
Brushite	$Ca(PO_3OH) \cdot 2H_2O$	
Churchite-(Y)	$YPO_4 \cdot 2H_2O$	
Collinsite	$Ca_2Mg(PO_4)_2 \cdot 2H_2O$	
Crandallite	$CaAl_3(PO_4)_2(PO_3OH)(OH)_6$	
Fluorapatite	$Ca_5(PO_4)_3F$	
Foggite	$CaAlPO_4(OH)_2 \cdot H_2O$	
Hydroxylapatite	$Ca_5(PO_4)_3(OH)$	
Leucophosphite	$K(Fe^{3+})_2(PO_4)_2(OH) \cdot 2H_2O$	
Monetite	$Ca(PO_3OH)$	
Taranakite	$K_3Al_5(PO_3OH)_6(PO_4)_2 \cdot 18H_2O$	
Tinsleyite	$KAl_2(PO_4)_2(OH) \cdot 2H_2O$	
Variscite	$AlPO_4 \cdot 2H_2O$	
Kaolinite	$Al_2Si_2O_5(OH)_4$	Silicates
Quartz	SiO_2	

[a]International Mineralogical Association—Commission on New Minerals, Nomenclature, and Classification
In **bold**, minerals first described in a cave environment worldwide; in *italics*, allochthonous minerals

numerous speleothems, and left behind large amounts of waste and debris.

Successive cave flooding events resulted in the accumulation of large quantities of sand, silt, and clay that were either inter-bedded with bat guano horizons or completely buried the organic sediment. In addition, thousands of bones are scattered throughout these sediments. The geochemical environment under which the phosphate-rich deposit accumulated is not yet fully understood. However, the 14 phosphate minerals identified throughout time (Table 2) point toward distinct stages of phosphatization processes that have affected to various degrees both the limestone bedrock and the siliciclastic sediments. Such examples are visible throughout the cave (Fig. 11a, b) at locations where the mining activities left behind good exposures, which also provided quality material for detailed mineralogical work.

Because of the multi-reactant system (limestone, sand, silt, clay, guano, and bones), depending on whether the percolating waters leaching guano/bones react with carbonate rocks or the allogenic sediments, common or exotic Ca^{-}, K^{-}, or Al-rich phosphates were deposited (Onac et al. 2009). Along with the ordinary cave minerals (i.e., calcite, aragonite, gypsum, hydroxylapatite), a number of rarities makes up an imposing list (Table 2), among which are ardealite, burbankite, atacamite, churchite-(Y), collinsite, crandallite, foggite, kröhnkite, leucophosphite, and tinsleyite (Constantinescu et al. 1999; Breban 2002; Marincea et al. 2002; Onac et al. 2002, 2005, 2009, 2011; Dumitraş et al. 2005, 2008).

In some parts of the cave, the weight of the overburden sediments was significant and the underlying material underwent excessive compaction; hence, the original textures and structures were changed (Fig. 11c). In restricted parts of the cave (e.g., Bivouac Room) owing to microbial processes, the temperature inside the buried guano increased until spontaneous ignition led to its combustion. This process produced berlinite and hydroxylellestadite, two rare and unusual high-temperature minerals for sedimentary environments (Onac and White 2003; Onac et al. 2006; Onac and Effenberger 2007; Cîntă-Pînzaru and Onac 2009).

Of the minerals listed in Table 2, detailed studies have been conducted on 14 of them, all representing rare occurrences in cave environments worldwide. The coexistence of some low- (hydroxylapatite) and high-temperature (berlinite) minerals generated discussions and controversies (Onac and White 2003; Marincea and Dumitraş 2005; Dumitraş et al. 2008; Onac and Effenberger 2009). All studies mentioned above emphasize that Cioclovina Uscată Cave is a world-class underground mineralogical museum of significant scientific value.

Cave Fauna

The area under investigations belongs to the second biospeleological province of the South Carpathians, more precisely, to the region between Olt River and Timiş–Cerna Couloir (Decu and Negrea 1969). Before the major mining operations commenced, the following cave invertebrate groups were documented: Araneida, Colembola, Diptera, and Isopoda (Gheorghiu et al. 2009) from samples collected between 1923 and 1924 by P.-A. Chappuis and R. Jeannel from Cioclovina Uscată and Cioclovina cu Apă caves (Jeannel and Racovitza 1929). Both caves are characterized by the presence of several endemic species (Dumitrescu et al. 1967; Gruia 2003; Nitzu et al. 2016).

Fig. 11 Phosphate minerals in Cioclovina Uscată Cave. **a** Sandy-silty sediments impregnated with ardealite and brushite, **b** Centimeter-thick layers of various minerals formed by the phosphatization process affecting limestone blocks, **c** Heavily compacted and "baked" guano phosphate sediments from which berlinite and hydroxylellestadite were described (photograph by B. P Onac)

Maternity colony of *Rhinolophus ferrumequinum* was observed in Ponorici–Cioclovina cu Apă Cave (Gheorghiu et al. 2009; Nagy and Postawa 2011). Bones of *Nyctalus noctula*, a species that is primarily adapted to life in wood shelters were discovered in this cave by Dumitrescu et al. (1955); nevertheless, this species is no longer present in the cave. Summer colonies of *R. ferrumequinum*, *Rhinolophus hipposideros*, *Myotis myotis/oxygnathus*, and *M. myotis* were documented from Cioclovina Uscată (Gheorghiu et al. 2009; Nagy and Postawa 2011). After a massive ecological rehabilitation of the Cioclovina Uscată Cave undertaken between 2004 and 2005, a far larger diversity of bats species have been reported by Gheorghiu et al. (2009). In addition to ones mentioned above, the list also includes *Myotis daubentonii, Myotis Blythii, Myotis capaccinii, Myotis nattereri, Plecotus auritus, Miniopterus schreibersii, Pipistrellus pipistrellus, Pipistrellus pygmaeus, Barbastella barbastellus*, etc. For a complete inventory of the bats inhabiting the caves discussed in this chapter, readers should consult the study of Dumitrescu et al. (1967) and Gheorghiu et al. (2009).

Paleontology

Cioclovina Uscată Cave is the most important site in the investigated area with respect to paleontological remains, both in terms of number and variety of fossil species, which include: *U. spelaeus, Felix spelaea, Capra ibex, Panthera (Leo) spelaea, Canis lupus*, and *Ursus arctos* (Simionescu 1942; Rădulescu and Samson 1959; Petrea 2009). The vast majority of the fossil repository consists of remains belonging to the *U. spelaeus*, the bones are no longer in anatomic connection, but they are transported by water and subsequently disturbed by the guano phosphate mining operations.

In 1985, cavers from the Proteus Hunedoara SpeleoClub discovered Cioclovina 2. Before the most recent phosphate mining operations started, the cave hosted beautifully preserved cave bear bioglyphs, including scratches and several well-preserved hibernation nests with skeletons in anatomic connection. In addition, numerous Bärenschliffe (Bear-polish), meaning wall surfaces polished to a mirror-like sheen by the passage of countless bears during hundreds or thousands of years, were also documented. In just four years, due to the survey conducted to evaluate the possibility of mining guano phosphate, this scientific resource was almost completely destroyed.

Anthropology and Archeology

The first systematic archeological excavations in Cioclovina Uscată Cave begun in 1911, but the results were published later in several papers by Roska (1912a, b, 1923, 1925) who describes pottery and vase fragments along with a very rich association of tools (blades, scrapers, spearheads, etc.) made of jasper, opal, limestone, quartz, chalcedony, and bones, assigned to cultures belonging to Mousterian and Aurignacian (Roska 1925). Unfortunately, the stratigraphic context of these discoveries can no longer be evaluated as the mining activities destroyed the original excavations. Materials assigned to the Coțofeni Culture were excavated in the 1950s by C. S. Nicolăescu-Plopșor, C. N. Mateescu, and L. Roșu (Boroneanț 2000). A major archeological discovery was made in 1955 in the Cioclovina cu Apă Cave, where T. Orghidan found over 2800 glass, faience, and amber beads, bronze tools and vases assigned to the Hallstatt Culture of Early Iron Age (Dumitrescu et al. 1955). These were later investigated in great detail by Comșa (1966) and Emödi (1978). The spectroscopic investigations on one of the amber beads indicated that the artifact was manufactured from amber occurring along the southern shore of the Baltic Sea. The presence of this archeological amber in Cioclovina cu Apă suggests that the amber trade route (from the Baltic region to Rome) once passed very close to this cave (Banerjee et al. 1999).

In terms of anthropological materials, prior to 1911 (unrecorded dates), M. Roska was aware that T. Györö and I. Lager from Călan found a human skull in the Big Room of Cioclovina Uscată Cave, which subsequently and under uncertain circumstances was destroyed (Roșu 2004). Sometimes before 1941, during the extraction of guano phosphate, miners discovered another fragment of a human skull within the Aurignacian horizon. This new finding was first mentioned in a thorough study that investigated the *U. spelaeus* remains from this cave (Simionescu 1942). Later that same year, Rainer and Simionescu described the skull with a considerable degree of detail, concluding that it belongs to a 30–40-year-old woman of Upper Paleolithic age.

Olariu et al. (2005) first dated the Cioclovina skull by means of accelerator mass spectrometry (AMS) radiocarbon (^{14}C) to 29,700 ± 700 years BP (33,726 ± 1132 cal yrs BP). Based on the obtained age, the authors considered the skull to be of a *H. sapiens fossilis* with strong Neanderthalian characters. To improve the large error value obtained for this first ^{14}C age, a second date was performed by Soficaru et al. (2007), who reported a slightly younger age of 33,212 ± 693 cal yrs BP. Altogether, these two ages firmly confirm that the Cioclovina specimen is one of the few early modern humans remains from Europe, which in fact represents a Neanderthal—early modern human hybrid (Harvati et al. 2007).

Cave Climatology

The average annual temperature of the area varies between 9 and 10 °C. The snow covers the land surface for up to 150 days beginning in October, and with the last snowfall normally occurring in the first half of April. The thickness of the snow layer can reach up to 2 m. The cave atmosphere mean annual temperature in the region is generally between 9 and 10.5 °C. Cave climate parameters were not monitored continuously in any of the cave, but spot measurements at various times of the year exist for all caves presented in this chapter. This information is available in Gheorghiu et al. (2009).

Cave Conservancy

All caves presented in this chapter are part of the Grădiştea Muncelului–Cioclovina Natural Park (GMC). Vale Stânii and Cioclovina Uscată caves are class A of protection, whereas Ponorici–Cioclovina cu Apă is B. To enter any of these caves, permission from the Romanian Speleological Heritage Commission and the GMC Natural Park Administration is required.

Acknowledgements We wish to thank all those who, over the years, have participated in the exploration, mapping, and scientific research of the caves within the 2063 karst basin. The Administration of the Grădiştea Muncelului–Cioclovina Natural Park is acknowledged for granting permission to collect samples for various studies. Part of this work was supported by CNCSIS grant 1696 to BPO.

References

Abafi Aigner L (1897) A ponoricsi barlangok Hunyadmegyében. Turisták Lapja IX:221–228

Balintoni I (1997) Geotectonica terenurilor metamorfice din România. Ed Carpatica, Cluj-Napoca, p 178

Bandrabur G, Bandrabur R (2010) Sebeş mountains. In: Orăşeanu I, Iurkiewicz A (eds) Karst hydrogeology of Romania. Ed Belvedere, Oradea, pp 77–88

Banerjee A, Ghiurcă V, Langer B, Wilhelm M (1999) Determination of the provenance of two archaeological amber beads from Romania by FTIR- and solid-state-carbon-13 NMR spectroscopy. Archäol Korrespondenzblatt 29:593–606

Bielz EA (1884) Beitrag zur Höhlenkunde Siebenbürgens. Jahrbuch des Siebenbürgischen Karpathen-Vereins IV:1–66

Boroneanţ V (2000) Arheologia peşterilor şi minelor din România. Ed cIMeC, Bucureşti, p 327

Bosch R, White WB (2004) Lithofacies and transport of clastic sediments in karstic aquifers. In: Sasowsky ID, Mylroie J (eds) Studies of cave sediments-physical and chemical records of paleoclimate. Kluwer Academic/Plenum Publishers, Dordrecht/New York, pp 1–21

Breban R (2002) Studiul fosfaţilor din Peştera Cioclovina Uscată. Unpublished BSc Thesis, Babeş-Bolyai Univ, Cluj-Napoca, p 92

Breban R, Şerban M, Viehmann I, Băicoană M (2003) Istoria exploatării de guano-fosfat şi a descoperirii omului fosil din Peştera Cioclovina Uscată. Hunedoara, Club Speleol Proteus, p 254

Cîntă-Pînzaru S, Onac BP (2009) Raman study of natural berlinite from a geological phosphate deposit. Vib Spectrosc 49:97–100

Comşa E (1966) Le dépôt en bronze de Cioclovina (Carpates Meridionales). Acta Archaeol Carpathica VIII:169–174

Constantinescu E, Marincea Ş, Crăciun C (1999) Crandallite in the phosphate association from Cioclovina cave (Şureanu Mts., Romania). In: Scientific works by Emil Constantinescu. Mineralogy in the system of Earth Sciences. Imperial College Press, London, pp 1–5

Decu V, Negrea Ş (1969) Aperçu zoogéographique sur la faune cavernicole terrestre de Roumanie. Acta Zool Cracoviensia 14:471–546

Dumitraş DG, Constantinescu E, Marincea Ş, Bourgier V (2005) New data on ardealite from the type locality (the "dry" Cioclovina Cave, Şureanu Mountains, Romania). In: Proceedings of Annual Scientific Session "Geo 2005" (Roşia Montană), pp 31–36

Dumitraş DG, Marincea Ş, Bilal E, Hatert F (2008) Apatite-(CaOH) in the fossil bat guano deposit from the "dry" Cioclovina Cave, Şureanu Mountains, Romania. Can Mineral 46:431–445

Dumitrescu M, Orghidan T, Tanasachi J (1955) Două descoperiri interesante în Peştera Cioclovina cu Apă. Bul Şt, Sec Şt Biol, Agronom, Geol Geogr VI I(2):359–368

Dumitrescu M, Orghidan N, Orghidan T, Puşcariu V, Tanasachi J, Georgescu M, Avram Ş (1967) Contribuţii la studiul peşterilor din regiunea Hunedoara. Lucr Inst Speol "Emil Racoviţă" VI:9–88

Emödi J (1978) Noi date privind depozitul de la Cioclovina. SCIVA 29 (4):481–495

Gheorghiu V, Murariu D, Borda D, Farcaş A, Chachula O (2009) Ecological rehabilitation and restoration of the hibernating colonies of chiroptera from Ciclovina Uscată Cave. In: Petculescu A, Murariu D (eds) The first ecological reconstruction of underground environment from Romania, Cioclovina Uscată Cave. Ed Universitară Bucureşti, pp 19–66

Gherman I (1932) Contribuţii la cunoaşterea regiunii carstice de la NE de Pui. Rev Muz Geol Univ Cluj V(1):119–133

Götzinger G (1919) Die Phosphathöhle von Csoklovina in Siebenbürgen. Mitt Geogr Ges 62:304–333

Gruia M (2003) Collembola from Romanian caves. Trav Mus Hist Nat "Grigore Antipa" 35:139–158

Halaváts G (1898) Az Ohába-Ponori Kretá-Terület. A magyar kir földt int Évijelentése, Budapest, pp 94–106

Harvati K, Gunz P, Grigorescu D (2007) Cioclovina (Romania): affinities of an early modern European. J Hum Evol 53:732–746

Häuselmann AD, Häuselmann P, Onac BP (2010) Speleogenesis and deposition of sediments in Cioclovina Uscată Cave, Şureanu Mountains, Romania. Environ Earth Sci 61:1561–1571

Horusitzky H (1918) Das Phosphorhältige Material der Höhle bei Csoklovina. Barlangkutatás 6(1–4):74–76

Horváth B (1916) Über die phosphorhaltigen Ablagerungen der Höhlen. Barlangkutatás 4(3–4):197–202

Jeannel R, Racovitza EG (1929) Énumération des grottes visitées, 1918–1927 (7e série). Arch Zool Exp Gén 68(2):293–608

Marincea Ş, Dumitraş D, Gibert R (2002) Tinsleyite in the "dry" Cioclovina Cave (Şureanu Mountains, Romania): the second occurrence. Eur J Mineral 14:157–164

Marincea Ş, Dumitraş D-G (2005) First reported sedimentary occurrence of berlinite ($AlPO_4$) in phosphate-bearing sediments from Cioclovina Cave, Romania–Comment. Am Mineral 90(7):1203–1208

Mitrofan H (1979) Observations sur le conditionnement tectonique du développement et de la répartition des creux du réseau Ponorici-Cioclovina cu Apă (Monts de Sebeş). Lucr Inst Speol "Emil Racoviţă" VIII:225–231

Mitrofan H (1988) Sistemul "Ponorici-Ciclovina cu Apă". Speleotelex 4(16):20–25

Nagy ZL, Postawa T (2011) Seasonal and geographical distribution of cave-dwelling bats in Romania: implications for conservation. Anim Conserv 14(1):74–86

Negrea Ş (1974) Explorarea sistemului hidrocarstic Ponorici–Cioclovina. Bul Club Speol "Emil Racoviţă" Bucureşti 3:25–35

Nitzu E, Giurginca A, Nae A, Popa I, Baba Ş, Meleg IN, Vlaicu M (2016) The catalogue of caves with endemic cavernicolous arthropod fauna of Romania. Trav Inst Spéol "Émile Racovitza" LV:3–62

Olariu A, Skog G, Hellborg R, Stenström K, Faarinen M, Persson P, Alexandrescu E (2005) Dating of two Paleolithic human fossil bones from Romania by accelerator mass spectrometry. In: Olariu A, Stenström K, Hellborg R (eds) Proc Int Conf Applic of high Precision Atomic and Nuclear Methods. Ed Acad Rom, Bucharest, pp 222–226

Onac BP, Effenberger HS (2007) Re-examination of berlinite ($AlPO_4$) from the Cioclovina Cave, Romania. Am Mineral 92:1998–2001

Onac BP, Effenberger HS (2009) Apatite-(CaOH) in the fossil bat guano deposit from the "Dry" Cioclovina Cave, Şureanu Mountains, Romania: Discussion. Can Mineral 47:207–208

Onac BP, White WB (2003) First reported sedimentary occurrence of berlinite ($AlPO_4$) in phosphate-bearing sediments from Cioclovina Cave, Romania. Am Mineral 88:1395–1397

Onac BP, Breban R, Kearns JB, Tămaş T (2002) Unusual minerals related to phosphate deposits in Cioclovina Cave, Şureanu Mts. (Romania). Theor Appl Karstol 15:27–34

Onac BP, Ettinger K, Kearns Balasz II (2005) A modern, guano-related occurrence of foggite, $CaAl(PO_4)(OH)_2{\cdot}H_2O$ and churchite-(Y), $YPO_4{\cdot}2H_2O$ in Cioclovina Cave, Romania. Mineral Petrol 85:291–302

Onac BP, Effenberger HS, Ettinger K, Cîntă-Pânzaru S (2006) Hydroxylellestadite from Cioclovina Cave (Romania): Microanalytical, structural, and vibrational spectroscopy data. Am Mineral 91:1927–1931

Onac BP, Effenberger HS, Breban RC (2007) High-temperature and "exotic" minerals from the Cioclovina Cave, Romania: a review. Studia UBB Geologia 52:3–10

Onac BP, Bernhardt H-J, Effenberger HS (2009) Authigenic burbankite in the Cioclovina Cave sediments (Romania). Eur J Mineral 21:507–514

Onac BP, Effenberger HS, Collins NC, Kearns JB, Breban RC (2011) Revisiting three minerals from Cioclovina Cave (Romania). Int J Speleol 40(2):99–108

Papiu VC, Mânzatu S, Iosof V, Udrescu C, Giuşcă R (1971) Alcătuirea chimico-mineralogică a formaţiunilor bauxitifere din bazinul Haţegului. CR Inst Geol 57(2):77–123

Petrea C (2009) Paleontological and taphonomic characters of *Ursus spelaeus fossilis* from Ciclovina Uscatã Cave. In: Petculescu A, Murariu D (eds) The first ecological reconstruction of underground environment from Romania, Cioclovina Uscatã Cave. Ed Univ Bucureşti, pp 67–88

Rădulescu C, Samson P (1959) Contribution à la connaissance de la dentition lactéale d'*Ursus spelaeus*. Eiszeitalter Gegenwart 10:205–216

Rainer F, Simionescu I (1942) Sur le premier crâne d'homme paléolithique trouvé en Roumanie. Anal Acad Rom, Mem Sec Şt (series 3) 17:489–503

Roska M (1912a) A csoklovinai Cholnoky-barlangban gyűjtött paleolit kőeszközök bemutatása. Dolgozatok az Erdélyi Nemzeti Múzeum 3 (2):201–237

Roska M (1912b) Traces de l'homme diluvien dans la caverne "Cholnoky" à Csoklovina. Dolgozatok az Erdélyi Nemzeti Múzeum 3(2):238–249

Roska M (1923) Săpăturile din peştera de la Cioclovina. Comisunea Monumentelor Istorice II:27–51

Roska M (1925) Recherches sur le Paléolithique en Transylvanie. Bul Soc Ştiinţe Cluj I I(4):183–192

Roşu L (2004) With C.S. Nicolăescu-Plopşor through the ages. Ed Mica Valahie, Bucureşti, p 139

Săndulescu M (1984) Geotectonica României. Ed Tehnică, Bucureşti

Savu H, Pavelescu M, Stancu I, Lupu D (1968) Geologic map of Romania, Orăştie Sheet (26), L-34-XXIV. Scale 1:200.000, Inst Geol Rom, Bucureşti

Schadler J (1932) Ardealit, ein neues Mineral $CaHPO_4{\cdot}CaSO_4 + 4H_2O$. Zentral-bl Miner A:40–41

Schréter Z (1917) A csoklovinai barlang (Die Höhle von Csoklovina). Barlangkutatás 5(3–4):175–191, 211–219

Simionescu I (1942) *Ursus spelaeus* din peştera Cioclovina (Hunedoara). Anal Acad Rom, Mem Sec Şt (series 3) 17(7):383–403

Soficaru A, Petrea C, Doboş A, Trinkaus E (2007) The human cranium from the Peştera Cioclovina Uscată, Romania: context, age, taphonomy, morphology and paleopathology, Curr Anthr 48:611–619

Stilla A (1972) Zona Pui şi bazinul Haţeg–unităţi structurale distincte în Carpaţii Meridionali. DS Inst Geol LVII I(5):173–176

Stilla A (1985) Géologie de la région de Haţeg-Cioclovina-Pui-Băniţa (Carpathes méridionales). An Inst Geol Geofiz 66:91–179

Téglás G (1878) A dél-keleti Kárpátok és ezek két új barlangja. Földt Közl VI:113–121

Tomuş RB (1999) Complexul caustic Ciclovina–Bazinul 2063. Hunedoara, Club Speleo Proteus, p 55

Tomuş RB (2011) Monografia carstului din Şureanu. Hunedoara, Club Speleo Proteus, p 244

Retezat Mountains: Jiul de Vest–Cernisoara Basins

Ioan Povară and Gheorghe M. L. Ponta

Abstract

The Jiul de Vest, Lăpuşnicul Mare, and Cernişoara rivers are in the western part of the Southern Carpathian Mountain Range. They form the natural boundary between the Retezat and Godeanu Mountains to the north and west, respectively, and the Vâlcan Mountains to the south. The Piule-Iorgovanu Mountains (Retezatul Mic Mountains) are between the Buta and Jiul de Vest Rivers and are an alpine karstic plateau (2000 m above sea level) of Jurassic limestones. On this plateau, an extensive network of dry valleys was developed and shaped by glaciers, which deeply eroded the carbonate (limestones) and noncarbonate rocks. Limestones are exposed from 710 m elevation at Cerna Spring (Izvorul Cernei) to 2080 m elevation in the Piule Mountain. Between the three rivers, hydrogeological connections have developed, supported by the continuity of the limestones and by structural-tectonic features. All three rivers contribute to the recharge of the most important karst spring in Romania, Cerna Spring.

Keywords

Karst spring • Dye studies • Carbonate rocks Hydrogeology

The text in this chapter is an update version of an earlier publication in Carbonates and Evaporites Volume 25 Number 2, pp. 117–126 by Povara and Ponta (2010). It is reproduced here by kind permission of the editors of the Carbonates and Evaporites.

I. Povară
"Emil Racoviţă" Institute of Speleology of the Romanian Academy, Calea 13 Septembrie 13, 050711 Bucharest 5, Romania
e-mail: ipov.iser@gmail.com

G. M. L. Ponta (✉)
Geological Survey of Alabama, 420 Hackberry Lane, Tuscaloosa, AL 35401, USA
e-mail: gponta@yahoo.com

Introduction

The topography is dominated by a karstic plateau located at 2000 m elevation, oriented northeast–southwest, between Piule, Stănuletele Mare, and Piatra Iorgovanului Mountains (Fig. 1). The Borăscu (Eocene) peneplain, present across the Carpathian Mountains, was identified at the highest elevation of this plateau (De Martonne 1906), with extensions toward Jiul de Vest (2–4 km long), and Buta and Lăpuşnicul Mare Rivers. The last two stages of the Alpine Ice Age (Riss and Würm) have shaped a topography consisting of glacial cirques and valleys at the edge of the platform (Soarbele, Iara, and Scorota Rivers). The Cernişoara Basin is dominated by limestone ridges, crossed by deep gorges created by tributaries of the Cernişoara River. The climate is continental, with ranges typical for high mountain areas. Between November 1981 and November 1982, the total rainfall was 1356 mm, slightly over the multi-annual average (1300 mm) for this area (Bulgăr et al. 1984). The snow lasts for more than 200 days above 1800 m elevation, and the maximum value of the water reserve in the snow layer is 300 mm. This was recorded in March 1981 and is characteristic for the 1300–1800 m elevation range, where over 65% of the Cerna Spring recharge area is located. The maximum snow melting takes place in April and May, while in early June, the snow recharge reaches zero. The maximum daily rainfall in the summer is 62.8 mm, while evapotranspiration is about 480 mm/year.

The hydrographic network is divergent, with the orientation of the channels of the most important rivers (Jiul de Vest and Cernişoara) influenced by the main tectonic structures in the area (the Cerna-Jiu Fault and the Getic Nappe) or by lithological contacts (Buta and Lăpuşnicul Mare). The surface runoff of Jiul de Vest and its northwest-side tributaries is temporary. During the dry season, more than 10 km of the Jiul de Vest is dry.

G. M. L. Ponta and B. P. Onac (eds.), *Cave and Karst Systems of Romania*,
Cave and Karst Systems of the World, https://doi.org/10.1007/978-3-319-90747-5_16

Fig. 1 Location of Jiul de vest–Cernisoara karst area within Romania

A Brief History

The geological investigation of the region began at the end of the nineteenth century. The first geological sketch was drawn by Inkey (1884), which was later revised by Mrazec (1897). Murgoci (1906, 1910) is the first geologist to recognize the crystalline structures in the Ţarcu-Godeanu Mountains as a large outlier belonging to the Getic Nappe; later, Codarcea (1940) completed a significant study of the geology and tectonic structure of the region. Additional contributions to the geology of the area were made by Gherasi (1937), Pop (1963), Codarcea and Năstăseanu (1964), Codarcea and Răileanu (1968), Năstăseanu (1967, 1980), Bercia et al. (1987), Conovici (1999), etc.

The Godeanu Mountains were described by Niculescu (1965) in a comprehensive geomorphological study. A general presentation of the Jiul de Vest karst landforms (the limestone area of the Retezat Mountains) was presented by Niculescu (1960) and Bădescu (1991), while a study of the most important caves, horizontal and vertical, of the Jiul de Vest Basin was completed by Ponta et al. (1984a, b), Ponta (1998), and Ponta and Terteleac (2006).

In 1967, T. Orghidan, M. Dumitrescu, and I. Vintilescu presented the theory that the recharge area of the Cerna Spring includes the upper part of the Lăpuşnicul Mare Basin, but this idea failed to be proved (Vintilescu 1972).[1] The first hydrogeological study was performed by Pascu (1968, unpublished). He suggested an underground connection between the sinking stream cave in the Turcineasa valley and Cerna Spring and brings forward the hypothesis of a large water transfer from Jiul de Vest River to Cerna Spring. Later studies (Povară 1976, 1980; Bulgăr and Munteanu 1982[2]; Bulgăr et al. 1984; Ponta et al. 1984a, b) pointed out that the

[1]In September 12, 1967, 10 kg of fluorescein was injected in Lăpuşnicul Mare, upstream of Lunca Berhinei. The dye was not detected at Cerna Spring.

[2]Bulgăr and Munteanu (1982) Evaluarea potentialului hidrologic al zonelor de carst din bazinele hidrografice Jiul de Vest, Cerna Superioara şi Dâmboviţa (unpublished).

carbonate succession from the Cerna-Jiu Syncline is part of the Cerna Spring recharge area.

The Geology and Hydrogeology of the Area

Geology

The geology of Jiul de Vest–Cernişoara area is comprised of Danubian Autochthonous units and of metamorphic formations of the Getic Nappe (Fig. 2). *The Godeanu Outlier represents The Getic Domain*, as it was defined by Murgoci (1906), with further details and information provided by Streckeisen (1934), Bercia et al. (1987), and Conovici (1999). Between Lăpuşnicul Mare and Cerna, this domain is represented by two structural units: the Godeanu Unit in the upper part (part of the Getic Nappe) and the Borăscu Unit (digitations) in the lower part.

- The Godeanu Unit consists of the Precambrian metamorphic rocks of the Sebeş-Lotru series in which lenses of crystalline limestones and skarns occur.
- The Borăscu Unit, identified on the south side of Lăpuşnicul Mare River, under the Godeanu Outlier, is a crystalline formation similar to the Getic Nappe, formed by a series of small nappes oriented east and southeast. These small nappes include Permian and Mesozoic rocks, calcareous sandstones and marly limestones, both with low hydraulic properties. The crystalline formations are overlain by Permian deposits, 20–30 m thick formed by conglomerates, sandstones, and siltstones.

The Danubian domain. The basement of the Cerna-Jiul de Vest area consists of crystalline schists of the Drăgşanu and Lainici–Păiuş series and granite massifs. The Vârful lui Stan–Curmătura Olteţului tectonic line (Berza et al. 1984) is the boundary between the two crystalline units. In the northwest portion of the Vâlcan Mountains, the lower part of the Danubian Domain consists of the remnants of an ancient sedimentary cover, affected by a metamorphic stage (probably Hercynian). Between the Oslea Mountain and Coada Oslei, this formation is overlain by the crystalline rocks of Drăgşanu and Lainici-Păiuş series and consists of metamorphosed sedimentary rocks, several hundreds of meters thick (Pop 1973). A sequence of 200–300 m recrystallized (Oslea limestones), white or gray limestones is underlain by layers of slates and sandstone-like slates.

In the upper part of Scorota Valley, between Piule, Piatra Iorgovanului Mountains, and the Lăpuşnicul Mare River, underlying the limestones, a succession of sedimentary rocks, 10–20 m thick (Lower Jurassic) consisting of sandstones and red and purple microconglomerates, was identified.

In the north, the metamorphic rocks and granites are overlain by Jurassic deposits consisting of Liassic sandstones and Middle Jurassic–Aptian limestones and dolomites (the Cerna–Jiu sedimentary zone). These deposits are exposed in a synclinal structure (Fig. 3). A fault transects the southern slope, which is the contact between limestones and crystalline rocks.

The Mesozoic sedimentary cover of the Jiul de Vest–Cerna area may be divided into two sectors, lithostratigraphically and structurally differentiated:

1. The area between the Lăpuşnicul Mare and Jiul de Vest Rivers (Piatra Iorgovanului, Scăriţa, Piule, and Pleşa Mountains), the sedimentary series outcrops in a northeast–southwest-oriented syncline (10.5 km long and 3–5 km wide), enlarged to the west, where it is covered by the crystalline schists of the Getic Nappe. The Liassic deposits (quartzitic conglomerates, arkoses, sandstones, and slates) unconformably overlay the Permian conglomerates, followed by calcareous sandstones (Dogger) and blackish, fractured limestones with chert (Malm) overlain by a limestone reef structure (Aptian), 500–600 m thick. The sedimentary cycle ends with Upper Cretaceous wildflysch, known as the Nadanova Formation.
2. In the area south of Obârşia Cernişoarei, the sedimentary sequence is incomplete, and the Liassic sandstones and the calcareous sandstones (Dogger) are missing.

The most frequent sedimentary cover consists of quaternary glacial and fluvio-glacial deposits.

The structure of this mountain is the result of multistage tectonics, affecting the crystalline-granite formations of the Precambrian basement. Transgressively and unconformably overlaying these are 1500-m-thick Mesozoic deposits, forming an asymmetric, northeast–southwest-oriented syncline. The syncline plunges to the southwest, while its thickness decreases. The Jurassic limestones (calcareous sandstone) and the Cretaceous deposits (arenitic, algal, and bioclastic limestones) are more than 1000 m thick and overlain by the Upper Cretaceous detrital sediments (clays, marls, and sandstones with Senonian conglomerates) (Pop 1963, 1973).

In the western sector of the area, the Getic's Nappe crystalline rocks overlie the Mesozoic sedimentary cover, while the limestones crop out only on a small area, gradually narrowing and decreasing in thickness toward Cerna Spring. The tectonic evolution of the Godeanu Outlier and of the Borăscu Digitation involved the "squeezing" of limestones and wildflysch against the "wall" of granites and Danubian crystalline elements in the eastern portion of the nappe's leading edge, along the Cerna Fault (Miocene: intra-Burdigalian). The wall acts as an impermeable barrier, directing the underground flow toward the south.

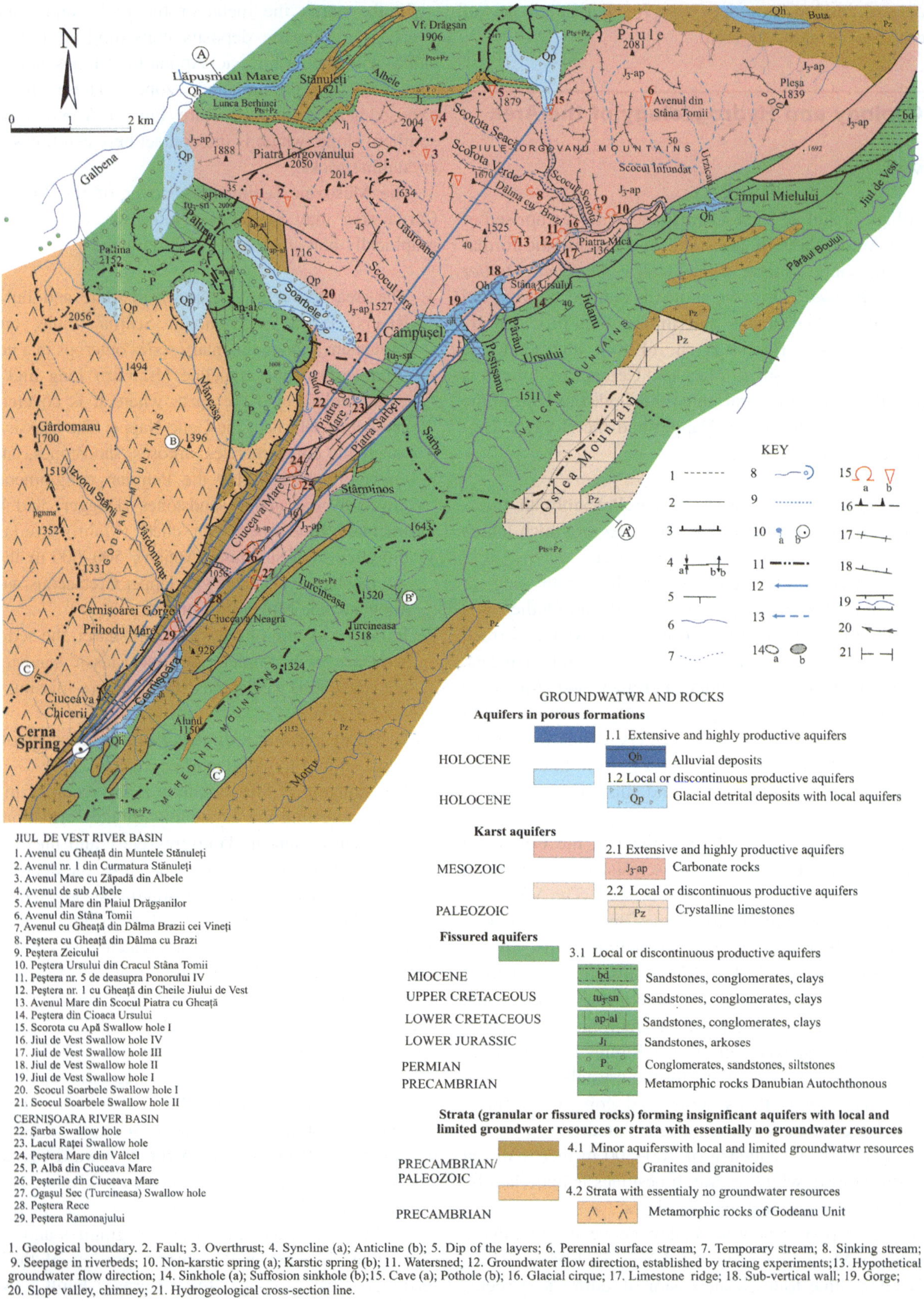

Fig. 2 Hydrogeological map of the upper part of Jiul de Vest and Cernişoara Rivers' Basins (after geological map Baia de Aramă Sheet (1:200,000) and Oslea Sheet 1:50,000, unpublished)

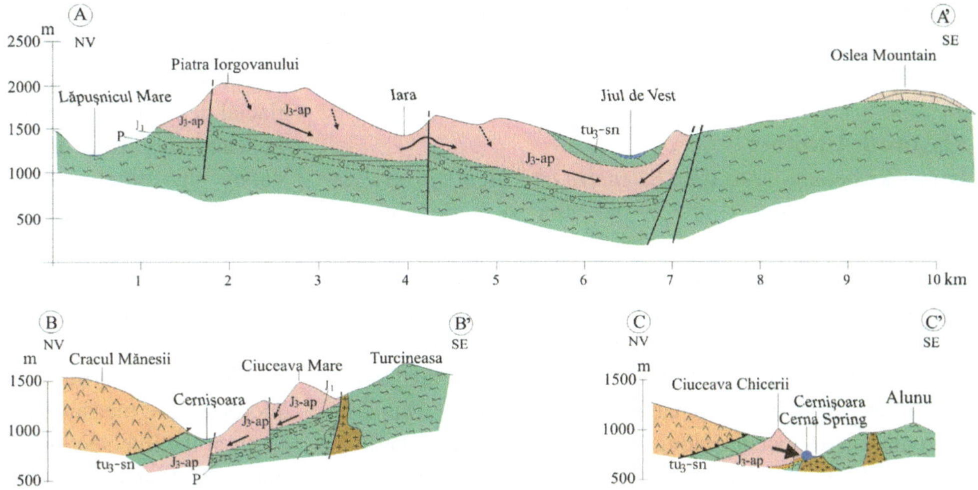

Fig. 3 Cross sections in the Jiul de Vest–Cernişoara Basins (for legend see Fig. 2)

Hydrogeology: Occurrence and Availability of Water

The availability of groundwater in the Jiul de Vest–Cernişoara varies widely, largely due to the geologic complexity of the area. The water-bearing characteristics of the aquifers are discussed based on the physical characteristics of the rocks or sediments in the area (porosity and permeability, granular or fractured) and the occurrence of springs and sinking streams. Dye studies, which were used to define the hydrogeologic watershed of each basin, are also discussed.

The hydrogeologic units on Fig. 2 are represented according to the International Standard Legend for General and Special Hydrogeological Maps (Struckmeier and Margat 1995), with the groundwater and rocks category subdivided into three main subcategories, with karst included in "Fissured aquifers, including karst aquifers," as shown below:

Aquifers in which flow is mainly intergranular

1. Extensive and highly productive aquifers
2. Local or discontinuous productive aquifers or extensive but only moderately productive aquifers

Fissured aquifers, including karst aquifers

1. Extensive and highly productive aquifers
2. Local or discontinuous productive aquifers

Strata (granular or fissured rocks) forming insignificant aquifers with local and limited groundwater resources or strata with essentially no groundwater resources

1. Minor aquifers with local and limited groundwater resources
2. Strata with essentially no groundwater resources
3. Where there is an extensive aquifer immediately underlying a thin cover.

Because of the large variation in the development of karst features and the volume of data related to ground/surface water in karst areas, data are presented on the map as a new subcategory, "Karst aquifers" using a range of pink colors to differentiate aquifers in karstic terrain, versus the traditional range of green colors, used to represent the fissured non-karstic rocks (Ponta 2003).

Aquifers in Porous Formations

Glacial detrital deposits with local aquifers and the alluvial deposits, with *extensive and highly productive aquifers*, are located along the Soarbele, Scorota, Jiul de Vest, and Cernişoara Rivers. Water is concentrated in open pore spaces between the grains, pebbles, and boulders. Some small springs and seeps are identified, and their flow depends on the precipitation and snow melting. The thickness of these deposits is between 2 and 8 m.

Karst Aquifers

The water in the crystalline limestones of the Oslea Mountains is concentrated in small fractures, and along bedding planes, with *locally productive karst aquifers*. Small springs (<1 L/s) can be present at the contact with noncalcareous rocks. The thickness of the crystalline limestone ranges between 200 and 300 m.

Extensive and highly productive karst aquifers occur in solution cavities, joints, fractures, and along the bedding planes of the Mesozoic carbonate rocks. This aquifer has a more extensive network of interconnected fractures than any other aquifer in the region. These interconnected fractures serve as a conduit, leading the waters from the Jiul de Vest

Basin toward Cerna Spring. Tracer studies demonstrate that the upper part of the Jiul de Vest River is recharging Cerna Spring.

Fissured Aquifers

No rocks belonging to the "*Extensive and highly productive aquifers*" subcategory are in the area. A wide range of sandstones, conglomerates, and siltstones from Lower Jurassic to Miocene representing the "*Local or Discontinuous productive aquifers*" subcategory are present. These units generally yield small quantities of water to springs, which at the interface with limestones sink underground.

Groundwater in the metamorphic rocks of the Danubian Autochthonous occurs along fractures, joints, and bedding planes. These metamorphic occasionally are capable of supplying moderate quantities of water to springs. These metamorphic rocks form an impervious layer under the limestone deposits that are up to 600 m thick.

Strata (granular or fissured rocks) forming insignificant aquifers with local and limited groundwater resources or strata with essentially no groundwater resources

The Precambrian deposits of the Godeanu Unit (Godeanu Outlier—part of the Getic Nappe) essentially do not represent a groundwater resource. Minor aquifers with local and limited groundwater resources are found in the Precambrian/Paleozoic granites and granitoides and Permian conglomerates, sandstones, and siltstones. The water is concentrated in fractures/joints.

Karst Landforms

The area underlain by limestones is extensive (Table 1). Rainfall is abundant, and the karst vertical potential exceeds 1000 m. Exokarst landforms are weakly represented. Various types of karrens, small-to medium-sized sinkholes, canyons (dry or with temporary flow), and gorges were identified.

The limestones of the Jiul de Vest Basin (23.8 km^2) form a karstic plateau oriented east–west (Piule, 2080 m—Piatra Iorgovanului, 2014 m), which continues with a narrow ridge to the southwest, toward Cerna Spring. The tectonically controlled karrens are present extensively across the site, with most identified in the lower part of the Soarbele and Iara valleys and on the southern slope of the Albele Mountain (Niculescu 1960). Small-sized sinkholes (10–15 m in diameter and 5–8 m deep) are found on Pleşa, Albele, and Stănuleţi Mountains. The most interesting are the subsidence sinkholes, formed on the west side of the Soarbele Valley, and the suffusion sinkholes formed on the glacial deposits of the same valley (Bădescu 1991). In two of the sinkholes from the upper glacial level of the Soarbele Valley, the presence of impermeable clay deposits favored the appearance of perennial lakes.

The endokarst of the Jiul de Vest Basin is represented by 328 caves. Only 6.4% are longer than 100 m, while 2.9% are vertical caves. The most karstified interval is between elevation 1100 and 1200 m, where 61% of the caves, horizontals and verticals, have developed Fig. 4). The microtectonic measurements revealed that most of the caves occur at the intersection of the Cerna–Jiu Fault with adjacent fractures (Ponta et al. 1984a, b) and are in the vicinity of sinking streams of the Jiu de Vest riverbed.

Five vertical caves, 55–92 m deep, between elevations 1750 and 2000 m, host perennial ice accumulations: Avenul Mare cu Zăpadă from Albele–Găuroane, Avenul cu Zăpadă from Scorota Seacă, Avenul cu Gheaţă from Dâlma Brazii cei Vineţi, Avenul cu Gheaţă from Piule, and Avenul cu Gheaţă from Stănuleţi. The perennial ice from these caves recharges the groundwater during periods with low precipitations. The water resulted from precipitations and snow melting infiltrates along vertical fractures and occasionally generates vertical caves. One of the most important vertical caves (shaft) is Avenul din Stâna Tomii with 114 m drop (Fig. 5).

Most of the northwest-side tributaries of the Jiul de Vest River are located on limestones (Scocul Iara, Găuroane, Scorota Verde, Scorota Seacă, and Urzicari) and have occasional surface runoff, following heavy rainfall or snowmelt. Only two tributaries (Soarbele and Scorota cu Apă) have an upper sector developed on nonkarstifiable rocks, where the flow is perennial. Once the streams enter the limestones, the waters disappear underground through sinking streams, or diffuse infiltration of the riverbed. The snow melts (April–June) and the rain storms (June–July) may provide a continuous flow to the confluence with Jiul de Vest River.

The tributaries on the southeast side of Jiul de Vest (Pârâul Rece, Pârâul Jidanului, Pârâul Ursului, and Peştişanu Rivers) have 99% of their basins developed on nonkarstifiable rocks of the northern slope of the Vâlcan Mountains. Before reaching Jiul de Vest River, they cross a limestone ridge, through short gorges, which represents the eastern flank of the Cernişoara syncline. Downstream, Jiul de Vest River sinks underground, and the riverbed is dry from the

Table 1 Area of the Cerna Spring hydrogeological basin

	Km^2	Cernişoara	Jiul de Vest	Lăpuşnicul Mare
Nonkarstifiable rocks	47.85	27.45	19.46	0.935
Limestones	32.77	4.30	23.820	4.650
Total	80.62	31.75	43.285	5.585

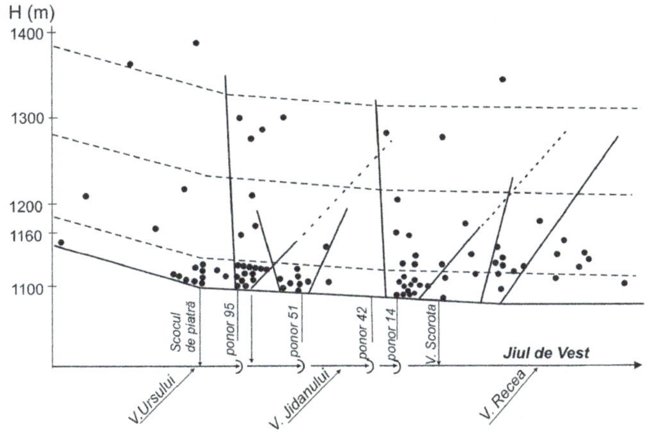

Fig. 4 Distribution of the caves in the Jiul de Vest Basin, according to elevation (Ponta et al. 1984b)

later part of July until October–November. The Jiul de Vest River sinks along a parallel fault upstream of Câmpul Mielului and recharges the Cerna Spring through underground pathways (Povară 1976).

The absence of the antithetic faults adjacent to the sinking streams along with Jiul de Vest River supports the hypothesis that the underground drainage toward Cerna Spring is recent, most probably in the Riss-Würm Interglacial Stage.

The limestones of the Cernişoara Basin outcrop on a small area (4.30 km^2) and are part of the eastern flank of the Cernişoara syncline, underlying the nonkarstifiable formations of the Getic Nappe. South of the saddle (water divide) with Jiul de Vest, a plateau oriented northeast–southwest, is at 1300–1400 m elevation and is crossed by several valleys (Sturu, Şarba, and Lacul Răţii). Along them, perennial sinking streams and small antithetic faults have been developed.

The increased fragmentation of the plateau by the Cernişoara tributaries led to the formation of an elongated limestone ridge oriented NE–SW, known as "*ciuceve.*" The ridge is about 1500 m wide in the upper part of the Cernişoara River, gradually narrowing and decreasing in thickness toward Cerna Spring. In the area of "ciuceve," the karren fields are the predominant karst feature.

The endokarst is represented by 294 caves, 79.9% are less than 50 m long (Fig. 6), and 58% are in the 800–1000 m elevation range (Dancău et al. 1968; Avram et al. 1964, 1966).

Underground Flow

The main groundwater flow within the karst aquifer is influenced by three structural-tectonic and physiographic features:

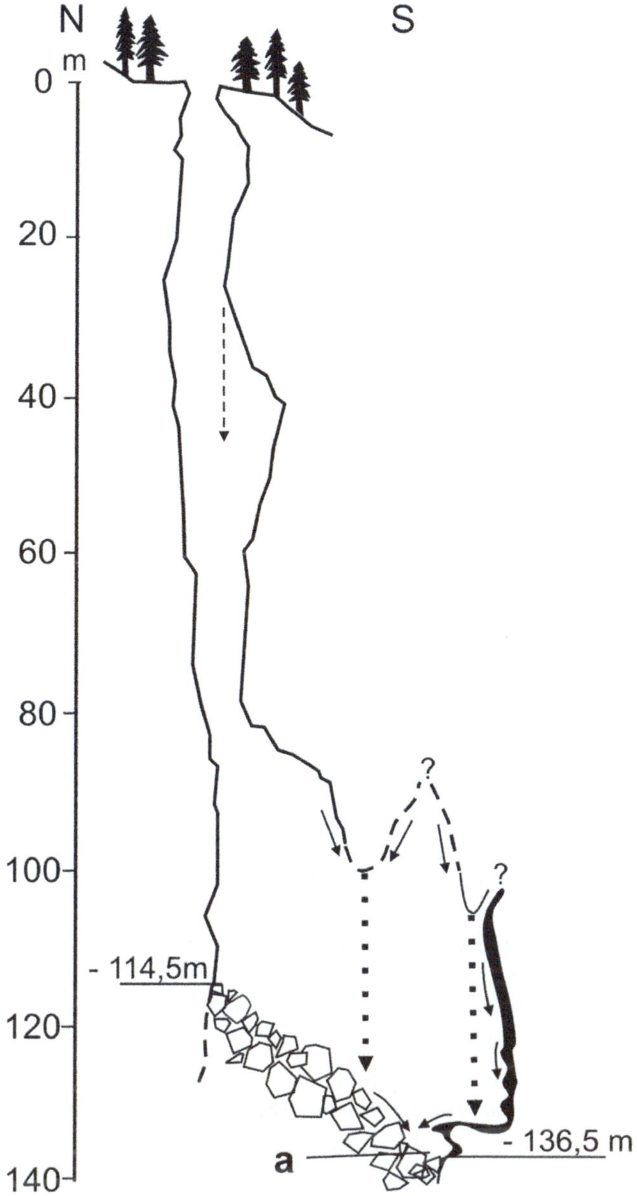

Fig. 5 Stâna Tomii Pit (Avenul din Stâna Tomii) (longest drop in Romania—114 m; Survey, Focul Viu Bucuresti Grotto) (Bleahu et al. 1976)

- The occurrence of the limestone in a wide, asymmetric synclinal structure, with the eastern flank cut by the Cerna–Jiu Fault, which is parallel to the leading edge of the Getic Nappe. East, the limestone deposits are in contact with impermeable crystalline formations.
- The gradual deeping of the syncline axis from the northeast to the southwest, under the nonkarstifiable formations of the Godeanu Outlier.
- The general slope of the karst surface from 2000 m in Jiul de Vest to only 700 m in the Cernişoara Basin.

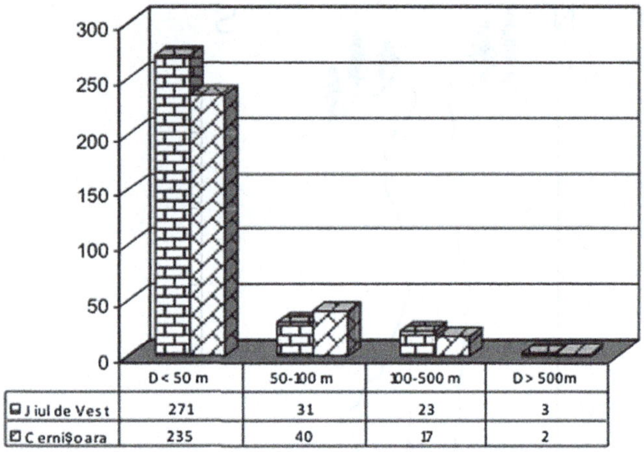

Fig. 6 Distribution of the karst cavities according to length (chart drawn after data from Goran 1982)

Consequently, the underground flow is driven from the northeast to the southwest, parallel to the Cerna–Jiu Fault. Groundwater in the entire structure is discharged through Cerna Spring.

Several dye studies using fluorescein and In-EDTA proved the connection between the sinking streams of the Jiul de Vest River and Cerna Spring, which confirmed that Cerna Spring's recharge area includes the entire karst area of Retezatul Mic Mountains (Pascu 1968, unpublished; Povară 1976, 1980; Bulgăr et al. 1984; Ponta et al. 1984a). The most important features are presented in Table 2.

The average theoretical flow velocity ranges between 32.5 and 55.5 m/h (0.886–1.33 km/day) and is typical for conduit flow. In 1982, the Geological and Geophysical Prospecting Company with the Institute of Nuclear Physics (IFIN), Bucureşti, completed a tracer study in the Scorota sinking stream on the plateau. The stream is situated at the contact between impermeable crystalline rocks and the Jurassic limestone at an elevation of 1390 and 13,350 m from Cerna Spring, which is at an elevation of 700 m above sea level (Ponta 1998).

On August 9, 1982, 100 g of Indium-EDTA was used in the Scorota sinking stream, which has a flow of 25 L/s. The travel time of the tracer to Cerna Spring, 13.3 km away and 700 m lower in elevation, was 28 days; the recorded velocity was 55.6 m/h. As much as 33% of the tracer was recovered (Ponta et al. 1984a). The low level of tracer recovery points to high dilution or lateral dispersion, from the main drain to the adjacent structures.

Cerna Spring Karst System

Cerna Spring is on a small, northwest-side tributary of the Cerna River, Ogaşul Obârşiei, and 100 m upstream of the confluence, at 710 m elevation. Its hydrogeological basin (watershed) is 80.6 km^2, of which 40.6% are limestone outcrops. Part of the flow is through karst conduits 20–35 cm in diameter. At the spring, air bubbles are released as result of the decompression process accompanying the accessional *Vauclusian* Spring.

The only systematic hydrometric records were performed by National Institute of Hydrology and Meteorology from Bucharest, Romania, during November 1, 1981, to December 31, 1982. Occasionally, Cerna Spring flow-rate data were recorded, the largest being of 10.5 m^3/s (May 1980). The hydrograph of the spring flow rate and rainfall recorded at Câmpul lui Neag highlights a connection between rainfall and flow rate, as well as a delay of the flow rate response to the rain event (Fig. 7).

The parameters based on the recession hydrograph analysis performed for the period August 25–September 30, 1982 (Fig. 8), following the method issued by Mangin (1975), point out certain drainage features confirmed by the field data:

- a poorly developed phreatic karst (low k);
- a quick discharge (high a = drainage coefficient);
- the significance of the surface runoff for the karst system recharge (high i);

Table 2 Dye studies completed in Jiul de Vest–Cernişoara basins

No.	Inlet	H (m)	Outlet	H (m)	L (km)	⊗H (m)	Tracer	T (hours)	V (km/d)	Year	References
1	Scorota Sinking Stream	1410	Cerna Spring	710	13.3	700	In	240	1.33	1982	Ponta et al. (1984a)
2	Jidanului (Caprei) Creek	1140			12.1	475	F	282	1029	1975	Povară (1980)
3	Ursului Creek	1180			10.95	470	F	228	1.15	1974	Povară (1980)
4	Ştirbu (Peştişanu) Creek	1170			9.9	460	F	268	0.886	1974	Povară (1976)

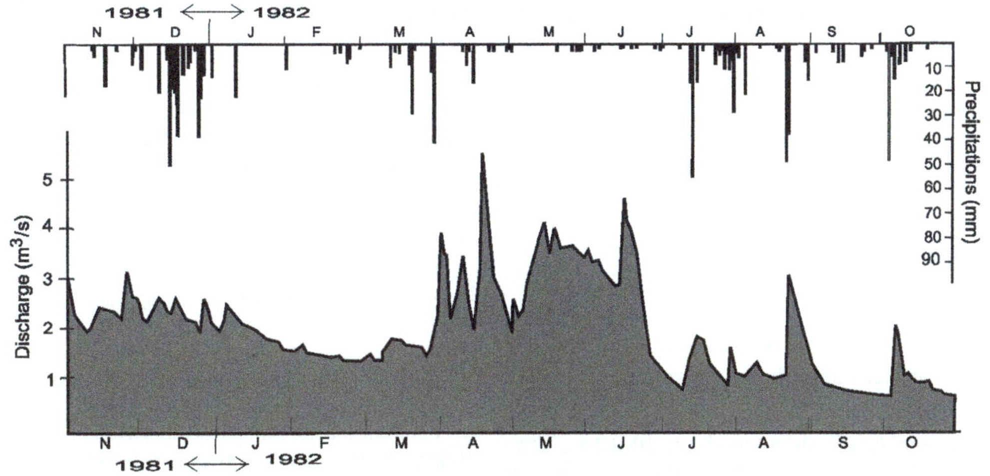

Fig. 7 Hydrograph of the Cerna Spring flow rate and rainfall diagram recorded at Câmpul lui Neag (after Bulgăr et al. 1984)

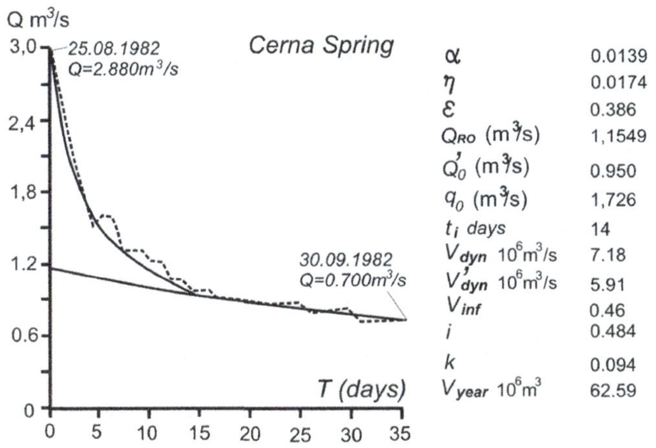

Fig. 8 Recession curve and parameters for the Cerna Spring flow rate during August 25–September 30, 1982

- a rather short flow-rate decrease time (t_i= 14 days), related to the time span of the depletion of infiltration.

Using the systemic analysis (Mangin 1975) of the Cerna Spring flow rate and of the rainfall recorded at Câmpul lui Neag, the following results were obtained (Table 3):

- The simple correlogram and the spectrum of the variance density for the rainfall denote a low structured character of it for intervals of 14, 26, and 40 days.
- The simple correlogram of the flow rate suggests a low decrease. The value r_k = 2 is reached after 43 days (MT), while the correlogram reaches 0 after 55 days. In this case, the memory effect of the basins developed on the nonkarstifiable formations on the southeast side of the Jiul de Vest River is also involved. These areas are mostly covered by superficial deposits and forests or pastures.
- The regulation time (RT) for this system is 36 days, and it highlights the duration of the rainfall influence. In this case, we can also mention the influence of the nonkarstifiable rocks.
- For Cerna Spring, the cross-correlogram for the year 1982 has a low inter-correlation coefficient (r_k), indicating a weak connection between rainfall and flow rate. This reaches a maximum value after 2 days (0.193) and becomes 0 after 18 days. The correlogram shape shows a significant supply from riverbeds (diffuse infiltration and swallow holes). The system has a low inertia, comprising a poorly developed phreatic subsystem. The low inter-correlation coefficient may be explained by the fact that the rainfall measured at Câmpul lui Neag is not significantly influencing Cerna Spring Basin.

Using the karst aquifer classification based on the results of the recession hydrograph analysis (Mangin 1975), the authors conclude that the Cerna Spring system is recharged by an aquifer that is more karstified upstream than

Table 3 Spring features for the 1981–1982 hydrologic cycle

	Q_{max} (l/s)	Q_{min} (l/s)	Q_{max}/Q_{min} (m^3/s)`	Q_{med} (l/s)	(	Dyn. vol. ($10^6 m^3$)	ME	TF	RT	M
Izvorul Cernei	5.700	0.600	9.5	1.985	0.0139	7.18	43	0.084	36	C

ME memory effect, *TF* truncation frequency, *RT* regulation time, (discharge coefficient

downstream, with a delay in the supply process (type III: $k < 0.5$; $0.25 < i < 0.5$).

The systemic analysis was based on a single hydrologic cycle, and the rainfall recorded at Câmpul lui Neag may not be representative for the Cerna Spring hydrogeological basin.

Conclusions

The Retezatul Mic Mountains are between the Buta and Jiul de Vest Rivers and are an alpine karstic plateau

- (2000 m above sea level) developed on Jurassic limestones. An extensive network of dry valleys was developed and shaped by glaciers, which deeply eroded the carbonate (limestones) and noncarbonate rocks on the plateau. The mountain area of the upper part of Jiul de Vest, Cernişoara, and Lăpuşnicul Mare Rivers is recharging the Cerna Spring. Cerna Spring is the most important karst spring of Romania, with a maximum recorded flow rate exceeding 10 m^3/s. The Cerna Spring recharge area is 80.6 km^2, of which 32.77 km^2 is covered by Mesozoic (J_3-ap) limestones. These carbonates crop out in a wide, asymmetrical syncline, oriented northeast–southwest. The syncline axis plunges toward the southwest. The groundwater flow is toward the southwest and is controlled by the syncline. South of Scocul Soarbele, the limestones are overlain by the Borăscu Nappe (lower part of the Getic Nappe).
- The limestones have been less or moderately influenced by karst processes. The exokarst is represented by karren fields, dissolution, and suffusion sinkholes. Both categories of sinkholes are relatively small (diameter 4–15 m, depth 2–10 m). There are many caves, horizontal and vertical (328 on Jiul de Vest and 294 on Cernişoara). Most are short, and more than 80% are less than 50 m long.
- Surface water flow in all northwest-side tributaries of Jiul de Vest is ephemeral. Therefore, an important role in the recharge of the Cerna Spring is played by the seepage losses from the hydrographic network. The groundwater flow velocities determined with chemical and activable tracers range between 0.78 and 1.33 km/day. The maximum runoff is recorded between May and June when spring rainfall is combined with snow melting. The minimum occurs between September and October.
- The Cerna Spring karst system is better developed upstream (Jiul de Vest Basin), while the water-filled karst is less represented. The system dynamic volume (V_{dyn}) is 7.18×10^6 m^3/s, while the annual runoff volume is 62.59×10^6 m^3/s.

References

Avram Ş, Dancău D, Şerban E (1964) Cercetări speologice în Bazinul Cernei. Nota I. Lucr Inst Speol "Emil Racoviţă" III:131–200

Avram Ş, Dancău D, Şerban E (1966) Cercetări speologice în Bazinul Cernei. Nota II. Lucr Inst Speol "Emil Racoviţă" V:11–20

Bădescu D (1991) Le karst alpin des Monts Retezat (Alpes de Transilvanie Roumanie). Karstologia 17:43–53

Bercia I, Bercia E, Năstăseanu S, Berza T, Iancu V, Stănoiu I, Hârtopanu I (1987) Geological map of Romania, Obârşia Cloşani Sheet. Scale 1: 50,000. Inst Geol Geofiz, Bucureşti

Berza T, Pop G, Seghedi A, Iancu V, Hârtopanu I, Hârtopanu P, Năstăseanu S, Moisescu V (1984) Geological map of Romania, Muntele Oslea Sheet. Scale 1:50,000. Inst Geol, Bucharest

Bleahu M, Decu V, Negrea Ş, Pleşa C, Povară I, Viehmann I (1976) Peşteri din România. Ed Ştiinţifică şi Enciclopedică, Bucureşti, p 415

Bulgăr A, Diaconu V, Oancea V (1984) Modern methods in karst hydrological research. Application to some principal karst systems from the Southern Carpathians. Theor Appl Karstol 1:215–225

Codarcea A (1940) Vues nouvelles sur la tectonique du Banat Meridional et du Plateau de Mehedinti. An Inst Geol Român XX 1936:74

Codarcea A, Năstăseanu S (1964) Contribuţii la cunoaşterea stratigrafiei depozitelor calcaroase din bazinul Văii Cerna şi de la Cazane (Dunărea). St cerc geol geofiz geogr, Seria geol 9(2):241–250

Codarcea A, Răileanu G (1968) Geological map of Romania, Baia de Aramă Sheet. Scale 1:200,000, Inst Geol, Bucureşti

Conovici M (1999) The geological study of the Getic-Supragetic Sector from the Southwestern part of the Southern Carpathians. Unpubl PhD Thesis, Babeş-Bolyai University, Cluj-Napoca

Dancău D, Danielopol D, J Ilinca, Şerban E (1968) Cercetări speologice în Bazinul Cernei. Nota III. Lucr Inst Speol "Emil Racoviţă" VII:45–54

De Martonne E (1906) Recherches sur l'Évolution morphologique des Alpes de Transylvanie (Karpates méridionales). Paris

Gherasi N (1937) Etude géologique et pétrografique dans les Monts Godeanu et Tarcu. Ann Inst Geol Rom XVIII, Bucharest

Goran C (1982) Catalogul sistematic al peşterilor din România. Ed CNEFS, Bucureşti, p 496

Inkey B (1884) Die Transylvanischen Alpen vom Rotenturmpasse bis zum Eisernen Tor. Math u. naturwiss, Berichte aus Ungarn IX

Mangin A (1975) Contribution a l'étude hydrodynamique des aquifere`s karstiques. Ann Speleol 30(1):21–124

Mrazec L (1897) Essai d'une classification des roches cristallines de la zone centrale des Carpathes roumaines. Bul Soc St Roum, Bucharest

Murgoci GM (1906) Sur l'age de la grande nappe de charriage des Carpathes Méridionales. Bul Soc St XVI:1–2

Murgoci (1910) The geological synthesis of the South Carpthians. In: CR Congress of Geology Institution, Stockholm, pp 871–880

Năstăseanu S (1967) Cretacicul superior din Valea Cernei si date noi privind tectonica Munţilor Cernei (Banat). D S Com Geol LII I (1):401–411

Năstăseanu S (1980) Géologie de Monts Cerna. Ann Inst Géol Géophys 54:153–280

Niculescu G (1960) Câteva observaţii asupra reliefului carstic din Munţii Piule - Iorgovanu. Natura XI I(4):82–86

Niculescu G (1965) Munţii Godeanu-Studiu geomorfologic. Ed Acad RPR, Bucureşti

Ponta G (1998) Karst of Romania. In: Yuan D, Liu Z (eds) Global karst correlation IGCP 299. Science Press, Beijing, pp 193–207

Ponta G (2003) Proposed modifications of the International standard legend for large scale hydrogeological maps in carbonate rocks. In: International conference on karst hydrogeology and ecosystems. Western Kentucky University, Bowling Green, Kentucky

Ponta G, Terteleac N (2006) Geology and dynamics of underground waters in Jiul de Vest–Cerna Valley, Băile Herculane (Romania). In: Onac BP, Tămaş T, Constantin S, Perşoiu A (eds) Archives of climate change in karst. Karst Waters Institute Special Publication, 10, pp 227–230

Ponta G, Strusiewicz R, Simion G, Gaspar E (1984a) Subterranean stream piracy in the Jiul de West-Cernisoara karst area–Romania. Theor Appl Karstol 1:235–238

Ponta G, Aldica G, Bădescu D, Panaiotu C, Solomon A (1984b) Cercetări speologice în perimetrul Jiul de West–Cernişoara. Bul Info 8:77–117

Pop G (1963) Contribuţii la cunoaşterea paleozoicului şi mezozoicului din părţile de est şi nord-est ale masivului Godeanu (Carpaţii Meridionali). Stud cerc Geol VIII/3:369–391

Pop G (1973) Depozitele mezozoice din Munţii Vâlcan. Ed Acad RPR, Bucureşti

Povară I (1976) Note sur la provenance des eaux de la résurgence "Izvoru Cernei"(Carpates Meridionales–Roumanie). Trav Inst Spéol "Émile Racovitza" XV:187–195

Povară I (1980) Note sur la circulation souterraine des eaux dans les calcaires du bassin de Cerna. Trav Inst Spéol "Émile Racovitza" XIX:237–241

Streckeisen A (1934) Sur la tectonique des Carpathes Me´ridionales. Ann Inst Geol Roum XVI, Bucharest

Struckmeier W, Margat J (1995) Hydrogeological maps: a guide and a standard legend. Int Assoc Hydrogeol 17:177

Vintilescu I (1972) Sur l'étendue du bassin de la Cerna dans sa partie supérioure. Trav Inst Spéol "Émile Racovitza" XI:349–354

Southern Carpathians: Pârgavului Cave

Doru Bădescu, Horaţiu Roman, and Iulia Bădescu

Abstract

The Pârgavului Cave is located in the Vâlcan Mountains (South Carpathians) and develops in Lower and Middle Jurassic limestones. Due to various carbonate facies, the dissolution acted selectively; thus, cave passages show different morphologies. The Mesozoic tectonics events combined with the hydrogeological conditions existing during the Quaternary, created a unique imprint on various geomorphological aspect of the cave. Speleothems are present in the sub-horizontal part of the cave. The progressive incision of the underground river led to the formation of several generations of shelfstones.

Keywords

Carbonate facies • Tectonics • Waterfalls
Differential dissolution

Geographic, Geologic, and Hydrologic Settings

The Pârgavului Cave (750 m alt.) develops in Lower, Middle, and Upper Jurassic limestones on the southern slopes of Vâlcan Mountains (Fig. 1) and is situated north of Topeşti and Vâlcele villages (Goran 1982). The entrance of the cave is located at the base of a cliff, 30 m above the resurgence that drains the subterranean spring. On the plateau above the cave, several sinkhole and shafts, which facilitated the infiltration of rainwater into the karst system, were identified. This area of Vâlcan Mountains is a gentle monocline, descending toward the SE. The elevation of Vâlcan Mountains ranges between 400 and 500 m in the southeast and 800–900 m in the center and northeast. The total length of the cave passages reaches ~3600 m and its vertical development is 119 m (+110 m; −9 m).

Metamorphic (Upper Proterozoic), magmatic, and sedimentary rocks of Marginal Dacides (Danubian Unit) tectonic unit outcrop in the region (Fig. 2) (Pop et al. 1975; Marinescu et al. 1989; Bădescu 2009). The Tismana Granite (Upper Paleozoic) penetrates the metamorphic rocks. The Mesozoic begins with the Lower Jurassic in Gresten facies, which consist of quartzite sandstones (with elements of feldspar, metaquartzite fragments, muscovite, and rarely, biotite) (Pop 1973; Bădescu 2009). On average, the rocks contain 80% quartz and approximately 16% feldspar. The Upper part of Lower Jurassic consists of limestones and marls. The Lower Jurassic is overlain by Middle Jurassic carbonate deposits (Carozzi 1960). These deposits present numerous lateral facies variations, which can easily be observed in the Pârgavul Cave. The limestones are represented by biosparite, pelsparite, and micrite, which occasionally contain remains of bivalves, gastropods, corals, and foraminifers. Dolomite rocks appear alongside the limestones. Differentiating between the Middle and Upper Jurassic deposits is difficult, as paleontological evidence are sparse or do not exist.

The Upper Jurassic is represented by limestones and dolomites (150–200 m thick), sometimes interbedded with chert layers, which frequently contain an argillaceous horizon (0.2–10 m thick) at the base. These deposits are overlain by 50 m of micritic limestone of Lower Cretaceous (Neocomian) age and 300 m of Barremian-Aptian limestones (Urgonian facies). The Upper Cretaceous is composed of arenite and lutite deposits in wildflysch facies.

D. Bădescu (✉)
Laboratory of Geology and Paleontology, Faculty of Geology and Geophysics, University of Bucharest, N. Bălcescu Blvd. 1, 010041 Bucharest, Romania
e-mail: dorubadescu@ymail.com

H. Roman
Faculty of Geology and Geophysics, University of Bucharest, Traian Vuia 6, Bucharest, Romania
e-mail: horace_the_horace@yahoo.com

I. Bădescu
Département d'Anthropologie, Université de Montréal, Pavillon Lionel-Groulx, 3150 Jean-Brillant, Montréal, QC H3T 1N8, Canada
e-mail: iulia.badescu@umontreal.ca

G. M. L. Ponta and B. P. Onac (eds.), *Cave and Karst Systems of Romania*,
Cave and Karst Systems of the World, https://doi.org/10.1007/978-3-319-90747-5_17

Fig. 1 Location of Pârgavului Cave

The Tertiary includes Neogene molasses of the Getic Depression, which also comprises deposits of Sarmatian, Meotian, and Pontian age. During the Alpine neo-Cretaceous tectogenesis (Laramide) or post-Cretaceous the sedimentary deposits experienced folding and faulting.

The hydrographic network is relatively well developed, with valleys oriented NNW-SSE, along which the landscape is fragmented in hills and ridges. The watercourses flowing over the carbonate deposits originate from the impermeable rocks (igneous and crystalline schists). The vicinity of Pârgavului Cave is dominated by Pârgavului Valley and its tributaries, as well as by Piscurilor and Pârgaşului valleys, which cross the impermeable unit of Tismana Granite, the metamorphic rocks, and other Lower to Upper Jurassic and Lower Cretaceous deposits. The main water source is Pârgavului resurgence, which collects its waters from the Şesul Pietrei Plateau and Padina Crovurilor (Iurkiewicz and Mangin 1994; Iurkiewicz 2010) (Fig. 2).

History of Explorations

In 1971 and 1972, Radu Petre along with "Focul Viu" Caving Club organized two summer camps in the Topeşti-Vâlcele karst area to identify and survey new caves and shafts. After one month of exploration, the entrance of the Pârgavului Cave was located, with a lake at the entrance making impossible the exploration of the cave without scuba diving gear. A team of cavers from the same club visited the area during the dry seasons of 1986 and 1987 and found low water levels in the entrance's lake. Thus, they were able to enter the cave crossing a 1m long and deep sump, and then explore and survey the cave to its present length. Several trips (1986 and 1987) were necessary to complete the exploration, which was slowed down by numerous waterfalls needed to be climb (Dudnic 1989). These were extremely difficult, since the cave walls around the waterfalls are heavily weathered, and

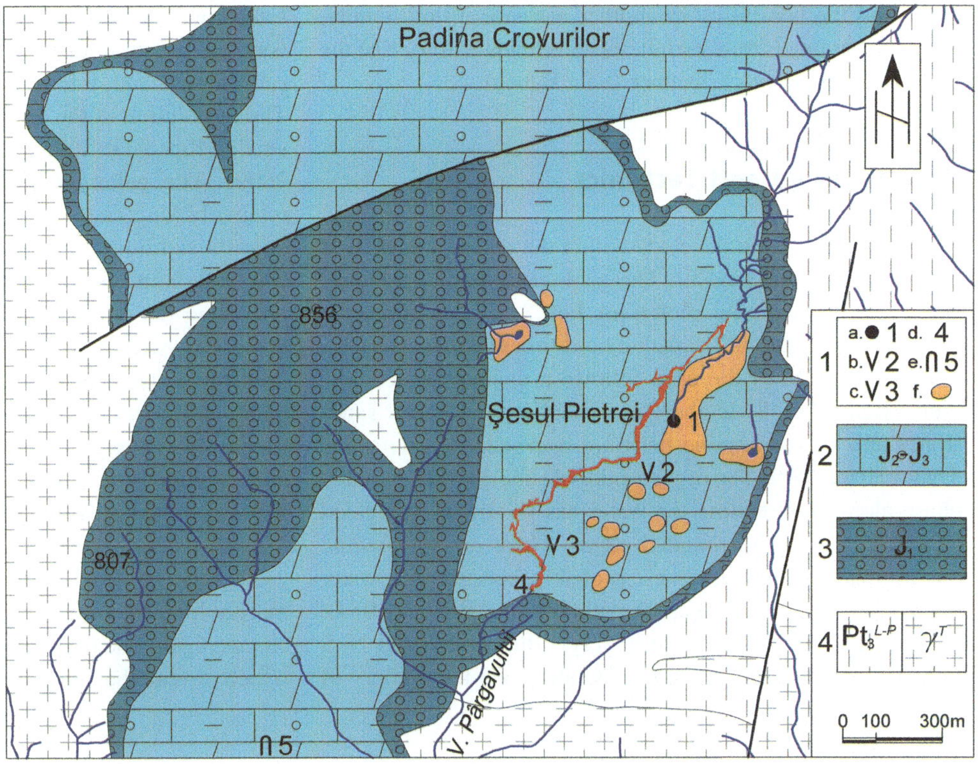

Fig. 2 Geologic map of Pârgavului drainage basin (after Pop 1973; Pop et al. 1975; Marinescu et al. 1989). *1a* Poiana Ponor (swallow hole), *b* Avenul Mare din Şesul Ponor (pit), *c* Peştera de la Cunună (pit), *d* Pârgavului Cave, *e* Şura Plaiului Cave, *f* sinkhole, *2* J_2–J_3 Middle to Upper Jurassic limestones, *3* J_1 Lower Jurassic sandstones, *4* Pt_3 Upper Neoproterozoic metamorphic series of Lainici-Păiuş, *γ* Tismana Granite (Paleozoic)

therefore, both free climbing and use of single rope technique equipment were required; in addition, waters were extremely cold.

Cave Morphology

The entrance is small, with a strong air current that suggests a vast subterranean cavern beyond that point. The entrance passage descends along the first 20 m to the lowest point in the cave, ending in front of a "pseudo-sump" that blocked the access further in the cave for a long time, being open only during dry periods. On the other side of the "sump" begins an ascending gallery named Focul Viu. Several high water level marks are visible on the cave walls. During the rainy season, occasionally, the passage is completely submerged (flooded) for several meters in height.

The first section of Focul Viu Gallery is dry, 2–3 m wide and 3–5 m high, and the stream flows at a lower impermeable level. Not far from the entrance, a relatively well-decorated and ascending passage to the left called Bear's Gallery (Galeria Ursului), houses a whole skeleton of *Ursus spelaeus*.

The main gallery continues south–north for 300 m and changes direction (east to west) in the vicinity of a small shaft. Immediately after this change in direction, the stream reappears and will be present throughout the rest of the cave. After 350 m, the Boat's Gallery (Galeria Bărcii) is reached; its profile and the presence of solutional notches indicate an initial phreatic phase in the development of the cave, subsequently followed by a vadose one. The cave continues almost horizontally through the Rimstones' Gallery (Galeria Gururilor) leading to a beautiful shelfstone nick-named Lavinia's Mystery (Taina Laviniei), which remained suspended above the ground. We interpreted the origin of this shelfstone as a stage in the cave evolution when this passage was partly filled with alluvia over which a horizontal carbonate flowstone precipitated. Following the reactivation of stream incision, and likely to a major flood event, the streamed sediments were removed and the shelfstone became suspended.

Further on, is the well-decorated Gallery of Lakes (Galeria Lacurilor), with some small canyons and rills carved in limestones as a result of differential weathering between calcareous sandstones and limestones. After passing several tight sections in which the ceiling is approximately 1 m high, the Sumps' Tributary (Afluentul Sifoanelor) is reached; this point marks the end of the gallery with sub-horizontal floor. From the cave entrance to the Sumps' Tributary, the vertical relief of the cave is only 10 m. Beyond this point, the cave has a completely different morphology, a change that is discussed in the next section.

The Waterfall's Gallery (Galeria Cascadelor) is a vadose canyon passage along which numerous waterfalls between 6 and 18 m in height are encountered. The gallery has a phreatic morphology, and the air current is very strong, indicating the possibility of another entrance it the vicinity. After climbing a 6 m waterfall, the cave ends with one of the largest rooms, Baboş Chamber (Sala Baboş).

Cave Geology and Tectonics

The Pârgavului Cave develops within the entire sequence of Jurassic carbonate deposits. The geological map of the cave and cross section are shown in Fig. 3. The entrance is located in gray and black spathic limestones (Bădescu 2009). The first portion of the cave follows a disjunctive tectonic fault oriented N–S. In the Bear's Gallery, quartzite sandstones`and silty limestone beds are interbedded, giving the cave walls a particular rugged outlook. During the phreatic phase along this section, the water has generated elongated ceiling grooves with rounded edges. There are no specific orientations of these grooves, probably because the water moved very gentle in an epiphreatic regime (Bleahu 1974).

The sudden change in the gallery's direction (from N–S to E–W, between the Rimstones' Gallery and the Gallery of Lake concurs with the presence of the vertical faults, which control the dissolution of micritic limestones, calcareous sandstones, and intensely dolomitized micritic limestone. The dissolution of the dolomites results in gray surfaces with small ellipsoidal stains or relict micritic zones (Bădescu 2009). The gallery continues in an E–W direction for a relatively long distance, crossing through arenitic limestone, dolomitized limestone, and micritic limestone. Affected by a NW–SE fault, the interbedded chert layers are revealed, providing the only stratigraphic reference in this section of the cave. After a well-developed section in finely grained limestones and dolomites, follows another change in the gallery's direction that occurs along a fault oriented N–S, at Lavinia's Mystery. The Gallery of Lakes develops NE–SW and remains the same until Sumps' Tributary. On this well-decorated passage, we cannot distinguish any tectonic feature that would explain the genesis of the gallery.

Beyond the intersection with the Sumps' Tributary, the morphological aspect and geology of the main gallery change again due to the presence of a series of normal faults, which reveal the impermeable layers within the carbonate

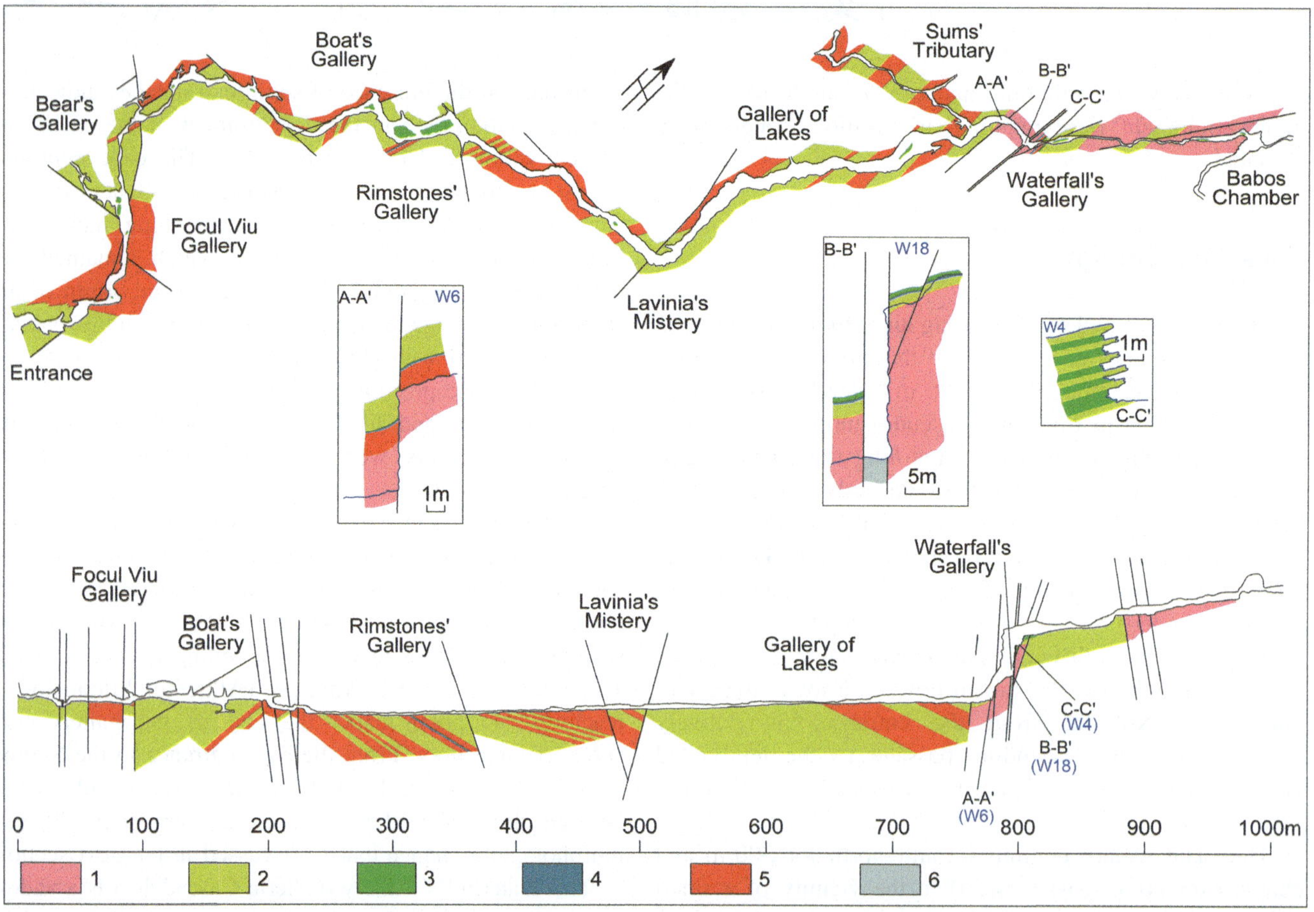

Fig. 3 Plan and cross section of the Pârgavului Cave. *1* Quartzitic sandstones and micro-conglomerates (Hettangian), *2* sequence of sandy limestones and/or grayish calcareous sandstones, black spathic limestones, and spotted grayish limestones (lower Jurassic), *3* grayish marly limestones (Bajocian), *4* black chert (Callovian-Oxfordian), *5* whitish and grayish fine limestones (Tithonian), *6* fault breccia/clay

deposits (i.e., Gresten facies arenites and micro-conglomerates). Up to this location, the underground river flows in a direction opposite to that of the dip of the strata (obsequent valley). Upstream from here, the dip of the bedrock strata changes and the water flow coincides with the inclination of the bedrock (consequent valley).

The direction of displacement for the fault's compartments is clearly highlighted by the fault pipes, which are easy to observe, as they affect a chert level showing interesting models of kink folds. These features are best view in front of the 6 m waterfall (W6, see A–A′ section in Fig. 3). Upstream from the 6 and 18 m waterfalls and all the way to the Baboş Chamber, the stream flows almost exclusively on impermeable deposits (see B–B′ cross sections in Fig. 3). About 40 m from the 4 m waterfall, a directional strike-slip fault that continues until the entrance to the Baboş Chamber is responsible for another waterfall.

The Genesis of the Waterfalls

It is easily evident that Pârgavului Cave has two distinct structural compartments (Fig. 3). The genesis of waterfalls at certain locations within the cave is associated with the presence of faults. These are responsible for the occurrence of staircase-type structures dipping downstream and the uplift to higher levels of the impermeable Lower Jurassic bedrock. The water has removed the thin layers of carbonate rocks through erosion and corrosion and intercepted the micro-conglomerate and quartzite sandstone rocks. Therefore, the incision of the stream bed was halted and the water forced to erode laterally along structural discontinuities (i.e., faults) generating waterfalls. The 18 m waterfall (W18) has at its bottom a room carved by the intense erosion that affected the 5–6 m thick clay layers (also known as fault breccia) present in the vicinity of the fault. In contrast to the W6 and W18 waterfalls, the W4 waterfall is developed on carbonate rocks. It is also a waterfall of natural tectonics but the displacement of the compartments was not sufficient to expose the impermeable bed. The particular morphology of this waterfall showing a succession of thin sandy limestone shelves is due to differential dissolution; the resistant ledges contains insoluble clay and quartz, whereas the softer interbedded marl layers are weaker, thus easier to weather.

References

Bădescu D (2009) Geologia Peşterii Pârgavului. In: Focul Viu 40 de ani de speologie. Ed RAO, Bucureşti, pp 134–138

Bleahu M (1974) Morfologia carstică. Ed Ştiinţifică, Bucureşti, p 590

Carozzi A (1960) Microscopic sedimentary petrography. Wiley, New York, p 485

Dudnic M (1989) Explorarea Peşterii Pârgavului. Bul Info 12:107–132

Goran C (1982) Catalogul sistematic al peşterilor din România. Ed CNEFS, Bucureşti, p 456

Iurkiewicz A (2010) Southern Vâlcan mountains. In: Orăşeanu I, Iurkiewicz A (eds) Karst hydrogeology of Romania. Ed Belvedere, Oradea, pp 105–120

Iurkiewicz A, Mangin A (1994) Utilisations de l'analyse systémique dans l'etude des aquifères karstiques des Monts Vâlcan. Theor Appl Karstol 7:9–96

Marinescu F, Pop G, Stan N, Gridan T (1989) Geological map of Romania, Pestişani Sheet. Scale 1:50.000. Inst Geol Geofiz, Bucureşti

Pop G (1973) Depozitele mezozoice din Munţii Vâlcan. Ed Acad RSR, Bucureşti, p 155

Pop G, Berza T, Marinescu F, Stănoiu I, Hârtopanu I (1975) Geological map of Romania, Tismana Sheet. Scale 1:50.000. Inst Geol Geofiz, Bucureşti

Mehedinti Mountains: Cioaca cu Brebenei and Closani Caves

Ioan Povară, Virgil Drăguşin, and Ionuţ Mirea

Abstract

Since both caves developed in similar geological and hydrogeological conditions, they are presented in this chapter side by side. Although very different in size, both caves are well-known for their wide diversity of speleothems. Cioaca cu Brebenei is an 85 m long cavity that hosts some of the largest and most spectacular helictites of the Romanian karst. The cave has been carved by the Izvorele Creek, a right-side tributary of the Motru River. Cloşani Cave is one of Romania's most thoroughly studied caves that hosts the only underground laboratory devoted to a variety of "in situ" cave studies. The cave has two major galleries (Laboratories and Crystals) summing up 1458 m of passages. Abundant and diverse speleothems, including shields, draperies, eccentrics, helictites, and gorgeous calcite pool spar crystals, are decorating the Crystals Passage. Cloşani appears to be a base-level cave, genetically connected with the flow of the Izvorele Creek and at a later stage of the Motru River. U-series analyses of speleothems indicate the cave is older than 600,000 years. Both caves are speleological reservations (Protection class A) located within the Domogled-Valea Cernei National Park and the ROSCI-0035 (Romanian Sites of Community Importance) "Natura 2000" Site.

Keywords

Base-level cave • Eccentric speleothems • Draperies
Calcite spar

I. Povară (✉) · V. Drăguşin · I. Mirea
"Emil Racoviţă" Institute of Speleology of the Romanian Academy, Calea 13 Septembrie 13, 050711 Bucharest 5, Romania
e-mail: ipov.iser@gmail.com

Geologic Settings

The central-eastern area of the Mehedinţi Mountains (Fig. 1) in which the two caves have developed belongs to the Danubian Autochthonous. The crystalline bedrock (Tismana granite) (Berza 1978) is overlain by a Mesozoic sedimentary cover consisting of Lower Jurassic successions (sandstones and microconglomerates), followed by Middle Jurassic rocks (gritty, sparry limestones) and by a massive Barremian-Aptian limestone unit, covered by the Upper Cretaceous flysch (Diaconu 1990; see Fig. 2). Both caves have developed in the lower Cretaceous limestones. Brown-reddish clay overlying the limestone could have resulted (as speculated by Diaconu 1984) following the alteration of feldspars and mica minerals derived from the crystalline schists and granitoids in adjacent areas that are situated at higher elevations than the limestone unit. That clay, carried by rainfall into the cave, is encountered on the floor of the Crystals Gallery (Galeria Cristalelor) and stains some of the crystals.

Cioaca cu Brebenei Cave

History of Exploration

A team of researchers (A. Burghele, D. Dancău, V. Decu) from the "Emil Racoviţă" Institute of Speleology first visited the cave in 1959 (Bleahu et al. 1976). The first map was published in a book describing the karst and caves from the southwestern Romania (Decou et al. 1967). Since then, the cave has been surveyed for fauna, and the speleothems have been sampled and radiometrically dated.

Cave Description

The entrance (1 × 1.2 m), located on the eastern (left) slope of the Izvorele Valley (at 518 m a.s.l.), is a 6.25 m deep

G. M. L. Ponta and B. P. Onac (eds.), *Cave and Karst Systems of Romania*,
Cave and Karst Systems of the World, https://doi.org/10.1007/978-3-319-90747-5_18

Fig. 1 Location of Cioaca cu Brebenei and Cloşani caves

shaft. At its base, soil, clay, along with vegetal and animal debris originating from outside the cave have piled up. From the bottom of the shaft, an 8 m long descending (30° slope) passage leads to a 15 × 35 × 8.5 m (width × length × height) chamber having its floor covered by large collapsed blocks, along with boulders and other debris, partly covered by coralloids (Fig. 3).

Two passages open from this chamber: an ascending one toward north, ending in a chimney, and a south-oriented descending one, which ends in a massive calcite flowstone. In the middle part of this corridor, on a surface of *ca*. 4 m^2, impressively large and transparent calcite helictites have developed (Fig. 4).

Cave Speleogenesis and Speleothems Dating

Based on surface geomorphology, location of the cave, and orientation of its passages, Diaconu (1990) suggested that it was formed by the Izvorele Creek, a right-side tributary of the Motru River. The underground stream that carved Cioaca cu Brebenei Cave presumably drained out along the Crystals Passage of Cloşani Cave; the distance between the extremities of these two caves is 400 m (Fig. 5).

Ten samples out of four stalagmites from Cioaca cu Brebenei Cave were dated using the U-series alpha spectrometry method. The oldest of these speleothems formed some 303 ka (Constantin 2003).

Cave Biology

The following troglobiont invertebrate species have been identified in Cioaca cu Brebenei Cave: *Trachysphaera orghidani, Trichoniscus aff. inferus, Centromerus europaeus, Lithobius decapolitus,* and *Duvalius spinifer*, also recorded in the Cloşani Cave, and *Saphrochaeta oltenica* (Decou et al. 1967; Giurginca et al. 2015). Among bats, two species (*Miniopterus schreibersii* and *Barbastella barbastellus*) roost in the cave.

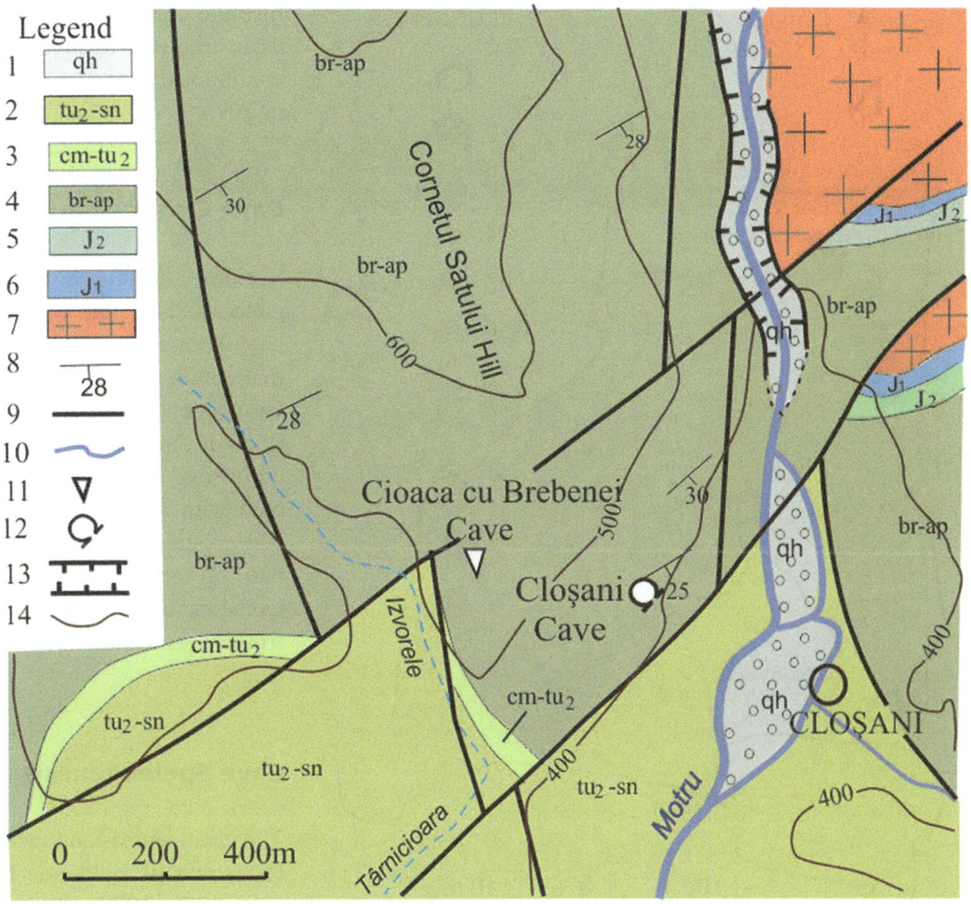

Fig. 2 Geological map of the Cloşani area (modified after Diaconu 1990). 1. Quaternary floodplain and terrace alluvia, 2. Upper Cretaceous wildflysch, 3. Upper Cretaceous marly limestones, 4. Lower Cretaceous massive limestones (Urgonian facies), 5. Middle Jurassic gritty, sparry limestones, 6. Lower Jurassic sandstones and microconglomerates, 7. Tismana Granite, 8. dip and strike of the limestone beds, 9. fault, 10. river, 11. shaft, 12. cave, 13. gorge, 14. gradient lines

Cave Protection

The cave has been declared a scientific (speleological) reserve (Class A) and included in both the Domogled-Valea Cernei National Park and the ROSCI-0035 "Natura 2000" Site.

Cloşani Cave

History of Exploration

Ionescu (1913) is the first to mention Cloşani Cave in his biospeleological study dedicated to the Southern Carpathians' caves. It was later cited by Chapuis and Jeannel (1951) who provided the first description of the Laboratories Passage (Galeria Laboratoarelor). In 1959, M. Ghica (a former geologist at the "Emil Racoviţă" Institute of Speleology) discovered the Crystals Passage. The map of the Laboratories Passage was published by Mitulescu (1960). Since then, numerous field campaigns have been undertaken by research teams of the "Emil Racoviţă" Institute of Speleology of the Romanian Academy. In 1962, a regional research center was built close to the cave. A detailed map and a thorough study were provided by Decou et al. (1967) and Bleahu et al. (1976). In 1978, Diaconu published data on the cave genesis and in 1990 he completed a monographic study on Cloşani Cave as part of his PhD thesis.

Cave Description

Cloşani Cave consists of Laboratories and the Crystals galleries (Fig. 6) that begin 30 m away from the entrance. To access both of them, one needs to crawl through a low and narrow corridor, part of the labyrinth nearby the entrance. The total length of the *Crystals Gallery* (initially named "Matei Ghica" Passage), including several side corridors is 827 m. Large, common speleothems have been well developed along it. The particular mineralogical and crystallographic value resides in the presence of several calcite aggregates, rare in cave environment (Diaconu 1990).

The Laboratories Gallery is slightly sinuous, 4–6 m in height, 4–5 m in width, and extends for 485 m at ~10–12 m below the Crystals Gallery. It hosts abundant and diverse speleothems, showing many narrow sectors, due to the presence of massive stalagmite domes and columns.

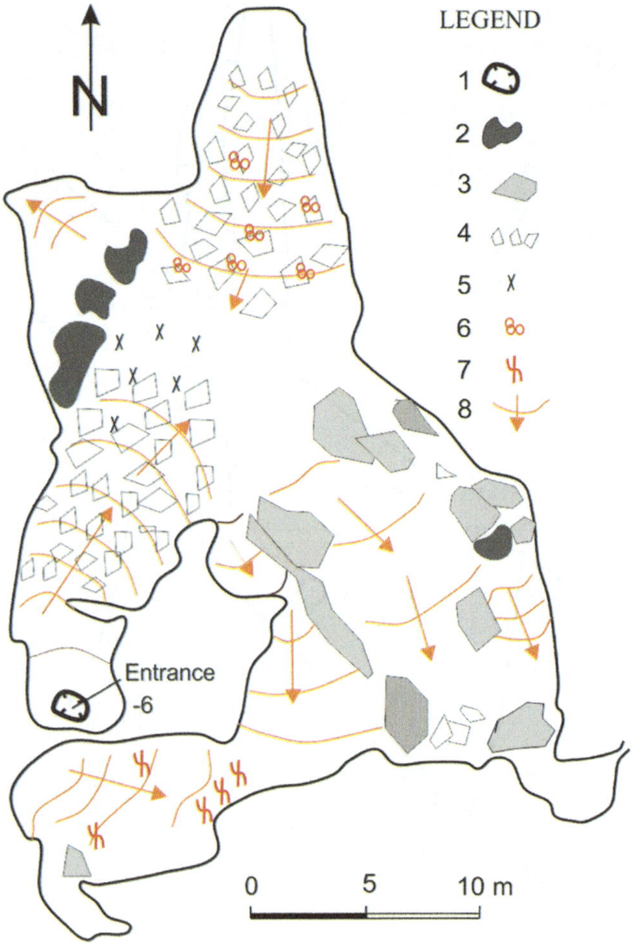

Fig. 3 Map of the Cioaca cu Brebenei Cave (modified after Decou et al. 1967). 1. shaft, 2. pillar, 3. collapsed blocks, 4. allochthonous debris, 5. bat guano, 6. coralloids, 7. helictites, 8. gradient lines

Fig. 4 Helictites in Cioaca cu Brebenei Cave (photograph courtesy of C Lascu)

Along this lower passage, at 30, 90, 125, and 155 m from the entrance, the Institute of Speleology constructed four ceiling-less cubicles (3 × 1.5 m), equipped with concrete tables to facilitate the "in situ" study of the troglobiont species behavior. At the northern end of the passage, the Institute of Geodynamics of the Romanian Academy set up its instruments to investigate terrestrial tides. After running continuously for almost a decade, the research was stopped in 1976.

Cave Clastic Sediments and Speleothems

Various clastic sediments were deposited along the two galleries. The floor of the oldest corridor (the Crystals Gallery) is mainly covered by reddish-brown clays, along with fine sands and gravels in its northern sector. Further, the clastic deposits are locally covered by calcite crusts. The floor in the middle section of this gallery is covered by a number of rimstone pools having their walls and floor covered with abundant and well-developed rhombohedral calcite crystals up to 15 cm in length (Fig. 7). In the terminal part of the Crystals Gallery, abundant and diverse speleothems decorate the entire space. Among these, calcite fringes, scalenohedron spar crystals, and candlesticks (Fig. 8).

Cave Speleogenesis and Speleothems Dating

The cave genesis and evolution are related to both the Motru River and its right-side tributary, the Izvorele Creek (Diaconu 1978, 1990). During the first stage of cave evolution, the upper level corresponding to the Crystals Gallery was carved by the Izvorele Creek. Next, in response to a drop in the base level caused by Motru River's erosional downcutting, the Laboratories Gallery has been formed.

More than 60 U–Th measurements were performed on several speleothems collected from Cloşani Cave (Constantin 2003; Constantin and Lauritzen 1999; Constantin et al. 2007). The obtained ages range between Middle Pleistocene (483 ka) and Holocene. Nevertheless, attempts to date two cores recovered from a large and thick flowstone accumulated right at the cave entrance failed because the samples were older than 600 ka, hence beyond the U-series limits (Constantin 2003).

Cave Climatology

The cave topoclimate is controlled by the distance from the entrance and the thickness of the limestone above the cave. The following particular features of the cave topoclimate have been highlighted by the results of a study performed between December 1987 and March 1990 (Racoviţă et al. 1993):

- bidirectional airflow within the entrance area and unidirectional airflow along the two passages;

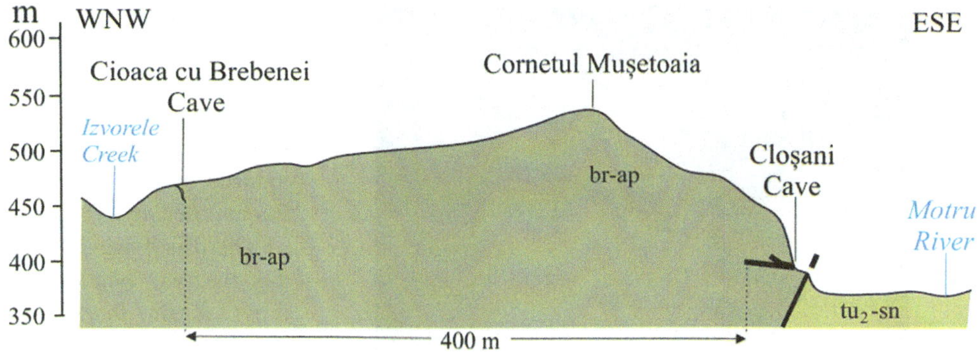

Fig. 5 Geological cross section through the Cornetul Muşetoaia between Cioaca cu Brebenei and Cloşani caves (geology after Diaconu 1990)

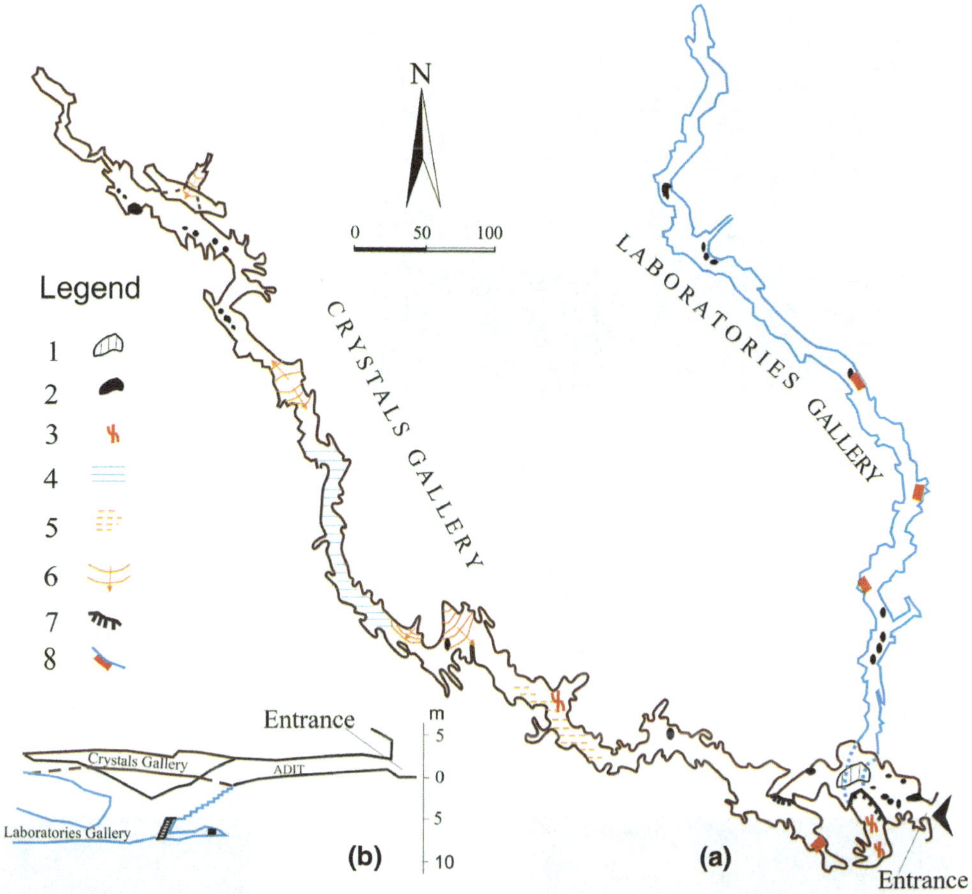

Fig. 6 **a** Map of the Cloşani Cave (modified after Diaconu 1990): 1. pillars, 2. columns, 3. helictites, 4. rimstone pools with rhombohedral calcite crystals, 5. clay, 6. gradient lines, 7. step, 8. laboratory cubicles; **b** section through the entrance zone

- unstable meroclimate close to the entrance, with large seasonal air temperature variations;
- stable meroclimate along the two passages, with a quasi-constant average air temperature of about 11 ± 0.5 °C;
- air relative humidity exceeding 98% all year long in both passages, whereas close to the entrance it varies significantly.

CO_2 Concentration

The CO_2 concentration of the cave atmosphere and drip water was measured at two points inside the cave: one on the Laboratories Passage and one on the Crystals Passage (Fig. 9). We used a Vaisala GMP 222 probe calibrated for the range 0–10,000 ppm, with an accuracy of (1.5% of range +2% of reading). For drip water concentration, we used the method described by Drăguşin et al. (2017), which relies on the measurement of the CO_2 concentration in a confined atmosphere at equilibrium with drip water. Using this CO_2 concentration, the method allows for direct calculation of drip water CO_2 using Henry's law (Sander 2015).

Our data show that cave air values have seasonal variability, with values as high as 9704 ppm recorded during the warm season, whereas values as low as 550 ppm were

Fig. 7 Rhombohedral and scalenohedral calcite pool spars in the Crystals Gallery (photograph courtesy of BP Onac)

measured at the end of the cold season (Fig. 9). This implies that CO_2 is produced in the soil above the cave by organic processes during summer (Lloyd and Taylor 1994; Breecker et al. 2012) and is transported into the karst system where its concentration depletes until the start of a new warm season, usually in March–April. The fact that in both passages the CO_2 concentrations have similar tendencies indicates that the cave's atmosphere is homogenized by air circulation. Nevertheless, the Laboratories Passage shows generally lower values (average ~3000 ppm) than Crystals Passage (average ~4000 ppm), possibly due to stronger ventilation induced by its closeness to the hillside, an area with higher fracture density.

The dissolved CO_2 signal is more intricate, as it appears that the two locations, although similar, act differently at times. Again, the Crystals Passage shows slightly higher average values when compared to Laboratories Passage (~7859 and ~7594 ppm, respectively). These differences could reflect ventilation events taking place at different times along the two flow paths. Further study could shed more light on CO_2 dynamics in this karst system and would allow for a better understanding of deposition versus corrosion phases affecting speleothems. Moreover, it would bring more insight into the degassing gradients between drip water and cave air, which bears great importance for the stable isotopes study of speleothems.

Cave Biology

More than 70 animal species have been identified and described from the Cloşani Cave, including 13 terrestrial troglobiont species (*Closania winkleri*, *Closania orghidani* (Decou 1959), *Trachysphaera orghidani*, *Polydesmus oltenicus*, *Trichopolydesmus eremites*, *Troglovitrea argintarui*, *Trichoniscus* aff. *inferus*, *Trichoniscus dancaui* (Giurginca et al. 2015), *Mesogastrura ojcoviensis*, *Onychiurus cloşanicus*, *Centromerus europaeus*, *Lithobius decapolitus*, *Duvalius spinifer*, *Neobisium cloşanicus*, *Nesticus ionescui*) and one aquatic troglobiont subspecies—

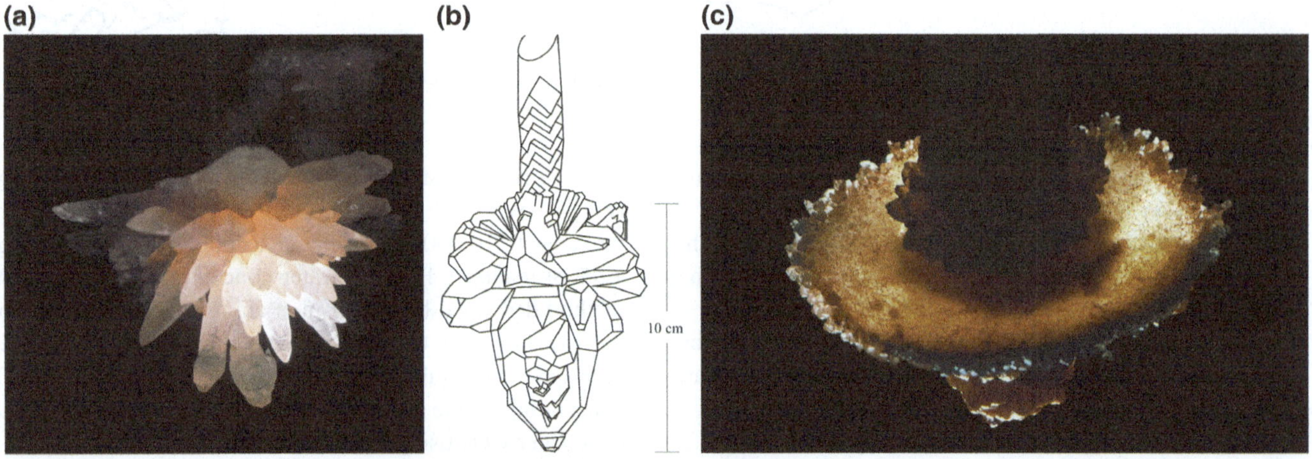

Fig. 8 Water lily flower (**a**) and its crystalline structure (**b**) composed of rhombohedron and scalenohedron crystals grown below the water level of a rimstone pool (drawing after Diaconu 1990), **c** calcite shelfstone (candlestick) precipitated around the tip of partly submerged stalactite in the end section of the Crystals Gallery (photograph courtesy of C. Lascu)

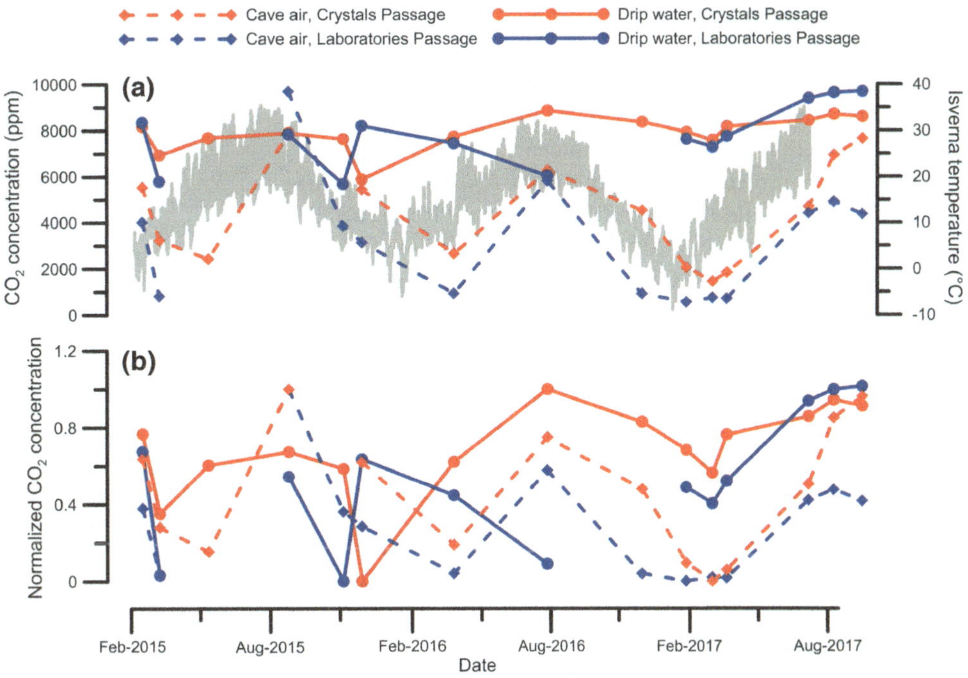

Fig. 9 Variability of CO_2 in cave air and drip water: **a** gray line represents the atmospheric temperature recorded at Isverna, **b** normalized values of CO_2 concentration in both cave air and drip water

Niphargus carpathicus variabilis (Gruia 1969; Decu et al. 1978; Nitzu et al. 2016). The cave invertebrate inventory also comprises troglophile and subtroglophile species, inhabiting the entrance area (e.g., *Typhloiulus mehedintzensis*; Tabacaru 1976). A detailed study of the cave ecology was undertaken by Decou and Herdlicka (1978).

Small bat colonies of *Rhinolophus ferrumequinum*, groups of *M. myotis*, *M. blythii*, *M. mystacinus*, *R. blasii*, and *Nyctalus noctula*, but also individuals of *R. hipposideros* and *M. emarginatus* were observed over the years in the cave (Gheorghiu et al. 2001). *Rhinolophus euryale* populates only the Crystals Gallery (Fig. 10).

Fig. 10 Hibernation colony of *Rhinolophus euryale* and *R. blasii* in the Cloşani Cave, during the winter of 2015 (photograph courtesy of C. Jére)

Cave Protection

The cave has been declared a scientific (speleological) reserve in 1951 by the Romanian Academy's Commission for the Protection of Natural Monuments. Since 1990, it is included as a protected area, within the Domogled-Valea Cernei National Park, the Piatra Cloşani Natural Reserve and the ROSCI-0035 "Natura 2000" Site. The "Emil Racoviţă" Institute of Speleology is officially in charge of the cave custody. The access into the cave is restricted for research activities only.

Conclusions

During an initial stage, a common evolution was likely experienced by the two caves, both of which were generated by Izvorăle stream. Ensuing to the subsequent, further entrenchment of the Izvorăle valley, the two cavities started to evolve independently from each other. Cioaca cu Brebenei entered a phase when filling was taking place by means of rock collapse and by speleothems deposition. Remarkable witnesses of this stage are coral shaped and eccentric speleothems. In Cloşani Cave, the stable cave climate together with the ground surface derived seepage that was oversaturated in calcium bicarbonate has favored a fast growth rate of the speleothems. Large single crystals and crystal aggregates, both of which are uncommon in the underground environment, developed within the Crystals Gallery. That passage can be considered, taken as a whole, as an enormous geode. The troglobiont fauna consists of 13 species of invertebrates and of 10 species of bats. Both caves are subject to a regime of scientific nature reserve under the "Emil Racoviţă" Institute of Speleology supervision; the access inside them being strictly controlled.

Acknowledgements The authors address special thanks to C. Lascu, C. Jére, R. Tafet, and V. Gheorghiu for the information, photographs, and references on the bat species inhabiting the Cloşani Cave.

References

Bleahu M, Decu V, Negrea Şt, Pleşa C, Povară I, Viehmann I (1976) Caves from Romania. Ed Ştiinţifică şi Enciclopedică, Bucureşti, p 416

Berza T (1978) Studiul mineralogic şi petrografic al masivului granitoid de Tismana. An Inst Geol Geofiz LIII:5–176

Breecker DO, Payne AE, Quade J, Banner JL, Ball CE, Meyer KW, Cowan BD (2012) The sources and sinks of CO_2 in caves under mixed woodland and grassland vegetation. Geochim Cosmochim Acta 96:230–246. https://doi.org/10.1016/j.gca.2012.08.023

Chapuis PA, Jeannel R (1951) Enumération des grottes visitées. Biospeologica LXXII:97–98

Constantin S (2003) Evoluţii paleoclimatice în Cuaternar pe baza speleotemelor din carstul Munţilor Banatului şi Mehedinţi. Unpublished PhD Thesis, Univ Bucureşti

Constantin S, Lauritzen SE (1999) Speleothem datings in SW Romania. Part 1: evidence for a continuous speleothem growth in Peştera Cloşani during oxygen isotope stages 5–3 and its paleoclimatic significance. Theor Appl Karstol 11–12:35–46

Constantin S, Lauritzen S-E, Lundberg J (2007) New data on the chronology of the Second Termination and the paleoclimate during MIS 5 based on the study of a stalagmite from Closani Cave (SW Romania). Quat Int 167–168:77

Decou V (1959) *Sophrochaeta motasi n. sp.* et *Closania orghidani n. sp.*, nouvelles espèces de Bathysciinae (Coloeoptera-Catopidae) des Carpathes Méridionales. Ann Spéléol 14(1–2):185–196

Decou V, Herdlicka W (1978) Recherches écologiques dans les grottes des Monts Mehedinţi (Carpates Méridionalés). Trav Inst Spéol "Émile Racovitza" XVII:95–116

Decou A, Decou V, Bleahu M (1967) Grottes d'Olténie explorées de 1959 á 1962 In: Botoşâneau L, Negrea A, Negrea Ş (eds) Recherches sur les grottes du Banat et d'Olténie, Roumanie (1959–1962). Centre National de la Recherche Scientifique, Paris, p 429

Decu V, Diaconu G, Povară I (1978) Peştera Cloşani. Ed Sport-Turism, Bucureşti, p 142

Diaconu G (1978) Peştera Cloşani. Genèse et évolution. Trav Inst Spéol "Émile Racovitza" XVII:185–191

Diaconu G (1984) Considérations concernant la genèse des argiles sur les calcaires de la "Zone Cloşani"—Monts Mehedinţi. Theor Appl Karstol 1:13–21

Diaconu G (1990) Cloşani Cave. Mineralogical and genetic study of carbonates and clays. Misc Speol Rom 2:190

Drăguşin V, Balan S, Blamart D, Forray FL, Marin C, Mirea I, Nagavciuc V, Orăşeanu I, Perşoiu A, Tîrlă L, Tudorache A, Vlaicu M (2017) Transfer of environmental signals from surface to the underground at Ascunsă Cave, Romania. Hydrol Earth Syst Sc 21:5357–5373. https://doi.org/10.5194/hess-21-5357-2017

Gheorghiu V, Petculescu A, Iavorschi V (2001) Contribution to the knowledge of the Chiroptera distribution from Romanian sector of the Carpathian Mountains. Stud Chiropterol 2:1746

Giurginca A, Munteanu CM, Vlaicu M, Tabacaru I (2015) Cavernicolous Oniscidea of Romania. Ed Semne, Bucureşti, p 165

Gruia M (1969) Date asupra răspândirii colembolelor în peşterile României. Trav Inst Spéol "Émile Racovitza" VIII:161–178

Ionescu CN (1913) Biospeologia Carpaţilor Maridionali. Bul Soc Reg Rom, Geogr XXXIV:79–98

Lloyd J, Taylor JA (1994) On the temperature dependence of soil respiration. Funct Ecol 8:315–323

Mitulescu N (1960) Peştera de la Cloşani. Ocrotirea Naturii 5:112–117

Nitzu E, Giurginca A, Nae A, Popa I, Baba Ş, Meleg IN, Vlaicu M (2016) The catalogue of caves with endemic cavernicolous arthropod fauna of Romania. Trav Inst Spéol "Émile Racovitza" LV:3–62

Racoviţă G, Boghean V, Petrescu M (1993) Le bilan topoclimatique de la grotte de Cloşani (Monts Mehedinţi, Roumanie). Theor Appl Karstol 6:7–86

Sander R (2015) Compilation of Henry's law constants (version 4.0) for water as solvent. Atmos Chem Phys 15:4399–4981. https://doi.org/10.5194/acp-15-4399-2015

Tabacaru I (1976) Sur un nouvel Iulide cavernicole de Roumanie: *Typhloiulus (Lamellotyphlus) mehedintzensisn sg, n.sp.* Trav Inst Spéol "Émile Racovitza" XV:63–75

Mehedinti Mountains: Martel and Lazului Caves

Gheorghe M. L. Ponta, Gheorghe V. Aldica, and Tiberiu Tulucan

Abstract

Martel and Lazului are active phreatic/epiphreatic caves with a superimposed vadose morphology. Martel is a branchwork cave developed parallel with the river as a left (north) side meander and is located at about 8 m below the thalweg. The water level in the cave rises and falls along the main gallery, where the lakes/streams at the lightest rain become sumps. The cave is relatively poor in speleothems, except the fossil gallery (CS FV Gallery of Memories) located at the upstream end of the cave, but numerous erosion and corrosion features are present. Lazului is also a branchwork cave, forming a large meander on the right (south) side of the Motru Sec River, where at the lowest points of the cave, five streams are disappearing underground. Erosion and corrosion features are present, and speleothem is found occasionally in the upper level of the cave.

Keywords

Karst • Branchwork cave • Phreatic • Epiphreatic

Geographic, Geologic, and Hydrologic Settings

Martel and Lazului caves are located along the Motru Sec Valley, Mehedinţi Mountains, in the central western part of Southern Carpathians. Both caves develop in Barremian-Aptian limestones of the Cerna-Jiu sedimentary zone of the Danubian Autochthonous (Mutihac and Ionesi 1974). The caves are located on the left (north) side and right (south) side, respectively, of the Motru Sec River at 4.2 km upstream from the confluence with Motru River (Fig. 1).

Cerna-Jiu sedimentary zone is characterized by the development of an extensive carbonate platform, which formed starting from Middle Jurassic and lasted until the end of Lower Cretaceous. It is bounded by Cerna Valley in the west, continues on the southern slopes of the Vâlcan Mountains, extending east of the Jiu Valley. Cerna-Jiu area functioned mostly as a marginal sedimentary area, where reef facies developed, especially in Jurassic and Early Cretaceous, ending with the accumulation of wildflysch-type deposits (Ponta et al. 2018).

The area geology consists of a crystalline basement (Lainici-Păiuş Terrane) and Tismana granite of the Danubian Autochthonous (Săndulescu 1984), covered by the Cerna-Jiu sedimentary deposits. From upstream to downstream along the Motru Sec River, the sedimentary sequence is formed by Lower Jurassic sandstones and conglomerates, which are overlain by Middle Jurassic–Lower Cretaceous (Neocomian) micritic and pelletal limestones sometimes dolomitized, followed by Barremian–Aptian Urgonian-type limestones. The limestones deposits are overlied by wildflysch deposits of Turonian–Senonian age. This sedimentary sequence forms the Nordic flank of Orzeşti-Godeanu Syncline.

The Barremian–Aptian limestones deposits in which both caves develop are thick bedded (30–90 cm), with well-defined primary and secondary porosities, permitting water to flow underground along the bedding planes and fractures, respectively (Fig. 2). Along the Motru Sec River, the flow diminishes due to the presence of numerous ponors in the river's thalweg, through which the water disappears totally during the dry season.

The southern flank of the syncline is exposed in the Brebina Baia de Aramă area where large springs are

G. M. L. Ponta (✉)
Geological Survey of Alabama, 420 Hackberry Lane, Tuscaloosa, AL 35401, USA
e-mail: gponta@yahoo.com

G. V. Aldica
National Institute of Materials Physics, 105 Bis Atomiştilor Street, P.O. Box 077125 Măgurele, Ilfov, Romania
e-mail: aldica2000@yahoo.com

T. Tulucan
Vişinului 77, 310197 Arad, Romania

G. M. L. Ponta and B. P. Onac (eds.), *Cave and Karst Systems of Romania*,
Cave and Karst Systems of the World, https://doi.org/10.1007/978-3-319-90747-5_19

Fig. 1 Location of Martel and Lazului caves within Romania

discharging from Barremian–Aptian limestones. The sedimentary sequence between Lupşa and Brebina Valley is overlay by the Sebeş Lotru Terrane of the Getic Domain (700–1000 m thick).

Several tracer studies performed in the early 1970s and late 1980s (Bandrabur et al. 2010) proved the continuity of Danubian Autochthonous limestones under the Getic Domain Nappe at a depth of 1500 m. The approximate distance between the ponors and the springs is 7–8 km. Based on a cross section realized by Pop et al. (1985), the crystalline deposits are about 1000 m thick and overlie the 500-m-thick wildflysch deposits. For the waters to travels at 58–73 m/h, some major faults/joints in the crystalline/wildflysch deposits must exist to justify these velocities.

In the late 1970s, when the dam at Ivan Valley on Cerna River (Ponta et al. 2013) was under construction, during the excavation of a tunnel through crystalline formation to divert the waters of Cerna to Motru River through Mehedinţi Mountains, several large underground streams were intercepted flowing along fractures, with no limestones in the area (Ţurcanu pers. comm.). Further detailed studies of the geology of the area, an updated karst inventory, and new dye studies are recommended.

Martel Cave

General Data

Martel Cave (named after the French geographer), also known as the Cave Nr. 6 from Motru Valley is located on the southern slope of the Piatra Mare a Cloşanilor Mountain, in the Motru River Basin. The cross section in Fig. 3 shows the location of the caves relatively to the Motru Sec River's thalweg.

The first cave exploration was made by Decu and Argintaru between 1960 and 1962, reaching a length of 2200 m; the description and cave map were published in 1967 (Decu and Bleahu 1967; Bleahu et al. 1976). In 1975, Focul Viu Caving Club (Bucharest) has resumed the cave exploration reaching 4133 m in length (Tulucan 1979). In the cave, the average temperature is 9.5–10 °C, with very high humidity. In the entrance area, a weak air current is present.

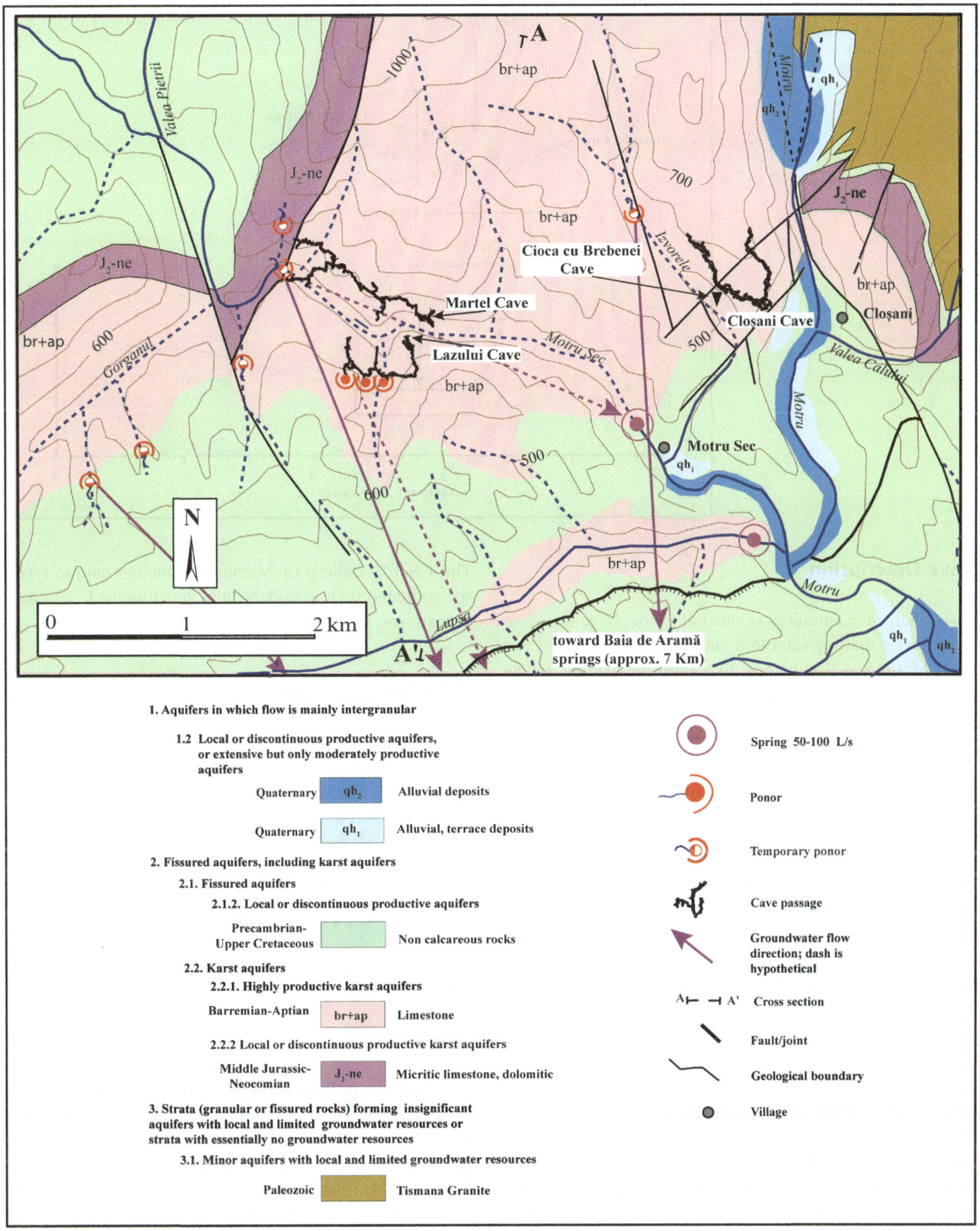

Fig. 2 Karst hydrogeologic map of Martel and Lazului area (geology modified after Pop et al. 1985)

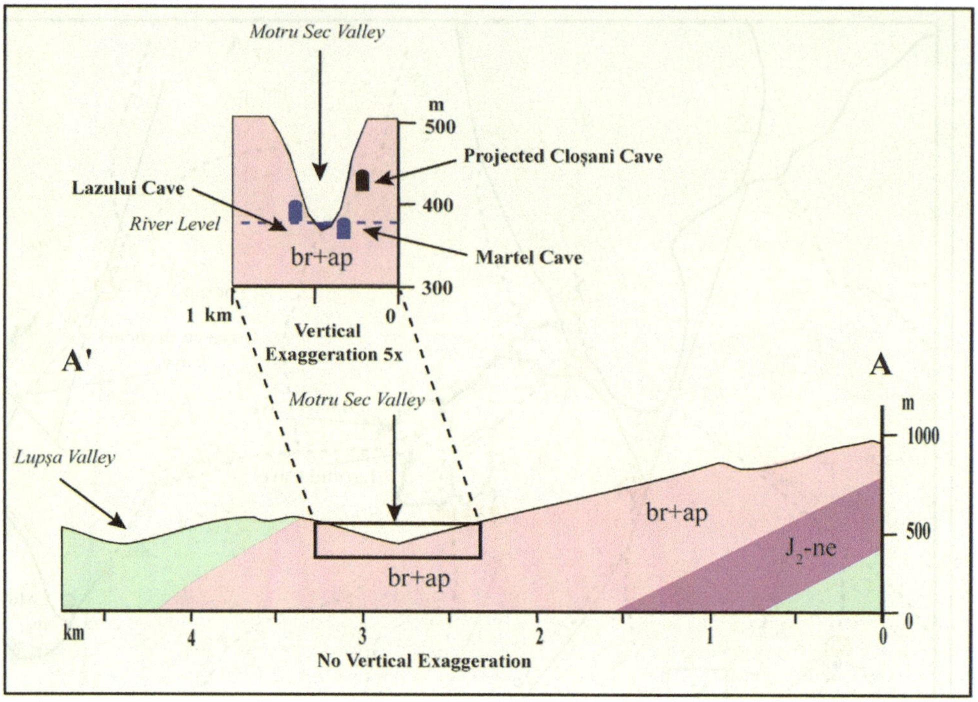

Fig. 3 Hydrogeological cross section showing the location of the caves relatively to the Motru Sec River's thalweg

Cave Description

Two small cave entrances situated at +1 m above the Motru Sec River's thalweg next to a culvert under the gravel road from Motru Sec Village lead to a short descending gallery, which opens in the main passage of the cave oriented NW-SE (Fig. 4). The cave continues downstream on two levels, both ending in deep lakes. Upstream, the main subfossil passage (−12 to −5 m below entrance elevation), after crossing four shafts with depths of 20 m (each) that give access to perennial stream passage and a canyon, opens in a large passage, 8–10 m high and 1.5–2 m wide, interrupted by low sections (Emissary/Emisar, Little/Mic, or terminal/final sumps) and three large chambers (one was named the Great Room/Sala Mare). A short distance beyond the Great Room, a 350-m-long side gallery oriented toward the valley slopes was surveyed (crawling passage/Galeria Târâşului). Above the terminal sump, a vertical passage opens in the 700-m-long fossil and well-decorated CS FV Gallery of Memories.

Cave Sediments and Speleothems

Sand and clay are the main sediments covering the undulating floor of the active and subfossil galleries. Limestones blocks appear in the larger areas of the cave. Speleothems (stalagmites, stalactites, etc.) occur in the fossil galleries. The CS FV Gallery of Memories contains unique type of speleothems such monocrystals, bi-trifurcated, trifurcated stalagmites, and soda straws (Tulucan 1979).

Cave Speleogenesis

The vadose morphology of Martel Cave characterizes the underground stream that is recharged by the sinking waters of the Motru Sec River and its north side tributary. Since the cave is located at about −8 m under the thalweg of Motru Sec River, it partly develops below the water table in phreatic conditions, resulting a unique irregular and undulatory longitudinal profile. At high water levels, parts of the cave enter in epiphreatic regime.

Lazului Cave

General Data

Lazului Cave is located on the right side of the Motru Sec River at about 20 m off the gravel road at 4.2 km upstream from the confluence with Motru River. Chappuis and Winkler described the first 170 m of the cave in 1928 (Chappuis and Jeannel 1951). The first survey of the cave was done by Decu and Bleahu (1957–1962) the map being published in 1967 and later in the Caves of Romania book (Bleahu et al. 1976).

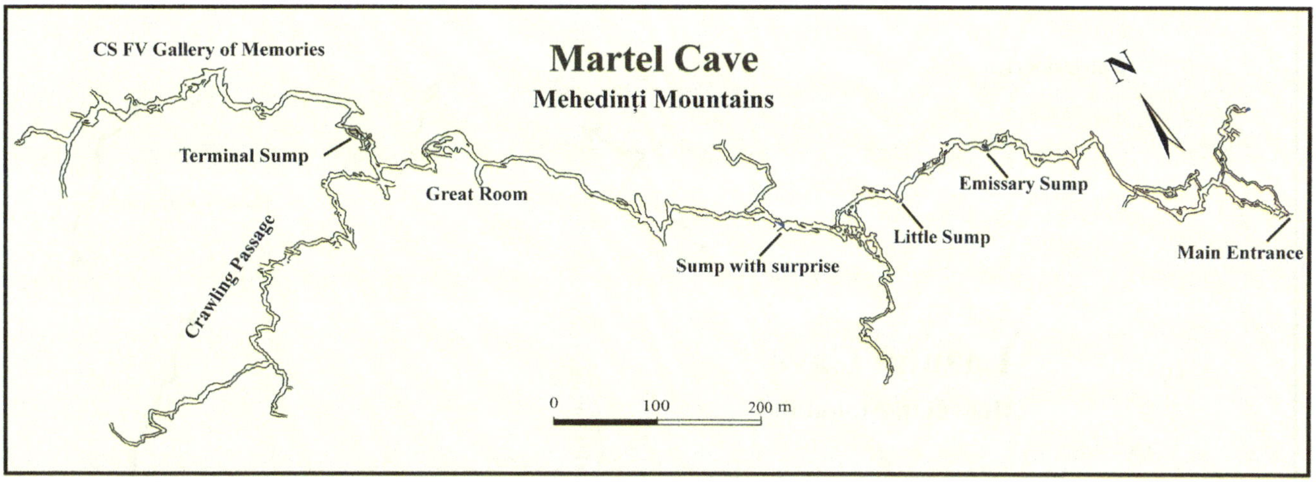

Fig. 4 Martel Cave (surveyed by Focul Viu Caving Club and Niphargus Praha)

At that time, the total length of the cave was 2200 m. In December 1975, Focul Viu Caving Club began the exploration and remapping of the cave, and by the end of 1976, it reached 3201 m in length and a vertical range of 30 m (−22 to +8 m) (Aldica and Ponta 1978; Aldica et al. 1980).

Cave Description

The cave is a branchwork type developed mainly along vertical or oblique joints and fractures with fossil, subfossil, and active (stream) galleries (Palmer 2007). The entrance, 7.5 m × 2.5 m, is located at 1.5 m relative altitude and is the largest of the four openings of the cave. Two distinct sections were identified in the cave. The first one near the entrance has a labyrinth pattern (Fig. 5) with numerous galleries, not higher than 1.5 m that make the connection with an upper maze located ~8 m above. The influence of exterior factors (temperature, humidity, airflow) has a strong effect on the speleothems and walls of the cave, resulting in extreme weathering. In the entrance area, up to 80 m inside the cave, during winter abundant ice speleothems form and frequent blocks are present on the floor as by-product of gelifraction. Guano is accumulated in one small room in the Labyrinth (Dumitraş et al. 2002). The Labyrinth ends after 200 m in a lake, which become a sump in the rainy season, followed by a 2.5 m drop, where the gallery narrows and marks the entrance in the second section of the cave.

This part is formed by a unique gallery oriented first south-north, with an average height of 2.5 m, and a few side stream passages and several short upper-level sections, which usually ends in chimneys. After approximately 250 m (from −2.5 m drop), the main gallery turns east, becoming parallel with the Motru Sec River in the so-called Whirlpools Gallery. As the name said, several whirlpools are present, full with water in the rainy season. In the cave were identified five streams, two delimiting the cave on the western side, which are disappearing underground in the vicinity of the gallery with sand (Fig. 5). Another stream is coming into the cave at the northern end of a south–north side passage named crawling with mud (Târâşul cu argilă), which after about 50 m is distempering. All of these streams are draining water from the Motru Sec River and recharging the springs from Brebina or Baia de Aramă areas.

Cave Sediments and Speleothems

The cave floor is covered by clay, sand, and gravel, and occasionally, in the larger areas, numerous breakdown blocks are found. On the walls quite frequently, a clay layer is present in areas that are flooded. The cave is poorly decorated, the presence of the speleothems being more frequent in the upper level of the entrance's labyrinth, where some stalactites, stalagmites, or flowstones are found.

Cave Speleogenesis

The Lazului Cave has a branchwork pattern and was partly created under phreatic conditions below the water table of the area. Later, these were remodeled in vadose regime by five streams recharged by the sinking water of the Motru Sec River.

Cave Climatology

From climatological point of view, two distinct sections were identified: (1) The Labyrinth section of the cave is influenced by the outside factors, thus both temperature and humidity

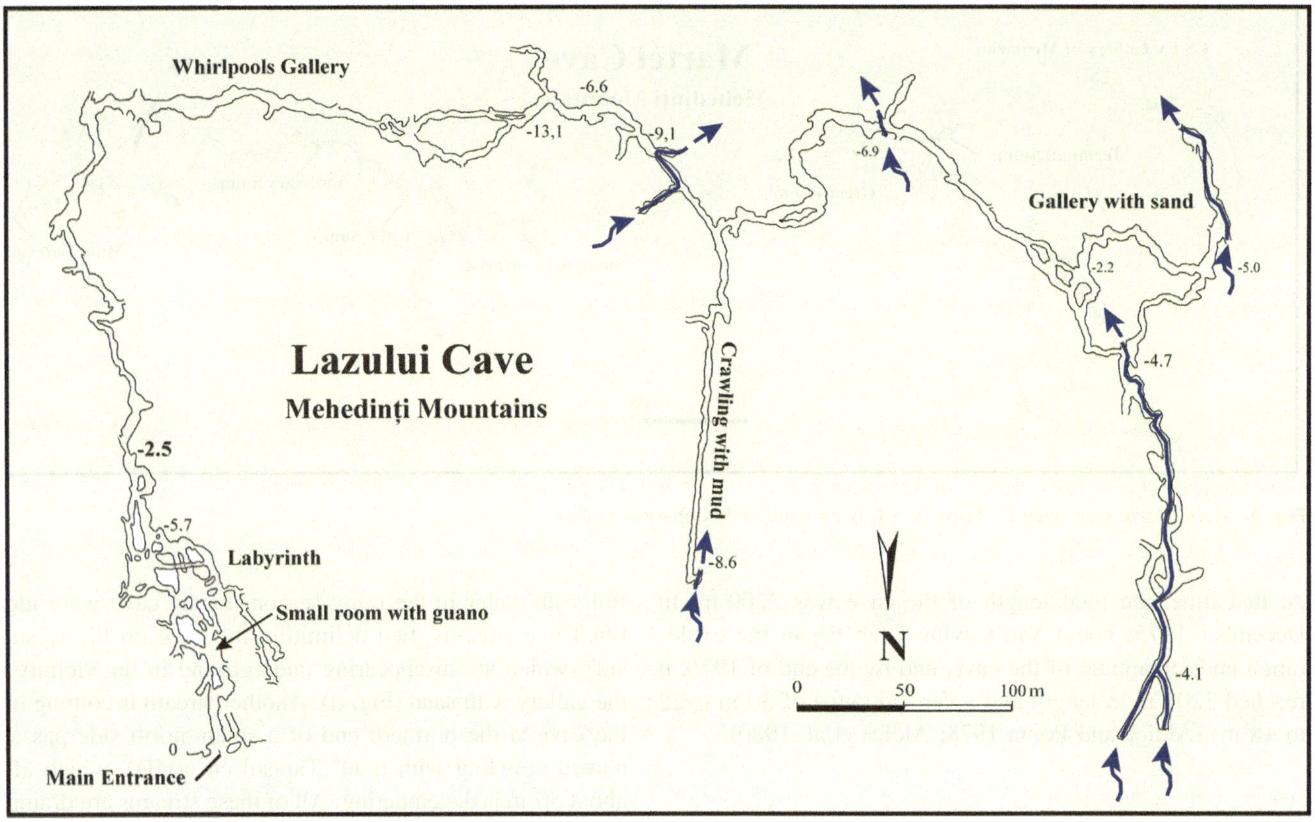

Fig. 5 Lazului Cave (surveyed by Focul Viu Caving Club and Niphargus Praha)

vary greatly. As far as the 2.5 m drop, there is a permanent thermic exchange between exterior and interior, perceptible through the air currents, which traverse the labyrinth, resulting in low air temperatures, dry walls, and the presence of ice in the winter; (2) in the second part of the cave, the air temperature ranges between 9 and 10.5 °C, and the relative humidity oscillates between 97 and 100%. The water temperature in the stream is around 10 °C, during the summer.

Acknowledgements We are thankful to Focul Viu Caving Club Bucharest and Nyphargus Praha for their support during the exploration and research activities at these two caves. We are also grateful to G. Diaconu (Senior Researcher with the "Emil Racoviţă" Institute of Speleology), the scientific coordinator of Focul Viu Caving Club in the 1970s and 1980s, for his advices and recommendations for fieldwork activities.

References

Aldica G, Ponta G (1978) Peştera Lazului. Bul Speol 2:30–35

Aldica G, Ponta G, Sluka M (1980) Jeskyne Lazului. Cesky Kras 5:76–79

Bleahu M, Decu V, Negrea Ş, Pleşa C, Povară I, Viehmann I (1976) Peşteri din România. Ed Ştiinţifică şi Enciclopedică, Bucureşti, p 415

Bandrabur G, Povară I, Bandrabur R (2010) Mehedinţi Mountains and Plateau. In: Orăşeanu I, Iurkiewicz A (eds) Karst hydrogeology of Romania. Ed Belvedere, Oradea, pp 121–136

Chappuis PA, Jeannel R (1951) Enumération des grottes visitées, 1927-1949 (8e série). Arch Zool exp gén 88:81–230

Decu V, Bleahu M (1967) Grottes d'Olténie explorées de 1959 à 1962. In: Decu V, Bleahu M (eds) Recherches sur les grottes du Banat et d'Olténie (Roumanie, 1959–1962). CNRS Paris, pp 229–239

Dumitraş D, Marincea S, Diaconu G (2002) Leocophosphite and taranakite in the bat-guano deposit from Lazului Cave (Mehedinţi Mountains). Rom J Miner Depos 80:23–25

Mutihac V, Ionesi L (1974) Geologia României. Ed Tehnică, Bucureşti, p 646

Palmer AN (2007) Cave geology. Cave Books, Dayton, p 454

Ponta GML, Aldica GV, Dumitru R (2018) Polovragi Cave - Oltetului Gorge karst area. In: Ponta GML, Onac BP (eds) Karst and cave systems of Romania. Springer International Publishing, Cham, pp 83–91

Ponta G, Povară I, Isverceanu G, Onac BP, Marin C, Tudorache A (2013) Geology and dynamics of underground waters in Cerna Valley/Băile Herculane (Romania). Carbonates Evaporites 28:31–39

Pop G, Berza T, Marinescu F, Stănoiu I, Hârtopanu I (1985) Geologic map of Romania, Tismana Sheet (106d; L-34–106-D). Scale 1:50.000, Inst Geol Rom, Bucureşti

Săndulescu M (1984) Geotectonica României. Ed Tehnică, Bucureşti

Tulucan T (1979) Peştera Martel. Bul Speol 1:14–24

Further Reading

Bleahu M (1972) Karst of Rumania. In: Herak M, Springfield VT (eds) Important karst regions of the Northern Hemisphere. Elsevier, Amsterdam, pp 341–353

Cocean P (1995) Peşterile României. Potenţial Turistic. Ed Dacia, Cluj-Napoca, p 264

Orghidan T, Negrea Ş, Racoviţă G, Lascu C (1984) Peşteri din Romania. Ghid Turistic. Ed Sport-Turism, Bucureşti, p 504

Useful Websites

www.speologie.org
www.karstgeology.com

Mehedinti Mountains: Isverna Cave

Mihai Baciu, Gabor Mogyorosi, Virgil Drăguşin, Zsolt Gyurka, and Szabolcs Storozynski

Abstract

Isverna Cave is situated on the eastern slope of the Mehedinţi Mountains, at an altitude of 450 m. It is over 4 km long, and it functions as the water collector for a karst area of over 20 km^2. It is traversed by a river with very clear water except during flood conditions. A series of sumps, of which the longest is the Black Sump ($\sim$700 m), are spread along the underground river. This makes the cave an excellent destination for cave diving activities. The first few hundred meters of galleries host hibernation and maternity bat colonies.

Keywords

Isverna Cave • Black Sump • Mehedinţi Mountains Romania

Introduction

Isverna Cave is remarkable due to the fact that the river flowing through it has very clear waters, a characteristic little seen in Europe. Its submerged passages are the longest in Romania, the Black Sump being almost 700 m long. Although the passages nearer to the entrance have been known for a long time, the exploration of the submerged galleries started only in the 1980s due to the high degree of difficulty imposed by the underwater passages. The cave is situated at the west end of Isverna Village, at the foot of the Mehedinţi Mountains (Fig. 1), and is the collector of a mostly underground hydrographic system that covers more than 20 km^2.

History of Exploration

The first 200 m of the cave were explored and described by C. N. Ionescu in 1914, followed by P. A. Chappuis and A. Winkler in 1951. In subsequent years, biologists from the "Emil Racoviţă" Institute of Speleology visited the cave and studied its fauna (Decu and Povară 1976). C. Goran surveyed the cave in 1973, and the Focul Viu Caving Club explored a few more passages in 1979.

In 1980, F. Păroiu begins the exploration of the submerged passages. In 1981, cave divers from GESS returned better prepared and with a larger team: F. Păroiu, Ş. Sârbu, C. Lascu, I. Povară, N. Grigore, C. Vânău, and their cave diving instructor, T. Iliffe. In 1982, C. Lascu and M. Oancea reach a depth of 42 m in the Black Sump (Fig. 2), a Romanian record at the time. After 1984, the GESS team begins a prolific collaboration with Hungarian cave divers from the Verticum Club. The explored length of the Black Sump grew each year, and in 1985, Ş. Sârbu and G. Mogyorosi completed the exploration of the first 200 m, pushing it to 260 m in 1986. In 1987, during a collaboration between Romanian and Czech divers, M. Piskula, T. Piskula, and F. Păroiu reach 365 m in length and 36 m of depth and the cave still continues!

The next few years witnessed a national effort of Romanian cave divers to further explore the cave. Among those, there were R. Geza and F. Baciu. In 1990, the latter reaches the end point of the existing guideline in a partly

M. Baciu (✉)
The Group for Underwater and Cave Exploration (GESS), Frumoasă 31, 010986 Bucharest, Romania
e-mail: mihai@SubAquaSport.com

G. Mogyorosi
Hungarian Speleological Society, Alkotas 5, Budapest, 1123, Hungary

V. Drăguşin
"Emil Racoviţă" Institute of Speleology, Romanian Academy, Frumoasă 31, 010986 Bucharest, Romania

Z. Gyurka
Hungarian Speleological Society, Bartok Bela 32, Isaszeg, 2117, Hungary

S. Storozynski
Hungarian Speleological Society, Szarvasmezo 19, Budapest, 2045, Hungary

G. M. L. Ponta and B. P. Onac (eds.), *Cave and Karst Systems of Romania*,
Cave and Karst Systems of the World, https://doi.org/10.1007/978-3-319-90747-5_20

Fig. 1 Location of Isverna Cave in SW Romania

Fig. 2 Map of Isverna Cave (map credits: Sz. Gyurka, Sz. Storozynski, I. Nemeth, J. Kovacs, V. Drăguşin)

solitary dive. The Franco-Swiss explorer J. J. Bolanz, during a solitary dive in 1994, reaches the same end point and advances for another 25 m.

The exploration enters a new phase when, in 1999, G. Mogyorosi and I. Brankovics reach the end of the guideline. At the same time, G. Mogyorosi and M. Baciu start a collaboration that would see the Black Sump finally crossed. One by one, the guidelines the other teams used in the sump are removed and the main guideline is correctly placed. On December 18, 2000, the two divers continue the exploration and reach an air bell after 495 m. Still, the Black Sump continues, and in 2001 the end is finally reached and it is surveyed for its full length of 695 m.

In 2002, the team reaches the Gravel Sump, in the main gallery, and partially explores it. After they cross it, in 2003 they begin exploring the Sand Sump but they abandon it shortly after discovering dry passages leading from the Throne Room toward the Boulder Room (Fig. 2). In 2005, G. Mogyorosi, A. Sari, Sz. Storozynszki, and M. Baciu reach another sump that M. Baciu explores solitary. Two years later the team returns to survey another 320 m; they stop in front of a large pile of boulders blocking the main passage. Exploration is still continuing!

Cave Description

Presently, the total length of the cave is 4055 m, of which more than half is completely submerged (Figs. 3, 4, and 5). The portion between the entrance and the Black Sump was described in more detail by Bleahu et al. (1976) and Orghidan et al. (1984). It formed in upper Jurassic–lower Cretaceous (Aptian) limestone (Codarcea et al. 1964) with stratification dipping toward the SE.

Fig. 3 Green Sump (photograph courtesy of A. Cohn)

Fig. 4 Exit of the Yellow Sump (photograph courtesy of A. Cohn)

Fig. 5 Black Sump (photograph by M. Baciu)

Although the underground river which surfaces just outside of the cave entrance, inside the cave it is first encountered at the Green Lake, some 200 m from the entrance. A small waterfall upstream of the lake connects it to the Green Sump (Fig. 3). From the lake room, an active and a dry passage both lead to the Yellow Sump (Fig. 4); beyond this point cave exploration is only possible using cave diving gear. Following an 80 m long passage, we arrive at the Black Sump, a wide gallery with a vast volume (Fig. 5), developed on a SW-NE direction. It first descends to −42 m then goes up to −12 m, only to descend again to −30 m, followed by another ascent to −7 m. Afterward, a vertical pit descends to −15 m and is followed by a passage running almost horizontally at −20 m for a length of 60 m. Again, the gallery descends to −44 m into a large chamber and then rises almost vertically to an air bell. From here, a high fracture that is partially aerated in the upper part runs for another 200 m with a maximum depth of −9 m up to the exit

Fig. 6 Throne Room (photograph by M. Baciu)

of the Black Sump. Beyond the Black Sump, the passage dimensions are smaller and the quantity of fine sediment is more substantial, hampering underwater exploration and survey.

From here, the active passage leads in a NW direction to the Gravel Sump (195 m long, −18.5 m deep) which is followed by the Throne Room (Fig. 6), formed at the intersection of two main fractures. From this chamber, there are two ways of progressing that run in parallel on a NE direction; one, through the Dry Passage with Formations (347 m) which is rich in speleothems but also presents important clay accumulations; the other is through the Sand Sump (~100 m long and a maximum depth of −39 m). The Sand Sump presents a difficult restriction at −20 m that made the survey of this long active passage very difficult. It is followed by two smaller sumps (maximum depth −20 m, total length approximately 250 m). At the convergence of the two parallel passages, we encounter the Boulder Room; this reaches heights of up to 20 m and is packed with large boulders, witnesses of an important collapse. In the middle of the Boulder Room, there is a major hurdle for the explorers, who have to negotiate their way through a very narrow passage in between the boulders. The main active passage continues in the same direction. Although the river here is at a low level, progress is slower due to the presence of several lakes and large accumulations of boulders.

At some point the main orientation of the passage becomes S-N for a distance of 150–200 m until the Final Sump where it returns to SW-NE. About 50 m before the Final Sump, the ceiling drops to a height of 2–3 m and the passage continues underwater. The Final Sump (30 m long, −5 m deep) surfaces in a tall passage (~8 m high) that is rich in speleothems. This last passage continues for a distance of about 400 m and is blocked by a large boulder collapse. During dry conditions, there are a few tributaries that emerge from the right side of the river, one after the exit from the Boulder Room and another one in the Side Chamber, right before the Final Sump. The last one had a temperature of 10 °C in the summer of 2007, compared to the 8 °C of the main river.

Cave Climatology

The cave is part of a monitoring program of the "Emil Racoviţă" Institute of Speleology. During 2014–2015, we monitored air temperature and CO_2 concentration. A Tinytag Plus 2 data logger was used to record temperature inside the cave since August 2014. The average temperature is 10.05 °C, showing little variability, between 10.02 and 10.10 °C. The CO_2 concentration in the cave atmosphere, which is an indicator of ventilation processes, was measured using a Vaisala GMP 222 probe. We observed a significant difference between summer and winter regimes. Summer values were found to be highest at the Green Lake (3680 ppm in August 2015), probably due to the large water surface of the lake and of the Green Sump that, together with the small waterfalls connecting the two, promote CO_2 degassing from the water. A similar situation is encountered at the Changing Room (3660 ppm), close to the Yellow Sump and close to the underground river. The lowest summer values were found closer to the cave entrance where, in August 2015, we measured concentrations of 1840 ppm. Winter values were as low as 300 ppm in December 2013 near the cave entrance, while at the Green Lake we recorded 1580 ppm. Nevertheless, at the Changing Room the values were as low as 720 ppm, revealing strong ventilation during winter at this site.

Cave Biology

The cave offers shelter to a bat colony of *Rhinolophus ferrumequinum*, *R. hipposideros*, *Myotis myotis,* and *M. emarginatus*. In their description of the cave, Bleahu et al. (1976) give a fauna list composed of *Mesachorutes ojcoviensis*, *Acherontia sp.*, *Quedius mesomelinus skoraszewski*, *Heteromyza atricornis*, *Trichoniscus* cf. *inferus*, *Trachysphaera jonescui isvernae*, *Lithobius decapolitus*, and *Sophrochaeta jeannelli*.

Acknowledgements We thank I. Nemeth and J. Kovacs for their contribution to the survey of the Black Sump. We also thank Sz. Bodony and the various team members that helped the exploration effort. VD was financially supported by the FREem project (IFA-CEA C4-08/2014). We thank A. Cohn and M. Piskula for providing the underwater photographs.

References

Codarcea A, Răileanu G, Năstăseanu S, Bercia I, Bercia E, Bişioanu C (1964) Geological map of Romania, Baia de Aramă Sheet. Scale 1:200.000, L-34-XXIX, Inst Geol, Bucureşti

Decu V, Povară I (1976) Peştera Isverna. In: Bleahu M, Decu V, Negrea Ş, Pleşa C, Povară I, Viehmann I (eds) Peşteri din Romania. Ed Ştiinţifică Enciclopedică, Bucureşti, pp 317–318

Orghidan T, Negrea Ş, Racoviţă G, Lascu C (eds) (1984) Peştera Izverna. Ed Sport-Turism, Bucureşti, pp 345–348

Mehedinti Mountains: The Cave from Mohilii Creek (Ascunsă Cave)

Virgil Drăguşin, Marius Vlaicu, and Emilian Isverceanu

Abstract

Ascunsă Cave is situated in the Mehedinţi Mountains and is part of the Isverna cave system. The cave was discovered in the late 1970s by members of the Focul Viu Caving Club, who surveyed the first 13.5 m of it. Since 2008, the "Emil Racoviţă" Institute of Speleology and the Underwater and Cave Exploration Group explored and surveyed 691 m of passages totaling 185 m of vertical development. Studies based on stalagmites from this cave revealed the climate evolutions during the last glacial cycle.

Keywords

Contact cave • CO_2 monitoring • Mehedinţi Mountains Romania

Geographic, Geologic, and Hydrologic Settings

The cave is located on the eastern side of the Mehedinţi Mountains (Fig. 1) and is part of the Isverna cave system, with its underground river being a tributary of Isverna Cave (Povară 2012). Between the closest points of the two caves, there are two kilometers in a straight line, implying a mature underground network. The cave is developed at the contact between Turonian–Senonian wildflysch (mélange) and the overlaying Upper Jurassic-Aptian limestone (Codarcea et al. 1964). Ascunsă is a temporary stream cave, functioning as a sinking stream (ponor) of the Mohilii Valley but also having its own underground river that appears in the White Chamber (Sala Albă) along the Tributary Passage (Galeria Afluentului). Drăguşin et al. (2017) showed that this tributary drains a mixed aquifer and is not infiltrating from the nearby creek.

History of Exploration

The cave was first reported by members of the Focul Viu Caving Club in the late 1970s, when it was surveyed on a total length of 13.5 m (Goran 1981). In August 2008, V. Drăguşin and E. Isverceanu blocked (redirected) the water of Mohilii Creek from entering the cave and in September of the same year E. Isverceanu, R. Băncilă, and A. Crânguş explored most of the new part of the cave. In October 2008, during the initial survey of the cave, E. Isverceanu reached a small sump on the Girls' Passage (Galeria Fetelor) at ~160 m below the entrance level. Because it was too small to dive a decision was taken to dig a hole through the wall to bypass it. This work lasted throughout 2008 and until August 2009 and was done mainly by V. Drăguşin and E. Isverceanu with the help of T. Marin, V. Voiculescu, M. Baciu, and M. Robu. After the bypass, E. Isverceanu and E. Buduran reached another end point in September 2009, where the passage was filled with sediments. Throughout 2009–2012, work was done in order to facilitate the access to the new end point. Located in a very narrow part of the cave, the excavated sediment could not be easily deposited, hindering progress. The work at this end point was completed with the help of E. Buduran, I. Axinte, I. Mirea, G. Ruică, and C. Cojocaru. Finally, in 2013 V. Drăguşin and A. Crânguş managed to pass through and crawled for another 6–7 m, up to a point where a large stalactite is blocking the passage. The exploration is still ongoing, with the goal to remove the stalactite and lower the water level beyond it, which stands at about 5 cm below the ceiling.

V. Drăguşin (✉)
"Emil Racoviţă" Institute of Speleology, Romanian Academy, Frumoasă 31, 010986 Bucharest, Romania
e-mail: virgil.dragusin@iser.ro

M. Vlaicu
"Emil Racoviţă" Institute of Speleology, Romanian Academy, Calea 13 Septembrie 13, 050711 Bucharest, Romania

E. Isverceanu
Underwater and Cave Exploration Group (GESS), Frumoasă 31, 010986 Bucharest, Romania

G. M. L. Ponta and B. P. Onac (eds.), *Cave and Karst Systems of Romania*,
Cave and Karst Systems of the World, https://doi.org/10.1007/978-3-319-90747-5_21

Fig. 1 Location of the Ascunsă Cave in the karst of Mehedinţi Mountains

Cave Description

The cave has two entrances, one in the form of a portal and the second one a pothole. After a series of tight crawls, the main passage opens up in the White Chamber (−25 m) and continues to have large dimensions for most of the explored parts of the cave. Passages developed entirely in limestone near the entrance and in the back of the cave are the tightest. The cave is at the moment 691.5 m and is 185 m deep (Fig. 2).

The cave morphology is relatively simple, a main gallery with only one short side passage, named The Tributary Passage. Because the wildflysch walls could not support massive limestone formations or were undermined by fluvial erosion, large collapsed blocks occupy most of the main passage. Calcite deposition cemented some of these blocks, generating new passages that run in between (Fig. 3).

Being a stream cave, there are a few fluvial sediment accumulations, mostly fine sand and silt, but only in restricted areas. The largest part of the sediments resulted through the collapse of the wildflysch walls, of the limestone ceiling, or even massive speleothems. A very large rock debris accumulation is located in the Mushrooms' Chamber (Sala Ciupercilor; −55 m), and it appears to have moved down through a small diameter chimney, now impenetrable. All along the cave, there are numerous speleothems, mostly stalactites, stalagmites, and flowstones, but other forms such as shields are also present.

Speleogenesis

The shaft entrance is probably a former sinking point (ponor) of Mohilii Valley, which was abandoned after the stream eroded into the subsurface and used the portal as its new

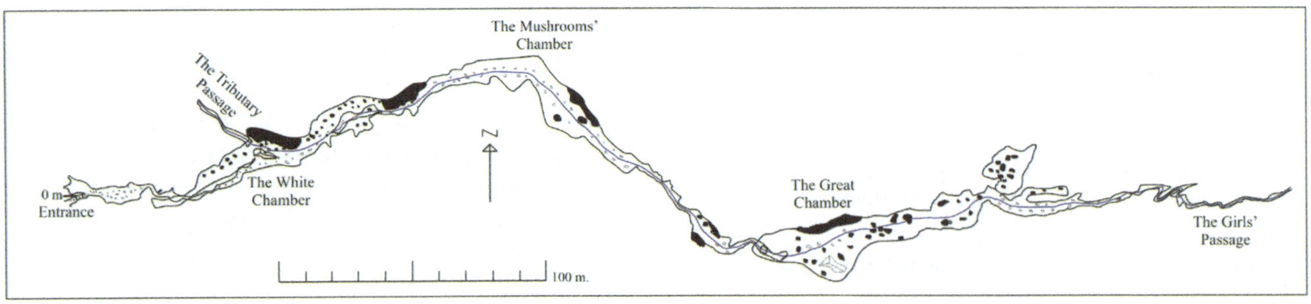

Fig. 2 Map of the Ascunsă Cave

Fig. 3 The Great Chamber. Please note the caver on the right (photograph courtesy of AP Iordache)

entrance/ponor. The presence of slickensides and ceiling anastomoses is showing that the initial development of the cave was along bedding planes. U–Th dating work (Drăguşin 2013; Drăguşin et al. 2014; J. Hellstrom pers. comm.) identified stalagmites with various ages, spanning the period since ~119,000 years ago until present. From field observations, much older speleothems are present, implying an even older age of the cave.

Cave Climatology

Ascunsă Cave is included in a monitoring program by the "Emil Racoviţă" Institute of Speleology (Drăguşin et al. 2017). The study aims to clarify the relationship between the underground climate and that from the outside, and the way in which climate signals are transferred from the surface into the underground via physical and chemical parameters of drip waters (isotopic composition, geochemistry, etc.). The mean cave temperature between February and August 2015 close to the White Chamber (−30 m) was 7.0 °C, whereas in the Great Chamber (Sala Mare; −112 m) it reaches 7.3 °C, reflecting the increase in values with depth.

For the period 2012–2015, CO_2 concentration in cave air varied between seasons, with the lowest levels recorded during March–May and the highest in November–January. These values do not rise above 3500 ppm although CO_2 concentrations around 9000–10,000 ppm were measured in drip water. This suggests that the cave is continuously ventilated, with an intensity that differs with the seasons. During the February–November 2015 period, when also drip water CO_2 levels were measured, these were higher in the White Chamber than in the Great Chamber (~9000 and ~7000 ppm, respectively). Nevertheless, due to stronger ventilation, air CO_2 concentration was almost equal at the two sites (~2200 ppm).

Cave Biology

During the cold season, a bat colony composed of several hundred individuals of four species: *Myotis myotis, M. emarginatus, Rhinolophus ferrumequinum,* and *R. hipposideros* congregate in the cave. Moreover, *Salamandra salamandra,* and species of the Coleoptera, Diptera, and Lepidoptera orders were observed inside the cave. Even in the deepest parts of the cave, at 100 m below the surface, rodents are active throughout the year.

Acknowledgements This work was financially supported by the FREem project (IFA-CEA C4-08/2014). It is also part of the research program of the "Emil Racoviţă" Institute of Speleology (Romanian Academy). We thank the administration of the Mehedinţi Plateau Geopark and all our friends and colleagues for their support during the exploration and research activities at the cave.

References

Codarcea A, Răileanu G, Năstăseanu S, Bercia I, Bercia E, Bişioanu C (1964) Geological map of Romania, scale 1:200000, L-34-XXIX, Baia de Aramă sheet, Institutul Geologic, Bucureşti

Drăguşin V (2013) Late Pleistocene climate variability recorded in stalagmites from Romania. Unpublished Ph.D. Thesis, Babeş-Bolyai University, Cluj-Napoca

Drăguşin V, Staubwasser M, Hoffmann DL, Onac BP, Ersek V, Vereş D (2014) Constraining Holocene hydrological changes in the Carpathian-Balkan region using speleothem $\delta^{18}O$ and pollen-based temperature reconstructions. Clim Past 10(4):1363–1380

Drăguşin V, Balan S, Blamart D, Forray FL, Marin C, Mirea I, Nagavciuc V, Orăşeanu I, Perşoiu A, Tîrlă L, Tudorache A, Vlaicu M (2017) Transfer of environmental signals from the surface to the underground at Ascunsă Cave, Romania. Hydrol Earth Syst Sci 21:5357–5373

Goran C (1981) Catalogul sistematic al peşterilor din România. Ed CNEFS, Bucureşti, p 456

Povară I (2012) Valea Cernei: morfologie, hidrologie, ape termominerale. Ed AGIR, Bucureşti

Mehedinti Plateau: The Zăton-Bulba Karst System

Ioan Povară and Cristian Lascu

Abstract

In the central part of the Mehedinţi Plateau, west of the Nadanova-Baia de Aramă tectonic corridor, Zăton and Ponoarele depressions, along with Bridge's and Bulba caves form a karst system developed in Mesozoic limestones. The Zăton Lake hosted by a karst marginal depression is located on Precambrian granitoides rocks, with its northeastern part in Mesozoic limestones. Dry for most part of the year, the lake fills after heavy rains or snow melting and is drained along the subterranean galleries of Bulba Cave (5360 m long). When the Zăton Lake is full, the water level rises and flow through the Bridge's Cave, flooding a large sector of the cave, forming a pond hydrodynamically connected to the Zăton Lake. Once the lake level recedes or dries out, the underground pond disappears. During years with heavy rainfall, a second temporary lake forms within the Ponoarele endorheic depression, which also drains along the underground network of the Bulba Cave. The Bridge's Cave, 734 m in length, presents an underground network consisting of dry galleries, galleries temporarily invaded by rising waters from below, and underlying—only partly accessible—galleries with perennial flow. Vertical shafts connect the passages with perennial flows, to overlying galleries. Bulba Cave is the local base level, which control the evolution of the entire karst system. In the central sector of the dry galleries, calcite draperies, soda straw, and candlestick stalagmites are common. Both caves host bat colonies, represented by the species: *Rhinolophus ferrumequinum*, *Rh. hipposideros*, *Rh. blasii*, *Miniopterus schreibersii*, and *Myotis dasycneme*.

Keywords

Karst system • Endorheic depression • Natural bridge Speleothems

Introduction

Zăton and Ponoarele endorheic depressions (Fig. 1), along with Bridge's and Bulba caves, are closely related as a result of sharing the same hydrographic network evolution and development of the subterranean drainage, which led to the present-day configuration of the karst system (Goran 1978). The Zăton Lake is disappearing underground at the interface between the non-karst rocks and the limestones. The low vertical range between the ponor of Zăton endorheic depression and the Bulba Cave outlet (only 20 m on a horizontal distance of 1230 m) has led to an extensive accumulation of clastic sediments (clays, sands, pebbles, cobbles) and organic debris (vegetal fragments), transported by tributary streams from non-karst catchment areas.

A complex description of the Zăton-Bulba karst system has been published by Bleahu et al. (1963) and Goran (1978). The latter study provides a map of the karst surface landforms and a cross section between Zăton and Bulba (Fig. 2), explaining how the hydrological karst system operates.

Geological and Hydrological Settings

The bedrock of the area consists of an igneous core (the Tismana granite), overlain by a Mesozoic sedimentary cover, assigned to the Danubian Autochthonous domain (Codarcea et al. 1968). It is comprised by Lower Jurassic sandstones, Middle Jurassic sandy limestones, and a thick Middle Jurassic to Lower Cretaceous limestone sequence, which is covered by transgressive marly limestones (Upper Cretaceous) and Upper Cretaceous Wild Flysch (black

I. Povară (✉) · C. Lascu
"Emil Racoviţă" Institute of Speleology, Romanian Academy, Calea 13 Septembrie 13, 050711 Bucharest 5, Romania
e-mail: ipov.iser@gmail.com

C. Lascu
e-mail: cristilascu48@gmail.com

G. M. L. Ponta and B. P. Onac (eds.), *Cave and Karst Systems of Romania*,
Cave and Karst Systems of the World, https://doi.org/10.1007/978-3-319-90747-5_22

Fig. 1 Location of Zăton-Bulba cave system

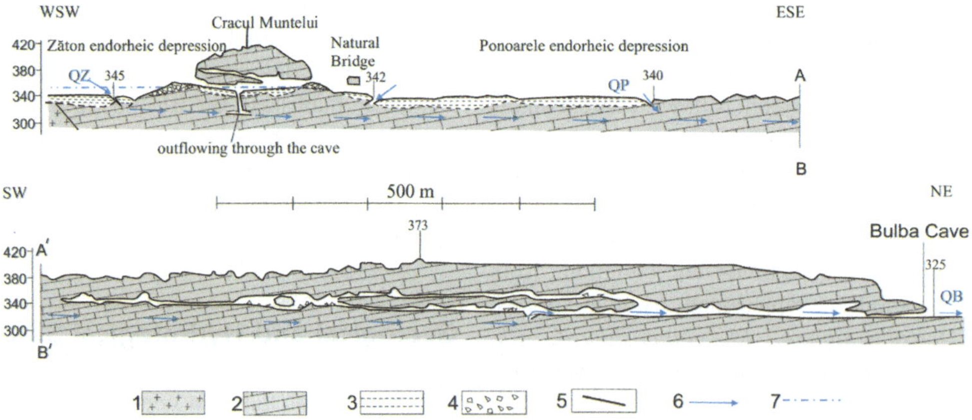

Fig. 2 Morphological and hydrogeological cross section between the Zăton endorheic depression and the Bulba Cave outlet (after Goran 1978, modified): 1. Precambrian granites and granodiorites; 2. Middle Jurassic to Lower Cretaceous limestones; 3 Quaternary alluvium; 4. Collapsed blocks; 5. Fault; 6. Groundwater flow path; 7. Highest water level in the Zăton Lake; QZ = discharge coming through the ponor of the Zaton endorheic basin; QP = discharge coming through the ponor of the Ponoarele endorheic basin; QB = total flow evacuated through Bulba Cave

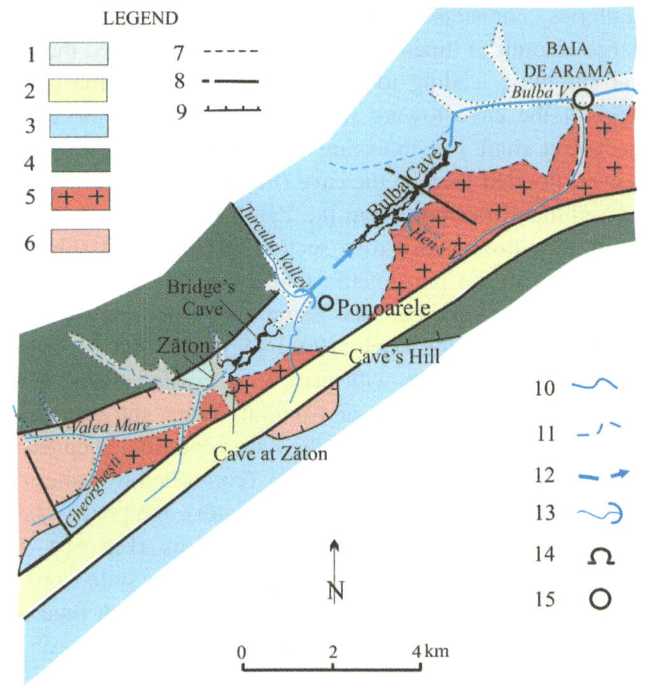

Fig. 3 Geological map of the Zăton-Bulba area (after Pop et al. 1975, modified): 1. Quaternary deposits; 2. Neogene gravels, sands, and clays; 3. Lower Jurassic to Lower Cretaceous rocks (sandstones, massive limestones, marls); 4. Upper Jurassic to Lower Cretaceous serpentines and argillites; 5. Precambrian granites and granodiorites; 6. Getic Nappe; 7. Geological boundary; 8. Fault; 9. Overthrust; 10. Perennial stream; 11. Temporary stream; 12. Groundwater flow path; 13. Swallow hole; 14. Cave; 15. Locality

argillites) (Fig. 3). The sedimentary sequence has been overthrusted by the Severin ophiolite Nappe and by the Getic Nappe (Drăghici 1962).

The area is an autochthonous outlier, exposed during the Miocene to disjunctive tectonic movements, which have led to the formation of the Balta-Baia de Aramă Graben, oriented NE-SW. To the west, the area is bordered by the Isverna-Ponoarele Fault. Located between these two major faults, the study area presents numerous fissures, joints, fractures, which facilitated the subaerial erosion, and the rainwater infiltration into the limestone rocks, to recharge the underground stream.

Zăton Endorheic Depression

Zăton Lake partly overlaps the catchments of two valleys (Valea Mare and Gheorghești) and their tributaries, which at the non-karstic/limestone interface are sinking underground through galleries located in the limestones of the Cracul Muntelui Hill. On the eastern edge of the depression, the Zăton Lake forms periodically. At the highest water level, the lake covers an area of ~ 1 km^2. Formed on the 6–8 m thick sediments overlaying the limestones, several funnel-like features (ponors) transfer the water of the surface streams, through karst conduits toward the Bulba Cave. When water reaches higher levels in the lake, the Zăton Cave entrance, situated 13 m above the depression bottom, becomes flooded.

Peștera Podului (Bridge's Cave)

The second component of the karst system, also known as Cracul Muntelui Cave, Ponoarele Cave, Podul lui Dumnezeu Cave, Podul de Piatră Cave, or Podului Cave, lies between the Zăton and Ponoarele depressions. The cave is 734 m in length and develops on three distinct hydrological levels: inactive, temporarily active, and active, all within a vertical range of 17 m (Goran 1982). The northern end of the Stream Gallery has collapsed, leading to the formation of a large doline and of the Natural Bridge (Fig. 4).

History of Exploration

Ionescu (1913) provided the first brief description of the cave. In 1928, Chappuis and Winkler explored the cave, but their findings were published only decades later (Chappuis and Jeannel 1951). A detailed presentation of the cave and its hydrogeological behavior as a part of the Zăton-Bulba karst system has been undertaken by Bleahu et al. (1962, 1963), Decou et al. (1967), and Goran (1978). The "Emil Racoviță" Institute of Speleology performed in 1973 a complex study with the intention of developing the cave for tourism.

Fig. 4 Natural Bridge from Ponoarele (photograph by I. Povară)

Cave Description

The *Stream Gallery* can be accessed for only 10 m, at the bottom of a 17.5 m deep shaft, opened in a short corridor oriented southeast, 60 m away from the Zăton Lake entry (Fig. 5). This short gallery is partially clogged at both ends, with very soft clay deposits. The *Stream Gallery* (Galeria Activă), 10–30 m wide and 4–10 m high, which temporarily hosts a pond generated by the flooding of the lower stream galleries, connects Zăton and Ponoarele depressions. The large amount of limestone breakdown has modified the floor morphology, leading to ascending slopes from the middle part of the cave toward the two entrances. In the central sector, a shaft is connecting the Stream Gallery with the lower, flooded level of the cave (Fig. 6a). As a consequence of sediment transport from the Zăton Lake, several sectors, with low flow path cross section, acting as flashboards, according to the piezometric tubes principle, formed along the stream passage between Zăton and Ponoarele endorheic depressions. Bleahu et al. (1963) suggested that the system features a main drain, with narrow and wide sectors, which control both water input and output. When the water flow rate measured on the edge of the Zăton endorheic depression is larger than the drainage capacity (controlled by galleries's diameter, which works as a regulator) temporary lakes (Zăton and the Bridge's Cave), will form (Fig. 6b). The same process may occur between the swallow hole from the Ponoarele endorheic depression and the upstream final section of the Bulba Cave, leading to the formation of the Ponoare temporary lake.

The Dry Gallery (Galeria Uscată) represents the first drainage pathway from the Zăton Lake to the Ponoarele endorheic depression. There are well-developed calcite crusts (on the floor) and other massive speleothems, which have blocked the galleries toward Zăton.

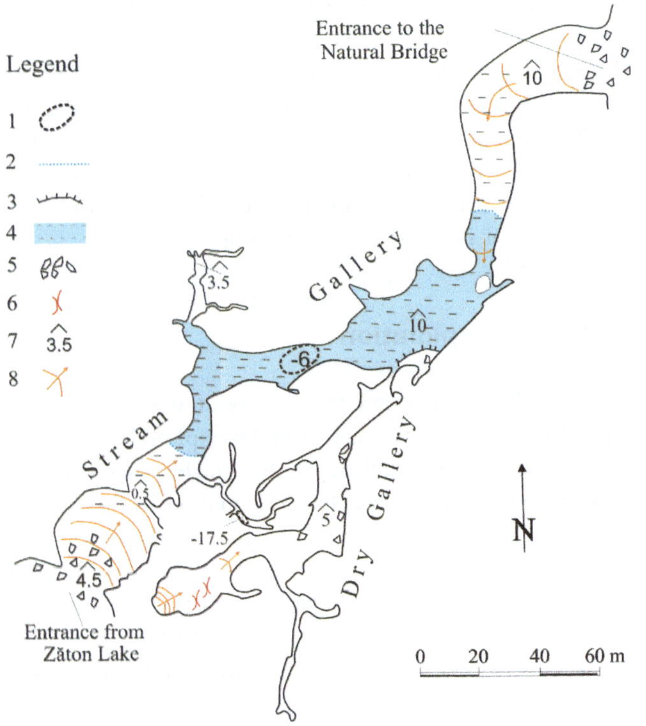

Fig. 5 Map of bridge's cave (after Bleahu et al. 1976, modified): 1. Shaft toward the lower, partly accessible stream gallery; 2. Underground pond boundary; 3. Terrace; 4. The maximum extent of the underground pond; 5. Breakdown; 6. Bat guano; 7. Ceiling height; 8. Contour lines with gradient arrows

Cave Clastic Sediments and Speleothems

The floor of the dry passages is covered by up to 6 m of clays, fine sands, and organic debris, washed in from the Zăton Lake. Breakdown deposits and boulders prevail in the vicinity of both entrances, along with other cryogenic-related sediments. The speleothems (stalagmitic domes, stalagmites, and stalactites) cover the most part of the walls and the ceiling of the Dry Gallery, whereas calcite crusts are present on its floor.

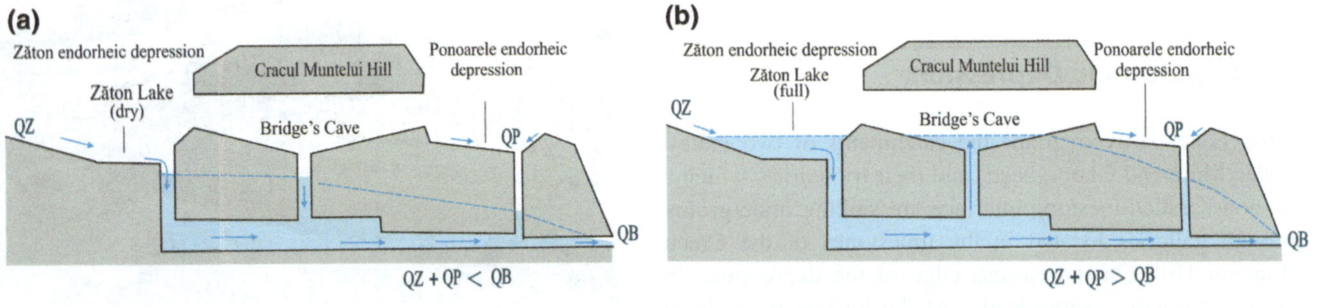

Fig. 6 Conceptual model of the hydrogeological functioning of the bridge's cave: $Q(Z + P)$ = cumulative flow Zăton + Ponoarele; QB = total flow evacuated through Bulba Cave

Cave Climatology

Temperature measurements performed in September 1961 and December 1973 indicated 10 °C in the final sector of the Dry Gallery, and 3 °C close to the shaft toward the stream gallery, respectively (Decu and Povară in Bleahu et al. 1976).

Cave Biology

Specimens of Gastropods, Isopoda, Aranaea, etc., have been recorded only along the Dry Gallery where favorable conditions are present. This gallery also hosts a bat hibernation colony (30–350 individuals) represented by the following species: *Rhinolophus ferrumequinum*, *Rh. hipposideros*, *Miniopterus schreibersii*, and *Myotis dasycneme*. During the summer, the same gallery shelters a few individuals of *Rh. ferrumequinum*.

Recreational Caving

Both entrances give access to the cave, but the one from Bridge's Cave is recommended. When the underground pond is full with water, the visitors will need an inflatable boat and rubber boots, the latter being useful even after the water drains away.

Bulba Cave (Baia de Aramă Cave, Peştera Mare)

History of Exploration

As with the previous cave, Ionescu (1913) provided a description of Bulba's first 50 m from the entrance. A general sketch of the cave was initially published by Bleahu et al. (1963), and later in Decou et al. (1967). A map of the underground network, showing 4860 m of galleries has been produced by the "Focul Viu" Caving Club, after successive surveying campaigns undertaken between 1974 and 1975 (see Bleahu et al. 1976). Later on, the members of the same caving club discovered and mapped other 300 m of galleries. In 1984, I. Grigore dove the sump at the upstream end of the river galleries. Beyond, he progressed for ca. 200 m upstream, until reaching a new sump (Şerban 1985). Along this newly explored cave section, he also discovered a dry gallery that reached into a previously known gallery. The survey of these latest explorations is not provided on the cave map in Fig. 7. A photograph album illustrating the morphology of Bulba's cave galleries and its speleothems was published by Lascu and Bleahu (1985).

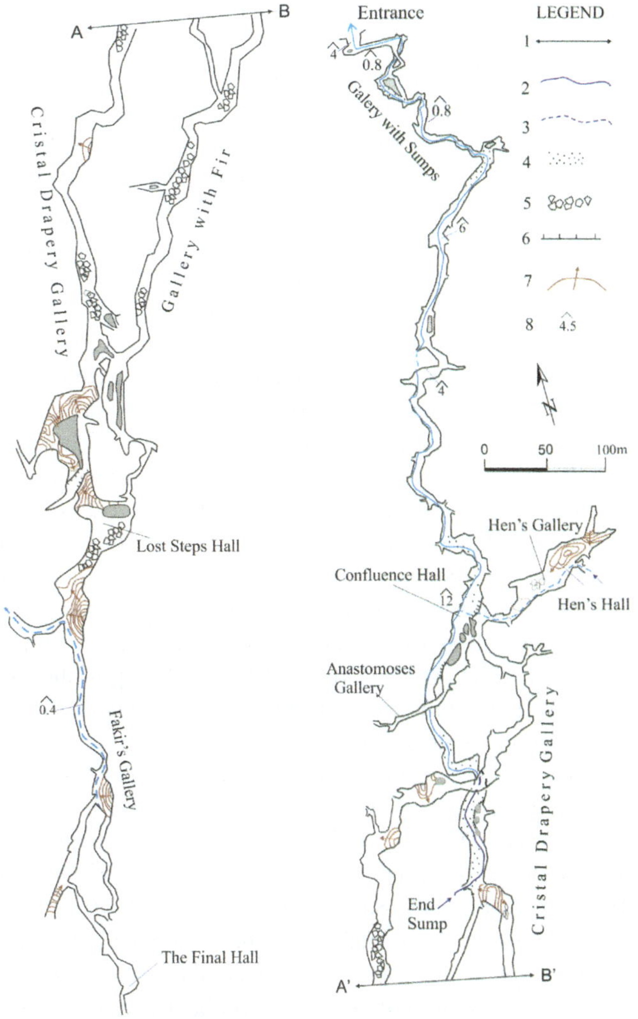

Fig. 7 Map of the Bulba Cave (after Povară et al. in Bleahu et al. 1976): 1. Connection line; 2. Perennial stream; 3. Temporary stream; 4. Alluvia; 5. Breakdown; 6. Terrace; 7. Contour lines with gradient arrows; 8. Ceiling height

Cave Description

The *perennial stream horizon* can be followed upstream from the cave entrance (Fig. 7). On the first 50 m, the passage is 2–4 m wide and 1.8 m high and continues for the next 180 m with a low (0.4 m high) corridor named *Gallery with Sumps* (Galeria cu Sifoane), which accommodates nine sumps, of which four are completely water-filled at high flow rates. The corridor continues with a 700 m long, 4–8 m wide and 5–12 m high gallery, ending in a sump. Given the amount of sediment carried and deposited by the underground river, the Gallery with Sumps flood rapidly causing upstream the rise of the water level. Fluorescein injected in the ponor of Zăton Lake has been detected at the Bulba Cave outlet after traveling 7 km in *ca.* 80 h, suggesting a velocity of 87.5 m/h (Unpublished data; A. Bulgăr pers. comm.). The drainage between the Ponoarele Depression and the Final Hall (Sala Finală) of Bulba Cave most likely occurs under phreatic conditions. From a hydrological point of view, Bulba Cave can be divided into two sectors: Between the Final Hall and the Confluence Hall, the flow is slow. Along this section, the stream meanders on a bed covered by clayey sand and sandy sediments, which may locally reach a thickness of 60 cm. In the Confluence Hall (Sala Confluenței), the tributary from the Hen's Hall (Sala Găinii) joins the river passage. Downstream from this confluence, gravels and large vegetal fragments, occasionally tree trunks, have been encountered. At exceptionally high flow rates, this debris may obstruct the Gallery with Sumps, flooding all upstream galleries to a height of *ca.* 7 m.

The temporary stream level includes the Gallery with Fir (Galeria cu Fir), which starts near the End Sump of the river galleries and continues for 1200 m on SSW direction. The only time the underground stream runs along this cave level is when the two endorheic depressions are water-filled.

The dry cave level is nearly 1 km long and formed by the Crystal Drapery (Galeria Valurilor de Cristal) and Anastomoses (Galeria cu Anastomoze) galleries. Close to the Lost Steps Hall (Sala Pașilor Pierduți; 80/30 m), several lateral corridors, make the connection with the temporary stream passages. A remarkable aspect of the underground morphology of the Bulba Cave is the evidence of structural control on the karst processes. The fractures pattern controlled the formation of parallelepiped limestone blocks ranging from a brick size (Fig. 8), up to 3 m. The collapse of limestone blocks and boulders due to weathering processes played a major role in the evolution of the Bulba's Cave underground network. The Lost Steps Hall and two-thirds of the temporary stream level (Gallery with Fir) have changed their volume as a result of these collapses.

The cave ceiling is separated from the ground surface by only a few tens of meters of limestone. This proves to be a mechanically labile setting, which favors the breakdown of limestone beds intercepted by the cave passages, as well as the enlargement of joints occurring in the cave walls. Passages which originally had formed under phreatic conditions have subsequently been altered (to a large extent) by such processes. Only certain mechanically favorable shapes (arch or cupola vaults) of the original passages remained unaffected.

Cave Sediments and Speleothems

There are a variety of sediments covering the floor of the galleries. Along the stream passages, clays, fine sand, and gravel deposits prevail (Fig. 9), whereas along the dry level, collapsed blocks and boulders have been accumulated. Throughout the cave, it can be observed how rectangular sections, with vertical walls and flat sub horizontal ceiling, alternate with elliptical and semicircular sections (Fig. 10).

The network of joints heavily influenced the type and distribution of speleothems. It can be noticed that dripstones mostly occur in fissure-rich sections of the galleries; moreover, there are alignments of stalactites (and corresponding stalagmites) along the joints opened in the ceiling (Fig. 11). Bulba Cave hosts (possibly) the most numerous and spectacular calcite veils (curtains) known in any Romanian cave. These formations abundantly occur along the Crystal Drapery Gallery, presenting red and white translucent calcite bands (Fig. 12), with a constant thickness of 5–8 mm and toothed like a saw blade (odontholites) edges.

Fig. 8 Parallelepiped blocks detached from the ceiling, partially coated with calcite (photograph by C. Lascu)

Fig. 9 Image from the upstream sector of the stream gallery (photograph by C. Lascu)

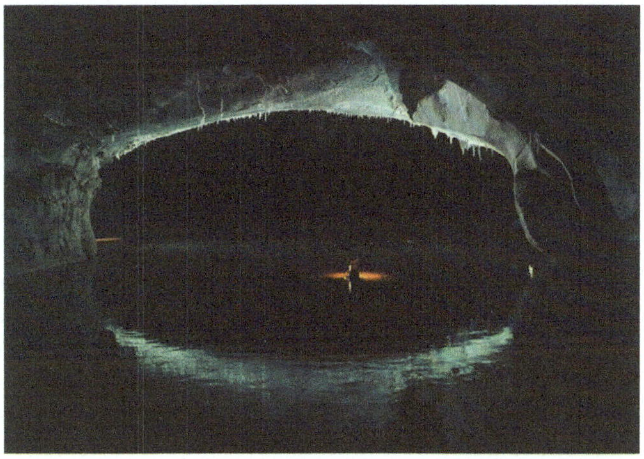

Fig. 10 Semicircular section in the dry passage (photograph by C. Lascu)

Fig. 11 Alignments of veils and soda straws developed along fissures (photograph by C. Lascu)

Fig. 12 Calcite veils in the crystal drapery gallery (photograph by C. Lascu)

Besides veils, Bulba Cave hosts other speleothems types, such as common conical stalactites, soda straws up to 40 cm long or bulbous stalactites, presenting in section calcite bands which alternate with layers of clays, related to exceptionally high floods. The most typical stalagmites (up to 2.5 m in height) are the "candlesticks". The "collars" (shelfstones) are witnesses of periods when the gallery was clogged with sediments, and the underground stream temporarily stopped flowing. One such speleothem documenting multiple clogging/reworking events is the "Small Fir" (Brăduțul), a stalagmite from the temporary stream galleries, which shows several successive circular calcite "collars" (Fig. 13). In the central sector of the upper level there are two rimstone dams, with fringed and meandering edges (Chinese walls).

Until the beginning of the Holocene, the Hen's Gallery was connected, through a large opening, with the homonymous valley; later on, the opening has been closed with debris related to collapses or originating from the slope (partially coated with flowstone), tree trunks and other vegetal fragments. This explains the presence in the Hen's Hall of several *Ursus spelaeus* skeletal fragments, along with a fossil bat guano deposit, with a volume of *ca*. 50 m^3, partially covered by present-day bat guano.

Cave Climatology

Only single measurements have been performed on the water and air temperature. During winter, when the Zăton Lake is frozen, the water temperature in the Gallery with Sumps reaches 4–5 °C, whereas in summer it rises up to 9–10 °C. In the Hen's Hall, the air temperature varies between 12 and 14 °C, likely as a result of bat guano fermentation (Decou et al. 1967).

Fig. 13 Stalagmite with successive circular calcite "collars"—a result of the alternate clogging/reworking events (photograph by C. Lascu)

Cave Biology

The cave trogloxene fauna is rich (Isopoda, Aranaea, Acari, Opiliones, Myriapoda, Diplura, Coleoptera), but, up to now, only two troglobious species have been identified, *Trichoniscus* cf. *inferus* and *Trachisphaera jonescui* (Decu in Bleahu et al. 1976). The cave hosts numerous bat species. A nursery colony has been observed in the Gallery with Sumps, in a hydrologically inactive, side corridor, located between sumps 1 and 2. The colony consists of specimens of *Myotis myotis*, *M. blythii* (*ca.* 200 individuals), and *M. schreibersii* (*ca.* 130 individuals). During the winter, small groups of *Rh. ferrumequinum* (the most frequent species), *Rh. blasii*, and *Rh. hipposideros* may be encountered in the same place. In the Hen's Hall, a maternity colony has been observed.

Cave Protection

The Zăton-Bulba karst system is a protected area included in the Mehedinţi Plateau Geopark, managed by the Mehedinţi County Council. The exokarst landforms (Zăton Lake, the karren field from the Cracul Muntelui Hill, and the Natural Bridge from Ponoarele) are well preserved and attract tourists. The natural bridge, partially damaged in 2010 by a natural process of physical–mechanical weathering, has been reinforced and restored. Moreover, a *ca.* 1 km long, detour road loop has been built in order to re-direct the heavy traffic. The Bulba Cave has been assigned to the maximum protection category ("A" class). Due to the difficult access, only a few visitors, mostly cavers and researchers enter the cave, thus is well preserved.

Conclusions

The Zăton-Bulba is one of the most representative karst systems in Romania. It has been developed in the Mesozoic limestones of the Mehedinţi Plateau. It comprises two endorheic depressions, Zăton and Ponoarele, as well as the Bridge's and Bulba caves. The outlet of the Bulba Cave is the local base level for the entire karst system. Following heavy rainfalls or snow melting, the Zăton and Ponoarele endorheic depressions are flooded, forming seasonal lakes at the contact between limestone and non-karst rocks. The filling of the two lakes and the flooding of several parts in both Bridge's and Bulba caves is controlled by a series of karst conduits having different diameters along the main drainage flow path from Zăton Lake toward Bulba Cave. The central section of the Stream Gallery within the Bridge's Cave can be temporarily invaded by rising waters from the partly accessible stream gallery below, allowing water derived from the Zăton Lake to form an underground pond inside the cave. The galleries of Bulba Cave reveal the relationship between the structural features of the limestones and the epiphreatic processes, rising a special interest for speleogenesis studies. On only 500 m, the Crystal Drapery Gallery hosts the largest concentration of flowstone veils known in any Romanian cave. The Natural Bridge from Ponoarele (also known as the God's Bridge) is an arch formed after the partial collapse of the ceiling next to the northern entrance of the Bridge's Cave. It is the best-known karst feature of this kind in Romania, and the only one that allows road traffic over it. Both caves host bat colonies that include six different species.

Acknowledgements The authors would like to thank Dr. C.-M. Munteanu for translating this chapter, Dr. M. Conovici for providing the geological information, as well as to Mr. V. Gheorghiu for the new data on bat species.

References

Bleahu MD, Decou A, Decou V (1962) Recherches spéologiques en Olténie en 1961. Spelunca 2:52–54

Bleahu MD, Decu A, Decu V (1963) Das hidrographiche System von Zăton-Ponoarele (Baia de Aramă). Rev Geol Geogr VII(1):147–156

Bleahu MD, Decu V, Negrea Ş, Pleşa C, Povară I, Viehmann I (1976) Peşteri din România. Ed Ştiinţifică Enciclopedică, Bucureşti, p 416

Chappuis PA, Jeannel R (1951) Enumération des grottes visitées, 1927–1949 (8-série). Biospeol LXXII(3):253–264

Codarcea A, Bercia I, Boldur C, Constantinof D, Maier O, Marinescu F, Mercus D, Năstăseanu S (1968) Geological structure of the Southwestern Carpathians. In: 23rd International geologigal congress, Prague, p 50 (Guide to excursion 49)

Decou A, Decou V, Bleahu M (1967) Recherches sur les grottes du Banat et D'Olténie (Roumanie, 1959–1962) Deuxiéme partie. Grottes D'Olténie explores de 1959 a 1962. Ed CNRS, Paris, pp 230–399

Drăghici C (1962) Structura geologică a Platoului Mehedinţi între Izverna—Cloşani—Padeş—Baia de Aramă—Ponoarele. DS Inst Geol XLVIII(1960–61):117–128

Goran C (1978) Le Karst du Plateau Mehedinţi, I. Partie nord. Trav Inst Speol "Émile Racovitza" XVII:217–226

Goran C (1982) Catalogue systematic al peşterilor din România. Ed CNEFS, Buxcureşti, p 496

Ionescu CN (1913) Biospeologia Carpaţilor Meridionali. Bul Soc Reg Rom Geogr XXXIV(2):74–98

Lascu C, Bleahu MD (1985) Peştera Bulba. Ed Sport-Turism, Bucureşti, p 128 (in Romanian with English summary)

Pop Gr, Berza T, Marinescu Fl, Stănoiu I, Hârtopanu I (1975) Geological map of Romania, Tismana Sheet. Scale 1:50.000, Inst Geol, Bucureşti

Şerban S (1985) Scurtă prezentare a sifoanelor din peşterilor României. Styx 2:71–83

Mehedinti Plateau: Epuran-Topolnita Karst System

Cristian Goran and Ioan Povară

Abstract

Epuran-Topolniţa karst system develops in the central-southern part of the Mehedinţi Plateau, in the vicinity of Jupâneşti and Marga villages, along a geological structure represented by a lower Cretaceous narrow limestone stripe oriented NNE-SSW. The underground drainage of the Topolniţa, Ponorăţ and Peţimea streams generated spectacular, large-sized caves developed on several levels, such as Epuran (3.6 km) and Topolniţa (22 km). The Epuran Cave is located on the western side of the limestone band, in the highly tectonized wall of a closed depression and presents two tiers of galleries. The upper, dry (fossil) level has a series of wide and narrow spaces, abundantly decorated by speleothems (stalagmites, flowstones, curtains, rimstone pools and cave pearls. The stream passage (lower level) is a labyrinth, characterized by breakdowns, clay deposits and marks indicating the water levels on the walls. Topolniţa Cave is the main collector in the area, the underground network being developed on five distinct levels, with large galleries and rooms, rich in speleothems and gigantic blocks collapsed from the ceiling. The cave presents multiple entrances, but the most spectacular one is Gura Prosăcului, a hydrogeological break into the limestone band by the Topolniţa River.

Keywords

Limestone bar • Tectonics • Speleothems
Archeology • Bats

Introduction

The Mehedinţi Plateau is a peneplain platform, 500–600 m above sea level (asl) cut by deep valleys. It is lower in elevation than the Mehedinţi Mountains in the east, which are part of the South Carpathian ridge. Even though the surface covered by limestones is small, the region is one of the most important karst areas of Romania due to its particular paleogeographic evolution and morphotectonic settings. There are over 200 known caves, some of them several kilometres long, such as Epuran-Topolniţa and Zăton-Bulba systems (Goran 2000). Due to its size and complex evolution of the karst network, the Epuran-Topolniţa system is the most important karst feature of the Mehedinţi Plateau.

The Epuran-Topolniţa system is located in the central-southern part of the region (the Topolniţa Basin) in a 3.5 km long, 50–200 m wide limestone band, oriented NNE-SSW, with an average altitude of 500 m (Fig. 1). The limestone ridge is partially covered by forest. Due to the presence of sinkhole valleys crossing the limestone bar (karst paleo-valleys), the ridge is divided in small closed depressions and plateaus, by saddle-like areas (Cornetul Jupâneştilor, 522 m, Cornetul Prosăcului, 527 m and Cornetul Sohodol, 475 m). The limestone stripe has steep slopes (100–150 m). Below them, the closed depressions Ponorăţ, Peţimea and the corridor of the Topolniţa Valley are formed along lithological contacts (Fig. 2).

Geologic and Hydrologic Settings

The geologic structure of the Mehedinţi Plateau comprises longitudinal ridges generally striking NNE-SSW. In the central part of the plateau, the crystalline bedrock is covered by Mesozoic sedimentary rocks (Iancu et al. 1986), whereas on its edges, two crystalline massifs are overthrusted on the sedimentary sequences forming a step-like landform. The Precambrian crystalline formations located on flanks outcrop

C. Goran (✉) · I. Povară
"Emil Racoviţă" Institute of Speleology, Romanian Academy, Calea 13 Septembrie 13, 050711 Bucharest 5, Romania
e-mail: cristian.goran@gmail.com

I. Povară
e-mail: ipov.iser@gmail.com

G. M. L. Ponta and B. P. Onac (eds.), *Cave and Karst Systems of Romania*,
Cave and Karst Systems of the World, https://doi.org/10.1007/978-3-319-90747-5_23

Fig. 1 Location of Epuran-Topolniţa cave system

at higher altitudes than the limestones and the Mesozoic flysch covering a central, quasi-continuous, tectono-erosional corridor.

Two Tithonian-Neocomian white-yellowish limestone ridges traversed by either transversal faults or paleo-hydrographic networks outcrop along the axis of the plateau's central corridor. These rocks form a line of high and prominent, conical or trapezoidal limestone erosional remnants (locally called *cornete*). The average width of the ridges is about 200 m and is covered by karren, closed depressions, dolines and doline valleys. On the edges, there are tectonic planes (100–150 m) suspended above closed depressions, which formed at lithological contacts. In the central corridor, the limestones are covered by upper Cretaceous siliciclastic deposits (sandstone–argillitic formation) developed in wildflysch facies (Fig. 2).

The tectonic structure in the region is complex. Besides the Severin Nappe and the crystalline of the Getic Nappe, which overthrust the autochthonous Mesozoic formations in Miocene as a result of several regional movements, a graben formed on the Baia de Aramă-Balta direction, amplifying the offset of the central corridor. South of Jupâneşti village, the graben continues as one major fault with numerous sub-parallel and transversal fractures, and only one limestone ridge is present. The hydrographic network is oriented NW-SE, perpendicular to the geological structure.

Several valleys show significant flow rates, being recharged by streams flowing over non-karst formations and karst springs originating in the Mehedinţi Mountains. The presence of the limestones in the central corridor and the opening of successive sinking points have concentrated and directed the subterranean drainages along the geological structure, towards outlet points located at both north (Zăton-Bulba system) and south (Epuran-Topolniţa system) (Goran et al. 2006).

The karst catchments feeding the Epuran-Topolniţa system are located on both sides of the ridge. These are old (pre-Quaternary) features that have been cyclically deepened. The remnants of the last surface drainage are now suspended on the plateau at 150 m whereas the cave entrances open at

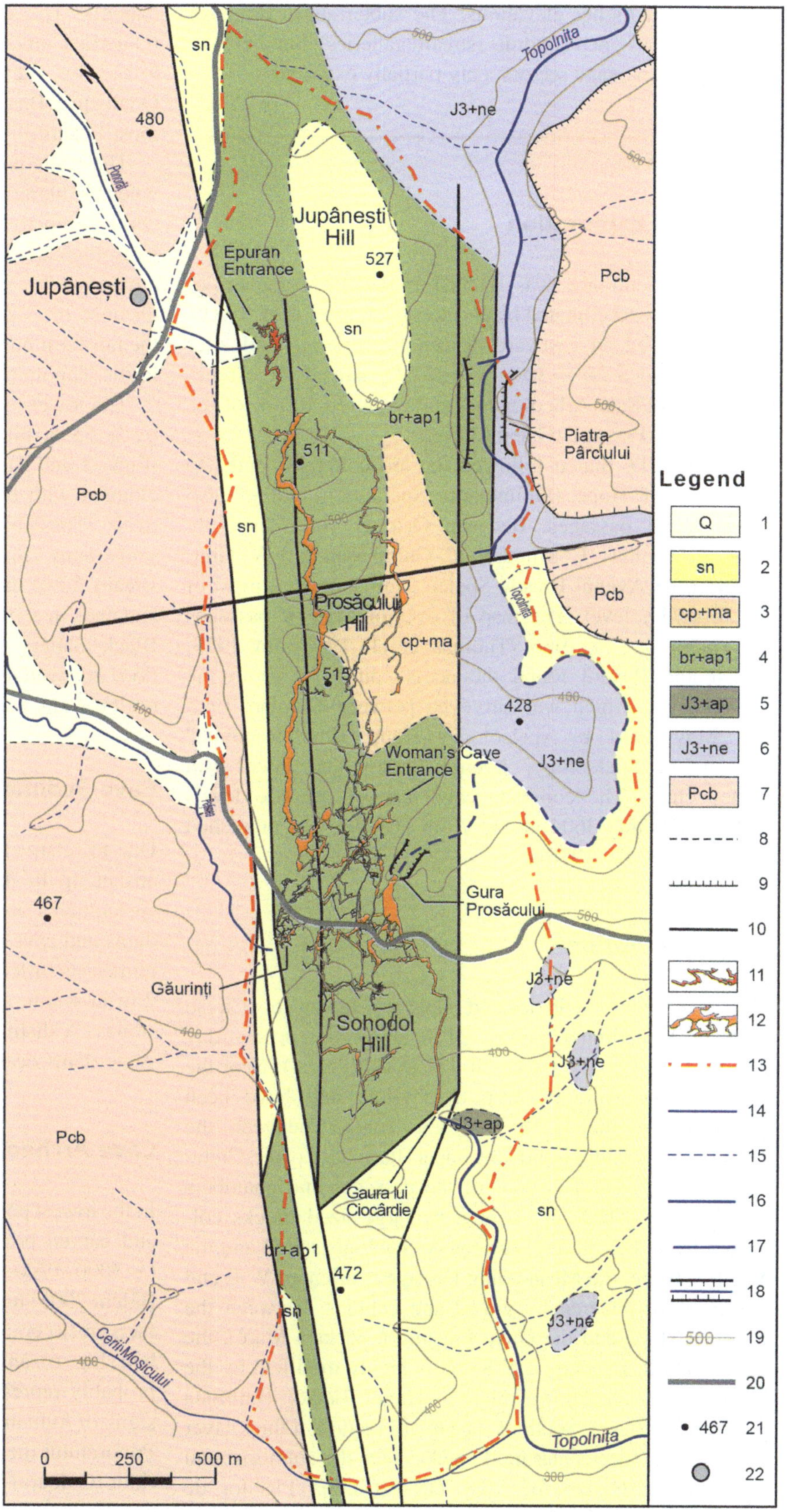

Fig. 2 Geomorphological map of the Epuran-Topolnița karst system (modified after Iancu et al. 1986; Goran and Vlaicu 2005). 1 Quaternary (clays, sands, gravels), 2 Senonian (clays, micro conglomerates), 3 Campanian-Maastrichtian (micaceous sandstones, clays, marls), 4 Barremian-Aptian (limestones in Urgonian facies), 5 Upper Jurassic-Aptian (reef limestones, nodules and chert), 6 Upper Jurassic-Neocomian (marls, marly limestones, calcirudites, calcarenites, cherty limestones, sandstones), 7 Precambrian (crystalline schists of the Getic Nappe), 8. Unconformity, 9 Overthrust, 10 Fault, 11 Epuran Cave, 12 Topolnița Cave, 13 Boundary of the karst system protection perimeter, 14 Perennial valley, 15 Temporary valley, 16 Ponor, 17 Outlet point, 18. Gorges, 19 Elevation line, 20 County road, 21 Elevation, 22 Village

different, but higher elevations. The subterranean drainage mainly follows open conduits (stream galleries), but there are also flooded passage sectors, only partially explored.

Epuran Cave

History of Exploration

The cave's entrance was opened in 1962 by a team of archeologists coordinated by N. Plopşor (Povară et al. 1981). They followed an eastward descending passage and after 12 m entered a 20 × 30 m room (later named the Archeologists Hall/Sala Arheologilor) and explored 100 m of the Meandering Passage (Galeria Cotită).

In 1964, D. Staicu followed the archeologists' paths to the Strait of Hope (Strâmtoarea Speranţei), finding also some lateral passages. Between October 1965 and July 1966, F. Thomas, L. Munthiu, C. Ghimbăşan and W. Gutt, all from the Avenul Braşov Speleo Club, surveyed most of the fossil/dry level. The map of over one km of passages was published in 1966 (Thomas et al. 1966). In 1973, Staicu and Povară found the access pit (−55 m) to the stream level, which was surveyed between October 1973 and May 1974 by speleologist from "Emil Racoviţă" Institute of Speleology, Avenul Braşov and Focul Viu speleo clubs. The complete cave map of the two levels (Fig. 3) totalling 3604 m and 75 m of relief was published in Bleahu et al. (1976).

Cave Description

The cave entrance is located close to Jupâneşti (Cireşu Village, Mehedinţi County), 24 m above the present-day ponor of the Ponorăţ Creek, at 425 m asl (Fig. 4). Near the entrance, the galleries form a labyrinth and have been developed parallel to the tectonic contact between the Danubian crystalline and the upper Jurassic–Lower Cretaceous limestones. The floor of the passages and chambers (Archeologists Hall) is covered by large-sized blocks collapsed from the ceiling. Between Archeologists Hall and the Strait of Hope, the Meandering Passage is oriented W-E and has its floor covered by gravel and cobbles. Between the Strait of Hope and the Great Fracture (Diaclaza Mare), the cave network is oriented N-S and is highly modified by the presence of breakdown blocks (Huge Blocks Hall/Sala Blocurilor Uriaşe). South of the Great Fracture, a thick layer of clay with spectacular mud cracks coats the subhorizontal floor of the Gallery of Mud Tiles (Galeria Plăcilor de Nămol). The cave continues with the Treasures Gallery (Galeria Comorilor) rich in a variety of speleothems, followed by the Basins Gallery (Galeria cu Bazine) and Odalisques Basins (Bazinele Cadânelor), which represent the most beautifully decorated part of the cave.

The walls preserve corrosion features and erosion levels, whereas along several sectors, there are gravel and sand deposits up to 2 m in thickness. The southern end of the hydrologically inactive passage (fossil) has been blocked by a massive flowstone formation. The survey indicates a linear distance of ~250 m between this end of Epuran Cave and the northern point of the Racoviţă Gallery (Galeria Racoviţă) in the Topolniţa Cave.

The lower level (stream passage) of the cave is accessible by a 55 m pit located in the vicinity of Great Fracture (Figs. 3 and 4). Two other pits found in the same area are clogged with boulders and clay at −9 and −23 m, respectively. The labyrinth like passage is 1300 m long and presents many chimneys, 15–20 m in height. The subterranean stream flows through the southern sector of the lower level and ends in a 10 m sump lake, the water heading towards the Black Gallery (Galeria Neagră) from Topolniţa Cave. The floor is covered by gravel, sand and large boulders, whereas on the walls corrosion features are present.

Cave Sediments and Speleothems

Due to relatively recent rock collapses, the speleothems are absent up to the Bears Hall (Sala Urşilor), where calcite speleothems are abundant, especially in terms of rimstone dams and cave pearls. South of the Great Fracture, there are numerous furled bacon draperies reaching 4 m in length and 2 m in width, soda straws, stalagmites, columns, and flowstones. A distinct feature is the 1–1.5 m high stepped rimstone dams occurring in the Odalisques Basins (Fig. 5).

Cave Archeology

In the lowest point of the Archeologists Hall, ceramic artifacts and carved bones dated to the beginning of the Iron Age (~3000–1700 BC), probably Getic (Lascu and Marinescu Bâlciu 2005) have been discovered. The most important artifacts are: a ceramic pot with three prominences and an alveolar belt (Fig. 6), an axle disk and nine bone points (Fig. 7). These probably represent a ritual "offering". Otherwise, there are signs of human habitation around the Cireşu Village, in the Ponorăţului meadow and at the entrance in the Hearths Corridor (Culoarul Vetrelor) from Topolniţa Cave.

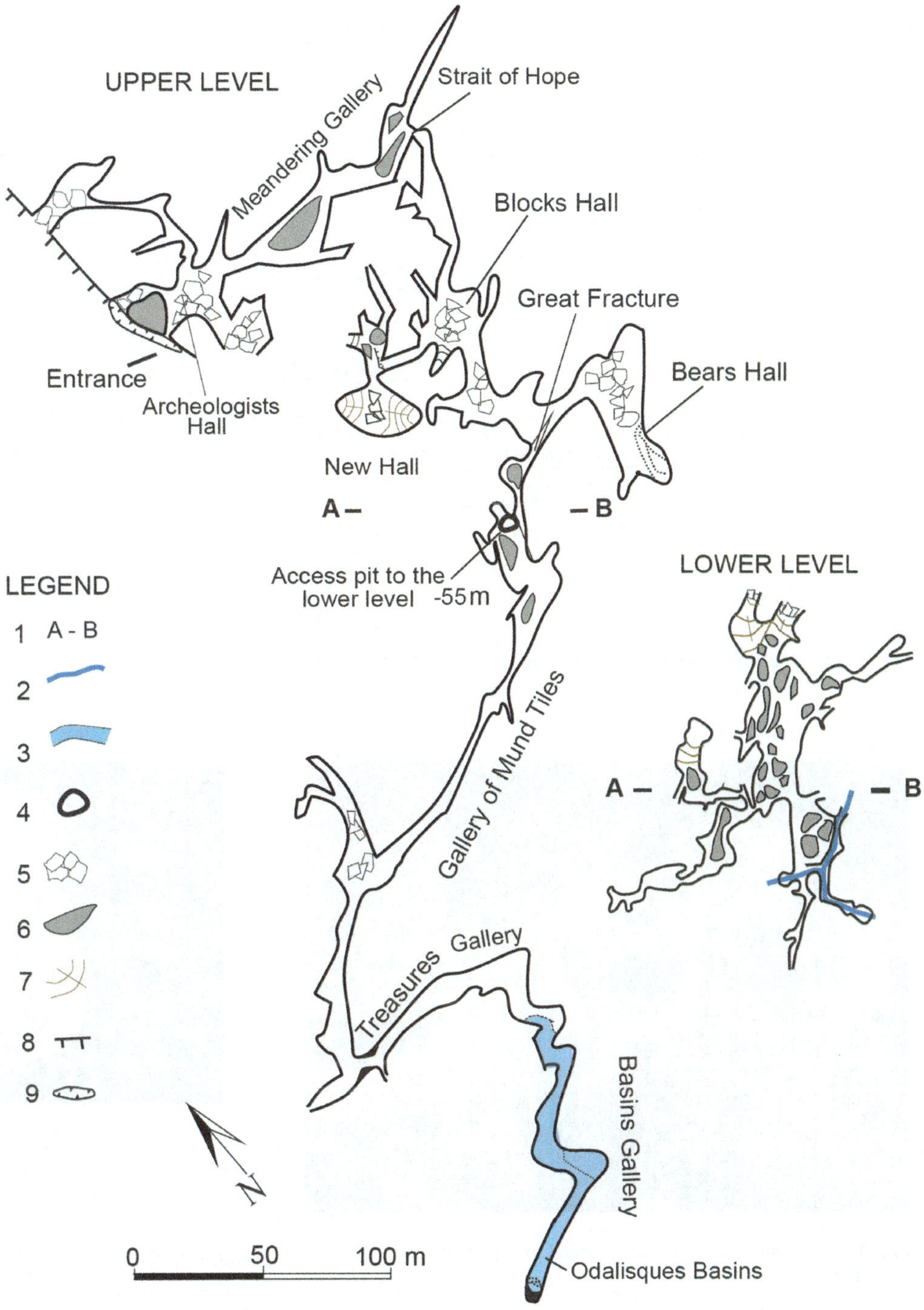

Fig. 3 Map of Epuran Cave (after Bleahu et al. 1976). 1 Connection line, 2 Perennial flow, 3 Temporary flow, 4 Pit, 5 Collapsed blocks, 6 Pillar, 7 Elevation line, 8 Limestone cliff, 9 Archaeological excavation

Cave Climatology

Because Epuran Cave has a single entrance, the predominant air current flows towards the exterior, and the air temperature varies between 5 °C in December and April and 7 °C in July (Thomas et al. 1966). Beyond the Strait of Hope, the air maintains a temperature of ±9 °C, a value close to the annual average at the exterior. There are no air currents in the stream passage, and the water temperature varies between 6 °C in winter and 8 °C in summer. At the entrance zone, the air temperature drop to −2 °C in the winter (e.g., December 2015), allowing for ice speleothems to form on the floor and at the base of the walls.

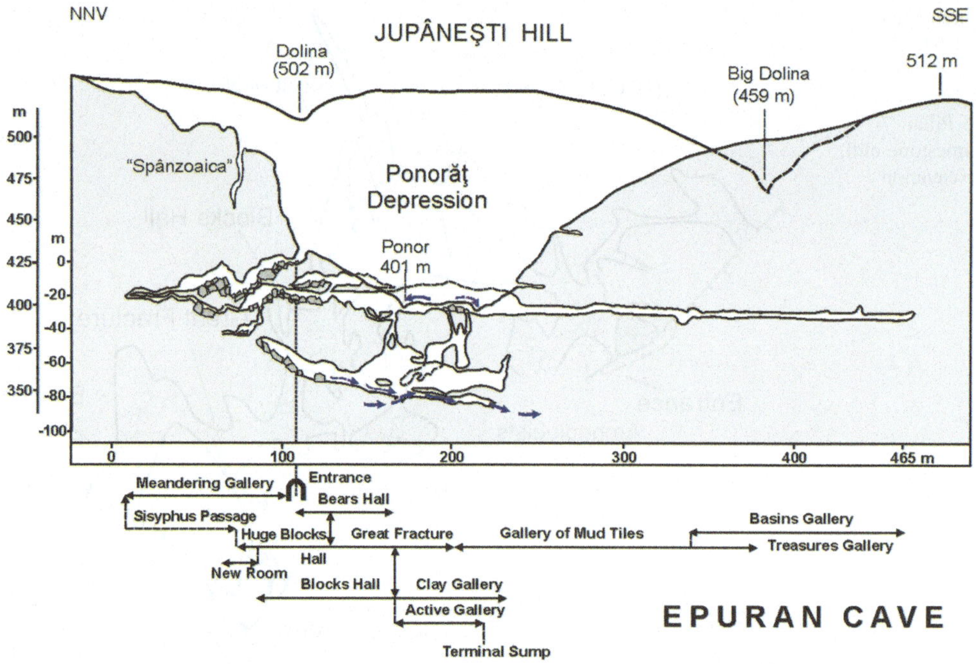

Fig. 4 Cross section showing the extension of Epuran Cave (after Goran 1998)

Fig. 5 The Basins Gallery in the Epuran Cave (photograph courtesy of R. Taffet)

Fig. 6 Ceramic vessel with three prominences and an alveolar belt (photograph courtesy of C. Lascu)

Cave Biology

The subterranean biotope of the cave favours the development of troglobitic fauna. The air temperature and humidity are relatively constant, and the trophic sources, for the most part, transported from the surface by infiltration waters, are abundant. The coleopteran *Cloşania orghidani*, the isopod *Trichoniscus inferus* and the amphipod *Niphargus variabilis* were found (Thomas et al. 1966).

In the winter of 2015, the species *Myotis myotis, M. blithii* (250 individuals), *Rhinolophus ferrumequinum* (50 individuals), *R. hipposideros* (15 individuals), *Myotis emarginatus* (10 individuals) have been recorded, but only in the Meandering Passage, up to the Strait of Hope (Vlaicu et al. 2013).

Fig. 7 Carved bone tools (photograph courtesy of C. Lascu)

Topolniţa Cave

History of Exploration

The first record of the cave belongs to Dimitrescu in 1880 (cited by Bleahu et al. 1976), followed by Murgoci (1898) and Vintilescu (1937–1938), the latter two mentioning the cave, but only describing karst landforms. Until 1961, just two cave entrances were known: Woman's Cave (Peştera Femeii; 21 in Fig. 8) and Gura Prosăcului, with only limited passages being investigated. The first underground exploration belongs to Prof. Sever Popescu, reaching through Gura Prosăcului [44], to the Otters Corridor (Culoarul Vidrelor; 46) and the first part of the Fractures Gallery (Galeria Diaclazelor; 18–42) through the Woman's Cave entrance (Bleahu et al. 1976). In 1961, A. Decu and V. Decu from the Institute of Speleology, joined in the following year by M. Bleahu, started the systematic investigation and mapping of Topolniţa, exploring all the entrances (Decu et al. 1967). The map of the cave with 10,330 m of subterranean passages was published in 1964 (Bleahu and Decu 1964). At the beginning of that year, D. Staicu, a native of the area, discovered three lateral passages near the Gura Prosăcului Entrance: Hidden Gallery (Galeria Ascunsă; 45), Suspended Gallery (Galeria Suspendată; 49–53) and Staicu Gallery [48–52].

From 1971 to 1975, I. Povară, G. Diaconu and C. Goran from the "Emil Racoviţă" Institute of Speleology, along with cavers from "Focul Viu" (Bucharest) and "Avenul" (Braşov) clubs resumed the exploration and mapping of Topolniţa Cave using precision survey equipment. Their survey extended the known development of the cave to 17 km. Since the hydrological link between the Epuran and Topolniţa caves was obvious, and the large stream passages ended upstream with sumps, there have been repeated attempts to connect the two caves by scuba divers. In 1975, F. Păroiu and C. Vânau made the connection between the Deep Sump (Sifonul Adânc; 11) from the upstream part of the Black Gallery (Galeria Neagră; 19) and the active sector of Gallery below the Karrens (Galeria de sub Lapiezuri; 17) (Halasi 1984). In 1980, C. Goran and four members of the "Avenul" (Braşov) discovered the entrance of an unknown passage system, developed parallel with the Prosăcului Gallery (Galeria Prosăcului) for 3500 m, which was named Topolniţa 2 (T2) (Goran 1982).

After 1990, Mehedinţi Speoalpine Club started a new campaign in the area. Beginning with 2001, German cavers led by R. Taffet join the investigation, achieving a series of important discoveries in the Epuran-Topolniţa System. These include several chimneys in the E. G. Racoviţă Gallery (followed by lateral passages) and a whole new level (~800 m long) connecting the Slope Gallery (Galeria cu Pripor; 28) with the Sohodol (Găurinţi) Cave. The partially mapped sector was named Topolniţa 3 (T3).

Cave Description

After the recent discoveries, Topolniţa cave system is almost 22 km long. Figure 8 shows 20.5 km of passages, cumulating with a vertical relief of 127 m (−109 and +18) and a total extension of 2.1 km. The survey benchmark (0 m elevation) is at the Women's Cave entrance, the highest elevation is in a chimney in Pripor's Gallery, whereas the lowest at Gaura lui Ciocârdie resurgence [51]. The cave system is developed along the limestone band occupying almost its entire width. The general development of the cave system follows a NNE-SSW direction, similar with that of the main geologic structure (Fig. 2). All five cave entrances are located in the southern half of the limestone stripe, on both sides of it, and at the boundary between limestone and non-karst rocks.

Because the junction between previously known openings has been achieved years after they were discovered, some of the entrances are still called "cave", whereas those crossed by the Topolniţa River are called "mouth". Also, the explorers called "galleries" the large, subhorizontal, longitudinal spaces and "corridors" the various connecting or secondary passages (Fig. 8). Three of the most important galleries of the cave are named after Romanian scientists: Emil G. Racoviţă, polar explorer, the founder of biospeleology, and of the world's first Institute of Speleology in Cluj (Romania) in 1920, Gheorghe Munteanu-Murgoci, a renowned geologist who studied the region and the first Romanian author of a paper dealing with karst issues and C.

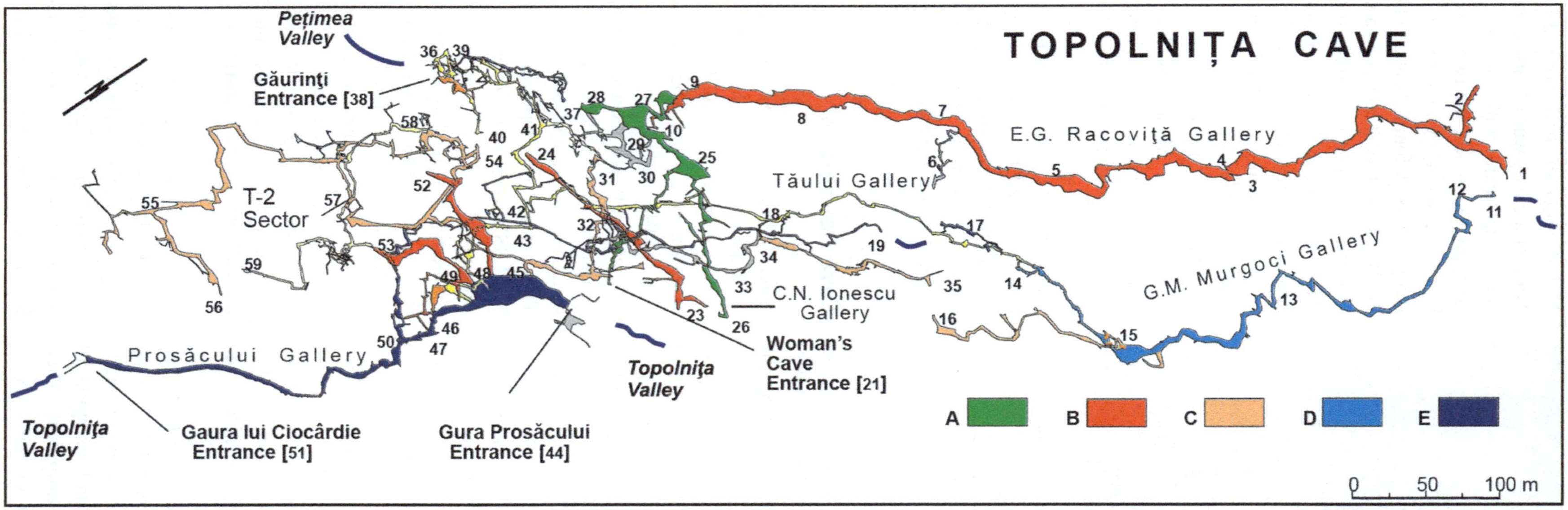

Fig. 8 Topolnița Cave—general map (surveyed by I. Povară, G. Diaconu, C. Goran, W. Gutt, O. Busuioceanu). A—upper fossil level, B—middle fossil level, C—lower fossil level, D—connecting passages, E—temporarily active level, F—permanent active level. See text: 1 Crystal Veil, 2 Red Dead-end, 3 Crystal Lake, 4 Great Dome, 5 White Palace, 6 Wonder Gallery, 7 Candles Forest, 8 Great Candle, 9 Quicksands, 10 The Rope ("La Coardă"), 11 Deep Sump, 12 Bended Sump, 13 Great Sump, 14 "350 m" Gallery, 15 branching Green Sump Gallery, 16 Green Sump, 17 Gallery below the Karrens, 18 Tăului Divide, 19 upstream sump on Black Gallery, 20 Tight Corridor, 21 Woman's Cave Entrance, 22 The Bridge, 23 end of Column Corridor, 24 end of Straight Corridor, 25 Guano Chamber, 26 end of Ionescu Gallery, 27 Great Chamber, 28 Slope Gallery, 29 Giants Gallery, 30 Confluences Chamber, 31 Three Maces, 32 Strait of Hope, 33 Basins Gallery, 34 Abyss Divide, 35 end of Abyss Gallery, 36 Pețimea Ponor, 37 Small Sump, 38 Găurinți, 39 Balcony Corridor, 40 end of Faults Corridor, 41 Blocks Chamber, 42 Sohodol Divide, 43 "Whirlpool" and "Roaring", 44 Gura Prosăcului, 45 Hidden Gallery, 46 Otters Gallery, 47 Aisle with Balcony Corridor, 48–52 Staicu Gallery, 49–53 Suspended Gallery, 50 Muddy Sump, 51 Gaura lui Ciocârdie, 54–55 Mineralogists Gallery, 55–56 Rimstone Dams Gallery, 57–58 "Upstream 45" Gallery, 57–59 "Downstream 45" Gallery

N. Ionescu, the first biospeleologist who explored caves in the South Carpathians (Bleahu et al. 1976).

One of the main entrances located on the eastern side of the limestone ridge is Gura Prosăcului, developed on a lithological contact at the end of the blind valley Prosăc Gorges. Gura Prosăcului is a 59 m high portal and the inlet point of the Topolnița River (Fig. 9). On the right slope of the gorges, some 85 m above the thalweg is the entrance to the Woman's Cave (400 m asl), which is closed by a metal gate. To the east, ~1 km south from Gura Prosăcului, the Topolnița River emerges through the 35 m high portal of Gaura lui Ciocârdie (282 m asl; [51]), and then continues along a short steephead valley.

In the western part of the limestone bar is the inlet point for the water of the Pețimea Valley (355 m asl) [36] as well as the "Găurinți" Entrance [38], which is located on the slope of a small limestone amphitheatre. Using any of the two entry points, cavers access what is known as the Sohodol Cave. Its galleries cross the limestone ridge towards the Prosăcului Gallery. At surface, between Gura Prosăcului and Găurinți, a saddle-like morphologic feature named La Varniță (Varniței Pass; 438 m asl) is located. It represents a remnant of the former Topolnița riverbed (Goran 1976).

Topolnița Cave presents a multilevel network developed on five main tiers, comprising large, mostly horizontal galleries parallel to the tectonic lines, and numerous side passages converging at different levels in chambers, or connecting passages and large collapsed voids towards the Prosăcului Gallery. Prosăc is the main collector of the karst system, parallel with the south-eastern edge of the limestone band (Fig. 8). From north to south, the cave network shows three different sectors, each with distinct morphology, speleogenesis and functionality.

Northern Sector. In the vicinity of the terminus passage of the Epuran Cave and the sinking point of the Topolnița River (Piatra Pârciului) originate three large, longitudinal drainages, corresponding to three galleries arranged vertically (Fig. 8) as follows: Racoviță (fossil), Murgoci-Tăului (temporarily active) and Black (stream passage). None of the three levels are directly connected to the surface, and their genesis is related to the water losses from the Ponorăț Depression through the Epuran Cave (NW), and in the case of the Black Gallery, by an additional input from the Piatra Pârciului Gorges (Bleahu and Decu 1964). All these galleries are sinuous, low-gradient single passages with a clear linear trend that follows the fracture networks. They host alluvial deposits and have relatively constant transversal profiles, but because belong to different levels, the morphological and sedimentological aspects are distinctive.

Racoviță Gallery [1–10] is the most important passage, both in terms of total length (1.7 km), size (widths and heights of 10–20 m), and by the abundance of some spectacularly beautiful speleothems. It has been developed parallel to the north-western edge of the limestone band. The passage is relatively linear, but it also presents a series of small bands related to the transfer from one fracture system to another. These changes in direction correspond to alternating sectors, some very high and others very wide. The first ones are rich in speleothems, whereas the others have the floor covered by massive collapsed blocks and more recent speleothems. The presence of the breakdowns led to important sedimentary deposits to accumulate behind them. In some areas, these alluvial deposits (sands, clays) are covered by calcite crusts on which various speleothems (domes, candlestick stalagmites, rimstones dams, etc.) have been precipitated (Figs. 10 and 11).

Suggestive names such as: the Crystal Veil (Vălul de Cristal; 1), the Crystal Lake (Lacul de Cleştar; 3), the Great Dome (Marele Dom; 4), the White Palace (Palatul Alb; 5), Candles Forest (Pădurea de Lumânări; 8), or the Great Candle (Marea Lumânare; 8) proposed by cavers for certain places along the gallery, provide a virtual image on the abundance and beauty of the speleothems (Bleahu and Lascu 1975).

The Racoviță Gallery has a few side passages; one of them is the Red Dead-end (Fundătura Roşie; 2), a clayey corridor with speleothems, chimneys, and rock fractures, similar to the Odalisques Basins in the Epuran Cave (Fig. 3). The other one is the Wonder Gallery (Galeria Minunată; 6), a descending corridor with a large guano deposit formed along a vertical fracture. At the downstream end of the gallery, in the area called Quicksands (Nisipurile Mişcătoare; 9), there is evidence of a flow bed, which downstream is split in two by a 10–12 m steep slope, called "The Rope" [10]. The upper level can be reached at the base of a massive breakdown in the Great Chamber (Sala Mare; 27), whereas the lower one connects under this chamber with the descending, transversal stream coming from the Giants Corridor (Culoarul Uriaşilor; 29), which then continues towards Gura Prosăcului.

Murgoci [11–14], ***Tăului*** [15–18], ***and Fractures*** [18–42] **galleries** develop along an alignment situated in the central part of the limestone ridge. They lack speleothems, but preserve many solution features formed by both slow-moving and turbulent water flow. The 1 km long and meandering Murgoci Gallery is located in the upstream section of the karst system. Its starting point is the Deep Sump [11] an insufficiently explored site. It continues towards south as a single stream passage showing a succession of high ceiling sectors, having their walls and floor covered by solution features (whirlpools, scallops, pendants, etc.), and low ceiling sections, which often are flooded or closed by temporary sumps: the Bended Sump (Sifonul Cotit; 12), the Great Sump (Sifonul Mare; 13) and the Green Sump (Sifonul Verde; 16). Initially, this gallery was considered to be at the base level represented by the Topolnița

Fig. 9 Gura Prosăcului Entrance (photograph by C. Goran)

Fig. 10 The Crystal Lake in the Racoviţă Gallery (photograph courtesy of R. Taffet)

River. However, because presently it is active (have running stream) only during very large flash floods, it is considered to be part of the epiphreatic (floodwater) zone.

The Murgoci Gallery ends downstream, in a short descending passage which hosts several whirlpools, leading to the Green Sump. An ascending branch located 10–12 m above the Green Sump connects with the Tăului Gallery [15–18], a fossil passage consisting of a succession of ascending and descending sectors, with several open fractures in the floor functioning as ponors (Fig. 12). Above the Tăului Gallery is the “350 m” Gallery [14], an inactive level accessible from the Murgoci Gallery, heading towards the upstream end of the Abyss Gallery (Galeria Prăpastiei; 35) and a lower one, which can be entered by a small pit. At the base of the pit is the Gallery below the Karrens [17], a continuation of the drainage coming from the Green Sump towards the upstream end of the Black Passage [19].

From the Tăului Gallery, the longitudinal drainage continues beyond Tăului Divide (Bifurcaţia Tăului; 18), with the Fractures Gallery [18–42], which is a high, narrow corridor with smooth walls, presenting a flow channel and small whirlpools on the floor. Its drainage capacity has been gradually reduced by the presence of pits on the south-eastern wall descending into the Black Gallery. The Fractures Gallery is initially linear, but afterwards presents a series of sharp turns following the rock fractures system and ends in the Sohodol Divide (Bifurcaţia Sohodol; 42). At this confluence it crosses the transversal, temporarily active drainage, coming from Găurinţi [38] along the Surprises Gallery (Galeria Surprizelor; 41–43), part of the middle sector of the cave, and continues downstream through the Otters Corridor [43] towards Prosăcului Gallery [46].

The Black Gallery [19–50] represents the lower, hydrologically active level of the northern sector, ending upstream in a sump [19]. It is much shorter than the other two longitudinal galleries and is located south of Murgoci Gallery and the Gallery below the Karrens. The water flowing along this passage is recharged from the Ponorăţ Depression (the stream from the Epuran Cave) and from Piatra Pârciului Gorges. Although the sump is unexplored, it is likely that a flooded passage parallel to the Murgoci Gallery or a branch beginning somewhere upstream of the Deep Sump [11] exist. The Black Gallery develops mainly along fractures and has sections of 1–2 m in width, but uneven ceiling. Its very low sectors are flooded during rainy seasons. The underground stream flows almost over the entire width of the

Fig. 11 Various speleothems along Racoviță Gallery (photograph courtesy of C. Lascu)

gallery, which host various flowstones and small amounts of alluvial deposit.

A number of chimneys drain the water flow from the upper-level galleries (Abyss, Tăului, and Fractures) and redirect them along the Black Gallery. The lower third part of this gallery is located in the middle section of the cave where it collects waters of different source: (1) a tributary originating in the Tight Corridor (Culoarul Strâmt; 20), (2) the stream from Găurinţi sector, through a series of conduits with a labyrinth pattern located in the eastern section and (3) from the diffuse losses in the Prosăcului Gorges.

In the final part of the gallery, the stream emerges from below the low ceiling of the Muddy Sump (Sifonul Înămolit; 50), on the right side of the Prosăcului Gallery. The upstream half of the Black Gallery is doubled by an upper fossil level represented by the Abyss Gallery [34–35]. This is located between the downstream end of the "350 m" Gallery and the Abyss Divide [34], close to the final part of the Basins Gallery (Galeria cu Bazine; 33), and only a few metres away from the Tăului Divide. The abyss Gallery is wide and contains speleothems and crusts deposited over thick clay deposits. Two large funnel-like passages descend to the stream level of the Black Gallery.

The Middle Sector. It is the second largest part of the Topolniţa cave system (Fig. 13), with passages oriented predominantly transversal to the main system of faults, many confluences and passages descending towards the lower level of the cave. The presence of faults between Prosăcului (N) and Sohodol (S) ridges is indicated at surface by a series of saddle-like depressions (Prosăcului, Varniţei and Găurinţi). Underground, this network of fractures controlled passages developed by draining the water at lithological contact on both sides of the limestone band. Two flow directions resulted based on the fracture pattern: (1) the first one is the result of losses of Topolniţa River water in the Prosăcului Gorges (E), which formed the section between the entrance in the Woman's Cave and Prosăcului Gallery and (2) the second is linked to Peţimea Creek water losses in the Găurinţilor wall (W), forming the section between the entrances in the Sohodol Cave and Prosăcului Gallery. In their final part (downstream from the Sohodol Divide), the two streams confluence.

The configuration of the cave patterns in the middle sector is more complicated compare to the other two due to numerous confluences, the presence of connecting passages between different levels, the presence of large areas

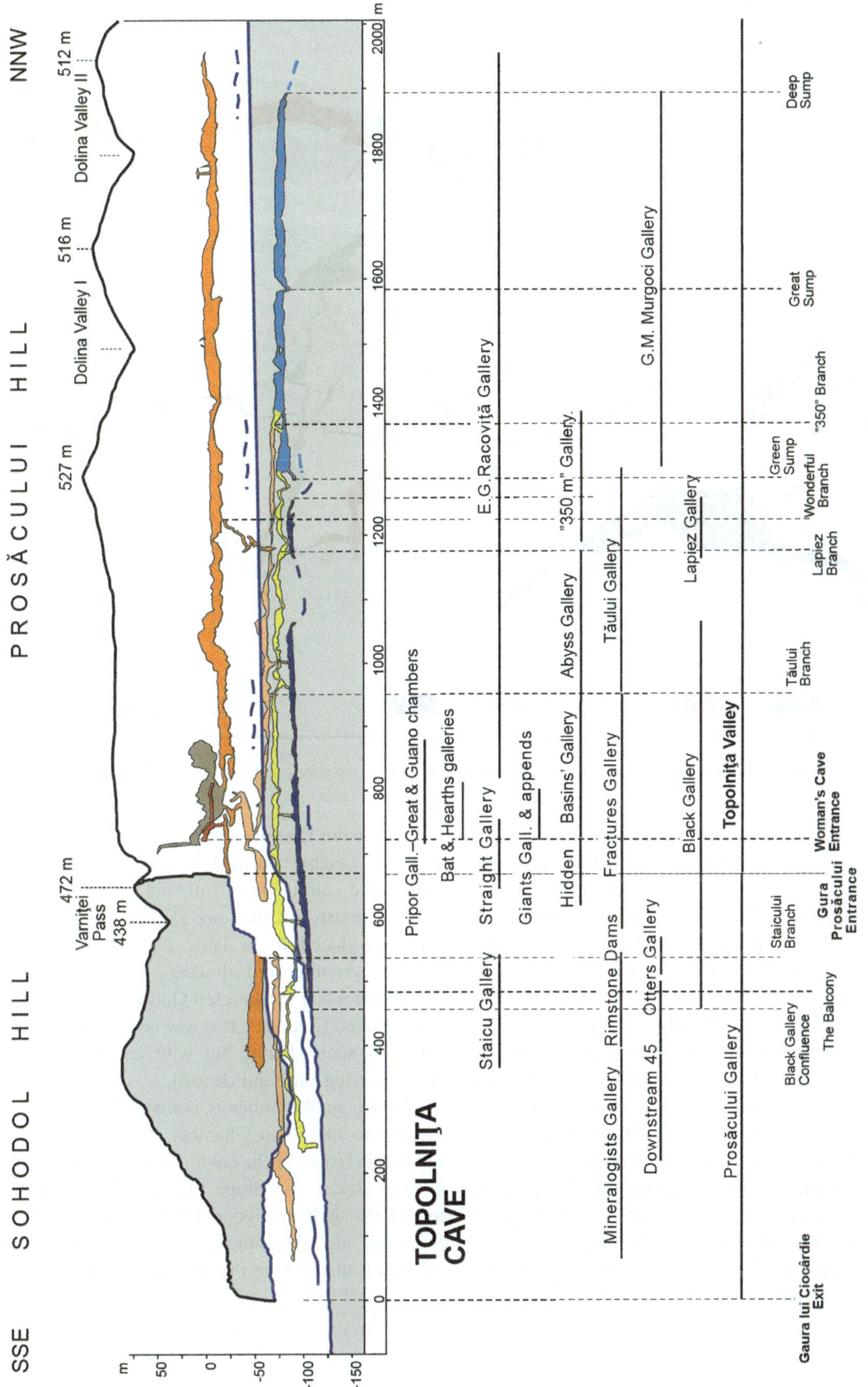

Fig. 12 Topolnița Cave: vertical projection on the extension direction (modified after Goran 1998)

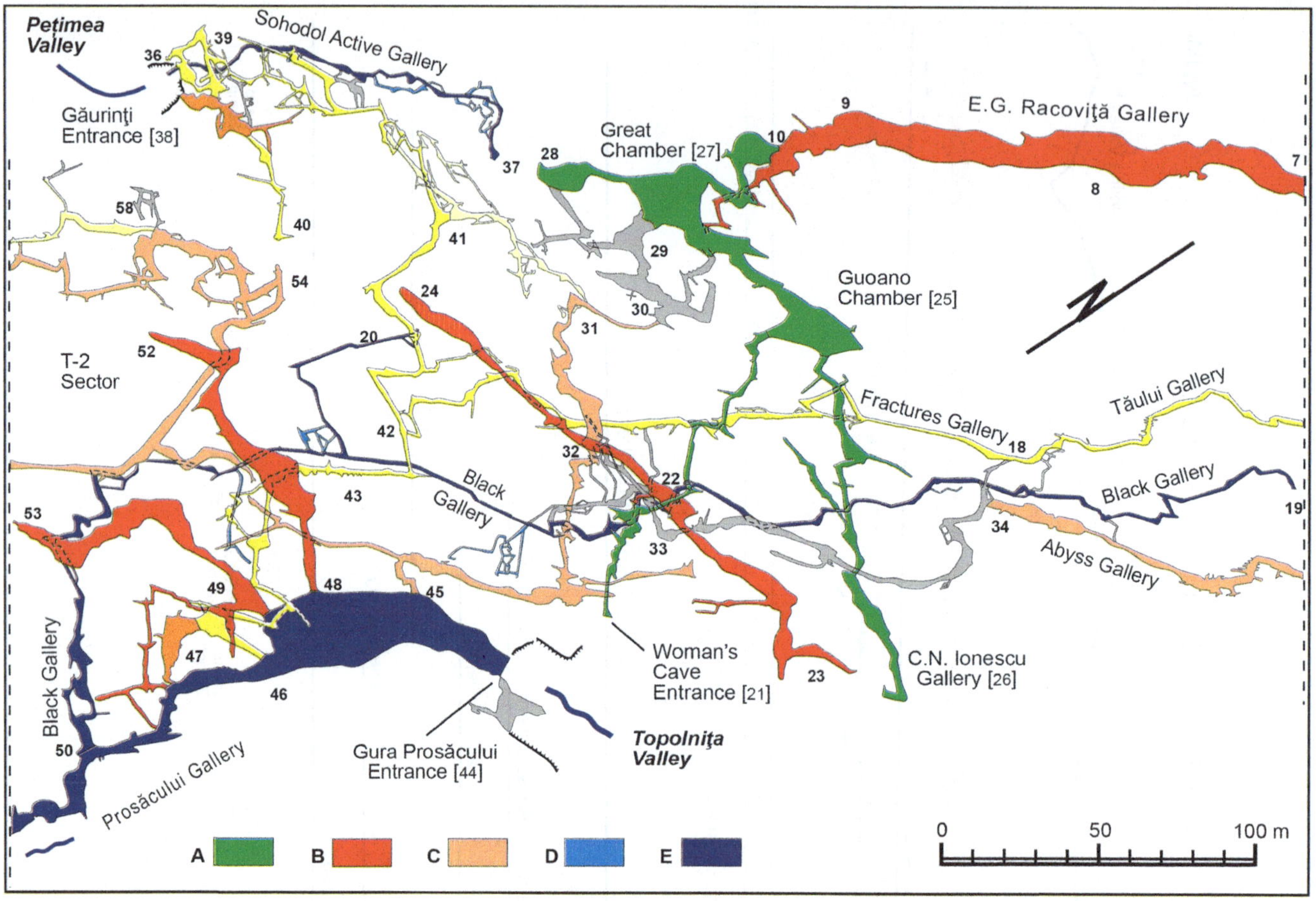

Fig. 13 Detail map of the middle sector of Topolniţa Cave (for legend and authors see caption of Fig. 8; after Goran 1998)

modified by collapsed rocks and large accumulations of boulders alternating with relatively smaller passages (Figs. 8 and 13).

The eastern drainage is the oldest and belongs to the upper fossil level. The gallery located at the highest elevation in the cave begins at the entrance in the Woman's Cave [21] with two relatively narrow (1–2 m) passages, Bats Corridor (Galeria Liliecilor; 21–22) and Hearths Corridor (Culoarul Vetrelor; 22–25). These lead to the Guano Chamber (Sala cu Guano; 25), the first big room of the cave. A metallic bridge [22] connects the Bats Gallery and the Hearths Corridor over a 10 m deep shaft, formed when the floor collapsed. The shaft leads to a lower level formed by the Columns Corridor (Culoarul Coloanelor; 22–23) and the Straight Corridor (Galeria Dreaptă; 22–24). Along these two lower-level galleries, probably part of the Topolniţa River was either lost or a drainage might exist from Găurinţi (Faults Corridor/Culoarul Faliilor; 40) towards the Prosăcului Gallery (Bleahu and Decu 1964).

The Guano Chamber is a large void (40 m in diameter, 25 m in height), located at an old confluence, with a cupola-like ceiling due to massive rock falls. Sizable speleothems exist, but they are extensively weathered due the guano produced by a large bat colony roosting at this location. Another old drainage coming from the gorge, functioned along the Ionescu Gallery [26], which also opens in the Guano Chamber. It is low and narrow towards its end, with less speleothems, but with an interesting stalagmitic floor covering the sand deposit.

The Guano Chamber is connected through a horizontal corridor to the Great Chamber (Sala Mare, 27; the largest and highest room of the cave). From here, at the upper end of a steep ascent, the Slope Gallery [28] begins, with large spaces and speleothems on the floor, and ends in a system of chimneys and labyrinth type passages leading to the Găurinţi sector of the cave (Taffet pers. comm.).

The longitudinal drainage coming from the Racoviţă Gallery (access is blocked by collapsed boulders) crosses at 15–20 m below the Great Chamber and continues on the other side with the Giants Gallery (Culoarul Uriaşilor; 29), towards the Confluences Chamber (Sala Confluenţelor; 30), the Tight Corridor [20] and Strait of Hope [32] (Fig. 13). In this part of the cave, an almost horizontal floor is covered with rimstones dams and massive stalagmites. At the place called the "Three Maces" (triplets stalagmites) ("La Trei Măciuci"; 31), the passage splits into two levels, both ultimately leading to the Prosăcului Gallery. The upper level continues through the ceiling of the Corridor of Hope [30–32] towards south through narrow sections filled with gravel deposits (access is restricted for protection reasons), to Hidden Gallery [30–45]. This is a wider, higher, relict passage, developed on a longitudinal fracture, ending in the right wall of the Prosăcului Gallery, 25 m above the stream.

The lower level belongs to the present day floor of the Corridor of Hope and continues, slightly descending, by a northward loop (in the opposite direction of the general drainage of the system). Following it, there is a narrow passage with beautiful speleothems—Basins Gallery [33–34]—leading towards the lower level of the Tăului and Fractures galleries, with which connects at the Tăului Divide [18]. The access in the Basins Gallery is also restricted for protection reasons.

The western drainage is named for historical reasons, Sohodol Cave. It represents a more recent tributary section of the Topolniţa cave system. Most of its galleries are smaller and narrower having a labyrinth pattern. This is due to many loops and niches opening in the lateral walls and numerous connecting passages between the fossil and stream levels. The sector presents three or four levels of galleries, decreasing in number downstream, which cannot be correlated with certainty with the levels from other sections of the cave system.

The water entering the cave through the Peţimea Valley ponor [36] forms the underground stream that flows along the Sohodol Active Gallery then through the Spiral Corridor (Culoarul cu Spirală) a passage with circular or elliptical morphology (1–2 m in diameter) ending in the Small Sump (Sifonul Mic; 37). This stream reappears in the Tight Corridor [20], a right side tributary of the Black Gallery. In the first part of the stream level, there are a series of conduits and narrow passages, which are connecting with the subfossil and fossil levels from above (Fig. 8).

The highest level of the Sohodol sector begins at the Găurinţi Entrance [38] with the Balcony Chamber (Sala cu Balcoane; 39), a void with unclear contour and many collapsed blocks, which opens towards the Step Gallery (Galeria cu Săritoare; 39–41), a high passage with straight walls, obvious corrosion traces and a series of speleothems towards its end. It developed along vertical fracture planes and becomes higher downstream. The name of the gallery comes from the Black Step (Săritoarea Neagră; 7 m), which preserves traces of past flooding events at its base. The water runs through several loops, but only during rainy periods. It flows from the side of the waterfall and leaves the main passage in the Blocks Chamber (Sala cu Blocuri; 41) through a gallery with a series of small waterfalls (Short Corridor/Culoarul Scurt), heading towards the Black Gallery. From the Blocks Chamber, the Sohodol drainage continues at the lower fossil level with the Surprises Gallery [41–43], a narrow, zigzagging corridor with alternating high and low sectors, which in the Sohodol Divide [42] joins the longitudinal drainage of the Fractures Gallery.

Downstream from the Surprises Gallery, the influence of tectonics on cave morphology is evident as the passages are very high and wide. The floor changes from bare rock in the "Whirlpool"/"La Marmită" and Roaring/"Vâjâitoare" (a pit towards the Black Gallery; 43) areas, to a floor covered by thick, sandy alluvial deposits. The last part of the drainage is a short corridor with several branches, the Otters Gallery [43–46], which ends in the Prosăcului Gallery [46].

Southern Sector. The southern sector of the Topolniţa Cave has been developed south of the tectonic line between the Găurinţi and Gura Prosăcului entrances, the karst network being located under the Varniţă and Cornetul Sohodol (Sohodol Hill; Fig. 12). It consists of multileveled, longitudinal conduits which reflect the drainage directions and the morphology of the limestone band. The Prosăcului Gallery, a typical major drainage, is the main system component, both with regard to its dimensions and by concentrating the entire drainage from the Epuran-Topolniţa karst system towards Gaura lui Ciocârdie outlet. Directly connected to the Prosăcului Gallery—developed very close to the southeastern boundary of the limestones unit—are a number of tributary passages located only on its west side, which have discharged in the past or continue to discharge the water draining from Hidden Gallery, Otters Corridor, and Black Gallery. Beside these, there are two other large, transversal galleries (Suspended and Staicu), which functioned as diverging various drainages towards former outlet points of the system. Northwest of the Prosăcului Gallery develops a parallel system known as Topolniţa 2, which consists of narrower and more winding corridors and diverticula (Fig. 12).

Prosăcului Gallery connects the inlet point of the Topolniţa River (Gura Prosăcului; 44) with the Gaura lui Ciocârdie [51] outlet through a 800 m long borehole, with a clear structural-tectonic configuration and very large passages. Its begins with a spectacular ogival portal (Fig. 9) developed on a vertical fracture, below which the riverbed is covered by huge limestone blocks. Topolniţa River flows through the Gura Prosăcului portal only at average and high water flow rates, because during drought periods its water entirely disappears underground at Piatra Pârciului or along the right side of the gorge. Downstream of the portal, the ceiling rises up to 80–90 m. The walls are vertical and present two ledges, representing remnants of old flow levels at 20–25 m and 50–60 m, respectively. The gallery remains very high through the first 250–300 m, in which all the side tributaries and passages opened on the right (west) wall. Downstream from this first sector, towards its final part, the gallery becomes more linear, the height decreases to 20–30 m, the collapsed blocks are smaller and fewer. Because the gallery floor is almost horizontal (very low gradient), the stream flow is slow; thus, it looks more like a lake than a river.

Another significant temporary tributary is the Hidden Gallery [45]. Its confluence hangs 20–25 m up the wall, only 40 m away from Gura Prosăcului entrance. About 100 m downstream is the confluence with the temporarily active Otters Gallery [46], which is connected through a steep slope to an upper meander of the Prosăcului Gallery that includes the Aisle with Balcony Corridor [47] (20–25 m level) and the Suspended Gallery [49–53], both opening in the Prosăcului Gallery a few tens of metres away from the cave entrance. The entrance into the Staicu [48–52] and Suspended is at the level of the ceiling (50–60 m) in the section between the Hidden and Otters galleries. This is where the Prosăcului Gallery reaches its maximum height. Both Staicu and Suspended have large passages (10–15 m in width and height), flat floors and interesting sediment deposits and speleothems. The Black Gallery is the only active tributary located 200 m downstream from the Gura Prosăcului entrance. The confluence is permanently clogged by alluvial deposits transported from the surface and form the Muddy Sump; 50). Downstream this point, Prosăcului Gallery has a perennial flow.

Topolniţa 2 is a well-individualized sector located northwest of the Prosăcului Gallery, oriented parallel with it, laying under the Cornetul Sohodol. Its 3500 m long passages are developed on two levels. Mineralogists Gallery (Galeria Mineralogilor; 54–55, Fig. 8) and the Rimstone Dams Gallery (Galeria cu Gururi; 55–56, Fig. 8) belong to the lower fossil level, whereas the Upstream Gallery 45 (Galeria Amonte 45; 57–58, Fig. 8) and the Downstream Gallery 45 (Galeria Aval 45; 57–59, Fig. 8) are temporarily active. Generally, they have narrow and high ceiling sections due to a higher degree of rock fracturing. These passages present frequent changes in direction and morphology and have lateral tributaries. The formation of this sector is linked to an old drainage (parallel with the Prosăcului Gallery) coming from the Găurinţi sector.

Cave Sediments, Speleothems, and Minerals

The most important sedimentary deposits reaching 1.2–3 m in thickness are located on the Racoviţă Gallery. Sedimentological, mineralogical and granulometric analyses emphasize the presence of two cycles, each including phases of alluvial accumulation, erosion and chemical precipitation (Horoi 1993). Significant alluvial deposits and collapsed boulders are known from the Staicu and Suspended passages and along the entire length of the Prosăcului Gallery. They have been transported and deposited inside the cave by the Topolniţa, Ponorăţ and Peţimea creeks.

Along the Hearths Corridor, Staicu Gallery, and in the Guano Chamber, there are up to 2.5 m thick guano deposits. Racoviţă Gallery has the most representative speleothems for the entire cave system: candlestick stalagmites (Fig. 14) with a diameter of 4–8 cm and 2.5 m height, banded draperies, soda straws and rimstone dams. Fine-floating calcite rafts form on the water's surface in these rimstone pools. The arenaceous stalagmites represent a distinct category in the central part of the Racoviţă Gallery. They are calcite-cemented sand features formed by water drops falling in the same spot from a height of 10–12 m. Sections of Staicu and Balcony galleries are exposed to periodic freezing; thus, bladed stalactites (Povară and Diaconu 1974) were noticed.

From a mineralogical point of view, Diaconu (1983) described several speleothem types composed of calcite, gypsum and aragonite, adding Topolniţa to a larger list of

Fig. 14 Candle stick forest in Topolniţa Cave (photograph courtesy of R. Taffet)

Romanian caves in which the gypsum–aragonite paragenesis has been identified. In the Suspended Room, brownish-to-black crusts deposited near guano deposits were investigated by means of X-ray diffraction, infrared, and thermal analysis. The results confirmed the presence of ardealite and brushite, two common cave phosphates (Diaconu and Dumitraş 2000–2001).

Cave Climatology

The size and complexity of the cave, along with the existence of five cave levels and five distinct entrances control the diurnal and seasonal topoclimate. During the warm period of the year, the underground air temperatures vary between 8.2 and 10.8 °C (Bleahu and Decu 1964). Several to tens of metres into the cave (adjacent to each of the entrances), there are ice speleothems. The air currents are generally unidirectional and vary diurnally and seasonally. In summer, the air gets into the cave through the upper entrances (Woman's Cave and Găurinţi) and exists through Gaura lui Ciocârdie. The flow reverses in the winter. In spring and autumn, there is also a diurnal circulation reversal, especially when temperatures vary significantly between day and night.

Cave Microbiology and Fauna

Hodorogea (1972) conducted analyses in order to detect microorganisms by their enzymatic activity (dehydrogenase, catalase, phosphatase, urease) on four different cave nutrient substrates (guano, moonmilk, marl and sandy clay). The results showed that the highest activity of the dehydrogenases, urease, phosphatase and catalase was recorded in the upper, fresh guano deposits. For all the other samples, the enzymatic activities were much lower. From a microbiological point of view, *Pseudomonas* and *Arthrobacter* were abundant in moonmilk and marls. In the latter one, *Bacillus circulars* was also identified. The greatest and most diverse population of microbes (*Arthrobacter, Bacillus circulars, B. megaterium, Actynomicetes* and a variety of *Cocci*) occurred in guano (Hodorogea 1972).

All types of subterranean biotopes and biocoenoses have been identified in the cave (Decu, in Bleahu et al. 1976). The following troglobitic species were recorded: *Closania orghidani, Trichoniscus cf. inferus, Orobainosoma hungaricum orientale, Neobisium maxbeieri* and *Niphargus sp.* (Dancău and Tabacaru 1964; Gruia 1976). There are numerous bat species; during winter, the Guano Chamber is occupied by a large colony of *Rhinolophus euryale, Myotis daubentonii, M. capaccinii*, whereas towards the entrance in the Women's Cave there are small colonies of *Rhinolophus ferrumequinum, R. blasii, R. hipposideros*, and *R. euryale* (Fig. 15). During summer, along the Hearths Corridor, there are several nursery colonies of *Myotis myotis* and *M. blasii* mixed with *M. schreibersii.* In the Staicu Gallery, during the entire year there is a colony of 4000–5000 individuals of *M. myotis, M. blithii, Nyctalus noctula,* and *N. leisleri* (Vlaicu et al. 2013).

Fig. 15 Nursery colony of *Rhinolophus euryale/blasii* in the Guano Chamber from Topolniţa Cave in the summer of 2015 (photograph courtesy of B. Szilárd)

Cave Conservancy

Epuran and Topolniţa caves as well as Topolniţa Gorges (the sector facing Gura Prosăcului) have been scientific reservation since 1965. Except for the Prosăcului Gallery, which is a B-class protected sector, the rest of the Epuran-Topolniţa cave system is A. Nowadays, this karst region is part of the protected area Mehedinţi Plateau Geopark. Epuran and Topolniţa caves have been classified as speleological preserves (protection class A). The protection of the speleological heritage has demanded important effort from the speleologists and custodians dealing with these caves. Over the years, walls limiting access to some sectors and gates at the entrances into the speleothem-rich passages have been built, especially with the contribution of the late speleologist W. Gutt from Avenul Braşov. Presently, access to the two caves requires authorization from the Speleological Heritage Commission and the presence of a ranger from the Mehedinţi Plateau Geopark.

Acknowledgements The authors express their appreciation to Avenul Braşov and Focul Viu Bucharest caving clubs, which took part in the 1972–1976 and 1980 exploration and mapping campaigns.

References

Bleahu M, Lascu C (1975) Peştera Topolniţa. Ed Sport-Turism, Bucureşti, p 75

Bleahu MD, Decu V (1964) Peştera Topolniţa. Ocrotirea Naturii 8:73–97

Bleahu MD, Decu V, Negrea Şt, Pleşa C, Povară I, Viehmann I (1976) Peşteri din Romania. Ed Ştiinţifică şi Enciclopedică, Bucureşti, p 416

Dancău D, Tabacaru I (1964) Observaţii zoogeografice asupra faunei cavernicole din Oltenia şi Banat. Lucr Inst Speol "Emil Racoviţă" I–II:509–575

Decu A, Decu V, Bleahu M (1967) Grottes d'Olténie explorées de 1959 á 1962. Recherches sur les grottes du Banat et d'Olténie, Roumanie (1959–1962). Centre National de la Recherche Scientifique, Paris, pp 239–242

Diaconu G (1983) Sur la paragenèse minéralogique gypse-aragonite rencontrée dans quelques grottes de Roumanie. Trav Inst Spéol "Émile Racovitza" 22:81–90

Diaconu G, Dumitraş D (2000–2001) Sur la présence de l'ardéalite dans la Grotte de Topolnita (Plateau de Mehedinti, Roumanie). Theor Appl Karstol 13–14:67–73

Goran C (1976) Evolution du reseau hidrographique dans la zone karstique Topolniţa-Epuran. Trav Inst Spéol "Émile Racovitza" XV:197–206

Goran C (1982) Catalogul sistematic al peşterilor din Romania. Ed CNEFS, Bucuresti, p 496

Goran C (1998) Mehedinţi Plateau. Carpatho-Balcanic Conf Geomorphol. Field Guidebook, Băile Herculane—Orşova—Drobeta Turnu Severin, 11–17 Oct 1998, pp 22–29

Goran C (2000) Carstul din Podişul Mehedinţi. Contextul regional şi subunităţile. St Cerc Geogr 47–48:131–149

Goran C, Constantin S, Horoi V (2006) South Mehedinţi Plateau and Epuran-Topolniţa cave system. In: Constantin S (ed) Field Trips in the karst of Romania. Selected routes from Southern Carpathians, Bucharest, pp 28–40

Goran C, Vlaicu M (2005) Planul de management al sistemului carstic Epuran-Topolniţa. APM Mehedinţi, Drobeta Turnu Severin, p 95

Gruia M (1976) Biotopes cavernicoles peuplés par des collemboles. Trav Inst Spéol "Émile Racovitza" XV:107–122

Halasi G (1984) Scurtă prezentare a sifoanelor din România. Bul Grup "Transilvania" Styx 1:21–38

Hodorogea P (1972) Recherches microbiologiques dans la grotte de Topolniţa (Roumanie). Première note. Trav Inst Spéol "Émile Racovitza" XI:335–342

Horoi V (1993) Sedimentological studies in Racoviţă gallery—Topolniţa cave (Mehedinţi Plateau, Romania). An evolutive interpretation. Theor Appl Karstol 6:129–144

Iancu V, Marinescu Fl, Hârtopanu M, Conovici M, Stănoiu I, Gridan T, Conovici N, Strusievicz R (1986) Geological map of României, Bâlvăneşti Sheet (41c, manuscript). Scale 1:50.000. Inst Geol Geofiz, Bucureşti

Lascu C, Marinescu Bâlciu S (2005) Noi date privind "depuneri" rituale în peşteri din Mehedinţi. St Preistorie 2:207–213

Murgoci GM (1898) Calcare şi fenomene de eroziune în Carpaţii Meridionali. Bul Soc Şt Bucureşti VIII(1):32

Povară I, Diaconu G (1974) Deroulement du processus de gelifraction dans le milieu souterraine. Trav Inst Spéol "Émile Racovitza" XIII:139–146

Povară I, Gutt W, Zakarias A (1981) Peştera Epuran. Ed Sport-Turism, Bucureşti, p 38

Thomas F, Munthiu L, Ghimbăşan C (1966) Peştera lui Epuran. Ocrotirea Naturii 10(2):177–198

Vintilescu I (1937–1938) Contribuţiuni la cunoaşterea carstului din Podişul Mehedinţilor. Cerc Stud Geogr, Ser II(1):123–140

Vlaicu M, Csaba J, Dragu A, Borda D, Goran C, Szodoray-Paradi F, Năstase-Bucur R, Niţu E, Murariu D (2013) Ghid pentru monitorizarea stării de conservare a peşterilor şi speciilor de lilieci de interes comunitar din Romania. Ed Advertising, Bucureşti, p 134

Caves of Cerna Valley

Gheorghe M. L. Ponta

Abstract

Over 300 dry (fossil) and active (stream) caves of thermal and hypogene origin are known in the upper Cerna Valley, between Naiba Valley and Pecinişca Village. The large variety of caves is sustained by a complex geology developed along the graben, with granites outcropping along the Cerna River's thalweg. The presence of this granitoid pluton is the source of a large geothermal anomaly in Băile Herculane area and a smaller one 21 km north, where two thermal springs are located. A synthesis of the Focul Viu Caving Club explorations, emphasizing the Great Cave of Şălitrari (the longest cave in the area) and Ogaşul Adânc, a stream cave that traverses the limestone strip, is presented in this chapter.

Keywords

Hypogene caves • Salpeter • Thermal springs

Geographic, Geologic, and Hydrologic Settings

In the Cerna Valley, the geological structure is composed of Danubian Autochthonous units (Cerna and Presacina sedimentary zones) and crystalline formations of the Getic Domain Nappe (Sebeş Lotru Series). The basement of the Cerna Valley consists of Precambrian gneisses (Neamţu Series), intruded by Cerna granite of Pan-African age (Liégeois et al. 1996), and covered partially by Permian red beds (conglomerates, sandstones, and shales) in Verrucano facies (Codarcea and Năstăseanu 1964; Ponta and Terteleac 2013) (Fig. 1).

Presacina sedimentary zone with deposits of Jurassic–Cretaceous age is developed on the right side (west) of Cerna Valley, in the Cerna Mountains being formed by a marginal facies/shallow waters deposits (Presacina facies) represented through marls, sandstones, and conglomerates of Lower to Middle Jurassic overlain by reefal facies limestones with chert nodules (Fig. 2). The entire limestone sequence is of Upper Jurassic (Malm)–Lower Cretaceous (Berriasian) age, with a thickness ranging between 100 and 200 m. These layers and some adjacent older deposits are overlaid by bedded limestones with marls interbeds, 50–75 m thick of Lower Cretaceous (Valanginian–Hauterivian), which gradually pass into limy marls that forms the so-called Iuta Layers (200–250 m thick) of Lower Cretaceous (Barremian–Aptian age) (Năstăseanu 1980).

On the left (East) side of Cerna Valley, in the Mehedinţi Mountains, is the Cerna–Coşustea sedimentary zone, where slightly recrystallized limestone sometimes oolitic or reefal in Urgonian facies with the same age as Iuta Layers directly overlay older Mesozoic deposits or the crystalline basement. In both sedimentary zones, discordant over the Iuta Layers or the limestones in Urgonian facies are disposed Turonian–Senonian (Upper Cretaceous) deposits in two facies: one predominantly formed by silty clays with olistoliths (wildflysch facies/Nadanova Formation) and the second one by conglomerate, sandstones (Mehedinţi flysch) (Simion et al. 1985; Bercia et al. 1987).

The Getic Domain Nappe is present in Godeanu or Bahna klippes, and along the Cerna Valley as a narrow strip. They are intense metamorphose formations of Sebeş Lotru Series and consist of gneisses pegmatites and micaschist with garnet and amphibolites. The present tectonic configuration is a result of the Alpine orogeny (Bercia et al. 1987).

Two types of tectonic elements, plicative and ruptural, are visible in the area. The plicative ones are represented through the Cerna syncline and anticline located in the right side of Cerna Valley (West) and Vârful lui Stan–Domogled anticline situated on the east side of the valley. The disjunctive elements as Cerna thrust sheet, the longitudinal fault which mark the Cerna Graben and also those, which cross transversal (East–West) the graben.

G. M. L. Ponta (✉)
Geological Survey of Alabama, 420 Hackberry Lane, Tuscaloosa, AL 35401, USA
e-mail: gponta@yahoo.com

G. M. L. Ponta and B. P. Onac (eds.), *Cave and Karst Systems of Romania*,
Cave and Karst Systems of the World, https://doi.org/10.1007/978-3-319-90747-5_24

Fig. 1 Location of Cerna Valley karst area within Romania

Cerna River is oriented northeast–southwest and flows along the graben. It represents a morphological boundary between Cerna and Godeanu to the west and Mehedinți Mountains in the east (Simion and Ponta 1983). Between these mountain units and downstream to the confluence with Belareca River, the Cerna Basin is about 412 km^2. The multiannual average discharge measured in 1978 by Meteorological and Hydrological Institute at Pecinișca, before building the dam at Valea lui Ivan was 15.8 m^3/s.

In 1979, G. Simion, G. Ponta with Geological and Geophysical Prospecting Company, and E. Gaspar with Institute of Physics and Nuclear Engineering Bucuresti performed a tracer study with I-131 (Iodine) in Poiana Țesna Ponor, located 13 km north of Băile Herculane, on the western banks of Cerna River. This sinking stream is located at the contact of noncalcareous rocks with the Jurassic limestones. Twenty days later, the tracer was intercepted in the Hercules I, Apollo II, and seven Izvoare Calde Dreapta thermal springs, extending the recharge of the thermal springs mentioned above at least 13 km upstream from Băile Herculane (Simion et al. 1985). The appearance of the tracer in 7 Izvoare Calde Dreapta situated 2 km upstream of Hercules I Spring confirms the presence of an underground diffluence along transversal faults, which determine a hydrodynamical connection of the Cerna syncline with the Cerna Graben Aquifer (Ponta 1998; Ponta and Terteleac 2006).

History of Exploration

The first speleological studies in the Cerna Valley were conducted in the second half of the nineteenth century and the beginning of the twentieth century, when the first 14 caves located in the Băile Herculane area were surveyed. In 1961, the "Emil Racoviță" Institute of Speleology (ISER) from București initiated an extensive study in the Cerna Valley, during which 62 caves located in the middle and upper part of the valley were surveyed and the findings published in three separate papers (Avram et al. 1964, 1966; Dancău et al. 1968).

The speleological and hydrogeological activities continued in the late seventies and early eighties under the leadership of

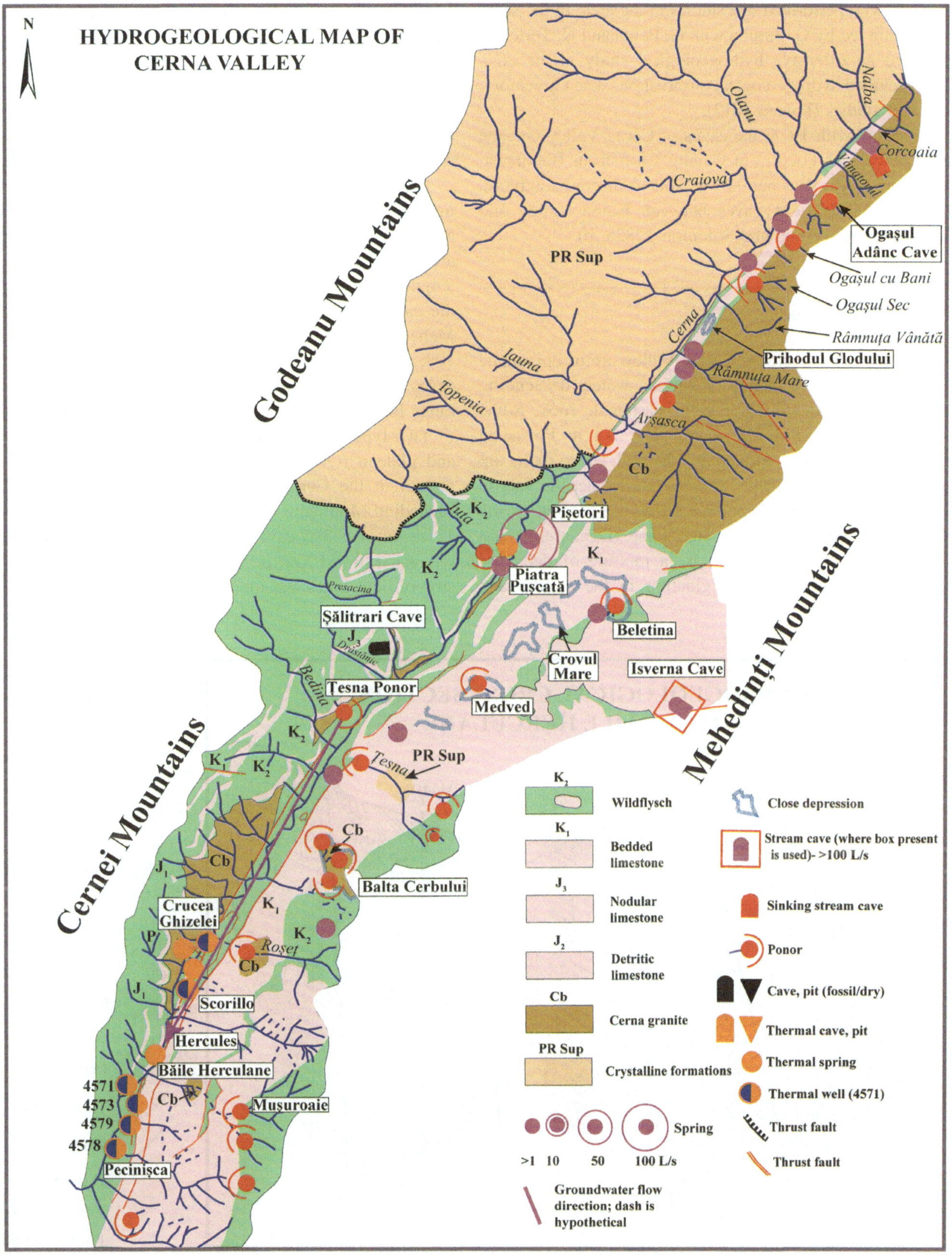

Fig. 2 Karst hydrogeologic map of the Cerna Valley, modified after Simion et al. 1985, with permission (geology after Năstăseanu (1980) and Bercia et al. (1987)). Key for the legend is available in the karst hydrogeology chapter

I. Povară, who performed dye studies in the area. In the same time, a team led by G. Simion with G. Ponta and N. Terteleac conducted an extensive hydrogeological study in the area, which included hydrogeological observations, karst inventory, and tracer studies (Povară 2012).

Beginning with 1978, the caves of Cerna Valley became the objectives of Focul Viu Caving Club from Bucharest, which over a two year period was able to discover, explore, and survey ~300 new caves between Naiba Valley and Băile Herculane (Ponta and Solomon 1982a, b).

Karst Morphology

Most of the caves along the Cerna Valley are of epigenetic origin, formed by the action of surface waters descending into the ground and dissolving the carbonate rock. Some caves located in Băile Herculane area and in Presacina Valley are hypogenic, formed by thermal water rising up from depth and mixing with surface waters; later the vadose phase prevailed. These hypogene caves are located in the geothermal anomaly of Băile Herculane (Demetrescu and Andreescu 1994) being carved by a mixture of deep hot waters (Fig. 3), which are originating in the granite and rainwater/river water. Some of the caves are developed along gravitational cliff fracture.

On the Mesozoic limestones of the Cerna Valley, a rich exo- and endokarstic landscape developed. Based on Table 1, the limestones outcrop in the upper Cerna Valley between Naiba Valley and Pecinişca are 79.69 km^2, which represent 19% from the total area in Cerna Basin (412 km^2).

From Table 1, it can be observed that the 83% of the limestones with 61.5% of the caves are located in the Mehedinţi Mountains and only about 10% of the limestones are in Cerna Mountains, with 38.5% of the caves. The total length of the cave passages is about the same, resulting that the caves in Cerna Mountains are longer than the ones in Mehedinţi Mountains. The total vertical range shows that 70% of vertical passages are in Mehedinţi Mountains. The caves from the plateau are fewer, but with a total vertical range higher than those developed in the bar type karst.

Two types of karst were identified in this area: karst bar and plateau (Onac and Goran 2018). The bar type karst prevail in the Cerna Mountains and north of Arşasca in Mehedinţi Mountains. The bar type karst with a total area of 12.18 km^2 has 295 caves versus 61 developed in the plateau type karst (67.51 km^2 or 85%). More than likely the narrow strip of limestones was more affected by the fracture systems

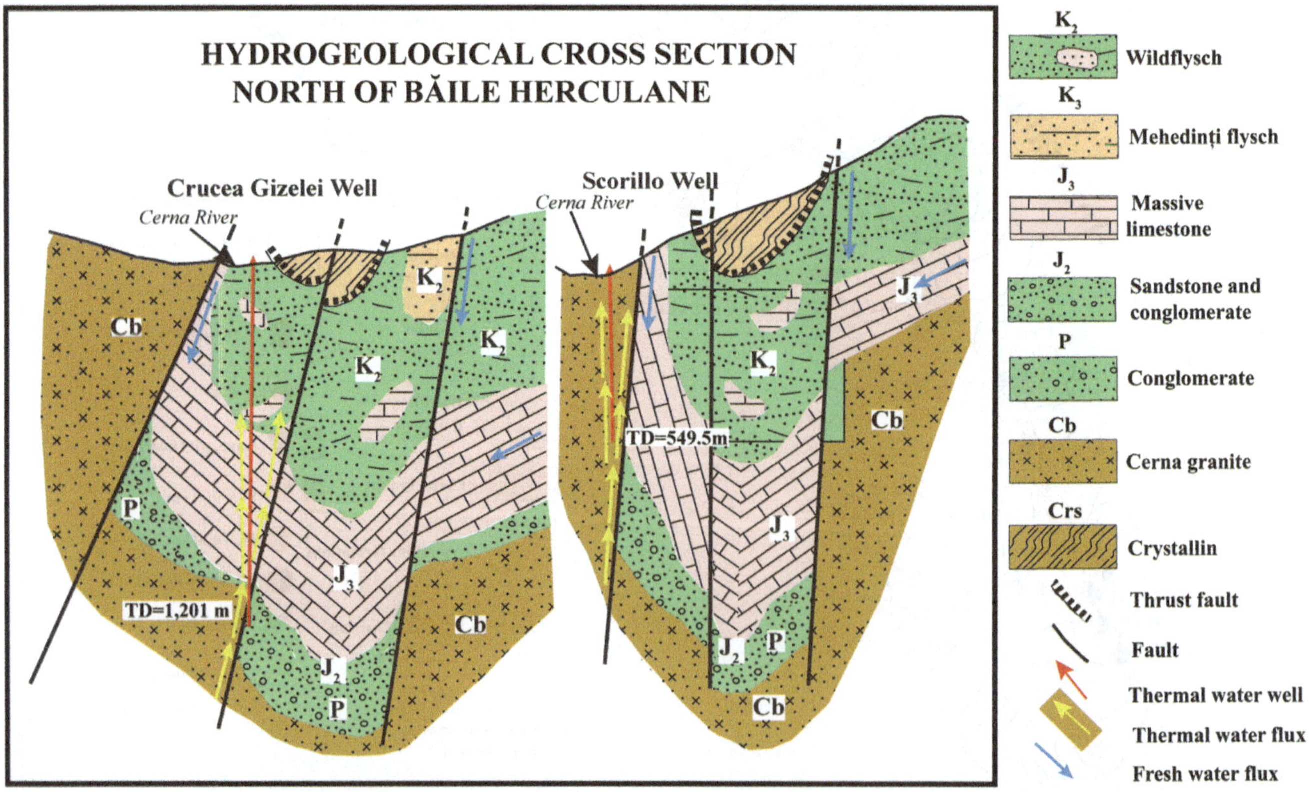

Fig. 3 Geologic cross section through Scorillo and Crucea Gizelei wells

Table 1 Area of the Cerna Valley between Naiba and Pecinişca

	km^2	km^2 (%)	Karst type	Age of limestones	Number of caves	Number of caves (%)	Total length	Vertical range	Vertical range (%)
Presacina sedimentary zone	8.5	17	Bar Olistoliths	J_3 Cr_1	137	38.5	8373	716	30
Cerna—Coşustea sedimentary zone	71.2	83	Plateau Bar	Cr_1	219	61.5	8499	1684	70
Total	79.7	100			356	100	16872	2400	100

of the graben, generating a more extended secondary porosity, resulting in the development a higher number of caves. Also, there is a possibility that in the near future, more caves that are still unknown, might be discovered.

The distribution of the caves from north to south is:

1. In the limestones between Naiba Valley and Corcoaiei Gorges are a narrow strip of limestones on the right/west side of Cerna, which host four small caves.
2. The limestones between Corcoaia Gorge and Arşasca Valley form a 15-km-long narrow strip (bar type) on the left side of Cerna Valley bordered by 200–400 m high cliffs interrupted by several gorges as Corcoaia (along Cerna), Râmnuţa Mare, and Râmnuţa Vânată. Some of the streams were not able to traverse the strip, forming suspended valleys, which disappears underground at the contact with the limestones. In this category are included Ogaşul Adânc and Ogaşul cu Bani. This bar is only 400–500 m wide and presents some discontinuities mainly in the northern part. In the southern part, the limestones form a suspended syncline filled with flysch deposits. The erosion modeled differently these two types of deposits generating elongated close depressions known as Poieni (Meadows). One of these meadows is located between two Râmnuţe rivers and is known as Prihodul Glodului.
3. The limestones between Arşasca and Pecinişca valleys are known as the Mehedinţi Karst Plateau, which is divided in the Northern and Southern sections by Ţesna River. The Northern section, located between Arşasca and Ţesna valleys, presents two morphological uneven levels developed parallel with Cerna River/Graben. A cliff separates them, which in this case divide also two types of karst.

The lower level is formed by a 250-m-wide strip/bar. Between the cliff line which is the western border of the upper level and the limestone strip of the lower level, a suspended syncline is located with Turonian–Senonian flysch deposits in the axis along which a meadow alignment was formed. Some springs originate in these deposits, which disappears underground at the contact with the lower/inferior bar/strip of limestones (e.g., Ogaşul Ţiganului, Ogaşul lui Şularu Gorges, Bobotului Gorges along Cerna Valley).

Springs originating in the flysch deposits and disappearing underground at the contact with limestones are recharging large springs like Pişetori. It is possible that the recharge zone of the Pişetori Spring is much larger, but future dye studies will define the recharge zone.

The upper level is formed by the Mehedinţi Plateau, which is characterized by a spectacular alignment of Meadows (Fig. 4). These are large endorheic basins/close depressions formed at the limestones/flysch contact, where almost always-dry ponors are present, which after heavy rains become active. These meadows have a subhorizontal floor (up to 1 km wide), surrounded by almost vertical cliffs (Poiana Beletina, Crovul Mare, Poiana Porcului, and Crovul Medved).

The Southern section has a plateau aspect with close endorheic basins with granite or flysch bottom (e.g., Balta Cerbului, Şaua Padina, and Poiana Muşuroaie). The other exokarst features are less obvious in this area, because 75% is covered with forest.

4. The limestones from the right (West) side of Cerna Valley are a bar type karst developed along Cerna River between Bobotului Gorge and Băile Herculane. In some section, the limestones present two parallel strips (Cerna syncline flanks). The western strip/flank presents very few caves versus the eastern one, in which the majority of the caves are located.

The main exokarst features in this area are the gorges/canyons. The Iuta Valley presents a 300-m-long gorge with 100–200 m high walls, most of the time being covered with vegetation. The slope of the Iuta thalweg is 100 m/km, the confluence with Cerna being marked by a waterfall with several steps.

Presacina Valley presents the longest gorge from Cerna Basin (2 km long). Their width is 10–12 m in the vicinity of confluence with Cerna River, upstream becoming 50–60 m wide. At 1.5 km from the Cerna River, a 400-m-long canyon (4–6 m wide), with a 5-m waterfall in its middle part exist. The downstream section from this waterfall has a steep slope

Fig. 4 Poienile/The Meadows (photograph by G. Ponta)

interrupted by another 1.5 m waterfall, while the upstream section presents several short waterfalls with whirlpools at the base. Drăstănicului Valley with 500 m of gorges and a slope of 222 m/km is marked by the presence of a 6-m waterfall in the central area and a 10 m one close to the confluence with Cerna.

Presacina–Drăstănic area presents the maximum exposure of the bar type karst in the Cerna Valley, resulting in hosting the longest caves in the area.

Great Şălitrari Cave (Peştera Mare din Şălitrari)

ISER Bucureşti first explored the cave in 1960 that mapped the Nitrate Passage. In 1979, Focul Viu Caving Club resurvey the main gallery, where T. Chiriţă dug at the end of the main passage and found the continuation into Focul Viu Passage. By the end of 1979, the cave was 1500 m long. In 1998, R. Puşcaş with Speotimis Caving Club resurveyed the cave, found a southeast continuation, resulting in 1350 m of passages, with 85.5 m vertical range, and 470 m extension (Fig. 5). The cave is located on the right side of the Presacina in a cluster of caves, close to the confluence with Cerna, at about 180-m-relative elevation (Ponta and Solomon 1982b).

The cave has two entrances, which end in the same gallery after 15 m, and three large sections united through narrow passages. The entrance section (Nitrate Passage) is oriented east–west and is 5–6 m in height. The floor is covered with silt, sand, boulders, and saltpeter, and funnel aspect holes resulted from the saltpeter mining. At the end of this section, a rimstone pool is present, from where the connecting gallery with the new sector of the cave begins. The second section (Focul Viu Passage begin with Chamber 1 which is bordered by a canyon-type gallery up to 30-m deep which function as a ponor. The upper end of the cave is a well-decorated Chamber 3, and 20–30-m-thick layer of boulders with a maze of small passages in between. The third section discovered in 1998 by R. Puşcaş, and his team is very well decorated, 10 m wide a 7 m high.

Great Şălitrari is a vadose cave with a branchwork pattern suspended at 180 m above Preasacina's thalweg, formed by a stream originated in the Drăstănic Valley. The new section

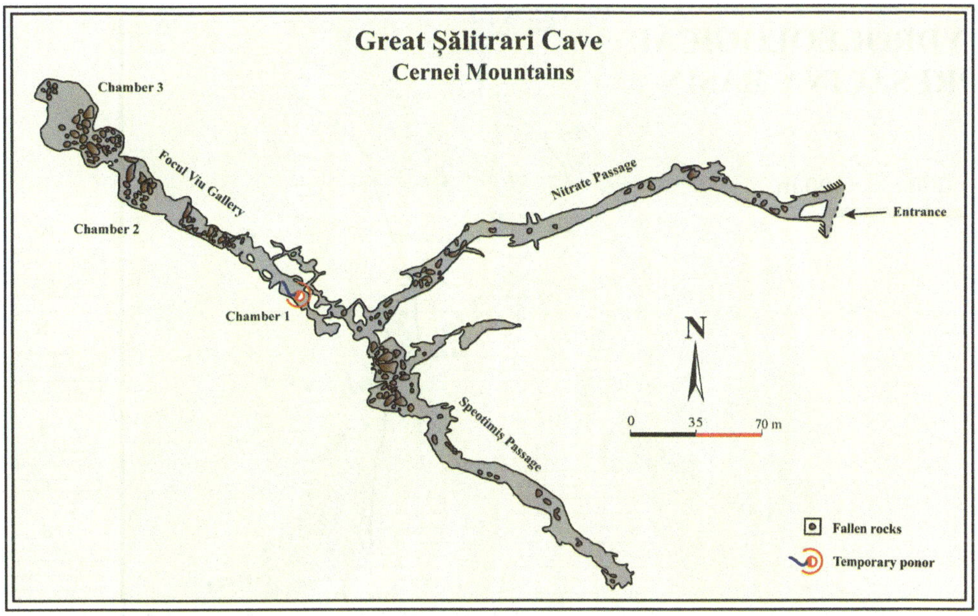

Fig. 5 Map of the Great Şălitrari Cave. Surveyed by R. Puşcaş and Speotimiş Caving Club

discovered by Speotimis although is developed on the same direction with Focul Viu Gallery, its slope is oriented toward northwest, suggesting that this section was formed by loses of downstream section of Drăstănic or Cerna River, the outlet being through the same Nitrate Passage. Chemical analyses of the saltpeter show a mineral association, which confirm at least a hypogene episode in the cave evolution, which overprinted the vadose morphology (Onac et al. 2011). As shown in Fig. 6, the Presacina River thalweg is formed partially by granite, which is the main factor of the geothermal anomalies along Cerna River. The small geothermal anomaly is located only a few hundred meters north of the cave, what justify the finding related to hypogene origin.

Ogaşul Adânc Cave

The Vârful lui Stan–Domogled anticline has a granitic core (Cerna Granite) covered by Mesozoic sedimentary deposits of the Danubian Autochthonous. The western flank is traversed by faults parallel with the Cerna Graben, with wildflysch deposits in between. The limestone deposits of Lower Cretaceous (Barremian–Aptian) age on eastern flank of the anticline covers the granite core of the anticline, south of Arşasca Valley. These deposits are overlaid by limestones of the same age as those covering the anticline granite core axis, which were thrusted over the flysch formation (Ponta et al. 2013). North of the Arşasca River only the western flack is present under a narrow limestone strip, which was traverse by Ogaşul Adânc cave with its 650 m of passages and 75-m-vertical range. The limestone bar/strip is bordered on both sides by vertical cliffs and sharp ridges. In the central part, the limestones present an intercalation of flysch deposited in a suspended syncline of Senonian age. Due the differential erosion between these two units, large close depressions are formed. Tectonically, the limestone strip is bordered by the Cerna Graben, and numerous perpendicular faults along which streams are draining underground at the limestone interface and reappear at the foot of the cliff.

The cave is located in the left side of the Cerna Valley opposite of Olanului Creek. At about 100 m, relative elevation can be observed the downstream entrance of the cave, the ponor entrance being 50 m above, next to a very large fractured boulder cross by a canyon-type passage, which belong to the same cave (Ponta 1983).

The cave was discovered by I. Povară (ISER Bucharest) in June 1979, the first partial exploration being performed in the same day by C. Lascu and G. Ponta. The stream part of the cave was surveyed during the 1979 National Speleological Convention (Speosport) held in Herculane,

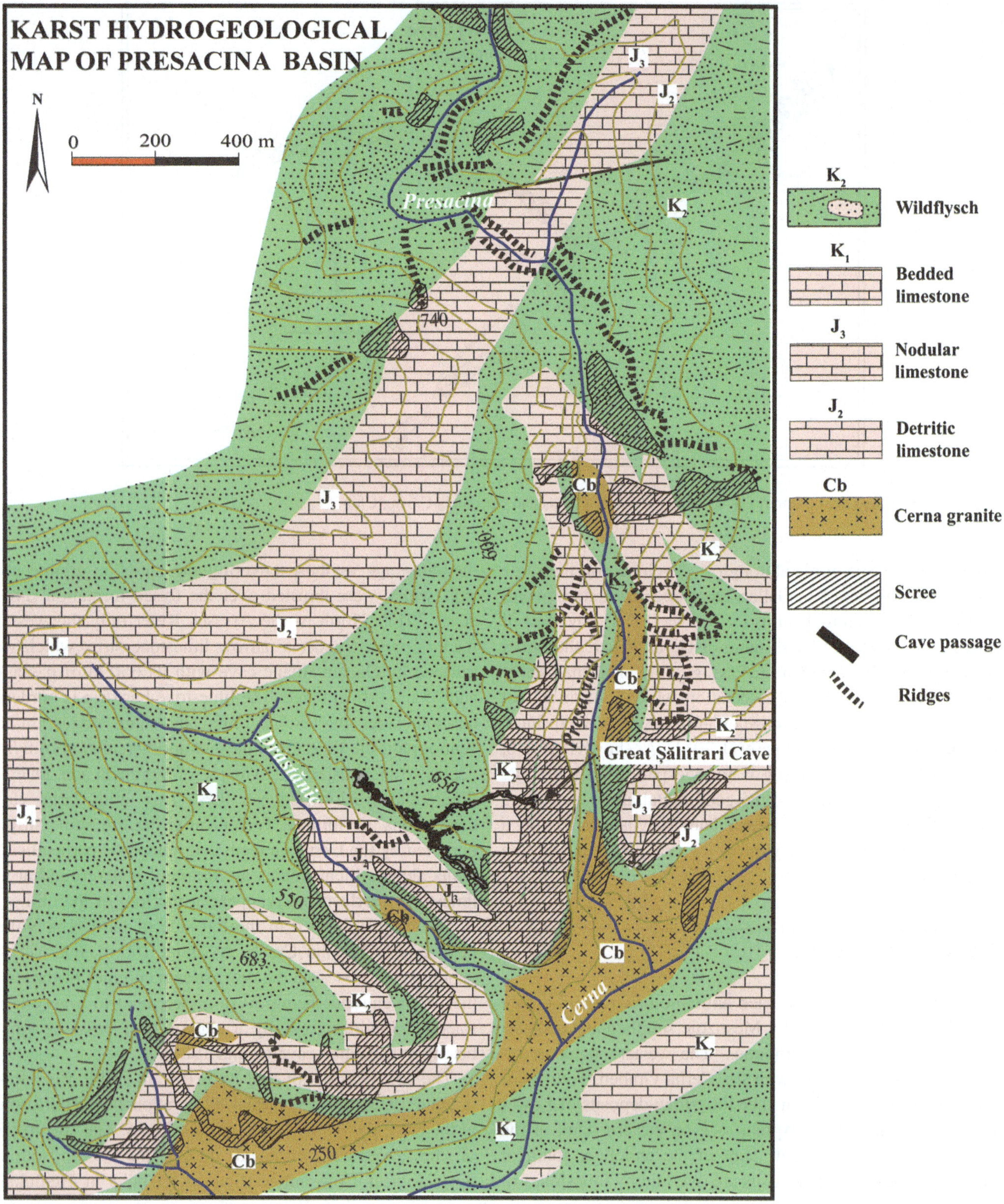

Fig. 6 Karst hydrogeologic map of Presacina Basin

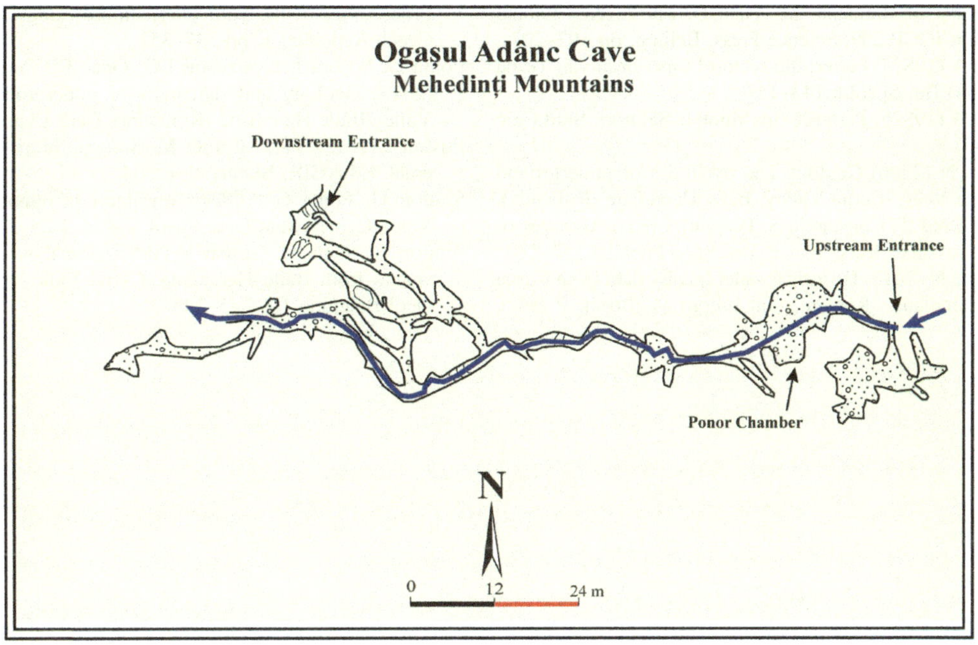

Fig. 7 Map of the Ogaşul Adânc Cave (Survey by G. Ponta, I. Povară, C. Lascu, S. Bulgăr, G. Ionescu, A. Constantin)

when S. Bulgăr climbs a 25 m waterfall. During this trip was found that the cave traverses the entire limestone bar/strip, and the vertical range was 55 m (Fig. 7).

In the winter of 1979, the cave was completed mapped, and additional passages were discovered through digging by G. Ionescu and A. Constantin, resulting a vertical range of 75 m and 650 m in length, 130 m extension and a ramification coefficient of 5. The upper entrance elevation is 595 m and relative elevation—95 m. The Upper entrance of the cave begins with a series of small waterfalls (until −17 m) with a Hall at their base from where two narrow passages begin which intercept the stream. The Stream Gallery is a canyon 5–8 m height. The upstream section is formed by a steep gallery, which ends at the base of a 25-m waterfall, from where through a squeeze the Ponor Chamber can be reached, traversed by the mainstream, which is fed by two additional ponors. The Active gallery descends on a slope, with a smaller height, ending in an impenetrable sump where the entrance in a subfossil section is located. The total length of the active passage is 125 m and is the longest stream in the Cerna Valley.

Acknowledgements We are thankful to Focul Viu Caving Club Bucharest, Liliacul Arad, and Nyphargus Praha, and all our friends and colleagues from home and abroad for their support during the exploration and research activities at these caves. Also, we are thankful to R. Puşcaş and Speotimiş Caving Club for letting us to include their map of the Great Şălitrari Cave in this chapter.

References

Al Codarcea, Năstăseanu S (1964) Contribuţii la cunoaşterea stratigrafiei depozitelor calcaroase din bazinul Văii Cerna şi de la Cazane (Dunărea). St cerc geol geofiz geogr, Seria geol 9(2):241–250

Avram Ş, Dancău D, Şerban E (1964) Cercetări speologice în Bazinul Cernei. Nota I. Lucr Inst Speol "Emil Racoviţă" III:131–200

Avram Ş, Dancău D, Şerban E (1966) Cercetări speologice în Bazinul Cernei. Nota II. Lucr Inst Speol "Emil Racoviţă" V:11–20

Bercia I, Bercia E, Năstăseanu S, Berza T, Iancu V, Stănoiu I, Hârtopanu I (1987) Foaia Obârşia Cloşani. Scale 1:50,000. Inst Geol Geofiz, Bucureşti

Dancău D, Danielopol D, Juvara I, Şerban E (1968) Cercetări speologice în Bazinul Cernei. Nota III. Lucr Inst Speol "Emil Racoviţă" VII:45–54

Demetrescu C, Andreescu M (1994) On the thermal regime of some tectonic units in a continental collision environment in Romania. Tectonophysics 230:265–276

Liégeois JP, Berza T, Tatu M, Duchesne JC (1996) The Neoproterozoic Pan-African basement from the Alpine lower Danubian nappe system (South Carpathians, Romania). Precam Res 80:281–301

Năstăseanu S (1980) Géologie de Monts Cerna. Ann Inst Géol Géophys 54:153–280

Onac BP, Goran C (2018) Karst and caves of Romania: a brief overview. In: Ponta GML, Onac BP (eds) Karst and cave systems of Romania. Springer International Publishing, Cham, pp 21–35

Onac BP, Wynn JG, Sumrall JB (2011) Tracing the sources of cave sulfates: a unique case from Cerna Valley, Romania. Chem Geol 288:105–114

Ponta G (1983) Peştera din Ogaşul Adânc. Carst 3:8–10

Ponta G (1998) Karst of Romania. In: Yuan D, Liu Z (eds) Global Karst Correlation IGCP 299. Science Press, Beijing, pp 193–207

Ponta G, Solomon A (1982a) Peşteri din sectorul superior al văii Ţesna (M-tii Mehedinţi) Bul Speol 6:143–157

Ponta G, Solomon A (1982b) Peşterile din Muntele Şălitrari. Stud Cerc Speol 2:39–50

Ponta G, Terteleac N (2006) Geology and dynamics of underground waters in Jiul de Vest—Cerna Valley, Băile Herculane (Romania). In: Onac BP, Tămaş T, Constantin S, Perşoiu A (eds) Archives of climate change in karst. Karst Waters Inst Spec Publ 10:227–230

Ponta GM, Terteleac N (2013) Historical water quality data from Cerna Valley/Băile Herculane—Romania. In: Filippi M, Bosák P (eds), Proceedings of the 16th International Congress Speleology, Brno, Czech Republic, 1, pp 347–352

Ponta G, Povară I, Isverceanu EG, Onac BP, Marin C, Tudorache A (2013) Geology and dynamics of underground waters in Cerna Valley/Băile Herculane (Romania). Carb Evap 28:31–39

Povara I (2012) Valea Cernei Morfologie, hidrologie, ape termominerale. Ed AGIR, Bucuresti, p 304

Simion G, Ponta G (1983) Contribuţii la cunoaşterea carstului din Valea Cernei. Stud Cerc Speol Vâlcea 3:29–34

Simion G, Ponta G, Gaşpar E (1985) The dynamics of underground waters from Băile Herculane, Cerna Valley (Romania). Ann Soc Géol Belgique 108:245–249

Cernei Mountains: Caves Conveying Geothermal Fluids at Băile Herculane

Ioan Povară, Horia Mitrofan, Bogdan P. Onac, Constantin Marin, Eugen Niţu, Daniela Ioniţă, Alin Tudorache, and Mădălina Vişan

Abstract

Several small caves located in the southwestern part of the South Carpathians discharge steam or/and chloride-rich hot groundwater derived from a geothermal reservoir situated nearby. The monitoring of natural tracers (heat, certain solutes, stable isotopes) documented uncommon levels for the fluids discharged within the Băile Herculane caves, thus revealing flow processes that ordinary meteoric-derived groundwater would be unable to depict. Furthermore, the exotic cave minerals and their isotopic composition suggest that the caves are the result of combination of hypogene and epigene processes. A rich and peculiar endemic troglobite fauna was identified in the Adam's Shaft, in which a guano accumulation dated to ~8425 cal yrs BP suggests it is one of Europe's oldest permanent bat colonies. The unusual warm/wet microclimate of the cave favored both an ideal roosting site for bats and a "tropical" biospeleological oasis.

Keywords

Geothermal fluids • Hot vapors • Hypogene speleogenesis • Cave minerals • Guano • Hydrogeology

I. Povară (✉) · C. Marin · E. Niţu · A. Tudorache
"Emil Racoviţă" Institute of Speleology of the Romanian Academy, Calea 13 Septembrie, 13, 050711 Bucharest 5, Romania
e-mail: ipov.iser@gmail.com

C. Marin
e-mail: constmarin@gmail.com

E. Niţu
e-mail: eunitu@yahoo.com

A. Tudorache
e-mail: altudorache@yahoo.co.uk

H. Mitrofan · M. Vişan
"Sabba Ştefănescu" Institute of Geodynamics of the Romanian Academy, 020032 Bucharest, Romania
e-mail: horia.mitrofan@geodin.ro

M. Vişan
e-mail: danamadalina@yahoo.com

B. P. Onac
Karst Research Group, School of Geosciences, University of South Florida, 4202 E. Fowler Ave, NES 107, Tampa, FL 33620, USA
e-mail: bonac@usf.edu

B. P. Onac
"Emil Racoviţă" Institute of Speleology of the Romanian Academy, Clinicilor 5, 400006 Cluj-Napoca, Romania

D. Ioniţă
Department of Analytical Chemistry, Faculty of Chemistry, University of Bucharest, 2-14 Bd. Regina Elisabeta, 030018 Bucharest, Romania
e-mail: dani17_82@yahoo.com

Introduction

A group of relatively small caves clustering within the boundaries of the Băile Herculane Spa located in the SW part of the Cernei Mountains, in the downstream section of Cerna Valley, transmit toward the ground surface atypical fluids: either steam or chloride-rich hot water, both derived from a nearby geothermal reservoir. Analogous "thermal caves" are not that common worldwide, although similar occurrences have been documented in Italy (Galdenzi et al. 2010; Apollaro et al. 2016), Hungary (Erőss et al. 2008), Mexico (Hose et al. 2000; Rosales Lagarde et al. 2014), France (Audra et al. 2007), and Greece (Lazaridis et al. 2011). For more recent studies on the topic, readers are directed to a recently published book on hypogene karst (Klimchouk et al. 2017). In Romania, only one other cavity is known (Movile Cave, in Dobrogea), which similarly, intercepts a thermal aquifer (e.g., Marin and Nicolescu 1993; Sarbu et al. 1996).

Despite the fact that none of them is longer than 200 m, the five caves discussed in this chapter stand out on several accounts:

1. They host "exotic" minerals precipitated following the interaction between thermal solutions and the bedrock.
2. Endemic troglobite fauna can inhabit them.

G. M. L. Ponta and B. P. Onac (eds.), *Cave and Karst Systems of Romania*, Cave and Karst Systems of the World, https://doi.org/10.1007/978-3-319-90747-5_25

3. The monitoring of natural tracers (temperature, certain solutes, stable isotopes) in the discharging fluids document abnormal values, which may reveal flow processes difficult to recognize by studying the more common, meteoric-derived groundwater.

Geographic, Geologic, and Hydrologic Settings

Cerna Valley represents the longest (22 km) limestone gorge encountered in the South Carpathians. This is a river that flows from NNE to the SSW, parallel to the mountain range main ridges. The examined caves open in the downstream section of the Cerna Valley, on its right (west) side (Fig. 1). The course of Cerna follows the regional Cerna-Jiu fault, a Tertiary age fracture system along which right (west) lateral displacements occurred (Fig. 2 inset). The latter movements accommodated (e.g., Linzer et al. 1998) the Oligocene–Lower Miocene clockwise rotation of the South Carpathians around the NW corner of the Moesian tectonic plate. This process also generated the trans-tensional Mehadia sedimentary basin, located only a few km to the NW of the Băile Herculane Spa, and which is filled with Upper Badenian to Lower Pannonian deposits (Mărunţeanu et al. 1994). Nowadays, sparse—yet occasionally significant—seismic events (instrumentally recorded magnitudes up to 5.6) indicate a persistent, roughly N-S oriented extensional stress (Radulian et al. 2000; Plăcintă et al. 2016).

That overall setting could contribute to explain the presence of a chloride-rich thermal water accumulation, hosted mostly (Povară 1992; Povară et al. 2008) by a Proterozoic age granitoid pluton (the so-called Cerna Granitoid). Worldwide, many occurrences of thermal groundwater located in fractured granite bodies were documented, e.g., Michard et al. (1986, 1989), Pauwels et al. (1997), Marques et al. (2006), Delgado-Outeiriño et al. (2009), and Chen et al. (2016). Yet in addition, as compared to the settings addressed by those authors, the aquifers that occupy the Carnmenellis granite pluton in England (Edmunds et al. 1985) and the Sila granite batholith in Italy (Apollaro et al. 2016) seem to also share certain common chemical signatures with fluids discharged from the Cerna Granitoid.

Fig. 1 Location of karst is in the lower Cerna Valley

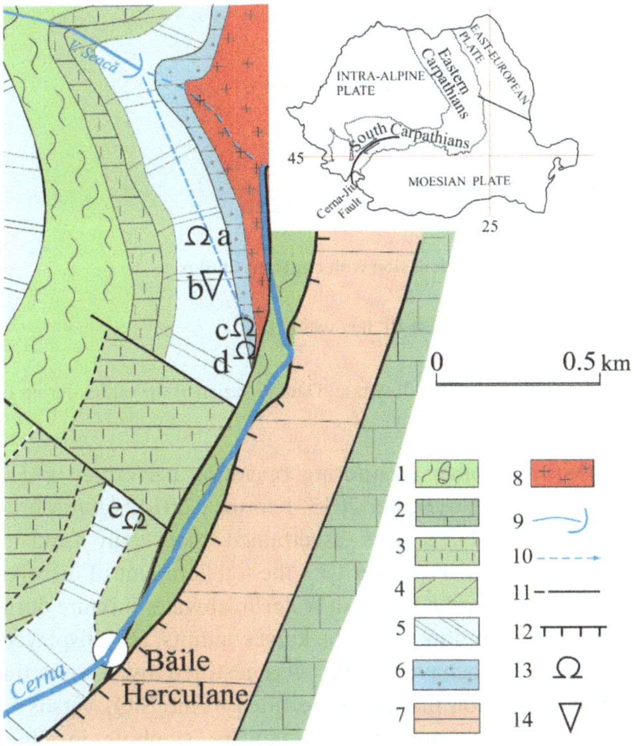

Fig. 2 Map illustrating the general geological and hydrological setting of the Băile Herculane caves which discharge geothermal fluids (geology after Năstăseanu 1980). *1* Upper Cretaceous wildflysch, *2* Lower Cretaceous (Urgonian) limestone, *3* Middle Cretaceous marly limestone ("Iuta Layers"), *4* Lower Cretaceous bedded limestone, *5* Middle–Upper Jurassic nodular limestone, *6* Lower Jurassic sandstone, *7* Proterozoic metamorphic rocks, *8* Cerna Granitoid (Proterozoic), *9* swallet, *10* underground flow path outlined by tracer tests, *11* fault, *12* nappe boundary, *13* cave, *14* shaft. The letters on the map represent: *a* Steam Cave, *b* Adam's Shaft, *c* Hercules Cave, *d* Despicătura Cave, and *e* Diana Cave. The inset map shows Romania's major tectonic features, outlining the role that the Cerna-Jiu strike-slip fault had played in the South Carpathians rotation around the Moesian Plate

The Middle–Upper Jurassic limestone deposits in which the caves are excavated belong to the South Carpathians Alpine nappes pile, whose emplacement has been completed in the Upper Cretaceous. Limestones do not directly overlie the Cerna Granitoid, but due to the complex structural setting of the area, those two lithologic units for granitoid pluton frequently come in direct contact to each other (Povară et al. 2008). The Jurassic limestone unit, in turn, is partly overlain by an aquiclude consisting of Middle Cretaceous clayey limestones, designated as the "Iuta Layers."

Both meteoric and chloride-rich thermal waters contribute to the underground flow occurring in Diana and Hercules outflow caves. Thermal waters having the same chloride-rich chemical characteristics also discharge to surface through several drill-holes and non-karst springs in the vicinity of Băile Herculane Spa. Overall, the investigated caves occur in two distinct hydrological settings (Fig. 2): one associated to Diana Cave, whereas the other to Hercules Cave, Despicătura Cave, Adam's Shaft, and Steam Cave (Grota cu Aburi), all four cavities belonging to a common karst network that conveys groundwater and steam fluxes.

The small (22 m long) Diana Cave hosts a couple of relatively low-yield (ca. 0.5–1.5 L/s) thermal springs, whose temperatures have been reported to range between 47.5 and 51 °C, whereas their total dissolved solids (TDS) vary between ~4.4 and 6.3 g/L (Povară et al. 2008). A distinctive characteristic of the discharged water is its elevated H_2S content (30–70 mg/L).

In contrast, Hercules Cave represents the main, perennial outflow of a typical karst drainage network, which still receives a significant contribution of chloride-rich thermal groundwater. In fact, both Diana and Hercules caves seem to be fed by a similar, hot brackish parent-water of Na–Ca–Cl type (Mitrofan et al. 2016), although H_2S is, nonetheless, completely absent in the Hercules spring flow.

Apart from direct seepage of cold meteoric water, the Hercules Cave underground stream is also fed (as proven by artificial tracer tests; Povară et al. 2008) by a series of creeks which sink in valleys located to the north of the cave (Fig. 2). As a function of the hydrometeorological inputs, the Hercules Spring discharge can fluctuate by a factor of 10 (in the 10.2–105 L/s range in the time interval 1984–1995; Povară et al. 2008), causing changes of the TDS content (in the 2.5–0.3 g/L range) and temperature of the outflow (in the 53.5–17.0 °C range) that are systematically of opposite sign.

The abundance of Cl^- supplied to the Hercules Spring by the Cl^--rich hot parent-water has been used for diagnosing a particular hydrological behavior observed in March–April 1981, when a couple of heavy flood events had quickly succeeded one after another (Mitrofan et al. 2015). The corresponding monitoring data (recorded on average, four times a day) were represented in a diagram in which the dissolved chloride flux (i.e., the dissolved Cl^- concentration, multiplied by the springwater discharge) was plotted against the simultaneously recorded water flow rate (Fig. 3). The plot outlines an outstanding behavior, part of which was hysteretic (cf. large arrows indicating a counterclockwise evolution). For flow rates smaller than about 45 L/s, the Cl^- flux (supplied entirely by the brackish hot fluid) remained remarkably constant (30.6 g/s on the average—as indicated by the orange horizontal line in Fig. 3). In contrast, for values in excess of about 68 L/s (recorded during the actual flood events), the flow rates displayed—paradoxically—only small-amplitude variations, while the Cl^- flux experienced successive depletions and enrichment with respect to the stable value of 30.6 g/s. The appropriate interpretation is summarized as follows (Mitrofan et al. 2015):

- At low-flow rates (less than <45 L/s), the stream passage was subject mostly to an open-channel regime and the

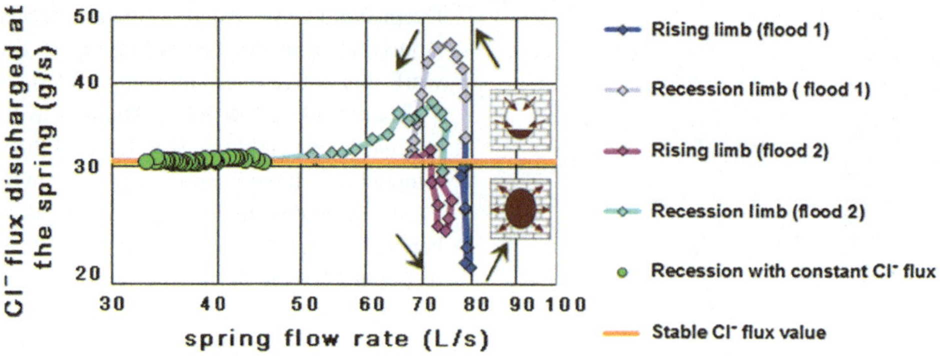

Fig. 3 Log–log plot of the chloride flux discharged by Hercules Spring versus the corresponding flow rate. Data recorded at the outlet during two successive flood events that occurred in March–April 1981

karst spring discharged—irrespective of the freshwater input—a constant flux of chloride: the latter was supplied, entirely, by the hot brackish parent-water.

- During a flood pulse rise, not only became the Hercules Cave passage completely filled with mixed (fresh + brackish) water, but a certain amount of this chloride-rich water was also forced and stored (as schematically depicted in the inset below the horizontal orange line of Fig. 3) into the fractured walls of the cave passage: therefore, a Cl^- flux deficit was recorded at the spring.
- Subsequently, as the flood pulse recedes and the open-channel flow regime progressively re-establishes, the chloride-rich waters previously stored in the fractured bedrock are released back into the stream passage (as illustrated by the inset above the orange line in Fig. 3): consequently, a Cl^- flux excess was recorded at the spring outlet.
- The actual flow rate variations actually induced by the flood were efficiently buffered by the water exchanges taking place between the cave passage and the surrounding fractured walls: that is why an almost constant flow rate was recorded at the spring outlet during the entire period of maximum flood intensity.

An analogous behavior has been speculated also for other investigated hydrokarst settings in France (Bailly-Comte et al. 2010), Germany and Austria (Goldscheider 2005), or the USA (Vesper and White 2004; Raeisi et al. 2007), although the most frequently mentioned in this respect is, by far, the Santa Fe underground river in Florida, USA (e.g., Martin and Dean 2001; Screaton et al. 2004; Martin et al. 2006; Bailly-Comte et al. 2010). Different, however, from all these cases where only meteoric-derived fresh groundwater is involved, the Hercules Spring setting also benefits from an abundant and quite stable flux of the conservative natural tracer Cl^-: this specific circumstance enabled more rigorous interpretations to be conducted in terms of fracture storage/release from storage.

During another monitoring operation performed at Hercules Spring (November 2013–February 2014; data recorded twice a week), it was ascertained that apart from the well-known natural tracer Cl^-, the total amount of dissolved SiO_2 carried by the brackish water/freshwater mixture during relatively high flow along the karst conduits, also displayed a conservative behavior. This was testified by a very tight linear regression that could be fitted to the SiO_2 versus Cl^- reciprocal concentration plot (square symbols in Fig. 4, where each data point corresponded to a different mixing ratio between cold freshwater and brackish hot water).

On the other hand, in each of the Băile Herculane thermal discharges there is a certain degree of freshwater dilution experienced by the hot and brackish common parent-water. For the most concentrated sample, however, the dilution should not be important. This most concentrated sample can be assumed, consequently, to have a Cl^- content that is also virtually equal to that of the brackish hot end-member involved in the two-component mixture discharged by the Hercules Spring. Based on this supposition, also the SiO_2 concentration of that hot brackish parent-water can be estimated: to this purpose, the regression in Fig. 4 has been extrapolated up to the maximum Cl^- concentration (4738 g/L) recorded in any of the Băile Herculane thermal discharges during the 2013–2014 monitoring work. Accordingly, the obtained SiO_2 concentration of 93.4 mg/L corresponds (cf. silica geothermometry model of Fournier 1977) to the temperature at which the hot brackish parent-water supplying the Hercules Spring had last equilibrated with a silica mineral species (either quartz or chalcedony) existing in the host rock. The silica minerals are assumed to originate in the Cerna Granitoid reservoir rock (e.g., Povară 1992). If assuming chemical equilibrium with chalcedony (Arnórsson et al. 1983) or with quartz (Fournier 1977), the temperature values which correspond to the inferred water–rock chemical equilibration are 105 or 130 °C.

Yet the temperature of the stream encountered along Hercules Cave never exceeded 55 °C. It hence results that the brackish hot parent-water should mix with the cold meteoric-derived freshwater somewhere beyond (i.e.,

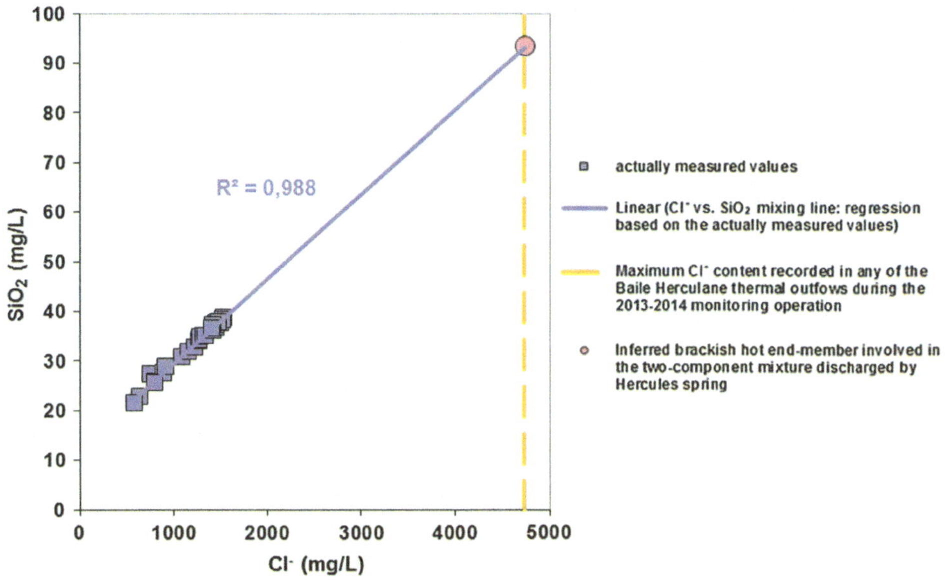

Fig. 4 SiO_2 versus Cl^- reciprocal concentration plot constructed for the Hercules Spring flow samples collected between 2013 and 2014

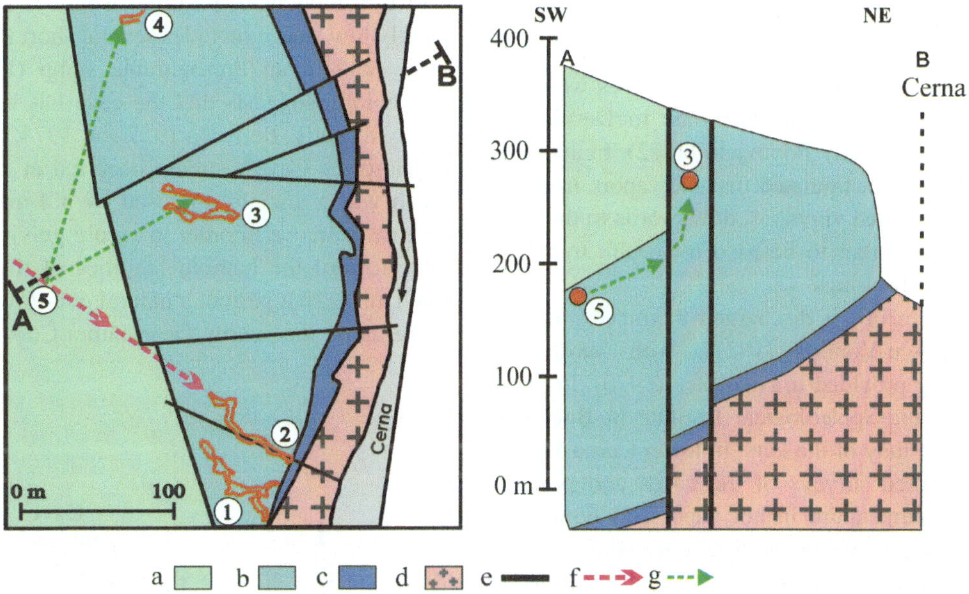

Fig. 5 Map (left) and cross section (right) illustrating the general setting of the Hercules cave system. The geological background is compiled, in a simplified form from Diaconu (1987) and Pop (in Terteleac et al. 1989). Letters in the legend indicate: *a* clayey limestone ("Iuta Layers"—Middle Cretaceous), *b* limestone (Middle–Upper Jurassic), *c* sandstone (Lower Jurassic), *d* Cerna Granitoid (Proterozoic), *e* fault, *f* presumed pathway (according to Mitrofan and Povară 1992) of the underground stream beyond the Hercules Cave current final sump, *g* inferred pathway followed by the hot vapor. Numbers on the map and cross section indicate: *1* Despicătura Cave; *2* Hercules Cave; *3* Adam's Shaft; *4* Steam Cave; *5* the hot vapor release, which is presumably derived from a liquid water inflow with temperature >100 °C that enters an aerated passage of the underground stream, beyond the current final sump of Hercules cave. Note that cave maps when view on horizontal plane are plotted in red, whereas on the cross section, they are indicated only by red dots

upstream) the final sump of Hercules Cave. And given the open-channel setting inferred (see above) to exist in that conduit situated upstream the sump, it is likely that when the liquid inflow of brackish hot parent-water with temperature (cf. silica geothermometry) in excess of 100 °C enters this aerated cave passage, it starts boiling. The accordingly released amount of steam could provide a reasonable explanation for the hot vapor discharges encountered in Adam's Shaft and in Steam Cave—two presently dry cavities that are located 140 and 230 m, respectively, above Hercules's Cave stream (Fig. 5). Such an inference is also supported by the existence of a direct underground flow pathway between Steam and Hercules caves—as proven by the fact that fluorescein injected (Povară et al. 2008) into a steaming vent of Steam Cave was subsequently detected in the Hercules Spring. Also, the 100 °C liquid brackish parent-water left behind after steam is released, when intercepting the underground stream mixes with fluctuating amounts of cold meteoric freshwater transported by that rivulet; eventually, the corresponding liquid mixture discharges through the Hercules Cave outlet.

History of Exploration

In the second century AD, during military expeditions in Dacia (nowadays Romania), Hercules Spring has been discovered by Romans, who subsequently built bathing facilities (Stoica de Haţeg 1984). Roman aristocracies and soldiers used Despicătura Cave and the entrance zone of Hercules Cave for bathing. Sources about rheumatic diseases having been healed at these locations can be found in votive tablets dedicated to Hercules, which are now preserved at the Art History Museum in Vienna, and at the local museum in Băile Herculane (Bălteanu 2007). No information is currently available on when Diana and Steam Caves were discovered. The Roman-times aqueduct next to Despicătura Cave has been mentioned by Popoviciu (1872). In addition, Popescu-Voiteşti (1921) stipulated that the vapors from the Steam Cave were directed via pipes, downwards to the level of Hercules Cave, in order to be used for health treatment and for heating the dwellings.

Chemical analyses of the discharged thermal waters are reported in Popescu-Voiteşti (1921), with subsequent investigations being provided in Nicolescu (1970). In 1972, the researchers of the Speleological Institute in Bucharest began a systematic study of the karst in lower Cerna Valley. This included detailed surveys of the caves and periodic measurements of temperature in hot vapor and liquid discharges, as well as gauging of the flow rates (Povară et al. 1972). Several of these caves have been actually addressed, over the last 50 years, by a large variety of investigations: artificial tracer tests (Pascu 1968; Povară 1973, 1980; Simion et al. 1985), mineralogy (Diaconu and Medeşan 1973; Diaconu 1974; Diaconu et al. 2010; Onac et al. 2009, 2011, 2013a, b; Puşcaş et al. 2013), biospeleology (Decou et al. 1974; Decou and Tufescu 1976; Tufescu and Decou 1977), hydro-geochemical (Bulgăr and Povară 1978; Marin 1984; Povară and Marin 1984; Povară et al. 2008, 2015; Wynn et al. 2010; Ponta and Terteleac 2013; Mitrofan et al. 2015, 2016), geotectonic (Diaconu 1987), geothermometric (Povară 1992; Mitrofan and Povară 1992), geoelectrical (Mitrofan et al. 1995, 2008), and radiocarbon dating (Carbonnel et al. 1999).

Caves Description

Diana Cave (situated at 11-m-relative elevation above Cerna streambed) was initially 14 m long (Fig. 6), but its present-day morphology has been severely altered by engineering works, which aimed to tap the two thermal springs discharging inside. The cave is partly developed at the contact between the main Middle–Upper Jurassic limestone and the clayey carbonates of the Iuta Layers. The clays present in the latter unit are responsible for the rich mineralogical association (Onac et al. 2009). In addition, they likely provide an explanation for the cation-exchange processes that control, to some extent, the ionic composition of the discharged water (Mitrofan et al. 2016).

Hercules Cave (at 5.5-m-relative elevation) has a ~90-m-long main passage, which is rather narrow, frequently low, and it includes several short flooded sections, to finally end in an impenetrable sump (Fig. 7). Its overall morphology suggests that the cave has been generated relatively recently. Between 1957 and 1974, the discharge from the Hercules Cave's stream occurred at a higher elevation, which was artificially imposed by a 4-m-high dam built at the cave entrance in order to secure gravitational flow along pipes toward the bathing facilities of the Băile Herculane Spa. During that period, a natural overflow was occasionally active along the nearby Despicătura Cave.

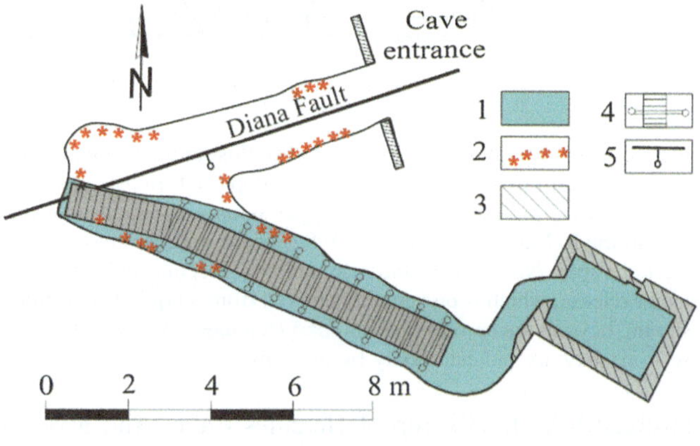

Fig. 6 Diana Cave map *1* thermo-mineral water, *2* gypsum crust, *3* concrete tank for water storage, *4* concrete pier, *5* fault

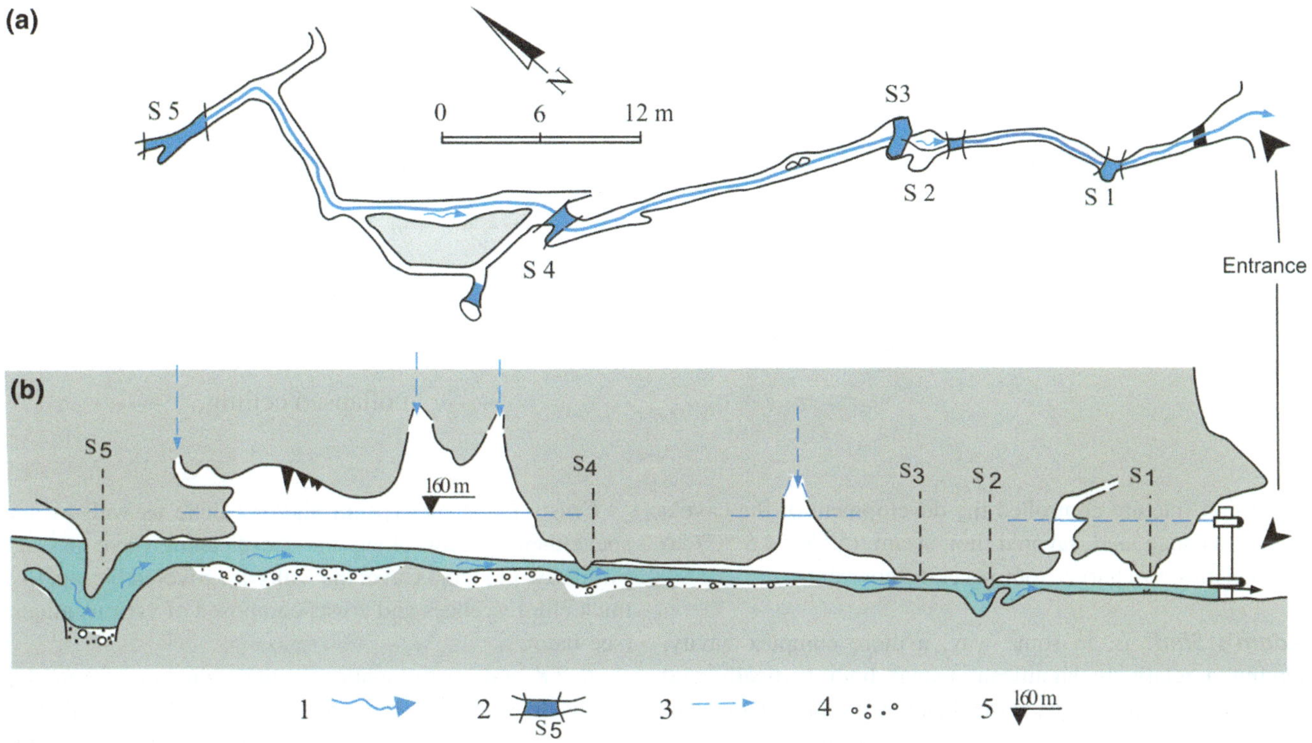

Fig. 7 Hercules Cave map (**a**) and cross section (**b**) *1* flow direction of the underground stream, *2* sumps (S1–S5), *3* dripping points, *4* alluvia, *5* the maximum elevation of the cave water level imposed by the artificially built dam

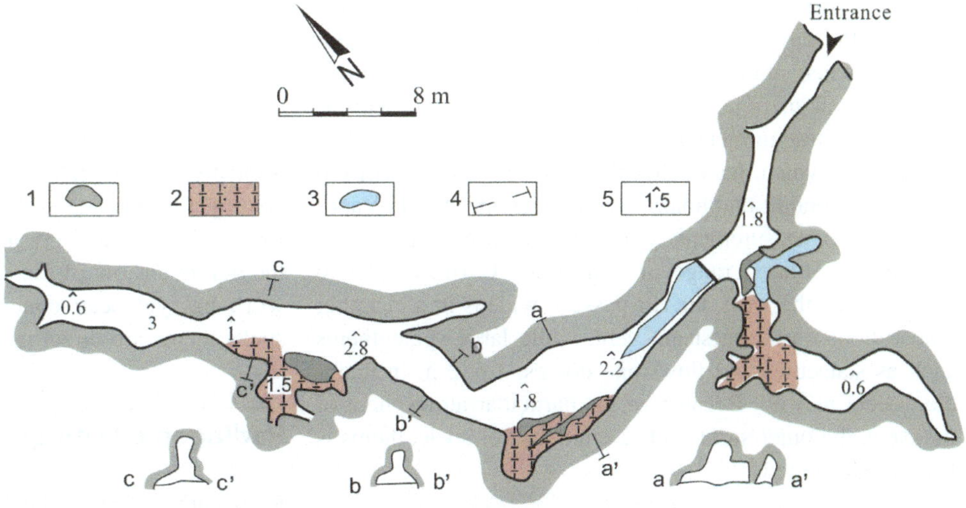

Fig. 8 Despicătura Cave map. *1* rock pillar, *2* gypsum deposits, *3* water pond, *4* passage cross section in selected sites, *5* passage height

Despicătura Cave is located very close to the Hercules Cave, acting as a temporary overflow for their common karst drainage system. The cave passages are rather narrow with the ceiling generally above 1.8 m. Very particular to this cave are the abundant gypsum deposits (wall crusts and floor blocks). Two distinct speleogenetic stages are suggested by the cross sections of the sub-horizontal, ~100-m-long passage of Despicătura Cave (Fig. 8): (i) an initial phreatic stage, during which ceiling pockets have been excavated and (ii) a subsequent phase, characterized by lateral evolution, contemporary, most probably, with the shaping of the Cerna Valley's first terrace. A small bat colony dwells toward the end of the cave.

Steam Cave is a 14-m-long cavity (Fig. 9) that opens at 383 m elevation (230 m above the Cerna streambed).

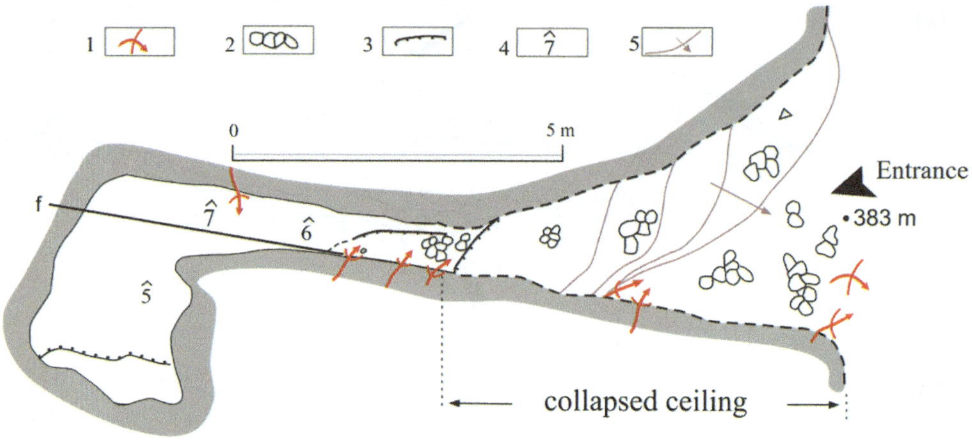

Fig. 9 Steam Cave map *1* vent, *2* collapsed boulders, *3* step on the floor, 4 passage height, *5* contour lines

A tension fracture controlled the development of the cave's single passage and favored hot steam (52.8–54.5 °C) to permanently discharge from several floor vents.

Adam's Shaft is, in some way, a more complex cavity, having a relatively significant extent both vertically and horizontally (Fig. 10). Its vertical entrance shaft (11 m deep) occurs on a limestone ridge, which separates two gullies located on the mountain slope behind the Roman Hotel, within the Băile Herculane Spa. The cavity is developed along three parallel tension fractures that exhibit approximately E–W strikes. On the northernmost positioned rift passage, it is located the steam vent which heats, to various degrees, the cave atmosphere and the cave walls. Exceptionally, when abundant rainfall amounts occur within a short interval of time, the vapor release may stop. This is because infiltrating meteoric cold waters flood the karst flow paths along which, habitually, the steam rises (Fig. 5) from the stream conduit that finally reaches Hercules Cave. Vapor temperature measurements conducted in different seasons indicate a variation range of almost 17 °C (from 29.8 to 46.5 °C). To the west, the rift passage widens in a chamber whose floor hosts the steam vent, while large speleothems such as stalactites and flowstone domes, 1.5–2 m in diameter, occur as well. Because of the particular atmosphere of the cave, the outer surface (1–2 cm thick) of speleothems is heavily weathered.

Cave Sediments and Speleothems

The stream in Hercules Cave flows on a bed that consists of finely stratified sediments (~1.4-m-maximum thickness), within which, 10–15-cm-thick layers of brown-reddish clay alternate with thinner gray, to gray-blackish, or even black horizons. Both types of horizons can include in their mass fragments of speleothems and poorly rounded gravels.

Crusts made of gypsum and/or calcite as well as small speleothems occur along the Despicătura Cave passages. Similarly, in Diana Cave, the walls are covered by mm to cm thick efflorescences and crusts composed of various minerals (see below).

A 2.8-m-deep excavation in the sediments accumulated on the floor of the southern passage in Adam's Shaft has revealed bat guano deposits, alternating with gray/reddish clay and partly rounded gravels (4–5 cm in diameter) made of quartzite, gneiss, and quartzitic sandstone. Fragments of highly weathered speleothems have also been found.

In the northernmost rift passage of Adam's Shaft, before reaching the passage section which hosts the steam vent, an unusual type of speleothems was encountered. Their distinctive feature is the significant involvement of some kind of gel in their constitution, with only minute amounts of $CaCO_3$ and clay particles embedded within them. Morphologically, these jelly-like deposits form flowstone crusts coating the passage walls and floor and snottites (4–8 cm long and less than 1 cm in diameter), which swing slowly in the gentle airflow of the cave. Although no microbiological work has yet been conducted, it is very likely that these speleothems are direct products of microbial activity.

Radiogenic Datings

Radiocarbon dating has been performed on two guano samples collected from the previously mentioned bat guano deposit in Adam's Shaft (Carbonnel et al. 1999). The first sample (1.2 m depth) has an age of 2858 cal year BP, whereas the other one, recovered at −2.5 m, returned an age of 8425 cal year BP. Considering that bat guano occurs throughout the entire depth of the excavated profile, it is reasonable to conclude that a significantly large bat colony permanently populated the cave beginning even before 8000 years ago. This could indicate that in the early

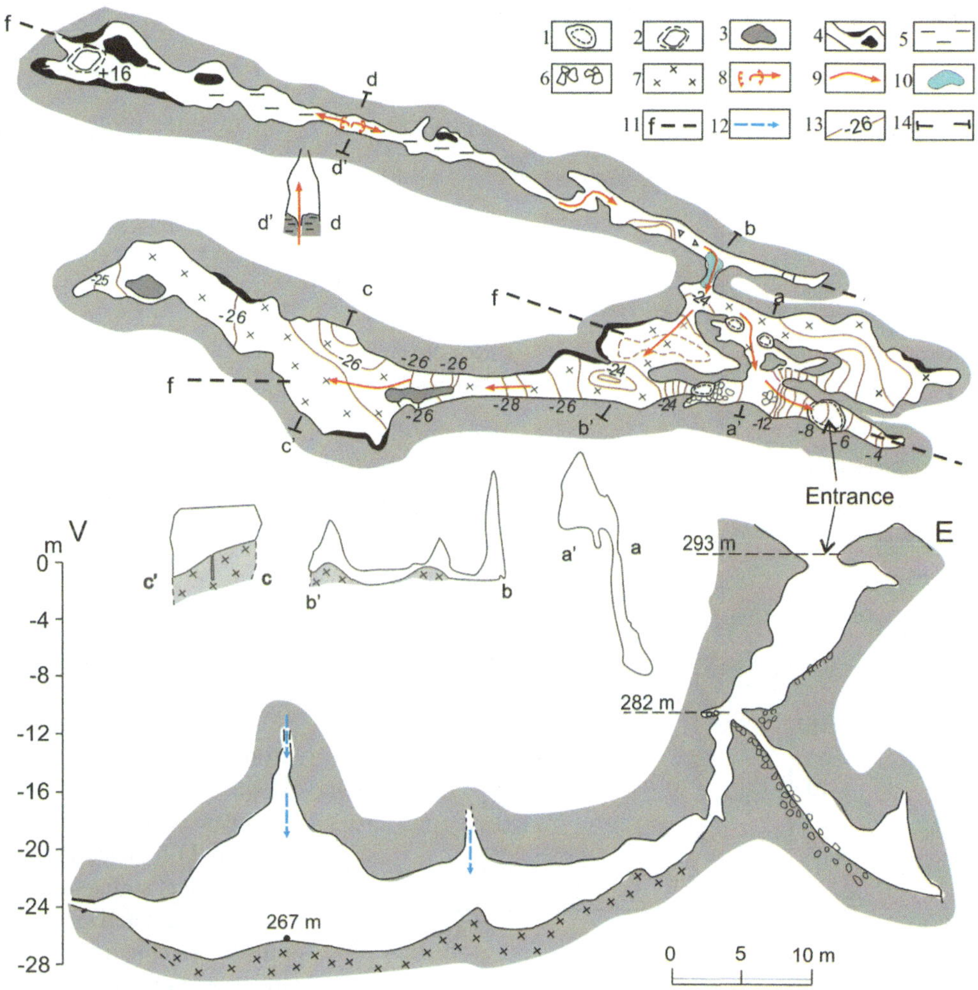

Fig. 10 Adam's Shaft map (above) and cross section (below) *1* entrance shaft, *2* chimney, *3* rock pillar, *4* speleothems, *5* clay deposit, *6* collapsed boulders, *7* guano deposit, *8* vent, *9* vapor flow direction, *10* pond, *11* fault, *12* dripping points, *13* contour lines, *14* passage cross section in selected sites

Holocene, the unusual warm microclimate established in Adam's Shaft attracted the bats, which formed probably one of Europe's oldest permanent bat roost site (Carbonnel et al. 1999).

Attempts were made to date gypsum samples from Despicătura Cave using U-series disequilibrium method. However, the δ^{234}U values were very low (<310‰), preventing us for calculating the initial δ^{234}U, altogether suggesting the gypsum deposits are relatively young.

Radon Content in the Cave Atmosphere

A survey of radon levels has been undertaken in Adam's Shaft and Despicătura Cave using 14 solid-state nuclear tracks detectors and 4 CR-89 passive detectors, respectively (Carbonnel et al. 1999; Cucoş Dinu et al. 2016). Between May 15 and July 15, 1998, the radon contents measured in 14 stations within the Adam's Shaft varied from 1754 to 5514 Bq/m^3. These values are extremely high, considering that in a recent European Union Directive for protection from ionizing radiation, the reference level was set at 300 Bq/m^3 (annual average). In light of these results, to prevent serious health issues in Adam's Shaft, cavers or researchers entering this cave should limit their stay to maximum 5 h/month. In contrast, the radon levels in Despicătura Cave (measured over 2-month period in both warm and cold season) are very low, never exceeding 455 Bq/m^3; thus, health hazards are minimal.

Mineralogy

A total of 15 minerals have been reported from the five cave presented in this chapter (Table 1). In terms of chemical groups, sulfates are the most abundant (8 minerals), followed by carbonates and phosphates (3 minerals each), and native elements (only sulfur). The only minerals that occur in all caves are calcite and gypsum. Diana Cave is by far the most outstanding given the presence of 11 minerals, 4 of which reported for the first time from a cave environment worldwide (Diaconu 1974; Onac et al. 2009, 2013a, b;

Table 1 Minerals identified in the investigated caves

Mineral	Formula	Cave
Sulfur	S	Diana
Aragonite	$CaCO_3$	Diana
Calcite	$CaCO_3$	All caves
Monohydrocalcite	$CaCO_3{\cdot}H_2O$	Hercules
Anhydrite	$CaSO_4$	Diana
Apjohnite	$Mn^{2+}Al_2(SO_4)_4{\cdot}22H_2O$	Diana
Epsomite	$MgSO_4{\cdot}7H_2O$	Diana
Gypsum	$CaSO_4{\cdot}2H_2O$	All caves
Halotrichite	$Fe^{2+}Al_2(SO_4)_4{\cdot}22H_2O$	Diana
Pickeringite	$MgAl_2(SO_4)_4{\cdot}22H_2O$	Diana
Rapidcreekite	$Ca_2(SO_4)(CO_3){\cdot}4H_2O$	Diana
Tamarugite	$NaAl(SO_4)_2{\cdot}6H_2O$	Diana
Brushite	$Ca(PO_3OH){\cdot}2H_2O$	Adam
Fluorapatite	$Ca_5(PO_4)_3F$	Adam
Hydroxylapatite	$Ca_5(PO_4)_3OH$	Adam, Steam

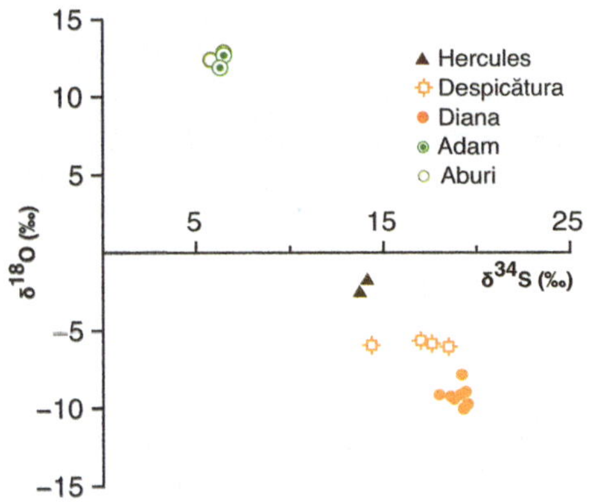

Fig. 11 $\delta^{18}O$ and $\delta^{34}S$ values of gypsum speleothems from Băile Herculane thermal caves

Puşcaş et al. 2013). The abundance and unicity of the mineral assemblages in Diana Cave illustrate its highly active steam condensate, strong acid-sulfate weathering environment. The presence of Diana Fault along, which a section of the cave develops (Fig. 6), played a critical role in the minerogenic processes because: (1) exposed the marly limestones of the Iuta Layers, a source of Na, Al, K, Fe, and Mn, and (2) provided paths for the thermo-mineral water to rise from depth and discharge in the cave. Next, the hydro-geochemical component (Na–Ca–Cl-type thermal waters rich in sulfate and H_2S) created the uniquely aggressive sulfate acid environment under which via key chemical reactions (above and below water surface) the sulfate-dominated mineral assemblage was precipitated (Onac et al. 2013a, b; Puşcaş et al. 2013).

The isotopic composition ($\delta^{18}O$ and $\delta^{34}S$) of gypsum from all five caves identifies two distinct populations (Fig. 11). The more positive values (~14–20‰) of the samples from Diana, Hercules, and Despicătura caves (Group 1) indicate an almost complete sulfate-limited thermochemical sulfate reduction (TSR) process that occurred in waters characterized by elevated H_2S/SO_4^{2-} ratio. Such values are consistent with an oxidation of dissolved sulfide in springs of the lower Cerna Valley as discussed by Wynn et al. (2010). Group 2 includes the two steam caves (Adam and Steam); the somehow lower $\delta^{34}S$ values (5.5–6.5‰) in their gypsum may reveal that the sulfuric acid which reacted with limestone resulted from the oxidation of dissolved sulfide produced during methane-limited TSR.

Since oxygen in gypsum is acquired during oxidation of sulfides, $\delta^{18}O$ values can shed light on conditions present during their formation. As such, the $\delta^{18}O$ values (0 to −10‰) of gypsum samples from Group 1 caves suggest highly anoxic conditions in the deep thermal water, with abundant ^{18}O exchange with liquid water but not with O_2. Instead, in the steam-heated environment of Group 2 caves, the very high $\delta^{18}O$ values (~12–14‰) advocate for the majority of sulfate O being derived from atmospheric O_2 in gas-phase oxidation prior to hydration (Onac et al. 2011).

Cave Fauna

Overall, the caves that discharge geothermal fluids at Băile Herculane are not very rich in terms of inhabiting species. In Despicătura, Hercules and Steam caves no invertebrate or vertebrate species were identified. Only a sub-troglophilic species (*Scutigera coleoptrata* Linnaeus 1758) (Chilopoda, Scutigeromorpha) was found in Diana Cave (Negrea 1994). The lack of species in these caves is a direct consequence of the relatively constant and elevated temperatures. It was observed (Delay 1974) that the annual "hetero-thermal" regime characteristics in karst zones represent an important factor in the ontogeny of many cave-dwelling species. The seasonal variations of temperature were proven to be an important factor in the development of many troglophile species (Lunghi et al. 2014). A situation similar with the caves at Băile Herculane was observed in the Acquasanta thermal caves in Italy, from where a non-specialized fauna was described (Galdenzi et al. 2010). On the other hand, in so-called warm caves (with average air temperatures of 14–15 °C, like Hoţilor and Movile in Romania, or Frasassi Cave in Italy), a rich troglophile or troglobite fauna was documented (Negrea and Negrea 1977; Bertolani et al. 1994; Negrea 1994; Nitzu 2001; Nitzu et al. 2016; Galassi et al. 2017).

The biodiversity in Adam's Shaft represents a particular case, where the species richness is conditioned by the combined influence of rich organic substrate and high temperature, on one hand, and by the cave morphology, on the other. According to Negrea and Negrea (1977), the morphology of this cave does not correspond to a type of shaft that simply closes in a "dead end," but to a shaft that is connected to subsequent passages at depth. A quasi-similar case was observed in Spain in *Alhama de Murcia en el Valle del Guadalentín* (Strinati 1953; Lisón et al. 2010). A large number of the invertebrate species inhabiting Adam's Shaft belongs to a rich guanophilous community (gunanocenosis) (Decou and Tufescu 1976).

In the southern passage of this cave, the elevated air temperature favors the existence of a maternity colony that includes several bat species: *Rhinolophus euryale* (dominant species), *Rh. ferrumequinum*, *Myotis myotis*, *Miniopterus schreibersii*, and *M. oxygnathus*) (Decou et al. 1974). The particular cave climate of that passage also supports a particular and rich troglobite fauna.

A total of 25 species of invertebrates were identified in the Adam's Shaft (Decou et al. 1974) (Fig. 12), most of them being trogloxene or troglophilous species with European or Palaearctic ranges. The most interesting species identified in Adam's pothole are *Neotrombicula adamensis* Feider 1974 (Acari, Trombidiformes), a troglobite endemic for this cave, *Trachelipus trilobatus* (Stein 1859) (Malacostraca, Isopoda), endemic for the Băile Herculane zone, and *Trichoniscus inferus* Verhoef 1908 (Isopoda), which is endemic for caves situated between Olt and Timiş—Cerna Basin. *Uroactinia (Chiropturopoda) cavernicola* (Hutzu 1997) (Acari, Uropodita), previously mentioned by Decou et al. (1974) as *U.* cf. *coprophila* Sellnick 1958 (a tropical species with African distribution range) is an endemic guanobiont species, dominant in the Adam's guanocenosis.

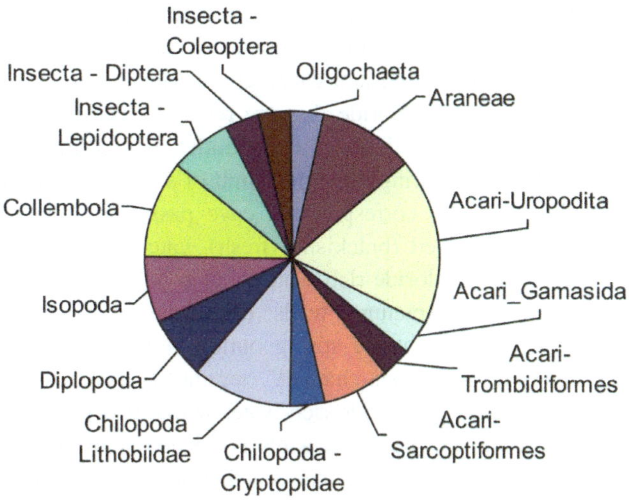

Fig. 12 Proportional distribution of 25 cave-dwellers (invertebrate species) identified in Adam's Shaft, distributed per 14 macrotaxa

Among the troglophilic species of springtails, besides the guanobiont species *Mesogastrura ojcoviensis* (Stach 1919) and *Heteromurus nitidus* (Templeton 1835), another troglophilic species—*Entomobrya pasaristei* (Denis 1936), known in Romania only from Dobrogea Caves, was identified by Gruia (2003).

Cave Conservancy and Recreational Caving

All five caves discussed in this chapter are within the Domogled-Valea Cernei National Park and Romanian Sites of Community Importance (ROSCI) 0069 "Natura 2000" Site. Despicătura and Hercules caves are gated, whereas Adam's Shaft is naturally protected both by its remote and largely unknown location (for tourists) and by its vertical entrance (11-m drop). Steam Cave is the only one that can be visited without a guide, in any season.

Conclusions

A group of small limestone caves discharge either steam or hot (up to ~55 °C), Na–Ca–Cl-type groundwater. Those warm fluids' initial source is the nearby geothermal reservoir hosted by a granitoid pluton. The geothermal (hot and brackish) parent-water is variably diluted by mixing with different amounts of meteoric-sourced karst freshwater. This rather atypical setting enabled plausible interpretations to be made about certain specific behaviors displayed by the resulting underground flows.

At low flows, the discharge of hot brackish parent-water secures a constant flux of the conservative natural tracer Cl^-: with respect to that constant value, alternating depletions and enrichment of the chloride flux are recorded, during heavy flood events, at the Hercules Cave outlet. It is accordingly conjectured that during the rising limb of a storm event, not only becomes the corresponding cave passage completely flooded with mixed (brackish + fresh) water, but a certain amount of this chloride-rich water is also forced and stored into the passage fractured walls: consequently, a Cl^- flux deficit is recorded at the spring outlet. Next, as the flood recedes and an "open-channel" regime is progressively re-established, the chloride-rich water which had previously been stored in the fractured walls is released back into the stream passage: a Cl^- flux excess is therefore recorded at the spring outlet.

In this karst setting, not only Cl^-, but also SiO_2 behaves as a conservative natural tracer. Consequently, using silica geothermometry, it is inferred that the involved liquid parent-water has a temperature in excess of 100 °C: therefore, upon its arrival in an aerated cave passage situated close to the local base level that liquid water starts to boil. The consequently released water vapors ascend through open cracks in the limestone rock, to eventually reach a couple of presently dry—yet "steaming"—cavities (Adam's Shaft and Steam Cave), which are perched 140–230 m higher up, whereas the 100 °C liquid water which was left as a result mixes with fluctuating amounts of meteoric origin cold freshwater, the corresponding mixture eventually discharging by the Hercules stream cave outlet.

In all investigated caves, "exotic" minerals precipitated following the interaction between thermal solutions and the bedrock. In Diana Cave, in particular, an active steam condensate, strong acid-sulfate weathering environment favored the occurrence of 11 such minerals, four of which reported for the first time from a cave environment worldwide.

Adam's Shaft hosts a rich and diverse endemic troglobite fauna. Within the same cave, the radiocarbon age of 8425 cal years BP obtained from the lower part of the guano accumulation, argues for one of Europe's oldest permanent bat colony.

Acknowledgements Parts of this work were supported by grants from the Romanian National Authority for Scientific Research, CNDI-UEFISCDI; PN-II-PT-PCCA-2011-3.1-1619 to MC and ID_544 to BPO). For their dedicated assistance provided during the field operations, Elisabeta Primejdie and Socrate Bucur are gratefully acknowledged. We also thank Floarea Răducă for her continuing support in the laboratory work and Dr. Victor Polyak for providing the U-series ages on gypsum speleothems.

References

Apollaro C, Vespasiano G, Muto F, De Rosa R, Barca D, Marini L (2016) Use of mean residence time of water, flowrate, and equilibrium temperature indicated by water geothermometers to rank geothermal resources. Application to the thermal water circuits of Northern Calabria. J Volc Geotherm Res 328:147–158

Arnórsson S, Gunnlaugsson E, Svavarsson H (1983) The chemistry of geothermal waters in Iceland. III. Chemical geothermometry in geothermal investigations. Geochim Cosmochim Acta 47:567–577

Audra P, Hoblea F, Bigot J-Y, Nobecourt J-C (2007) The role of condensation-corrosion in thermal speleogenesis: study of a hypogenic sulfidic cave in Aix-les-Bains, France. Acta Carsologica 36:185–194

Bailly-Comte V, Martin JB, Jourde H, Screaton EJ, Pistre S, Langston A (2010) Water exchange and pressure transfer between conduits and matrix and their influence on hydrodynamics of two karst aquifers with sinking streams. J Hydrol 386:55–66

Bălteanu D (2007) Herculane—Arc peste timp. Ed Info, p 138

Bertolani R, Manicardi GC, Rebecchi L (1994) Faunistic study in the karst complex of Frasassi (Genga, Italy). Int J Speleol 23(1–2):61–77

Bulgăr A, Povară I (1978) Separation of karstic thermal springs discharge components as based on the analysis of discharge and temperature variations measured at exsurgence. Trav Inst Spéol "Émile Racovitza" 17:209–214

Carbonnel J-P, Olive P, Decu VG, Klein D (1999) Datations d'un dépôt de guano holocène dans les Carpates méridionales (Roumanie). Implications tectoniques. CR Acad Sci Paris, Sciences de la terre et des planètes 328:367–370

Cucoş Dinu A, Călugăr MI, Burghele BD, Dumitru OA, Constantin C, Onac BP (2016) Radon levels in Romanian caves: an occupational exposure survey. Environ Geochem Health. https://doi.org/10.1007/s10653-016-9878-1

Chen L, Ma T, Du Y, Xiao C, Chen X, Liu C, Wang Y (2016) Hydrochemical and isotopic (^2H, ^{18}O and ^{37}Cl) constraints on evolution of geothermal water in coastal plain of Southwestern Guangdong Province, China. J Volc Geotherm Res 318:45–54

Decou VG, Negrea A, Negrea Ş (1974) Une oasis biospéologique tropicale développeé dans une région tempereé: "Peştera lui Adam" de Băile Herculane (Carpates Méridionales, Roumanie). Trav Inst Spéol "Émile Racovitza" 13:81–103

Decou VG, Tufescu M (1976) Sur l'organisation d'une biocénose extrème: la biocénose du guano de la grotte "Peştera lui Adam" de Băile Herculane (Carpates Méridionales, Roumanie). Trav Inst Spéol "Émile Racovitza" 15:113–132

Delay B (1974) Les conditions thermiques des milieux terrestres dans la zone d'hétérothemie des massifs calcaires et leurs influences sur le développement de *Speonomus longicornis* (Coléoptère, Bathyscinae). Ann Spéléol 29(1):121–136

Delgado-Outeiriño I, Araujo-Nespereira P, Cid-Fernández JA, Mejuto JC, Martínez-Carballo E, Simal-Gándara J (2009) Behaviour of thermal waters through granite rocks based on residence time and inorganic pattern. J Hydrol 373:329–336

Diaconu G (1974) Quelques considérations sur le présence de l'anhydrite dans la grotte "Peştera Diana" (Băile Herculane, Roumanie). Trav Inst Spéol "Émile Racovitza" 13:191–194

Diaconu G (1987) La géologie et la tectonique du périmètre d'influence la source thermo-minérale "Hercule", Baile Herculane. Theor Appl Karstol 3:109–116

Diaconu G, Medeşan A (1973) Sur la présence du pickeringite dans la grotte de Diana (Băile Herculane, Roumanie). Trav Inst Spéol "Émile Racovitza" 12:303–309

Diaconu G, Dumitraş D-G, Marincea Ş (2010) Mineralogical analyses in various caves from the Băile Herculane area, the Cerna Passage. Trav Inst Spéol "Émile Racovitza" 49:135–148

Edmunds WM, Kay RLF, McCartney RA (1985) Origin of saline groundwaters in the Carnmenellis Granite (Cornwall, England): natural processes and reaction during Hot Dry Rock reservoir circulation. Chem Geol 49:287–301

Erőss A, Mádl-Szőnyi J, Csoma AÉ (2008) Characteristics of discharge at Rose and Gellért Hills, Budapest, Hungary. Central Eur Geol 51:267–281

Fournier RO (1977) Chemical geothermometers and mixing models for geothermal systems. Geothermics 5:41–50

Galassi DMP, Fiasca B, Di Lorenzo T, Montanari A, Porfirio S, Fattorini S (2017) Groundwater biodiversity in a chemoautotrophic cave ecosystem: how geochemistry regulates microcrustacean community structure. Aquatic Ecol 51(1):75–90

Galdenzi S, Cocchioni F, Filipponi G, Morichetti L, Scuri S, Selvaggio R, Cocchioni M (2010) The sulfidic thermal caves of Acquasanta Terme (central Italy). J Cave Karst Stud 72:43–58

Goldscheider N (2005) Fold structure and underground drainage pattern in the alpine karst system Hochifen-Gottesacker. Eclogae Geol Helv 98:1–17

Gruia M (2003) Collembola from Romanian Caves. Trav Mus Nat Hist Naturelle "Grigore Antipa" 45:39–158

Hose LD, Palmer AN, Palmer MV, Northup DE, Boston PJ, DuChene HR (2000) Microbiology and geochemistry in a hydrogen-sulphide-rich karst environment. Chem Geol 169:399–423

Hutzu M (1997) Chiropturopoda cavernicola nov. Spec., eine neue höhlenbewohnende Uropodiden—Art (Actinotrichida: Uropodina). Mém Biospéol 24:163–168

Klimchouk A, Palmer AN, De Waele J, Auler AS, Audra P (eds) (2017) Hypogene karst regions and caves of the world. Springer International Publishing

Lazaridis G, Melfos V, Papadopoulou L (2011) The first cave occurrence of orpiment (As_2O_3) from the sulphuric acid caves of Aghia Paraskevi (Kassandra Peninsula, N. Greece). Int J Speleol 40 (2):133–139

Linzer H-G, Frisch W, Zweigel P, Girbacea R, Hann H-P, Moser F (1998) Kinematic evolution of the Romanian Carpathians. Tectonophysics 297:133–156

Lisón F, Yelo ND, Haz Á, Calvo JF (2010) Contribución al concimiento de la distribución de la fauna quiropterológica de la región de Murcia. Galemys 22(1):11–28

Lunghi E, Manenti R, Ficetola GF (2014) Do cave features affect underground habitat exploitation by non-troglobite species? Acta Oecol 55:29–35

Marin C (1984) Hydrochemical considerations in the lower Cerna river basin. Theor Appl Karstol 1:173–182

Marin C, Nicolescu T (1993) The geochemistry of groundwater from southeastern Dobrogea, Romania. Trav Inst Spéol "Émile Racovitza" 32:229–247

Marques JM, Andrade M, Carreira PM, Eggenkamp HGM, Graça RC, Aires-Barros L, Antunes da Silva M (2006) Chemical and isotopic signatures of $Na/HCO_3/CO_2$-rich geofluids, North Portugal. Geofluids 6:273–287

Martin JB, Dean RW (2001) Exchange of water between conduits and matrix in the Floridan aquifer. Chem Geol 179:145–165

Martin JM, Screaton EJ, Martin JB (2006) Monitoring well responses to karst conduit head fluctuations: Implications for fluid exchange and matrix transmissivity in the Floridan aquifer. Geol Soc Am Special Paper 404:209–217

Mărunţeanu M, Şerban E, Rusu A (1994) Neogene of Caransebeş-Mehadia Basin. Rom J Stratigraphy 76:79–87

Michard G, Sanjuan B, Criaud A, Fouillac C, Pentcheva EN, Petrov PS, Alexieva R (1986) Equilibria and geothermometry in hot alkaline waters from granites of S.W. Bulgaria. Geochem J 20:159–171

Michard G, Grimaud D, D'Amore F, Fancelli R (1989) Influence of mobile ion concentrations on the chemical composition of geothermal waters in granitic areas. Example of hot springs from Piemonte (Italy). Geothermics 18:729–741

Mitrofan H, Povară I (1992) Delineation of a thermal water carrying karstic conduit by means of thermometric measurements in the Băile Herculane area (Romania). Theor Appl Karstol 5:139–144

Mitrofan H, Mafteiu M, Povară I, Mitruţiu M (1995) Electrometric investigations on the supply channels of Hercules spring (Romania). Theor Appl Karstol 8:129–136

Mitrofan H, Povară I, Mafteiu M (2008) Geoelectrical investigations by means of resistivity methods in karst areas in Romania. Env Geol 55:405–413

Mitrofan H, Marin C, Povară I (2015) Possible conduit-matrix water exchange signatures outlined at a karst spring. Groundwater 53 (Suppl 1):113–122

Mitrofan H, Marin C, Povară I, Onac BP (2016) Hercules and Diana hypogene caves (Herculane Spa, Romania): dissimilar chemical evolutions documented by their present-day thermal water discharges. In: Chavez T, Reehling P (eds) Proceedings of DeepKarst 2016: origins, resources, and management of Hypogene Karst. Carlsbad, New Mexico, pp 145–150

Năstăseanu S (1980) Géologie de Monts Cerna. Annuaire Inst Géol Géophys Bucureşti 54:153–280

Negrea Ş, Negrea A (1977) Sur les associations de plancher des grottes du Banat (Roumanie). Trav Inst Spéol "Émile Racovitza" 16:99–139

Negrea Ş (1994) Chilopodes (Chilopoda) cavernicoles de Roumanie connus juqu'à present. Trav Mus Hist Naturelle "Grigore Antipa" 34:265–283

Nicolescu M (1970) General overview on the geology and the mineral waters of Caraş-Severin county. In: Apele minerale şi nămolurile terapeutice din Republica Socialistă România, vol 3, Ed Medicală, Bucureşti, pp 123–150

Nitzu E (2001) Edaphic and subterranean Coleoptera from the Dobrogean karstic areas Romania. A zoogeographic approach. Mitt Hamb Zool Mus Inst 38:131–169

Nitzu E, Giurginca A, Nae A, Popa I, Baba S, Meleg IN, Vlaicu M (2016) The catalogue of caves with endemic cavernicolous Arthropoda of Romania. Trav Inst Spéol "Émile Racovitza" 55:3–62

Onac BP, Sumrall J, Tămaş T, Povară I, Kearns J, Dârmiceanu V, Veres D, Lascu C (2009) The relationship between cave minerals and H_2S-rich thermal waters along the Cerna Valley (SW Romania). Acta Carsologica 38:27–39

Onac BP, Wynn JG, Sumrall JB (2011) Tracing the sources of cave sulfates: a unique case from Cerna Valley, Romania. Chem Geol 288:105–114

Onac BP, Effenberger HS, Wynn JG, Povară I (2013a) Rapidcreekite in the sulfuric acid weathering environment of Diana Cave, Romania. Am Mineral 98:1302–1309

Onac BP, Puşcaş CM, Effenberger HS, Povară I, Wynn JG (2013b) The hypogene origin of Diana Cave (Romania) and its sulfuric acid weathering environment. In: Filippi M, Bosák P (eds), Proceedings of the 16th International Congress of Speleology, Brno, Czech Republic, 3, pp 470–473

Pascu M (1968) Mitigation of cold waters infiltrations at the thermal spring Hercules I. Hidrotehnica, Gospodărirea Apelor, Meteorologia 14:33–37

Pauwels H, Fouillac C, Goff F, Vuataz F-D (1997) The isotopic and chemical composition of CO_2-rich thermal waters in the Mont-Dore region (Massif-Central, France). Appl Geochem 12:411–427

Placinta AO, Popescu E, Borleanu F, Radulian M, Popa M (2016) Analysis of source properties for the earthquake sequences in the south-western Carpathians (Romania). Rom Rep Phys 68:1240–1258

Ponta GM, Terteleac N (2013) Historical water quality data from Cerna Valley/Băile Herculane—Romania. In: Filippi M, Bosák P (eds), Proceedings of the 16th International Congress of Speleology, Brno, Czech Republic, 1, pp 347–352

Popescu-Voiteşti I (1921) Étude géologique sur les sources minérales des Bains d'Hercule. Ann Mines Roumanie 4:115–135

Popoviciu A (1872) Băile lui Ercule sau scaldele de la Meedia. Ed Albatros, Pesta, p 125

Povară I (1973) Contributions á la connaissance des sources thermo-minérales de Băile Herculane. Trav Inst Spéol "Émile Racovitza" 12:337–348

Povară I (1980) Note sur la circulation souterraine des eaux dans les calcaires du bassin de Cerna. Trav Inst Spéol "Émile Racovitza" 19:237–241

Povară I (1992) New data on the Hercule thermal aquifer, obtained by temperature measurements (Băile Herculane, Romania). Theor Appl Karstol 5:127–138

Povară I, Marin C (1984) Hercule thermomineral spring. Hydrogeological and hydrochemical considerations. Theor Appl Karstol 1:183–193

Povară I, Diaconu G, Goran C (1972) Observations préliminaires sur les grottes influencées par les eaux thermo-minérales de la zone Băile Herculane. Trav Inst Spéol "Émile Racovitza" 11:355–365

Povară I, Simion G, Marin C (2008) Thermomineral waters from the Cerna Valley Basin (Romania). Studia UBB Geologia 53:41–54

Povară I, Conovici M, Munteanu C-M, Marin C, Ioniţă ED (2015) Karst systems within the Southern Carpathians structure (Romania). Carpathian J Earth Env Sci 10(2):5–17

Puşcaş CM, Onac BP, Effenberger HS, Povară I (2013) Tamarugite-bearing paragenesis formed by sulphate acid alteration in Diana Cave, Romania. Eur J Mineral 25:479–486

Radulian M, Mândrescu N, Panza GF, Popescu E, Utale A (2000) Characterization of seismogenic zones of Romania. Pure Appl Geophys 157:57–77

Raeisi E, Groves C, Meiman J (2007) Effects of partial and full pipe flow on hydrochemographs of Logsdon River, Mammoth Cave Kentucky USA. J Hydrol 337:1–10

Rosales Lagarde L, Boston PJ, Campbell AR, Hose LD, Axen G, Stafford KW (2014) Hydrogeology of northern Sierra de Chiapas, Mexico: a conceptual model based on a geochemical characterization of sulfide-rich karst brackish springs. Hydrogeol J 22:1447–1467

Sarbu SM, Kane TC, Kinkle BK (1996) A chemoautotrophically based cave ecosystem. Science 272:1953–1955

Screaton E, Martin JB, Ginn B, Smith L (2004) Conduit properties and karstification in the unconfined Floridan aquifer. Ground Water 42:338–346

Simion G, Ponta G, Gaşpar E (1985) The dynamics of underground waters from Baile Herculane, Cerna Valley (Romania). Ann Soc Géol Belgique 108:245–249

Stoica de Haţeg N (1984) The Chronicle of Mehadia and of Băille Herculane. Ed. Facla, Timişoara, p 232 (in Romanian)

Strinati P (1953) Une grotte chaude près d'Alhama de Murcia. Speleon 4(2):95–104

Terteleac N, Veliciu Ş, Pop G (1989) Complex geological and geophysical research concerning the hydrostructure of the Hercules Spring. Unpublished report, IPGG Archive, Bucharest (in Romanian), Băile Herculane

Tufescu M, Decou VG (1977) Modèles préliminaires de la dynamique des populations principales de la biocénose du guano de la grotte "Pestera lui Adam" de Baile Herculane (Roumanie). Trav Inst Spéol "Émile Racovitza" 16:85–98

Vesper DJ, White WB (2004) Storm pulse chemographs of saturation index and carbon dioxide pressure: implications for shifting recharge sources during storm events in the karst aquifer at Fort Campbell, Kentucky/Tennessee, USA. Hydrogeol J 12:135–143

Wynn JG, Sumrall JB, Onac BP (2010) Sulfur isotopic composition and the source of dissolved sulfur species in thermo-mineral springs of the Cerna Valley, Romania. Chem Geol 271:31–43

Aninei Mountains: Captare Cave (Pestera de la Captare)

Petr Barák, Tomáš Svoboda, and Gheorghe M. L. Ponta

Abstract

The Captare Cave (Peştera de la Captare) is located in the western part of Romania (Banat), in the sedimentary zone Reşiţa-Moldova Nouă, which host the largest single exposure of limestones in Romania ($\sim$800 km^2). The stream from the Captare Cave is a right (south) side tributary of Miniş River that crosses from northwest to southeast the northern/upper section of the carbonate rocks. This chapter focuses on the explorations of the cave, including the recent (since 2013) scuba diving activities, which resulted in the discovery of three major galleries (Chámovod, Never-ending, and Broken Lines Corridors), which increased the length of the cave from 961 to 2900 m.

Keywords

Doline • Ponor • Cave diving • Spring

Introduction

Geographic, Geologic, and Hydrologic Settings

Captare Cave is located in the Anina Mountains, on the right side of Miniş Valley, at km 9.2 on the Anina–Bozovici highway (Goran 1982). The cave entrance is located 85 m above the river's thalweg and 60 m above the spring (Fig. 1).

The Reşiţa-Moldova Nouă sedimentary zone is formed by a sequence of anticlines and synclines separated by major fractures/faults parallel with the general northeast–southwest direction/strike of the structure (Năstăseanu and Savu 1970). The cave is hosted in the Valea Aninei limestones of Upper Oxfordian–Lower Kimmeridgian age, which are thin- to medium-bedded limestones intercalated with thin-bedded chert layers (Bucur 1997). The limestones are part of the eastern flank of Pitulaţi Syncline, which dips west 45°–70°. The cave's development is controlled by fractures parallel and perpendicular to the longitudinal axis of the syncline. The Valea Aninei and Marila limestones (Upper Tithonic-Berriasian) form a karst plateau, bordered to the west by Păuleasa stream and ridge, Golumbului Valley to the east, and to the north by the Miniş Gorge.

The spring at the Captare Cave has an estimated flow of 100 L/s, and the water has a pH of 6.5 and a temperature of 8.5 °C (September 3, 1978). The spring is the main outlet/discharge point for the waters of the Padina–Cracul Roşu Plateau. Due to its high flow, the spring was developed (dammed) as a public water supply for the Miniş Settlement and the Cerbul Chalet.

Overlaying the cave survey on a topographic map shows that between the final sump and the nearest known ponor there is still a considerable distance. The N-S branch of the cave is heading toward a large doline formed on the Valea Aninei limestones. This doline is the head of a temporary/dry valley, which flows toward north parallel with the main N-S cave passage. The western branch of the cave presently ends under this valley, close to the interface between Valea Aninei and Marila limestones.

History of Exploration

Construction workers discovered at the end of 1970s the fossil entrance of the cave during the development (confining) of the spring as public water supply for Miniş Settlement and Cerbul Chalet. Considering the widespread occurrence of karst rocks, beginning with 1977 Focul Viu Caving Club from Bucharest organized several trips to find and map the caves and pits in the area limited at north and

P. Barák (✉) · T. Svoboda
Czech Speleological Society (Devon Caving Club), Lhota Rapotina 82, 679 01, Czech Republic
e-mail: barak.petr@post.cz

G. M. L. Ponta
Geological Survey of Alabama, 420 Hackberry Lane, Tuscaloosa, AL 35401, USA
e-mail: gponta@yahoo.com

G. M. L. Ponta and B. P. Onac (eds.), *Cave and Karst Systems of Romania*, Cave and Karst Systems of the World, https://doi.org/10.1007/978-3-319-90747-5_26

Fig. 1 Karst regions of Romania with location of the Captare Cave

east by Miniş Valley, Nera Valley to the south, and Beiul Valley at west.

In the summer of 1978, Focul Viu organized an international caving camp with the participation of Speleogroup No 22, Čachtice of the Slovak Speleological Society. During this camp, a team of cavers notices along the Miniş River (right/southern side) a 20–25 m thick travertine deposit in which they identified three caves (about 10 m long each), and on the top, they found the fossil entrance in the Captare Cave. By the end of 1978, the first map totaling 961 m of passages was completed (Ponta 1979; Sluka 1982). At that time, the 3rd sump was the end of the cave (Fig. 2).

In the 1980s, Czech Speleological Society, with Speleoaquanaut, Labyrinth, and Jiří Hovorka, dived the first two sumps (M. Piškula) of the cave. In 2013, T. Svoboda (Czech Speleological Society, Devon Caving Club) passed the 3rd sump and the remapping of the cave began under the supervision of O. Burneci with Exploratorii Reşiţa Speological Association (Barák et al. 2014).

In 2014, the exploration is continued beyond the 3rd sump by the Czech cave diving team (P. Barák, T. Svoboda, Czech Speleological Society, Devon Caving Club, and V. Kaman, Czech Speleological Society, Pustý žleb Caving Club), with the survey of the Chámovod Corridor, Inclined Dome, and Choco-voko Dome, adding 663 m of new passages, including sumps 4th, 5th, and 6th (Barák 2014). In 2015, P. Barák and T. Svoboda continued the explorations beyond the 4th and 5th sumps (Fig. 2), surveying the following new passages: Broken Lines Corridor with Reşiţa's, Anina's, and Like Brno domes, and the Never-ending Story Corridor, totaling 1123 m and increasing the number of sumps to 14 (Barák and Svoboda 2016). The activities in 2016 added 153 m of new passages and included explorations of the Never-ending Story Corridor (P. Barák, T. Svoboda) and a climb before the 3rd sump performed by P. Anelt and P. Kubálek of the Czech Speleological Society, Speleoklub Praha.

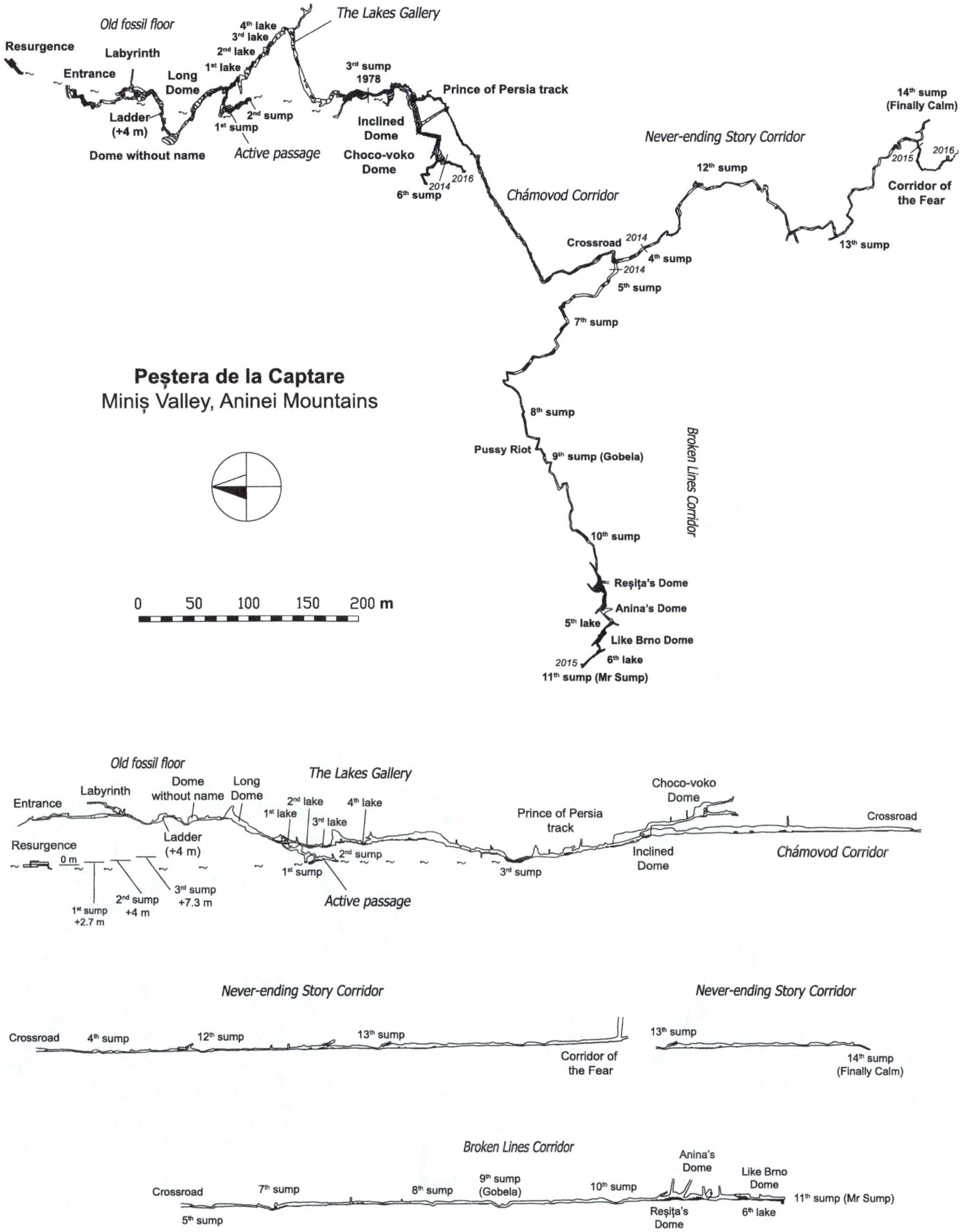

Fig. 2 Map of the Captare Cave

Cave Description

The caves began with a narrow gallery which ends in a room packed with limestone blocks fallen from the ceiling. On the right-side wall, there is a spectacular outcrop where the alternation between limestones and chert layers can be seen. Beyond the entrance room, the main passage is higher and larger, and the first speleothems appears. The most impressive one, a 2 m tall stalagmite formed in a side passage called the Labyrinth. Vandals caused some damages in these entrance galleries, but their impact disappears after climbing a 4 m wall, where three signatures dated 1857 were found. From this point, a descending gallery ends into a small lake. On the other side of the lake, there are two rooms: "Dome without name" and Long Dome, which continues with a well-decorated passage that descends in steps to a junction.

To the right (west) of the mentioned junction, the Active Passage is reached after a few drops (the deepest is 7 m) and a vertical shaft, 3 m in diameter, with a lake on the bottom. On the opposite wall of the shaft, after a delicate traverse to a window, a climb down intercepts a stream with sumps at both ends (1st and 2nd sumps; Fig. 2).

The passage to the left (southeast) is named The Lakes Gallery, which consists of four lakes. In fact, the number of lakes varies depending on the season. The upstream section ends in a large passage, 7 m wide and 15 m high, with numerous boulders on the floor through which the main stream of the cave is flowing. This is the second location where the fossil passage intercepts the stream level. Downstream, the gallery continues along a fracture plane as a 0.4 m wide and 4–5 m high, tight passage that ends in a pit with the walls covered with clay and sand. Upstream, the cave continues for another 70 m, ending in a lake where the 1978s final 3rd sump is located. Up to this hydro-morphological feature, the cave measures 961 m in length with −50 m vertical development.

Immediately beyond the 3rd sump, a junction is intercepted. To the right (west), an upper level begins with the Inclined and Choco-voko domes (large amount of clay on the floor), but ends shortly in the 6th sump. To the left (south–southwest), the cave continues upstream along the main streamlet that traverses the Chámovod Corridor (Fig. 3) coming from the Crossroad where the 4th and the 5th sumps are located. On the other side of the 5th sump, the Broken Lines Corridor develops (Fig. 4), where four more sumps [7th, 8th, 9th (Fig. 5), and 10th] were encountered before reaching the final one (11th sump) that represents the end point of this part of the cave. Between sumps 10th and 11th, Reşiţa's (Fig. 6), Anina's, and Like Brno domes

Fig. 3 Chámovod Corridor (photograph by P. Barák)

Fig. 4 Broken Lines Corridor (photograph by P. Barák)

Fig. 5 Ninth sump "Gobela" (photograph by P. Barák)

(Fig. 7) are located, with large boulders and slab breakdown accumulated on the floor.

Southeast of the Crossroad, beyond the 4th sump, the "Never-ending Story" Corridor began, which hosts two sumps along the main passage and ends in the 14th sump. The stream bed has no alluvial sediments, suggesting that large amounts of water flow along this gallery during rainy season. This section lies at the interface between the phreatic and vadose zones of this hydrokarst system.

In summary, the cave develops on three levels: The upper one (**Fossil/Dry level**) includes the entrance passages, the Lakes Gallery, and the section between Choco-voko Dome and the 6th sump (Fig. 2). In the Lakes Gallery, four lakes are located; the water level depends on the amount of the percolating water dripping through the fractured limestone. Occasionally (during the rainy season), temporary streams are flowing in these areas as well as along the Chámovod Corridor (Fig. 4), which is the result of the confluence between the Broken Lines and Never-ending Story Corridors. The longitudinal cross section (Fig. 2) points out the connection between the 1st, 2nd, and 3rd sumps and the active phreatic resurgence, through underwater conduits (**Phreatic level**). The 3rd sump ends in a corridor with water disappearing between collapsed blocks and flowing along an unknown phreatic path underneath the Active Passage to wards the resurgence. A perennial flow in the cave was observed upstream of the 3rd sump (**Active level**).

Fig. 6 Crystal formations in Reşiţa's Dome (photograph by P. Barák)

Fig. 7 Like Brno Dome (photograph by P. Barák)

Conclusions

Captare Cave explored and surveyed between 1970s and 2016 is an important karst system with one permanent active flow and two important temporary tributaries. The recharge area is in the Padina–Cracul Roşu Karst Plateau. Cave passages are developed on three levels: the fossil part with various types of speleothems, the active (stream) passages, and the permanent sumps (phreatic zone). The most widespread cave level is located at the interface between phreatic and stream zone, with periodical active flow and flooded sumps. In a few areas of the cave, chert layers are present. The most spectacular corridors are hosted in limestones with no fluvial sediments. The total extent of the cave passages by the end of 1978 was 961 m, whereas by mid-2016, it reached a length of 2900 m with 60 m vertical range (+19.5 m, −40.5 m). These achievements are the result of a great international cooperation between Czech and Romanian cavers and environmentalists.

Acknowledgements The 2013–2016 Czech team acknowledges Asociaţia Speologică Exploratorii Reşiţa for cooperation in the cave activities and assistance in obtaining the necessary permits to work in the cave. M. Sluka is thanked for providing information from the initial research in the 1970s.

References

Barák P, Kaman V, Svoboda T, Crîsta R (2014) Banát 2013 - úspěšný výzkum krasu Aninských a Lokvejských hor. Speleofórum 33:97–100

Barák P (2014) Explorări speologice ceheşti în vestul României. Speomond 17:25–27

Barák P, Svoboda T (2016) Peştera de la Captare – linie prolomeny! Speleofórum 35:88–92

Bucur II (1997) Formaţiunile mezozoice din zona Reşiţa-Moldova Nouă (Munţii Aninei şi estul Munţilor Locvei). Ed Presa Univ Clujeană, Cluj-Napoca, p 214

Goran C (1982) Catalogul sistematic al peşterilor din Romania. Ed CNEFS, Bucureşti, p 496

Năstăseanu S, Savu H (1970) Geological map of Romania, Anina Sheet. Scale 1:50.000, Inst Geol Geofiz, Bucureşti

Ponta G (1979) Peştera de la Captare - Valea Minişului. Bul Speol 1:1–5

Sluka M (1982) Peştera de la Captare. Slovensky Kras XX:167–173

Useful Websites

http://6-22.speleo.cz
http://www.exploratorii.ro

Aninei Mountains: The Ponor-Plopa Cave System

Silviu Constantin

Abstract

Ponor-Plopa is a tiered karst system developed along a small underground stream that includes five sumps. While part of the system was known and visited for a long time, the sections located beyond the sumps were only explored in the last decades. These sections became known for their rich paleontological deposit, including remains of cave bear and other large carnivores and herbivores that have been accumulated within some passages c. 50–30 ka ago. The cave is internationally known for yielding the oldest remains of a modern human (c. 40 ka) that has shown proof of admixture with Neandertalians.

Keywords

Oase Cave • Banat • Early modern humans
Neandertal-sapiens admixture • Fluviokarst evolution

Introduction

The Ponor-Plopa cave system is a relatively extensive network of underground passages located in SW Romania that became internationally famous in 2003, under the name of *Peştera cu Oase*, following the discovery of the oldest *Homo sapiens* remains in Europe (Trinkaus et al. 2003a). This triggered immediate attention and a series of scientific campaigns started, leading to extensive documentation of one of the most rich and better preserved paleontological cave sites in the country.

S. Constantin (✉)
"Emil Racoviţă" Institute of Speleology, Romanian Academy, Frumoasă 31, 010986 Bucharest, Romania
e-mail: silviu.constantin@iser.ro

Geographic, Geologic, and Hydrologic Settings

The Ponor-Plopa cave system is located in the central part of the Aninei Mountains, southwest of Steierdorf, a small neighborhood of Anina, at elevations ranging between 575 m (the Ponor) and 545 m (the Plopa resurgence) for the main river cave, and ~600 m for the dry caves on the Plopa Plateau. The system includes a total of six caves of which two, Ponor and Plopa, were connected by cave diving. The remaining four are relatively small, dry (inactive) cavities that were connected to the system at various stages during its evolution (Fig. 1).

The cave system develops mainly within massive Barremian (Lower Cretaceous) reef limestone (Plopa limestones) on the western flank of a local syncline (the Central Syncline), which forms a small karst plateau. To the west, the plateau is bordered by a rocky escarpment formed at the contact between the Plopa limestones and a Hauterivian mudstone formation (the "Lower Plopa" formation); to the north, it is bordered by a secondary ridge of the Culmea Frumoasă, whereas to the south and east it is delimited by the Miniş Gorge (Fig. 2).

The cave system was formed along the Ponor stream, a tributary of the Miniş River. This small stream, with a flow rate of ~10 l/s (Iurkiewicz et al. 1996), gathers its waters from the impervious Jurassic deposits of the Anina Anticline. When reaching the Cretaceous marls and limestones, it sinks in the underground at the bottom of a ~30-m high cliff. The sinking point and corresponding escarpment were formed at the interface of different permeabilities of the Plopa limestones and Hauterivian mudstones as well as along the Miniş strike-slip fault. The Ponor stream resurfaces through the high porch of Plopa cave, at ~100 m from the Miniş River.

Exploration History

The Ponor-Plopa system was first described by the geographer Vasile Sencu (Senco 1973; Sencu 1964, 1977) who performed water tracing experiments that revealed the local

G. M. L. Ponta and B. P. Onac (eds.), *Cave and Karst Systems of Romania*,
Cave and Karst Systems of the World, https://doi.org/10.1007/978-3-319-90747-5_27

Fig. 1 Location of the Ponor-Plopa cave system in the karst region of Aninei Mountains

underground drainages of Ponor-Plopa and Uteriş-Irma, respectively. He also published the first maps of the Plopa resurgence and the Ponor Uscată (dry) Cave (1964). Both caves have been frequently visited by tourists, which led to an almost total destruction of the speleothems from the Ponor Uscată Cave. Due to the presence of some deepwater pools, Plopa Cave was less affected by uncontrolled tourism until the 1980s when a diesel dump reaches the Ponor stream, and the fumes have reached the underground pools and covered speleothems, sediments, and walls with a thick black coating that is still present.

The exploration of the unknown passages located between the Ponor sink and the Plopa Cave started in 2001 by a team of Pro Acva Group in Timişoara (Ş. Milotă, L. Sarcina) and A. Bîlgăr from Drobeta-Turnu Severin (Bîlgăr 2004). The team was able to unplug the Ponor and dive the narrow entrance sump to reach the Big Chamber (Sala Mare). Further on, they were able to dive several other small sumps and ultimately reach the Plopa resurgence. In February 2002, the three explorers were able to climb up a shaft located beyond Sump 6 and get access to a network of passages located above it. By de-clogging a narrow passages filled with bones and clay ("The Gate"), they reached a series of pristine passages filled with a fossil bone bed and cave bear activity marks such as hibernation nests. Within the first large-sized chamber (subsequently named "The Mandible Chamber"), they discovered and photographed a human mandible. The discovery was announced to Dr. O. Moldovan (OM), at that time Head of the Cluj-Napoca Branch of the "Emil Racoviţă" Institute of Speleology. Dr. Moldovan subsequently visited the cave together with the Pro Acva Group team and retrieved the mandible. Further on, she contacted professor E. Trinkaus (ET) of the Washington University in St. Louis who analyzed the remains and have it radiocarbon dated. The ensuing age of >35,200 ^{14}C BP points toward the potentially oldest *Homo sapiens* is discovered in Europe and justified the start of an extensive study of the sector that was now known as *Peştera cu Oase*. The first scientific reconnaissance campaign of 2003 yielded yet another discovery: a human facial skeleton (Oase 2), as well as the complete map of the post-sump sector. Systematic excavations by an international team were carried out in 2004 under the leadership of O. Moldovan, E. Trinkaus, and J. Zilhão and in

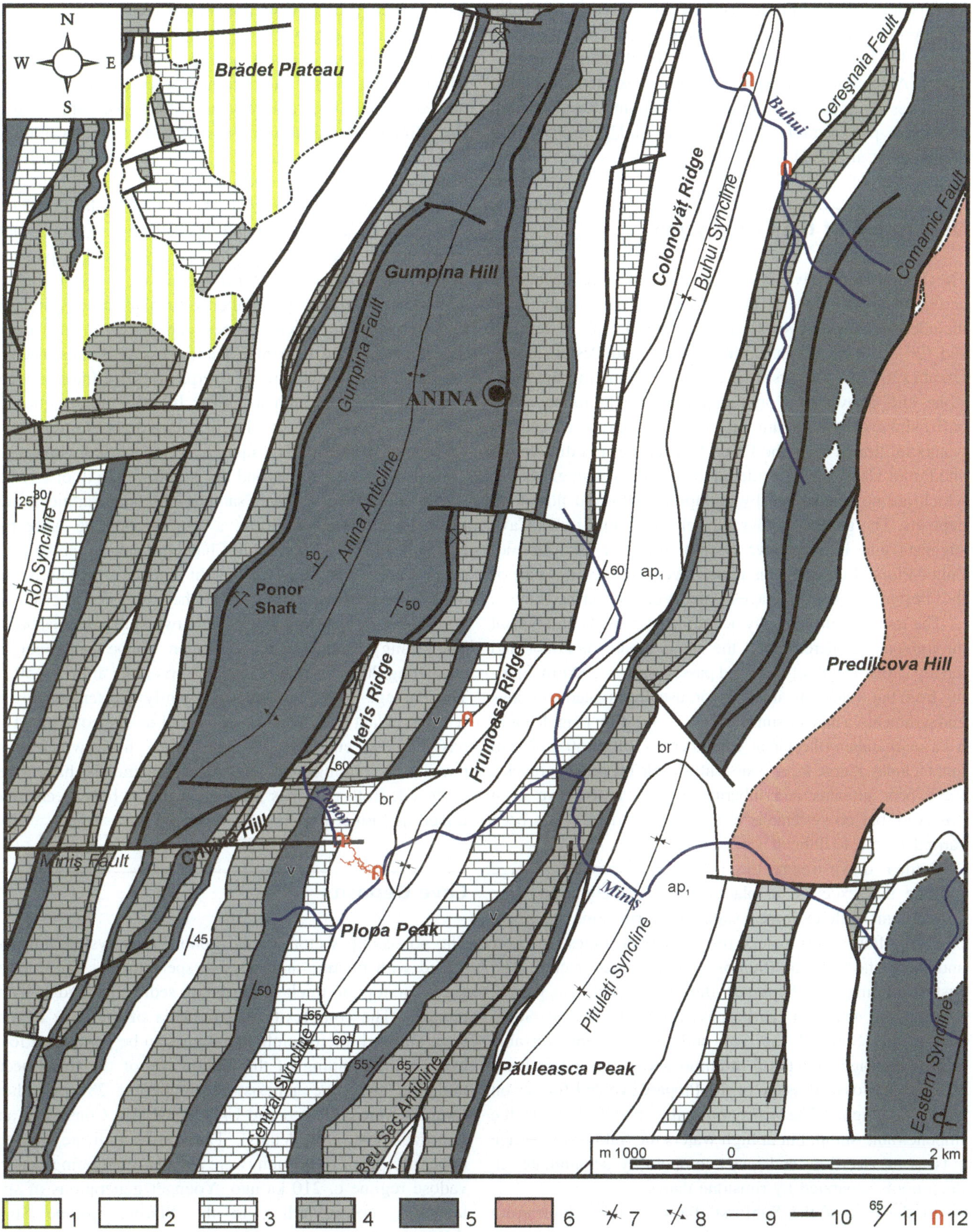

Fig. 2 Simplified geological map of the Anina region (Constantin 2013). Key: 1. Quaternary deposits; 2. karst-forming deposits (Valea Aninei, Plopa, and Valea Minişului limestones); 3. Lower Cretaceous limestones (non-karst or with limited karst features); 4. Jurassic limestones (non-karst or with limited karst features); 5. impervious sedimentary rocks; 6. crystalline schists and or granite basement; 7. syncline; 8. anticline; 9. geologic limit; 10. fault; 11. layers strike and dip; 12. cave entrance

2005 (with S. Constantin replacing O. Moldovan). A precise remapping of the surface features (Fig. 3) was performed by a team led by B. Bădescu from the "Exploratorii" Speleological Association in Reşiţa, in 2006. Subsequent analytical work was done between 2006 and 2010, and a monograph book was published in 2013 by Oxford University Press (Trinkaus et al. 2013).

Description of the Cave System

The Ponor-Plopa cave system is a tiered cave developed across two main levels (Figs. 3 and 4). The upper level includes four small caves: Peştera Hoţilor (35 m), Peştera din Zid (20 m), Peştera din Dolină (25 m), and Peştera Ponor Uscată (265 m). The first two caves are located in the limestone cliff above the Ponor and are thought to be either paleo-sinks of the stream or side-valley caves carved by waters infiltrated from the Plopa Plateau. Peştera din Dolină and Ponor Uscată are considered to be parts of a single cavity which are now separated by collapsed rocks and flowstone deposits. The general topography of the Ponor Uscată Cave suggests that it may have been a tributary of a former Ponor-Plopa drainage. This upper level also includes part of the "Peştera cu Oase" passages which will be described below.

The lower (hydrologically active) level consists of a typical through-cave centered along the small stream of Ponor and having a total length of surveyed passages of ~2000 m. Starting from the water sink at the Ponor, usually clogged with logs and sediments, a narrow sump (Sump 1, ~100 m long) leads to a large chamber (the Great Chamber, ~100 × 50 × 25 m) that includes a large scree cone with sizable breakdown blocks. Cave bear remains and hibernation bear nests have been observed, but no systematic research of the deposit has been carried out due to difficult access.

Further downstream, the river flows through a large passage that exceeds, in places, 20 m in height. Massive speleothems (calcite domes, stalagmites, rimstone dams) are present. After ~500 meters, the passage becomes narrower and three short sumps (Sumps 2–4) are encountered. The first sump may be bypassed through a narrow squeeze, whereas the next two can be passed only by diving. Some 150 m after Sump 4, the river sinks into an untraversable narrow passage, but a 1.5 m-climb above this point leads into the Subfossil Passage of the Plopa Cave and further on to the daylight. The river normally flows through the impenetrable sump, but at high waters the Subfossil Passage acts as an overflow and water temporarily accumulates in deep pools separated by rimstone dams.

At the end of the Subfossil Passage, the subterranean stream appears again at the surface after only a couple of hundred meters of large passage. Following the watercourse upstream in about 100 m, another sump (Sump 5, 20 m long) is reached. Diving this sump leads to the so-called Cloakroom—a relatively large chamber where the stream continues upstream toward the impenetrable sump. From the Cloakroom, a short climb of a muddy scree cone reaches the base of a 17-m high chimney (the Shaft). The Shaft can be free climbed along its northwestern side, along a drainage channel that appears to have been carved by a former waterfall. On top of the shaft, a large-sized subhorizontal passage is reached—this is the beginning of the Peştera cu Oase sector which corresponds, genetically and topographically, to the upper level of the cave system. The passages are relatively wide (10–15 m) and up to 6–8 m in height, and include pristine speleothems such as candlestick stalagmites, massive calcite domes, and large rimstone dams. After ~50 m, the Rimstone Dams Passage seems to come to a dead end, which can be overpassed by climbing a 2-m step and going through a narrow blowing hole (squeeze), which was enlarged by the first explorers, to reach the Passage of the Nests. The squeeze is now closed with a gate. The passages located beyond the gate have the floor covered with a thick layer of gravel, sand, speleothem fragments, and fossil bones. After c. 30 m, a steep slope (the Ancestors Ramp) leads to the Mandible Chamber (Sala Mandibulei), a room exceeding 15 m in height and hosting numerous massive speleothems, which are either encrusted in calcite or lying loose on the floor. From here, two side passages begin. The Long Passage has a canyon shape with heights exceeding 20 m and ends into a debris cone after ~150 m. The Passage of Three Skulls is slightly shorter and lower, with abundant cave bear nests as well as speleothems. This passage also ends into a debris cone from which bone remains belonging to more recent mammals have been retrieved. Both passages are considered to have been functioned as former ponors (swallets).

Cave Deposits

The post-sump sectors of the Ponor-Plopa system are decorated with pristine and massive speleothems. These often grew on top of thick deposits of sediments and/or bone remains. The complexity and richness of the cave deposit have allowed interdisciplinary studies to be performed for a better understanding of the taphonomy of the fossil accumulation and cave genesis. A number of 28 speleothem samples were collected and U/Th dated (see Constantin and Lauritzen 2013 for details). The results indicate that the upper sectors of the system were already functioning under a vadose regime c. 210 ka ago. Younger generations of speleothems, corresponding to the Marine Isotope Stage (MIS) 5e, have grown within the Long Passage and have been found as broken fragments in the Passage of the Nests. Speleothem datings provided crucial indirect proof for the

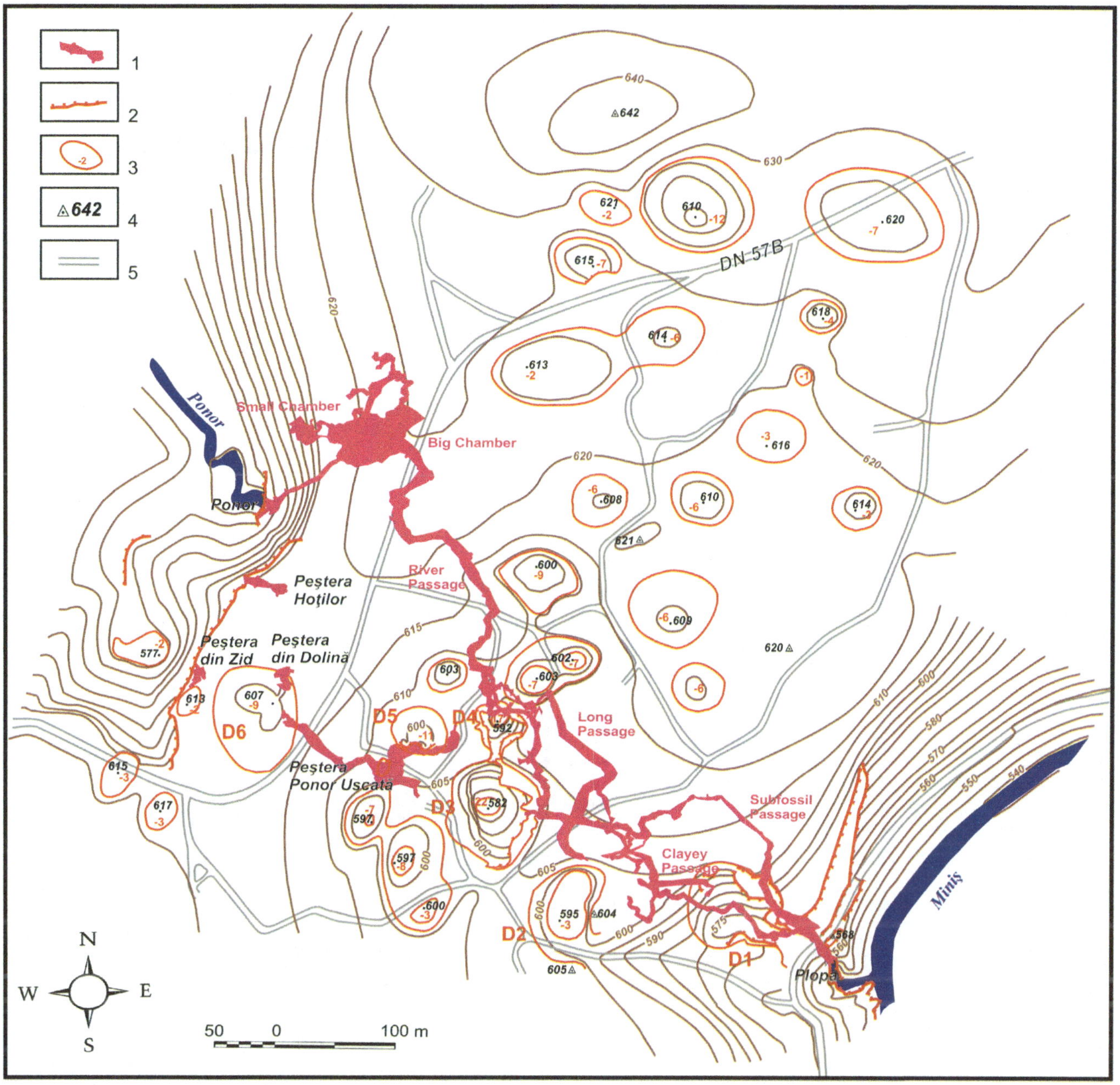

Fig. 3 Map of the Ponor-Plopa karst area and simplified projection of surveyed caves. Topographic survey by B. Bădescu and Asociaţia Speologică "Exploratorii" Reşiţa (contours at 5 m). Key: 1. Cave projection; 2. limestone cliff; 3. doline and depth; 4. topographic points and absolute elevation; 5. roads. D1 through D6 are arbitrary indicatives of several dolines significant for the genesis of the system. Note that the superimposition of the underground and overground maps is based on the only two points shared by both the Ponor and the spring. As the underground map was drawn using compass and clinometer measurements made across several sumps, it must be borne in mind that the locational relationships between features of the endo- and the exokarst apparent in the combined map may be affected by the minor errors inherent in the procedure (from Constantin et al. 2013)

age of the fossil bone accumulation (including human ones) along the Ramp of Ancestors. Finally, dating of two speleothems collected from the debris cones that clog the former swallets (ponors) in Peştera cu Oase allowed to establish a minimum age of the obstruction at c. 20.6 ka for the Long Passage and c. 8 ka for the Passage of Three Skulls, respectively.

Cave sediments were studied both from the excavation site and from the southern wall of the Shaft where they have a total thickness of more than 15 m. A 13.4-m long sampling trench was excavated, and samples were collected for

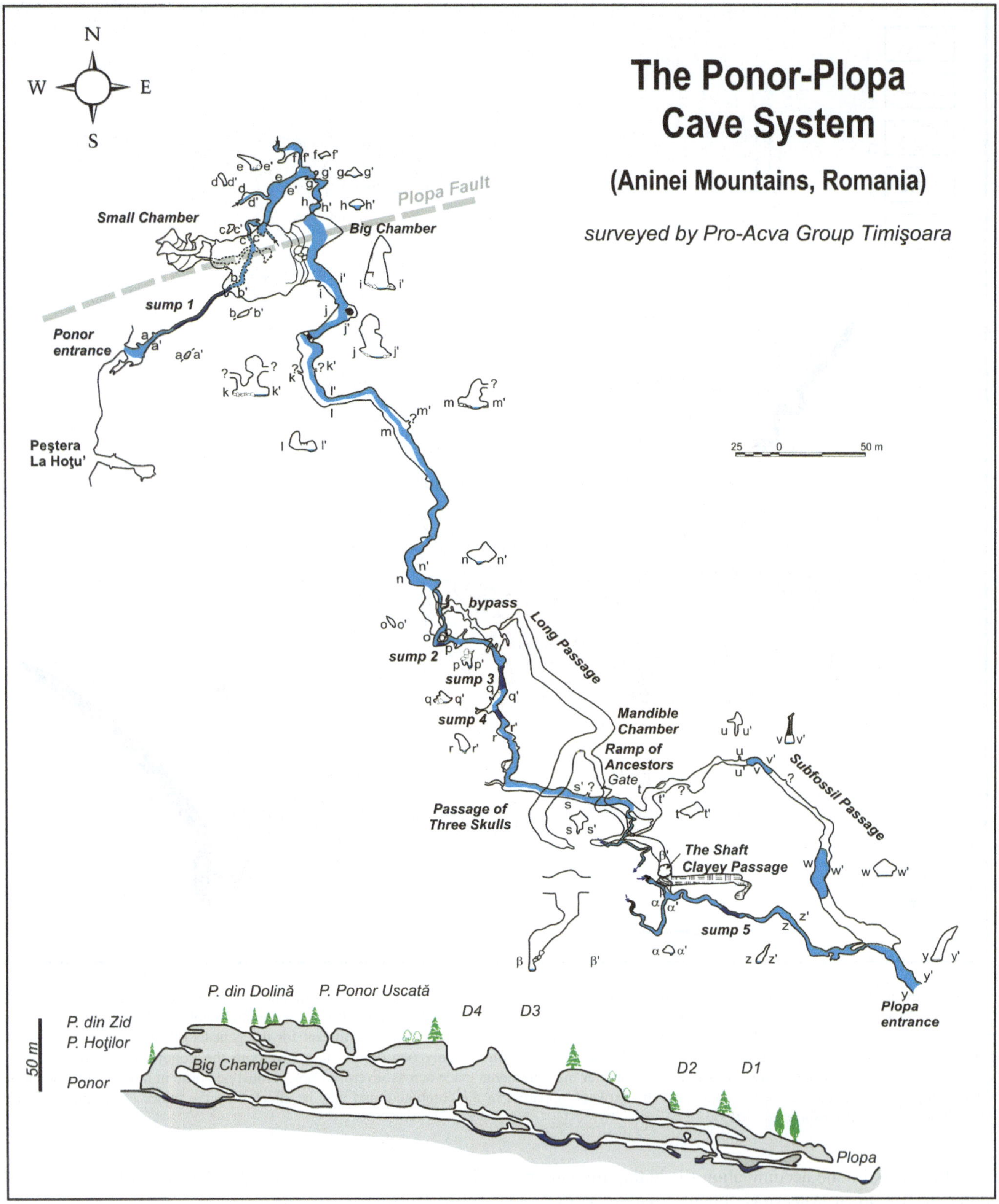

Fig. 4 Simplified map of the Ponor-Plopa cave system (from Constantin et al. 2013)

sedimentological analysis and measurements of magnetic properties. In addition, fossil bone fragments retrieved from the trench were Electron Spin Resonance (ESR) dated (see Panaiotu et al. 2013 for details). The sedimentological study suggests that the Shaft may have been flooded all the way to the top by the rising Ponor waters with potential inflow of coarser sediments and bones from the upper levels.

In the excavation from the Ramp of Ancestors, three stratigraphic levels were described: "Surface," "Level 1," and "Level 2." The Surface unit refers to bones lying on the calcitic crust; the other two levels include numerous bone remains. The grain size analysis indicates different deposition regimes, Level 2 corresponding to a rapid flow, and Level 1 to a much slower one (Constantin et al. 2013).

Genesis of the Cave System

The Ponor-Plopa cave system is a typical fluviokarstic drainage toward the main river, established through a small karst formation that has deepened over time leaving above a tiered network of dry passages. Constantin et al. (2013) have reconstructed the history of the cave system in detail by piecing together information on overall topography, cave morphology, underground stratigraphy, and radiometric datings. Currently, the system is considered to have undergone three major evolutionary stages. In the first one, drainages from the Ponor Valley toward the Miniş were established along the upper-level caves, i.e., Peştera din Zid—Peştera Ascunsă—Peştera Ponor Uscată—Peştera cu Oase. The overall direction of the drainage was WNW-ESE starting from Ponor Cliff and ending to a couple of small "blind valleys" located at c. 40 m above the Plopa entrance (Fig. 5a). During this stage, the underground stream probably followed a route passing through Ponor Uscată Cave, then through the Long Passage, and re-emerging at surface through the Passage of Three Skulls. Subsequently, the stream may have eroded to a lower level, creating the Passage of the Nests and emerging at surface through the Clayey Passage.

In Stage II, the deepening of the Ponor stream has left the upper level largely dry with the exception of occasional inflow of the waters accumulating within the small catchment area of Plopa. The system was divided by collapses and deep sinkholes. The collapsed sinkholes separated the caves in the upper level from each other, while the Long Passages and the Passage of Three Skulls started to act as temporary ponors. At this stage, the main drainage was rapidly established between the Ponor swallet and the Plopa entrance acting as a local base level for the temporary streams in the Peştera cu Oase sector. Eventually, these were captured by the main drain through the Shaft (Fig. 5b). Numerous radiometric datings and stratigraphic studies suggest that the flash flooding repeatedly affected the upper passages during MIS 3, between c. 45 and c. 35 ka, washing out debris and animal remains from the bottom of the collapsed sinkholes, as well as cave bears that have used the cave. All this fills material accumulated along the Ramp of Ancestors and the Passage of the Nests, behind the narrow squeeze (at the gate). After the Last Glacial Maximum (Stage III), fluvial inflow into the upper levels decreased and finally ceased, the ponors being clogged with debris and flowstone (Fig. 5c).

Paleontology

Ponor-Plopa is one of the most important paleontological cave sites of Romania. Almost 5000 cave bear (*Ursus spelaeus*) remains were found across only 9 m^2 of excavation, with a minimum number of individuals of 41 adult and 37 juvenile cave bears. Cave bear remains constitute c. 94% of the findings, the remainder being shared by large carnivores (cave hyena, wolves, and foxes) and herbivores such as red deer (*Cervus elaphus*), the giant deer (*Megaloceros giganteus*), and ibex. There is an extensive bibliography dedicated to the fossil accumulation at Oase, and the reader is referred to the work of Trinkaus et al. (2013) and references therein.

Anthropology

The discovery of the two early modern human remains in Peştera cu Oase is one of the most important findings in Romanian speleological research. While it was initially believed that Oase Cave could be an early archaeological site, it soon became obvious that no archaeological remains will be found in the cave, the research being focused toward paleontological and anthropological work. The cave contains the remains of two early *Homo sapiens* (Oase 1 and 2) along with a temporal bone that was initially believed to belong to a different individual, but now is considered as part of Oase 2. Works by Trinkaus et al. (2003b, 2006) and Rougier et al. (2007) indicated that these individuals share traits typical for both modern humans and Neanderthals. These studies were suggesting a potential in-breeding between the modern humans arriving in Europe and their local "cousins," as in many similar cases the findings initiated vivid arguments. The first independent proof came recently (2015), when Fu et al. have analyzed the ancient DNA of Oase 1 and found that it contains 6–9% Neanderthal genome, more than any other modern humans known to date.

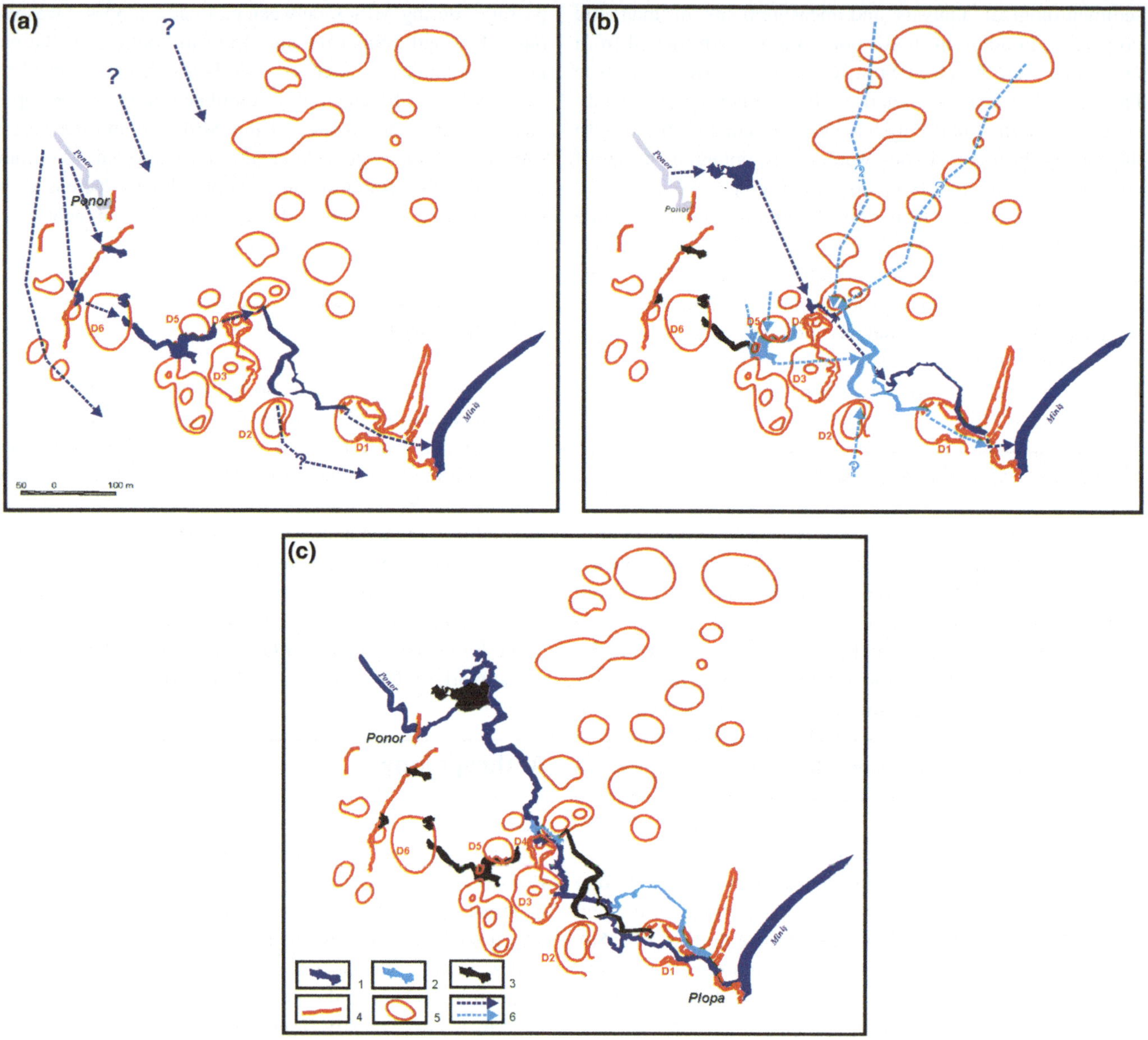

Fig. 5 Tentative scenario of the evolution of the cave passages in the Ponor-Plopa area. **a** Inception stage: successive karst captures along the Ponor Cliff toward the Peştera Ponor Uscată-Long Passage-Passage of Three Skulls-D2 (D1); **b** deepening stage: The Ponor stream creates the Big Chamber sector and flows toward the Miniş via the Subfossil Passage. Some of the upper levels acted as temporary drainages for the endorheic basin of Plopa. **c** Current stage: the only main drain of Ponor-Plopa, a few temporarily flooded passages and all upper levels being hydraulically inactive. Legend: 1. underground passage, permanent flow; 2. temporary flow; 3. dry passage; 4. limestone cliff; 5. doline; 6. possible drainages pathways: dark blue—permanent; light blue—temporary. Note that drainages indicated in **a** and **b** are not synchronous so the suggested inputs/outputs of the system indicate possible reorientations of the drainages during the same stage (from Constantin et al. 2013)

Cave Conservancy

The Ponor-Plopa cave system is protected as a scientific reserve ("class A" cavity), with the Subfossil Passage and the Ponor Uscată being assigned class C and cave tourism being permitted. In fact, these sectors are well known to tourists and frequently visited, while the rest of the system is naturally protected by the various sumps. The Peştera cu Oase sector is also protected by a gate, being the most pristine and scientifically valuable part of the system.

Conclusions

The Ponor-Plopa is an example on how relatively modest fluviokarstic system may yield invaluable scientific information to better understand the geological past, environmental history, and the human lineage. The discovery of the new passages and fossil deposits by a team of mindful cavers, followed by the work of equally aware professionals, has set an example on how international cooperation may lead to spectacular and valuable results. This is perhaps one of the best studied cave bear sites in the world, and further research may still be done as soon as investigation techniques will advance.

Acknowledgements The Ponor-Plopa research is the result of hard work of more than 50 individuals from all over the world. All of them and their corresponding grants are acknowledged in the monograph book of Trinkaus et al. (2013). I would like to thank all colleagues involved, for their help and insightful discussions, either in the field or in the laboratory. I am especially grateful to O. Moldovan for inviting me to take over the geology and dating of the site in the early stage of the study.

References

Bîlgăr A (2004) Ion—file de poveste. Lumea Peşterilor 1:19–22

Constantin S (2013) The karst geology of the Ponor-Plopa system in its regional context. In: Trinkaus E, Constantin S, Zilhão J (eds) Life and death at the Peştera cu Oase. A setting for modern human emergence in Europe. Oxford University Press, Oxford, pp 33–40

Constantin S, Lauritzen S-E (2013) Uranium-series dating on speleothems from the Pestera cu Oase. In: Trinkaus E, Constantin S, Zilhão J (eds) Life and death at the Peştera cu Oase. A setting for modern human emergence in Europe. Oxford University Press, Oxford, pp 56–72

Constantin S, Munteanu C-M, Milota S, Sarcină L, Gherase J, Zilhão J (2013) The Ponor-Plopa Cave System: description, sediments, and genesis. In: Trinkaus E, Constantin S, Zilhão J (eds) Life and death at the Peştera cu Oase. A setting for modern human emergence in Europe. Oxford University Press, Oxford, pp 41–55

Fu Q, Hajdanjak M, Moldovan OT, Constantin S, Mallik S, Skoglud P, Patterson N, Rohland N, Lazaridis I, Nickel B, Viola B, Prufer K, Meyer M, Kelso J, Reich D, Paabo S (2015) An early modern human from Romania with a recent Neanderthal ancestor. Nature 524:216–219

Iurkiewicz A, Constantin S, Bădescu B (1996) Sisteme carstice majore din zona Reşiţa-Moldova Nouă. A.S. "Exploratorii" Reşiţa, p 70

Panaiotu C, Constantin S, Petrea C, Horoi V, Panaiotu C-E (2013) Rock magnetic data of Late Pleistocene sediments from the Peştera cu Oase and their paleoclimatic significance. In: Trinkaus E, Constantin S, Zilhão J (eds) Life and death at the Pestera cu Oase. A setting for modern human emergence in Europe. Oxford University Press, Oxford, pp 86–99

Rougier H, Milota Ş, Rodrigo R, Gherase M, Sarcină L, Moldovan O, Zilhão J, Constantin S, Franciscus RG, Zollikofer CPE, Ponce-de-León M, Trinkaus E (2007) Peştera cu Oase 2 and the cranial morphology of early modern Europeans. PNAS 104:1164–1170

Sencu V (1964) Cercetări asupra carstului din partea sudică a localităţii Anina (Banat). Peşterile din bazinele pîraielor Steierdorf şi Ponor. St Cerc Geol Geofiz Geogr, Ser Geogr 11:149–162

Senco V (1973) Les eaux karstiques de la zone minière de la ville d'Anina (Banat). Livre du cinquantenaire de l'Institut de Spéologie "Émile Racovitza", pp 591–604

Sencu V (1977) Carstul din cîmpul minier Anina. St Cerc Geol Geofiz Geogr XXIV:199–212

Trinkaus E (2006) Modern human versus Neanderthal evolutionary distinctiveness. Cur Anthropol 47:597–620

Trinkaus E, Moldovan O, Milota S, Bîlgar A, Sarcină L, Athreya S, Bailey SE, Rodrigo R, Mircea G, Higham T, Bronk Ramsey C, van der Plicht J (2003a) An early modern human from the Pestera cu Oase Romania. Proc Nat Acad Sci USA 100(20): 11231–11236

Trinkaus E, Milota Ş, Rodrigo R, Gherase M, Moldovan O (2003b) Early modern human cranial remains from the Peştera cu Oase, Romania. J Hum Evol 45:245–253

Trinkaus E, Constantin S, Zilhão J (eds) (2013) Life and death at the Peştera cu Oase. A setting for modern human emergence in Europe. Oxford University Press, Oxford, p 438

Aninei Mountains: Buhui Cave

Eugen Crețu

Abstract
Buhui Cave in the central part of the Reşiţa-Moldova Nouă karst region (Aninei Mountains) has a total length of 7429 m, of which 2100 m are represented by the main meandering stream passage that hosts spectacular vadose and phreatic morphologies. The cave has multiple entrances and includes a number of well-decorated chambers, in one of which remains of *Ursus spelaeus* and *Capra ibex* (Steinbock) were identified.

Keywords
Vadose/Phreatic morphology • *Ursus spelaeus*
Buhui Cave • Romania

Introduction

Buhui Cave is located in the Anina Mountains (Banatului Mountains), 4 km west of Anina–Steierdorf, in the middle of the Reşiţa-Moldova Nouă karst area (Fig. 1). From a geomorphological point of view, Buhui Basin is situated in the Colonovăţ Plateau, which is rich in surface and underground karst features. The cave itself develops in the Plopa limestone of lower Cretaceous age (Năstăseanu and Savu 1970; Bucur 1997). For additional information on the hydrogeology of the cave and surrounding areas, the reader is directed to Sencu (1986), Iurkiewicz et al. (1996a), and Iurkiewicz (2010).

History of Exploration

Buhui Cave has been discovered in 1773, when the Steierdorf colony was established. The cave is very important for the community, because its stream represents the main source of drinking water (Orghidan et al. 1984). With increasing economic importance of the area due to coal mining activities, the Buhui Lake was built, being the first water reservoir in Romania.

Between 1875 and 1884, geologist R. Hoernes and paleontologist G. Téglas studied the cave for the first time, whereas T. Ottlik explored and took pictures in the cave (1934–1935). Survey of the cave and hydrogeological observations were conducted by Sencu (1963) between 1956 and 1959, followed by thorough biological studies led by Negrea and Botoşăneanu between 1961 and 1972 (Negrea et al. 1965; Negrea and Negrea 1972). During this period, 3217 m of cave passages was surveyed. In 1981, Speotimis and Cristal caving clubs from Timişoara resurveyed the cave, finishing the exploration campaigns with a total of 6547 m of galleries. In 1996, L. Kalitzky remapped the main gallery using a theodolite, confirming the 1981 survey (Goran 1982). Even so, the length reported in some subsequent publications (e.g., Orghidan et al. 1984; Cocean 1995) is mistakenly using the old 3217 m. The most recent discoveries extended the cave length to 7429 m (Iurkiewicz et al. 1996b).

Cave Description

Downstream of the Buhui Lake, the Buhui stream sinks underground through ponors and flows through a 2100 m-long main gallery (Fig. 2). It is the longest underground river in the Banat region and one of the longest in Romania. Between these ponors and the entrance called "Buhui Cavern," the cave has two more access points. The first one is "Certej Entrance," named after the homonymous tributary that enters the cave at this location. The second one is called "Doline Entrance" and is in fact a 15 m-deep shaft likely formed due to ceiling collapse caused by erosion–corrosion processes.

The size (width and height) of the main, meandering gallery varies notably and has many short side passages, and only a few big chambers. The cave is developed along

E. Crețu (✉)
Terassond, Am Kalkwerk 1, 06447 Bad Bibra, Germany
e-mail: Eugen.Cretu@terrasond.de

G. M. L. Ponta and B. P. Onac (eds.), *Cave and Karst Systems of Romania*,
Cave and Karst Systems of the World, https://doi.org/10.1007/978-3-319-90747-5_28

Fig. 1 Location of Buhui Cave in the karst area of Aninei Mountains

bedding plane (mainly in the first part) and two main fractures systems.

The first sector of the cave upstream from the Certej Entrance is characterized by low and large galleries. The stream has low velocity flow, mainly in the N–S direction, filling the main gallery almost entirely with sand sediments. Further, downstream "Certej Entrance," the morphology of the Buhui Cave changes drastically. The waters of Buhui and Certej streams converge, and as a result, high galleries (Fig. 3) with waterfalls, whirlpools, and canyon-like passages formed. Downstream "Doline Entrance," the fracture system involved in the genesis of the cave is W–E-oriented. Large passages and chambers beautifully are decorated (Fig. 4), and chimneys become more and more frequent. The main sediments are still sands and gravel, which along with the clear water contribute to the beauty of the subterranean landscape. Bleahu et al. (1976) and Orghidan et al. (1984) provide more detailed descriptions of the cave.

It is worth mentioning that traces and remains of *Ursus spelaeus* and *Capra ibex* (Steinbock) were found in the Bear's Chamber (Sala Urșilor; Fig. 5) adding to the Buhui Cave heritage (Bleahu et al. 1976).

Cave Climate

The temperature in the inner parts of the Buhui Cave ranges between 8 and 9.5 °C (cooler near entrances), whereas the relative humidity is above 98% year around (Bleahu et al. 1976).

Cave Fauna

The caves of the Aninei Mountains are well investigated from a biospeleological point of view, given the extensive work carried out by various researchers of the "Emil Racoviță" Institute of Speleology (Botoșăneanu et al. 1967; Botoșăneanu 1971; Negrea and Negrea 1972, 1977). The terrestrial species are represented by *Banatosoma*

Fig. 2 Map of the Buhui Cave

Fig. 3 Image from the river passage downstream the Certej Entrance (photograph courtesy of A. Posmoşanu)

Fig. 4 Stream passage near the Great Waterfall (photograph courtesy of A. Posmoşanu)

Fig. 5 Large and well-decorated Bear's Chamber (photograph courtesy of A. Posmoşanu)

ocellatum, Polydesmus sp., *Trachysphaera* sp., and *Trochus* sp. (Myriapoda, Diplopoda), *Paranemastoma* sp., *Centromerus jacksoni* (Arachnida), and *Hyloniscus dacicus* (Malacostraca) (Tabacaru 1972; Tabacaru et al. 2002–2003). Among the aquatic species, *Niphargus* sp. (Amphipoda), *Acanthocyclops milotai* (Iepure and Defaye 2008), *Acanthocyclops propinquus* (Pleşa 1969), *Elaphoidella romanica, Elaphoidella phreatica* (Copeopoda) (Nitzu et al. 2016), and *Pseudocandona* sp. (Ostracoda) are the most common.

Only three species of bats, namely *Myotis oxygenatus*, *Rhinolophus ferrumequinum*, and *Rhinolophus mehelyi* (Dumitrescu et al. 1963; Bleahu et al. 1976; Csősz et al. 2015) were identified in Buhui Cave.

Cave Conservancy and Recreational Caving

Buhui Cave is not open for mass tourism, and except for the V. Sencu Room, which is classified as a B-protected sector, the rest of the galleries are ranked as C (local importance, see the *Cave Protection* chapter in this book). Caving and diving gear are needed to visit the cave, which normally lasts for 6–8 h.

References

Bleahu MD, Decu V, Negrea Ş, Pleşa C, Povară I, Viehmann I (1976) Peşteri din Romania. Ed Ştiinţifică şi Enciclopedică, Bucureşti, p 416

Botoşăneanu L (1971) Observations sur la faune aquatique hypogée des Monts du Banat (Roumanie). Trav Inst Spéol "Émile Racovitza" 10:123–166

Botoşăneanu L, Negrea A, Negrea Ş (eds) (1967) Recherches sur les grottes du Banat et l'Oltenie, Roumanie (1959–1962). Centre National de la Recherche Scientifique, Paris, p 392

Bucur II (1997) Formaţiunile mezozoice din zona Reşiţa-Moldova Nouă (Munţii Aninei şi estul Munţilor Locvei). Ed Presa Univ Clujeană, Cluj-Napoca, p 214

Cocean P (1995) Peşterile României. Ed Dacia, Cluj-Napoca

Csősz I, Jére C, Bücs S, Bartha C, Barti L, Szodoray-Parádi F (2015) The presence of Mehely's horseshoe bat *Rhinolophus mehelyi* in South-Western Romania. North-Western J Zool 11(2):351–356

Dumitrescu M, Tanasachi J, Orghidan T (1963) Răspândirea chiropterelor în R.P. Română. Trav Inst Spéol "Émile Racovitza" 34:509–575

Goran C (1982) Catalogul sistematic al peşterilor din Romania. Ed CNEFS, Bucureşti, p 496

Iepure S, Defaye D (2008) The *Acanthocyclops kieferi* complex (Crustacea: Copepoda) from South-Eastern Europe, with description of a new species. Crustaceana 81:611–633

Iurkiewicz A (2010) Banat Mountains (Reşiţa-Moldova Nouă Synclinorium). In: Orăşeanu I, Iurkiewicz A (eds) Karst hydrogeology of Romania. Ed Belvedere, Oradea, pp 137-168

Iurkiewicz A, Constantin S, Bădescu B (1996a) Sisteme carstice majore din zona Resita-Moldova Nouă. Ed AS Exploratorii Reşiţa, p 72

Iurkiewicz A, Dragomir G, Rotaru A, Bădescu B (1996b) Karst systems in Banat Mountains (Reşiţa-Nera zone). Theor Appl Karstol 9:121–140

Năstăseanu S, Savu H (1970) Geological map of Romania, Anina Sheet. Scale 1:50000, Inst Geol Geofiz, Bucureşti

Negrea Ş, Negrea A (1972) Recherches sur l'association pariétale des grottes du Banat (Roumanie). Acta Zool Cracov 17(3):45–96

Negrea Ş, Negrea A (1977) Sur les association de plancher des grottes du Banat (Roumanie). Trav Inst Spéol "Emile Racovitza" 16:99–139

Negrea Ş, Botoşăneanu L, Negrea A, Sencu V (1965) Caves of Banat (Romania) explored in 1963. Int J Speleol 1(4):397–439

Nitzu E, Giurginca A, Nae A, Popa I, Baba Ş, Meleg IN, Vlaicu M (2016) The catalogue of caves with endemic cavernicolous arthropod fauna of Romania. Trav Inst Spéol "Émile Racovitza" LV:3–62

Orghidan T, Negrea Ş, Racoviţă G, Lascu C (1984) Peşteri din România. Ed Sport-Turism, Bucureşti, pp 170–177

Pleşa C (1969) Fauna cavernicolă acvatică din peşterile calcaroase. In: Călinescu R (ed) Zoogeografia Romîniei. Ed Ştiinţifică şi Enciclopedică, Bucureşti, pp 354-358

Sencu V (1963) Cercetări asupra carstului din jurul localităţii Anina (Banat) - Peşterile din bazinele pâraielor Anina şi Buhui. Prob Geogr 10:155–180

Sencu V (1986) Field measurements relating to the subterranean runoff of karst waters in the Aninei Mountains. Z Geomorph 30(1):59–68

Tabacaru I (1972) Contribution a l'etude du genre *Hyloniscus* Verhoeff (Crustacea, Isopoda) 1. Deux nouvelles especes du groupe flammula. Trav Inst Spéol "Émile Racovitza" 11:233–246

Tabacaru I, Giurginca A, Vănoaica L (2002–2003). Cavernicolous Diplopoda of Romania. Trav Inst Spéol "Émile Racovitza" 41–42:121–148

Domanului Mountains: Comarnic Cave (Pestera de la Cantonul Comarnic, Pestera din Ogasul Ponicovei)

Ioan Povară

Abstract

Comarnic Cave is situated in southwestern Romania, in a region in which the Mesozoic limestones cover an area of ~ 600 km^2 and form parallel ridges with high plateaus that are fragmented by deep gorges. The karst plateaus are dotted by sinkholes and sinking streams that recharge permanent or temporary karst springs. Comarnic Cave (6201 m) represents a hydrological penetration of the Ponicova Creek across the Upper Jurassic–Lower Cretaceous limestone bar. Its passages are mostly descending and develop on three levels (inactive, temporarily active, and active), with an offset of 101 m between the access points. The upper (inactive) level is rich in speleothems, such as draperies and stair-stepped rimstone pools, well developed in the Virgin Chamber. Along the lower level, erosional and/or solutional features are abundant. Comarnic Cave is well known for its spectacular chert layers (in the Zebras' Chamber) protruding from the limestone along the bedding planes. Adventure caving trips can be arranged at the Semenic-Cheile Caraşului National Park Headquarter.

Keywords

Chert layers • Rimstone dams • Anemolites
Comarnic Cave • Romania

Introduction

Comarnic Cave is situated in the Domanului Mountains (southwestern Romania), part of the Banatului Mountains (Fig. 1). Its 6201 m of passages and 101 m total relief develop on three karstification levels: inactive/fossil/dry, temporarily active/subfossil, and active/stream. The water originating in the Ponicova Creek flows through the lowermost gallery. The cave can be reached from the town of Reşiţa, by following for 9 km the road DJ 58 B toward Anina, up to Iabalcea village; subsequently, follow for another 7 km a secondary road toward Canton Comarnic, the starting point of the tourist trail toward the Comarnic entrance into the cave. The dry section of the cave, rich in speleothems, is open for tourism.

History of Discovery and Research

The cave was discovered in 1856 and was first explored by Telegdi in 1893 (Bădescu and Vlaicu 2013). The first map of the cave, with over 4 km of passages, was completed by Balogh in 1933 and published by Piţu in 1945 (Balogh 1969). The same author made a series of remarks on the cave geology and mineralogy (Bleahu et al. 1976). Detailed studies on the geology and morphology of the area belong to Mateescu (1961), Puşcariu et al. (1964), and Sencu (1964). Sencu (1972) finishes the map of the fossil and subfossil levels, with observations on the tectonic control on the cave development and the alluvial deposits found within the Comarnic Cave. Detailed descriptions of the cave were published by Protopopescu (1934), Piţu (1947), and Orghidan et al. (1984) in touristic guides. After 1981, Exploratorii Reşiţa Speleological Club explored and map over 2 km of new passages. Between 1987 and 1988, the researchers of the "Emil Racoviţă" Institute of Speleology remapped the cave using a theodolite. This new survey has been part of a project aiming to develop a touristic path in the dry section of the cave.

Geologic, Geographic, and Hydrologic Settings

Paleozoic (conglomerates, sandstones, shales) and Mesozoic rocks that outcrop in folded and faulted synclines and anticlines, transgressively and unconformably, cover the

I. Povară (✉)
"Emil Racoviţă" Institute of Speleology, Romanian Academy, Frumoasă 31, 010986 Bucharest, Romania
e-mail: ipov.iser@gmail.com

G. M. L. Ponta and B. P. Onac (eds.), *Cave and Karst Systems of Romania*,
Cave and Karst Systems of the World, https://doi.org/10.1007/978-3-319-90747-5_29

Fig. 1 Location of Comarnic Cave in the Domanului Mountains

crystalline bedrock (Răileanu et al. 1957, 1964). The major folds are oriented NNE–SSW. The Comarnic Cave develops within the Lower Cretaceous limestones (Fig. 2). The Plopa limestone, in which Comarnic Cave has developed, has two distinct horizons: a lower one with abundant siliceous layers (Upper Valanginian–Lower Barremian) and an upper one made of reef limestone (Barremian–Aptian). The Plopa limestones present interbedded chert layers along the bedding planes and lenticular black chert nodules (Fig. 3). The yellowish white massive limestone of Lower Cretaceous age (Barremian–Aptian) is known as the Aninei Valley limestones (Bucur 1997). The major landforms are the limestone ridges overlying many hogbacks over 200-m-high overlooking the Caraş, Comarnic, and Buhui gorges (Mateescu 1961). The rather steep stream gradient along the Caraş Gorge (elevation drops from 600 to 400 m) caused the development of numerous whirlpools. The karst landforms are mainly represented by sinkholes (up to 200 m in diameter and 5–25 m in depth), uvalas, and sinkhole valleys (Rusu 1967; Sencu 1970). Among those 187 caves known in the Caraş Gorge and Socolovăţ Plateau, 165 are less than 500 m in length, 3 are longer than 500 m, but only 2 exceed 1 km (Goran 1982). All streams within the region are tributary to the Caraş River (Fig. 2). Permanent springs (2–5 l/s) recharged by the sinkholes and the sinkhole valleys emerge at the base of the gorge' cliffs (Rusu 1967). Fluorescein injected into the Ponicova Creek has been detected at the Comarnic sources after traveling 1.8 km in *ca.* 20 h, suggesting a velocity of 30 m/h (Iurkiewicz 2010).

Cave Description

Comarnic Cave has two main entrances: Ponicova and Comarnic (Fig. 4). The large chambers and passages of the inactive level spread over 2.5 km and include: Sala Virgină (Virgin Chamber), Sala Domului (Dome Chamber), Orga Mare (Great Organ), Sala Mare cu Blocuri (Great Chamber with Blocks). Within the northern half of this level, limestone blocks collapsed from the ceiling are common occurrence. Abundant and various speleothems decorate the southern half of the dry level. Within this sector, spectacular

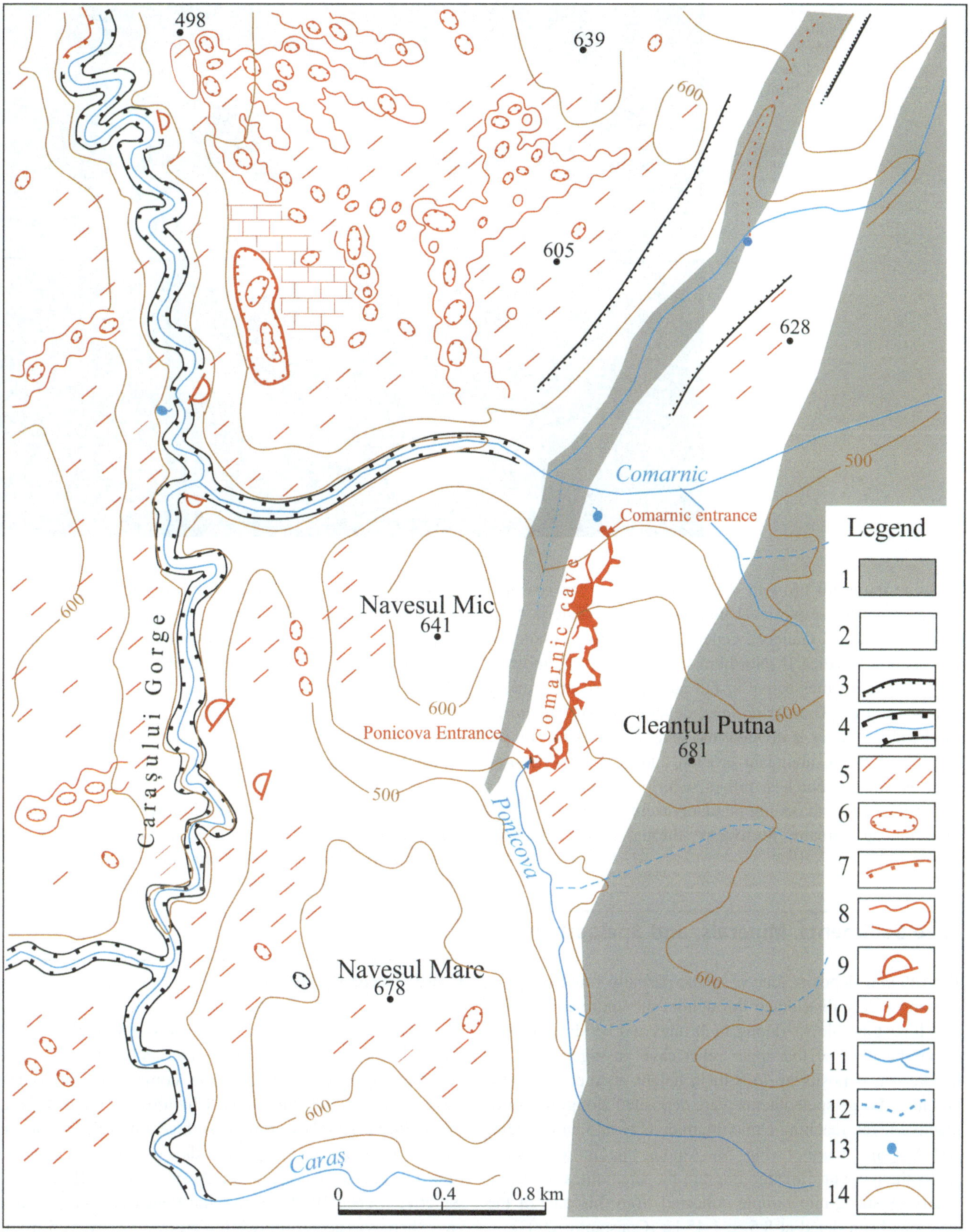

Fig. 2 Karst morphology of the Comarnic Cave area (after Sencu 1972): 1 non-karst lithology; 2 Lower Cretaceous limestones; 3 ridge; 4 gorge; 5 karren; 6 sinkhole; 7 uvala; 8 sinkhole valley; 9 cave entrance; 10 comarnic Cave; 11 permanent stream; 12 temporary stream; 13 spring; 14 contour lines

Fig. 3 Chert layers and lenses interbedded within limestone beds in Zebras Chamber (photograph courtesy of Asociaţia Speologică Exploratorii Reşiţa)

rimstone dams fill up with water only during wet seasons. The temporarily active level that includes the Galeria Nordică I (Northern I Gallery), Galeria Arcuită (Bend Gallery), and Galeria Nordică II (Northern II Gallery) is dry most of the year. It becomes active in fall and spring or after major summer storms. The stream level developed along fractures and bedding planes is divided into several sectors by sumps. In each of them, abundant erosion and corrosion features such as alluvial notches, pillars, whirlpools, scallops, corrosion cupolas, and ceiling pendants were documented. The subterranean stream resurfaces through the Comarnic Spring.

Cave Sediments, Minerals, and Speleothems

The alluvial deposits (clays, sands, partially rounded gravels) are widely distributed throughout the entire cave. Massive limestone blocks (along fault lines and bedding planes) cover the floor of the upper relict cave passages, forming breakdown deposits up to 8 m in height covered with stalagmites. Large speleothems were deposited along the entire inactive level, reaching a maximum density in Sala Domului and Virgin Chamber (Fig. 5). Among the more specific speleothems, worth mentioning are the anemolites and cave pearls. A stalagmite sample collected from this part of the cave has been dated at 8.5 ± 1.45 ka (Constantin 2003).

A mineralogical investigation conducted by Zaharia (2006) revealed the presence of two very common cave phosphates: hydroxylapatite and brushite. They occur as dark brown to black and white crusts, respectively. The reaction between phosphate-rich solutions leaching out from the guano deposits and the underlying limestone is responsible for their precipitation.

Paleontology

To date, only incomplete skeleton remains belonging to small mammals *Clethrionomys glareolus*, *Rhinolophus ferrumequinum*, *Rhinolophus hipposideros*, *Myotis myotis*, *Myotis oxygnathus*, as well as *Ursus spelaeus* have been identified (Bleahu et al. 1976).

Cave Speleogenesis

The cave was generated by the action of Ponicova stream, which formed between the Ponicova and Comarnic entrances the upper level of the cave comprising the Bended Passage, the Virgin and Dome Chambers, the Great Organ, etc. Following the lowering of surface's base level, the cave stream downcuts to the present-day horizon where it continues to enlarge passages like Northern I and II, and Bend galleries (Sencu 1972). Above the Ponicova entrance of the cave there was accordingly left a protruding, over 35-m-high limestone wall. Subsequently, in the region of the Sala Domului and Southern Gallery (Galeria Sudică), the

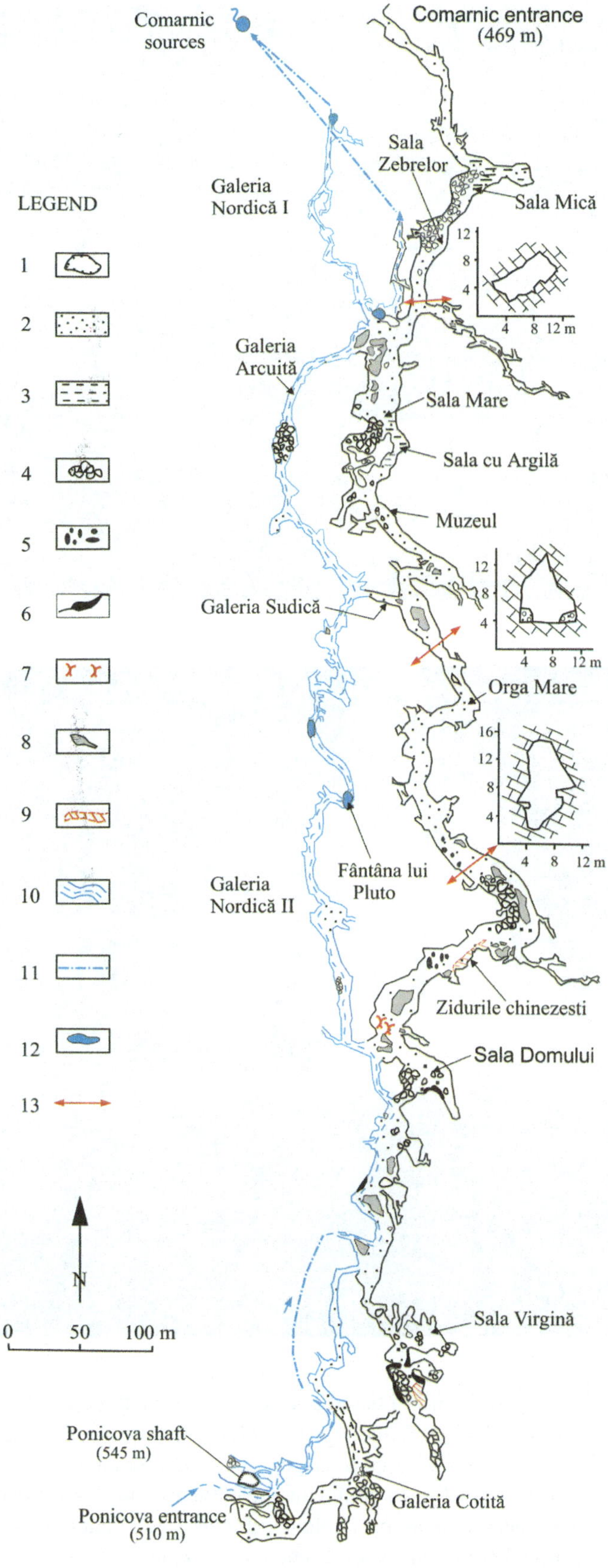

Fig. 4 Comarnic Cave map (after Sencu 1972, completed by Bădescu and Vlaicu 2013): 1 shaft; 2 alluvia; 3 clays; 4 blocks; 5 speleothems; 6 crusts; 7 guano; 8 pillars; 9 rimstone dams; 10 temporarily active and active level; 11 unexplored subterranean stream passage; 12 lake; 13 passage cross-section

limestone bedding planes favored the excavation of connecting passages toward the underlying, temporary stream passage level (Northern II Gallery, the Bend Gallery Arcuită, and the Northern I Gallery).

Cave Climatology

Due to its morphology (two entrances), Comarnic Cave posses a unidirectional ventilation that changes seasonally. In the entrance area, the air temperature ranges between 6 and 7.5 °C in summer, while in winter, values are below freezing causing ice speleothems to form (Fig. 6). Along the stream passage, relatively constant values (9–9.5 °C) were recorded.

Cave Fauna

The subterranean fauna was first studied by Jeannel (1931) and Botoşăneanu et al. (1976). Negrea and Negrea (1983) presented in detail the cave's biocoenoses, emphasizing the high number of troglophile and troglobitic species. In the entrance area, the presence of the opilionid *Nemastoma silli* and the isopod *Mesoniscus graniger* is recorded, along with other species attracted by the abundant organic matter deposited on the floor. Among the endemic troglobitic species, *Lithobius dacicus*, *Onychiurus romanicus*, and *Duvalius milleri* were identified (Bleahu et al. 1976). The guanophilous dipteran *Heteromyza atricornis* and various Nycteribiidae (bat parasites) are recorded in the guano deposit under the bat colony (located at approx. 40 m from the Dome Chamber). The aquatic fauna is relatively poor, only the genus *Nyphargus* can be found in the rimstone pools. Other invertebrate species are the isopod *Ligidium hypnorum* and the diplopod species *Trachysphaera costata*, *Strongylosoma stigmatosum*, *Brachydesmus troglobius*, and *Dyocerasoma lignivorum* (Tabacaru et al. 2004; Giurginca et al. 2015). The following species of Chiroptera have also been identified in the Dome Chamber: *R. ferrumequinum, R. hipposideros, M. myotis, M. oxygnathus,* and *Miniopterus schreibersii* (Bleahu et al. 1976).

Cave Conservancy

Comarnic Cave (an area of 0.1 ha) was declared a natural reservation by the Romanian Academy in 1947 and later included in the Semenic-Cheile Caraşului National Park. The Virgin Chamber within the Comarnic Cave is a class A protected area. The Semenic-Cheile Caraşului National Park Administration, in cooperation with the Speleological Association "Exploratorii" in Reşiţa, is in charge of the management of the protected area.

Fig. 5 Rimstone dams in the Virgin Chamber (photograph courtesy of Asociația Speologică Exploratorii Reșița)

Fig. 6 Ice formations located at 80 m downstream from the Ponicova entrance (photograph courtesy of Asociația Speologică Exploratorii Reșița)

Conclusions

Comarnic Cave, the longest cave in southwestern Romania (6201 m), represents the drainage of the Ponicova Creek, which crosses from north to south the Jurassic–Cretaceous limestone ridge, belonging to the Reșița-Moldova Nouă Synclinorium. It is a tiered cave developed on three levels as a result of discontinuous downcutting within the drainage basin. A series of large passages and grandiose chambers (in the upper level) host abundant speleothems (rimstone pools, cave pearls, massive calcite domes, and anemolites). Although some of its speleothems are rather young as documented by a single U/Th age determination

(8.5 ± 1.45 ka), the cave itself is certainly much older considering the total relief between the upper and lower entrances. Likely the cave had a slow vertical evolution, the temporarily active level coalescing with the stream passage. The latter one can be visited only at low water levels on discontinuous sectors, which ends in sumps. Along the stream passage, the presence of continuous chert layers within the limestone highlights its bedding. In the upper, dry level the air temperature remains constant at 9–9.5 °C all year around, creating favorable conditions for three endemic troglobites and five bat species.

Acknowledgements The author is grateful to his colleagues Dr. A. Giurginca and V. Gheorghiu for the data concerning the subterranean fauna and to Asociaţia Speologică Exploratorii Reşiţa for the photographs.

References

Balogh E (1969) Csepkkö világ. Tineretului, Bucureşti, p 203

Bădescu B, Vlaicu M (2013) Atlasul habitatului 8310 din situl Natura 2000 Semenic-Cheile Caraşului: peşteri şi lilieci. BrumarR, Timişoara, p 71

Bleahu MD, Decu V, Negrea Ş, Pleşa C, Povară I, Viehmann I (1976) Peşteri din România. Ed Ştiinţifică şi Enciclopedică, Bucureşti, p 416

Botoşăneanu L, Negrea A, Negrea, Ş (eds) (1976) Grottes du Banat explorées de 1960 à 1962. Recherches sur le grottes du Banat et d'Olténie (Roumanie, 1959–1962), CNRS, Paris

Bucur II (1997) Formaţiunile mezozoice din zona Reşiţa-Moldova Nouă. Ed Presa Univ Clujeană, Cluj-Napoca, p 280

Constantin S (2003) Evoluţii paleoclimatice în Cuaternar pe baza speleotemelor din carstul Munţilor Banatului şi Mehedinţi. Unpubl Ph.D. thesis, Universitatea Bucuresti

Giurginca A, Munteanu C-M, Vlaicu M, Tabacaru I (2015) Cavernicolous Oniscidea of Romania. Editura Semne, p 165

Goran C (1982) Catalogul sistematic al peşterilor din România. Ed CNEFS, Bucureşti, p 496

Iurkiewicz A (2010) Banat Mountains (Reşiţa-Moldova Nouă Synclinorium). In: Orăşeanu I, Iurkiewicz A (eds) Karst hydrogeology of Romania. Ed Belvedere, Oradea, pp 137-168

Jeannel R (1931) Origine et evolution de la faune cavernicole du Bihar et des carpates du Banat. Arch Zool Ital 16:47–60

Mateescu F (1961) Influenţe structurale în relieful Munţilor Caraşului. Probleme de geografie VIII:205–220

Negrea A, Negrea Ş (1983) Considération sur les écosystèmes des grottes Comarnic et Popovăţ (Banat, Roumanie). Trav Inst Spéol "Émile Racovitza" XXII:47–56

Orghidan T, Negrea Ş, Racoviţă G, Lascu C (1984) Peşteri din Romania. Ed Sport-Turism, Bucureşti

Piţu A (1947) Peştera Comarnic. Enciclopedia turistică românească XIV:48–57

Protopopescu N (1934) Peştera Comarnic. Enciclopedia turistică românească I:25–42

Puşcariu V, Rusu T, Viehmann I (1964) Peşterile din carstul Caraşovean. Lucr Inst Speol "Emil Racoviţă" III:83–129

Răileanu G, Năstăseanu S, Mutihac V (1957) Cercetări geologice in regiunea dintre Anina şi Doman (zona Reşiţa-Moldova Nouă) Bul Acad RPR Geol-Geogr II(2):289–310

Răileanu G, Năstăseanu S, Boldur C (1964) Le sédimentaire paléozoïque et mésozoïques du domaine gétique de la partie SW des Carpates Méridionales. Ann Com Geol Inst Geol 34(2):5–58

Rusu T (1967) Cheile Caraşului. Ocrotirea Naturii 11(1):37–50

Sencu V (1964) Cercetări asupra carstului din partea sudică a localităţii Anina (Banat). St Cerc geol geofiz geogr, Ser geogr 11:149–163

Sencu V (1970) Văile de doline din carstul munţilor Banatului. St Cerc geol geofiz geogr, Ser geogr XVI(2):177–187

Sencu V (1972) Peşterile Comarnic şi Popovăţ. Ed Acad RSR, Bucureşti, p 168

Tăbăcaru I, Giurginca A, Vănoaica L (2004) Cavernicolous Diplopoda of Romania. Trav Inst Spéol "Émile Racovitza" XLI–XLII:121–148

Zaharia L (2006) Phosphatic minerals in some caves from south-western Romania. Proc XVIIIth Cong Carpatho-Balkan Geol Assoc 678–680

Domanului Mountains: Exploratorii 85 Cave (Pestera Exploratorii ′85)

Günther Karban

Abstract

The Exploratorii 85 Cave with 5120 m of passages is one of the largest underground cavities of the Bănăţean karst, which is located in the western part of the Southern Carpathians. Discovered in 1985 by I. Marius, with Exploratorii Reşiţa Speleological Association, it was intensively explored and surveyed until 1987. The cave is formed by upper-level well-decorated fossil galleries and a lower level with 5 underground streams that come out in the Caraş River, near Racoviţă Cave. Dye studies show that the cave potential is about 10,000 m. The cave is in custody of the Exploratorii Reşiţa Speleological Association.

Keywords

Karst • Cave • Spring • Dye studies

Introduction

The Exploratorii 85 Cave is located in the Domanului Mountains, Iabalcea village, on the right side of the Comarnic Creek (Fig. 1). The cave is 5172 m long, with 54 m (+29 m, −25 m) vertical range. The extension of the cave is 700 m and is located between 0.5 and 388 m relative and absolute altitude, respectively (Nania 1996).

Geographic, Geologic and Hydrologic Settings

The cave is accessible by foot from Iabalcea, following a blue marked tourist trail or by an all-terrain vehicle on a gravel road to the Comarnic Rangers Station. The entrance in the cave is located 600 m downstream from the station, on the right side of the Comarnic Creek. The entrance has a metal gate mounted on a concrete slab, to prevent flood-waters to reach it (Fig. 2). The cave is located in the Semenic-Caraş Gorge National Park, is protected by law and is in the custody of the Exploratorii Reşiţa Speleological Association (Asociaţia Speologică Exploratorii Reşiţa).

The area where the cave is formed belongs to the Getic Domain and is composed of an impervious crystalline basement (quartzitic schist and mica schist), transgressively overlain by sedimentary deposits (Năstăseanu et al. 1985). The calcareous rocks in which most of the caves develop consist of Anina Valley limestones of Upper Jurassic (Oxfordian–Kimmeridgian) age. The limestone is grayish-yellow, finely crystalline, with 20–40 cm thick beds, and numerous chert layers and nodules. The Emil Racoviţă and the Exploratorii galleries, located at the west end of the cave, are carved in the Poplar limestones of Lower Cretaceous (Barremian–Lower Aptian) age. This lithological unit is a massive yellowish white, finely to medium crystalline (Năstăseanu et al. 1985).

In the right side of the Comarnic Creek, over 30 caves and potholes were found at different elevations; their development does not exceed 100 m in length. Exploratorii 85 Cave is a notable exception. Its passages parallel the Comarnic Creek were formed by successive water losses along Comarnic and Topliţa valleys, located upstream and downstream of the present entrance.

Presently, the swallets along the Comarnic Creek are temporary active, most of the time the waters disappearing underground through a new sinking point formed about 200 m from the limestone/non-karst rocks interface. In the dry season, this swallet drains the entire flow of Comarnic Creek through a lower still unknown level of the cave. Future dye studies will clarify the underground water path, more than likely heading towards the spring located under Racoviţă Cave, in the Caraş River (Iurkiewicz et al. 1996).

The hydrological network of the cave consists of five underground streams with a total length of 292 m (courses were active/flowing during exploration), representing

G. Karban (✉)
Munich, Germany
e-mail: guenti_k@yahoo.de

G. M. L. Ponta and B. P. Onac (eds.), *Cave and Karst Systems of Romania*,
Cave and Karst Systems of the World, https://doi.org/10.1007/978-3-319-90747-5_30

Fig. 1 Map of the Romanian karst, showing location of Exploratorii 85 Cave

Fig. 2 Exploratorii 85 Cave. Graphics: R. Bala, G. Karban, M. Ţigla. Survey and exploration: B. Bădescu, A. Florea, M. Iucu, V. Nania, and C. Varga. Survey simplified by G. Karban and M. Ţigla

5.7% of cave's total length. The dye studies proved that three of these courses, namely Streams I, II and III (Activul I, II and III), originate in Comarnic Valley. For Streams IV and V (Activul IV and V), water origin is not yet known.

History of Exploration

The Exploratorii 85 Cave was discovered in the fall of 1985, during a study on the genesis of Racoviță Cave that is located in the Caraş Gorge. In 1984 fluorescein was injected in a sinking stream located in the Comarnic Creek, 500-m upstream from the confluence with Caraş River (Q = 27 L/s). The dye came out 12 h later in a spring located under the Racoviță Cave, in the Caraş Gorge (Fig. 3). This dye study proved the hypothesis that the Racoviță Cave was generated by the sinking waters of the Comarnic River and not of Caraş River (Orghidan et al. 1984). Thus, in 1985, as a result of systematic research of Comarnic Valley, a dry ponor was discovered, which represents the current entrance in the Exploratorii 85 Cave (Fig. 3).

The swallet was discovered by I. Marius, who was helped to remove about 2 m^3 of gravel, boulders, and clastic sediments by V. Radu, and B. Nania, both members of the Exploratorii Reşiţa Speleological Association.

Cave Description

The current entrance of the cave (Fig. 4) was opened by removing over 2 m^3 of limestone rock and alluvial material; the narrow entrance passage descends 8 m at 55° slope along the bedding plane, where crawling is necessary due to severely reduced height of the gallery. The end of the gallery intercepts the first underground stream of the cave, which disappears after 12 m in an impenetrable swallet. On the left side, through an ascending passage, a small well-decorated chamber was discovered that opens in a 15 m high and 1.5 m wide canyon, which ends in the Restaurant Hall (Sala Restaurant) where the floor is covered with limestone blocks detached from the inclined (45° SE) ceiling.

From the Restaurant Hall, the gallery split into three passages. Towards east-southeast is the Iucu Marius Passage, which is rich in speleothems, sizable halls (Mushrooms Hall/Sala cu Ciuperci) and large passages developed by widening initial canyon-type galleries. In this passage, Stream II (Activul II) is intercepted; it drains the waters from Swallets I and II located upstream of the cave entrance, along Comarnic Creek's riverbed. The hydrologic connection was determined by dye studies conducted with fluorescein (Fig. 3). The end of the gallery is a labyrinth, with many breakdowns and a room (Bones Cemetery/Cimitirul de Oase) in which cave bear bones and skulls

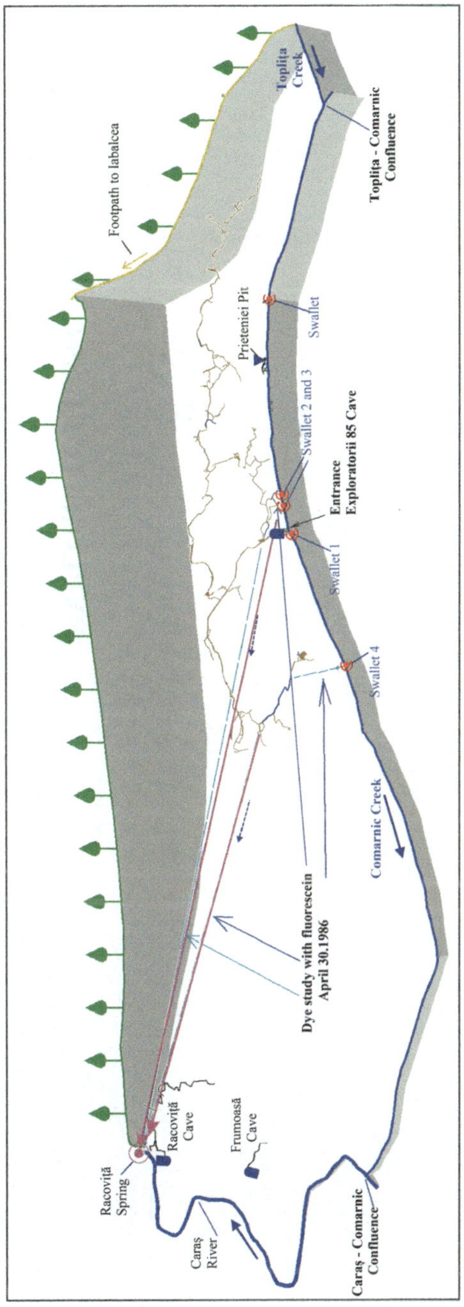

Fig. 3 Dye studies performed in the Exploratorii 85 Cave area

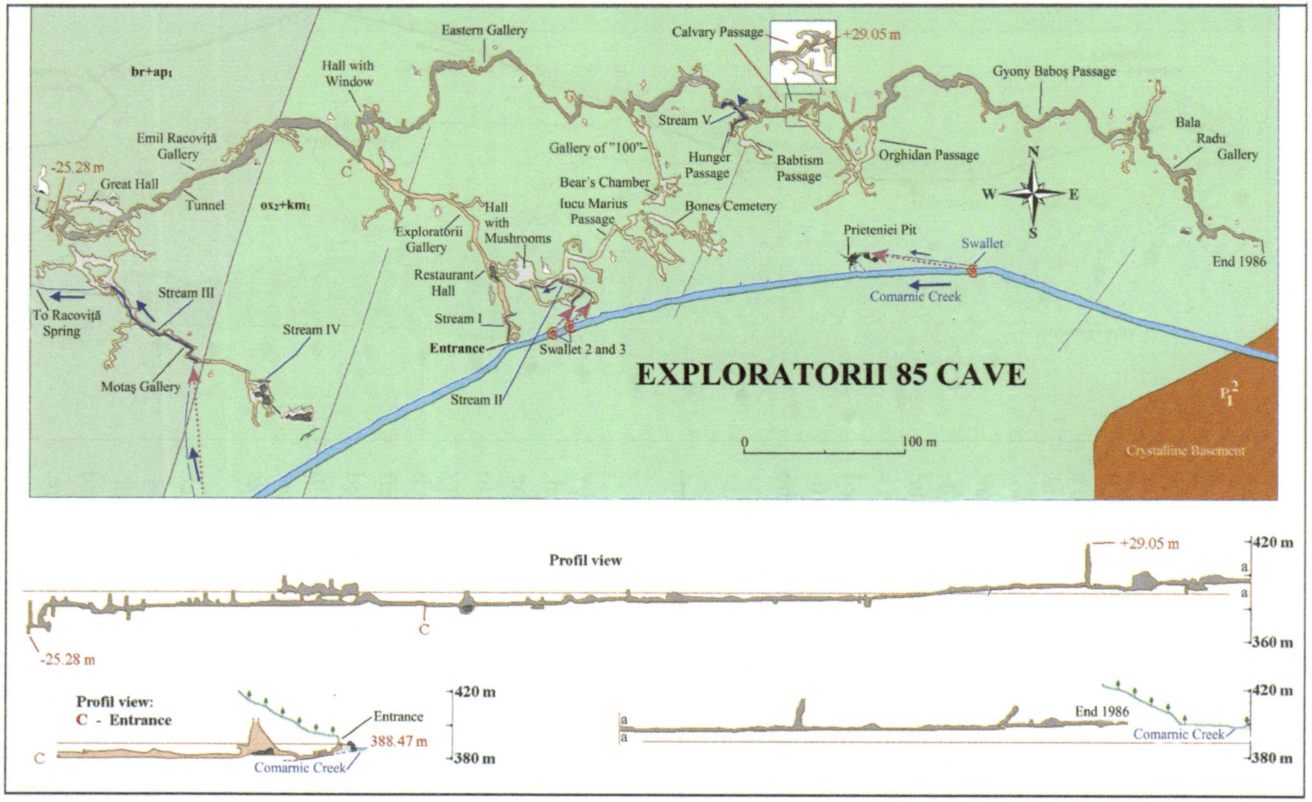

Fig. 4 Cave entrance (photograph by G Karban)

Fig. 5 Various dripping speleothems (**a–b**) and the Corkscrew (**c**) in Exploratorii 85 Cave (**a–b** photographs by M. Iucu, **c** by G. Karban)

were found, indicating that these galleries were once directly opened to the land surface.

Two other galleries heading north-northwest originate in the Restaurant Hall: a short (15 m) ascending one and the Explorers Gallery (Galeria Exploratorilor). The latter develops along bedding planes, first descends to a swallet and then continues slightly upwards until intersecting another gallery on the right. Here, the galleries forks east and west. The western part comprises the large and high Emil Racoviţă Gallery (Galeria Emil Racoviţă), which continues with Motaş Gallery (Galeria Motaş) on a south-eastern direction. The Emil Racoviţă Gallery ends in a large room with many breakdowns, shafts, and chimneys, with labyrinthic aspect, where the lowest point of the cave (−25 m) is located. The connection between Emil Racoviţă Gallery and this room is through a very narrow passage named the Tunnel (Tunelul) (0.40 m/0.35 m).

Motaş Gallery intercepts two streams: Stream III, whose origin is a sink along Comarnic Creek located downstream of the entrance, and Stream IV, which was only 4 m long during the exploration. These two streams drain their waters to the resurgence located under Racoviţă Cave This section of the cave is developed in Barremian–Aptian limestones and crosses the fault line into the Oxfordian–Kimmeridgian limestones located in eastern part of the cave. From the junction of the Exploratorii Gallery with Emil Racoviţă Gallery, another passage continues north for about 20 m, then turns eastwards and crosses the Hall with Window (Sala cu Fereastră) and continues with the Eastern Gallery (Galeria Estică) that leads into the easternmost end of the Radu Bala Gallery.

Along the way, four large galleries are intercepted on the right side: Gallery of "100" (Galeria celor "100") ending with the Bear's Chamber (Sala Ursului), where skeletal remains of *Ursus spelaeus* were found; Hunger Passage (Galeria Foamei) through which Stream V flows; Baptism Passage (Galeria Botezului); and Orghidan Passage (Galeria Orghidan). After the intersection of Orghidan Passage with the Eastern Passage, the cave continues with Gyony Baboş Passage that opens into the final Radu Bala Gallery. Between the Baptism Passage and Orghidan Passage is Calvary Passage (Galeria Calvarului), which ends in a chimney, where the highest (current) elevation in the cave (+29 m) is reached.

Cave Sediments and Speleothems

Active/stream galleries or temporary active sections are characterized by the presence of erosion features (erosion levels, whirlpools, meanders, pillars, etc.) and corrosion features (echinoliths, scallops, nodules and chert ledges).

The sedimentary deposits have an allochthonous origin and comprise gravel, coarse or fine sand, which alternates with vegetal detritus (leaves, tree branches). The floor of a few galleries is covered with clay. On some galleries, the sedimentary deposits were washed out, followed by a new phase of deposition. The fossil/dry passages of the cave are well decorated with stalactites, stalagmites, flowstones, draperies and a specular stalactite ending with a corkscrew (Fig. 5). Also, some rare speleothems are present as shelfstones, helictites and crystalictites.

Cave Climatology

It is a relatively warm cave, with temperatures ranging between 3 and 11 °C. Even though the cave has only one entrance, it has a dynamic thermocirculation with air currents present especially in the sectors with breakdowns.

Cave Fauna

From a biospeleological point of view, the cave has been investigated in several stages during the month of October 1985 and March through May 1986. Although there are five underground streams, the water is filtered in the cave entrance area through boulders, gravel and sand that retain most of the plant debris (leaves, twigs, etc.), which leads to poor vegetal resources inside the cave. The only abundant matter is the black humus deposited on the riverbed during low flow periods.

Fauna types identified in the cave are: trogloxenes and troglophiles (Negrea et al. 1993). The cave wall fauna is best represented by: butterflies (*Triphosa* sp.), flies (*Limonia* sp., *Culex* sp., etc.), snails (*Oxychilus* sp., *Spelaeodiscus triaria,* etc.), spiders (*Meta menardi*), opilionids (*Paranemastoma* sp.) and springtails.

The floor fauna is represented by springtails (various species), Crustaceans (*Niphargus* sp., isopods), oligochets, flies, trichopteran larvae and beetles. Among bats (Chiroptera), only one species of *Myotis* was identified (isolated individuals). Inside the cave, other three troglophiles were identified: the snails *Orcula jetschini* and *S. triaria* (and the isopod *Trichoniscus transsilvanicus.* On the upper floor of the "Iucu Marius" Passage, and at the end of the "100" Passage skeletal remains of cave bears were collected by Dr. Ş. Negrea with the "Emil Racoviţă" Institute of Speleology, and on the other side of the "Gastropods" Passage (Galeria cu Gasteropode), *Felis silvestris*, mandibles of *Lepus europaeus* and a jaw fragment of a juvenile *Ovis* were identified.

Archaeology

Although the cave had no natural opening when discovered, some archaeological researches were undertaken, revealing in the sedimentary deposits at the entrance in the Motaş Gallery (−21 m elevation), two pieces of ceramics dated to the eighteenth century. It is assumed they were transported into the cave by Stream III.

Cave Conservancy

The cave is part of the Semenic–Caraş Gorge National Park, and it was declared a Speleological Reserve category B. Therefore, it is closed with a metal gate and its custody belongs to the Exploratorii Reşiţa Speleological Association.

Acknowledgements Many thanks to M. Ţigla for producing the graphics for this paper.

References

Iurkiewicz A, Constantin S, Bădescu B (1996) Sisteme carstice majore din zona Reşiţa-Moldova Nouă. A.S. "Exploratorii" Reşiţa, p 70

Nania V (1996) Peştera Exploratorii 85. Cerc Speol pp 38–40

Năstăseanu S, Iancu V, Savu H, Săndulescu DR (1985) Geological map of Romania, Reşiţa Sheet. Scale 1:50.000, Inst Geol Geofiz, Bucureşti

Negrea Ş. Negrea A, Karban G (1993) Grottes explorèes dans vallèes Luca, Comarnic et Topliţa (Banat, Roumanie) et leur faune. Trav Inst Spéol "Émile Racovitza" XXXII:141–142

Orghidan T, Negrea Ş, Racoviţă G, Lascu C (1984) Peşteri din România. Ed Sport-Turism, Bucureşti, p 504

Domanului Mountains: Poiana Gropii Pit (Avenul din Poiana Gropii)

Günther Karban

Abstract

Poiana Gropii with 1900 m in length and −249 m depth is the deepest vertical cave in Banat region formed by vadose water flowing along fractures and bedding planes. Speleothems are present only below −130 m, but spectacular morphologies due to erosion and corrosion occur throughout the cave.

Keywords

Karst • Pit • Vertical cave • Blind valley

Introduction

Poiana Gropii Pit with a depth of −249 m is located in the Domanului Mountains, Southern Carpathians, at an elevation of 746 m, in the northeastern part of Reşiţa–Moldova Nouă Syncline (Fig. 1), which represents the largest, unfragmented limestone area in the country (ca. 800 km^2). It was the first deep vertical cave explored in Romania in the sixties by the Exploratorii Reşiţa Speleological Association using ropes and ladders (Bleahu et al. 1976).

The cave opens in the Semenic—Caraş Gorge National Park near the town of Cuptoare, which administratively is under the jurisdiction of Reşiţa (Caraş—Severin County). The pit can be accessed from the Poiana Bichi ranger's station located on the Reşiţa-Văliug highway following for 45 min an uphill touristic trail marked with a blue circle (route to Crivadia—Semenic Touristic Complex) to Frantz Meadow/Spring. From here a dirt road began which in a few hundred meters intersects a blind valley; the pit entrance is at its upstream end.

The vertical cavity is developed at the contact between Anina's limestones (Barremian–Lower Aptian; Upper Oxfordian–Lower Kimmeridgian) and non-karst rocks (conglomerates and sandstones) (Bucur 1997).

History of Exploration

The Poiana Gropii Pit was known to the locals for a long time and as the story goes it was said to be bottomless (Fig. 2). This is because after the First World War, a logging area was developed in the vicinity of the pit. The workers were paid by how many carts were loaded and left the site, not by how many reached their final destination. Thus, most of the logs ended up in the "bottomless" pit, fact that fed local myths and legends. Truth is that during that time it was inaccessible due to its 80 m vertical shaft located at the entrance.

In 1961, a group of three climbers from the Reşiţa "Metal Association," H. Horst, J. Francis, and G. Karban, learned about the existence of Poiana Gropii Pit and decided to explore it (Karban 2014). Before taking on this new challenge, they discovered and explored many caves, but a venture of this magnitude has never occurred. The first attempt to explore the cave was in September 1961, with a 25-m-long rope. The rope being too short to explore the first drop, a pulley was used to lower the late H. Kopetzky to −18 m, where a small platform/ledge was identified. This ledge was later named "Heinz's step" (Fig. 3).

Based on the echo after throwing a few rocks into the pit, the conclusion was that the pothole continues with a succession of pits of unknown depths. It became clear that any further exploration will require significantly more caving gear. During that time, cave explorations were accomplished by rappelling in "S" or "Dülffer" (body rappelling) method and climbing short sections employing the "Prusik" knot.

The very first attempts were made by rappelling and climbing the rope using the so-called firefighter's style, feet/leg propped against the wall and pulling the rope with hands—that demanded additional effort and a great physical condition. Because we only had one climbing rope borrowed from the Reşiţa's firefighters and our own one, at each ledge/drop

G. Karban (✉)
Munich, Germany
e-mail: g_karban@yahoo.com

G. M. L. Ponta and B. P. Onac (eds.), *Cave and Karst Systems of Romania*,
Cave and Karst Systems of the World, https://doi.org/10.1007/978-3-319-90747-5_31

Fig. 1 Location of Poiana Gropii Pit in the main karst area of Domanului Mountains

a caver has to stay to pull up the rope. With this method, which was time-consuming, the first 45 m were descended.

In 1962, a rope ladder with wooden steps, 12 m long and weighing 25 kg was borrowed from a manufacturing facility in Reşiţa. Helped by this new gear, the team reached the depth of −62 m, in 15–18 h. After these rather unsuccessful attempts and challenging physical escapades, the conclusion was that we had to manufacture our ladders, using 5 mm steel cables with 1.5 cm diameter steel steps.

In the summer of 1963, a 17 m ladder was finished, and a second one, 21 m long, was completed by fall of that year. Also, our personal equipment was improved, replacing hats with helmets on which lights were attached. In this way in October 1963, the depth of −125 m was reached, the exploration ending at a narrow passage, later named the "Mouse hole" (Gaura de Şoarece). In 1964, new metal pitons (spikes), steel carabiner (the one's borrowed from the fire-fighters were too heavy), and ladders were manufactured.

On a Saturday night (to be able to get out Sunday and make it in time to work on Monday), in the fall of 1965, four people went back to the cave. Benefiting from the improved equipment, the previous −125 m (Mouse hole) was reached much quicker. Here, H. Neff, who was a well-built athlete failed to cross the narrow passage, thus only the remaining three continued the exploration. After descending a 10 m shaft, a steep gallery developed along conglomerate/limestone interface (Hell's Gallery) led to the depth of −215 m, where it narrowed again in an impenetrable sump. In the vicinity, a side gallery with beautiful rimestone pools (gours) and stalagmites, called "Corner of Heaven" (Colţ de Rai), is located (Fig. 4). With the depth of −215 m, Poiana Gropii Pit became the deepest cave in Romania, a record that will be exceeded nine years later by Izvorul Tăuşoarelor Cave. The depth of −215 m was surpassed in 1977.

During mapping and systematic exploration, new passages and pits were discovered. In September 1966, a base camp was established at −110 m, with a phone line making the connection to the support team at the surface. In November 1966, a new branch beginning at Heinz's step/ledge is explored. After several drops, at −130 m, a steep passage (Scrambling Gallery/Galeria Ramonajelor)

Fig. 2 Entrance in the Poiana Gropii Pit (photograph by G. Karban)

developed along a vertical fracture was explored down to the current terminus of the cave (−249 m), where it ends in a sump. In the coming years, the junction between the right (System I) and the left (System II) branches is discovered. The survey completed in 1971 by G. Karban and R. Pauler shows a vertical development of −236 and 1050 m in length.

In 1977, during the annual meeting of Romanian cavers (Speosport), organized by the Exploratorii Reşiţa Speleological Association, a field trip was organized in the Poiana Gropii Pit. With this occasion, a member of the Club "Z" Oradea (N. Sasu) successfully passed the terminus sump at −235 m, adding an additional one meter to the vertical development of the cave. In 2012, a team led by R. Puşcaş (with a major contribution from I. Parvulescu) resurveyed the pit with modern instruments/methods and discovered new galleries, traversed the sump at −236 m, and reached a depth of −244 m and a length of 1760 m (Puşcaş 2012). More recently, the pit acquired 5 m more in depth and 140 m in length so that its present-day size is 1900 m long and −249 m in depth.

This cave was the central point that led to the foundation of the Speogroup Reşiţa (1961), which in 1975 will change its name to Exploratorii Reşiţa Caving Club, and more recently to Exploratorii Reşiţa Speleological Association.

Fig. 3 Heinz's step, the place where the real adventure in the Poiana Gropii Pit begins (photograph by G. Karban)

Cave Description

Poiana Gropii Pit is situated at the end of small blind valleys, which is collecting its waters on non-calcareous rocks and disappears underground at the base of the limestone cliff through a 6 × 4 m entrance (Fig. 2). The cave consists of two branches (System I and II), which are interconnected at different points via horizontal and descending galleries or pits. Both systems begin with a series of short drops down to −125 or −130 m, respectively, and then continue with steep downstream galleries with small waterfalls. System I hosts

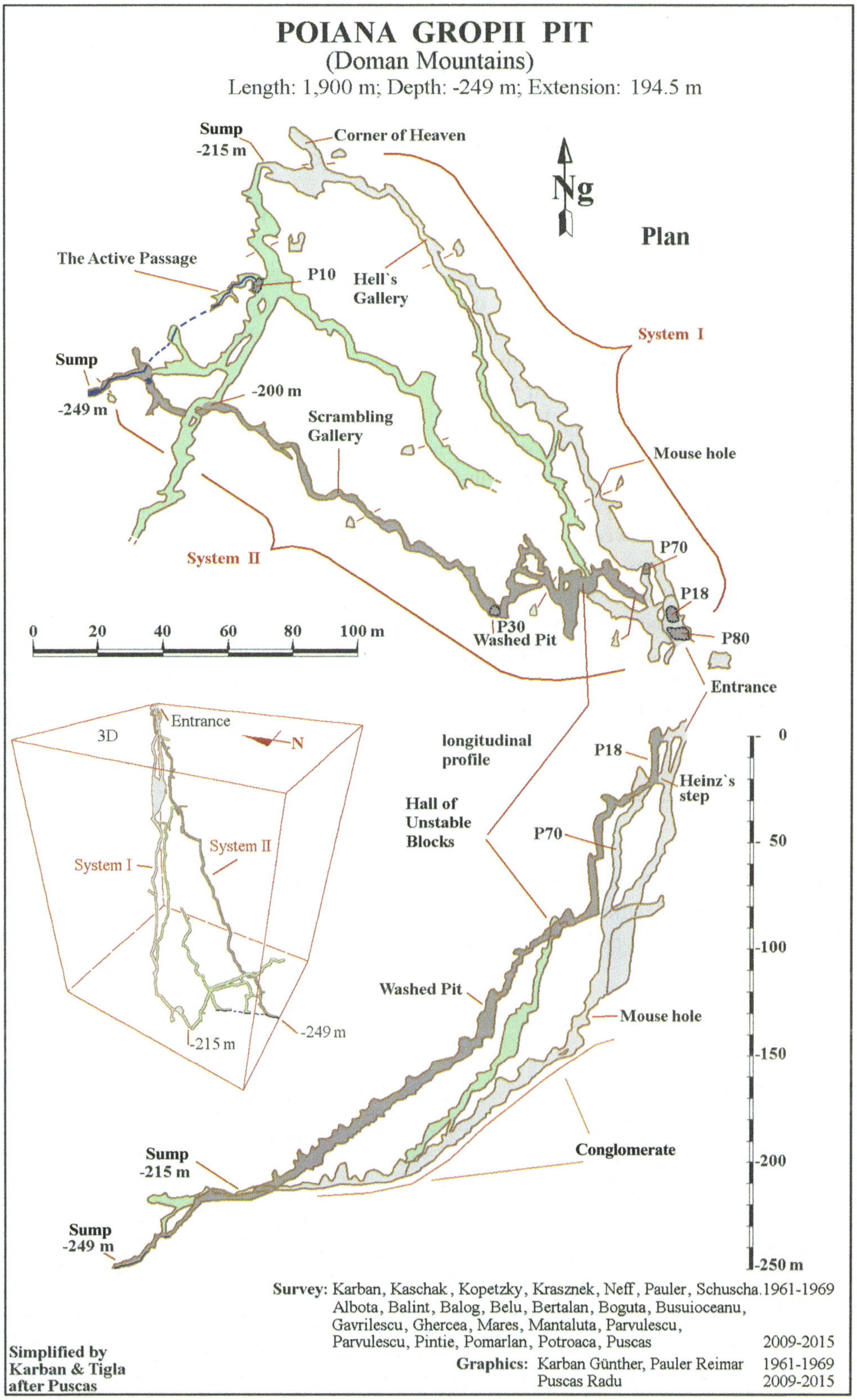

Fig. 4 Map and profile of the Poiana Gropii Pit

the longest drop of 70 m, which can be bypassed through a succession of parallel, shorter shafts. Right at the entrance, two parallel shafts of 70 and 18 m in depth were identified (Bădescu and Vlaicu 2011).

The cave exploration trips are made by rappelling the 18 m shaft to reach the convenient platform/ledge known as Heinz's step (Fig. 3) from where one can continue straight ahead through the 70 m drop, which is communicating with the room ending in the "Mouse hole" and beyond to −215 m sump. An alternative pathway continues to the left through a descending gallery followed by a succession of drops through the "Hall of Unstable Blocks" and the "Washed Pit" (Fig. 4) to a depth of −130 m. From here, a fracture-type gallery with many drops is intercepted. At −244 m is located the final sump, which was explored in 2015 by R. Puşcaş to −249 m. Between the lowest points of the two systems, a stream flowing from System I to System II and ending in the final sump was intercepted (Iurkiewicz et al. 1996).

Speleothems are not present between entrance and −130 m, but spectacular morphologies due to erosion and corrosion occur. The presence of chert nodules and layers interbedded in limestone makes some passages very attractive. Below −130 m, stalagmites, stalactites, and rimestone pools are only occasionally present. Also, along the stream passage connecting System I and II, some "pockets" of crystals were observed.

Cave Fauna

Part of the fauna recovered from Poiana Gropii Pit was examined by Negrea (1977) who identified bats and invertebrates (Arachnida, Crustacean, Isopoda, Oligochaeta, Insecta, Gastropoda), but other samples are still under investigations. Temperature in the cave ranges between 5 and 8 °C.

Cave Conservancy

Poiana Gropii Pit is an unprotected cavity, but because it is part of the 8310 habitats (Bat Protection) Natura Semenic-Caraş Gorge, it can be visited only with the prior approval issued by the Semenic-Caraş Gorge National Park. Visiting the pit requires very good knowledge of single-rope technique.

References

Bădescu B, Vlaicu M (2011) Atlasul habitatului 8310 din situl Natura 2000 Semenic-Cheile Caraşului: peşteri şi lilieci. Ed BrumarR, Timişoara, p 71

Bleahu M, Decu V, Negrea Ş, Pleşa C, Povară I, Viehmann I (1976) Peşteri din Romania. Ed Ştiinţifică şi Enciclopedică, Bucureşti, p 415

Bucur II (1997) Formaţiunile mezozoice din zona Reşiţa-Moldova Nouă. Ed Presa Univ Clujeană, Cluj-Napoca, p 214

Iurkiewicz A, Constantin S, Bădescu B (1996) Sisteme carstice majore din zona Reşiţa-Moldova Nouă. Ed Asoc Speol Exploratorii Reşiţa, Reşiţa, p 72

Karban G (1971) Avenul din Poiana Gropii. Trav Inst Spéol "Émile Racovitza" 10:339–343

Karban G (2014) Amintiri din viaţa unui speolog. Ed Asoc Speol Exploratorii, Reşiţa, p 179

Negrea Ş (1977) Sur les associations of plancher des grottes du Banat (Roumanie). Trav Inst Spéol "Émile Racovitza" 16:99–139

Puşcaş R (2012) Avenul din Poiana Gropii. Speomond 16:27–29

Metaliferi Mountains: Zidită Cave

Ferenc L. Forray

Abstract
Zidită Cave distinguishes itself from other caves in the region due to the archaeological artifacts discovered in the sediments accumulated at the cave entrance, the paleoclimatic information recorded in the cave's sediment, and longtime hibernacula and nursery for several bat species.

Keywords
Bat guano • Stable isotopes • Archeology
Speleothems • Cave • Romania

Introduction

Metaliferi Mountains are famous for their gold–silver (Au–Ag) and lead–zinc–copper (Pb–Zn–Cu) ore deposits, but small patches of limestone (mainly of Tithonian age) (Bleahu et al. 1967; Lupu et al. 1967), carbonate metamorphic rocks (e.g., the Rapolt crystalline) (Savu et al. 1968; Bordea et al. 1978), or Quaternary carbonate tufa/travertine (Lupu et al. 1982) (e.g., South of Rapolt crystalline formation) host interesting karst landscapes. According to the Romanian Cave Inventory (Goran 1982), there are 318 caves reported from these karst areas. Their maximum length is ~1200 m, whereas the vertical development never exceeds 150 m, but, however, the vast majority of these caves are small. With its 547 m, the Zidită Cave is the fourth longest cave in the Metaliferi Mountains (Fig. 1).

Geographic, Geologic, and Hydrologic Settings

Geoagiu River (a right-side tributary of Mureş River) flows over Mesozoic ophiolites (Borcoş et al. 1981), traverses the ~3.7 km^2 limestone block forming the Mada Gorge (~3 km long), and before reaching the Mureş River, it passes over sandstone/shale bedrock (Maastrichtian flysch) (Bordea et al. 1978).

The hydrogeology of the upper watershed of Geoagiu River and the karst nearby is described by Orăşeanu (2010). The cave is located (Lat/Long: N46.00746°, E23.12869°) on the southwestern side of the limestone massif called Pleşa Mare Hill at 410 m above sea level.

History of Exploration

Due to its large entrance, the cave was known to local people for a very long time. The first document mentioning the cave was published by Téglás (1887, 1902) and Floca (1957). Archaeological studies have revealed the use of the cave entrance as temporary shelter from Chalcolithic–Early Bronze Age up to modern times (Căstăian 2014). Despite this long history, the cave was mapped only in 1980 by M. Nedopaca, D. Avramescu, and A. Cor from Zarand Caving Club (Brad, Hunedoara County). The first survey revealed a 402-m-long cave with a +18 m vertical development (Nedopaca 1989).

In 1982–1983, the explorations conducted by the Proteus Caving Club (Hunedoara, Romania) extended the cave with 145 m to the present-day total length of 547 m (data from Proteus Caving Club Archive).

Cave Description

The Zidită Cave's relatively large entrance was fortified through a stone wall, which was built for defense purpose; remnants of this wall are style visible. Behind the wall, small phreatic tubes lead to the Great Hall (Fig. 2) and a maze of

F. L. Forray (✉)
Department of Geology, Babeş-Bolyai University,
M. Kogălniceanu 1, 400084 Cluj-Napoca, Romania
e-mail: ferenc.forray@ubbcluj.ro

G. M. L. Ponta and B. P. Onac (eds.), *Cave and Karst Systems of Romania*,
Cave and Karst Systems of the World, https://doi.org/10.1007/978-3-319-90747-5_32

Fig. 1 Location of the Zidită Cave in Metaliferi Mountains

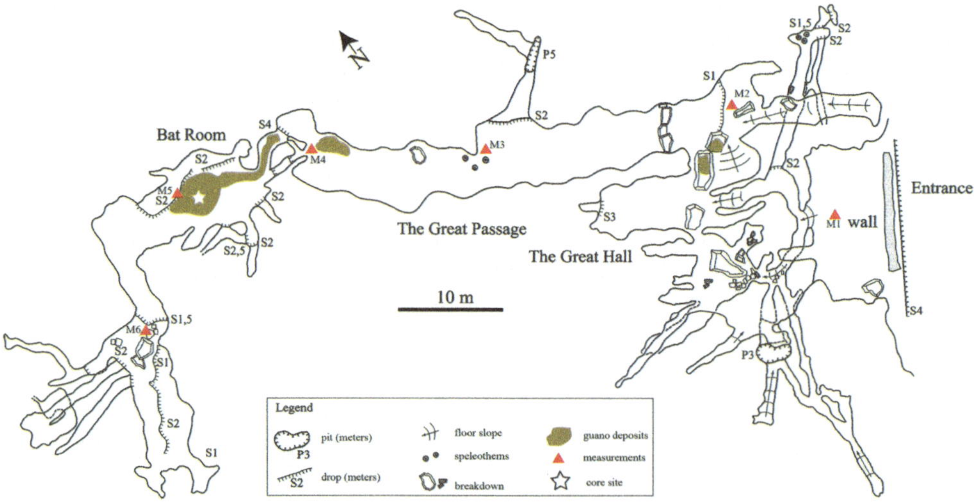

Fig. 2 Map of Zidită Cave (reprinted from Forray et al. 2015, with permission from Elsevier). The measurement points (triangle) are for temperature and humidity. The location of the guano core site is marked by the white star

small passages. From this hall, a large East–West oriented Great Passage connects with the Bat Room through a chimney. Beyond this fairly large chamber, the cave continues toward south and east with several narrow passages dominated by large breakdown blocks.

Cave Sediments, Speleothems, and Minerals

The cave is dominated by breakdown blocks, clay, and scattered guano on the floor. Although occurring throughout the cave, the largest guano deposits are in the Bat Room (~1.5 m thick) and Great Hall (~0.6 m thick), respectively. Speleothems are less frequent throughout the cave (the Great Passage) and are represented by calcite flowstones, stalactites, stalagmites, columns, gours (rimstone pools), calcite rafts, soda straws, and draperies.

Carbon stable isotopes (Fig. 3) and pollen analyses were carried out, at 1-cm interval along the 1.5 m long core recovered from the guano heap located in the Bat Room. Based on twelve radiocarbon (^{14}C) ages, we inferred that the guano accumulated continuously from around AD 1000 until present. This suggests that the fortification wall never fully closed the cave entrance, thus allowing bats to roost inside the cave.

The $\delta^{13}C$ values revealed a gradual onset of the Little Ice Age (LIA) after the Medieval Warm Period (MWP). During the LIA, there were two major cold/wet periods, one around AD 1500 and the other around AD 1865 (Fig. 3). In the twentieth century, the $\delta^{13}C$ values indicate a steady change toward a warmer and drier climate (the so-called global warming). Additional information on the climate and environmental changes inferred from studies on the guano core is available in Forray et al. (2015) and Cleary et al. (2016).

Calcite is the most common mineral in the composition of cave speleothems; rarely, aragonite was also identified. The mineralogical studies conducted by Giurgiu and Tămaş (2013) on crusts formed on limestone fragments in contact with guano identified two calcium phosphate minerals: brushite and hydroxylapatite.

Cave Genesis

The cave develops along two major faults (oriented approximately E–W and NW–SE), which are responsible for many breakdown blocks. Later, the passages were enlarged by water, possible drained from Ardeu creek. In several places, flow was under phreatic condition, creating solution pockets in the ceiling (or cupolas) and echinoliths. Higher fracture density leads to the creation of labyrinth phreatic tubes in the southern part of the cave.

To date, there are no radiometric dates (e.g., U/Th) on speleothems available from this cave, which would help establishing the minimum age of it and allow to estimate the incision rate of the Geoagiu River that cut the Mada Gorge.

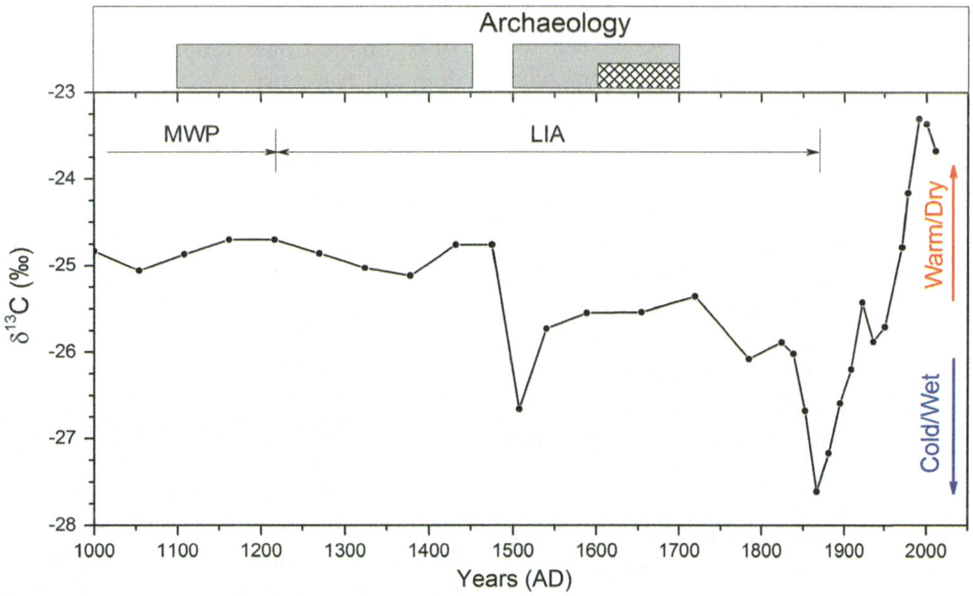

Fig. 3 $\delta^{13}C$ profile through the guano core. The values and age-depth model are according to Forray et al. (2015). To give roughly equal sampling through the whole guano sequence, ten-year average values are represented after 1950. Archaeological data are from Căstăian (2014). Gray area represents ceramic fragments/fireplaces, pottery, and the cross-hash pattern represents the possible period in which the fortification wall had defense purpose (LIA: Little Ice Age; MWP: Medieval Warm Period)

Cave Climatology

The only cave climate parameters documented in the cave survey report form (Nedopaca 1989) are a pair of temperature (6.7 °C) and relative humidity (76%) recorded in the far end of the cave in April 13, 1980, while mapping the cave. On that day, the surface temperature was 14.4 °C. Five measurements collected in October 2014 along the major cave passages (Fig. 2) indicate that the average temperature was ~13.5 °C (outside: 15.3 °C) whereas the relative humidity ranged between 87 and 90%. Fluctuations of these two parameters are controlled by the permanent bidirectional ventilation that operates between the cave entrance and the end of the Great Passage (Forray et al. 2015).

Paleontology

Below a prehistoric horizon, archaeological excavations revealed the existence of reddish, yellowish limestone pebbles with traces of Pleistocene fauna (Căstăian 2014). No further information is available on what type of fauna is or how the author determined its age. In addition, Nedopaca (1989) reported the presence of *Ursus spelaeus* bones in the Great Passage.

Cave Fauna

Conducted as part of a national campaign, the first observations on bats roosting in Zidită Cave were collected in 1952. The synthesis was published 11 years later and reported a large colony of *Miniopterus schreibersii* (common bent-wing bat or Schreibers' bat) and isolated individuals of *Rhinolophus blasii* (Blasius' horseshoe bat), *Rhinolophus ferrumequinum* (greater horseshoe bat), *Rhinolophus hipposideros* (lesser horseshoe bat), and *Myotis myotis* (greater mouse-eared bat) (Dumitrescu et al. 1963).

In 1980 when Zidita was included in the Romanian Caves Database, the cave survey mentioned the presence of several bat colonies (tens of individuals each) in the Great Hall (Nedopaca 1989); however, the report offered no details about bat species. Between 2012 and 2014, the maternity colony was composed of *Rhinolophus euryale* (116 individuals), but the cave represents hibernacula for other bat species like *Rh. ferrumequinum, Rh. hipposideros, and M. schreibersii* (Forray et al. 2015). In 2015, the cave hosted a vigorous maternity colony of *M. myotis* (Coroiu pers. comm.).

Anthropology/Archaeology

Archaeological excavations were carried out in 1996, followed by two other campaigns in 1998 and 2000 led by Pinter (Pinter et al. 2001; Căstăian 2014). The results are summarized below, but readers are encouraged to consult the original publication for more details. The archaeological investigations unearthed three human mandible fragments and a piece of a vertebra. The bones were found in a site with artifacts mixtures of different ages. For the time being, this makes the bone fragments undated, unless in the future ^{14}C dating will be performed.

The flint flakes, stone axes, and axe–hammer are the earliest archaeological objects, possibly of Neolithic Age. Evidence of fireplace from Coţofeni Culture (Chalcolithic and Early Bronze), ceramic fragments and fireplace from Wietenberg Culture (Middle Bronze Age), and ceramic fragments from Hallstatt Culture (Early Iron Age) followed by medieval objects (ceramic pot, razor, and the fortification's iron key) completes the archaeological inventory.

Cave Conservancy

Even though Zidită is not a show cave, it is frequently visited due to touristic advertising campaigns, founded through a European Union project. Most of these visits end in the entrance chamber behind the stone wall, but sometimes tourist wanders further inside. The cave represents an important habitat for bats, and all species documented are listed in the European Council Directive 92/43/EEC (EEC 1992) on the Conservation of Natural Habitats and of wild flora and fauna list.

Conclusions

The Zidită Cave is not particularly outstanding in terms of size, morphology, or speleothems. What makes it special is the amount of guano accumulated especially in the Bat Room. Using multi-proxy analyses (stable isotopes, pollen, and microcharcoal), these organic sediments allowed scientists to decipher the changes in climate and environment that occurred over the past 1000 years. At the same time, the archeological and anthropological materials preserved within the sediment layers near the cave entrance reveal a long human occupation starting probably in Neolithic/Early Bronze Age and continued up to the twentieth century.

Acknowledgements We are thankful to Proteus Caving Club (Hunedoara, Romania) and to R. Breban for kindly providing the map of Zidită Cave and information on the cave exploration. Dr. T. Tămaş, A. Giurgiu, and I. Vişan are acknowledged for logistical help during some of the field activities. This research was partly funded through grant PN-II-ID-PCE-2011-3-0588 from Romanian National Research Council (CNCS) to B. P. Onac. We are also thankful to M. C. Căstăian for the archaeological information and to Dr. I. Coroiu for bat species identification during several field works.

References

Bleahu M, Savu H, Borcoş M (1967) Geological map of Romania, Brad Sheet (17). Scale 1:200.000, Inst Geol Geofiz, Bucureşti

Borcoş M, Berbeleac I, Bordea S, Bordea J, Mantea G, Boştinescu S (1981) Geological map of Romania, Zlatan Sheet (74c). Scale 1:50.000, Inst Geol Geofiz, Bucureşti

Bordea J, Berbeleac I, Borcoş M, Mantea G, Stancu J, Rogge-Ţăranu E (1978) Geological map of Romania, Geoagiu Sheet (90a). Scale 1:50.000, Inst Geol Geofiz, Bucureşti

Căstăian MC (2014) The archaeology of caves in the Apuseni Mountains. Unpubl Ph.D. Thesis, Lucian Blaga University, Sibiu, p 304 (in Romanian)

Cleary DM, Onac BP, Forray FL, Wynn JG (2016) Controls on $\delta^{15}N$ values of cave bat guano: an assessment for reconstructing paleoclimate and environmental changes. Palaeogeogr Palaeoclimatol Palaeoecol 461:87–97. https://doi.org/10.1016/j.palaeo.2016.08.012

Dumitrescu M, Tanasachi J, Orghidan T (1963) Răspândirea chiropterelor în R. P. Română. Lucr Inst Speol "Emil Racoviţă" 1–2:509–575

EEC (1992) Council Directive 92/43/EEC of 21 May 1992 on the conservation of natural habitats and of wild fauna and flora, p 44

Floca O (1957) Regiunea Hunedoara. Ghid turistic. Tipografia Judeţeană, Deva, p 594

Forray FL, Onac BP, Tanţău I, Wynn JG, Tămaş T, Coroiu I, Giurgiu A (2015) A Late Holocene environmental history of a bat guano deposit from Romania: an isotopic, pollen and microcharcoal study. Quat Sci Rev 127:141–154. https://doi.org/10.1016/j.quascirev.2015.05.022

Giurgiu A, Tămaş T (2013) Mineralogical data on bat guano deposits from three Romanian caves. Studia UBB Geol 58:13–18. https://doi.org/10.5038/1937-8602.58.2.2

Goran C (1982) Catalogul sistematic al peşterilor din România. Ed CNEFS, p 496

Lupu M, Borcoş M, Dimian M, Lupu D, Dimitrescu R (1967) Geological map of Romania, Turda Sheet (18). Scale 1:200.000, Inst Geol Geofiz, Bucureşti

Lupu M, Kräutner HG, Ţicleanu N, Boştinescu S, Bandrabur T, Kräutner F, Horvath AR, Nicolae I (1982) Geological map of Romania, Deva Sheet (89b). Scale 1:50.000, Inst Geol Geofiz, Bucureşti

Nedopaca M (1989) Pestera Zidită, Nr. 321/29 August 1989. Romanian Caves Database. Institutul de Speologie "Emil Racoviţă", Bucureşti

Orăşeanu I (2010) Upper watershed of Geoagiu River (The Metaliferi Mountains). In: Orăşeanu I, Iurkiewicz A (eds) Karst hydrogeology of Romania. Ed Belvedere, Oradea, pp 311–315

Pinter ZK, Căstăian M, Ţuţuianu D (2001) Mada, com. Balşa, jud. Hunedoara. Punct: Peştera Zidită, în Cronica cercetărilor arheologice din România. Campania 2000, p 139

Savu H, Pavelescu M, Stanciu I, Lupu D (1968) Geological map of Romania, Orăştie Sheet (26). Scale 1:200.000, Inst Geol Geofiz, Bucureşti

Téglás G (1887) Ujabb barlangok az erdélyrészi érczhegység délkeleti övéből. Matematikai Természettudományi Értesitő 5:180–186

Téglás G (1902) Hunyadvármegyei kalauz. Erdélyi Kárpát-Egyesület Kiadása, Kolozsvár. Magyar Polgár Könyvnyomda, p 214

Bihor Mountains: Dârninii Cave (Pestera din Peretele Dârninii)

Felix Papiu and Bogdan P. Onac

Abstract

Some of Romania's largest and most interesting cave systems are located in the Bihor Mountains. Among these, Dârninii Cave is one of them and excels through its length (5645 m) and size of the passages/rooms, as well as some of the speleothems (columns, domes). To date, the scientific research comprises some studies on cave minerals, fauna, underground climatology, and microbiology. Although scanty, all these investigations point out to some very intriguing cave resources that await future detailed studies.

Keywords

Speleothems • Mineralogy • Microbiology
Climatology • Cave fauna

Introduction

Dârninii Cave opens at an absolute altitude of 1220 m on the upper section of the Albac Valley (~30 m above valley's thalweg), downstream from the place called Fleiu (belonging to the Sforţea Hamlet) in the central part of the Apuseni Mountains (Bleahu et al. 1976; Vălenaş et al. 1977; Papiu 2002; Onac et al. 2010) (Fig. 1). The cave is accessible by car from Huedin via Transursoaia road (Beliş-Poiana Horea-Ursoaia Pass; ~70 km) or along the Arieş Valley, on the route Albac-Horea-Mătişeşti-Albac Valley (~20 km). Hikers can follow the trail that links Scărişoara Ice Cave to the Albac Valley (12 km) passing through some scenic hamlets such as Gheţar, Ocoale, and Sforţea.

The cave develops in lower Triassic (Anisian) dolomitic limestones, and its passages form an anastomotic maze that spreads on two main karstification levels. The upper one is dry (fossil) and is generally composed of rather rectilinear galleries of various heights (0.5–15 m) and large rooms, which are abundantly decorated by a variety of speleothems (Sântămărian 1992). The water looses in the riverbed of the Albac Valley feed an underground stream that flows along the lower level of the cave. A dye tracing experiment conducted in 1982 by the "Emil Racoviţă" Speleological Club concluded that the cave waters surface in the Mătişesti Spring located 4 km downstream from the cave. The underground course has some minor tributaries originating in the dolines and ponors of the Mununa Plateau, fact established through dye tracing. However, considering the volume of water discharged from the cave through the main spring, the amount of recharge supplied by these tributaries is insignificant.

History of Exploration

The cave was discovered in 1963 by Paşca Ispas, a forester that contacted the "Emil Racoviţă" Institute of Speleology in Cluj. In the same year, I. Viehmann and G. Racoviţă explored and surveyed the first 212 m of galleries, pointing out the fact that they heard the sound of flowing water somewhere at depth. It was only in March of 1981 that the "Emil Racoviţă" Speleological Club (CSER) in Cluj-Napoca began the systematic exploration of the Dârninii Cave. A team composed of D. Soriţău, M. Someşan, and D. Selişcan visited the cave and discovered a small hole in the cave floor through which they felt a very strong air current. Over the summer, CSER members (M. Crăciun, D. Soriţău,

F. Papiu (✉)
"Omega"/CSER Speleological Club, N. Iorga 1A, Cluj-Napoca, Romania
e-mail: papiufelix@yahoo.com

B. P. Onac
Karst Research Group, School of Geosciences, University of South Florida, 4202 E. Fowler Ave., NES 107, Tampa, FL 33620, USA
e-mail: bonac@usf.edu

B. P. Onac
"Emil Racoviţă" Institute of Speleology, Romanian Academy, Clinicilor 5, 400006 Cluj-Napoca, Romania

G. M. L. Ponta and B. P. Onac (eds.), *Cave and Karst Systems of Romania*,
Cave and Karst Systems of the World, https://doi.org/10.1007/978-3-319-90747-5_33

Fig. 1 Location of Dârninii Cave

V. Lungu, D. Purcaru, M. Almăşan, R. Silberg, Ş. Şchiau, I Ghindriş, and E. Rujdea) returned to the cave and started a 6 day hard work of enlarging and removing the filling from the blowing hole. Each team was assigned a 10–12 h digging shift. By the end of the 6th day, M. Crăciun, M. Almăşan, and D. Purcaru succeeded to pass beyond the narrow passage and discovered the Forest Room (Sala Pădurii) and continued the exploration all the way to the Great Abyss (Marea Prăpastie) (Fig. 2). Two 4 day bivouacs gathered 11 cavers (A. Caba, I. Ghindriş, M. Găvruş, D. Ridici, S. Dumitraşcu, A. Moldovan, M. Wittenberg, I. Viehmann, M. Almăşan, C. Popa, and I. Vlad) who discovered, explored, and surveyed the Mammoths Gallery (Galeria Mamuţilor).

In February of 1982, the Labyrinth (Galeria Labirintul), the Active (Galeria Activă) galleries, and a series of other small side passages, pits, and chimneys were discovered during another bivouac organized by D. Soriţău who led a team of 13 cavers. After this last very successful underground camp, for protection reasons, CSER decided to suspend any further bivouacs in Dârninii Cave.

The joined effort of 52 cavers from various speleoclubs in Romania: CSER Cluj, "Avenul" Braşov, "Silex" Braşov, C.S.A. Cluj, "Emil Racoviţă" Institute of Speleology (both Cluj-Napoca and Bucharest), and "*Ursus spelaeus*" Baraolt, led to new discoveries and surveys in August 1982. In September of the same year, A. Moldovan, M. Almăşan, D. Soriţău, and M. Wittemberg explored a series of pits from the Mammoths Gallery. In one of them, they intercepted at −52 m the cave stream but were able to follow it for only 20 m (Moldovan 1984). Nevertheless, during this cave trip, the team performed a dye study with fluorescein to trace the underground path of the river, whose water appeared in the above-mentioned spring after 23 h. The explorations slowed down somehow between 1983 and 1988, but F. Papiu along with M. Crăciun and M. Wittemberg continued to add small passages, bringing the total length of the cave to 5645 m, with a vertical relief of −112 m. To better protect the cave, in 1985 a solid metallic gate was put in place (D. Selişcan) at ~50 m south of the Big Room (Sala Mare) in a narrower section of the cave passage (Fig. 2).

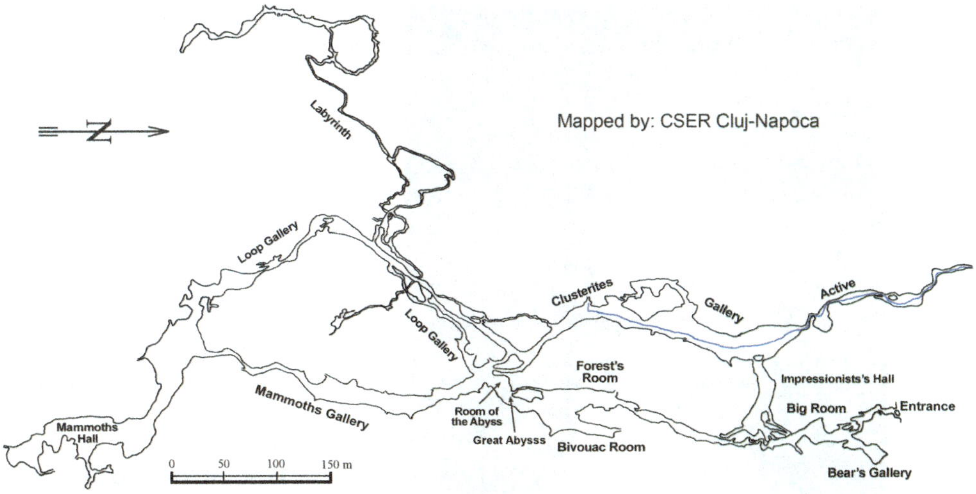

Fig. 2 Map of Dârninii Cave

Cave Description

After passing the cave entrance (2 × 3 m), the main gallery is relatively horizontal for ~70 m, with the floor covered with limestone breakdowns, which in turn are coated with various speleothems. At its end, a 6 m drop marks the entrance in the Big Room (Sala Mare; 35 × 15 × 10 m), which continues to the northeast with the Bear's Gallery (Galeria Ursului) and to the south, through a narrow gallery (where the gate is emplaced) connects with the Forest Room (Sala Pădurii; 10 × 5 × 5 m), east of which is the Bivouac Room (Sala Bivuacului) (Fig. 2). From here, a relatively large (15–30 m in height), heavily decorated (stalactites, stalagmites, pools, etc.) gallery continues toward southwest, suddenly ending at the Great Abyss with a 25 m vertical drop. At the bottom of this descent, we find ourselves in the Room of the Abyss (Sala Prăpastiei), a big chamber (100 × 30 × 50 m) from where three galleries begin: the Clusterites Gallery (Galeria Clusteritelor), the Mammoths Gallery, and the Loop Gallery (Galeria Buclei).

In plan view, the Clusterites Gallery (~600 m) resembles the letter S (Fig. 2) and is trending toward north. A small stream flows along this gallery and disappears in an impenetrable fracture. South from the Great Abyss, the Mammoths Gallery connects with the Mammoths Hall (100 × 50 × 50 m), which is richly decorated by a variety of speleothems (e.g., domes, rimstone pools) and hosts large deposits of moonmilk (Papiu 2002). Northwest from this big room, the cave continues with the Loop Gallery that returns back at the Great Abyss. West from the Loop Gallery is the entrance in another long meandering gallery called the Labyrinth, along which opens several side passages, some of them connecting with the stream in the lowest part of the cave (Figs. 3 and 4).

Cave Climate

The climatological study performed by Caba and Găvruş (1984) proved that the cave presents unidirectional ventilation; the air current moves from the lowest entrance toward some inaccessible (likely fracture) upper one during the winter and reverses on summer. In this last period, the main air current is doubled in the entrance zone by a convection cell. Regardless of the season, the thermal gradient corresponds to a logarithmic function. The relative humidity (RH) measured in different parts of the cave (between 89 and 99%) indicates an irregular spatial distribution. On the base of the ventilation regime and considering the distribution of the thermic and RH values, two meroclimatic zones were distinguished: (1) *perturbation*, extending ~400 m from the cave entrance, and (2) *stability*, which was documented in the rest of the cave.

Speleothems and Minerals

Dârninii Cave is known for its rich and diverse calcite and aragonite speleothems, especially large stalagmites, columns, domes, and moonmilk (Sântămărian 1992). Coralloids (also known as clusterites in Romanian) are well developed in the Clusterites Gallery, where they abundantly cover the cave walls. A unique aspect about some of these coralloids is that they are in fact multi-aggregates composed of a variety of minerals that were precipitated in a very particular sequence (Onac and Kearns 2000). Directly on the limestone bedrock or calcite crusts covering cave walls are first formed calcite knobs, which are then covered by a submillimeter layer of high Mg-calcite from which transparent and delicate needle-like crystals of aragonite grow outward. Toward their

Fig. 3 A massive column in the Mammoths Hall (photograph courtesy of C. Ciubotărescu)

apexes, these crystals are covered by patches of shiny, white powdery material (huntite), which in turn, during summertime, is either replaced or hidden under some crumbly, dull-white globular deposits composed of hydromagnesite. This sequence of minerals precipitation was interpreted to reflect specific cave climate conditions (temperature, RH, and partial pressure of CO_2), which control the stability of various Ca and Mg carbonates.

A rare and yet not fully understood speleothem reported from Dârninii Cave is folia, which forms on the cave walls and resembles an inverted rimestone. The samples collected from the Great Abyss and in the Mammoth Gallery were X-rayed and showed to be made of oxides and hydroxides of iron and manganese and a substantial content (100–400 ppm) of rare earth elements (Onac and Kearns 2000).

Geochronology

A stalagmite was collected near the Impressionists's Hall (Sala Impresioniştilor) and dated by means U-series disequilibrium method. The lower part of the stalagmite grew during the last interglacial period (between 123.2 ± 0.8 and 118.8 ± 0.8 ka) and was followed by a long, 109 kyr hiatus before the rest of it formed during a depositional event that lasted only ~1400 years (11.8 ± 0.1 to 10.4 ± 0.2 ka) and took place immediately after the last Ice Age (Tămaş 2003).

Microbiology

Microbiological and enzymological investigations were conducted on six cave sediments comprising moonmilk and clays from floors and walls collected between the cave

Fig. 4 Image from the Forest Room (photograph courtesy of C. Ciubotărescu)

entrance and Impressionists's Hall. Qualitatively, the microbial community in Dârninii Cave includes oligotrophic bacteria, aerobic mesophilic heterotrophs, denitrifiers, ammonifiers, and nitrifiers (Carpa and Butiuc-Keul 2009). The presence of the last three types in some of the samples may indicate the existence of a food chain (likely guano). The enzymological analysis documented dehydrogenase, catalase, and phosphatase activities. The first one denotes the presence of living microorganisms at the sampling time, whereas the last two signal their persistence in time. Altogether, the results point out the microbiological potential of the cave environment and that the findings are unrelated with the distance from the cave entrance (Carpa and Butiuc-Keul 2009).

Paleontology/Cave Fauna

The cave hosts life traces of the cave bear, *U. spelaeus* (claw marks, resting sites). With respect to the invertebrate cave fauna, the following groups/species were documented: Archnida: *Ixodes vespertilionis*; Coleoptera Leptodirinae: *Pholeuon* (s. str.) *knirschi albacensis* (troglobite endemic); Diptera; Lepidoptera: *Triphosa dubitata*, *Scoliopteryx libatrix*; Trichoptera (Racoviță 2004–2005). The cave also hosts three species of bats (Chiroptera): *Myotis emarginatus, Myotis brandtii,* and *Plecotus auritus*.

Cave Conservancy

Given the exceptional value and uniqueness of its resources, Dârninii is an A class protected cave (scientific reserve), meaning access is completely prohibited, except for exploration/survey and scientific purposes. This kind of activities, however, requires special permissions from the Romanian Speleological Heritage Commission and the Apuseni Natural Park Administration.

References

Bleahu M, Decu V, Negrea Ş, Pleşa C, Povară I, Viehmann I (1976) Peşteri din România. Ed Ştiinţifică şi Enciclopedică, Bucureşti, p 415

Caba E, Găvruş M (1984) Recherches topoclimatiques dans le grotte de Peretele Dârninii (Valée d'Albac, Massif du Bihor). Theor Appl Karstol 1:131–138

Carpa R, Butiuc-Keul A (2009) Microbial activity in the subterranean environment of Dârninii Cave, Bihor Mountains. ELBA Bioflux 1(1):13–22

Moldovan A (1984) Memoriile unei echipe de explorare (Dârninii-Izbândiş-Stanu Foncii). Peştera 1:113–117

Onac BP, Kearns J (2000) Crystallography and mineralogy of multi-aggregate speleothems from Peretele Dârninii Cave (Bihor Mts., Romania). In: Onac BP, Tămaş T (eds) Proc of the joint meeting "Friends of Karst" and "Theoretical and Applied Karstology". Ed Presa Univ Clujeană, Cluj-Napoca, pp 87–91

Onac BP, Racoviță G, Brad T (2010) Atlasul peşterilor din Munții Apuseni. 1. Bazinul Arieşului Mare. The University of South Florida Karst Studies Series 2, p 90

Papiu F (2002) Peştera din Peretele Dârninii. Ecocarst 3:26–27

Racoviță G (2004–2005) Révision systématique des Leptodirinae souterrains des Monts Apuseni. IV. Le sous-genre *Pholeuon* (s. str.) du bassin de l'Arieş (Monts du Bihor). Trav Inst Spéol "Émile Racovitza" XLIII–XLIV:165–191

Sântămărian A (1992) Mineralogia speleotemelor din Munții Bihorului. Unpubl B.Sc. Dissertation, Babeş-Bolyai University, Cluj-Napoca

Tămaş T (2003) Mineralogy and geochemistry of speleothems from caves in Bihor Mountains. Unpubl Ph.D. Thesis, Babeş-Bolyai University, Cluj-Napoca, p 168

Vălenaş L, Bleahu M, Brijan P, Halasi G (1977) Inventarul speologic al Munților Bihor. Nymphaea 5:209–335

Bihor Mountains: Zgurăsti–Poarta Lui Ionele Cave System

Christian Ciubotărescu

Abstract

Zgurăşti–Poarta lui Ionele (Gate of Ionele) cave system totals 5809 m (5424 + 385 m) of passages and is located in the Bihor Mountains in NW Romania. It includes the Zgurăşti Cave, renown for a series of lakes perched at ~118 m above Ordâncuşa River and Poarta lui Ionele Cave, a part of which is open for tourists. The cave system has been formed by an underground river that was once recharged by waters of the Mununa limestone plateau under which it is formed. The final part of the Zgurăşti Cave is clogged by rock debris and clay, generating the largest underground lake in Romania. Presently, the water appears in Poarta lui Ionele Cave and flows into Ordâncuşa River. Poarta lui Ionele Cave houses a bat nursery of *Miniopterus schreibersii.*

Keywords

Karst system • Shaft • Perched lake • Radionuclides

Introduction

Zgurăşti–Poarta lui Ionele cave system is located in the Bihor Massif (Fig. 1), part of the Apuseni Mountains, on the right side of Ordâncuşa Valley (Fig. 2), a tributary (via Gârda Seacă River) of Arieşul Mare River. Both caves are easily accessible along a modern road that departs from the center of Gârda de Sus village and follows upstream Gârda Seacă and then Ordâncuşa valleys.

The geological data provided by Bleahu et al. (1976) indicates that the cave system develops in middle Triassic limestone (Ladinian), but considering the revision of this unit by Bucur and Onac (2000), it is likely that the carbonate sequence belongs to the Upper Jurassic (Kimmeridgian).

At present, the cave system discharges the diffuse losses of water from the internal drainage area of the Munună Karst Plateau situated above them (Orăşeanu 2010). Following a dye (fluorescein) experiment conducted in Zgurăşti Cave, the tracer appeared in Poarta lui Ionele after 12 h, with maximum concentrations at 14 or 42 h, respectively. The gap between the two maxims indicates the existence of two different drainage pathways, one probably involving the existence of some lakes (Damm 1997; Onac et al. 2010; Orăşeanu 2016).

Brief History of Exploration

Poarta lui Ionele Cave (3403/83; Goran 1982) was mentioned in the nineteenth century by Vass (1857) and Bielz (1884). It was first studied between 1921 and 1923 by Jeannel and Racoviţă (1929). Viehmann and Racoviţă produced the first survey of the cave in 1964. Until 1988, only the lower part of the cave was known and accessible to visitors. In 1988, members of "Polaris" Blaj Caving Club and cavers from Czechoslovakia discovered an upper level, where a bat colony roosts. At that time, wood ladders were built for easing the access over the large flowstone in the lower section of the cave. Also, the connection with the upper galleries was enlarged allowing tourists to visit a part of the upper level. Since 1992, Sfinx Caving Club from Gârda de Sus explored new passages, resurveyed the cave, and rebuilt and maintained the touristic facilities. In 2003, in partnership with the Gârda de Sus Town Hall and sponsored by the Eximtur Travel Agency and with approvals from "Emil Racoviţă" Institute of Speleology (ERIS) and the Commission for Natural Monuments of the Romanian Academy, Sfinx Caving Club installed for the first time electricity in the cave.

Due to the increased number of visitors between 1988 and 1992, the in-cave touristic facilities (ladder, platform, etc.) had degraded and the bat colonies disappeared. Following the advices received from bat/cave specialists

C. Ciubotărescu (✉)
Speleological Association for Environmental Protection and Karst "Sfinx" – Gârda, Gârda de Sus 129A, 517310 jud. Alba, Romania
e-mail: cristigarda@yahoo.com

G. M. L. Ponta and B. P. Onac (eds.), *Cave and Karst Systems of Romania*,
Cave and Karst Systems of the World, https://doi.org/10.1007/978-3-319-90747-5_34

Fig. 1 Location of Zgurăști-Poarta lui Ionele cave system

Fig. 2 Limestone cliffs in the Ordâncușa Gorge (photograph by C Ciubotărescu)

(Drs Coroiu and Viehmann), the members of Sfinx Caving Club performed an ecological reconstruction that led to the repopulation of the cave in spring 2008 with a bat nursery of *Miniopterus schreibersii* (Fig. 3) Basically, what cavers did was to obstruct the connection between the lower and the upper cave passages, thus bringing the cave to its original condition.

Even though the ecological reconstruction was a success, the Administration of Apuseni Natural Park disagreed with the procedure of closing the man-made enlarged passages and the dispute ended in court. The lawsuit extended over two years, time in which the cave was studied (climatology, bat observations), and the results were presented at two scientific meetings (Ciubotărescu 2008; Borda et al. 2009). Finally, the Park Administration lost the legal contest and in the meantime the bat nursery grew significantly. Since 2011, Poarta lui Ionele became a modern show cave administered by the Gârda de Sus Town Hall, who rebuilt the touristic path leading to the cave as well as the trail and stairs (solid steel structures) within the cave in partnership with the Romanian Ministry for Development and Tourism (see *Show caves of Romania* in this volume). All touristic developments strictly followed the recommendations contained in the above-mentioned studies as well as the cave management plan realized by ERIS. This way, the cave environment and the bat roosts are now effectively protected.

Zgurăşti Cave (3403/92; Goran 1982) was first explored and surveyed by Jeannel and Racoviţă in 1921 and published their results in 1929 in the well-known collection *Énumération des grottes visitées*. Between 1976–1978 and 1987–1989, cavers from various speleological clubs (e.g., "Z" and "Cristal" Oradea, "Polaris" Blaj, "Osiris" from Czechoslovakia, and "Focul Viu" Bucharest) continued to explore and resurvey the cave to a temporary sump ("Polaris Sump") (Musil 1987). Another major campaign of cave explorations and surveys took place between 1993 and 1995 when cavers of the above-mentioned clubs fruitfully collaborated with members of the "Politehnica" Cluj-Napoca, "Sfinx" Gârda de Sus, as well as cavers from France and Great Britain (Buşu 2009; Ciubotărescu et al. 1997; Ciubotărescu 2009) (Fig. 4). The cave system is still being explored.

Fig. 3 *Miniopterus schreibersii* colony in the Poarta lui Ionele Cave (photograph by C Ciubotărescu)

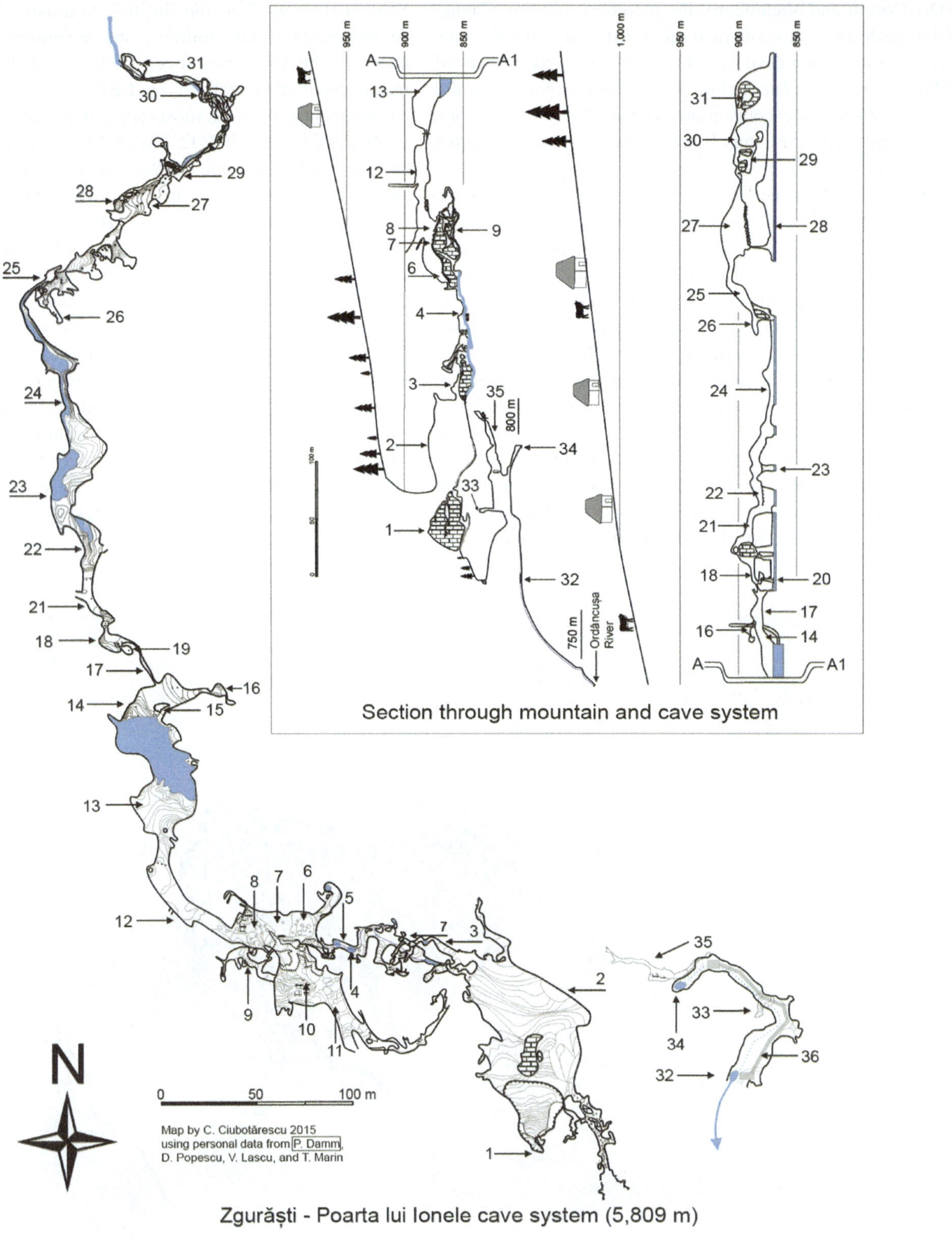

Fig. 4 Map and profile of the Zgurăști-Poarta lui Ionele cave system (data from C. Ciubptărescu and P. Damm). The numbers correspond to locations described in the text as follow: *1* Main Entrance, *2* Entrance Hall, *3* X Gallery, *4* Polaris Sump, *5* Czech Gallery, *6* Hall of the Fallen Boulders, *7* Dining Room, *8* Hall of Metal Horror, *9* The Muddy Inferno, *10* Osiris Room, *11* J.P. Gallery, *12* Nostrils Gallery, *13* Hall of the Lake, *14* Hall of Hope, *15* Rain Chimney, *16* C.T.K. System, *17* R.P. Gallery, *18* Hall of the Water Lilies, *19* R.E.C.K. Room, *20* Unknown Lake, *21* Roru Gallery, *22* Hall of the Mud Baths, *23* Lake Velența, *24* Long Lake Room, *25* Waiting Room, *26* V.A.P. System, *27* Hall of the Collapsed Temple, *28* Lake Styx, *29* Thierry-Simina System, *30* The Tyrolienne, *31* P.V.P. System, *32* Poarta lui Ionele Cave entrance, *33* Bats Gallery, *34* Temporary lake, *35* Gallery of Nursery Colony, *36* Touristic path

Description of Zgurăşti–Poarta lui Ionele Cave System

The access to Zgurăşti Cave is a vertical shaft with two entrances, the main one (labeled 1 in Fig. 4; hereafter, for simplicity we only use 2, 3, etc.) is 40 m in diameter and 45 m deep (Fig. 5a). The access to the cave is a short walk downhill along the rock scree, which ends in the "Entrance Hall" (Sala Intrării, 2), where a temporary lake forms (Fig. 5b). Continuing along the X Gallery (3) and descending a 14 m drop takes us to a small stream that ends in a temporary sump ("Polaris" Sump; 4). The cave continues beyond the sump with the so-called Czech Gallery (Galeria Cehilor; 5) at the end of which and after an 11 m climb, we enter the "Hall of Fallen Boulders" (Sala Prăbuşirilor; 6). This is a large square chamber (25 × 25 m) with abundant breakdown blocks on its floor, all covered with a thick clay layer (Fig. 6). Some of the boulders are a

(a)

(b)

Fig. 5 **a** Entrance shaft in the Zgurăşti Cave. **b** Temporary Lake in the Zgurăşti's Entrance Hall (photograph by C Ciubotărescu)

Fig. 6 Image from the Hall of Fallen Boulders (photograph by C Ciubotărescu)

few cubic meters in size. The cave continues toward southeast with the "Osiris Room" (10) and "J.P." Gallery (11), respectively, both well decorated with calcite speleothems (Ciubotărescu 2003).

After crossing the "Hall of Fallen Boulders" and the "Dining Room" ("Sala de Mese"; 7), the "Hall of Metal Horror" (Sala Ororii Metalice; 8) is reached; this is another room that hosts boulders, large broken stalagmites, domes, and rimstone pools, which are chaotically disposed. In this area, the collapsed floor over the galleries below formed a labyrinth named "The Muddy Inferno" (Infernul Nămolos; 9). Large rounded pebbles (up to 15 cm in diameter) are cemented in calcite, suggesting that once upon a time, a mighty river flew through this cave.

From the "Hall of Metal Horror," the main cave passage heads northwest through the "Nostrils Gallery" (Galeria Nărilor; 12), which displays a variety of speleothems; solution pockets/cupolas are visible in the ceiling, and two chimneys up to 23 m in height were also documented. These features attest the evolution of the cave under both phreatic and vadose conditions. "Galeria Nărilor" (ends in the "Hall of the Lake" (Sala Lacului; 13), a large passage having the floor covered completely with mud in its first part, whereas the rest of it accommodates the "Big Lake" (Lacul Mare) of which depth is $\sim$12 m (Lascu 2009) under normal conditions (Fig. 7). Also, underwater passages with small stalactites were discovered on the north-eastern side of the lake at about 6 m depth, clearly indicating that the lake was formed after the deposition of these calcite speleothems. A large stalactite (4 m long and 0.5 m in diameter) was found submerged in the north-western side of the lake (Lascu 2009).

The analysis of a 41 cm long core of a very fine clayey material recovered from the bottom of the Big Lake ($\sim$250 m from the entrance), revealed high concentrations of ^{210}Pb, ^{134}Cs, and ^{137}Cs (Pourchet et al. 1996). Of these radioisotopes, only the first one is natural, the other two are artificial derived from nuclear tests or accidents (in this case, the Chernobyl disaster from 1985). Since there are no streams sinking on the karst plateau above the cave, the presence of the radionuclides in the clayey sediment of Zgurăşti confirms that they were rapidly transported underground by the percolation waters. At the same time, they allowed to estimate the sedimentation rate, a process that apparently started not earlier than 250 years ago.

On the northern side of the lake, after a short climb on a muddy slope, the "Hall of Hope" (Sala Speranţei; 14) is reached; this is well-decorated chamber with stalagmites up to 2.5 m in height. From here, the cave continues upstream (north) through "R.P." Gallery (17), a canyon-type passage that leads to the "Hall of the Water Lilies" (Sala Nuferilor; 18), where lily pads (calcite encrusted stalagmites in standing pools of water) were discovered.

Fig. 7 The Big Lake (photograph by C Ciubotărescu)

From the "Hall of Water Lilies," the cave continues through Roru Gallery" (21), accessible after climbing a 6 m wall and crawling a short narrow passage, which descend into the "Hall of the Mud Baths" (Sala Băilor de Nămol; 22). After crossing the muddy area and traversing the "Lake Velenţa" (23), we enter the Long Lake Room (Sala Lacului Lung; 24), a narrow canyon-type passage, which ends at the bottom of a 20 m high wall. Next major chamber is the "Waiting Room" (Sala de Aşteptare; 25), located at the top of the wall and which gives access to the "Hall of the Collapsed Temple" (Sala Templului Prăbuşit; 27). As suggested by its name, cubic meter-size boulders and large broken stalagmites (up to 1.5 m diameter) were found in this room, thus resembling the "Hall of Metal Horror". Thirty meters upstream, "Lake Styx" (28) forms during wet periods in a narrow canyon. During the dry season, the lake turns into a small stream, which at low water levels can be followed upstream for another 155 m. The ceiling of the cave drops from 30 to 5 m after the first 105 m and continues to lowers until the gallery ends in an impenetrable muddy water sump.

Poarta lui Ionele is a show cave with an impressive entrance (32), 22 m high and 12 m wide. The main spring where the dye injected in the "Polaris" Sump of Zgurăşti Cave resurfaced is located at the base of the wall close to the cave entrance (Damm et al. 1999). Tourists can visit the cave on the gray marked path (36), which ends above a temporary lake (34). During flooding events (snow melting, heavy rains), the water from this lake overflow creating a spectacular waterfall over the massive flowstone (Fig. 8).

Cave Climatology

Cave temperature values were continuously recorded in Poarta lui Ionele Cave using five Gemini data loggers. The results indicate that the large gallery that forms the lower level is under a permanent bidirectional ventilation effect, whereas the upper gallery is characterized by an obvious stability meroclimate (Borda and Racoviţă 2011). According to the dynamics of air exchanges, the outside influences are strong enough, so that to the cave temperature varies during a year period with an evident periodicity, not only seasonally, but also daily. In spite of the lack of a similar periodicity, the daily amplitudes are quite important, exceeding 1 °C in the innermost part of the lower gallery. These facts suggest that, besides the convection phenomena, the outside influences can also propagate in the cave by conduction.

Cave Fauna

Zgurăşti–Poarta lui Ionele cave system fauna include gastropods: *Clausilia dubia*; oligochetes; pseudoscorpinoides: *Neobisium leruthi*; acariens: *Poecilophysis spelaea, Rhagidia spelaea*; coleopteres leptodirines: *Drimeotus* sp., *Pholeuon knirschi christiani* (Racoviţă 2004–2005); coleopteres trechine: *Duvalius hickeri*; lepidopteres: *Triphosa dubiata*; trichopteres: *Stenophylax permistus*, *Stenophylax vibex speluncarum*; miriapodes; copepodes ciclopoides: *Diacylops*

Fig. 8 Waterfall in Poarta lui Ionele Cave (photograph by C Ciubotărescu)

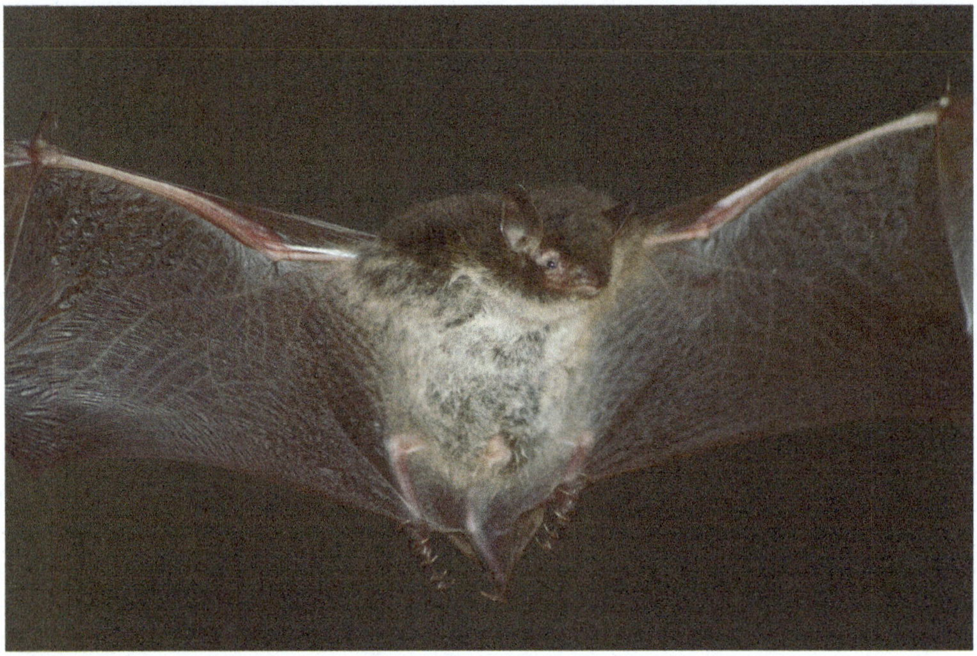

Fig. 9 Close-up of a flying *Myotis daubentonii* (photograph by C Ciubotărescu)

stygius deminutus; araneides: *Neticus spelaeus*; dipteres; chilopodes: *Cryptops hortensis*) chiropteres: *Myotis myotis, Myotis oxygnathus, Pipistrellus nathusii, Barbastella barbastellus, Plecotus* (spp), *Rhinolophus ferrumequinum, Rhinolophus hipposideros, Rhinolophus euryale,* and *M. schreibersii* (Borda et al. 2009; Onac et al. 2010).

In Poarta lui Ionele Cave, yearly since 2008, a nursery colony of *Miniopterus screibersii* forms, and during the winter, *R. ferrumequinum, R. hipposideros, R. euryale, M. myotis, M. oxignatus, M. daubentonii* (Fig. 9), and *M. schreibersii* use the upper level of the cave for hibernation (Borda et al. 2009).

Cave Conservancy

Zgurăşti and Poarta lui Ionele caves were declared protected areas (Nr 20/27.09.1995; Nr 27/11.07.1999) by the Alba County's Council and later by the Romanian Parliament (Law 5/6.03.2000). The entrance section of the first one is class B, whereas its inner parts are A. The lower part of Poarta lui Ionele Cave is developed for tourism. The caves were in the custody of Sfinx Caving Club from 1995 until 2003 when the Apuseni Nature Park was established.

References

Bielz E-A (1884) Beitrag zur Höhlenkunde Siebenbürgens als Vorarbeit einer Anthrologie dises Landes. Jarb Siebenbrg Karpathanvereins, Hermannstadt IV:1–66

Bleahu MD, Decu V, Negrea Ş, Pleşa C, Povară I, Viehmann I (1976) Peşteri din Romania. Ed Ştiinţifică şi Enciclopedică, Bucureşti, p 415

Borda D, Racoviţă G (2011) Données thermométriques relatives à l'atmosphère de la Grotte Poarta lui Ionel (Monts du Bihor, Roumanie). Trav Inst Spéol "Émile Racovitza" L:77–86

Borda D, Racoviţă G, Năstase-Bucur R, Ciubotărescu C (2009) Ecological reconstruction of a bat cave roost in Western Carpathians. Sustainability of the karst environment. Plitvice Lakes, Croatia, pp 17–24

Bucur II, Onac BP (2000) New data concerning the age of the Mesozoic limestone from Scărişoara (Bihor Mountains). Studia UBB Geologia XLV:13–20

Buşu L (2009) Avenul de sub Zgurăşti. Scurte episoade ale lungilor ture în Zgurăşti. In: Dumitru R (ed) Focul Viu 40 de ani de speologie. Ed RAO, Bucureşti, pp 86–95

Ciubotărescu C (2003) Peşera Zgurăşti. Ecocarst 4:27–29

Ciubotărescu CA (2008) Vulnerabilitatea liliecilor din peştera Poarta lui Ionele. In: Workshop interdisciplinar "Vulnerabilitatea elementelor de patrimoniu carstic în ariile naturale protejate", Sinaia, 23–26 Oct 2008, p 57

Ciubotărescu C (2009) Zgurăştii pagini de istorie. In: Dumitru R (ed) Focul Viu 40 de ani de speologie. RAO, Bucureşti, pp 83–84

Ciubotărescu C, Damm P, Popescu DP (1997) Peştera Zgurăşti. Bul Speol Gârda 1:33–36

Damm P (1997) Hidrogeologia zonei Gârda—Ocoale. Bul Speol Gârda 1:15–16

Damm P, Lascu V, Ciubotărescu C (1999) Condiţionarea tectonică şi hidrogeologică a sistemelor hidrocarstice din Platoul Ocoale (Munţii Bihor). Speomond 4:14–18

Goran C (1982) Catalogul sistematic al peşterilor din România. Ed CNEFS, Bucureşti, p 496

Jeannel R, Racoviţă EG (1929) Énumération des grottes visitées, 1918–1927 (7e série). Arch Zool Exp Gén 68(2):293–608

Lascu C (2009) "Lacul de sub Munună". In: Dumitru R (ed) Focul Viu 40 de ani de speologie. Ed RAO, Bucureşti, p 85

Musil J (1987) Nouvel Zgurăşti. Spelunca 28:14

Onac BP, Racoviţă G, Brad T (eds) (2010) Atlasul peşterilor din Munţii Apuseni. Munţii Bihor 1. Bazinul Arieşului Mare. Univ South Florida Karst Stud Ser 2:1–90

Orăşeanu I (2010) Karst hydrogeology of Apuseni Mountains. In: Orăşeanu I, Iurkiewicz A (eds) Karst hydrogeology of Romania. Ed Belvedere, Oradea, pp 181–324

Orăşeanu I (2016) Hidrogeologia carstului din Munţii Apuseni. Ed Belvedere, Oradea, pp 119–121

Pourchet M, Melieres MA, Silvestru E, Rajka G, Candaudap F, Carbonnel JP (1996) Radionuclides in a cave lake sediment core from Gheţarul de sub Zgurăşti (Romania). Theor Appl Karstol 9:23–28

Racoviţă G (2004–2005) Révision systématique des Leptodirinae souterrains des Monts Apuseni, IV. Le sous-genre *Pholeuon* (s. str.) du bassin de l'Arieş (Monts du Bihor). Trav Inst Spéol "Émile Racovitza" XLIII–XLIV:165–191

Vass I (1857) Eine Wanderung nach der Eishöhle bei Skerisoara. Verh u Mitt des Sieb für Naturwiss zu Hermannstadt VIII:162–170

Bihor Mountains: Coiba Mică–Coiba Mare–Tăuz Cave System

Christian Ciubotărescu and Bogdan P. Onac

Abstract

One of the main underground drainages in the Gârda area (Bihor Mountains) is Coiba Mică–Coiba Mare–Tăuz cave system, which holds several unique records in the Romanian karst: (1) Coiba Mare has the widest cave entrance (74 m) and the deepest dived sump (Lake of Death, −92.5 m), (2) it includes the first underwater cave junction made between Coiba Mare and Coiba Mică, and (3) Tăuz is regarded as one of the most beautiful karst springs and held the depth record for many years.

Keywords

Karst system • Underwater passage • Cave diving
Spring • Ponor • Vermiculations

Geographic, Geologic, and Hydrologic Settings

The Coiba Mică–Coiba Mare–Tăuz cave system is located in the southeastern part of the Bihor Mountains (Fig. 1) and is accessible by car along the Arieş Valley on the route Turda-Câmpeni-Gârda de Sus or from Ştei, over the Vârtop Pass and down to Gârda de Sus. From the center of this village, a secondary road follows upstream the Gârda Seacă River for 12 km up to a small settlement called Casa de Piatră. About 2 km before reaching the village, a short walk on a pathway on the right (west) side of the Gârda Seacă River takes visitors to the Tăuz Spring (10 m wide and 8 m deep) located at the foot of a 100-m high cliff. It is arguably the most beautiful karst spring from Romania, which reveals the turquoise waters resurfacing from Coiba Mică and Coiba Mare caves.

In the northern part of this village, a short hike (50 m) on a tourist path that goes down below the road leads to the widest cave entrance in Romania, which is the main access point in the Coiba Mare Cave. Just 600 m north of Casa de Piatră, upstream on the road parallel to the Gârdişoara Valley lies the entrance in Coiba Mică Cave. The access in this ponor cave (20 m wide and 2 m high) is now completely blocked with logs, mud, and debris accumulated during a major flush flood. In 2000, a lake formed outside the cave entrance (Fig. 2).

The cave system develops in well-bedded biogenic reefal and oncoidic limestones of Upper Jurassic (Kimmeridgian–Lower Tithonic) to Lower Cretaceous (Barremian–Lower Aptian) age (Bleahu et al. 1980; Turi et al. 2011). From a hydrologic point of view, the water of Gârdişoara River sinks in Coiba Mică (1027 m altitude) and appears in the Confluence Room (Sala Confluenţei) of Coiba Mare (Fig. 3), from where it flows along the cave's main passage into the Lake of Death (Lacul Morţii, also known as Final Sump, 970.5 m). From here, the water follows a largely unknown ~2.6 km long underground path over a vertical range of 120.5 m to resurface in the Tăuz Spring (850 m) (Fig. 4). Morphologic and hydrogeologic consideration regarding the underground karst conduits in the Gârda Valley area were presented by Vălenaş et al. (1982–1983), whereas the subterranean drainage was demonstrated by a dye tracer test conducted in 1985 (Orăşeanu 1996, 2010).

History of Exploration

Coiba Mică Cave The first documented visit in this cave belongs to R. Jeannel and A. Winkler (Jeannel and Racoviţă 1929). The cave was fully explored by L. Vălenaş, M. Oncu,

C. Ciubotărescu (✉)
Speleological Association for Environmental Protection and Karst "Sfinx"—Gârda, Gârda de Sus 129A, Jud., 517310 Alba, Romania
e-mail: cristigarda@yahoo.com

B. P. Onac
Karst Research Group, School of Geosciences, University of South Florida, 4202 E. Fowler Ave., NES 107, Tampa, FL 33620, USA
e-mail: bonac@usf.edu

B. P. Onac
"Emil Racoviţă" Institute of Speleology, Romanian Academy, Clinicilor 5, 400006 Cluj-Napoca, Romania

G. M. L. Ponta and B. P. Onac (eds.), *Cave and Karst Systems of Romania*,
Cave and Karst Systems of the World, https://doi.org/10.1007/978-3-319-90747-5_35

Fig. 1 Location of the Coiba Mică–Coiba Mare–Tăuz cave system in the karst of Bihor Mountains

Fig. 2 Lake formed in 2000 outside the entrance in the Coiba Mică Cave, which was clogged by sediments (photograph by C. Ciubotărescu)

and I. Bele ("Emil Racoviță" Caving Club from Cluj-Napoca) in February 1973 and mapped a month later (Vălenaş 1976, 1978).

Coiba Mare Cave R. Jeannel and A. Winkler explored and described the entrance part of this cave in 1922 (Jeannel and Racoviță 1929). Then, I. Viehmann, D. Coman, and M. Bleahu surveyed the first 760 m in 1953–1956 producing a map that was published by Bleahu in 1957. Between October 1975 and January 1978, cavers (L. Vălenaş, G. Drîmba, E. Vălenaş, G. Halasi, N. Sasu, D. Pop, H. Mitrofan, P. Brijan, E. Silvestru, O. Cuc, and G. Kajtor) from several caving clubs (Z Oradea, Speodava Ştei, Liliacul Arad) have explored and resurveyed the entire cave including the complex system of fossil galleries developed in the upper level of the cave using topofil, compass, and clinometer (Vălenaş 1976, 1978). The junction between the two caves was made in 1978 by F. Păroiu (Focul Viu Caving Club Bucharest), who dived a 60 m long sump connecting Coiba Mare with Coiba Mică (red dashed lines in Fig. 3). This was for the first time in Romania when two caves were connected via diving (Vălenaş and Păroiu 1978). After performing this junction, the total length of the cave system reached 5680 m with a vertical range of 121 m (Vălenaş et al. 1977; Vălenaş 1978; Orghidan et al. 1984).

Focul Viu Caving Club Bucharest remapped in 1989 the Coiba Mare Cave using measuring tape, mining compass, and clinometer, but never published a map. The total length was only 3700 m despite exploring and surveying a new 200 m long gallery (Aldica 2009). Members of the Sfinx Gârda Speleological Association discovered in the early 1990s new galleries, climbed a few chimneys in Coiba Mare, and entered Coiba Mică, which was once again accessible. Thus, they remapped the cave between 1993 and 2003 using measuring tape, clinometer, and compass with optical viewer. The survey pointed out that the system (without the sump in Coiba Mică which was not mapped) is 4186 m long. After the results of this last survey came out, cavers realized that each of the three surveys produced completely different results. The mystery over the real length of the cave remained (or increased) after a 4th team from Austria and remapped the main stream passage of the Coiba Mare Cave in 2011 using a Leica DISTO™ D3a surveying equipment (measuring range of 100 m with an accuracy of ± 1 mm; tilt: accuracy of the laser beam $\pm 0.3°$) and obtaining 924.4 m (Huber 2011). The team reported a total cave system length of 5042 m, but since they only remapped the stream passages, it remains unclear from where this number comes. They also obtained slightly different elevations from those reported by Vălenaş (1978) and the authors of this chapter.

In 1998, I. Novak and E. Zaruba dove down to −35 m in the Lake of Death, whereas in 2014 S. Paakkarinen and P. Gronqwist from Finland dived in the same lake to a depth of 92.5 m, currently the second deepest sump in Romania. Over the past 3 years, the gallery was blocked by tree trunks and any further exploration had to be halted.

Tăuz Spring (Sump)

The first attempt to dive Tăuz Spring was undertaken by F. Păroiu in 1979. One year later, G. Halasi and L. Vălenaş also dived the sump. In 1982, L. Czako from Hungary passed the first sump and reached −47 m. In 1988, L. Benysek from Czechoslovakia dives under −60 m (Pereț and Drăgan 2016). Between 2001 and 2006, cave divers from Wrolaw (Poland) dove in Tăuz Spring, with W. Bolek reaching −85 m in sump 2. R. Garki drowned in Tăuz during a dive in 2002. In January 2005, W. Szymanowski (Poland) passes the lowest point of the sump (−85 m) and starts on the ascending branch arriving to −30 m (Fig. 4) but had to return due to rebreather limitations (Pereț and Drăgan 2016). Two Finnish divers (S. Paakkarinen and P. Gronqwist) succeeded to cross in 2014 the entire Wlodek Branch of the sump and to explore more than 500 m of new passages including a new sump in which they reached −60 m and continue (Paakkarinen pers. comm.) (Fig. 4).

Description of Coiba Mică–Coiba Mare–Tăuz Cave System

Coiba Mică is a ponor cave situated at the altitude of 1027 m at the base of a 30 m high cliff and has a 20 m wide by 2 m high entrance (Fig. 2). After a descent on a slope covered by logs and debris, the gallery narrows to about 1 m wide. At 23 m from the entrance, a difficult 12 m waterfall ends in a 9 m long lake, from where a high canyon-like gallery (4 × 5 m) turns southeast and becomes wider as we enter in the Big Chamber (Sala Mare), the largest void of the cave (15 × 20 × 15 m). The vegetal remains on the walls indicate that during major flooding, the water level may reach even the ceiling. Moving downstream toward the southeastern end of the Big Chamber, after a dangerous run between unstable logs, follows another large and high room (Chimney Hall/Sala Hornurilor) in which the entrance is marked by a 6.5 m waterfall. After passing the Chimney Hall, the cave continues for another 50 m and ends in a room that is occupied by the terminal sump filled with large logs both in depth and surface (Fig. 3).

Coiba Mare Cave has three entrances, the main one being the largest in Romania (74 m wide and 50 m high; Fig. 5). It is situated at the altitude of 1007 m, at the southern end of a blind valley. The entrance gives access to the Big Chamber

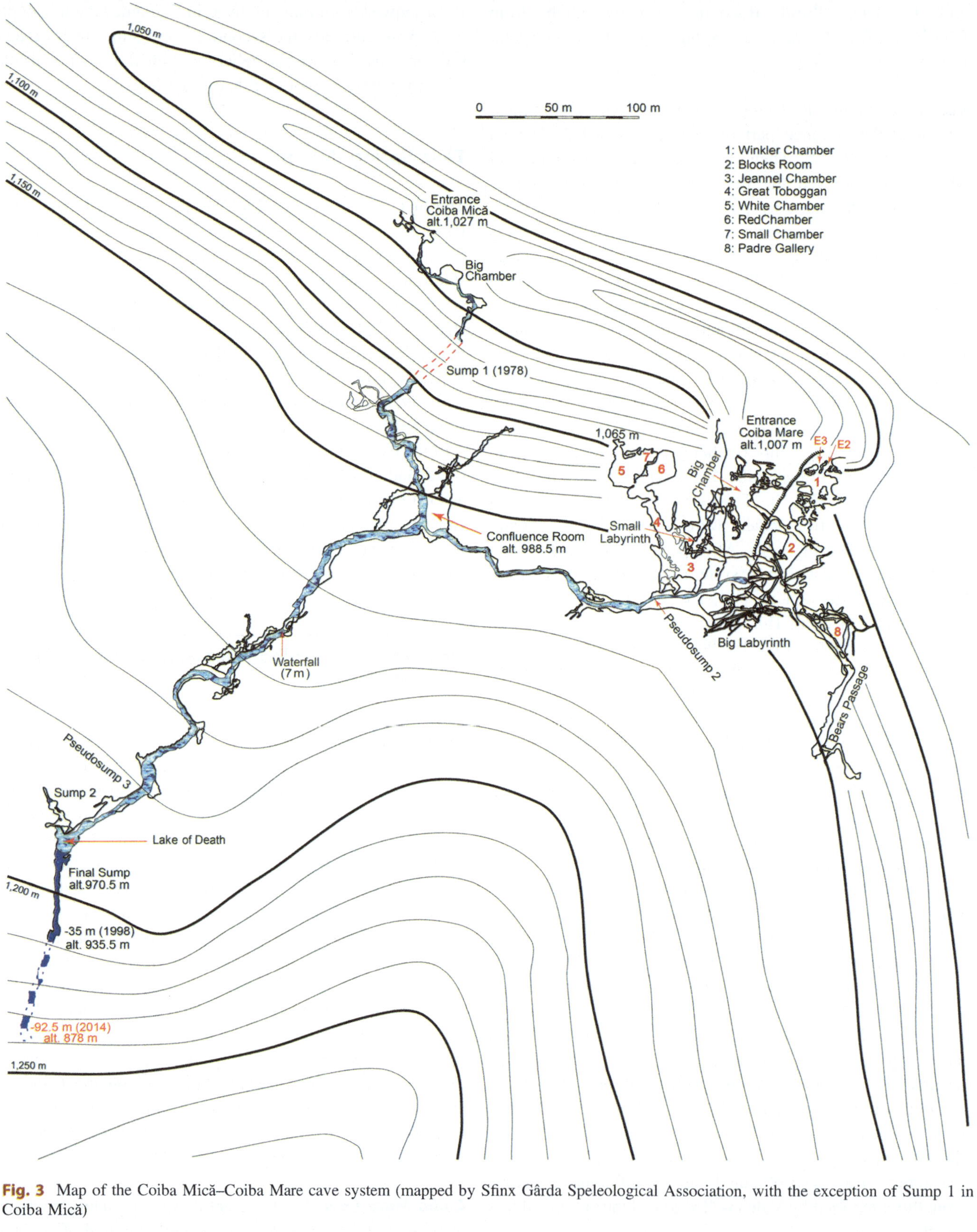

Fig. 3 Map of the Coiba Mică–Coiba Mare cave system (mapped by Sfinx Gârda Speleological Association, with the exception of Sump 1 in Coiba Mică)

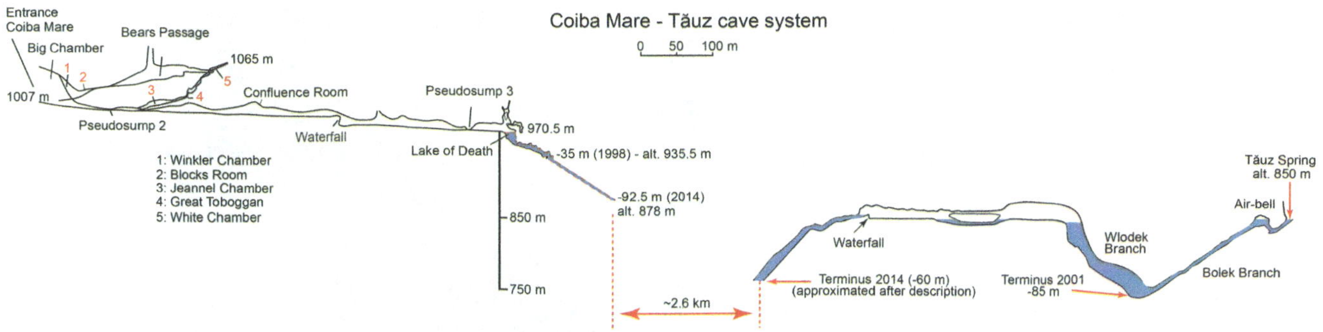

Fig. 4 Extended longitudinal profile through the Coiba Mare Cave and Tăuz Spring passages. The profile of the final sump (Lake of Death) is based on explorations and descriptions from I. Novak and E. Zaruba (down to −35 m) and S. Paakkarinen and P. Grönkvist (between −35 and −92.5 m). The longitudinal profile through the Tăuz Spring was realized by A. Peret (used with permission)

Fig. 5 Coiba Mare Cave entrance portal (photograph by C. Ciubotărescu)

(Sala Mare), a funnel-shaped hall, which communicates with Winkler Chamber situated at southeast, only 7 m above the floor. Two secondary cave entrances are accessible from Winkler Chamber after passing through a network of low galleries (0.4–2 m high). Advancing south of Winkler Chamber through the Blocks Room (Sala Blocurilor) and after overcoming a crawl between boulders is the entrance in the Bears Passage (Galeria Urşilor), an ascending gallery that hosts the most beautiful vermiculations from Romania that look like leopard print (Fig. 6). Below this section of the cave is the Padre Gallery, accessible from the Bears Passage after passing another crawl.

A complex network of passages and halls is reachable from the Big Chamber. Among these are: the Great Labyrinth (Marele Labirint) to southeast, Jeannel Chamber to the west from where climbing the Great Toboggan (Marele Tobogan) gives access to the Red/Roşie, White/Albă, and Small/Mică chambers. The Small Labyrinth (Micul Labirint) is also located in this western part of the Big Chamber.

An underground stream appears on the southern end of the Big Chamber and flows into the Confluence Room which joins the river coming from Coiba Mică (Fig. 3). Between these two points, there is only one low passage (Pseudosump 2), formed after a major flood event in 2000, when ~1.5 m thick

Fig. 6 Vermiculations (leopard prints) on the walls of Bears Gallery in Coiba Mare Cave (photograph by C. Ciubotărescu)

alluvial sediment was washed away from the place once called Pseudosump 1 that is now completely open. The next major obstacle along the river passage after continuing toward southwest is a 7 m drop (Fig. 7a) where water disappears between logs and boulders forming a nice waterfall (Fig. 7b). The active gallery continuing downstream with the roof dropping, from 5 to 0.5 m above the water surface at Pseudosump 3 (Fig. 7c). After this low passage, the ceiling of the cave rises to 47 m above the Lake of Death (Fig. 7d), which represents the final sump, presently completely covered with logs that have to be moved away to be able to dive the lake. Climbing the wall upstream of the lake, a team managed to overcome the sump along a 23 m dry gallery, but it connects to another impenetrable sump (Vălenaş 1978). The exploration of this section was completed by AS Sfinx Gârda, and the map is available in Onac et al. (2010).

Izbucul Tăuz is the most beautiful karst spring from Romania (Fig. 8). The drawing of the Tăuz Spring is made after the map by M. Stajszozyk, M. Czyerda, W. Bolek, W. Szymanowski (Wlodek), and the sketch and description of S. Paakkarinen and P. Gronqwist. Tăuz is now 945 m long out of which 220 m are air-filled galleries and 725 m of underwater passages. From the circular lake at the entrance, there is a −6 m drop under the archway of the sump. After diving 50 m to a depth of −12 m, an air-bell with a diameter of about 6 m is reached (Fig. 4). The mud covering the walls suggests that at high water levels this space is flooded. Based on the explorations performed by the Wroczlaw diving team (Poland), the cave continues with the Bolek Sump, which has a 200 m long downslope (30°) passage with some parts of low cross section, with sand and gravel on the bottom. Partly buried in this sediment, a 1.5 m long log (about 0.4 m in diameter) was found suggesting that it may come from the Lake of Death. If so, there is hope that divers too could get through this flooded karst conduit. W. Bolek reached the maximum depth of 85 m in 2001 after a 200 m dive from the cave entrance, noticing a narrow passage at about −80 m, which was explored only in 2006 by W. Szymanowsky, who reached after 375 m a depth of −30 m on the ascending branch of the Wlodek Sump (Fig. 4).

In 2014, S. Paakkarinen and P. Gronqwist successfully crossed on the other side of the sump, where they found a 200 m long and large gallery similar to the one in Coiba Mare, which is on the same direction with Sump 2 of Tăuz. This gallery is blocked by a new sump (5 m deep and 20 m long) that can be overpassed via a dry passage. On the other side of the sump, the gallery continues with a 4 m waterfall whose water comes from Sump 4, which they dived for 300 m to a depth of 60 m, and keeps going (Pereţ and Drăgan 2016).

Cave Sediments and Speleothems

Calcite speleothems (stalactites, stalagmites, columns, curtains, domes, flowstones, etc.), as well as abundant moonmilk deposits and spectacular vermiculations (leopard skin), are present in galleries and rooms on both sides of the Big Chamber (Fig. 6). This last type of speleothem was the

Fig. 7 Selection of images from the stream passage of Coiba Mare. **a** Rigging a traverse line at the 7-m drop, **b** waterfall, **c** Pseudosump 3, **d** the final sump, also known as the Lake of Death (photograph by C. Ciubotărescu)

subject of a recent investigation conducted by Bojar et al. (2015). The authors reported that the composition of vermiculations in Coiba Mare is almost entirely calcite (subordinately silicates and organics). This is in striking contrast to previous studies, which only documented mud or clay vermiculations (Hill and Forti 1997).

Two phosphate minerals, hydroxylapatite, $Ca_5(PO_4)_3OH$ and brushite, $Ca(PO_3OH){\cdot}2H_2O$, occur in areas where bats congregate. They were identified by means of X-ray diffraction analyses in the composition of millimeter-sized brownish to black crusts covering the cave wall and floor.

Fig. 8 Tăuz Spring (photograph by C. Ciubotărescu)

Cave Speleogenesis

The first speleogenetic hypothesis was put forward by Bleahu (1957), who considered the upstream migration of the Gârdişoara Valley water losses being the sole responsible for the genesis of this cave system. He considered that the piracy at the level of Coiba Mare Cave was active for a long period, which has led to the creation of large galleries and halls. Water losses through the Coiba Mică Cave instead are much more recent, which is why the dimensions of the cave are much modest.

In a later study, Vălenaş (1978) agrees with Bleahu's (1957) overall speleogenetic interpretation, but he expanded it by discussing the origin of the network mazes existing in the entrance zone. His take on the origin of the phreatic tubes is that they form at a time when Gârdişoara Valley was flowing at a higher elevation. Under this scenario, the recharge type of the limestone below was diffuse and the fractures and bedding planes were enlarged under phreatic conditions. In a later stage when the water of Gârdişoara Valley found a new ponor (the present entrance in Coiba Mare) to drain underground, the system begun to function under vadose conditions. At the time the paper by Vălenaş (1978) was published, the author was unable to clearly specify whether the speleogenetic model is bathyphreatic or epiphreatic, opting however, for the latter one.

Even though Vălenaş (1978) suggested a preference for the epiphreatic model, we note that this is not reflecting the current understanding of the epiphreatic cave development as described by Audra and Palmer (2013). A large number of karst water geochemical analyses have shown that flood-waters entering air-filled cave passages are far more aggressive (corrosion and erosion) than phreatic waters (Palmer 2007). Considering both the surface karst landscape and cave morphology, we believe that a full phreatic stage in Coiba Mare evolution was not a necessary condition for the genesis of the tubes. As a matter of fact, many cave passages and features of apparent phreatic origin may easily be explained by cave development in the epiphreatic (flood-water) zone, and this might have been the case for Coiba Mare. Ever since the stream piracy occurred at the level of the present-day entrance in Coiba Mare, the cave developed under epiphreatic conditions, resulting in a single-passage stream cave, with network diversions along bedding planes and fissures, and flood water injection features (horizontal and vertical enlarged fractures, anastomoses, blind terminations, size of scallops, etc.).

Cave Fauna

A rich cave fauna represented by: Gastropoda: *Discus ruderatus*, *Oxychilus glaber*, *Paladillia* (*Paladilliopsis*) *leruthi* (endemic stygobiont); Opiliones: *Gyan annulatus*; Pseudoscorpionide: *Neobisium* (*Blothrus*) *leruthi* (endemic troglobite); acarieni; Araneidae: *Nesticus spelaeus* (troglobiont); copepods ciclopoide: *Diacyclops languidoides*

languidoides, *D. nanus*, *Paracyclops fimbriatus*; collembole: *Isotomus alticola*; coleoptera leptodirine: *Drimeotus* (*Bihorites*) sp., *Pholeuon* (s. str.) *knirschi brevicule* (endemic troglobite); coleoptere trechine: *Duvalius* (*Duvaliotes*) sp.; Diptera; Trichoptera: *Micropterna testacea*; Chiroptera: *Plecotus austriacus*, *Myotis myotis*, *Myotis oxygnathus*, *Myotis dasycneme* (skeletal remains) was documented by Jeannel and Racoviță (1929), Damian-Georgescu (1963), Botoşăneanu (1966), Negrea (1966), Borda (1998–1999), and Racoviță (2004–2005).

Cave Conservancy

From a protection point of view, Coiba Mică is an unclassified cave, but currently is in self-conservation mode since its entrance is blocked with sediments and logs. Coiba Mare Cave is assigned to class B of protection, and thus permits from the Romanian Speleological Heritage Commission and the Apuseni Natural Park Administration are required for any cave activities.

References

Aldica G (2009) Peştera Coiba Mare. In: Dumitru R (ed) Focul Viu 40 de ani de speologie. Ed RAO, Bucureşti, p 98

Audra P, Palmer AN (2013) The vertical dimension of karst: Controls of vertical cave pattern. In: Schröder JF (ed) Treatise on geomorphology, vol 6. Academic Press, San Diego, pp 186–206

Bleahu M (1957) Captarea carstică şi rolul ei pentru evoluția morpfologică a regiunilor carstice. Prob Geogr V:55–99

Bleahu M, Bordea S, Bordea J, Mantea G, Dimitrescu R (1980) Geological map of Romania, Poiana Horea Sheet (56b). Scale 1:50.000, Inst Geol, Bucureşti

Bojar A-N, Guja O, Ştefănescu D (2015) Vermiculation patterns in Coiba Mare Cave, Bihor Mountains, Romania. Quat Int 357: 212–219

Borda D (1998–1999) The distribution of bats in Romanian caves (II). Trav Inst Spéol "Émile Racovitza" 37–38:223–230

Botoşăneanu L (1966) Genurile *Stenophylax* Kol. şi *Micropterna* Stein (*Trichoptera*) în România. Lucr Inst Speol "Emil Racoviță" V: 99–114

Damian-Georgescu A (1963) Fauna R.P.R., Crustacea. Copepoda. Fam. Cyclopidae. Forme de apă dulce. Ed Acad RPR, Bucureşti

Hill CA, Forti P (1997) Cave minerals of the world (2nd ed). Natl Speleol Soc, Huntsville, Alabama, pp 221–223

Huber W (2011) In the depths of the Coiba Mare. Global Mag Leica Geosyst, Reporter 67:14–17

Jeannel R, Racovitza EG (1929) Énumération des grottes visitées, 1918-1927 (7e série). Arch Zool Exp Gén 68(2):293–608

Negrea A (1966) Gasteropodele (Mollusca-Gasteropoda) din peşterile României. Lucr Inst Speol "Emil Racoviță" V:125–139

Onac BP, Racoviță G, Brad T (eds) (2010) Atlasul peşterilor din Munții Apuseni. Munții Bihor 1. Bazinul Arieşului Mare. Univ South Florida Karst Stud Ser 2:1–90

Orăşeanu I (1996) Contribution to the hydrogeology of the karst areas of the Bihor-Vlădeasa Mountains. Theor Appl Karstol 9:185–214

Orăşeanu I (2010) Karst hydrogeology of Apuseni Mountains. In: Orăşeanu I, Iurkiewicz A (eds) Karst hydrogeology of Romania. Ed Belvedere, Oradea, pp 181–324

Orghidan T, Negrea Ş, Racoviță G, Lascu C (1984) Peşteri din România. Ed Sport-Turism, Bucureşti, pp 170–177

Palmer AN (2007) Cave geology. Cave Books, Dayton, p 454

Pereț A, Drăgan C (2016) Explorarea Izbucului Tăuz. Diving România 3:27–29

Racoviță G (2004–2005) Révision systématique des Leptodirinae souterrains des Monts Apuseni. IV. Le sous-genre *Pholeuon* (s. str.) du bassin de l'Arieş (Monts du Bihor). Trav Inst Spéol "Émile Racovitza" 43–44:165–191

Turi V, Săsăran E, Bucur II (2011) New data on the Upper Jurassic–Lower Cretaceous limestone from Bihor Mountains: case study of Gârda Seacă–Hodobana region Romania. Studia UBB Geol 56 (1):11–27

Vălenaş L (1976) Carstul de la Casa de Piatră (Munții Bihorului. Bul CSER Bucureşti 4:150–168

Vălenaş L (1978) Morfologia sistemului Coiba Mare–Coiba Mică–Izbucul Tăuz. Nymphaea 6:329–362

Vălenaş L, Păroiu F (1978) Joncțiunea dintre Coiba Mare şi Coiba Mică. Bul Speol 2:36–41

Vălenaş L, Halasi G, Czakó L (1982–1983) La morphologie et l'hydrologie des conduits submergés du bassin de la Vallée Gîrda (Monts Bihor). Nymphaea X:195–205

Vălenaş L, Bleahu M, Brijan P, Halasi G (1977) Inventarul speologic al Munților Bihor. Nymphaea V:209–335

Bihor Mountains

Gheorghe M. L. Ponta

Abstract
One of the most important karst regions of western Romania is Bihor Mountains, which form the central part of the Apuseni Mountains. The centerpiece of Bihor Mountains is Padiş Plateau developed at about 1000–1200 m altitude and characterized by numerous karst features as dolines, ponors, springs, and caves. Cetăţile Ponorului is traversed by a stream that disappears in a sump at the end of the cave to resurface in Galbena Spring. Vărăşoaia cave system (V5) with −653 m is the deepest, whereas Humpleu-Poieniţa is the second longest (36 km) in the country. Another highlight of the Bihor Mountains karst is world's largest and oldest perennial ice deposit in Scărişoara Cave. Some of the oldest cave paintings in Europe have been discovered in Coliboaia Cave which along with the Secăturii Hill form a cave system. In Vârtop Cave, three Neanderthalian footprints were preserved in a hardened calcite moonmilk deposit, while the Bears Cave (Peştera Urşilor) is known for several cave bear skeletons in anatomic connection. Some of these caves are presented below or in separate chapters of this book.

Keywords
Karst plateau • Caves • Polje • Dye/tracer studies

Geographic, Geologic, and Hydrologic Settings

Bihor Mountains with 1299 cavities has one of the greatest concentrations of caves in Romania and the largest number of any massif in the country (Bălteanu et al. 2011). The Bihor Mountains presents large peneplains/leveling or erosional surfaces, Mărişel and Râmeţi, with their correspondent karstoplains Bătrâna and Scărişoara developed between 1200 and 1300 m elevation (Fig. 1). The most frequent karst features on the limestone plateaus are dolines (sinkholes), caves, and closed depressions crossed by short rivers, which in their lowest point are sinking underground, through either a cave or an impenetrable ponor. In the Padiş Plateau is Poiana Ponor, the only polje in Romania; this is a large close depression traversed by a stream, which originates in Poiana Ponor Spring and after about 700 m is disappearing underground (Fig. 2).

In the area, the Bihor unit and post-tectonic cover formations consist of Paleozoic, Mesozoic, Cenozoic, and Quaternary sedimentary formations, which are underlain by metamorphic rocks (Mantea 1985). In the Mesozoic deposits, several Triassic, Jurassic, and Cretaceous lithostratigraphic units have been identified. Quartzitic sandstones and a sequence of shale of Werfenian age (about 70 m thick) or Werfenian—Lower Anisian, and Permian deposits or crystalline formations form the main impervious units below the limestones (Fig. 3).

The Middle Triassic deposits are represented almost exclusively by carbonates, which appeared as a succession of Gutenstein limestones (whose thickness does not exceed 150 m), dolomites of Anisian age (60 m thick), and Wetterstein limestones that can reach 300 m in thickness. At the end of Triassic, the territory of Bihor Mountains was uplifted, and new sources of clastic sediments appeared, Hettangian—Lower Sinemurian clastic formations reach 160 m in thickness and include quartzitic conglomerates, sandstones, and shale. These deposits create an impervious layer between the Triassic and the Jurassic limestones, which are composed of calcareous units up to 45 m thick, considered the equivalent of the Gresten limestones (Upper Sinemurian–Carixian age), and Middle Jurassic deposits, which outcrop on restricted areas, and form a sequence of gray-blackish limestones and encrinitic limestones.

The Upper Jurassic formations cover remarkable large areas and are represented exclusively by limestones, recording various thicknesses, but do not exceed 300 m. When the Upper Jurassic sedimentation ended, favorable conditions for

G. M. L. Ponta (✉)
Geological Survey of Alabama, 420 Hackberry Lane, Tuscaloosa, AL 35401, USA
e-mail: gponta@yahoo.com

G. M. L. Ponta and B. P. Onac (eds.), *Cave and Karst Systems of Romania*,
Cave and Karst Systems of the World, https://doi.org/10.1007/978-3-319-90747-5_36

Fig. 1 Location of Bihor Mountains within Romania

the bauxite ore emplacements were created (during Neocomian), which are overlain by Early Cretaceous limestones.

The Padiș—Cetățile Ponorului Karst Plateau

The Padiș—Cetățile Ponorului karst plateau, with a 54 km^2 of carbonate rocks outcrop, is formed by 23.5 km^2 Triassic and 30.5 km^2 Jurassic and Cretaceous limestones, respectively (Fig. 3). In this area, 26 ponors and 18 karst springs have been identified. Most of the disappearing streams are situated on Triassic limestones. Tracing experiments indicate that the preferential flow directions are toward the west and south in the central part of the plateau (Orășeanu 1996, 2010, 2016).

The presence of the watershed is controlled by the outcrop of the Hettangian–Sinemurian deposits situated east of the Oșelu Spring and continues under the Quaternary deposits. The flow path reveals the strong dependence on the tectonic structures. In the Padiș-Trînghieşti area, a disappearing stream of 15 L/s was tested with 2 kg of fluorescein in June 18, 1991 (Ponta 1997, 1998). The first appearance of the tracer in the Poiana Ponor Spring was after 14 h. This stream sinks again after 700 m of subaerial course along a polje and reappears in the Cetățile Ponorului Cave at the place known as Nostrils (Nările) as a subterranean stream. From here, the river flows along 1700 m of passages before ending in a sump, and finally resurfacing through the Galbena Spring, after 40 h since injection in Trînghieşti Ponor. At the west end of Poiana Ponor, the stream is disappearing in a fracture zone, which allowed the waters to go through Hettangian-Sinemurian sandstones (Fig. 4). The short travel time is indicative of a vadose flow, with phreatic phases at different periods during the year (Fig. 5) (Ponta et al. 1992).

In 1985, Orășeanu conducted two dye studies in Arsuri and Renghii ponors (Orășeanu 2010, 2016). The one in Arsuri had the same path like the one presented above in Trînghieşti Ponor, and the one in Renghii ended up in Boga Spring, which was later confirmed by the topography of V5. The distance between these two ponors is only a few tens of meters, but in between is the watershed of two major karst systems, V5 and Cetățile Ponorului. No impermeable barrier exists, at least at the land surface between these two sinking points.

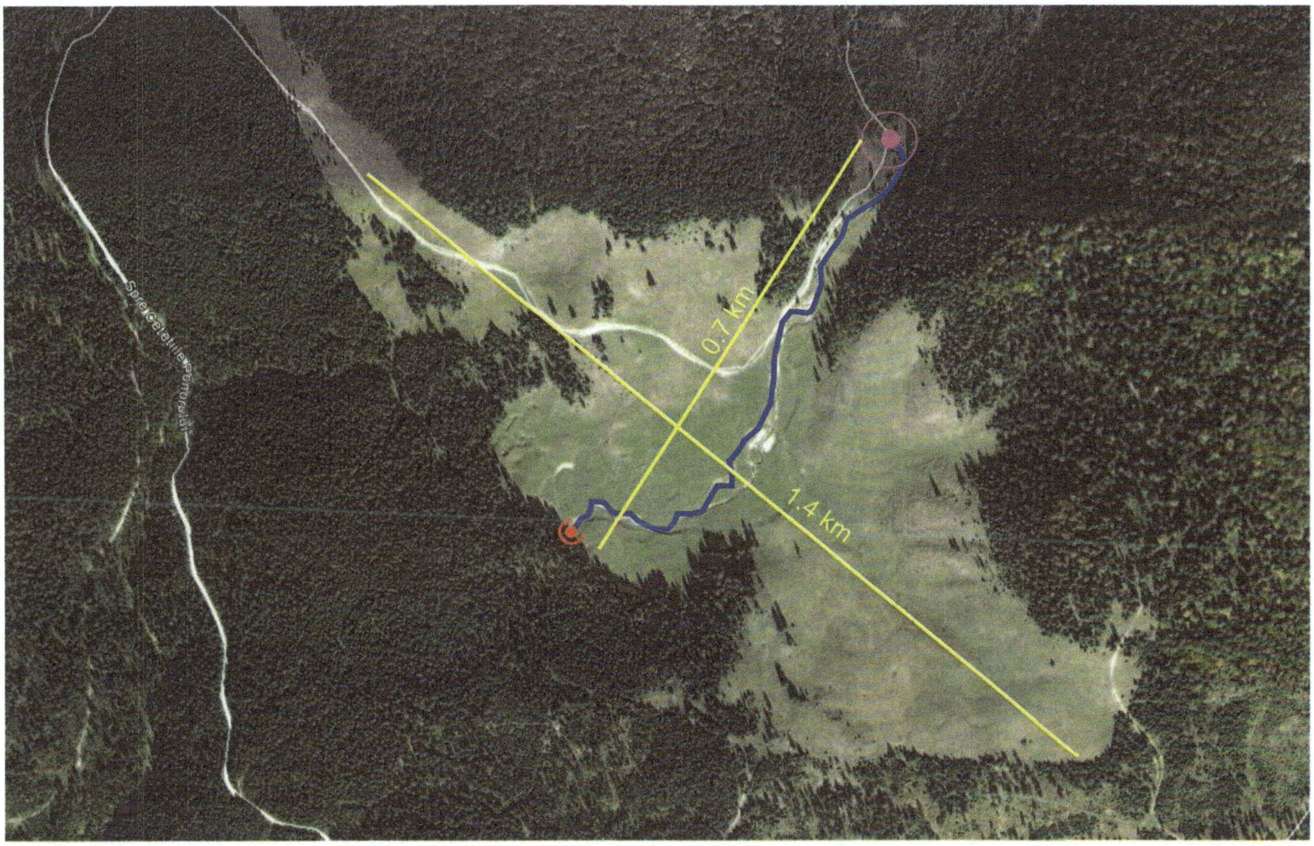

Fig. 2 Poiana Ponor Polje

Cetățile Ponorului Cave

Gheorghe M. L. Ponta
Geological Survey of Alabama
420 Hackberry Lane, Tuscaloosa, AL 35401, USA
e-mail: gponta@yahoo.com

Mihai Besesek
Speowest Speleological Association Arad, Speodava Ştei Caving Club, Arad Cozia 13, Apt. 2, Arad, 310101, Romania
e-mail: mihaibesesek@yahoo.com

History of Exploration

Despite its size and importance, the Cetăţile Ponorului Cave was explored relatively late. The first written citations date from 1885 by S. Nagy and Gy. Czaran in 1903. The magnificent entrance portal (Fig. 6 inset) was mentioned by Jeannel and Racoviţă (1929).

In 1949, a team from the "Emil Racoviţă" Institute of Speleology, Cluj-Napoca, explored the first 300 m of the cave. In 1951, a series of expeditions attended by climbers begun. In 1952 is reached the present end of the cave, but only in 1957, the terminal sump is explored, and the first attempt to climb the final wall to overpass the sump was performed by a team of A. S. Armata Braşov led by E. Cristea (Şerban et al. 1961; Bleahu et al. 1976). In 1971, a French–Romanian expedition reached the same point, and in January and July 1972 succeed to survey 1700 m of passages, and climb above the sump over 100 m, without finding a continuation (Vălenaş et al. 1977; Vălenaş 1984).

In the late seventies (1976–1978) I. Viehmann with CSER Cluj-Napoca, G. Brijan, and O. Cuc from C. S. Speodava Ştei (formerly Dr. Petru Groza) Caving Club organized several expeditions to explore, resurvey, and rig the cave (Bosdoc 2014), which by this time reached a total length of 3800 m. In preparation for a larger resurvey/exploration expedition, in 1984 and 1985, Cepromin Cluj-Napoca (R. Baboş) and Focul Viu Bucureşti Caving Club (G. Ponta, D. Cozea, and A. Ponta) with German and Slovak cavers organize two trips into the cave to locate new leads to overpass the sump.

Due to the 1989 political events in Romania, only in 1990 a large group of cavers from Focul Viu Bucureşti Caving Club with support of Zărand Brad Caving Club and French, British, and German cavers resurvey the cave and explore several new passages in the ceiling of the main borehole,

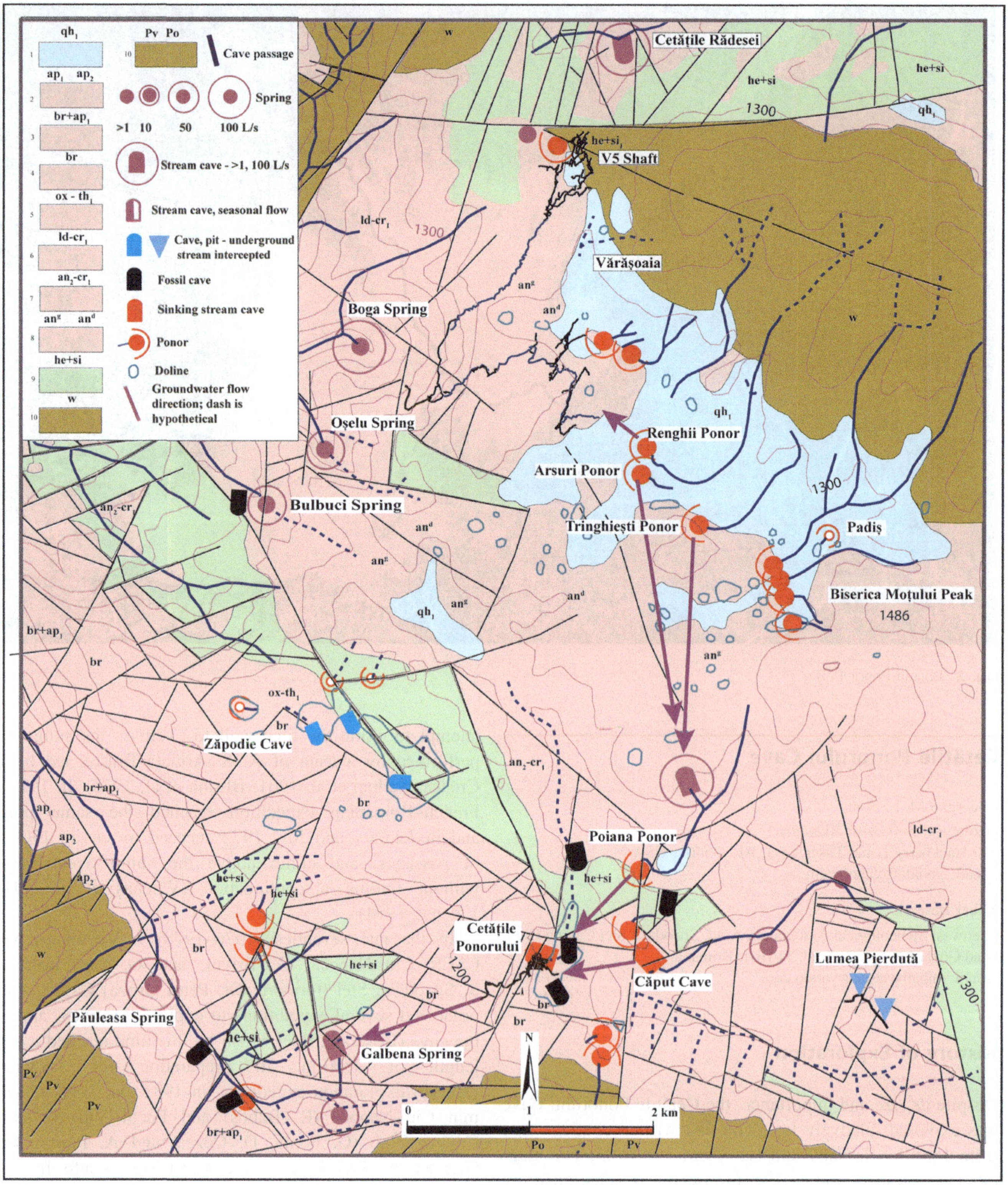

Fig. 3 Karst hydrogeologic map of Padiș—Cetățile Ponorului karst area, modified after Ponta (1997), with permission (geology modified after Mantea 1985). Key for the legend is available in the karst hydrogeology chapter

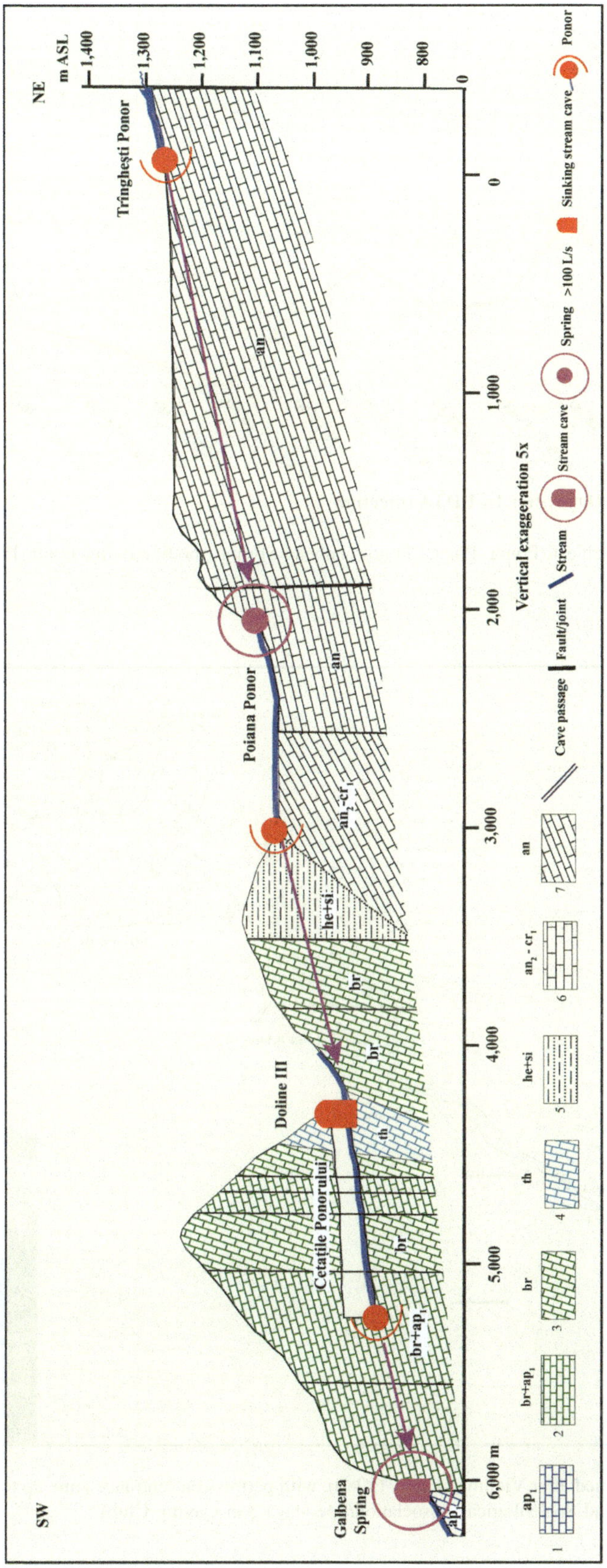

Fig. 4 Geologic cross section through Padiş—Cetăţile Ponorului karst area

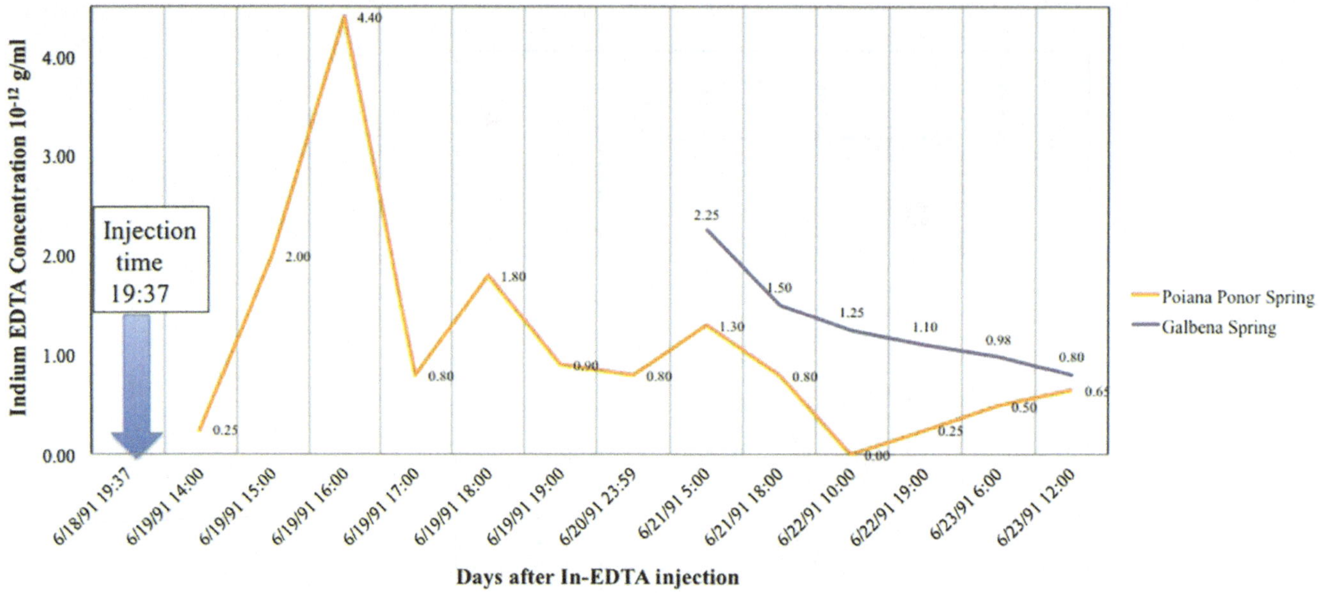

Fig. 5 Fluorescein breakthrough time curve Trînghieşti Ponor–Poiana Ponor/Galbena Spring (modified after Ponta 1997, with permission)

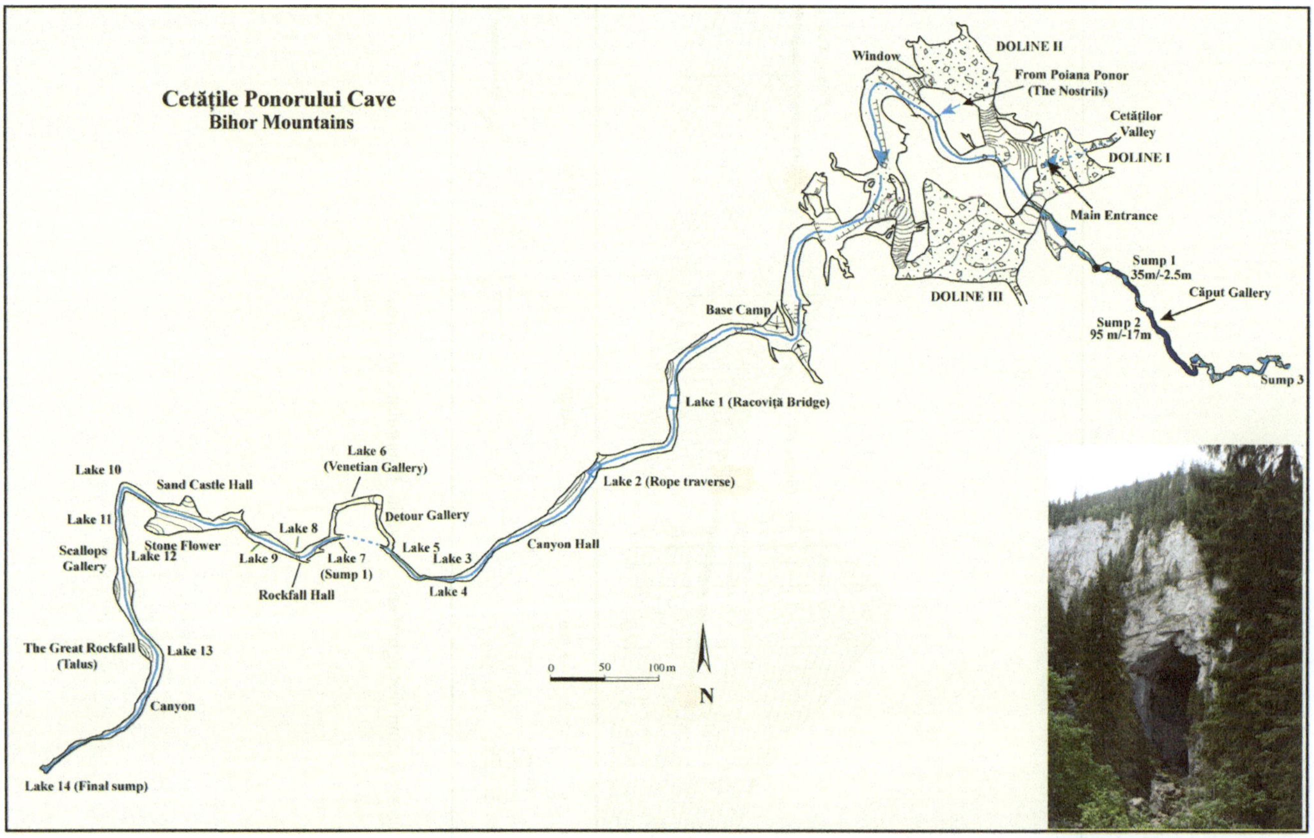

Fig. 6 Map of the Cetăţile Ponorului Cave, modified after Viehmann et al. (1980), with permission (entrance zone up to the Base Camp surveyed by Pitic/Cristal Oradea and Besesek/Speowest Arad Speleological Association/Speodava Ştei Caving Club)

but none was able to overpass the final sump (Aldica 2009). Recently, a team of cavers led by T. Rus (Speodava Ştei Caving Club), D. Pitic (Cristal Oradea Caving Club), and M. Besesek (Speowest Arad Speleological Association/ Speodava Ştei Caving Club) began to resurvey and explore the cave, process that is ongoing at the time of writing this book (entrance zone up to the Base Camp). Beyond the Base Camp, the map shows the 1978 survey of Viehmann and his team (Viehmann et al. 1980). Recently, M. Besesek with a team of divers explored the stream coming from Căput Cave.

Cave Description

Cetăţile Ponorului Cave is located at the downstream end of the Cetăţilor Valley, the lowest point (950 m) of the entire Padiş—Cetăţile Ponorului closed basin. The cave is a borehole type gallery, with few side passages located mostly close to the ceiling of the cave. It is 4275 m long, considerable in size, traversed by one of the largest underground rivers in the country (Fig. 6).

The entrance area is composed of three large collapse dolines (Dolines I, II, and III), with a combined diameter of approximate 0.5 km. The Doline I (100 m deep) is the end of a wild canyon created by the Cetăţilor Valley, with its temporary waters disappearing underground at the south end (left) of a large portal located at the base of the northern wall, 70 m high and 30 m wide. Thirty meters downstream the entrance, the river disappeared in Căput Cave resurfaces in the cave (Căput Gallery) as a left side tributary followed by the waters from Poiana Ponor, which are coming in strong through the Nostrils on the right side of the main stream passage, 118 m from the entrance. The Căput Gallery was surveyed on 475 m, with Sump 1 of 35 m long and −2.5 m deep, and Sump 2, 95 m long and −17 m deep. The exploration ended in Sump 3.

On the right (north) side of the portal, a large uphill passage begins (72 m long and 70 m high) with the floor formed by boulders (talus), the upper end being in the Doline II (60 m in diameter), which is an almost perfect circle with vertical walls. At the northern end of this doline, a downhill passage makes the junction with the main stream (window/fereastra). About 170 m downstream, the underground river resurfaces for about 20 m at the base of the Doline III (karst window at 930 m elevation asl) to continue its path for the next 1700 m (Fig. 7).

Doline III is the largest, with walls about 120 m high. It is accessible from Doline I, crossing the saddle that marks the boundaries between these two collapse dolines. Just 180 m

Fig. 7 Cetăţile Ponorului—main gallery (photograph courtesy of Christian Ciubotărescu)

downstream from Doline III, the Base Camp Hall is located, place used as a bivouac site in the early fifties and sixties. Beyond this relative large hall, the cave continues its path through an 8 m wide and 20 m high gallery (occasionally up to 70 m high or more), dimensions that will remain about the same until the final sump. Boulders often created dams behind which 14 lakes of varying lengths and depths formed, which requires wetsuits to be cross. Between Lake 3 and 4, the borehole develops along a major fault, the passage having canyon morphology.

The stream ends in Sump 1 at Lake 5, which can be avoided through a gallery that opens on the right (detour gallery/Galeria de Ocol), continue with the Venetian Gallery, the end of which is back into the mainstream passage on the other side of the sump. After Lake 7, the underground river reappears and the height of the gallery is over 100 m in the Rockfall Hall (Sala Grohotişului) area. From here on, follows a linear gallery that hosts Lakes 8 and 9, and finally, the Sand Castle Hall (Sala Castelului de Nisip), a large room where there are interesting stalagmites, a stone flower on the left and sand deposits on the right. Downstream, the gallery shows spectacular forms of erosion and accommodates lakes 10 and 11. The stream passage continues with the Scallops Gallery (Galeria Linguriţelor), which is about 80 m in length, and leads to a room with a big talus on the right, and then, gallery narrows (Canyon) and is completely flooded by Lake 14, with the final sump covered with numerous tree trunks, making it almost impossible to dive. At this point, several climbing tentatives (as high as 120 m) were performed over the years, but none succeeded to descend beyond the calcite dam of the final sump. Cetăţile Ponorului by its magnitude is one of the most important karst systems in the country.

Hydrogeology of the Upper Someşul Cald Basin

Another representative karst area in the Bihor Mountains is the upper basin of the Someşul Cald Valley, a region covering 20 km^2 and made up of Triassic (2.5 km^2) and Jurassic/Cretaceous (17.5 km^2) limestones (Fig. 8). In this area, 14 sinking streams and 15 springs have been identified. The main sinking points are situated at the contact between the Triassic limestones and the Cretaceous ones or on the Jurassic carbonates rocks. The watershed occurs in the central part of the area and was established by tracing experiments, which show a preferential flow direction to the south, toward the Alunul Mic Spring (Gligan 1987; Orăşeanu 2010, 2016). The flow path reveals the strong dependence on the tectonic structures.

The Alunul Mic Spring (also known as Peştera cu Oase/the cave with bones) with a flow ranging between 100 and 500 L/s drains the waters of the Valea Ponorului sinking stream, which has a total discharge of 70 L/s (Orăşeanu 2010, 2016). The waters originate on impervious formations and Triassic limestones (Fig. 8).

On October 8, 1991, 1 kg of fluorescein was injected into the Valea Ponorului sinking stream/swallet. The rainfall of those days is reflected in the general form of the Alunul Mic Spring graph (Fig. 9). The first appearance of the dye was 17 hours after injection, the graph presenting one pulse of high amplitude (Ponta et al. 1993). A flatter graph shows the reappearance of the tracer in the Humpleu Spring, after 27 h. The karstic diffluence controlled the formation of two important cave systems: Humpleu-Poieniţa cave system, 36 km long (see Chapter "Bihor Mountains: Humpleu-Poieniţa Cave System"), and Piatra Ponorului-Alunul Mic cave system, with 4 km of passages.

Secăturii Hill—Coliboaia Cave System

Mihai Besesek
Speowest Speleological Association Arad,
Speodava Ştei Caving Club, Arad, Cozia 13,
Apt. 2, Arad 310101, Romania
e-mail: mihaibesesek@yahoo.com

Valentin Alexandru Radu
Speleological Association Speowest Arad, Vladimirescu 39, Arad, Romania
e-mail: radugulea@yahoo.com

Gheorghe M. L. Ponta
Geological Survey of Alabama, 420 Hackberry Lane, Tuscaloosa, AL 35401, USA
e-mail: gponta@yahoo.com

Abstract

With a vertical range of −390 m, Avenul din Dealul Secăturii (Secăturii Hill Cave) ranks among the deepest caves in Romania and stands out from other cavities in the region primarily due the beauty of the speleothems and diverse morphology. The stream forming this cave resurfaces in Coliboaia Cave, which hosts Paleolithic parietal art dating back to ~36,000 years BP. Tracing experiments demonstrated the connection between these two caves having a total vertical range of 410 m.

Introduction

With a karstifiable area of ~15 km^2 and over 196 caves explored so far, Sighiştelului Valley has the highest density/km^2 in Romania. Many caves host archaeological

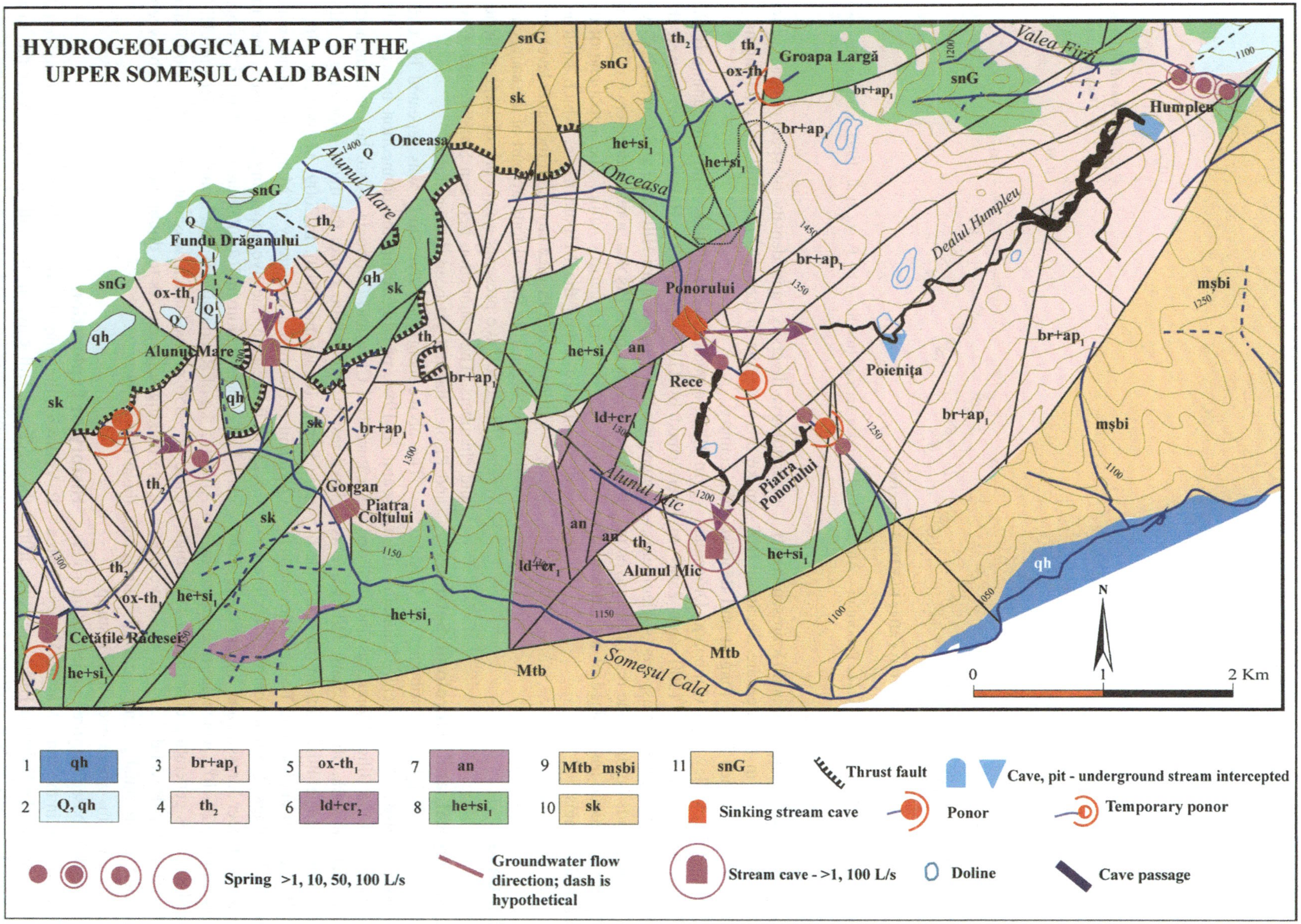

Fig. 8 Hydrogeologic map of the Upper Somesul Cald Basin, modified after Ponta et al. (1993), with permission (geology modified after Mantea 1985). Key for the legend is available in the karst hydrogeology chapter

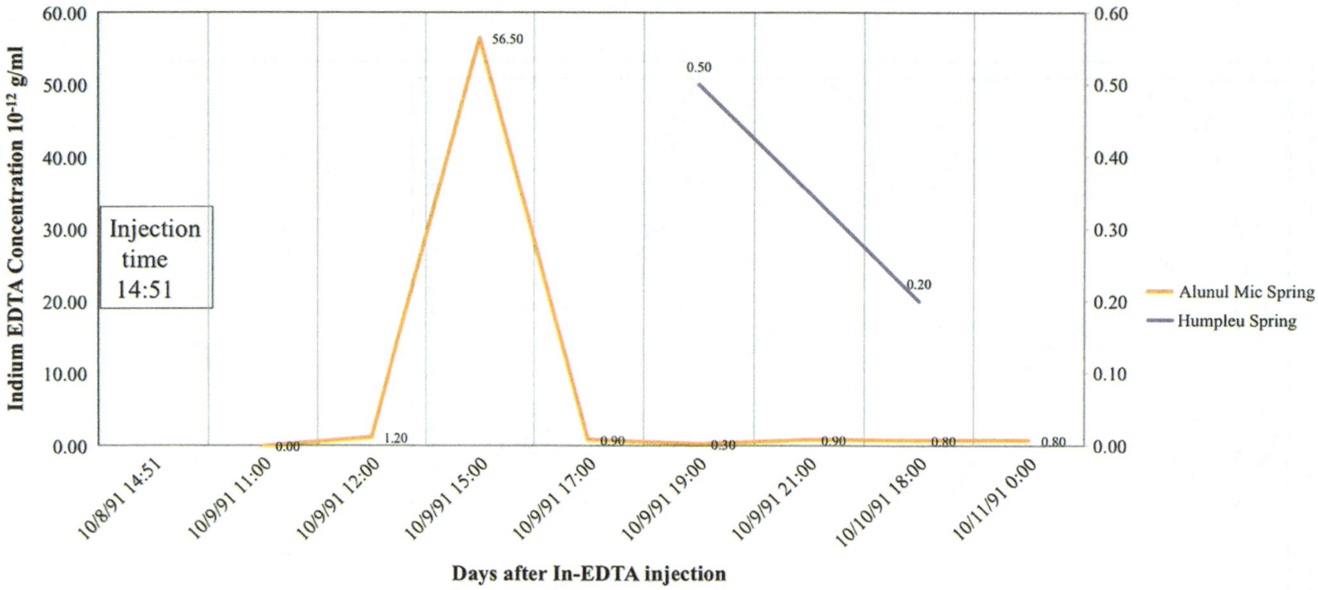

Fig. 9 Fluorescein breakthrough time curve Valea Ponorului—Alunul Mic/Humpleu springs (modified by Ponta et al. 1993, with permission)

artifacts of Paleolithic age (Cârciumaru and Dobrescu 2010). The most important cavities in the area are Secăturii Hill Cave (2990 m, −390 m), which is the longest and deepest cave in the Sighiştelului Valley, and Coliboaia, Drăcoaia, and Măgura with lengths between 1500 and 2500 m (Fig. 10). Besides its hydrogeological aspects, speleothems, and fauna, Coliboaia stands out from the rest of the caves in the region through its unique assemblage of archaeological and paleontological remains.

Secăturii Hill Cave

Geographic, Geological, and Hydrologic Settings

The Sighiştel Valley is a right side tributary of Crişului Negru River, which crosses the Apuseni Mountains from east to West. Situated in the Bihor Mountains and included in the Apuseni Natural Park, the Secăturii Hill Cave is located in the Valley Sighiştelului basin, on the right side of the Măgura Valley, at 970 m elevation. The cave develops to −354 m along the lithological contact between Oxfordian–Tithonic limestones (Upper Jurassic) and impermeable shale and marl layers dipping 20–40°, which are interrupted by a series of vertical fractures generating drops up to 15 m high (Halasi and Ponta 1984). Beyond −354 m, the cave continues with a series of horizontal galleries, carved entirely in limestone units to −390 m, where ends with a sump.

The cave network is based on three underground rivers, which have their sources in the same Măgura Valley. With a vertical range of −390 m the Secăturii Hill Cave is the fifth deepest cave in Romania, characterized by complex morphology, diversity, and density of speleothems, which make this cave one of the most beautiful and challenging in the country.

In 1981, a qualitative dye study with Rhodamine B was performed by G. Ponta and G. Halasi, which proved the connection between Secăturii Hill Cave and Coliboaia Cave. In the same area opens Pişolca, a resurgence cave of 290 m in length that ends in a narrow sump. The actual end of the cave is situated below the Măgura Cave, which stretches between Coliboaia and Pişolca caves, on the same level with the first one suggesting that all four are part of the same cave system. Măgura Cave presents a network of large galleries and chambers. The connection between Secăturii Hill Cave and Pişolca has not yet been demonstrated.

The above-mentioned dye study was repeated by Orăşeanu in 1984, proving again the connection between the Secăturii Hill Cave and Coliboaia. Also, he made an additional test in Muncelului Cave, the tracer appearing in Blidaru Spring (Orăşeanu 2010). In 2009, M. Besesek with I. Orăşeanu performed a dye study in Muncelului Ponor with 0.5 kg of fluorescein, the dye appearing eight days later in very low concentration in Răsuflătoarea Blidarului Spring. Also, 0.5 kg of Rhodamine B was launched in Secăturii Hill Cave at −350 m dye which was not detected in Coliboaia Cave making some cavers to doubt the connection between these two caves.

History of Exploration

P. Voitovici, then member of CSA Liliacul Arad, discovered the Secăturii Hill Cave in the fall of 1980. After the first explorations, Voitovici and other colleagues founded a new

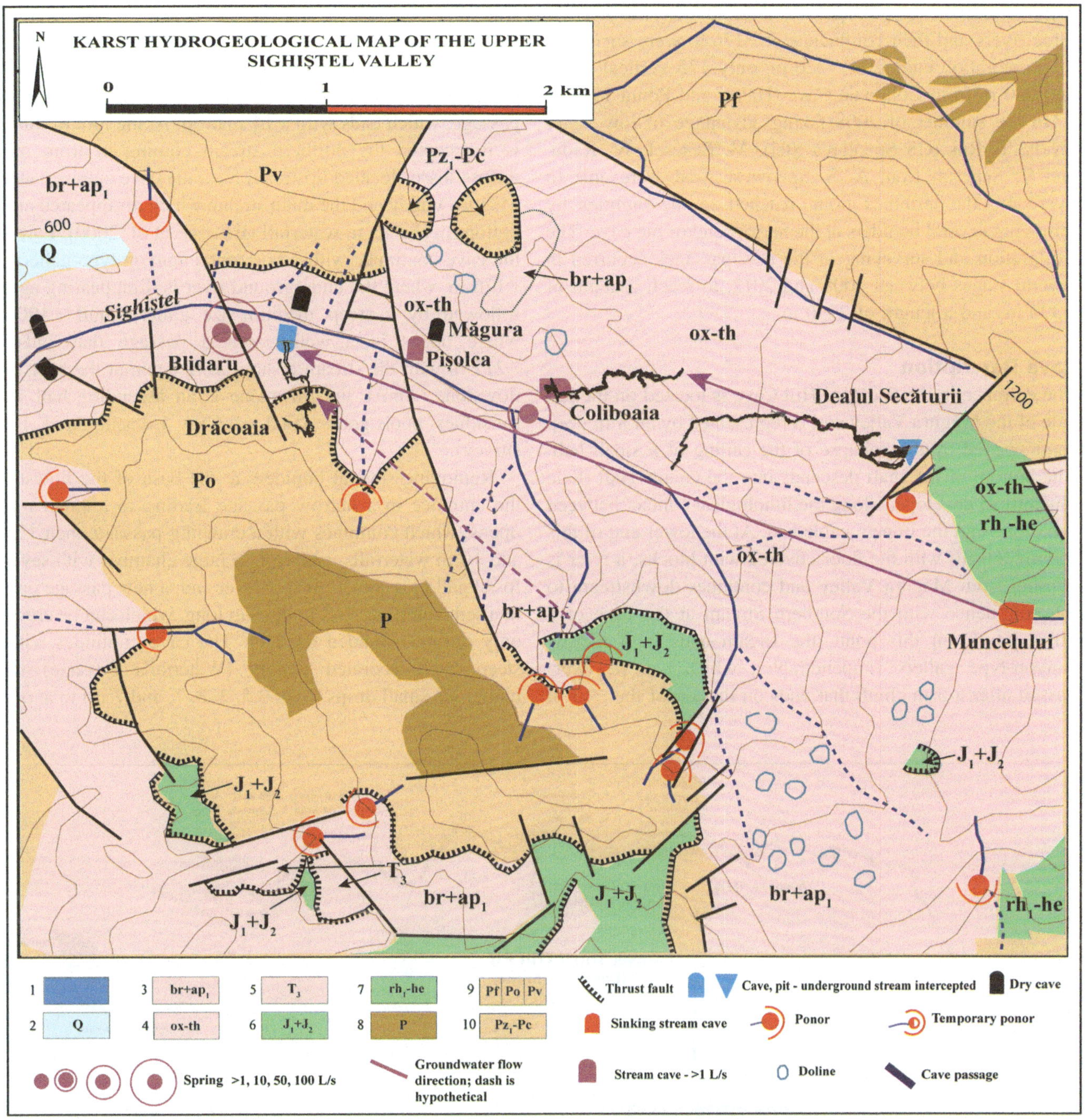

Fig. 10 Karst hydrogeologic map of the Upper Sighiştel Valley (geology modified after Mantea 1985). Key for the legend is available in the karst hydrogeology chapter

speleo club named Arago Arad and dedicate most of their time to the exploration of this cave. By the end of 1980, they have explored the cave to a depth of −130 m and stopped at the base of a 13 m waterfall (Marandiuc 1981). In order to protect the cave, they installed a gate at the entrance. Later, P. Voitovici and P. Brijan begin exploring possible other leads in the cave. They rigged the traverse over the 13 m waterfall for easier access to the confluence at −176 m and beyond.

In September 1981, G. Halasi and G. Ponta explored the cave to the bottom of the 13 m waterfall (which is a dead end) and discovered the continuation on the other side of the pit. A month later, members of CSA Liliacul Arad, G. Halasi, G. Ponta, R. Ermesz, and V. Brînzaş explored and mapped 1450 m of passages, to a depth of −220 m. The three rivers and their confluence at −176 m were surveyed, the exploration ending at −220 m, only 175 vertical range and 560 m from Coliboaia Cave (Halasi and Ponta 1984).

In the summer of 2008 taking advantage of low water levels, T. Rus (CS Speodava Ştei), M. Besesek, V. Radu, and C. Sebeşan from A. S. Speowest Arad were able to overcome the terminus point reached in the eighties by removing several boulders at the narrow end of the cave. The exploration and surveying of the new passages occurred in several stages between 2008 and 2012, to reach a depth of −390 m, and a length of 2990 m.

Cave Description

The entrance in the Secăturii Hill Cave is located on the left side of the Măgura Valley and is represented by an 8 m deep shaft opened by the collapse of the ceiling of a small hall. The bottom of the shaft descends into the Giants Hall (Sala Giganţilor) developed along the lithological contact between limestone and the impermeable bed. At the lower end of this room (−60 m) with the floor covered with blocks, a river is coming from Măgura Valley and continues downstream to the confluence with the Southern Stream at the base of a 20 m pit. From this point, the stream ends in a narrow canyon-type gallery (impenetrable), which can be over-passed after a 5 m climb that ends on the top of the −13 m waterfall. The base of the waterfall is at −130 m, where the stream disappears through breakdown blocks (Fig. 11).

From the top of −13 m waterfall, the cave continues on the other side of the pit (30 m long traverse) with a fossil passage, which ends with a 14 m drop. At the foot of this pit is intercepted the northern stream coming in from other diffuse losses located upstream from the cave entrance along Măgura Valley and the main stream which disappeared at the bottom of the 13 m waterfall (Besesek et al. 2009). Further, the cave continues with a horizontal canyon-type gallery to −176 m, where the underground river is disappearing again, followed by a steep medium-size gallery until −190 m, where opens in a well-decorated passage that ends at −217 m with the stream disappearing again at the base of a flowstone. Above the flowstone is an ascending hall with rimstones pools/gours (End of CSA Liliacul Arad 1981 survey).

Removing several boulders at the base of the flowstone the entrance in a narrow passage leading to a small room opens, which continues with a crawling passage, then 12, 6, and 10 m waterfalls, ending in a large chamber with several fossil galleries above. At the lower end, a new passage opens in a semisump at −268 m, 4–5 m long, which can be crossed only during extended drought. The cave continues with a steep, well-decorated gallery (Waterfalls Passage) with numerous small drops (6, 4, 3.5, 3, 8, 7, and 7 m) to another

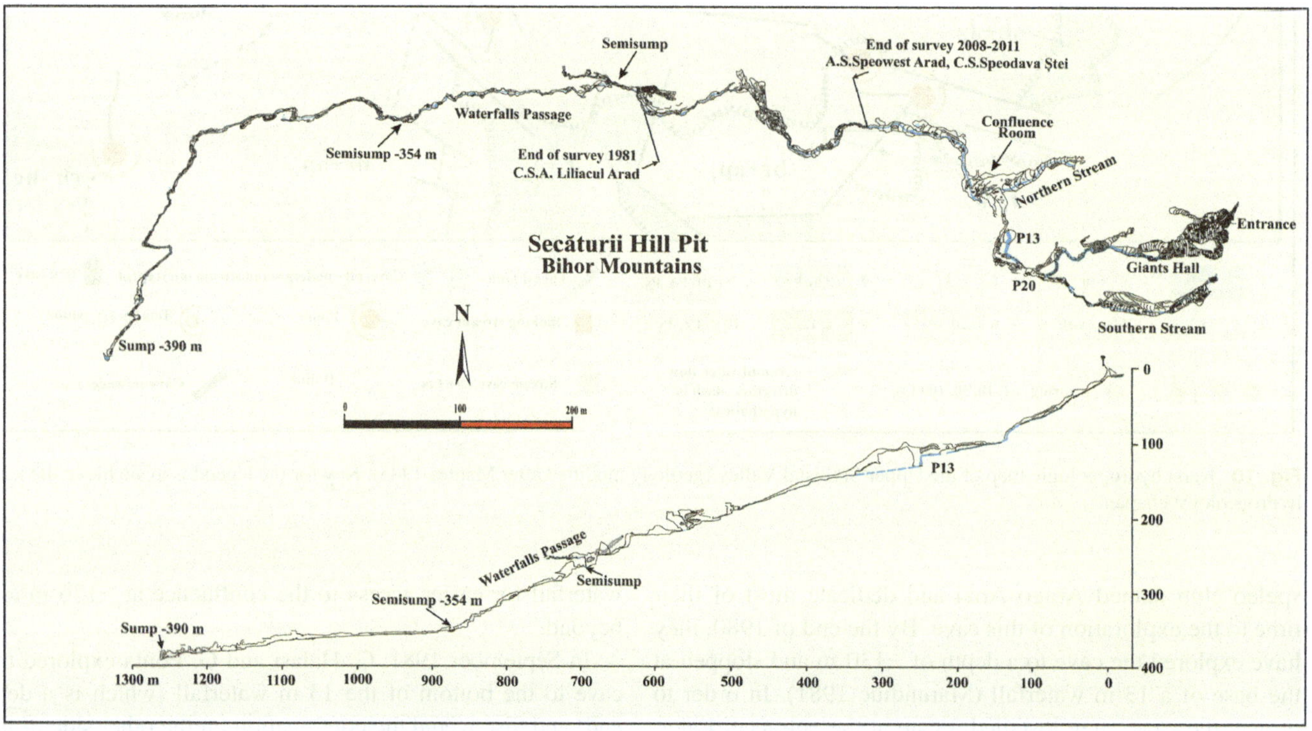

Fig. 11 Map of Secăturii Hill Cave (surveyed by: G. Halasi, G. Ponta, R. Ermesz, V. Brinzas (1981) and M. Besesek, V.A. Radu, R. Sarkozi, A. C. Oncu, S. Sebeşan, V. Şiclovan, R. Lazăr, A. Suciu, M. Velin, L.R. Ţociu, C. Bâc, F. Vida, T. Rus, P. Brijan, P. Brănescu, D. Lup, V. Gonceaov, A. Micula, A. Tocuţ, S. Ene, and D. Inăşel (2008–2012))

semisump located at −354 m. Beyond this squeeze, the cave leaves the contact with impermeable rocks and continues with a gallery totally carved in limestones. First horizontal portion is a spacious gallery, then alternate narrow and wider passages, starting to show signs of flooding and clastic sediments. In this sector, depending on the rainy/dry season, the stream may be temporary, when it disappears into the sediment or various impenetrable fractures. The filling is mostly clay sediment, pebbles, and collapse material. The final part is represented by a narrow canyon-type passage with an active stream, which opens in a gallery with chaotic aspect, where water disappears into the sediment, to reappear just before the final sump at −390 m elevation.

Cave Sediments and Speleothems

Between the entrance and −354 m, the galleries are developed mainly on the lithological contact between Upper Jurassic limestones and impermeable clay and marl of Upper Triassic–Lower Jurassic age. This sector is characterized by the presence of numerous and divers speleothems, with occasional spectacular white flowstones. Rocks and small quantities of clastic sediments cover the cave floor. Between −354 m and the terminal sump, the sediments appear in significant quantities, this section showing clear traces of flooding.

Cave Conservancy

The cave was gated shortly after its discovery limiting the underground access, thus is presently in a good state of preservation. Between entrance and −176 m, the access trail shows slight traces of degradation due to repeat visits in the cave. Southern and northern stream passages including the galleries to the final sump are very well conserved.

Coliboaia Cave

Geographic, Geological, and Hydrologic Settings

Located in the Apuseni Natural Park (Bihor Mountains, Câmpani village), the entrance in the Coliboaia Cave is on the right side (northern) of the Sighiştel Valley, at 560 m altitude, ~4 km upstream of Sighiştel village, near the entrance of the Măgura Cave (Fig. 10).

History of Exploration

Coliboaia Cave is mentioned in speleological literature for the first time by the Austrian geographer Schmidl (1863), later being visited by Jeannel and Racoviţă (1929), and scientists from the "Emil Racoviţă" Institute of Speleology in Cluj-Napoca. In the early eighties, G. Halasi investigated the active gallery reaching the fourth sump, mapped the Art Gallery (Galeria de Artă), but never noticed the paintings at that time (Halasi and Ponta 1984).

Coliboaia was initially explored only to Sump 3, represented by a narrow flooded passage with the ceiling descending to the water level. R. Ermesz (CSA Liliacul Arad) crossed this passage in 1981. Later that year, G. Halasi and D. Coloji explored and mapped 400 m of galleries behind the Sump 3, including the Art Gallery. Another trip in this part of the cave was accomplished by P. Brijan and D. Mateuţă (C. S. Speodava Ştei), who reached the Grifade Hall (Sala Grifadelor), located above the Sump 4, where they observed a significant bone deposit and numerous scratches on the walls.

In 2009, A. S. Speowest Arad (M. Besesek, V. A. Radu, R. Ţociu) with C. S. Speodava Ştei (T. Rus) and M. Kenesz from A. Speomontană Zărand Brad rediscovers the fossil gallery located behind Sump 3 and notices the Paleolithic paintings for the first time (Besesek et al. 2010, 2010–2011). The team also managed to pass Sump 4 through a narrow passage and explored over 200 m of an ascending active gallery to the base of a waterfall (8 m high). On the way back, they crossed the sump located south of Art Gallery on a side passage, exploring an additional 100 m to a large chamber with breakdowns and a massive flowstone. In 2012, M. Besesek, V. Radu, and R. Sarkozi continued the explorations behind Sump 4 climbing several waterfalls but stopped due to lack of equipment in a room at the base of another 7 m high waterfall.

Cave Description

Coliboaia has a big entrance, from where the main gallery continues through an opening of ~1 m high to a large hall, with a stream at the lower end. During dry seasons, the stream disappears gradually in a series of impenetrable ponors in the entry zone, to reappear in the Sighiştelului Canyon at bottom of a cone of scree beneath the cave portal. When ponors cannot drain all the water, a lake occupies almost the entire width of the main gallery on a length of 150 m (Fig. 12). Upstream, the water flows through wide galleries and chambers connected by narrow passages (three sumps), flooded nearly all year long. The fourth sump is at 1066 m from the entrance and a vertical range of +15 m.

Beyond the third sump, along the mainstream passage, the Art Gallery is reached by climbing a flowstone. The gallery is 7 meters above the actual stream level of the cave and hosts multiple Paleolithic paintings. Above Sump 4 was discovered a fossil hall about 10 × 10 m with the floor covered with a thick layer of sediment, collapses blocks, a large bones deposit and grifade on the walls (Grifade Hall). From this room, through a narrow, 40 m long passage with only 5 cm airspace, the cave continues along an ascending gallery developed on the lithological contact with non-calcareous rocks. After 400 m, the exploration stopped in 2012 at the base of a 7 m waterfall.

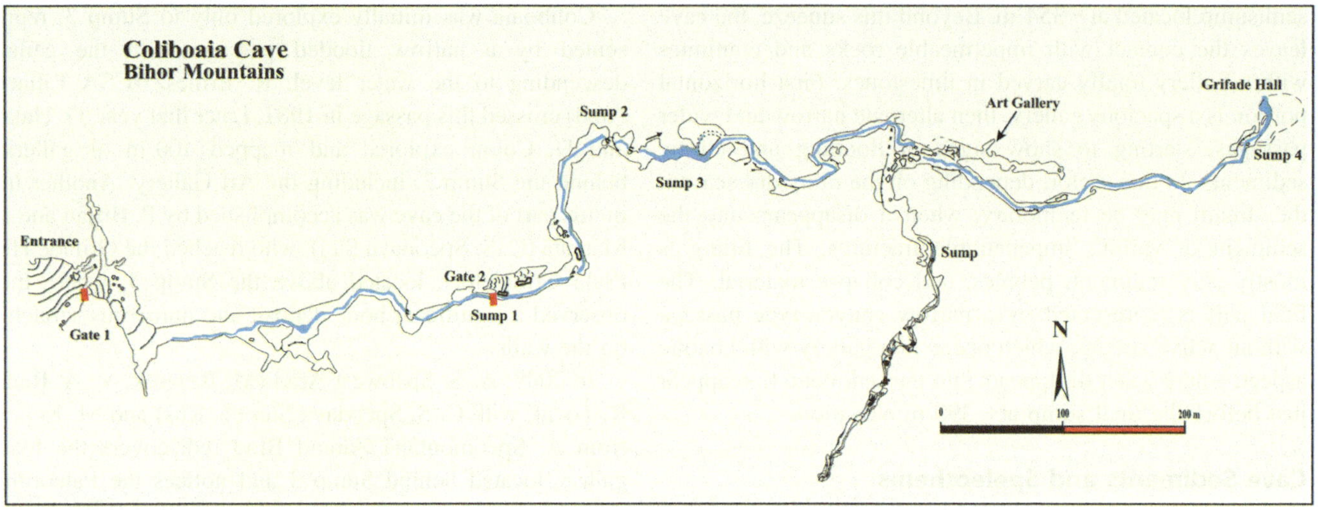

Fig. 12 Map of Coliboaia Cave (surveyed by: R. Ermesz, G. Halasi, D. Coloji, G. Ponta, P. Brijan, I. Mateuţă, L. R. Ţociu, T. Rus, M. I. Kenesz, V. T. Lascu, R. V. Sarkozi)

Cave Climate

Coliboaia shows a unidirectional airflow, during the summer air moves into the cave and opposite in the winter. The measurements recorded in the Art Gallery from June 2010 to February 2011 indicate that the air temperature fluctuated between 9.8 and 10 °C, whereas the relative humidity remained 100%.

Anthropology and Paleontology

In several caves along the Sighiştelului Valley, including Coliboaia, Paleolithic to the Middle Ages, artifacts were discovered. Coliboaia stands out through the Paleolithic cave drawings and one engraving discovered in 2009 behind Sump 3. The drawings (9) are in black pigment, probably made with charcoal (Clottes et al. 2011a, b). There are representations of animals: bison, horses, undetermined ones (horse or felid), one or two bear heads, a possible mammoth, and rhinoceros heads. At least one engraving is also visible (Ghemiş et al. 2011a, b). Based on radiocarbon testing of a sample collected from one of the painted animal and a piece of charcoal found within the drawing, the following ages were obtained:

- Painting: 27870 ± 250 (GIFA 11002), calibrated to 31450/32820 years BP;
- Charcoal: 31640 ± 390 (GIFA 11001), calibrated to 35120/36780 years BP.

These data confirm the very old age of the Coliboaia drawings, the only site in Central and Eastern Europe with parietal art. The oldest date (~36,000 years Cal BP) is similar to those obtained on the cave paintings of Chauvet Cave (Ardèche, France) (Clottes et al. 2011b).

Cave Conservancy

Coliboaia Cave is closed with two gates and is classified as a category a historic monument, being administered by the "Ţării Crişului" Museum in Oradea and the Apuseni Natural Park.

Acknowledgements Many thanks to all the cavers who explored and surveyed the Coliboaia Cave, *Survey team*: R. Ermesz, G. Halasi, D. Coloji, G. Ponta, P. Brijan, I. Mateuţă, L. R. Ţociu, T. Rus, M. I. Kenesz, V. T. Lascu, and R. V. Sarkozi; *Scientific team*: J. Clottes, B. Gely, F. Prud'Homme, M. Meyssonnier, V. Plichon, M. Venczel, E. Posmoşanu, A. Posmoşanu, I. Meleg, and A. Pitic; *Cave conservancy and support team*: M. Teuşdea, A. Dorotovics, A. C. Oncu, C. Bâc, M. Velin, V. Goncearov, A. Baltă, S. C. Sebeşan, A. Szilvagyi, A. Lăzău, L. Frenţiu, A. Buşe, A. Micula, and A. Ardelean.

References

Aldica G (2009) Peştera Cetatile Ponorului. In: Dumitru R (ed) Focul Viu 40 de ani de speologie. Ed RAO, Bucureşti, pp 96–97

Bălteanu D, Jurchescu M, Surdeanu V, Ionita I, Goran C, Urdea P, Rădoane M, Rădoane N, Sima M (2012) Recent landform evolution in the Romanian Carpathians and Pericarpathian regions. In: Lóczy D et al. (eds.) The carpatho–balkan–dinaric region, geography, vol 10. Springer, pp 249–286

Besesek M, Brijan P, Radu VA, Tulucan ST (2009) Sistemul Peştera din Dealul Secăturii - Peştera Coliboaia. Speomond 13:19–22

Besesek M, Gely B, Lascu V T, Radu VA (2010) Découverte d'une nouvelle grotte ornée paléolithique (Pestera Coliboaia), Roumaine, Département du Bihor. Int Newslett Rock Art 57:8–11

Besesek M, Radu VA, Rus T (2010–2011) Arta parietală paleolitică din România - Peştera Coliboaia. Speomond 15:28–34

Bleahu M, Decu V, Negrea Ş, Pleşa C, Povară I, Viehmann I (1976) Peşteri din România. Ed Ştiinţifică şi Enciclopedică, Bucureşti, p 415

Bosdoc T (2014) Amintiri de pe Râul fără Stele. Speomond 17:4

Cârciumaru M, Dobrescu R (2010) Recherches sur le terrain et sondages archeologiques dans la Vallée du Sighiştel, commune de Câmpani, departament de Bihor. Ann Univ Valahia Târgovişte, Archeol Hist XII(1):185–189
Clottes J, Besesek M, Gely B, Ghemiş C, Kenesz MJ, Lascu VT, Meyssonnier M, Philippe M, Plichon V, Prud'Homme F, Radu VA, Rus T, Ţociu RL (2011a) La grotte ornée Coliboaia - une découverte majeure par des spéléologues roumains. Spelunca 124:35–40
Clottes J, Gely B, Ghemiş C, Kaltnecker E, Lascu VT, Moreau C, Philippe M, Prud'Homme F, Valladas H (2011b) Un art très ancien en Roumanie. Les dates de Coliboaia. Int Newslett Rock Art 61:1–3
Czaran J (1903) Kalauz Biharfuredi Kirandulasokra, Beiuş, p 262
Ghemiş C, Clottes J, Gely B, Prud'Homme F (2011a) An exceptional archaeological discovery – the Gallery with Drawings from Coliboaia Cave, Apuseni Mountains, Romania. Acta Arch Carph XLVI:5–18
Ghemiş C, Clottes J, Gely B, Prud'Homme F (2011b) O descoperire arheologică de excepţie - Galeria desenelor din Peştera Coliboaia, Munţii Apuseni (România). Crisia XLI:51–64
Gligan M (1987) Drenaje subterane în zona de izvoare a Someşului Cald (M. Bihor). Bul Speol 11:25–32
Halasi G, Ponta G (1984) Subterranean drainage in the upper part of the Sighiştel Valley (Apuseni Mountains). Theor Appl Karstol 1:239–242
Jeannel R, Racoviţă EG (1929) Enumération des grottes visitées, 1918-1927 (7e série). Arch Zool Exp gGén 68(2):293–608
Mantea G (1985) Geological studies in the upper basin of the Someşul Cald Valley and the Valea Seacă Valley Region (Bihor-Vlădeasa Mountains). An lnst Geol Geofiz 66:1–91
Marandiuc C (1981) Raport de activitate pe perioada Nov. 1979 - Oct. 1980. Grupul Geo-Speologic "Arago" Arad. Bul Speol 5:128–130
Nagy S (1885) Biharország I–II–III, Nagyvárad
Orăşeanu I (1996) Contribution to the hydrogeology of the karst areas of the Bihor-Vlădeasa Mountains. Theor Appl Karstol 9:185–214
Orăşeanu I (2010) Karst hydrogeology of Apuseni Mountains. In: Orăşeanu I, Iurkiewicz A (eds) Karst hydrogeology of Romania. Ed Belvedere, Oradea, pp 181–324
Orăşeanu I (2016) Hidrogeologia carstului din Munţii Apuseni. Ed Belvedere, Oradea, p 289
Ponta G (1992) Hydrogeological map of Pietroasa (unpublished). Scale 1:50.000, Inst Geol Geofiz, Bucureşti
Ponta G (1997) Karst research in Romania: geological and hydrogeological setting of Bihor Mountains. In: Beck FB, Stephenson JB (eds) The engineering geology and hydrogeology of karst terranes, Proceedings of 6th multidisciplinary conference sinkholes engineering and environmental impacts of karst, Springfield, Missouri, 6–9 April, 1997, pp 97–103
Ponta G (1998) Karst of Romania. In: Yuan D, Liu Z (eds) Global karst correlation IGCP 299. Science Press, Beijing, pp 193–207
Ponta G, Selişcan R, Şteflea D (1992) The hydrogeology of Padiş karstic plateau - Bihor Mountains, Rumania: In: Hotzl H, Werner A (eds) Proceedings of 6th International Symposium on Water Tracing, Karlsruhe, Germany, pp 283–286
Ponta G, Selişcan R, Şteflea D (1993) Subterranean stream piracy in the upper basin of the Someşul Cald Valley Area, Bihor Mountains-Rumania. Proc Eur Speleol Conf, Liège, pp 73–77
Schmidl A (1863) Das Bihar-Gebirde an der Grenze von Ungar und Siebenburgen. Verl Forster u Bartelmus, Wien
Şerban M, Viehmann I, Coman D (1961) Grottes de Roumanie. Ed Meridiane, Bucharest
Vălenaş L (1984) Studiul complex al carstului din zona Izvorul Ursului-Pîrîul Sec (Munţii Bihor). Crisia XIV:559–580
Vălenaş L, Bleahu M, Brijan P, Halasi G (1977) Inventarul speologic al Munţilor Bihor. Nymphaea 5:209–335
Viehmann I, Cristea E, Şerban M, Cuc O, Ghiţea S (1980) La morphologie du complexe karstiques "Cetatile Ponorului" (Les Monts Apuseni, Roumanie). Trav Inst Spéol "Émile Racovitza" XIX:261–274

Useful Websites

https://www.facebook.com/Pe%C5%9Ftera-Coliboaia-603487983057479/?ref=hl
http://www.Speowest.ro/index2.php?p=coliboaia
https://ro.wikipedia.org/wiki/Pe%C8%99tera_Coliboaia

Bihor Mountains: Drăcoaia Cave

Jozef Psotka and Vladimír Papáč

Abstract

Drăcoaia Cave was known and visited by local people and tourists for a long time. The cave has an impressive entrance (15 × 12 m, width × height), which continues with a slightly ascending passage that ends after 150 m in a massive flowstone obstruction. The exploration activities conducted by Slovak cavers between 2002 and 2015 led to significant new discoveries in the cave, which now reaches 1963 m in length and a total vertical relief of 107 m. Cave consists of two levels of vadose canyon passages with sub-horizontal gradient, which are interconnected by steeply inclined meandering passages. Cave was formed by sinking streams draining from overlying non-karst Permian rocks. It hosts an endemic species of springtail, *Megalothorax draco*.

Keywords

Drăcoaia Cave • Bihor Mountains • Speleology Biospeleology

Introduction

Drăcoaia Cave is located in the Sighiştel Valley (western part of the Bihor Mountains, Fig. 1), a well-known karst area with one of the greatest density of caves in Romania. The majority of known caves are dry, but their morphology and sediment deposits suggest they functioned as active stream passages in the past. On the northern side (right) of the Sighiştel Valley, caves such as Măgura or Coliboaia are longer than 1 km. However, none of the caves on the southern side exceeded 500 m in length before the breakthrough in Drăcoaia Cave.

J. Psotka (✉)
Slovak Museum of Nature Protection and Speleology, Školská 4, Liptovský Mikuláš, Slovakia
e-mail: jozef.psotka@gmail.com

V. Papáč
Speleoclub Drienka, Kirejevská 37, Rimavská Sobota, Slovakia
e-mail: vladimir.papac@ssj.sk

Geographic, Geologic, and Hydrologic Settings

The Sighiştel Valley is a right-side tributary of the Crişul Negru River, which excavated a deep fluviokarst valley with narrow gorges in its upper section. The Sighiştel Valley karst area is about 15 km^2 (Halasi and Ponta 1984) and host around 200 caves. According to the geological map, the karst rocks in the Sighiştel Valley belong to the Vălani Nappe of the Codru Nappe system and comprise mainly Upper Jurassic and Lower Cretaceous limestones (Bleahu et al. 1985). Violaceous continental deposits; sandstones, conglomerates and green, weakly metamorphosed marine deposits of Permian age attributed to Arieşeni Nappe overlie the Mesozoic limestones. The contact between the Mesozoic and Permian rocks is a thrust plane, disrupted by N/NW–S/SE, NW/SE, and W/SW–E/NE faults. Granitic dykes of Upper Cretaceous age penetrated Mesozoic limestones in the SE part of the karst area (Balintoni 2001).

On both sides of the Sighiştel Valley, small surface streams formed on Permian rocks flow until they enter the limestone where quickly disappear underground. There are several perennial (one under Coliboaia Cave, Pişolca Cave, Blidaru, and Hidra Spring) and temporary karst springs in the valley. The most significant is Blidaru Spring located in front of the cave Răsuflătoarea Blidarului (about 300 m long), on the left side of the valley, 200 m downstream from Drăcoaia Cave. Numerous tracing experiments documented hydrological connection between a number of ponors from the NW slopes of Prislop Peak and the western slopes of Ţapu–Pietrele Negre ridge and Blidaru Spring, as well as other smaller sources along the valley (Orăşeanu et al. 1991; Oraseanu 1996, 2010). The longest proved hydrological connection is between Dosu Muncelului Pit (1100 m asl) and Blidaru Spring (435 m asl)—665 m vertical

G. M. L. Ponta and B. P. Onac (eds.), *Cave and Karst Systems of Romania*,
Cave and Karst Systems of the World, https://doi.org/10.1007/978-3-319-90747-5_37

Fig. 1 Location of Drăcoaia Cave

development and almost 4 km apart. According to results of the tracing experiments, the water disappearing underground in Pietrele Roşii Creek located in the Crăiasa River basin is connected with all main karst springs in Sighiştel Valley with most of the waters feeding the Blidaru Spring (Orăşeanu 2010).

History of Exploration

The remarkable large entrance passage of Drăcoaia Cave (Fig. 2) has been known for a long time. The first detailed description of the cave was included in Schmidl's (1863) book about the Bihor Mountains. Probably, explorers in the past thought about the cave continuation behind the massive flowstone obstructing the main passage, but no one ever tried to search for a way forward.

In July 2002, Slovak cavers noticed a small, fist-sized hole with a strong air current at the end of the cave. After a few intensive days of digging gravels out from the riverbed, the narrow passage led to a five-meter chimney ending with an inaccessible narrow passage (Horčík and Papáč 2002). Later, the cavers from Speodava Ştei Club tried without success to enlarge it. The entrance passage was resurveyed (Fig. 3) and published along with a short note (Máté 2007). In 2007, cavers continued to drill, and finally, in September 2008 more than 550 m of new corridors were discovered (Papáč and Psotka 2008, 2009). After surveying the new passages and exhausting all possibilities of continuation, we begun looking for further promising sites for exploration. In October 2009, a strong air current in a narrow meander at the bottom of the *Column Shaft* was noticed. Digging in this flooded location was very difficult; therefore, a manual water pump was used to remove the water. The 80-m-long *Cruel Passage* (the most difficult place in the cave) was finally passed in August 2010 and led to the discovery of 900 m of new galleries, the *Canyon, Gypsum Gallery,* and two sumps (Papáč 2013). In May 2012, cavers felt an air current at the end of the *Gypsum Gallery*; they started to dig again, since this is the most promising place to continue the exploration in Drăcoaia Cave. In August 2017, both sumps in *Canyon* were explored by cave divers D. Hutňan and K. Kýška,

Fig. 2 Entrance portal to Drăcoaia Cave (photograph by V. Papáč)

upstream sump ended in narrow constriction and downstream sump continued with dry gallery continuing with unexplored 5-m-deep shaft. Including sump, they surveyed 115 m of new passage.

Cave Description

The cave has an impressive entrance, which continues with a slightly ascending passage that ends after 150 m in a flowstone barrier with breccia and cemented fluvial sediments (Fig. 4). The first 100 m of the gallery is 10 × 10 m, becoming smaller in the vicinity of the flowstone obstruction (2–4 m wide and up to 6 m high). The floor of the entrance passage (ca. 50 m) is covered by angular debris and fallen blocks. Muddy rimstone dams and small lakes are covering the floor in the upstream part. Relics of ceiling channels and meanders, and remnants of large stalactites are visible on the cave ceiling.

Beyond the *First squeeze* (Fig. 5), climbing a 5-m-high chimney leads to the narrow *Discovery Meander,* which was artificially enlarged for over 16 m. The meander ends in a small room, 27 m above the squeeze, with anastomoses in the ceiling.

From here, after passing another tight spot and some fallen blocks, we approach the *Crossroad Passage*, from where the cave continues in several directions. A horizontal canyon-like, 8-m height gallery (*Crystal Passage*) develops toward northwest. It has an interesting morphology (cross section), and hosts crystals and fine-grained deposits. The *Crystal Passage* (Fig. 6) is terminated by sediment infilling all the way to the ceiling, where a fault is visible as several tens of cm wide zone of deformed rocks exposed on the wall. Another direction to follow from the *Crossroad* is along a steeply ascending corridor named *Meander Echo,* which continues the *Discovery Meander* as 15-m-high inflow chimney along which drips abundantly and narrows upwards.

Directly above the *Crossroad* is a large 40-m-high chimney, which branches in two directions. The largest passage yet discovered is the canyon-like *Channel Passage*, 8 m high and 2–4 m wide (Fig. 7a, b) with a flat ceiling and wall channels filled with sediments. Fluvial deposits, as gravel, sand, and clay, are present on the floor. Further on, the passage becomes narrow and carries a small temporary stream flowing directly on the bedrock. The cavers advanced slowly digging their way along this passage until stopped by blocks and boulders of Permian rocks. Before this section of the *Channel Passage,* it turns a 60-m-long and 10-m-high meander that ultimately leads to the *Column Shaft* (Fig. 7c). The shaft is 18 m deep and connects to the 80-m-long very tight meander called the *Cruel Passage*, which opens into an impressive 30-m-high *Canyon* (Fig. 7d). This is a vadose sinuous passage with a small active stream, originating from the *Inflow Sump*. In the upper parts of the passage, the former phreatic tube is preserved, whereas in the lower part, several levels of sub-horizontal wall notches indicate gradual vadose incision. The *Canyon Passage*'s floor is bare bedrock, but in the vicinity of the sump, gravel sediments and thick clay deposits occur.

From the *Canyon*, an ascending meandering passage with gravel deposits on the floor leads into the upper fossil level, which includes the *Nicolae Groza Hall* and the *Gypsum Gallery.* The first one is terminated by breakdown and fluvial sediments infilling all the way to the roof, whereas the *Gypsum Gallery* is heading toward NW as a nearly horizontal passage with abundant speleothems and traces of fluvial deposits. Anastomoses and relics of cemented gravels on the walls indicate periods of significant sediment accumulation. In fact, the downstream part of the *Gypsum Gallery* (its present terminus) is blocked by fluvial sediments in

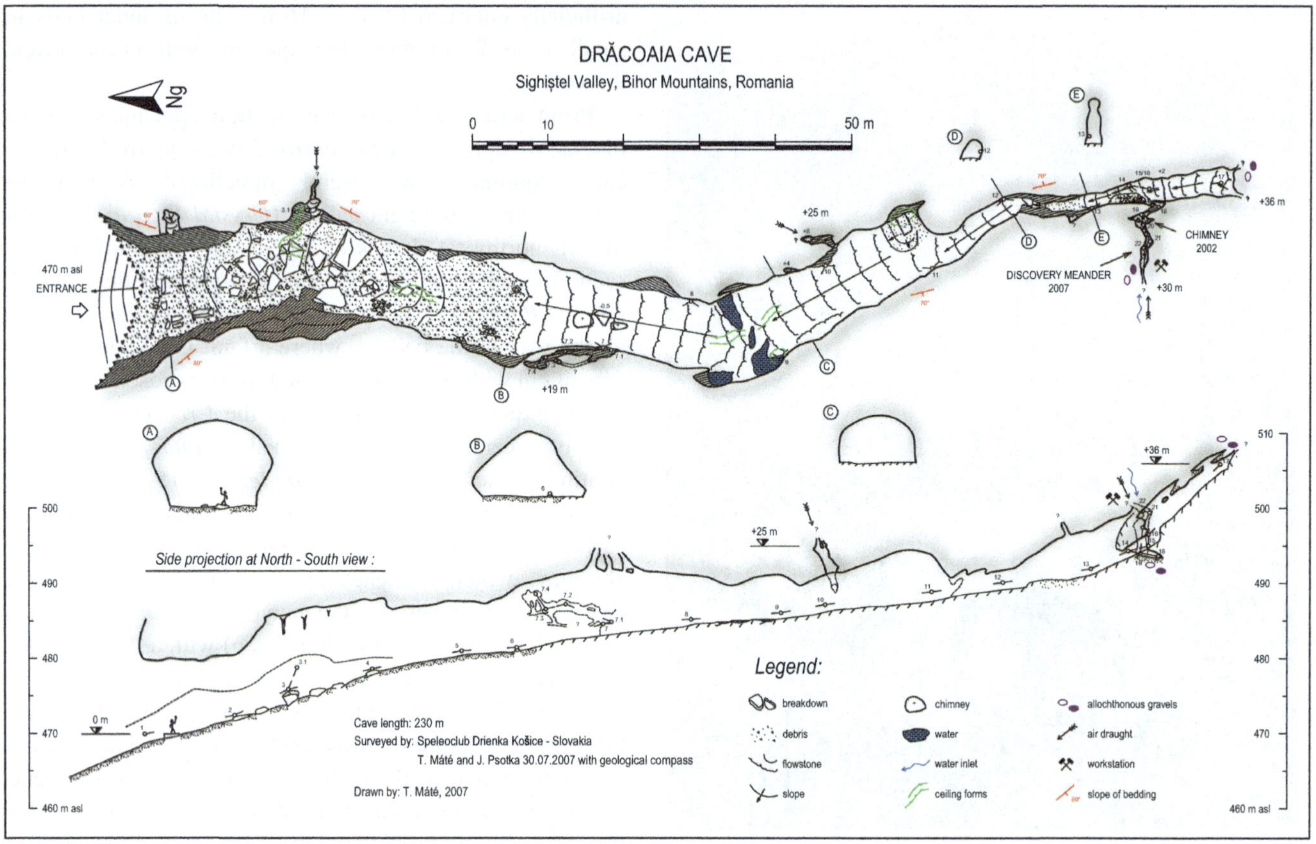

Fig. 3 Map of the entrance section in Drăcoaia Cave showing the location of the breakthrough chimney

which cavers dug a 20-m-long crawlway stimulated by the presence of a strong air current.

Cave Sediments and Speleothems

Autochthonous clastic cave deposits include limestone blocks and debris derived from collapses of the ceiling and cave walls. In the uppermost passages, large breakdown along fault line terminated former tributary passages. Most common deposits in the cave are allochthonous fluvial clastic sediments derived from the overlying Permian siliciclastic rocks (violet sandstones and conglomerates). Grain size ranges from clay, silt, to boulder. The majority of these deposits are preserved in the fossil passages in the form of past sedimentary infill. Modern active stream deposits are mostly thick muddy sediments located near the *Outflow Sump*. Chemical deposits include common types of speleothems (stalactites, soda straws, stalagmites, and flowstones). Speleothems and minerals of this cave have not been yet studied.

Cave Speleogenesis

Based on the preserved fluvial sediments, the cave was formed by aggressive allogenic waters that flowed over non-karst Permian sedimentary rocks. From the cave survey and surface topographic map, it is obvious that most of the sub-horizontal passages were formed by sinking streams originating in the nearby Frăpsineasa Valley. According to the cave survey, there are two distinct levels of sub-horizontal stream passages. They are probably related to a former local base level and each developed as vadose passages by independent streams. Later, they were interconnected through narrow steep meanders with vertical drops. Ascending parts of these meanders become vertical shafts/chimneys that end in fluvial deposits mixed with angular boulders or blocks of sandstone, from which water trickles into the cave. On the land surface above the cave, along the Frăpsineasa Valley, several small gullies in sandstone acting as temporarily active streams were identified. It is assumed that water from these gullies is percolating to the

Fig. 4 Photograph of the large gallery near the cave entrance (photograph by V. Papáč)

underlying limestone and created narrow, steeply inclined vadose passages with vertical drops.

Cave Climatology

Temperature and relative humidity (RH) were recorded during one trip (December 21, 2008) with a thermo hygrometer COMET 3120. At this date, the outside temperature was 2.5 °C, whereas at the cave entrance, temperature and RH reached 4.3 °C and 74%, respectively. In the *Channel Passage,* both temperature and RH increased to 9.1 °C and 94.5%; in the *Crossroad Passage,* the temperature was 8.1 °C and the relative humidity 93.2%. The differences between the *Channel* and *Crossroad* passages were probably caused by intense air current blowing from the *Discovery Meander* and the presence of a temporary active stream on the *Crossroad.* A very strong air current blows out from the *First squeeze* in the summer months and reverses in the winter. This suggests that Drăcoaia is the lower entrance of a much larger cave network.

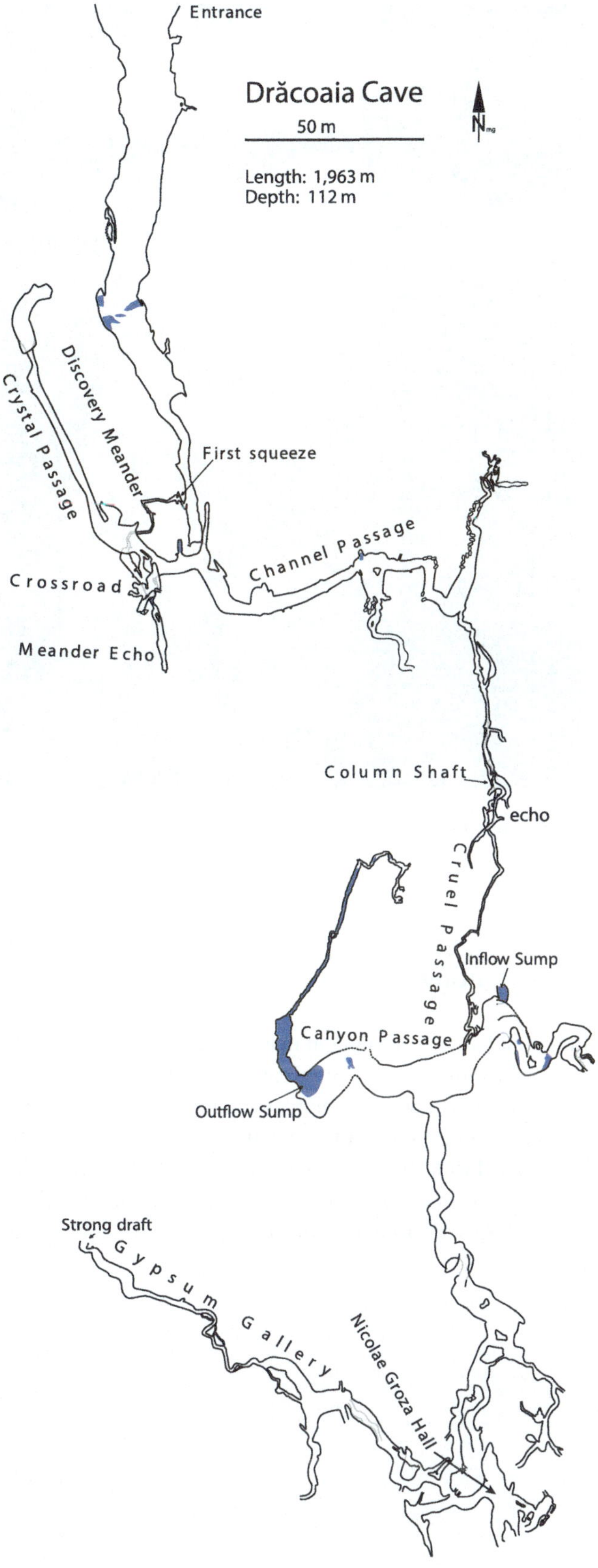

Fig. 5 Map of Drăcoaia Cave (drawn by M. Horčík, D. Hutňan, P. Imrich, T. Máté 2008–2017, completed by P. Imrich)

Fig. 6 Entry to the Channel Passage (photograph courtesy of Photo M. Horčík)

Cave Biology

Endemic fauna was found in many caves of Sighiştel Valley. The early biospeleological research focused on invertebrates and revealed a new troglobiotic springtail, *Onychiurus ancae* (Gruia 1971) in Măgura Cave and in other six caves nearby Drăcoaia. From this, last cave was described a troglophilic springtail (*Lepidocyrtus serbicus*), which is one of the most frequent species found in the Romanian caves (Gruia and Illie 2000). Endemic and highly troglomorphic species of springtail *Megalothorax draco* was discovered in the stream and pools in Drăcoaia Cave and remains the only known location (Papáč and Kováč 2013). During a sustained zoological campaign conducted in caves of the Sighiştel Valley, a new isopod, *Mesoniscus graniger dragani* was discovered (Giurginca 2003). This is a blind and depigmented troglophilous species, which occur mainly in caves with organic material like guano or rotten wood. These are endemic subspecies restricted to caves of the Sighiştel and Crăiasa valleys (Giurginca et al. 2015). According to our observations (since 2012), every summer more than 1000 individuals of Schreiber's bat (*Miniopterus schreibersii*) congregate to form a nursery colony.

Conclusions

Drăcoaia Cave (1963 m long) is a stream cave located on the left (south) side of Sighiştel Valley at an altitude of 470 m asl. The huge entrance allows access to a wide corridor ending in a massive flowstone. In 2002, enlarging a small tributary chimney located in the entrance chamber led to a complex network of mostly dry meandering passages. Between 2008 and 2012, Slovak cavers from Speleoclub Drienka and Šariš discovered more than 1700 m of new corridors, making Drăcoaia one of the longest caves in the Sighiştel Valley. The largest passages in the cave were formed by allogenic sinking streams from the nearby Frăpsineasa Valley, whereas the narrow steep meanders and chimneys were formed by aggressive water percolating from overlaying non-karst Permian siliciclastic rocks. This has led to the complex morphology of the cave with two levels of independently formed stream passages later interconnected by steep meanders and shafts. Although the cave is not impressively long, it clearly offers possibilities for further discoveries, especially if taking into consideration the results of the water tracer tests. The cave hosts an endemic species of springtail, *M. draco* and a sizeable maternity colony of *M. schreibersii*.

Fig. 7 **a** Morphologically interesting Crystal Passage with occurrence of crystals, **b** Channel Passage with flowstone decoration, **c** Column Shaft, **d** image from the Canyon Passage (photographs by V. Papáč)

Acknowledgements We thank to Romanian cavers from Speodava Ştei: P. Brănescu, T. Rus, M. Petrescu, and M. Inăşel for their permanent help during exploration in Sighiştel Valley. We are also grateful to the Apuseni Natural Park for granting permission to conduct speleological research in Drăcoaia Cave. Special thanks to all expeditions participants: M. Gaško, V. Papáč, J. Psotka, T. Máté, M. Miškov, Š. Kyšela, F. Horčík, M. Horčík (Speleoclub Drienka), P. Imrich, F. Majerníčková, M. Hajduk (Speleoclub Šariš), J. Šmoll (Speleoclub Červené vrchy) and I. Balciar (Speleological group Rimavská Sobota).

References

Balintoni I (2001) Short outlook on the structure of the Apuseni Mountains. In: Bucur II, Filipescu S, Săsăran E (eds) Fourth regional meeting of IFAA, Field Trip Guide, Cluj, Romania, pp 9–17

Bleahu M, Bordea S, Bordea J, Mantea G, Popescu A, Marinescu F, Cioflica F, Ştefan A (1985) Harta geologică a României, Pietroasa Sheet. Scale 1:50.000. Inst Geol Geofiz, Bucureşti

Giurginca A (2003) *Mesoniscus graniger dragani* n. ssp. from the caves of Valea Sighiştelului and Valea Crăiasa. The granulation at *Mesoniscus graniger*. Trav Inst Spéol "Émile Racovitza" 39–40: 45–56

Giurginca A, Munteanu C-M, Vlaicu M, Tăbăcaru I-G (2015) Cavernicolous Oniscidea of Romania. Ed Semne, Bucureşti, p 165

Gruia M (1971) Sur certaines espèces d'Onychiuridae (Collembola) des gorttes de Roumanie. Acta Zool Cracov 26(4):283–288

Gruia M, Illie V (2000) Collembola from the karstic system of Romania (II). Trav Inst Spéol "Émile Racovitza" 39–40:59–114

Halasi G, Ponta G (1984) Subterranean drainage in the upper parts of the Sighiştel valley (Apuseni Mountains). Theor Appl Karstol 1:239–242

Horčík F, Papáč V (2002) Rumunsko 2002. Spravodaj SSS 33(4):22–24

Máté T (2007) Bazinul Văii Sighiştel. Speomond 12:46

Orăşeanu I (1996) Contributions to the hydrogeology of the karst areas of the Bihor-Vlădeasa Mountains (Romania). Theor Appl Karstol 9:185–214

Orăşeanu I (2010) Bihor Vlădeasa mountains. In: Orăşeanu I, Iurkiewicz A (eds) Karst hydrogeology of Romania. Ed Belvedere, Oradea, pp 217–243

Orăşeanu I, Gaşpar E, Pop I, Tănase T (1991) Tracers experiments in the karst area of Bihor Mountains (Romania). Theor Appl Karstol 4:159–172

Papáč V (2013) Exploration of Drăcoaia Cave in Romania. Bull Slovak Speleol Soc (Issued for 16th Int Cong Speleol), Brno 2013, pp 93–96

Papáč V, Psotka J (2008) Peştera Drăcoaia. Speomond 13:39–42

Papáč V, Psotka J (2009) Jaskyňa Drăcoaia v doline Sighiştel, pohorie Bihor, Rumunsko. Spravodaj SSS 1:66–72

Papáč V, Kováč Ľ (2013) Four new troglobiotic species of the genus *Megalothorax* Willem, 1900 (Collembola: Neelipleona) from the Carpathian Mountains (Slovakia, Romania). Zootaxa 5:545–575

Schmidl A (1863) Das Bihar-Gebirge an der grenze von Ungarn und Siebenburgen. Verlag Förster und Bertekmus, Wien, p 442

Bihor Mountains: Vărăsoaia Cave System (V5)

József Zih, Katalin Zih-Perényi, and Ludovic Mátyási

Abstract

Vărăşoaia cave system (VCS) was discovered in 1986 in the Bihor Mountains of Western Romania, under the Padiş Plateau. After countless rounds of in-cave blasting aiming to overcome some very tight passages, the cavers managed to intercept a complex karst system with several very large active (carrying underground streams) and dry galleries and reached the −653 m in depth. To date, VCS is the deepest explored cave of Romania. The network currently has over 25 km of mapped passages, but explorations are still underway. Due to floods risk and high collapse possibility in some cave sections, visiting permits are granted only to experienced cavers.

Keywords

Karst • Pothole • Padiş Plateau • Morphology

Geographic, Geologic, and Hydrologic Settings

The Vărăşoaia cave system is located in the central-northern part of the Bihor Mountains, on the northern edge of the Padiş-Cetăţile Ponorului closed karst catchment (Fig. 1). The highest of the V5 pothole entrances is located at 1367 m above sea level (asl). The ponors in this region are located between 1250 and 1150 m altitude and at the western end of the streams disappearing in them, thus suggesting flow towards Boga Spring (663 m asl). Geologically, the Vărăşoaia-Şesul Padişului-Boga region that hosts VCS is part of the Bihor Unit, which includes in this area a lower impervious layer of Triassic sandstones, covered by a middle Triassic carbonate sequence (Bleahu et al. 1981). VCS develops in the latter unit composed of dolomite and limestone of Anisian age (Damm 1996). The strike of the entire lower and middle Triassic sequence is NNW–SSE and dips about 35° towards WSW. From a structural point of view, the region is characterized by a gentle monocline with the axis oriented NNW–SSE. The tectonic events associated with the Alpine Orogeny that uplifted the Carpathians are responsible for the network of fault systems, which had a significant role in the speleogenesis of VCS. The main fault directions are generally NE–SW and NNW–SSE.

From a hydrological point of view, the lithostratigraphy and tectonics of the region control the patterns of surface river network. The underground drainages in this area were evaluated via a number of tracer tests conducted by a team led by I. Orăşeanu between 1983 and 1986 and synthesized later in a number of publications (Orăşeanu et al. 1991; Orăşeanu 1996, 2010). The dye and tracer studies performed in 1985 in three ponors (Vărăşoaia, Cuţilor, and Renghii) indicated that all waters reappear in Boga Spring (300 L/s and mean annual temperature of 6.8 °C), after flowing with a velocity of up to 145 m/h (Orăşeanu 2010, 2016). Such short travel times are indicative of some very well-developed karst conduits, thus suggesting the existence of a large cave system situated somewhere between Vărăşoaia and Boga Spring.

History of Exploration

"*I only say this: for all our efforts, hardship, and also successes while exploring V5, there is only one man to be hold responsible, and this gentlemen is I. Orăşeanu*" (Damm 1999), were the words that the late P. Damm used to begin his speech at the opening of the V5 Cave exploration camp

J. Zih (✉)
Ponor Ltd., Maros u. 198, Budapest, 1237, Hungary
e-mail: ponors@yahoo.com

K. Zih-Perényi
Institute of Chemistry, Eötvös Loránd University, Pázmány P. sétány 1/a, Budapest, 1117, Hungary
e-mail: zihperkat@gmail.com

L. Mátyási
Geo Prospect, 13 Septembrie 30/1, 415600 Ştei, Romania
e-mail: matyasilajos@gmail.com

G. M. L. Ponta and B. P. Onac (eds.), *Cave and Karst Systems of Romania*,
Cave and Karst Systems of the World, https://doi.org/10.1007/978-3-319-90747-5_38

Fig. 1 Location of Vărăşoaia (V5) cave system in the Bihor Mountains

in 1992. The meaning of the quote is that after documenting the connection between the water losses in the northern part of the Padiş Plateau and Boga Spring, I. Orăşeanu constantly pushed on the caving community to substantiate its efforts to enter and explore a presumably large cave system in the area. That year, about 50 cavers begin to fight with the forces of nature in a summer camp to reach the long-awaited 600 m depth to surpass Grind Pit (−540 m). However, the 1992 expedition in V5 only reached the depth of −280 m.

The cave entrance was found by P. Damm and his father during a 1986 field trip in the Padiş Plateau karst area. The 1 m diameter hole filled by shepherds with twigs and branches was one of Vărăşoaia's countless similar woodpiles. A draft of air blowing out of this hole was felt, encouraging Paul to remove the litter under which he found a 3 m deep pit. This lucky discovery marks the beginning of the V5 cave system exploration (Fig. 2).

After passing through numerous squeezes, the team led by Damm and Matoş reached in 1988 the depth of −100 m where they stopped in front of a 10 cm wide and at least 2 m long squeeze. The technology available at that time was inadequate to overcome this narrow passage, thus they pushed leads in other parts of the cave. In 1991, after enlarging a squeeze at −70 m (Mousehole Nr. 2) they were able to reach −120 m. Here, they encountered another squeeze (12 m long, 15–20 cm wide) in the form of narrow fissure partly filled with mud, which only R. Pop was able to overcome. On the other side, he found himself on the edge of a pit. Opposite, a waterfall was rumbling into the pit, through its noise he shouted "THE CAVE CONTINUES!" The following year a few other cavers passed this squeeze and reached a depth of −280 m, where a narrow sump halted any forward exploration. Attempts of reaching below this depth failed until 2003, when a squeeze at −125 m was widen to be "accessible" and the exploration of lower levels could resume. At −200 m the cavers were able to enter in a meander, at the end of which the strong wind vanished in a pile of debris. Another few metres of digging and on 30 May 2004, K. Perényi and J. Zih were sitting in front of the Matos Hall's endless darkness. This huge 400 m × 50 m × 20 m hall continues in a labyrinthine that so far extends 6 km to the north along a fault system. Through a number of

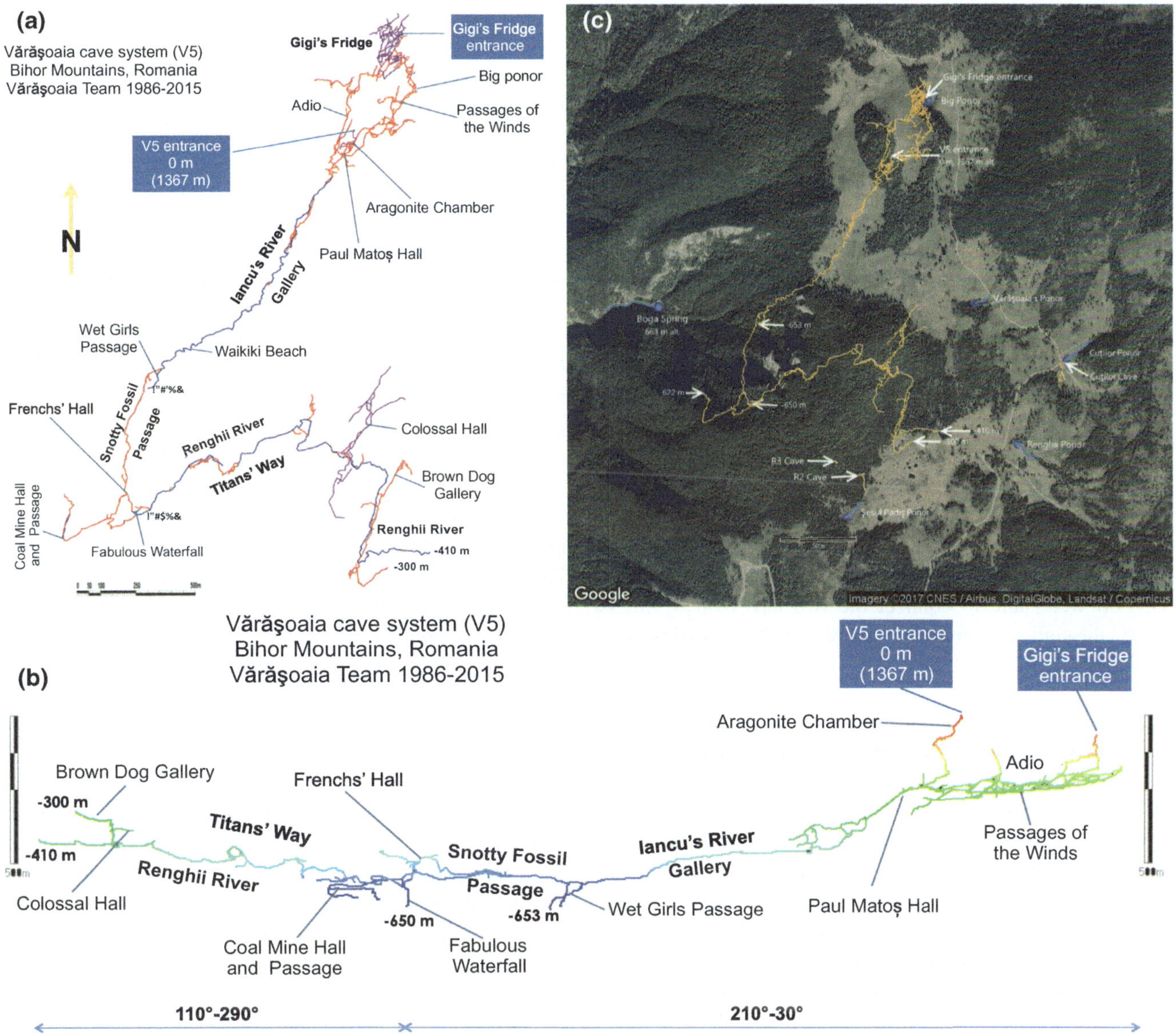

Fig. 2 Plan (**a**) and profile (**b**) of the Vărăşoaia cave system, showing its relationship with the topography of Padiş Plateau (**c**)

fracture-type passages the explorers almost reached to the surface, but the majority of them ended in large collapses.

Advancing towards south from the Paul Matoş Hall was mostly unobstructed, fact that allowed several 2–3-day exploration shifts until cavers reached the depth of −653 on 1 November 2004 (Damm et al. 2005). Although Iancu's River, which showed the way to this point, disappears in a sump, the cave continues along a dry gallery called the Snotty Fossil Passage (Fosilul Mucos). Further explorations in this section were slowed down for a while by finding and rigging the way. From here on, because of the dangerous summer floods, the exploration continued with weeklong winter camps, the first of which took place in January 2006 and gathered 13 cavers. For the first three days, the exploration targeted the Matoş Hall's northern section, followed by three more days in the −600 m deep tunnels. Here, after climbing a 15 m high wall, the team entered a 100 × 50 m dry hall (The French Hall/Sala Francezilor), which ends after a large breakdown zone with a 70 m drop in the Fabulous Waterfall (Cascada Fabuloasă) having a sump at its bottom (−650 m) (Damm et al. 2007). Upstream from the waterfall on this new hydrological system, there are 8–10 km of galleries to go through. Explored step-by-step between 2007 and 2010, the 10–30 m wide dry Titans' Way Gallery goes above the subterranean continuation of Renghii Creek, named Renghii River (Fig. 2c) (Perényi and Zih 2009; Damm et al. 2011). The upmost end of the Titans' Way is blocked by alluvial sediments, which fills up completely the gallery situated ~150 m below the surface. On the active part, the uppermost ending point is also a breakdown chamber, with a chimney of at least 60 m in height. In this section, between 2010 and 2014, three parallel enormous dry galleries were discovered, one of which hosts the large Colosală Hall (Sala Colosală; 300 × 50 × 5 m). All these galleries extend vertically almost to the surface. With a bit of luck, a convenient entrance to the VCS could be dug in this zone, which otherwise is at 15–20 h away from the V5's main entrance.

The second access point into the system is Gigi's Fridge (Frigiderul lui Gigi) entrance that opens from the side of the Groapa doline. This was originally a windy narrow opening, in which for many years the teams worked on enlarging it. During the 2006 summer camp, after 10 m of dismantling bedrock, the team reached an accessible pit. From here a depth of −170 m is reachable through larger pits. At this depth, a thick muddy layer covers everything, and the closest known part of the VCS is inaccessible. However, at −100 m from Gigi's Fridge's entrance, numerous parallel vertical fracture-type passages exist; in one of them, after confronting dangerous debris and squeezes the explorers led by O. Pop were able to connect the two caves on 8 August 2008, through an extremely narrow, hardly accessible passage filled with debris.

Description of the Vărăşoaia Cave System

At present, there are two known entrances to the system (Fig. 2a, b). The main 1986 entrance (V5), which is 200 m NE from the Vărăşoaia Peak on a grassy slope, and begins with a modest 1 m diameter pit. The other entrance (Gigi's Fridge) is at the foot of a rock wall, 50 m NW from the Groapa doline. This second entrance connects with the northern section of the VCS. Beyond this second entrance, about ten parallel wet vertical passages with total relief ranging between 50 and 120 m are known; they are generally safely accessed only by using single rope techniques (SRT). These shafts are connected between them through either narrow passages or galleries with significant breakdown deposits. The lower and upper ends of the pits are boulder choked or the way is too narrow. In one crevice, a very narrow section goes into the cave system's Farewell Gallery (Galeria Adio), which is a 10–15 m high and 4–5 m wide, NE–SW trending passage along which an underground stream flows. Both direction and size of the galleries resemble those of the northern section of the system, such as Passages of the Winds (Galeriile Vânturilor), which drains the Vărăşoaia North surface waters. The millions of cracks, which drain the surface waters, all lead to the Paul Matoş Hall, including the first discovered part of the cave.

The 1986 entrance begins with a −3 m pit, then another 3 m vertical drop, which is followed by a small chamber. From here, an extremely narrow section is passed, with several short drops that ultimately takes us at −40 m where the first rope is reached, which facilitates a 16 m rappel to First Chamber (Prima Sală). Then comes the Shower Room (Sala cu Duşuri), the Aragonite Chamber (Sala cu Aragonit) with its squeeze in Mouse Hole no. 2 followed by the Balustrade Hall (Sala cu Balustradă).[1] Further on opens a tight passage, the 20 m deep Romeo and Juliet Pit, and two 5 m drops after which a 12 m long passage is reached at the depth of −125 m. After successfully passing this section, the gallery becomes larger (average 5 m in diameter) and an 80 m vertical pit follows. The descent of multiple vertical sections in this part of the cave requires SRT. From the bottom, the cave continues along a short meandering passage that is interrupted by a 10 m drop. Following the small meander, it is reached the 1992 cave terminus at −280 m, after a series of pits of −10, −15, and −30 m.

From the −10 m pit forward, the cave continues for ~50 m along an approximately horizontal dry meandering passage with two squeezes, after which the gallery becomes larger, ending in the ceiling of the Paul Matoş Hall. A 15 m descent ends in the middle of the roughly 400 m long, 15 × 15 m diameter hall, filled with giant blocks and debris. To the north, the above-mentioned parallel crevasses' zone can be found, as well as Passages of the Winds and Gigi's Fridge sector (Fig. 2a, c). The southern part of the hall becomes progressively steeper and ends in blocks at −80 m. The cave continues through a 40–50 m long fissure that is in average 2 m wide. At the bottom of this fissure, all the waters accumulated in the north side of the cave come out from the collapse zone situated at −400 m. From here on, the stream passage is called Iancu's River Gallery (Galeria Râul lui Iancu), which until the sump at −653 m is not receiving any significant tributaries. The underground river at −600 m flows into the Wet Girls Passage, which is a steep, canyon-like gallery that ends in a 4 m diameter sump lake located at the lowest known point in the cave (−653 m). The Iancu's River Gallery has an average width of 5–6 m and along with it; there are a couple of small waterfalls and short, 10–20 m long lakes. In some places, flowstones decorate the walls, but generally there are few speleothems on this part of the cave. The tectonics played an important role in the morphology of the passages, which develops along 15–20 m in high fractures.

Some 200 m before reaching the endpoint, a 4–5 m wide dry gallery begins at −600 m right above the river passage. After walking 150 m along this conduit a 50 m deep pit is encountered, at the bottom of which is reached again the −650 m level through a tight sump. At the top of the pit, the dry gallery (3–4 m wide and 10–20 m high) continues.

[1]Not all the chambers, rooms, passages are shown on the map due to scale restriction.

Because of countless 5–10 m up and down climbs, the Snotty Fossil Passage resembles a roller coaster. Flowstones and helictites decorate this section and moonmilk also occurs in large quantities on ca. 50 m of the passage. After 800 m, a 35 m climb-up ends in the French Hall, a chamber of impressive size (150 × 40 × 30 m) filled with blocks. Abundant aragonite speleothems occur in two short passages starting out from this chamber (Fig. 3).

To the south, the chamber continues with a 40 m pit, which actually reveals that the sediment deposit in this hall is at least 40 m thick; its bottom is blocked with debris. This part has huge dimensions: the free space is about 100 m long, 30–40 m wide and at least 80 m high but appears chaotic due to the gigantic blocks, which make the walk difficult. In between the blocks after descending 20 more metres, the Fabulous Waterfall (70 m) is reached. At its bottom is a sump lake located at 650 m below the cave entrance. Another lake at the same level can be accessed using a different route through the breakdown.

Westward from the top of the boulder pile, the Coal Mine Passage (Mina de Cărbune) begins. This gallery develops along a fault line, which is 1–2 m wide and 15–20 m high. After ~150 m the black-walled Coal Mine Hall (50 × 30 40 m) is reached, from where the cave continues with a gallery similar in size and morphology to the Coal Mine Passage which ends in a sump. All passages of this cave section are covered by lavish sparkling white aragonite (Fig. 4).

Eastward from the debris zone of the breakdown begins the most spacious section of the cave system, The Titans' Way. For 2.5 km this passage is 20–30 m wide and high, and in many places it widens out to 40–50 m but these sections cannot be considered distinct chambers. The floor is covered by enormous boulders (some with thick flowstones) and blocks, forming piles up to 10–20 m high (Fig. 5).

Under the boulder, debris flows an underground stream whose water volume is about twice as large as Iancu's River. Since this water is coming from the direction of Renghii ponor, it was named Renghii River. The access to the river along the Titans' Way is difficult due to breakdowns and can only be reached in a few places. Above the dry passage another discontinuous fossil level exists, which partly collapsed into the Titans' Way. After 2 km, the upper dry/fossil and the stream passages depart from each other; the first one, rich in flowstones and helictites, takes an ascending trend (Fig. 6a) ending in a pebble clogs about 100 m above the underground river, whereas the latter one continues as a 5 to 10 m wide meandering stream passage (Fig. 6b). Morphologically, its last 100 m are narrower (2–5 m wide and 30–40 m high) and likewise the upper passage, this one too is blocked by a boulder-choke from under which the water rises.

Above this terminus, there is a chimney at least 60 m high. The water rising from under the boulder-choke originates in the Renghii Ponor, which is located only 200 m away in straight line. It represents V5's main sinking point, accounting for most of the water entering in the system, which is being increased in the Titans' Way by four tributaries. Out of these, three are of great significance, because

Fig. 3 One of the aragonite chambers above the French Hall (photograph courtesy of A. Piri)

Fig. 4 Aragonite in the Coal Mine Passage (photograph by K. Perényi)

they carved stream passages along which occur large size chambers such as the Colossal Hall or they include sections covered with flowstones and aragonite speleothems (Chocolate River/Râul de Ciocolată and Brown Dog Gallery/Galeria Câinelui Maro). All these stream passages develop along main faults and joints that eased the transfer of the waters from the central region of the Vărăşoaia Plateau to the Renghii underground river. All passages presented in this chapter were surveyed to a total length of 24.5 km; the maximum depth of the cave is −653 m (Fig. 2).

Speleothems and Sediments

Almost all passages host some sort of cave deposits. In the oldest, upper dry sections, one can find inactive flowstones and corroded stalactites, some covered with cauliflower-type calcite. Along the middle level and in the river passage there are actively forming stalactites, stalagmites, and flowstones. In the large chambers and in the Titans' Way, there are huge amounts of blocks and debris, in between which there are abundant gypsum flowers and anthodites. In some places, these are 10–15 cm in length, but generally the gypsum occurs as crusts or 1–2 cm long flowers. The most beautiful speleothems throughout the cave system are composed of aragonite (Figs. 2 and 3). These are found in large quantities,

Fig. 5 Old flowstone in the Titans' Way (photograph by K. Perényi)

Fig. 6 **a** Slickenside (left wall) near the upstream end of Titans' Way (photograph courtesy of Z. Németh), **b** Meander in the active stream near upstream end of Renghii River (photograph by K. Perényi)

various morphologies, and sizes in countless side passages starting out from the Titans' Way. The longest ones documented so far are about 1 m in length, and usually occur as tree-like or resemble large hedgehogs. Their colour can be anything from snow-white to rusty reddish-brown. In certain cave sections, abundant and large aragonite helictites were noted. Overwhelmingly, their size ranges between 20 and 30 cm in length, are either white or brown in colour when covered in clay.

The clastic sediments in the dry (upper level) sections consist of quartz and limestone pebbles, and clay. Often, the latter ones contain great quantities of bat bones. The clastic sediment layer in some places along the cave's middle level can reach up to 60–70 m. Such accumulations are only seen in at limited locations as from the other ones the waters carried them away. At the same time, on this level the passage floor houses enormous limestone blocks and debris. Most of the galleries developed along faults and joints and those, which aim towards surface, are filled in their upper parts (just below the Padiş Plateau) with alluvium; these are mainly composed of well-rounded quartz pebbles.

Speleogenesis of the Vărăşoaia Cave System

From a speleogenetic point of view, V5 raises numerous questions, just like any other major cave system. Among these, perhaps the most intriguing is how old is this cave. The morphology of the explored passages definitely points to a multistage cave, which evolved for a very long time. There is an ancient section of the cave accessible along some discontinuous galleries (Colossal Hall, upper parts of Titans' Way), which are either filled with sediments or have collapsed into the younger passage sections beneath. Their age can be linked to the times of the paleokarst development outside the cave. The paleokarst surface is presently filled with alluvial sediment, but the ancient karst surface shows a slope towards the oldest cave galleries, fact documented by means of geophysical measurements and drilling data. So far, the 3–4 dry levels explored (upper end of Titans' Way, Colossal Hall, and Brown Dog Gallery) can be linked with the paleokarst surface. However, in order to prove this, cosmogenic datings needs to be conducted on these sections.

The morphology of the cave's easiest-to-pass section, the Titans' Way fossil passage, is most likely the result of physicochemical processes acting since the last ice age. The spring elevation in this period probably changed drastically many times, fact that can be traced both at the surface and within the cave. Above the present-day spring location, several fossil resurgences can be found, the highest one identified is 150 m above the present one. Strong air current coming out from between the rocks can be felt at each of these paleo-springs. The ancient passages are collapsed and covered with debris from the foothills. Inside the cave, the Titans' Way at the French Hall ends with an 80–100 m drop. From this point on, three distinct passages could be followed towards the spring, but their exploration is in progress. One such passage starting right at the level of the French Hall, the Titans' Way, connects to the Iancu's River Passage; at this location a roughly 60 m high wall segment needs to be further explored.

In summary, the exploration and research activities in the Vărăşoaia cave system are far from being completed. The authors are confident that future of this unique cave system is bright and, it has the potential not only to expand (horizontal and vertical), but also to shed light on pre-Quaternary speleogenetic processes.

Acknowledgements The cavers of various speleo clubs are thanked for their contribution to the exploration and survey of the cave. We are grateful to I. Szlány for his kind assistance. The authors would like to thank B. P. Onac for his careful and helpful review.

References

Bleahu M, Lupu M, Patrulius D, Bordea S, Ştefan A, Panin Ş (1981) The structure of the Apuseni Mountains. Guide to Excursion B3. XII Congr Carpato-Balkan Geol Assoc, Inst Geol Geofiz, Bucharest, p 107

Damm PE (1996) Avenul V5 Sistemul carstic Vărăşoaia-Boga. Nymphaea XXII:29–35

Damm PE (1999) Vărăşoaia. Speomond 4:9–13

Damm PE, Lascu VT, Ursu V (2011) Karst and speleology in Romania. Turk J Speleol 2:19–35

Damm PE, Mátyási L, Zih J (2007) Le Gouffre V5. Spelunca 108:36–38

Damm PE, Perényi K, Pop C, Zih J (2005) Avenul V5. Istoria unui record naţional de adâncime. Speomond 9:3–7

Orăşeanu I (1996) Contribution to the hydrogeology of karst areas of the Bihor-Vlădeasa Mountains (Romania). Theor Appl Karst 9:185–214

Orăşeanu I (2010) Karst hydrogeology of Apuseni Mountains. In: Orăşeanu I, Iurkiewicz A (eds) Karst hydrogeology of Romania. Ed Belvedere, Oradea, pp 181–324

Orăşeanu I (2016) Hidrogeologia carstului din Munţii Apuseni. Ed Belvedere, Oradea, pp 63–136

Orăşeanu I, Gaşpar E, Pop I, Tănase T (1991) Tracers experiments in the karst area of Bihor Mountains (Romania). Theor Appl Karstol 4:159–172

Perényi K, Zih J (2009) A Titánok útja. HTM V(3):20–23

Useful Websites

http://www.speozet.ro
http://www.v5.atw.hu
http://www.varasoaia.org

Bihor Mountains: Valea Rea Cave

Katalin Zih-Perényi, József Zih, Paul-Erik Damm, Szabolcs Szűcs, and Bogdan P. Onac

Abstract

Valea Rea Cave in the Bihor Mountains of Western Romania contains large chambers, passages with distinctive morphology, and abundant and beautiful speleothems in which 33 minerals were identified. We found evidences for both hypogenic and epigenic speleogenetic processes contributing to the origin of the cave. With a known length of over 16 km and a depth of 384 m, Valea Rea overshadows most of the other caves in the country, and emerges as one of the mineralogical treasure of the world.

Keywords

Cave • Mineralogy • Maze • Hydrothermal Speleogenesis

Introduction

Although carbonate rocks make up only ca. 2.3% of country's land surface, Romania has several remarkable caves that host valuable mineralogical, paleontological, and archeological vestiges, as well as rich fauna. Valea Rea Cave in northwestern Romania is one of them hosting unique morphological features, a very rich assemblage of minerals, and has a polygenetic origin. The aim of this chapter is to highlight the most notable features of the Valea Rea Cave along with its 30-year history of exploration as well as a detailed description of the scientific work accomplished to date.

Geographic, Geologic, and Hydrological Settings

From a geomorphological point of view, the entrance into the Valea Rea Cave is located in a very spectacular scenery nearby the main ridge of the Bihor Mountains (Fig. 1), below the Bedeleu-Ciumărna (Fărcaşa-Cârligaţi) erosional platform (Bleahu et al. 1976). The landscape in this region does not reveal the existence of such an extraordinary cave. Considering the highly diverse petrography and the tectono-structural setting, Valea Rea Cave should simply not be there! Nevertheless, it is and develops almost entirely within the Middle Triassic dolomitic (mainly) rocks of the "Sebişel" tectonic sliver (Cornu Nappe; Bleahu et al. 1985; Balintoni 1997). Immediately below this unit occurs sandstones of Permian to Lower Triassic age known to host important concentrations of uranium (Matyasi 1998), whereas the topset beds continue with an Upper Triassic carbonate unit. In some areas, Upper Cretaceous post-tectonic sedimentary formations also appear. North of the cave, a large andesitic block has had a strong thermal and metasomatic effect on the whole carbonate succession (Ştefan et al. 1988). Relevant to the cave geology is the presence of granodiorite in

Paul-Erik Damm—Deceased February 17, 2012

K. Zih-Perényi
Institute of Chemistry, Eötvös Loránd University, Pázmány P. sétány 1/A, Budapest, 1117, Hungary
e-mail: zihperkat@gmail.com

J. Zih
Ponor Ltd., Maros u. 198, Budapest, 1237, Hungary
e-mail: ponors@yahoo.com

S. Szűcs
Evosoft Hungary Ltd., Arany János tér 1, Miskolc, 3526, Hungary
e-mail: szucs.szabolcs.laszlo@gmail.com

B. P. Onac (✉)
Karst Research Group, School of Geosciences, University of South Florida, 4202 E. Fowler Ave., NES 107, Tampa, FL 33620, USA
e-mail: bonac@usf.edu

B. P. Onac
"Emil Racoviţă" Institute of Speleology, Romanian Academy, Clinicilor 5, 400006 Cluj-Napoca, Romania

G. M. L. Ponta and B. P. Onac (eds.), *Cave and Karst Systems of Romania*,
Cave and Karst Systems of the World, https://doi.org/10.1007/978-3-319-90747-5_39

Fig. 1 Location of Valea Rea Cave in the karst of Bihor Mountains

the basement and a number of granite dykes crossing the sedimentary rocks. Worth noting is that the structure of the region is complicated by many faults and overthrusts, which likely played a significant role in the early speleogenetic phases and later, on cave's rich mineral assemblages.

All sedimentary units occur as narrow stripes oriented NE–SW, dipping 25–30°SE. The general direction of the cave overlaps that of the carbonate stripe of which maximum width is only 1 km. In addition, the vast majority of cave passages developed along the dip of the beds. The motion of the blocks along the faults oriented NW–SE has practically zigzagged the route of the cave and the trajectory of the watercourse. This might denote that the formation of the cave likely took place when these faults were still active.

History of Exploration

The entrance shaft into the Valea Rea Cave was discovered in 1986 by Adrian and Liviu Vălenaş. However, the first exploration started only in August 1987, when L. Vălenaş and P. Damm descended and explored the cave to a squeeze located at −29 m. During a collaborative effort (led by P. Damm) made by cavers of several clubs, the exploration continued to an impenetrable squeeze at −118 m, which was reached in 1990 (Damm 1991).

Finally, on Easter of 1993, J. Zih found a narrow route higher above the endpoint. Through that hole in the summer of 1993, the Mainstream Passage (Colector) was reached. The exploration stopped at the Sand Hall (Sala cu Nisip) and the S2 sump but it continued in autumn to the "T" point (Fig. 2). In April 1994 a bivouac (named Bivouac 1) was installed for the first time not far from the Ventilator. The Mud Sea at the "T" point was overcome and the way was again open to the Mainstream Passage, which finally reached the End of the World (Capătul Lumii). At the same time, the biggest halls of the cave were identified as a result of the accurate survey; the Giant Hall (Sala Giganţilor) and the Shower Chamber (Sala cu Duşuri) (Adamkó et al. 2009).

The long and narrow entrance zone dramatically reduced the number of potential explorers and has strongly limited the time remained for further exploration. Nevertheless, at the end of summer 1993 the Hidden Labyrinth (Labirintul

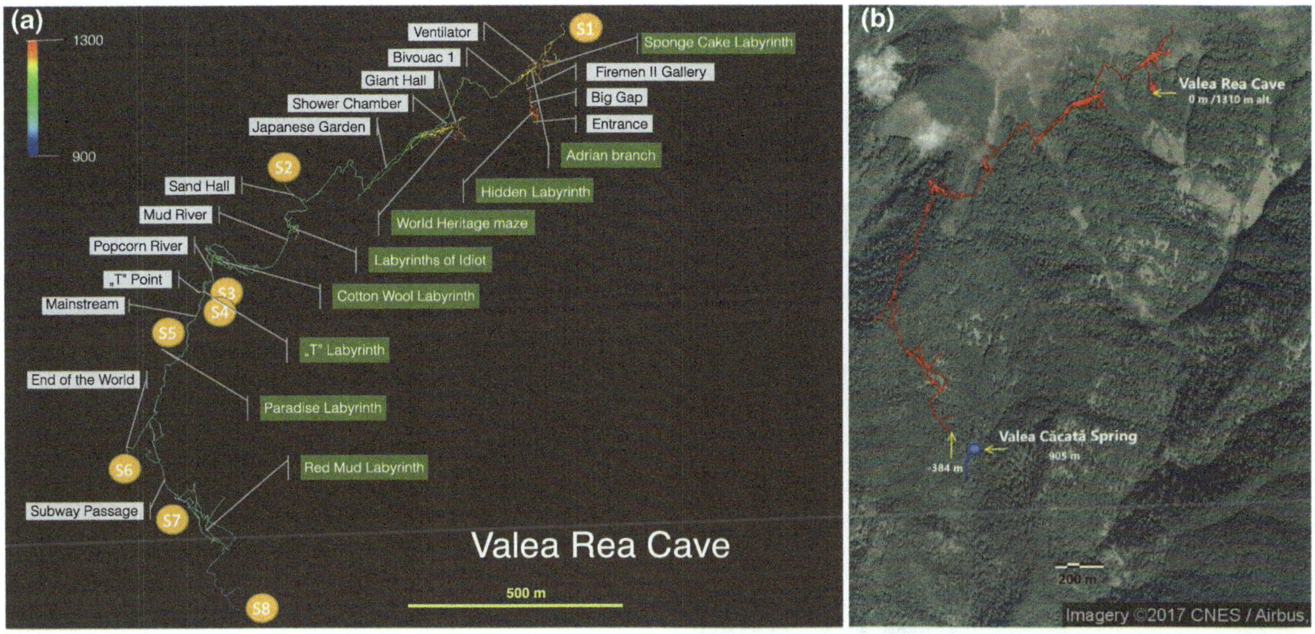

Fig. 2 **a** Plan of the Valea Rea Cave with the most important locations discussed in text, **b** Projection of the Valea Rea Cave map on a Google image of the area showing the entrance in the cave, its terminus, and the Spring from Valea Căcată

Ascuns) was identified in the entrance zone, and step by step, the Big Gap (Marea Prăpastie) was overcome, and a new fossil passage called Firemen II Gallery (Pompieri II) was discovered, allowing for a more comfortable access to the upper, previously unknown part of the Mainstream Passage. In this section of the cave the upmost sump (S1) was reached, downstream from a dry passage through which the connection to the known part of Mainstream Passage was established. The survey of the cave revealed that the Hidden Labyrinth approaches very close the surface. In May 1994, J. Zih localized its terminus on the surface and opened a new, wider entrance. These new explorations made the access to the Mainstream Passage much more comfortable along dry passages, instead of the original Explorer-branch (Adrian branch) (Adamkó et al. 2009). In 1995–96 were explored the mazes Sponge Cake Labyrinth (Fosilul Pişcot), Cotton Wool Labyrinth (Labirintul de la Vată) and the "T" Labyrinth and in 1998 the World Heritage maze. In 1996, the Paradise Labyrinth was explored and a short excavation opened the access into the part named Beyond the End of the World, all the way to the terminal sump S8 (Fig. 2a).

Due to lack of hydrological data, there were no indications regarding the origin of the water in the Mainstream Passage or where this stream was heading. In summer 1995, P. Damm found a spring discharging a significant amount of water into the Valea Căcată, a tributary to the Bad Valley. He also discovered a 50 m long cave behind the temporary active flood spring, which reaches the water level. Four years later (1999), the link between the final sump of the cave (S8) and the Valea Căcată Spring (Fig. 2b) has been finally confirmed, when the dye tracer (fluorescein) launched in S8 appeared at surface in 20 min (Adamkó et al. 2009).

Over the years, attempts were directed to find the upper part of the system. During the very dry summer of 1999, S1 opened and an additional 100 m of gallery was discovered passing two other intermittent sumps, but the advance stopped at a presumably permanent sump. In June 2003, two scuba divers (T. Dianovszky and P. Zsoldos) dived in S2 in order to explore the part beyond. It was for the first time a sump at this depth (−280 m) was dived in Romania. They reached a maximum depth of −7 m into the sump and extended its length with the addition of 20 more meters, after which a collapse blocked the way.

Until 2011, P. Damm and J. Zih organized and led the explorations, in which among many other cavers, K. Perényi, S. Szűcs, M. Botez, D. Pitic, and C. Pop played the most important role.

Cave Description

Valea Rea Cave has presently a surveyed length of 16,357 m and 384 m vertical range with a horizontal extension of less than 2 km (Fig. 2). The two entrances are located at 1310 and 1300 m asl, respectively, 40 m apart from each other. The entrance zone, Firemen's II Gallery, and Sponge Cake Labyrinth (Fig. 3) are indisputable evidences of the role of hydrothermal activity played in the cave genesis. The 1 km long, 2–5 m in diameter tubular passages have rounded morphologies, quartz–goethite bearing sediments, and

Fig. 3 Characteristic Gallery in the Sponge Cake Labyrinth located in the upper part of the cave. The tubular passages have rounded morphologies, the sediment contains hydrothermal quartz and goethite (photograph courtesy of A. Berentés)

frequently intersect fractures filled up with hydrothermal material, such as quartz, pyrite, and barite. As a result of long-term slope processes, one of the galleries has intersected and opened, thus providing one of the entrances. Nowadays, the galleries exhibit dual morphologies since subsequent vadose flows modified the original hydrothermal imprint. Morphological imprints of the vadose flows are the presently dry waterfalls and meanders in the descendent Firemen's II Gallery, as well as the wet, 30–100 m deep shafts. The latter ones penetrate the dry, sandy (paleo) galleries of Sponge Cake Labyrinth connecting it to the cave stream at numerous points.

The cave stream is reached in the Mainstream Passage at −110 m, just 150 m away from the most upstream sump (S1), which is in the upper third part of the entire drainage system. S1 has only 20 cm in diameter and practically function as a well filled with water. During flood, the overflow comes from a huge collapse located 30 m upward. Throughout the cave, the Mainstream Passage is 3–5 m wide (Fig. 4) and 5–10 m high (occasionally 30–70), often divided by breakdowns into pseudolevels of hundreds of meters long. It meanders toward SW with a small slope, whereas beyond the End of the World it changes the direction to SE.

Along the first 1.2 km, except for two huge collapses, the stream can be easily followed by using (if necessary) bypasses located close to the water. Next to the Sand Hall, the stream disappears at −280 m in sump S2, but 40 m above it begins a 5–15 m wide, 1-km-long inactive bypass. In many places, this gallery has collapsed generating sizable halls, such as Cotton Wool Chamber (Sala cu Vată) filled with debris and blocks. The dripping water forms various dripstone and spectacular mud formations: Mud River

Fig. 4 View of the Mainstream Passage close to the Japanese Garden (photograph courtesy of S. Kotarba)

(Râul de Nămol), Popcorn River (Râul cu Coralite). This gallery joins the Mainstream Passage after 700 m at the "T" point, but both its ramifications are closed by sumps S3 and S4, respectively. After another 300 m in this inactive gallery, the underground stream (emerging from sump S5) is reached again and can be followed across small lakes to the End of the World (S6), which is situated at −310 m. At 25 m above this point, begins the strongly ventilated, 3 m in diameter Subway Passage that after 1 km ultimately leads to the stream. The Mainstream Passage may be followed all the way to the lowermost sump (S8), which is about 50 m away from the surface. The underground stream discharges its water in the upper part of the Valea Rea Valley basin, through the spring of Valea Căcată, located at 905 m asl (Fig. 2b) (Orăşeanu 2010).

The underground river has two major tributaries, both in the upstream section of the cave. One of them is the Adrian branch, which means the young, vadose entrance to the cave. Along the second tributary appears the largest chamber of the cave, the Giant Hall (90 × 30 × 30 m) followed by the Shower Chamber; both have their floors covered with large blocks and debris. At least three recharge points on the surface feed the mainstream. Smaller galleries, generally of three-dimensional maze character, can be found above the Mainstream Passage and the bypasses. The Hidden Labyrinth, World Heritage, Labyrinths of Idiot above Sand Hall, Cotton Wool Chamber and "T" labyrinths, Paradise Gallery, and other dry passages Beyond the End of the World, sum up altogether about 6 km of phreatic passages resulting in locally, exceptionally high density of galleries (Damm et al. 2011).

Fig. 5 Gypsum flower in the Gypsum Cemetery, in the lower part of the Sponge Cake Labyrinth (photograph courtesy of L. Gyúró)

Cave Sediments and Speleothems

Valea Rea is famous for its richness and great diversity of speleothems. Most of them consist of the cave's common minerals, gypsum, aragonite, and calcite. Sparkling white gypsum crusts (up to 4–5 cm in thickness) cover large sections of the walls and ceilings. In some places, sections up to 0.5 m^2 of the gypsum crusts are detached and suspended away from the wall by the gypsum crystals growing behind. Branching and curving gypsum flowers (Fig. 5) and even fist-size blisters/balloons often protrude from the crusts.

Associated to gypsum, in some places, thick antler-type helictites form strange morphologies (Fig. 6). Gypsum cotton, beard, and snow, as well as needle-like acicular crystals complete the diverse morphology under which gypsum occurs in Valea Rea Cave. Two short corridors called Gypsum Cemeteries offer incredible high density of gypsum flowers, having their "leafs" as long as 40 cm. Starburst gypsum crust was also described by Onac et al. (1995). The so-called "gypsum transistors" can be found on the walls, as well as on gravels and sediments (Fig. 7).

Along the stream, anemolites are angled opposite to the direction of the airflow and are in general partly covered by coralloids (Fig. 8). At the entrance of the Giant Hall and next to the Japanese Garden, tangled mass of joining and branching helictites and quill anthodites deserve attention. Quartz boxworks are common at places where the hydrothermal quartz veins were exposed by weathering processes. The most beautiful flowstones can be found in the Japanese Gardens below the Shower Hall, and in the Paradise Labyrinth above the stream, between S5 and S6. They exhibit actively forming draperies, canopies, gours, etc.

The following delicate speleothems can only be found in certain parts of the cave. Curly, fluffy gypsum hair and ropes as long as 10 cm were found in the Cotton Wool Chamber, which develops in black fractured calcareous

Fig. 6 **a** Helictite from the entrance part of Giant Hall (photograph by K. Perényi), **b** Helictite "hands" next to the Japanese Garden (photograph courtesy of S. Kotarba)

Fig. 7 Acicular crystals of gypsum ("transistors") growing on the inner part of crust that is pushed away from the cave wall (photograph by K. Perényi)

dolomite. Neither selenite nor celestine can be found anywhere else in the cave. In the maze above "T" Point, popcorns overgrowing on speleothems resemble the spectacular tinder fungus. The passage before Cotton Wool Chamber is characterized by abundant mud formations.

A rare but interesting feature is the so-called bat scratches. The most obvious location is at one of the Paradise Labyrinth's endpoint, where hundreds of square meters of the wall and ceiling are covered with loose red material showing abundant 2–3 cm long parallel scratches in groups of four.

Fig. 8 Anemolites along the Mainstream Passage, between the Japanese Garden and Sand Hall (photograph courtesy of C. Egri)

Cave Speleogenesis

According to Damm (1999, 2000) and Onac and Drăguşin (2017), the cave developed in two distinct stages: a hydrothermal one (due to rising of post-magmatic hot fluids) and a vadose one (related to descending low-temperature meteoric waters). In the hydrothermal phase, a network of ascending phreatic conduits formed, which were lined with a variety of unusual minerals (see below). When the saturation and temperature of the hydrothermal solution decreased, the karstification process continued under epithermal conditions. As a consequence of the uplift of the Northern Bihor block and lowering of the base level, surface waters drained along fractures, faults, and hydrothermal veins, creating the present day typical vadose morphology.

Mineralogy

Gypsum is the most abundant mineral in the cave, associating often with anhydrite. It forms amazing euhedral crystals, crusts, fibrous speleothems, and anthodites. Selenite needles as long as 10 cm and less than 1 mm wide (Fig. 9) are radiating outward from the gypsum sand. They can be found only near sump S1, about 1 m above the water level (Onac et al. 1995; Damm et al. 1996).

In Cotton Wool Labyrinth above the Cotton Wool Chamber (ca. 30–50 m), the walls are covered by snow-white gypsum crust sprinkled with small (3–5 mm) light blue or grayish blue crystals (Fig. 10). Investigations by Onac et al. (1995) showed that celestine and anhydrite

Fig. 9 Selenite crystals embedded in gypsum powder in the Mainstream Passage, 1 m above the water level, next to Sump S1 (photograph courtesy of L. Gyúró)

Fig. 10 Celestite crystals embedded in gypsum crust in Cotton Wool Labyrinth. (photograph by K. Perényi)

crystals are intermixed and cover more than 10 m^2 of the walls; it is the only part of the cave where these minerals occur.

Due to the presence of dolomitic rocks, beside calcite and aragonite, hydromagnesite, periclase, and dolomite were also found to form speleothems (Onac et al. 1995). Periclase appears as light-yellow spots mainly on dolomitic rock or covering small surfaces of magnesium-rich calcitic crusts. Millimeter-size dolomite crystals were found growing outward directly from calcite coating. Moonmilk consists of hydromagnesite and occurs as a dry powder in many parts of the cave or sticky pure white blobs associated with acicular calcite (Onac et al. 1995).

Huntite occurrence was described by Ghergari and Tămaş (1999) in the Firemen's II Gallery. It appears as hard, creamy-white earthy masses filling some fissures in a ceiling pocket at $\sim$20 m after entering the passage. It is precipitated along with low-magnesium calcite on aragonite crystals and is also associated with hydromagnesite, each forming separate aggregates. Lansfordite appears as fine white powdery masses in close association with hydromagnesite (Onac 2003). Aluminite and meta-aluminite were identified by Feier (2003).

Brushite is one of the most common phosphate cave minerals, in Valea Rea occurs as white-yellowish earthy mass material where guano deposits covered gypsum floor crust. Four other phosphate minerals (bobierrite, vivianite, wavellite, and barrandite) identified by Onac et al. (1995) in nodules and small lenses appear in a particular setting where light brown and white clay layers are interbedded in guano. Since its original description, barrandite (a mineral of the strengite-variscite series; Hill and Forti 1997) has been redefined and the specimen from Valea Rea Cave it is actually an Al-rich strengite. Magnesium and iron originate from the bedrock or cave soil, aluminum derives from the minerals of white clay and phosphate is supplied by the guano. It was the first description of these minerals from a Romanian cave.

Quartz crystals up to 1 cm in length exist in many places, all of them are related to the hydrothermal activity (Onac et al. 1995; Damm et al. 1996). Veins, fissures, and tubes of hydrothermal origin are filled mostly with quartz (Ghergari et al. 1997). In the Sponge CakeLabyrinth it covers the entire wall of a 6–8 m deep shaft with 1.5–2 m diameter (Quartz Shaft). Fluid inclusion studies on some quartz crystals indicated that the quartz grew subaqueously at 270 °C (Onac et al. 1995). Malachite forms green earthy nodules on a millimeter layer of quartz. Sometimes rhodochrosite occurs as a black crust. Among the silicate minerals, illite in association with kaolinite was identified in the residual clays, whereas in alluvial deposits dickite and nacrite are more frequent (Onac et al. 1995). Millimeter-size clinochlore aggregates were also found in the entrance zone (Damm and Zih-Perenyi 2000).

Ghergari et al. (1997) studied two hydrothermal veins found at the depth of −115 m on the cave's stream passage. They have a columnar shape and oval section, with diameters between 30 and 50 cm. Their hydrothermal paleokarst filling consists of quartz, calcite, ferroan dolomite, barite, dickite, smectite/dioctahedral chlorite, illite/montmorillonite, montmorillonite, and from the metallic minerals, gold, and pyrite. Subsequently, the latter partially transformed into goethite and lepidocrocite. Gold was found in a very small quantity and appears as flakes of 2–6 μm associated with gray quartz. Barite was identified as 20–50 μm tabular crystals. Pyrite is either euhedral (single crystals or small groups) or as a fine powder disseminated in quartz. The authors concluded that the association above is typical epithermal, in strengthening the findings of Onac et al. (1995).

Metatyuyamunite, a uranium and vanadium-bearing mineral occurs as canary yellow sub-millimeter size plate-like crystals and forms yellow patches of a few cm^2 that cover delicate needle-like aragonite crystals altogether over several square meters on both walls of a narrow side passage (20 m in length), 300 m downstream from the Ventilator Waterfall (Onac et al. 2000, 2001). Uranium originates from the Permian sandstones occurring below the cave, in which it concentrated due to the reducing conditions existing at the time the siliciclastic deposits accumulated. The authors supposed that uranium was mobilized by hydrothermal solutions moving upward through the sandstones, transported by oxidizing carbonate-rich vadose waters and precipitated in presence of vanadium.

According to the compilation of all cave minerals made by Onac and Forti (2011), among the 319 minerals described worldwide, as many as 6 were reported for the first time from Valea Rea Cave. These are gold, periclase, lansfordite, meta-aluminite, wavellite, and nacrite. A list of all minerals positively identified in this cave is tabulated in Table 1.

Other Investigations

Chemical water analysis was used to study spatial and temporal distribution of major components and trace elements. It includes the sampling of possible recharges (sandstone and andesite), long-term irregular sampling (for 6 years) of S1, short-term (every 12 h, 5 times) sampling of S2, as well as occasionally other different sites like Adrian Tributary, dripstones and tributaries, S6, S8, and the surface discharge points. Some of the main conclusions are as follows.

Main Components Mean values of the main components are listed in Table 2. As expected, during wet seasons, as a consequence of higher recharge, conductivity, Ca^{2+}, and bicarbonate concentration of water in sump S1 is lower than in the dry period. The investigations suggest that the short-term chemical composition of the water in the mainstream is constant between rain events. The water entering the cave along the Adrian Branch becomes quickly (already at −10 m) saturated with respect to Ca^{2+} and bicarbonate. In the cave stream between S1 and S8, the concentration of Ca^{2+} and bicarbonate increase indicating active corrosion of the riverbed; however, in spite of the dolomitic bedrock, the concentration of Mg^{2+} shows practically neither temporal nor spatial variation along the stream, thus, Mg^{2+} should have a constant source above S1 (Zih-Perényi et al. 2000).

Spatial and Temporal Distribution of Trace Elements Concentration of some elements (Sr, Mo, Ti, and U) is higher in the upper sump (S1) compared to any of the surface waters potentially recharging the underground stream. This fact and the good correlation (correlation coefficient higher than 0.85) between Ca^{2+} concentration and these elements indicate that their origin is related to corrosion process in the upper, undiscovered part of the cave. Interestingly, the increase in V concentration occurs between S1 and S2, where metatyuyamunite aggregates were precipitated. The only element having higher concentration in the surface waters flowing over andesites was rubidium (Zihné Perényi et al. 2003).

On the basis of short time (48 h) change in their concentration, the traces elements were classified into two groups (Lásztity et al. 2002): stable-in-time and variable (Table 3). The elements of the latter group are sensitive to small change in drainage system; therefore, they could be indicators for the superimposed flows of different aquifers.

Chemical Form of Trace Elements The mobility, transport, reactivity, and fate of trace elements in natural waters highly depend on the chemical form of the element, which is controlled by the physicochemical and biological characteristics of the system. An operationally defined on-site fractionation scheme was developed that was based on the difference in the thermodynamic and kinetic behavior of the species (Zih-Perényi et al. 2005, 2008). Four categories were established: (1) Colloidal form of metals: oxo-hydroxides of the metal or adsorbed on the surface of the colloidal Fe/Al/Mn oxo-hydroxides; (2) Reactive form of metals (free ionic or complex form) of low stability and fast ion-exchange, such as labile organic or hydroxo complexes; (3) Exchangeable form of metals. Metals bound in unstable complexes with slow ligand-exchange, presumable a part of humic complexes; and unfilterable (<0.22 μm) colloidal forms; (4) Nonreactive form of metals (Fig. 11).

About half of Ti ions are in the particulate matter, as colloidal fraction (presumably as oxo-hydroxides) and the other half part is in nonreactive form. All the other elements studied are mostly in reactive, dissolved forms. This pattern is characteristic for pristine water. In a surface flow, due to higher concentration of organic and inorganic components as well as trace metals, the significant part of

Table 1 List of minerals described from Valea Rea Cave

Mineral	Chemical formula	Occurrence
Aluminite	$Al_2(SO_4)(OH)_4{\cdot}7H_2O$	Minute crystals
Anhydrite	$CaSO_4$	Crystals, wall crusts
Aragonite	$CaCO_3$	Wall crusts, helictites
Barite	$BaSO_4$	Tabular crystals
Bobierrite	$Mg_3(PO_4)_3{\cdot}8H_2O$	Earthy masses
Brushite	$CaHPO_4{\cdot}2H_2O$	Earthy masses
Calcite	$CaCO_3$	Most type of speleothems
Celestine	$SrSO_4$	Crystals, wall crusts
Clinochlore	$(Mg,Al)_6(Si,Al)_4O_{10}(OH)_8$	Aggregates, micro-crystals
Dickite	$Al_2Si_2O_5(OH)_4$	Earthy masses
Dolomite	$CaMg(CO_3)_2$	Spar, crusts
Epidote	$Ca_2Al_2(Fe^{3+},Al)Si_3O_{12}(OH)$	Micro-crystals
Epsomite	$MgSO_4{\cdot}7H_2O$	Aggregates, crystals
Gypsum	$CaSO_4{\cdot}2H_2O$	Crystals, crusts, flowers
Goethite	$FeO(OH)$	Crystals
Gold	Au	Micro-crystals (flakes)
Huntite	$CaMg_3(CO_3)_4$	Moonmilk, earthy masses
Hydromagnesite	$Mg_5(CO_3)_4(OH)_2{\cdot}4H_2O$	Moonmilk, crusts
Kaolinite	$Al_2Si_2O_5(OH)_4$	Earthy masses
Lansfordite	$MgCO_3{\cdot}5H_2O$	Small stalactites, crystals
Lepidocrocite	$Fe^{3+}O(OH)$	Crystals
Malachite	$Cu_2^{2+}(CO_3)(OH)_2$	Nodular speleothems, crusts
Meta-aluminite	$Al_2SO_4(OH)_4{\cdot}5H_2O$	Crystalline aggregates
Metatyuyamunite	$Ca(UO_2)_2(VO_4)_2{\cdot}3H_2O$	Plate-like crystals
Montmorillonite	$(Na,Ca)_{0.3}(Al,Mg)_2)Si_4O_{10})(OH)_2{\cdot}nH_2O$	Earthy masses
Nacrite	$Al_2Si_2O_5(OH)_4$	Small aggregates
Periclase	MgO	Crusts (limited extent)
Pyrite	FeS_2	Crystals
Quartz	SiO_2	Aggregates, crusts, crystals
Rhodochrosite	$Mn^{2+}CO_3$	Minute crusts
Strengite	$Fe^{3+}PO_4{\cdot}2H_2O$	Earthy masses, nodules
Vivianite	$Fe_3^{2+}(PO_4)_2{\cdot}8H_2O$	Earthy masses, nodules
Wavellite	$Al_3(PO_4)_2(OH,F)_3{\cdot}5H_2O$	Nodules

Table 2 Mean concentration of the main components

	pH	Conductivity (μS/cm)	HCO_3^- (mg/L)	SO_4^{2-} (mg/L)	Ca^{2+} (mg/L)	Mg^{2+} (mg/L)	K^+ (mg/L)	Na^+ (mg/L)	Mg/Ca molar
S1 (n = 16)	7.86 (0.41)	86 (26)	62 (19)	6.7 (1.8)	15.2 (3.0)	4.3 (0.8)	0.29 (0.10)	1.6 (0.2)	0.46 (0.05)
S2 (n = 9)	7.6 (0.4)	94 (15)	72 (18)	7.4 (0.7)	18.0 (4.1)	4.4 (0.5)	0.23 (0.03)	1.6 (0.2)	0.42 (0.06)
Dripstones (n = 5)	8.08 (0.35)	282 (66)	191 (38)	8.7 (3.3)	63 (21)	13.0 (7.4)	0.72 (0.33)	0.6 (0.4)	0.35 (0.19)
Surface recharges (n = 8)	7.68 (0.35)	32 (16)	14 (10)	2.5 (0.3)	3.7 (2.2)	0.49 (0.28)	0.32 (0.18)	1.0 (0.6)	0.18 (0.06)

Standard deviations are in brackets

Table 3 Trace element concentrations in S2

Stable-in-time elements									
	Fe	Mn	Al	Cr	Sr	Ba	Mo	U	V
μgL^{-1}	8	0.5	7	0.6	35	5	0.6	0.3	0.4
Variable elements									
	Ag		Cu		Pb			Sb	
μgL^{-1}	0.02–1		0.1–0.3		0.02–0.6			0.4–1.3	

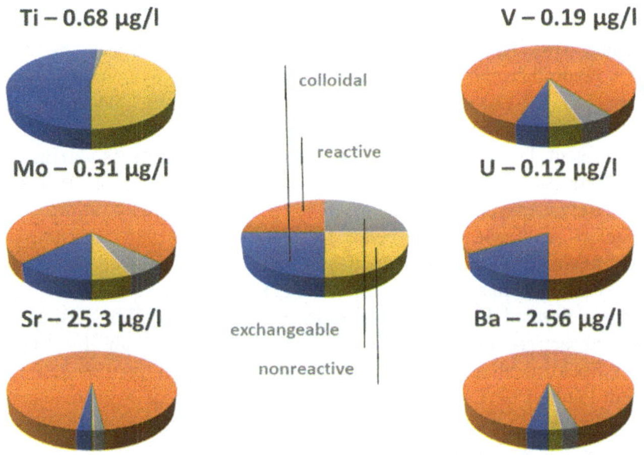

Fig. 11 Distribution of trace elements in sump water S1 according to an operationally defined on-site fractionation scheme, showing the thermodynamic and kinetic behavior of the species

the ions are bonded to coarse colloidal fraction or humic substances (Zih-Perényi and Lásztity 2005).

Discharge Rate A V-shaped dam was built 20 m downstream from sump S1 in 2004. Results of a 2.5-year monitoring period using a water level data logger are summarized in Table 4. The stream responded to rain events in 6–8 h. The yearly pattern of water temperature shows an almost monotone increasing tendency from spring to end of winter. It is only disturbed by heavy rain events.

Cave Conservancy

Vale Rea Cave is a scientific reserve, thus access requires authorization from the Speleological Heritage Commission and the Administration of the Apuseni Natural Park.

Conclusion

Far from an exhaustive covering, this chapter assembled the most significant aspects regarding Valea Rea Cave. Its speleogenesis includes a hypogene stage documented by quartz, pyrite, gold, and barite deposited from rising post-magmatic hot fluids and an epigenic (vadose) phase when several low-temperature minerals were deposited. The cave hosts Romanian's largest diversity of gypsum speleothems and 33 other minerals, of which 6 were first documented in a cave environment from this location. Through its length, depth, polygenetic origin, complex morphology, and mineral diversity, Valea Rea Cave is a flagship of the Romanian karst.

Acknowledgements We are grateful to L. Mátyási for his kind assistance and advices. The cavers of various speleo clubs are thanked for their contribution to the exploration and survey of the cave.

Table 4 Statistics on flow rate and temperature

Flow		Water temperature (°C)	
Mean (m^3/min)	1.37	Mean	6.45
Max. (m^3/min)	8.90	Standard deviation	0.21
Median (m^3/min)	0.99	Minimum	5.76
Sum (m^3/year)	7.2×10^5	Maximum	6.78
		Median	6.51

References

Adamkó P, Perényi K, Polacsek Z, Szilágyi PP, Zih J (2009) Valea Rea. In: Kucsera M, Lieber T (eds) Barlangtúrák a fold alatt a Föld körül. Kornétás Kiadó, Budapest, pp 49–57

Balintoni I (1997) Geologia terenurilor metamorfice din România. Ed Carpatica, Cluj-Napoca, p 176

Bleahu M, Decu V, Negrea Ş, Pleşa C, Povară I, Viehmann I (1976) Peşteri din România. Ed Ştiinţifică şi Enciclopedică, Bucureşti, p 416

Bleahu M, Bordea S, Bordea J, Mantea G, Popescu A, Marinescu F, Cioflica G, Ştefan A (1985) Geological map of Romania, Pietroasa Sheet. Scale: 1:50.000, Ins Geol Geofiz, Bucuresti

Damm P (1991) Avenul Adrian de sub Cornu Munţilor. Speotelex 7:14–17

Damm P (1999) Studiu geologic si mineralogic al zonei Valea Rea-Cornu Munţilor (Munţii Bihor), BSc Thesis, Babeş-Bolyai Univ, Cluj-Napoca

Damm PE (2000): Evolution of karst in the Valea Rea—Cornu Munţilor (Bihor Mountains, NW Romania). In: Onac BP, Tamas T (eds) Karst studies and problems: 2000 and beyond. Ed Presa Univ Clujeană, Cluj-Napoca, pp 22–25

Damm P, Zih-Perenyi K (2000) Mineralogia peşterii din Valea Rea. Nymphaea XXVIII:15–26

Damm P, Perenyi K, Pop C, Szucs S, Zih J (1996) Consideraţii asupra Peşterii din Valea Rea (Munţii Bihor). Cerc Speol 4:21–23

Damm P, Lascu VT, Ursu V (2011) Karst and speleology in Romania. Turkish J Speleol 2:19–35

Feier N (2003) Date noi privind mineralogia Peşterii Valea Rea, Munţii Bihor. Ecocarst 4:22–24

Ghergari L, Tămaş T (1999) Huntite—mineral formed in supergene conditions (Valea Rea Cave, Bihor Mountains). Rom J Miner 79:151–158

Ghergari L, Tămaş T, Damm P, Forray F (1997) Hydrothermal paleokarst in Peştera din Valea Rea (Bihor Mountains, Romania). Theor Appl Karst 10:115–125

Hill CA, Forti P (1997) Cave minerals of the world (2nd ed). Natl Speleol Soc, Huntsville AL, p 463

Lásztity A, Zih-Perényi K, Varga I, Bertalan É (2002) Groundwater analysis by atomspectrometric methods. ICP Inf Newslet 27:358

Matyasi S (1998) Studiul mineralogic, geochimic şi genetic al mineralizaţiilor de uraniudin extremitatea de NV a Munţilor Bihor. Unpubl PhD Thesis, Universitatea din Bucureşti, pp 146

Onac BP (2003) Minerals of the Carpathians: first update. Acta Miner Petrogr 44:31–34

Onac BP, Drăguşin V (2017) Hypogene caves of Romania. In: Klimchouk A, Palmer AN, De Waele J, Auler AS, Audra P (eds) Hypogene karst regions and caves of the world. Springer International Publishing, Cham, pp 257–265

Onac BP, Forti P (2011) State of the art and challenges in cave minerals studies. Studia UBB Geol 56:33–42

Onac BP, Bengeanu M, Botez M, Zih J (1995) Preliminary report on the mineralogy of Peştera din Valea Rea (Bihor Mountains, Romania). Theor Appl Karstol 8:75–79

Onac BP, Kearns J, Damm P, White WB, Matyasi S (2000) The first reported occurrence of metatyuyamunite $Ca(UO_2)_2(VO_4)_2 \cdot 3\text{-}5H_2O$ in Romania. Rom J Mineral 80:5–10

Onac BP, Kearns J, Damm P, White WB, Matyasi S (2001) Hydrothermal genesis of metatyuyamunite, $Ca(UO_2)_2(VO_4)_2 \cdot 3\text{-}5H_2O$ in the Valea Rea Cave, Romania. Proc 13th Int Congr Speleol 1:25–29. Brasilia

Orăşeanu I (2010) Karst hydrogeology of Apuseni Mountains. In: Orăşeanu I, Iurkiewicz A (eds) Karst hydrogeology of Romania. Ed Belvedere, Oradea, pp 181–324

Ştefan A, Lazăr C, Berbeleac I, Udubaşa G (1988) Evolution of magmatism in the Apuseni Mountains and associated metallogenesis. D S Inst Geol Geofiz 72–73(2):195–213

Zihné Perényi K, Lásztity A, Bertalan É, Varga I (2003) Atomspektrometriás nyomanalízis szerepe a karsztkutatásban. In: Magyar Spektrokémiai Vándorgyűlés Előadásai, Szeged, pp 128–131

Zih-Perényi K, Lásztity A (2005) On-site classification of manganese forms in natural waters by membrane filtration and chelating exchange. Spectrochim Acta, Part B 60:385–392

Zih-Perényi K, Kelkó-Lévai Á, Varga I, Lásztity A (2000) Trace analysis of groundwater samples by FI-GFAAS and other atomspectrometric methods. Proc 4th Eur Furnace Symp and XVth Slovak Spectros Conf, Košice-High Tatras, Slovakia, p 91

Zih-Perényi K, Lásztity A, Sugár É (2005) On-site fractionation of trace elements in karstic groundwater by membrane filtration and chelating exchange. 2005 Winter Conf Plasma Spectrochemistry, Budapest, Hungary, M 53

Zih-Perényi K, Dávid B, Sugár É, Mátyási S, Lásztity A (2008) On-site classification of trace elements in karstic groundwater. Pro Eur Symp Atomic Spectrom, ESAS 2008 Weimar, Germany

Bihor Mountains: Humpleu–Poienita Cave System

Philipp Häuselmann

Abstract

Humpleu Cave is currently Romania's second longest cave. Its highlights are the beauty of its fossil and water passages, the alpine-type narrowness of the second, vertical, entrance, and by the still exciting possibility of finding new passages. Except for a number of mineralogical studies, Humpleu Cave remains largely under-investigated. A variety of features (morphology, tectonics, speleothems, speleogenesis) are awaiting to be studied in order to better understand this huge cave and its surroundings. The present chapter gives a very brief and condensed overview of the cave and hopes to resuscitate interest to study it more thoroughly.

Keywords

Speleogenesis • Hydrogeology • Paragenesis
Sediments • Mineralogy

Introduction

Originally named "Peştera Mare din Valea Firii" or "Peştera din Dealul Humpleu," the cave was renamed to "Humpleu–Poieniţa Cave system" when other cavities were connected to it. The Humpleu Cave system consists of several caves and springs which are either physically or hydrologically connected. Together, they form a system, which excels both by its dimensions (~35 km in length) as well as by the size of its passages, which are among the largest in Europe and the world. Furthermore, the presumably very old age of the cave system allowed for the deposition of an abundance of different speleothems, which are remarkable in size and aspect. Finally, a 4.6 km long, almost horizontal, underground river passage showing remarkable dissolution morphologies, speleothems, and a clear water, adds to the beauty of the cave. So it is no wonder that the cave was officially protected and gated. Today, its location within the Apuseni Natural Park ensures an effective protection that sadly enough keeps the cavers outside but allows for paying spelunkers to enter. Because Humpleu Cave was too rapidly explored since its discovery in 1984, the original maps are mostly lost, and therefore, a Romanian-Swiss remapping project is underway since 2001 and hopefully will be completed by the end of 2018.

Geographic, Geologic, and Hydrologic Settings

The Humpleu system lies in the Apuseni Mountains, in Valea Firii, a left-hand tributary to the Someşul Cald River (Fig. 1). The system is basically the "first" encountered when traveling toward the famous Padiş area, where many significant karst features of Romania are found. In its close neighborhood lies the more famous Piatra Altarului Cave, and rumors go that there could have been a hydrologic link between Humpleu and Piatra Altarului in the past (Papiu 1988). However, proof for this link has not been found yet. Humpleu Cave draws its name from the hill in which it develops; the meaning of the word Humpleu itself is not clear.

According to the geology of the region proposed by Mantea (1985), the cave develops in Lower Cretaceous (Barremian–Aptian) limestones. A base level of more marly/shaley rocks is not observed in the near vicinity. Although no really prominent faults are directly seen in the cave or at surface nearby, the rather rectilinear pattern of Humpleu Cave from the entrance to the junction with Poieniţa Shaft (Fig. 2) suggests the presence of fracture pairs trending SSE and ESE. After the junction, the cave shows more bends and seems less influenced by fracture geometry. The whole area directly above Humpleu Cave is holokarstic,

P. Häuselmann (✉)
Swiss Speleological Society, Giebelweg 6, 3323 Bäriswil, Switzerland
e-mail: praezis@speleo.ch

G. M. L. Ponta and B. P. Onac (eds.), *Cave and Karst Systems of Romania*,
Cave and Karst Systems of the World, https://doi.org/10.1007/978-3-319-90747-5_40

Fig. 1 Location of Humpleu Cave in the Apuseni Mountains

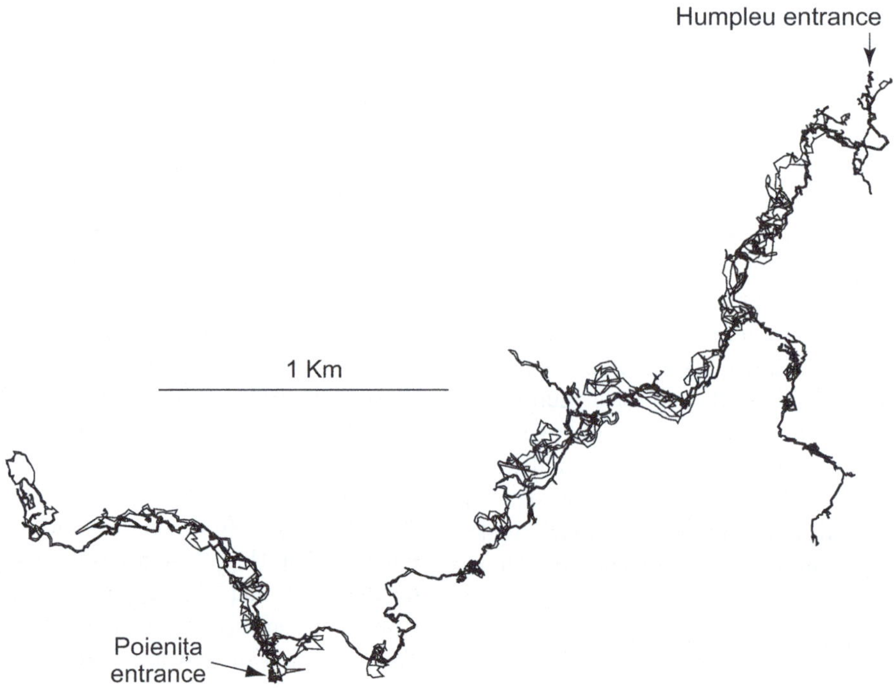

Fig. 2 New map of Humpleu Cave (95% complete; the remapping project goes on)

but only few longer caves are known (Poienița—linked with Humpleu, Poienița 2, and Ghețarul din Poiana Vârtop).

The hydrology of Humpleu Cave itself is rather simple, with an upstream sump recharging a long underground river, flowing along most of the present cave, before sinking in a downstream sink and reappearing through a number of springs along the Firii Valley, just north-east of the main cave entrance. However, looking at the riverbed on the cave map, one can see that it is rather winding, and tracer experiments (Ponta et al. 1993; Orășeanu 1996) reveal that waters are to come both from Groapa Largă and Valea Ponorului (Orășeanu 2010). This fact opens numerous questions on present and past hydrologic links, which for the moment cannot be answered, due to the (too) scattered and partially missing information on virtually all important cave systems nearby Humpleu. The spring cave of Humpleu, i.e., the "Peștera cu Apă din Valea Firii" (Water Cave from Firii Valley, hereafter Water Cave) is known for a very long time. It is an overflow spring which only discharges at very high waters; otherwise, the springs to the north-east of the cave are successively discharging when cave water level is rising.

History of Exploration

Due to the existence of the Water Cave, the presence of a large cave system was expected, but it was only in September 1984 when A. Cuș, a local forestman led A. Bulboacă (then a member of the Speleo Club "E. Racoviță" (CSER) in Cluj-Napoca) to a hill side opening from which cold air was exhaled. During 1984 and 1985, most of the fossil parts and long stretches of the river to the difficult semi-sumps were found and partially mapped by CSER and Speleo Club "Z" Oradea (Papiu 2001). However, already by the end of that year, a remapping project started by C. Popa with the aim of achieving a higher precision map, which ended in 1989 (Popa 1990a; Papiu et al. 1993). Between 1990 and 1994, the fossil rooms behind the first semi-sumps (Grenoblois, Charentais) were found and mapped, whereas the last chambers (Bingo, Amphitheater) were discovered in 1995–1996. In parallel, the explorations in the Poienița Shaft, discovered by "Z" Oradea in 1978, were continued by the Speleo Club "Politehnica Cluj" and led to the junction with Humpleu Cave in 1989. In the same year, the more distant rooms: Helictites (Helictitelor), Pagodas (Pagodelor), and Room of the Giants (Sala Giganților) were located and mapped.

In the period when Humpleu Cave was discovered and explored, but also later on, speleology in Romania, as in other east European countries, was not so much a matter of science, but mostly regarded as a competition. Who maps most? Who reached deeper? Who discovers most passages? In this respect, mapping precision was not high on the list, and communication with fellow cavers was not at all on the menu. Publication of the maps did not happen, since they were considered local secrets. This was one of the reasons why (with some few exceptions) most of the old maps and data of the earlier survey projects were lost or simply not available. Due to the eminent scientific interest of the cave, a Romanian-Swiss collaborative project was initiated in 2001 with the aim to remap at least the entrance parts of the cave. The main reason was to produce a reliable cave map and long profiles that will allow a three-dimensional view of the cave. This project was later expanded to cover the entire cave. Because most activities were carried out as summer camps (some years weather prevented any work), the project is still underway; however, it is our hope that it will be achieved soon after this chapter will be printed.

Cave Description

The overall general aspect of Humpleu Cave is relatively simple. It is a horizontal system that extends over roughly 80 m in vertical range, with an underground river running along its lower level. Added to this general picture is Poienița Shaft, an alpine-type cave with medium- to small-sized pits connected by meanders with a total relief of 280 m (Fig. 3).

The present entrance, which was dug out artificially, leads immediately into a vast room, called Entrance Hall (Sala de Intrare). However, along the way, it becomes obvious that this large volume is not actually a room, but a huge passage, whose roof still partially shows the initial corrosion features,

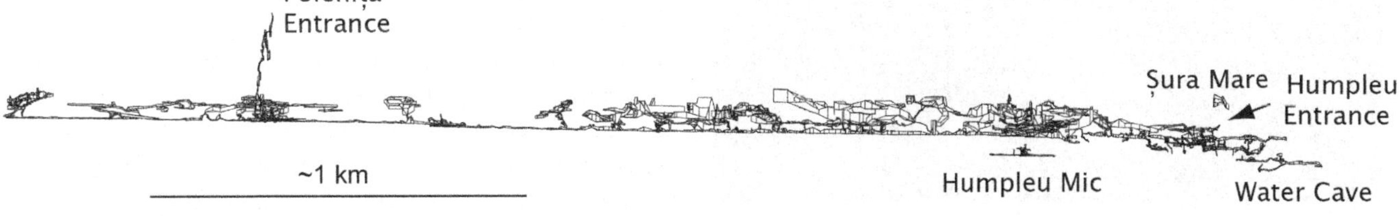

Fig. 3 Projected section of the Humpleu Cave, Șura Mare, and Water Caves, as well as the Humpleu Mic Cave, which is hydrologically linked to the same spring. Width of figure: ~3.5 km

despite abundant breakdown. The vastness of Humpleu Cave is one of the distinctive features (Onac and Papiu 1995). With heights between 10 and 20 m and widths of 30–50 m, it truly belongs to some of the largest cave passages on Earth. Occasionally, breakdown or massive flowstone domes narrow its dimensions, but in general, the passage continues more or less unchanged into the deeper parts of the cave.

In the Dancing Room (Sala de Dans), the ceiling of the passage collapsed and thus enabled the link with an upper gallery, which is equally large and also continuous (Fig. 4). A small opening in the bottom of the lower passage of the Dancing Room leads to a meandering canyon, where the underground river flows. It cascades down in some waterfalls and quickly disappears in the downstream sump, whose continuation was not found yet. Upstream, some cascades lead to a more horizontal part. The meander then continues peacefully, with a width between 1 and 8 m and heights ranging from 2 to 15 m. From place to place, it connects with the big rooms above. After roughly 1 km, the river is nearly blocked by flowstone and forms the first of the 11 semi-sumps (in other publications the number of them varies up to 22, depending on how they are counted). These semi-sumps occur where flowstone is blocking the canyon above, breakdown is abundant, or the river undercutting the walls of the canyon creates shortcuts with low ceiling, but wide passages.

Fig. 4 Dancing Room in Humpleu Cave system (photograph courtesy of C. Lascu)

In some places, the presence of breakdowns allow to ascent into the upper levels of the cave dominated by very large rooms; however, behind Wolf's Hall (Sala Lupului), which was discovered only in 2012 by a French–Romanian expedition, there is a rather long stretch where no rooms were located as yet. They do exist without any doubt, as documented by the existence of massive ceiling collapses in several places along the river. Only in the Balcony's Chamber (Sala Balcoanelor), the rooms are intercepted again. The largest underground voids are in this area of Humpleu Cave. Room of the Giants (540 × 111 × 35 m) with its estimated total volume in excess of 2 million m^3 is probably one of the largest room passage known in Europe.

The cave ends abruptly at an upstream sump, which after only 50 m is too narrow to dive, and a last large room, the Amphitheater, where breakdown and flowstone hinder further progress. Remapping this part of the cave is still in progress.

If analyzing the passages formed between the Water Cave and Dancing Room, we notice that successively the meander became free of water, which was drained toward the lower-level passages, which transformed into phreatic tubes downstream (e.g., Subway (Metrou), Fig. 5). Thus, we have evidences for a total of five subhorizontal passages, compared to only three identified above the underground stream. Although there was no scientific study conducted so far, it is safe to assume that the successive piracy of the waters was triggered by the deepening of the Valea Firii.

Cave Sediments and Speleothems

Although Humpleu Cave is rich in sediments and speleothems, in the entrance area only breakdown blocks and some giant calcite scalenohedrons (Fig. 6) filling a fracture in a small side chamber are present (Onac 1986; Onac et al. 2006).

After only 80 m into the cave, many blocks are lying on the cave floor, which are broken flowstone parts originating from a huge dome that is presently entirely dismantled. Such domes were also noticed in many other parts of the cave. Moonmilk and coralloids (popcorn) are abundant, mostly in corners and on passage walls, but also between and below breakdown (Chirienco 2002). Further inside the cave, well-developed and beautiful draperies,

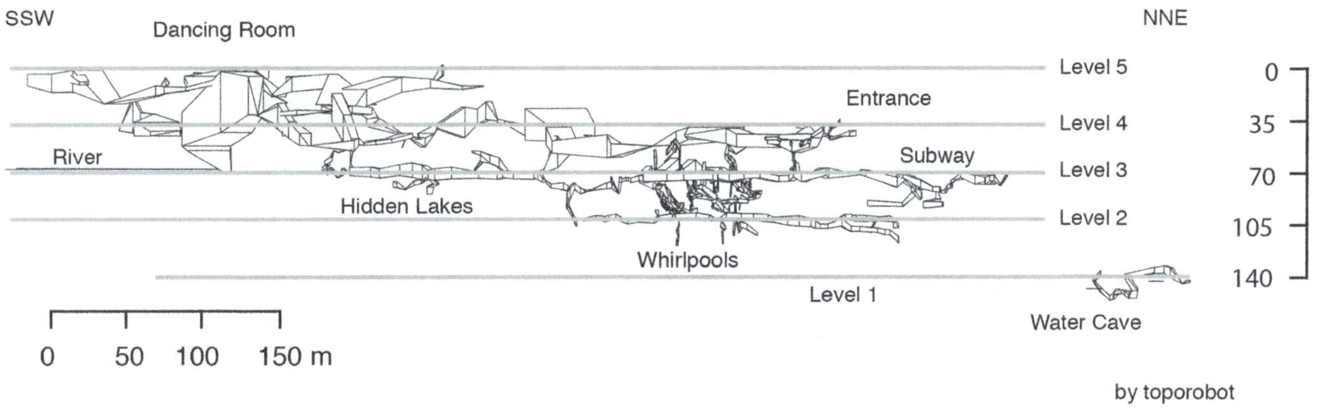

Fig. 5 Projection of the entrance part of Humpleu Cave. The horizontal meandering canyon (river) transforms into an epiphreatic tube downstream (subway)

stalagmites, shelfstones, rafts, calcite spar, pearls, giant columns, and flowstones occur. Helictites, monocrystalline stalagmites, aragonite frostwork, and water lilies are well hidden in the inner parts of the cave, all of exquisite beauty (Fig. 7).

Flowstone and massive stalagmitic domes can attain impressively sizes; one of the largest flowstone walls seen in the cave is the far end of Halasi Room (Sala Halasi), where a relief of 40 m is filled by such a broken dome.

Mârza and Selişcan (1987) conducted an interdisciplinary study (tectonics, hydrogeology, and karst) aiming at explaining the phenomena of the collapse of the speleothems in this emblematic cave of Romania. The authors conclude that neotectonic events seem to be episodically reactivated by seismic activities, which are felt with greater intensity in the Someşul Cald graben, a region with an extremely complicated tectonic structure. Along this line, Onac et al. (1998) documented at least two such major tectonic events over the last 250,000 years by dating with the U-series disequilibrium method multiple speleothems growing on broken ones.

Besides the omnipresent breakdown blocks, another common sediment are pebbles (Fig. 8). Although they are not easily visible near the entrance, there are some patches, which were not eroded by the river. Further into the cave, the remaining deposits grow in volume. The largest pebble deposits are found from Bivouac Room (Sala Bivuacului) to Room of the Giants, where such accumulations can reach up

Fig. 6 Giant monocrystalline calcite scalenohedrons (photograph courtesy of C. Lascu)

Fig. 7 Close-up image from the Helictite room: fragile, but well preserved due to the distance to the entrance (photograph courtesy of D. Sanz)

Fig. 8 Pebbles in the subfossil part; their size varies and in many places, violet sediment pebbles were noticed (photograph courtesy of M. Achtman)

Fig. 9 Photograph of a "pagoda," mixture of calcite crystals and clay; image width ~5 cm (photograph courtesy of D. Sanz)

to 30 m in thickness. The pebbles are clearly allogenic since they are composed of fragments of non-karst rocks (sandstones, igneous, and quartz veins) outcropping in the catchment area of the underground river; rounded limestone pebbles are virtually inexistent. A sedimentological study (Şoit 2009) revealed that there is no large difference between pebbles along the fossil and active passages, indicating that the catchment area remained unchanged during the evolution of the cave.

Sands, clays, and clay minerals are omnipresent in the cave in minor quantities compared to the pebbles. Often, they form an upper deposit covering pebbles. The presence of "pagodas" (Fig. 9) was at the origin of the appropriately named "Pagodas Room." These pagodas are not yet described in the speleological literature.

Speleogenesis

The size of the large passages and the presence of corrosion features inside them led to the assumption that once there was a huge river with a gargantuan discharge flowing through the cave. Changes in climate and/or diminution of the catchment area successively would lead to the present-day discharge and conditions. Today, however, the ideas have changed and the first hints to possible paragenesis came from Ph. Audra (unpublished report). Another recently completed study (Şoit 2009) revealed the presence of corrosion notches along the walls of the large passages, some of them still containing pebbles, indicating a formation by a river flowing atop of the sediment. Further on, the ceiling of a large part of the Room of the Giants is completely flat, and laterally, large pebble deposits are 50 cm distant from the ceiling, clearly indicating a paragenetic origin of the large room.

Next, the question to answer is the presence of only three levels above the stream passage compared to five between the entrance and Dancing Room. As a matter of fact, these levels are almost never separated by bedrock, but by breakdown or massive flowstones. It therefore seems that the whole cave has been traversed by the same river, and in two distinct phases, fill-up of pebbles caused the paragenetic generation of the large rooms. Figure 10 deciphers the speleogenetic hypothesis: After an initial phreatic genesis (a), a meandering canyon incises (b). An excess of sediment causes the watercourse to begin winding and undercutting the canyon walls (c); increased sediment deposition infers paragenetic raising of the ceiling (d). Less sediment load causes the river to empty the cave for the most part (e) and incises into bedrock subsequently (f). The process is then repeated (g–i). Today, the river seems to be incising again (j). The hypothesis that Humpleu Cave is mostly caused by paragenesis is still unconfirmed and, thus, needs to be taken with caution.

Mineralogy

Humpleu Cave became famous for its huge calcite scalenohedrons found near the entrance. They are among the largest found in natural caves (80 cm in length, more than 30 kg; Onac 1986; Ghergari et al. 1992). Sadly enough, the geode that was truncated by the present-day cave was devastated by mineral collectors, and only some rather unspectacular crystals remained in the cave.

A description of minerals in the Apuseni karst area (Onac 1992) also lists Humpleu Cave, where along with the ubiquitous calcite speleothems, among them a unique 4 m high moonmilk waterfall, immaculately preserved, also aragonite (dendritic aggregates) is found. Later, the moonmilk deposits from Humpleu Cave were analyzed in greater details allowing the identification of calcite, monohydrocalcite, and vaterite in their composition (Onac and Ghergari 1993; Chirienco 2002).

An investigation of the crusts and efflorescences sampled from various parts of the Humpleu cave system identified the following minerals: calcite, aragonite, gypsum, brushite, and hydroxylapatite (Ghergari et al. 1997). Recently, the finding of a greenish clay arose the curiosity of the speleologists. An X-ray analysis revealed the presence of crandallite, $CaAl_3(PO_4)(PO_3OH)(OH)_6$ (Exel 2013; Exel and Ottner 2015). A more thorough analysis of the same material yielded the presence of crandallite, hydroxylapatite, variscite, montgomeryite, and leucophosphite (Moldovan et al. 2015).

Cave Climatology

The cave has not yet received a thorough climatic study. Based on its morphology (entrances at different elevations) and seasonal observations, it belongs to the category of

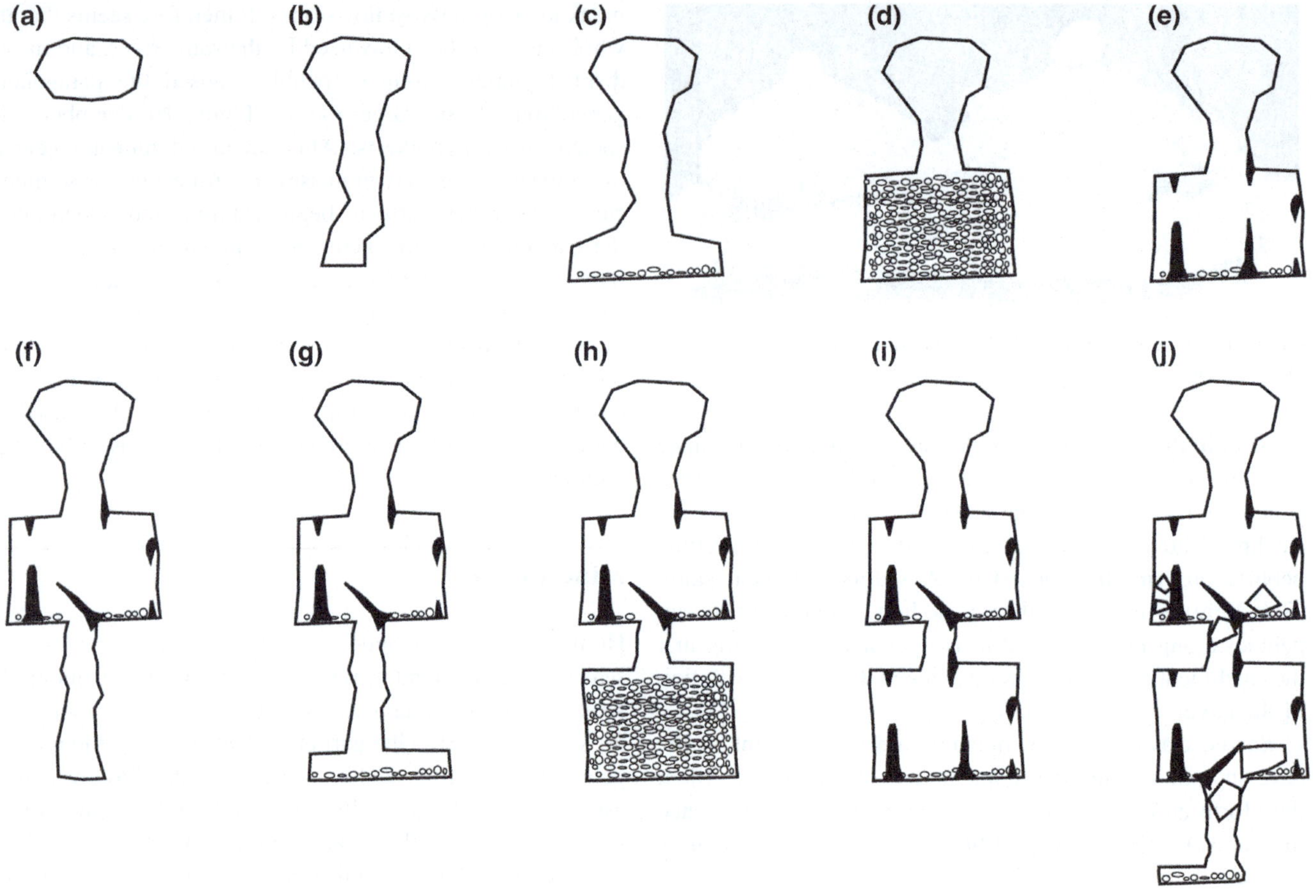

Fig. 10 Hypothetical stages in the genesis of Humpleu Cave by canyon formation, paragenesis, and subsequent deepening. The same process occurred twice; see text for more explanations

caves with unidirectional ventilation. During summer, airflow is from the higher entrance toward the lower one, whereas in winter the air current changes direction. A topoclimatic study conducted by Popa (1990b) between the cave entrance and the Dancing Room reveals a mean annual temperature of 4.9 °C in the inner sections of the cave. However, the study fails to mention the precision and accuracy of the instruments and reports only a handful of data.

Paleontology

Near the cave entrance in the Rol Gallery (Galeria Rol), but apparently also in Dan Coman Hall, bones of cave bear were found (Domşa and Popa 1988). At the latter location, however, the discoveries are still unverified. In the Entrance Room there is a goat skeleton in anatomic connection. Throughout the fossil part until the end of Charentais and Grenoblois rooms, bat bones are found, but they lack an investigation so far.

Cave Fauna

Humpleu Cave seems never to have been a main focus for cave biologists, although fauna does exist. The limited amount of information exists on the following species: *Myotis myotis*, *Myotis blythii* (Chiroptera); *Acanthocyclops balcanicus bisaetosus* (Cyclopoida; Iepure 2001); *Oncopodura pegyi*, *Paronychiurus boghieni* (Collembola; Gruia 2003); and *Pholeuon angusticolle arpadi* (Coleoptera; Racoviță 2011).

Conclusions

Humpleu Cave is the second longest cave system of Romania, but also a very interesting cave with respect to its genesis, mineralogy, and morphology. When remapping of the cave is completed in 2018 (hopefully), more thorough scientific investigations may begin. We are sure that this large cave system has many surprises to reveal.

Acknowledgements The large remapping project is not done single-handedly. More than 50 persons from all over the world participated in this project, from logistic help, organizational aid, to prolonged field work and continuous help throughout these years. It would lead too far to enumerate all the helping hands here, but the cave mappers, which did not contribute directly to the article, are mentioned here in alphabetical order: J. Achtman, M. Achtman, R. Crăiţa, G. Fraţilă, T. Kesselring, F. Maurer, F. Papiu, L. Plan, O. Pop, D. Sanz, late S. Schmassmann, R. Siegenthaler, T. Tămaş, and B. Wielander. Last but not least, the remapping project was initiated by B. P. Onac, who has been instrumental in organizing the first camps.

References

Chirienco M (2002) Calcite moonmilk in the Humpleu Cave system (Romania): the relationship between crystal morphology and cave topoclimate. Cave Karst Sci 29(3):101–104

Domşa M, Popa C (1988) Noi resturi de urside descoperite în peşterile României. Peştera 2:156–183

Exel T (2013) Tonmineralogische Untersuchungen von Höhlenlehmen in Hinblick auf die Entstehung von Höhlensystemen. Unpublished B.Sc. Thesis, University of Life Sciences, Vienna

Exel T, Ottner F (2015) Nachweis von Crandallit in rumänischem Höhlensediment. Mitt Österr Miner Ges 161:147–149

Ghergari L, Onac BP, Frăţilă G (1997) Mineralogy of crusts and efflorescences from Humpleu cave system. Proceedings of the 12th Int Congr Speleol, La Chaux-de-Fonds 1:231–234

Ghergari L, Mârza I, Bodolea A, Sehiau S (1992) Observations cristallographiques et génétiques sur les mégascalenoèdres de calcite de la grotte de Valea a Firii (Monts de Bihor, Roumanie). Karstologia 20:49–53

Gruia M (2003) Collembola from Romanian caves. Trav Mus Hist Nat "Grigore Antipa" 35:139–158

Iepure S (2001) Ciclopides du Parc National des Monts Apuseni. I. *Acanthocyclops plesai* n. sp. et *A. balcanicus bisaetosus* n. ssp. (Crustacea, Copepoda). Stud UBB Biol 46:31–41

Mantea G (1985) Geological studies in the upper basin of the Someşul Cald Valley and the Valea Seacă region (Bihor-Vlădeasa Mountains). An Inst Geogr Geofiz 66:5–85

Mârza I, Selişcan D (1987) Phénomenes d'éboulements massifs dans la Grotte de Valea Firei (Bassin du Somesul Cald), dus a la néotectonique épisodiquement réactivée. Stud UBB Geol 32 (2):81–86

Moldovan C, Tămaş T, Haeuselmann Ph (2015) The mineralogy of some particular phosphate deposits from the Humpleu cave system, Romania. Acta Miner Petrogr Abs Ser 9:43

Onac B (1986) Die grosse Höhle vom Fadental. Reflektor 4:9

Onac BP (1992) Mineralogy of the Apuseni Mountains caves. Theor Appl Karstol 5:193–201

Onac BP, Ghergari L (1993) Moonmilk mineralogy in some Romanian and Norvegian caves. Cave Sci 20(3):107–111

Onac BP, Papiu F (1995) Peştera Mare din Dealul Humpleu. In: Theor Appl Karstol (Prog Abs and Field Trip Guide), pp 52–54

Onac BP, Orăşeanu I, Perşoiu A, Tămaş T (2006) Karst landforms and caves in north-western Romania. Field guide for the Climate change: the karst record Conference (30 May–3 June 2006), pp 48–50

Onac BP, Papiu F, Aanei M, Bujor O, Irimie D, Filotie O (1998): Neotectonic features in the Humpleu cave system (Bihor Mountains, Romania). Speleocronos hors-série, Livre des contributions au colloque "Karst & Tectonics", Han-sur-Lesse, 9–12 mars 1998, pp 135–136

Orăşeanu I (1996) Contributions to the hydrogeology of the karst areas of the Bihor-Vladeasa Mountains, Romania. Theor Appl Karstol 9:185–214

Orăşeanu I (2010) Karst hydrogeology of Apuseni Mountains. In: Orăşeanu I, Iurkiewicz A (eds) Karst hydrogeology of Romania. Ed Belvedere, Oradea, pp 181–324

Papiu F (1988) Notă asupra ponorului din Valea Ponorului. Peştera 2:71–75

Papiu F (2001) Sistemul Humpleu. Speomond 6:26–29

Papiu F, Selişcan D, London JC (1993) La Pestera Mare din Valea Firii. Regards 12:1–4

Ponta G, Selişcan R, Steflea D (1993) Subterranean stream piracy in the upper basin of the Someşul Cald Valley area, Bihor Mountains, Romania. Bull Soc Géogr Liège 29:73–77

Popa C (1990a) Peştera Mare din Valea Firei (notă preliminară). Peştera Suppl 1:4–5

Popa C (1990b) Studiu topoclimatic din Peştera Mare din Valea Firii. Peştera Suppl 1:1–3

Racoviţă G (2011) Révision systématique des Leptodirinae souterrains des Monts Apuseni. VIII. Aperçu synthétique sur le genre *Pholeuon*. Trav Inst Spéol "Émile Racovitza" 50:37–60

Şoit VG (2009) Studiul evenimentelor erozionale, depoziţionale şi al ariei sursaă a sedimentelor din Peştera Humpleu—Peştera Mare din Valea Firei (Muntii Bihor). Unpublished BS Thesis, Babeş-Bolyai University, Cluj-Napoca

Useful Websites

https://sghbern.ch/newsghbern/romania/ (current remapping project)
http://pestera-humpleu.ro

Codru Moma Mountains: Vascău Karst Plateau

Gheorghe M. L. Ponta, Mihai Besesek, and Valentin Alexandru Radu

Abstract

Vaşcău Plateau hosts one of the most spectacular doline fields with karst lakes and ponors in the Romanian karst. The karst landscape is completed by over 70 short caves and shafts, including Câmpeneasca Cave with its spectacular waterfall at the entrance, the Iliei Pit the deepest (−165 m) cavity in Codru Moma Mountains, and last but not least, the Călugări intermittent spring.

Keywords

Karst • Dolines • Intermittent spring

Geographic, Geologic, and Hydrologic Settings

The Vaşcău Karst Plateau (90 km^2) is situated in the Codru Moma Mountains (part of the Apuseni Mountains) in the western Romania (Fig. 1). Its northern border follows the Briheni River, in the east the Crişul Negru Valley, to the south the boundary follows the line between Momuţa Peak and Călugări Spring, whereas to the west is Zugău Valley. The morphology of the karst plateau reflects the geological structure of the massif. The peaks and ridges correspond to hard rocks as rhyolites, Permian basalts, and Werfenian (Lower Triassic) quartzites, whereas the plateau consists of carbonate formations. The maximum elevation is reached in the southern part of the study area, in the Momuţa Peak (930 m). The mean annual temperature in the region is 8 °C, while the annual average rainfall is 1100 mm.

The dominant feature of the Codru Moma Mountains is the Vaşcău Plateau (500–600 m elevation), which shows an impressive karst topography generated by numerous dolines and closed depressions (Fig. 2). In most cases, the dolines form alignments or valleys of dolines, suggesting their spatial distribution is tectonically controlled. These alignments complete the tectonic image of the area, which initially was identified by geological survey. The closed depressions are major exokarst forms generated by karstic capture processes of the perennial/temporary streams in various stages of organization. Some are situated at the contact between limestones and other rocks/deposits, such as Pociovalişte, Recea, and Ponoraş or are develop exclusively on karst (Fântâna Ponor). Karst lakes occur in some of the dolines with their bottom covered by an impermeable clay layer (terra rossa) as a consequence of decalcification processes.

The karstic capture processes are also responsible for the genesis of dry valley known as sohodol. The best example is the Ţarina/Sohodol Valley, of which waters enter in the Câmpeneasca Cave, the stream bed downstream becoming dry. Ţarina, initially a tributary of the Briheni Valley, has a 5-km subaerial course developed on carbonate rocks before disappearing in Câmpeneasca Cave. It is worth mentioning that in the summer of 1984, 90% of the flow disappeared underground through a ponor situated 3 km upstream from the cave entrance. Vaşcău Plateau has only a few perennial rivers, most of them disappearing underground after a short subaerial course at the contact between impervious rocks and limestone (Fig. 3).

Codru Moma Mountains have a typical nappe structure (Codru Nappe system), in the area outcropping Moma, Vaşcău, and Coleşti nappes with a few small islands of Dieva Nappe on the top (Bleahu et al. 1979). The oldest carbonate deposits belong to the Moma Nappe and are

G. M. L. Ponta (✉)
Geological Survey of Alabama, 420 Hackberry Lane, Tuscaloosa, AL 35401, USA
e-mail: gponta@yahoo.com

M. Besesek
Speowest Speleological Association, Speleology Club Speodava Ştei, Cozia 13, Apt. 2, 310101 Arad, Romania
e-mail: mihaibesesek@yahoo.com

V. A. Radu
Speowest Speleological Association, Gării 39, Vladimirescu, 317405, Romania
e-mail: radugulea@yahoo.com

G. M. L. Ponta and B. P. Onac (eds.), *Cave and Karst Systems of Romania*,
Cave and Karst Systems of the World, https://doi.org/10.1007/978-3-319-90747-5_41

Fig. 1 Location of Vaşcău Plateau within Romania

represented by black dolomites of Lower-to-Medium Anisian age followed by dolomitic and gray limestones (Upper Anisian–Upper Carnian), and reef limestones (Upper Carnian–Norian). These deposits outcrop on 38.5 km^2 of which 80% are represented by black dolomites and dolomitic limestones. Within this formation, 13 ponors and 11 springs are known, the cumulated yield of the ponors being about 24% of the resurgences' yield (Ponta et al. 1986). One of the most interesting was the Călugări spring, which up to early 1990s was an intermittent spring (flowing about 5 min every 20–30 min/2–10 L/s). Presently, the spring is used by the monks of the Călugări Monastery and is running only at high flow (Fig. 4).

Dolomitic limestones, light-to-dark gray dolomites and limestones and shales in the upper part form the carbonates formations of the Vaşcău Nappe. The largest extend in the area are the Middle and Upper Anisian age limestones outcropping on 32.37 km^2. Within this formation, three ponors and three small springs were identified. The cumulated yield of the ponors represents only 4.5% of the resurgences' flow (Ponta and Gaşpar 1986).

The carbonate formations of Coleşti Nappe outcrop on 20.75 km^2 and are mostly of Carnian–Rhaetian age (Upper Triassic) formed by reef limestones, limestones with birdseyes structure, and red shales and host five ponors and nine springs, the largest one being Boiu discharging ca. 200–300 L/s. The cumulated flow of the ponors from this formation is 5–7% of all resurgences' flow. The three carbonate series belonging to different tectonic units act as a single hydrokarst structure.

Vaşcău Plateau was extensively studied from a hydrogeological point of view in the late 1970s by a team led by I. Orăşeanu, with G. Ponta, and N. Terteleac who took part in most of the dye/tracer studies performed in the area. (Orăşeanu 2010, 2016). In the mid-eighties, the Institute of Geology and Geophysics started a karst hydrogeological mapping program, the Vaşcău Sheet being the first one to be completed (Ponta and Gaşpar 1986; Ponta et al. 1986). With this occasion, an additional tracer study was performed to better define the recharge area of the Boiu Spring. In August 1985, Peştereli Valley Ponor was tested with In-EDTA, the tracer appearing after 12 h in Boiu Spring. The transfer

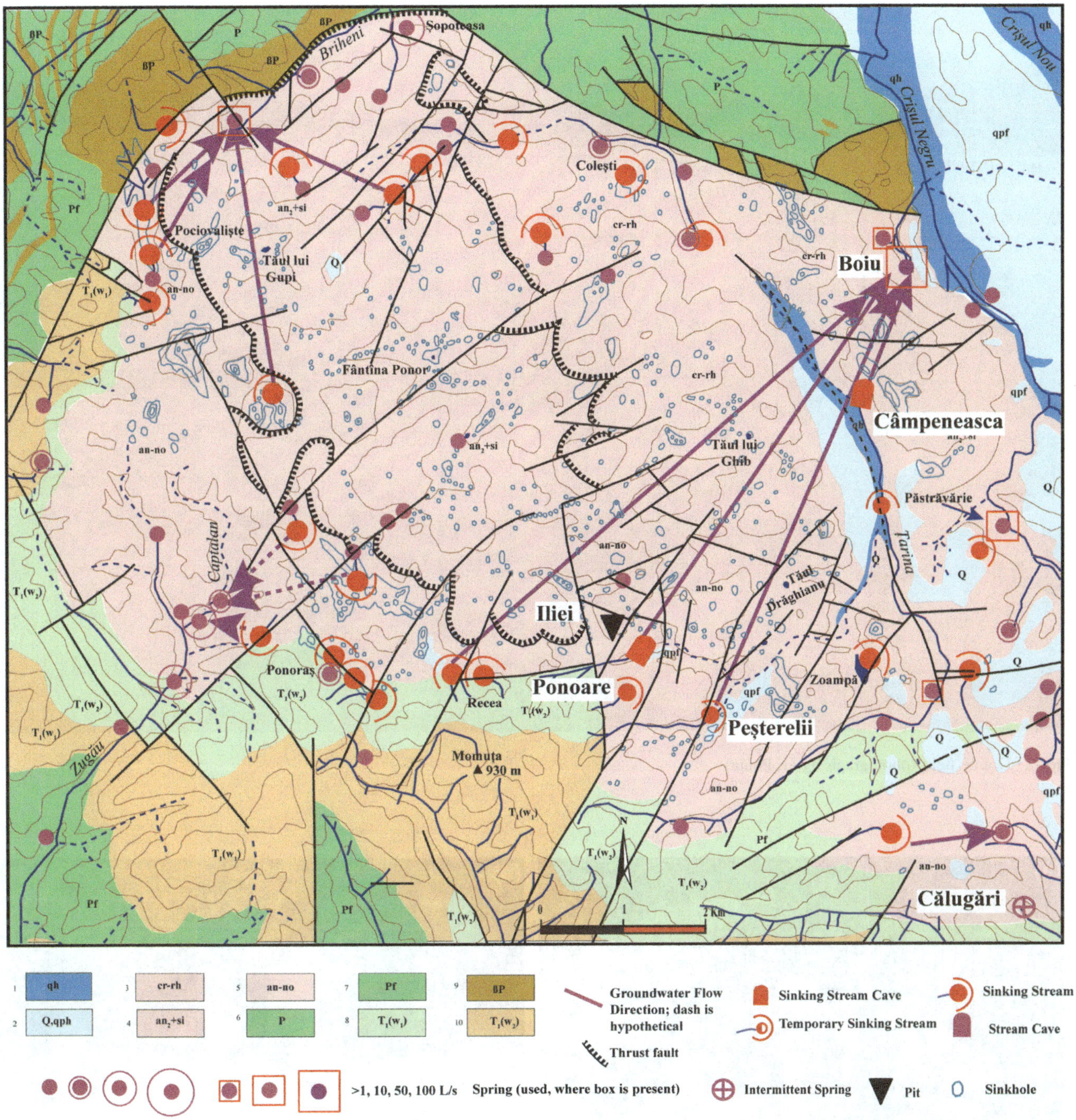

Fig. 2 Karst hydrogeological map of Vaşcău Plateau (modified after Ponta 1986, with permission) (geology modified after Bleahu et al. 1979). Key for the legend is available in the karst hydrogeology chapter

Fig. 3 Doline field (photograph by G. Ponta)

Fig. 4 Călugări intermittent spring in 1984 (**a**) and 2010 (**b**) (photographs by G. Ponta)

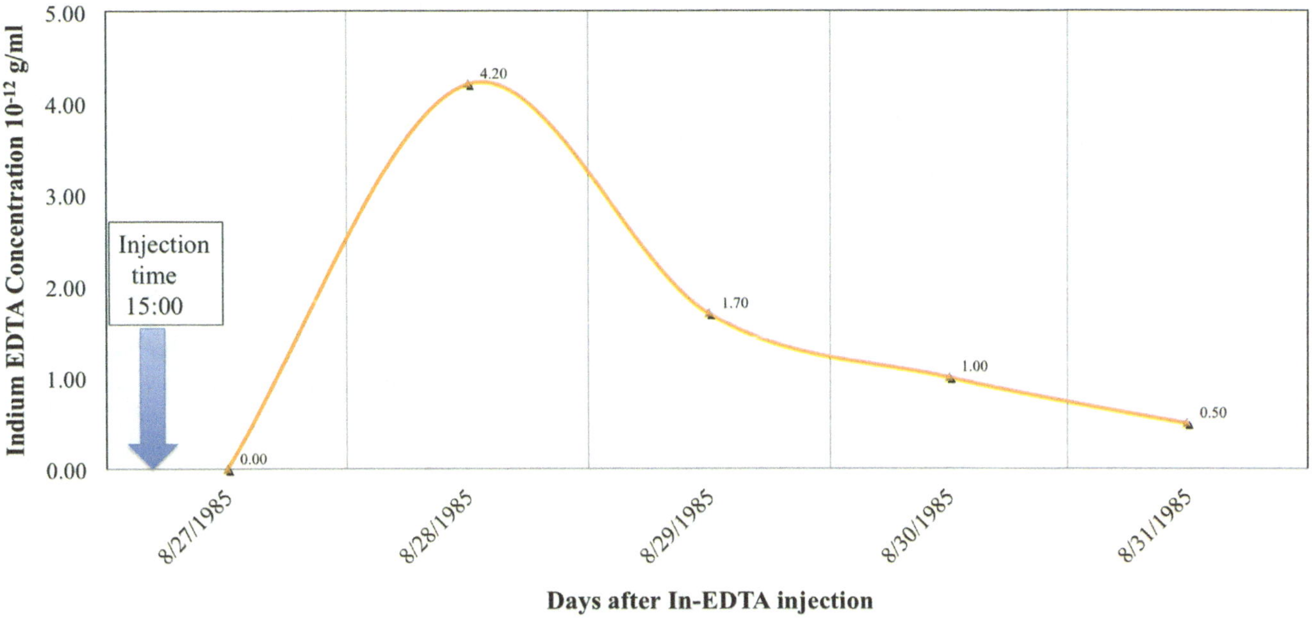

Fig. 5 Tracer (In-EDTA) breakthrough time curve Peştereli Valley Ponor—Boiu Spring (modified after Ponta 1986, with permission)

Fig. 6 Waterfall in Câmpeneasca Cave (photograph courtesy of A. Posmoşanu)

curve (in concentration versus time) is presented in Fig. 5. The transit time was very fast, unusual for a conduit flow (500 m/h), almost double than other tracer tests performed in the area. The heavy rain occurred during the test is partially responsible for this short travel time, but is recommended this test to be repeated in the near future with fluorescein.

Câmpeneasca Cave

The Câmpeneasca Cave is located at the end of the longest perennial valley of the plateau, Ţarina, in which waters are disappearing underground through a 20 × 15-m portal in a spectacular 40-m waterfall (Halasi 1979).

History of Exploration

The first reference of the Câmpeneasca Cave was made in 1858 in a publication by M. Jokai who visited the Vaşcău area and its surroundings, including the cave entrance. From reliable sources, he documented that 30 years earlier the Ţarina stream was sinking through a small entrance (in Halasi 1979). In 1863, A. Schmidl described the entrance of the cave to have the most beautiful waterfall in Europe (Fig. 6). Mihuţia performed the first investigation of the Boiu Spring in 1901 using coal dust as a tracer, which proved the connection between Câmpeneasca and Boiu (Mihuţia 1904). The tracer needed 4 h to travel the 1885 m aerial distance and 95 m vertical range. This was the first recorded tracer test performed in Romania. In 1978, I.

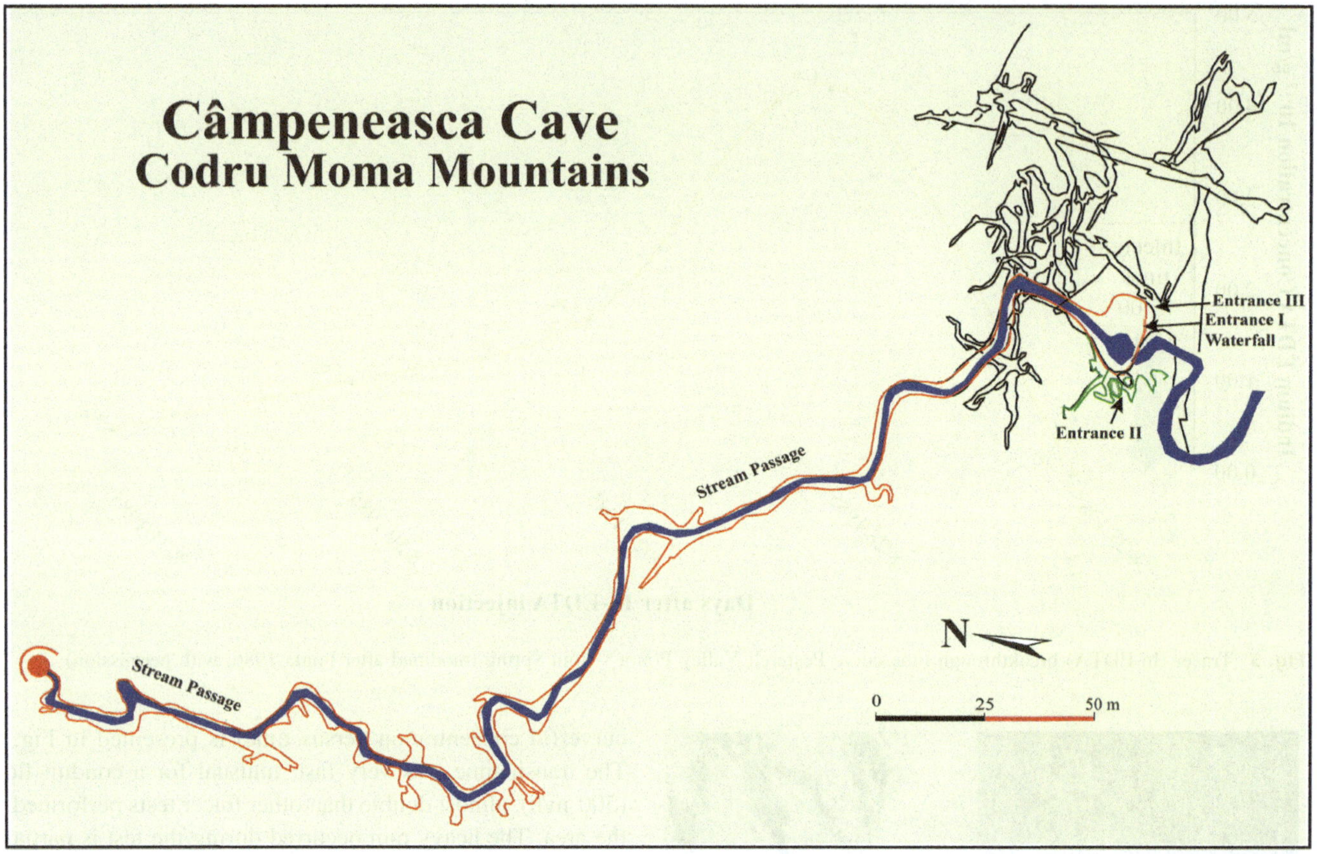

Fig. 7 Simplified map of Câmpeneasca Cave (modified with permission from Halasi 1979). Surveyed by G. Halasi, G. Ponta, K. Toth, M. Sluka (Red—Level IV, Green—Level III, Black—Level I and II)

Orăşeanu repeated the test with Rhodamine B at a low flow; this time it took 10 h for the tracer to cover the same distance (Orăşeanu 2010, 2016).

Only in 1958, the 40-m waterfall was descended for the first time by a team of researchers from the "Emil Racoviţă" Institute of Speleology (I. Viehmann, T. Rusu, M. Şerban, and M. Alb), when the active gallery was surveyed to the final sump (534 m) (Bleahu et al. 1976). In 1975, CS Liliacul Arad and Speodava Ştei resumed the exploration of the cave in several stages; by the end of 1979, it reached 1636 m and a vertical range of 68 m.

Cave Description

The cave has three entrances, the main one being at 395 m elevation and is a 20 × 15 m portal located on the edge of a 40-m waterfall (Fig. 7). To the left at 415 m elevation[1] is a second entrance, a 10-m deep pit that is connected with the main gallery through a few short passages. To the right of the portal is located the third, 2 × 3 m entrance, which ends in a dry ascendant gallery independent from the rest of the cave. At the base of the waterfall (Level IV), a stream is flowing through a 465-m-long meandering gallery (Stream Passage), which ends in a sump. The vertical range of this section is only 8 m. The third level of the cave is formed by a relatively horizontal gallery, which crosses the Stream Passage 40 m above. During high waters, this area is flooded. The second level is developed at the main entrance elevation and is formed by an assemblage of horizontal chambers and passages. Through two entrances disposed +15 m on the opposite wall of the cave, the first level of the cave is developed, which is formed by small galleries and a room with breakdowns, which communicate with the second level and opens in the 40-m waterfall at the portal elevation.

Câmpeneasca Cave is the most important ponor in the Vaşcău Plateau, draining about 60% of the Ţarina stream flow. The other 40% is lost along the 5-km-long flow on limestone deposits, mainly through a ponor located about 3 km upstream from the cave entrance. This ponor is more than likely to recharge a side gallery of the main collector, which it will be discovered in the future will show the branchwork type of the cave. The waters lost in the vicinity

[1]425 m elevation is the 0 m bench mark for the vertical range.

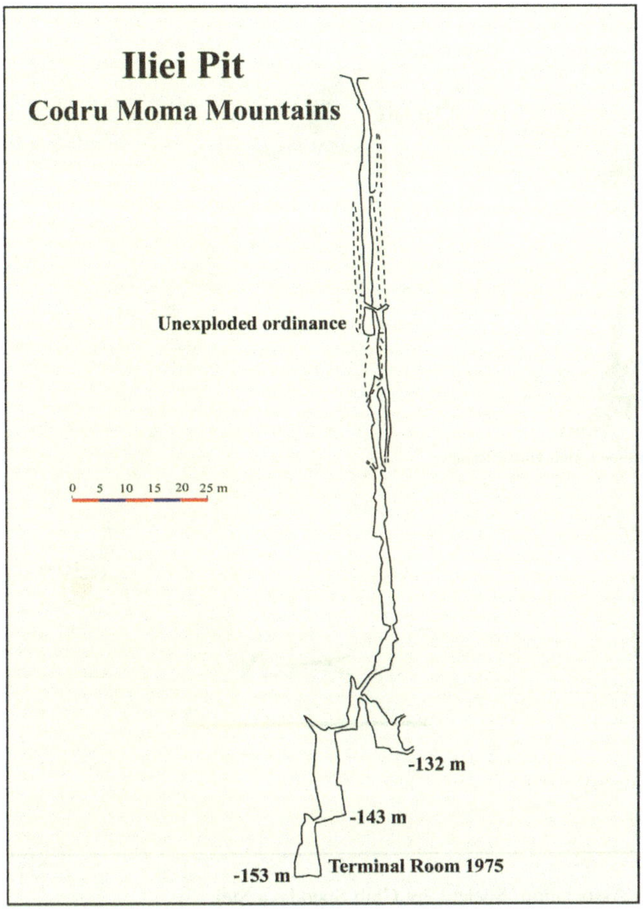

Fig. 8 Map of Iliei Pit. Surveyed by CS Liliacul Arad and CS Speodava Ştei

of the main entrance are reappearing underground on the fourth level as a small side tributary.

The distance between the Câmpeneasca Cave and the Boiu Spring is 1885 m with 95 m relief (considered from the Level I), distance traveled by the Rhodamine dye in 10 h (Orăşeanu 2016), resulting a velocity characteristic to a vadose flow. From the end sump to Boiu Spring is 1500 m and only 47 m vertical range. Hopefully, the next generation of cavers/divers will have the opportunity to clean the cave and then discover the rest of the karst system.

Cave Conservation

Because the stream entering the cave traverses the community of Izbuc and Călugări, with almost all the houses built along the valley, a large amount of trash is piling up at the base of the 40-m waterfall and further clogging the entire active gallery all the way to the sump. The cavers cleaned the area in 2008 removing a large quantity of trash, but it is all back into the cave, making the Boiu Spring not suitable for drinking. Although the cave is listed as a category C protected karst feature, it is certainly the most polluted cave in Romania (Besesek and Radu 2007).

Iliei Pit (Avenul Iliei)

Located at 675 m elevation, the Iliei Pit is just above the Ponoare Valley Cave, approximately 200 m away in straight line, what make us to consider that they are part of the same karst system.

History of Exploration

The explorations of Iliei Pit—Valley Ponoare Cave took place in the early 1970s by camps organized on Vaşcău Plateau by CSA Liliacul Arad and CS Speodava Ştei. The discovery of Iliei Pit occurred in the summer of 1974, whereas the first exploration was in December 27, 1974 by cavers from CSA Liliacul (L. Groh, G. Halasi, I. Marina, O. Honigesz, and G. Ponta). G. Ponta, a student in the Geology Department at the University of Bucharest, joined the Focul Viu Caving Club from where he borrowed a brand new 12 mm × 80-m-long Dederon rope (manufactured in the former East Germany), which was used to rappel the 50-m entrance shaft. In this 12-h long trip, the depth of −132 m was reached, using 160 m (1 × 80 m, 2 × 40 m) ropes and 10 (10 × 8) of homemade electron ladders. To be able to reach this depth, the surface team lowered the ladders with the rope to −50 m, to be able to explore the cave to −132 m.

In the summer of 1975, CSA Liliacul Arad (G. Halasi, and L. Groh) teamed with Speodava Ştei (D. Săsînă, G. Brijan, and O. Cuc) and using 140 m of ladders they reached −153 m. Also, for the first time in Romania, a winch was used for descending the entrance pit and a phone line connected the surface base camp with the one at −50 m. By the end of this trip, Iliei Pit with −153 m and 268 m in length became the deepest cavity in Codru Moma Mountains. During the explorations in the Ponoare Valley Cave in 2007, members of Speowest Arad and Speodava Ştei clubs began widening some of the squeezes at the end of the Iliei Pit in order to connect the two caves, reaching a 10-m drop followed by another 15 m to a depth of −165 m, located just a short distance from the Ponorului Valley Cave.

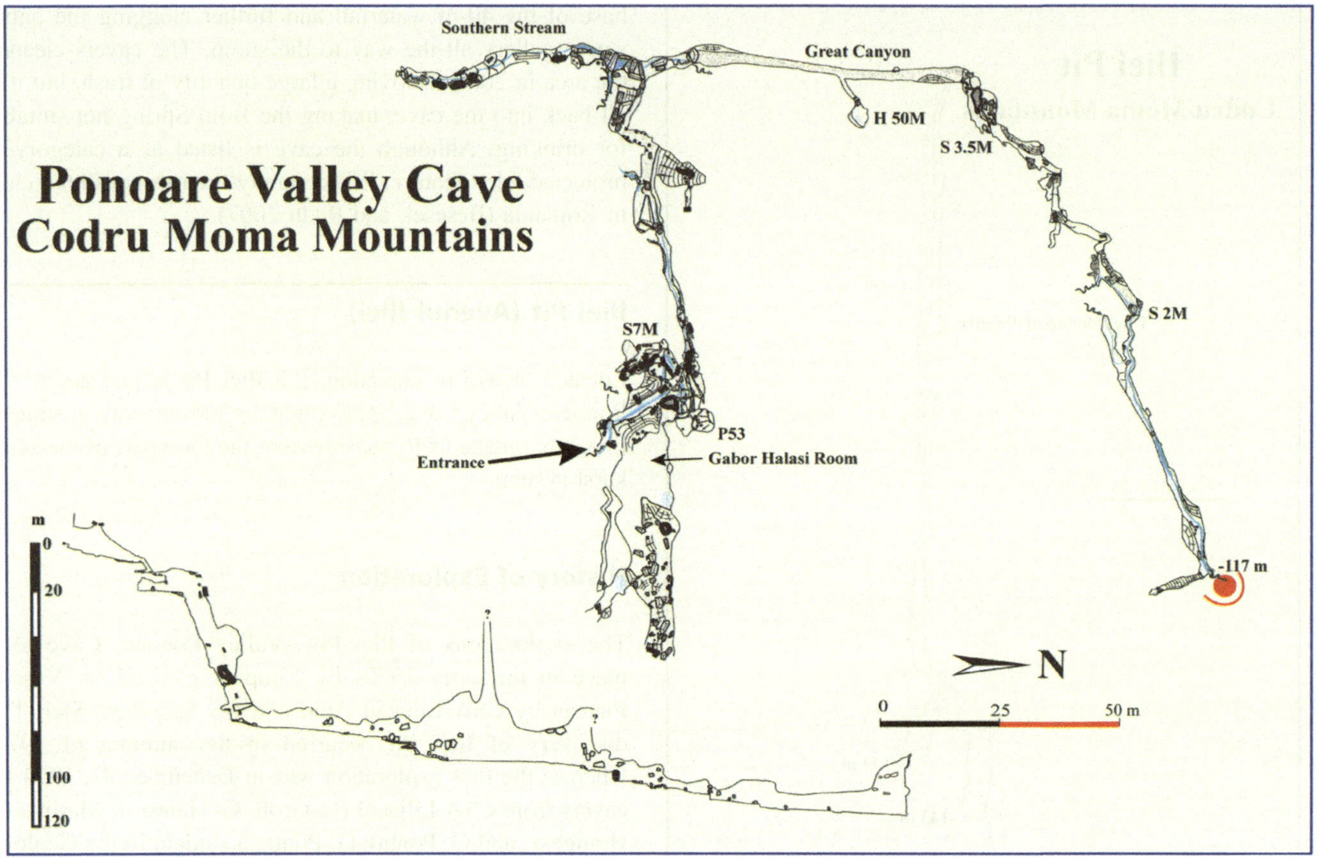

Fig. 9 Map of Ponor Valley Cave. Surveyed by Speowest Speleological Association, Speleology Club Speodava Ştei

Iliei Pit Description

Iliei Pit (Fig. 8) is located on the south side of the hill above Ponoare Valley, the opening being about 1 × 1 m. The first shaft is 50 m deep and ends in a room about 4 m in diameter. At the bottom, a second drop begins, which is caved in after 12 m with rocks (Halasi 1977). The cave continues after a +7-m climb to a ledge at −43 m where a system of parallel shafts is reached, with two penetrable openings. The one to the right is about 5–6 m deep, with an unexploded ordinance (World War II unexploded artillery missile) at the bottom. The narrow entrance into the second pit was enlarged to make it accessible, and the cave continues with a 47-m drop to a new ledge located at -90 m. Beginning from −105 m, several large ledges are found. At −105 m, the upper edge of another 20-m-long vertical drop is reached, where the entrances of two galleries occur. From this platform, one branch ends at −132 m, whereas the second one, after a series of 15, 8, and 20 m drops, with narrow passages in between, ends at −153 m in the Terminus Room 1975. As mentioned above, in 2007 two more drops were found, bringing the total depth of the cave to −165 m. Iliei Pit is of tectonic origin, being developed along vertical fractures resulting in a system of parallel shafts and chimneys with diameter ranging from 0.6 to 2 m at the top, and 4–5 m at the base.

Ponoare Valley Cave (Peştera Din Valea Ponoare)

Mihai Besesek · Valentin Alexandru Radu

Located at an altitude of 580 m, the Ponoare Valley Cave (700 m long and −117 relief) is one of the further ponors recharging the Boiu Spring. Between the cave and the spring, there is an aerial distance of 5900 and 279 m vertical range. The tracer test performed by I. Orăşeanu in 1978 with ^{82}Br indicated a flow of 630 m/day. At 210 m and +95 m vertical range is located Iliei Pit, the deepest in the Codru Moma Mountains, which is part of the same system, although they are not yet connected.

History of Exploration

The first exploration of Ponoare Valley Cave was made in 1978 by G. Halasi and G. Kajtor (22 m development). At that time, the cave ended with a lake/sump formed behind debris accumulated near the gallery entrance. In 1983, G. Halasi and J. Kokes found the entrance open and explore the cave. Later G. Halasi with G. Brijan from CS Speodava, and other cavers return for further exploration discovering 300 m of passages with a −70 m vertical range, the cave ending in a semi sump, which was not crossed.

In spring 2007, cavers from Speowest Arad and CS Speodava Ştei explored the semi sump, and M. Besesek, T. Rus, and V. Radu found a new stream gallery behind it. The gallery which begins at the base of the vertical drop in the entrance area of the cave, opened during a strong flood, making it possible to explore a system of active galleries with a tributary to the base of the wall where the underground river is disappearing again.

Cave Description

After a short distance inside the cave, a restriction is reached which was blocked by flood debris several times (Besesek 2007). Beyond this point, the cave continues with a series of vertical drops 3, 7, and 53 m, interrupted by short horizontal passages (Fig. 9). From the base of last vertical, at −73 m elevation, the cave keeps going in two directions. The first one requires a climbing a +6-m vertical wall leading to a section with dry galleries, where the explorations ended in the 1980s. The second one continues downstream (West) along the underground river through a canyon-type passage to a large room (Gabor Halasi Room) with its floor covered with boulders. At the lower end of this room, a south–north tributary exists (Southern Tributary), which upstream ends in narrow galleries.

Downstream (Southern Stream) from the confluence the stream disappears shortly and the cave continue toward north for about 100 m (Great Canyon), before turning east, and forming a parallel system with the entrance gallery. This sector has the floor covered with breakdowns and flood-related sediments. After descending a 2-m vertical drop, the junction with a short stream gallery is reached, from where it continues on a rectilinear gallery to the present end of the cave (−117.2 m).

Acknowledgements The authors would like to acknowledge the cavers of CS Liliacul Arad, Speowest Speleological Association, and Speleology Club Speodava Ştei that made this work possible. In addition, I would like to acknowledge all the other Romanian and foreign cavers who provided support for exploring and mapping Câmpeneasca Cave and Iliei Pit.

References

Besesek M (2007) Peştera din Valea Ponoare. Speomond 12:3–5
Besesek M, Radu M (2007) Dezastru ecologic în Peştera Câmpeneasca! Speomond 12:2
Bleahu M, Decu V, Negrea Ş, Pleşa C, Povară I, Viehmann I (1976) Peşteri din România. Ed Ştiinţifică şi Enciclopedică, Bucureşti, p 415
Bleahu M, Panin S, Tomescu C, Ştefan A, Istrate G, Ştefănescu M, Stan N (1979) Geological map of Romania, Vaşcău sheet. Scale 1:50.000. Inst Geol Geofiz, Bucureşti
Halasi G (1977) Avenul Iliei. Bul Speol 1:24
Halasi G (1979) Morfologia peşterii Câmpeneasca (Munţii Codru Moma). Nymphaea 7:265–271
Mihuţia S (1904): A vaskohi meszko-fensik hidrographiai viszonyai. Foldr kozlemenyek XXXII:1–31
Orăşeanu I (2010) Karst hydrogeology of Codru Moma Mountains. In: Orăşeanu I, Iurkiewicz A (eds) Karst hydrogeology of Romania. Ed Belvedere, Oradea, pp 245–281
Orăşeanu I (2016) Hidrogeologia carstului din Munţii Apuseni. Ed Belvedere, Oradea, p 289
Ponta G, Gaşpar E (1986) New tracing experiences with In-EDTA in the Vaşcău karstic plateau, Codru Moma Mountains (Rumania). Atti e Mem Comm Grotte "E. Boegan" 25:113–121
Ponta G, Bleahu M, Panin S, Orăşeanu I (1986) Hydrogeological map of Vaşcău. Scale 1:50.000. Inst Geol Geofiz, Bucureşti
Schmidl A (1863) Das Bihar-Gebirde an der Grenze von Ungar und Siebenburgen. Verl Forster u Bartelmus, Wien

Further Reading

Cocean P (1995) Peşterile României. Potenţial Turistic. Ed Dacia, Cluj-Napoca, p 264
Goran C (1982) Catalogul sistematic al peşterilor din România. Ed CNEFS, Bucuresti, p 496
Groh L, Halasi G, Oszker H (1978) Zonele carstice Briheni-Coleşti (Munţii Codru Moma). Bul CSER Bucureşti 4:123–139
Groh L, Heredi Z, Tulucan T, Honigesz O (1976) Fenomene endocarstice în Platoul Vaşcău. Bul CSER Bucureşti 4:82–100
Orghidan T, Negrea Ş, Racoviţă G, Lascu C (1984) Peşteri din Romania. Ghid Turistic. Ed Sport-Turism, Bucureşti, p 504

Useful Websites

www.speologie.org
www.karstgeology.com

Pădurea Craiului Mountains: Meziad Cave

Ioana N. Meleg, Marius Robu, Daniela R. Borda, Ludovic Mátyási, Călin Ghemiş, and Viorel T. Lascu

Abstract

Meziad Cave is known and studied since the late nineteenth century and has been developed for tourism at the beginning of the twentieth century. Today, part of its natural and cultural heritage revealed by scientists and cavers can be visited by tourists along a 1500 m path. Most of its chambers and passages are restricted to research for protection and conservation purposes. Meziad Cave is a landmark of the Pădurea Craiului Mountains (Apuseni Mountains, northwestern Romania) due to its diversity of scientific, aesthetic, and recreational values, such as: particular morphology characterized by huge voids, rare phosphate minerals related to the guano deposits, massive wind-controlled stalactites, some of the largest bat colonies in Romania, diverse and endemic subterranean fauna, and valuable paleontological and archeological remains.

Keywords

Show cave • Speleothems • Bats • Archeology Paleontology

Introduction

Meziad Cave is located in Pădurea Craiului Mountains (NW Romania; Fig. 1), in a limestone hill called Gruiul Peşterii (Cave's Hillock). The cave was home to big mammals and predators that went extinct about 25,000 years ago, and to humans who inhabited the cave during Neolithic and Middle Age. Today, the cave shelters unique subterranean fauna and a high diversity of bats. Because of the diversity of fauna and artifacts present in the cave, it attracted the attention of explorers, researchers, and cavers ever since the nineteenth century to study its geology, biology, paleontology, hydrology, and archeology. Most of the chambers and passages in Meziad Cave are restricted for scientific and preservation purposes.

Geographic, Geologic, and Hydrologic Settings

Meziad Cave is located at the border between Bihor and Pădurea Craiului Mountains, along the Peştera (Cave) Valley, in the village of Meziad, Remetea (Fig. 1). It occurs in an exceptional geo-structural context, with a rare petrographic complexity. On an area of 20 km^2, different geological formations are encountered: Bihor Unit found in the lower part of the basin, overlaid by Codru Nappe System (Vălani, Ferice and Arieşeni) mixed with Cretaceous volcano-sedimentary deposits, magmatic deposits (Vlădeasa type), fossiliferous deposits in the Gosau facies, and Miocene, Pliocene, and Quaternary sedimentary deposits belonging to the Beiuşului Depression (Mátyási 2015). The cave is developed in Urgonian facies limestone of Barremian

I. N. Meleg (✉) · M. Robu
"Emil Racoviţă" Institute of Speleology, Romanian Academy, Calea 13 Septembrie 13, 050711 Bucureşti 5, Romania
e-mail: ioana.meleg@iser.ro

M. Robu
e-mail: marius.robu@iser.ro

D. R. Borda
"Emil Racoviţă" Institute of Speleology, Romanian Academy, Clinicilor 5, 400006 Cluj-Napoca, Romania
e-mail: daniela.borda@academia-cj.ro

L. Mátyási
Geo Prospect, 13 Septembrie 30/1, 415600 Ştei, Romania
e-mail: geoprospect@geoprospect.ro

C. Ghemiş
Muzeul Ţării Crişurilor, Dacia Bldv. 1-3, 410464 Oradea, Romania
e-mail: ereshu@yahoo.com

V. T. Lascu
Centrul pentru Arii Protejate şi Dezvoltare Durabilă Bihor, Piaţa 1 Decembrie 4-6, Suite 8, 410068 Oradea, Romania
e-mail: violascu@spelemat.ro

G. M. L. Ponta and B. P. Onac (eds.), *Cave and Karst Systems of Romania*, Cave and Karst Systems of the World, https://doi.org/10.1007/978-3-319-90747-5_42

Fig. 1 Cave location in Pădurea Craiului Mountains

Age, and its genesis and evolution are related to the accentuated tectonic activity within the karst massif (Rusu et al. 1974).

Peştera Valley basin drains an area of 4.56 km^2, which extends from the Meziad Valley to the south and Măgura Meziadului Peak (782 m above sea level; asl) to the north. The cave was formed by two tributaries (Bradului and Gropilor) to the Peştera Valley that independently formed two karstification levels. The Bradului and Gropilor creeks enter underground at the Collapse Chamber and Poste Chamber, respectively (Fig. 2) and flow toward south until merging with the Peştera Valley (Rusu et al. 1974).

History of Exploration

The Austrian geologist K. Peters first documented Meziad Cave in 1861. One year later, the Austrian geographer A. Schmidl mapped the cave to a length of 1150 m, becoming the longest cave of Romania until 1934. Between 1896 and 1903, G. Czaran promoted and developed the cave as tourist attraction, building the first facilities and trained D. Negruţ as guide. The cave has been listed in the first tourist guide dedicated to the Bihor Mountains (Czárán 1903).

Further explorations involved biospeleological surveys. A new species of cave beetle has been discovered by E. Csiki in 1913, while in 1921 entomologist E. Bokor started to conduct research studies on cave insects. E. Racoviţă, the founder of biospeleology and of the world's first Institute of Speleology (1920), along with R. Jeannel and P. A. Chappuis led further research studies in 1921, 1922, and 1926 (Jeannel and Racovitza 1929).

The cave was surveyed several times by: V. Mátyás in 1930, E. Balogh, and the Tourist Club "Bihorul" Oradea in 1932; T. Rusu, G. Racoviţă, and V. Crăciun ("Emil Racoviţă" Institute of Speleology-ISER, Cluj Department) between 1968 and 1972 (Rusu et al. 1974); and O. Mărcuş and P. Damm (Cristal and Z Oradea caving clubs) between 1983 and 2002. All these cave exploration and mapping activities increased the length of the cave to 6292 m.

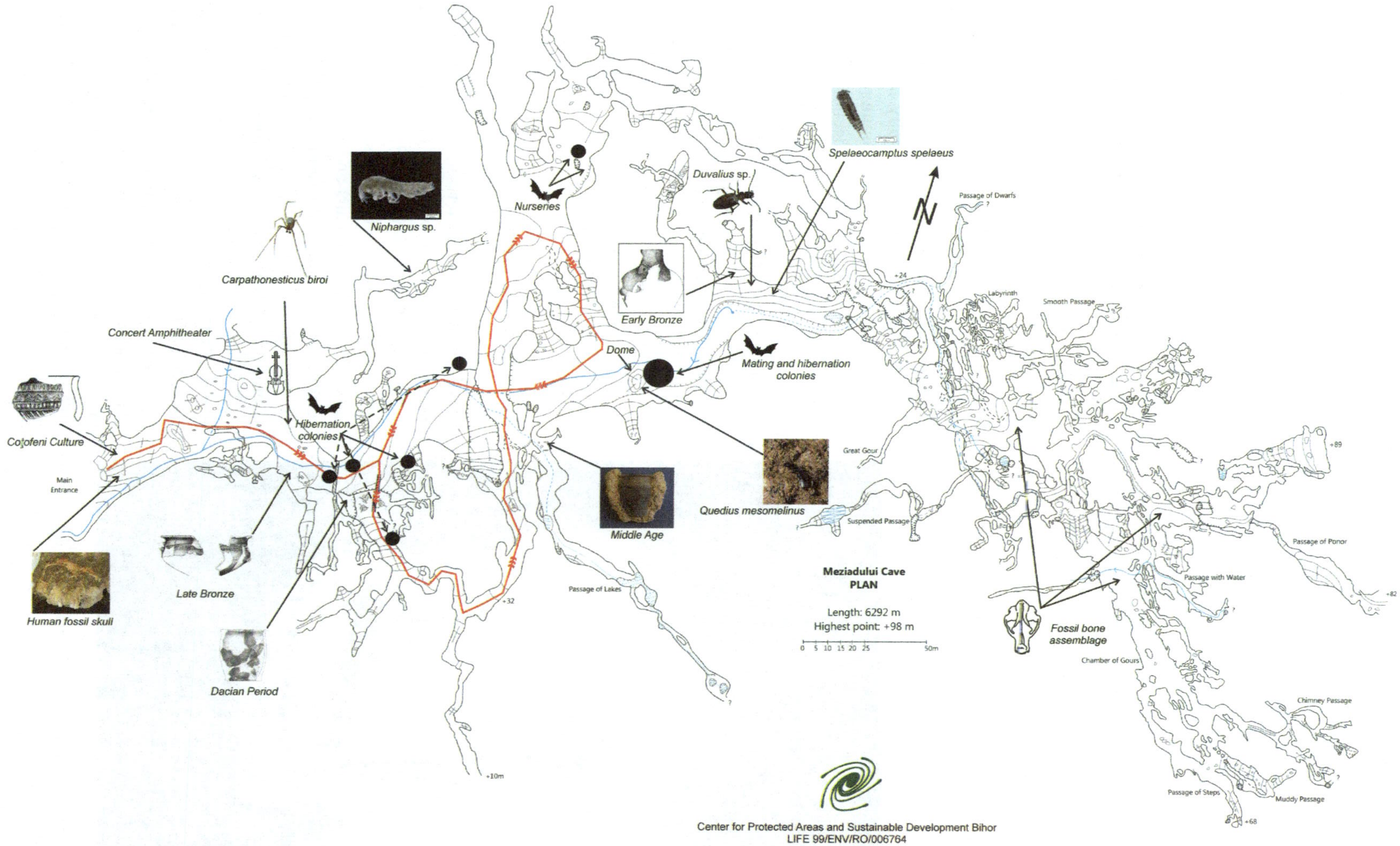

Fig. 2 Meziad Cave map showing natural and cultural heritage assets of the cave. Red trail: current show path. Small black circles: hundreds of bats. Big black circle: thousands of bats. Dotted line: occasional bat colonies (map courtesy of CAPDD Bihor)

Cave Description

The entrance of Meziad Cave (19 m wide and 15 m high; Fig. 3), located at the base of a 30 m cliff, leads to a large chamber developed along a fracture through which a subterranean river is flowing. During rainy times, it receives additional water from a surface river which is entering the cave through a window that opens on the right not far from the entrance.

The cave is developed on three levels: a 3.4 km fossil upper horizon, followed by a 1.8 km temporary active sub-fossil level along which the largest chambers of the cave are developed, and a lower level composed of a labyrinth of passages, almost entirely flooded, with only few dry/accessible ones (Fig. 2). The upper fossil level lacks an organized drainage, and its labyrinth of passages is well decorated with speleothems. Fallen blocks and pronounced corrosion marks are present. The sub-fossil level has a more linear development, a temporary stream drains it, and it is less decorated than the upper level. In turn, it has large chambers, some with distinct triangular sections with concave side aspects. Movement of water in both vadose and phreatic regimes shaped cave's galleries with various and beautiful corrosion/erosion levels and notches and alluvial terraces. Sumps, subterranean springs, and lakes are also common (Rusu 1988). In some places, large stalagmitic domes and fallen blocks can be found, while in side passages cave deposits are encountered; these are described below.

Fig. 3 Impressive entrance arch of the Meziad Cave (photograph courtesy of A. Posmoşanu)

Cave Sediments and Speleothems

In the sub-fossil level and some parts of the upper level, cave sediments are mainly represented by allochthonous detrital deposits (sands and gravel). Petrographic studies revealed that rhyolite and quartz-sandstone boulders are often found in Meziad Cave, while limestones are mainly represented by breakdowns. The thickness of the detrital deposit ranges between 1.5 and 3 m, occasionally reaching 5–7 m. Small volume of clay deposits is common [i.e., Sala Piramidei (Pyramid's Chamber), Sala Liliecilor (Bats Chamber), and Pasajul Oaselor (Bones Passage)]. Large guano accumulations are present in Bat and Pyramid chambers underneath and around bat colonies and paleontological sites, respectively (Fig. 2). In association with these deposits, a variety of phosphate minerals were documented (see Mineralogy entry). Meziad Cave is well known for the diversity and beauty of its speleothems like gours, calcite crusts (in the upper level), cave pearls, and gigantic anemolites (wind-controlled stalactites).

Cave Climatology

Meziad Cave is characterized by a constant and relatively intense air exchange with the outside. With two openings located at different elevation (82 m apart), Meziad is a unidirectional ventilated cave system, with a notable winter (ascending currents of air) and summer regime (descending currents of air). The temperature ranges between 8 and 11 °C, whereas the relative humidity (RH) reaches levels near saturation (97%). During summer, data loggers indicated 12.2 °C in Bats Chamber in the upper level, and 11.8 °C in Pyramid's Chamber in the lower level. Both places showed a saturated atmosphere (100% RH), and a constant temperature with very small amplitude (less than 0.15 °C), which corresponds to the stability climatic zone. In the same time, the outside temperature ranged between 15.46 and 41.86 °C, and humidity between 19.9 and 88.18% RH (unpublished data).

Mineralogy

A complex association of phosphate minerals occurs in Meziad Cave, including common phosphates (brushite, hydroxylapatite) and taranakite, a phosphate rarely encountered in caves of Romania. The genesis of these phosphate minerals is all related to the presence of bat guano deposits (Onac and Bengeanu 1992; Gafencu and Onac 2000).

Paleontology

In 1960, cave bear bone accumulations of Upper Pleistocene age were discovered (Rusu et al. 1974). In the upper level of the cave (Bones Passage) (Fig. 2), scientific reports and the photographic material from this site indicate the existence of a diversified paleofauna including carnivores, cervids, birds, and rodents. Similar to the majority of other caves with bone deposits from the Apuseni Mountains, the cave bear (*Ursus spelaeus*) is a dominant species. Moreover, as we know at the moment, the paleoecological peculiarities of this bear population from Meziad Cave (size of skeleton, diet, predators, period of extinction) seem to be in the range of the cave bears from the Carpathian Mountains (Robu 2015). During Upper Pleistocene in Pădurea Craiului Mountains is recorded the highest peak of cave bear population, and their predators as cave lions (*Panthera spelaea*), hyenas (*Crocuta crocuta spelaea*), and wolfs (*Canis lupus*) (Robu 2015). Around 30,000 years BP, cave bear population records a dramatic decrease, as well as all the associated fauna, which led finally to their extinction (Robu 2015).

Fauna

Meziad Cave is one of the most thoroughly investigated caves in terms of fauna (see Racoviță et al. 2002). Over time, various terrestrial and aquatic cave habitats acted as shelter, resting places, or natural traps to 39 epigean and hypogean invertebrate species belonging to 14 major taxa (Oligochaeta, Opiliones, Pseudoscorpiones, Acari, Araneae, Amphipoda, Isopoda, Bathynellacea, Cyclopoida, Harpacticoida, Collembola, Coleoptera, Diptera, and Lepidoptera).

Nowadays, some of these species are hardly observed, due to intensive and uncontrolled tourism in the cave over the past decades that affected their distribution and community dynamics along the cave. Among the species encountered in Meziad Cave, the hypogean species are represented by: *Carpathonesticus biroi* (Aranea), *Niphargus bihorensis* (Amphipoda), *Bathynella chappuisi* (Bathynellacea), *Spelaeocamptus spelaeus* (Harpacticoida), *Onychiurus meziadicus* (Collembola), *Duvalius (Biharotrechus) redtenbacheri bihariensis* (Coleoptera) (Giurginca 2000–2001; Gruia and Ilie 2001–2002; Moldovan 2002; Meleg 2013).

Meziad Cave is known as the home to one of the Romania's largest bat colonies. Thirteen species of bats usually roost in the cave in different seasons: barbastelle (*Barbastella barbastellus*), Schreiber's bent-winged bat (*Miniopterus schreibersii*), lesser mouse-eared bat (*Myotis blythii*), Geoffroy's bat (*Myotis emarginatus*), whiskered bat (*Myotis mystacinus*), greater mouse-eared bat (*Myotis myotis*), noctule (*Nyctalus noctula*), common pipistrelle (*Pipistrellus pipistrellus*), soprano pipistrelle (*Pipistrellus pygmaeus*), long-eared bat (*Plecotus auritus*), greater horseshoe bat (*Rhinolophus ferrumequinum*), a medium horseshoe bat (*Rhinolophus euryale/blasii*), and lesser horseshoe bat (*Rhinolophus hipposideros*) (Dumitrescu et al. 1963; Borda 2002; Coroiu et al. 2007). Some of the species, mainly *Miniopterus schreibersii*, *Rhinolophus ferrumequinum*, and *Pipistrellus pipistrellus,* establish winter colonies of thousands of bats between October and April. Also, big mixed nurseries of *Myotis myotis/Myotis blythii* and *Miniopterus schreibersii* are constantly present in the cave between May and August. From September till October, the cave is known as a swarming site for bats. The large colonies of bats provide huge accumulation of guano, which sustain a large community of guano invertebrates, as *Xenylla spelaea* (Collembola), *Quedius mesomelinus* (Coleoptera), various insects' larvae, and many rainworms (Rusu et al. 1981; Gruia and Ilie 2001–2002). Due to the aerosolization from guano, elevated concentration of bacteria, including pathogenic enterobacteria, and also fungi were detected in the air, which leads to biohazard around of the guano accumulations (Borda et al. 2014).

Archeology

Preliminary archeological discoveries unearthed valuable information about various periods when man used the cave as a temporary shelter (Middle Neolithic), as a sacral place (Copper Age-Coțofeni Culture, Early and Late Bronze Age, and Dacian Period-La Tene), or as a place for refuge between the tenth and twelfth century (Early Middle Age) as revealed by a recently discovered pottery vessel (Ghemiş 2015).

The archeology of Meziad Cave is better known thanks to the surface researches made in the last decade, and from several archeological diggings (2012, 2015). The first information about the use of this cave dates back to the Middle Neolithic. Research on discoveries belonging to this time span (i.e., shelter, pottery shards, and ritual deposition) is currently in progress.

In the Copper Age (Coţofeni Culture), the cave was probably used as a sacred place since a human skull showing evidence of violence was discovered in the summer of 2012 (Fig. 2). A typical amphora belonging to the Roşia Group (Early Bronze Age) was discovered in 2002 among a pile of rocks located upstream from the dome, on the right side of the cave (Fig. 2). In the Late Bronze Age, the cave was used for ritual purposes. Further archeological investigations provided fragments of an Igriţa group typical vessel, and more importantly, the La Tene/Dacian period pottery found near the entrance of the cave (Fig. 2).

The cave was also used in the Middle Age as a refuge place, fact suggested by a ceramic pot discovered in the "Main Gallery" (Fig. 2). The researches in progress will shed light on the character of the archeological complexes found until now in Meziad Cave.

Show Cave

Around 1900, Secession artists found inspiration in the beautiful landscape of Meziad Cave. Inspiring the architectonics of cities as Vienna (Austria) and Oradea (Romania), Meziad Cave impresses through its vaulted arches, natural bridges, various morphology of overlapping galleries, and gigantic yet delicate speleothems.

Meziad Cave has been developed for tourism in 1903. It has been refitted in 1970 and 2005, when LED electric lights replaced carbide lighting. In 2012, a radical restoration took place. Now the cave is integrated in the Romanian's show cave network. A short walk along the gorges leads from the parking lot located 200 m away from the cave to the resurgence draining the water of Meziad Cave and immediately after to the impressive arch of the cave (Fig. 3). Near the entrance, the guide welcomes tourists in a rustic, 150-year-old house that preserves the traditional woodcraft of local artisans.

The touristic infrastructure within the cave has been constructed using composite materials. The hidden lighting system highlights the mystery and the unique landscape of the cave, while handrails are used for safety. The entrance chamber is naturally illuminated through the arch and a window that opens on the left side of the entrance. This chamber is beautifully decorated and hosts a small concert amphitheater. The current show path follows an 8-shaped trail. Immediately after the historic gate, the path goes up toward the fossil level where very old speleothems, some of them shaped by air currents (anemolites), are covered by recent calcite deposits. The trail crosses a natural bridge that provides a breathtaking view of the cave scenery, then enters the passages near where bats congregate, and descends under the bridge until it crosses again the entrance chamber along the river. The visit is animated by the guide who names the speleothems after Romanian science fiction folk characters or animals (Fig. 4). Adventure caving is organized for groups of maximum 10 people and requires caving gear and specialized caving guide. It also has a via *ferrata* allowing the access into the giant entrance chamber through a natural window up to the belvedere above the cave.

Fig. 4 Part of the touristic trail in the Meziad Cave (photograph courtesy of A. Posmoşanu)

Cave Conservancy

The cave is designated as a natural monument (IUCN Category III) and archeological site. It is also included in the Natura 2000 site ROSCI 0062 Defileul Crişului Repede—Pădurea Craiului. Since 2011, it is jointly administered by the local authority and the NGO Center for Protected Areas and Sustainable Development (CAPDD), Bihor County. Both structures have a significant contribution in protecting and preserving the cave and its heritage, while finding the best solutions to promote the development of sustainable tourism in Meziad Cave and other karst areas.

Conclusions

Meziad Cave is an important landmark in the evolution of research and preservation of show caves in Romania. It provides fundamental practices concerning multidisciplinary research projects and methods for cave protection and preservation in other karst areas.

Acknowledgements We are very grateful to Center for Protected Areas and Sustainable Development (CAPDD), Bihor, and A. Posmoşanu for proving the Meziad Cave map and pictures, respectively. We thank the following people from Cristal and Z (Oradea) caving clubs for contributing (exploration and survey) to the production of Meziad Cave map: O. Mărcuş, R. Mărcuş, late P. Damm, A. Mărcuş, B. Mărcuş, L. Pusztai, late B. Horvath, D. Pitic, E. Zamfor, J. Kokes, A. Kondacs, late A. Tonţ, V. Mărcuş, B. Tamaş. We are thankful to both editors and reviewers for critical comments and thoughtful suggestions that improved the manuscript quality and clarity.

References

Borda D (2002) Chiropterele. In: Racoviţă G, Moldovan O, Onac BP (eds) Monografia carstului din Munţii Pădurea Craiului. Ed Presa Univ Clujeană, Cluj-Napoca, pp 172–182

Borda DR, Năstase-Bucur RM, Spînu M, Uricariu R, Mulec J (2014) Aerosolized microbes from organic rich materials: case study of bat guano from caves in Romania. J Cave Karst Stud 76:114–126

Coroiu I, David A, Borda D (2007) Colonii de hibernare a chiropterelor Peştera Meziad (Munţii Bihor). Misc Chiropterol 2:40–42

Czárán G (1903) Kalauz Biharfüredi Kirándulásokra. Sussmann Lázár, Beiuş

Dumitrescu M, Tanasachi J, Orghidan T (1963) Răspîndirea chiropterelor in R.P. Română. Lucr Inst Speol "Emil Racoviţă" 1–2:507–575

Gafencu M, Onac BP (2000) The mineralogy of phosphate speleothems in selected caves from Pădurea Craiului Mountains (Romania). Nymphaea 28:5–13

Ghemiş C (2015) Steps through time in Meziad Cave and surroundings. In: Meleg IN (ed) The book of Meziad Cave. CAPDD Bihor, Oradea, pp 9–16

Giurginca A (2000–2001) The spreading of the genus Mesoniscus in the Romanian Carpathians. Trav Inst Spéol "Émile Racovitza" 39–40:11–22

Gruia M, Ilie V (2000–2001) Collembola from the karstic system of Romania. Trav Inst Spéol "Emile Racovitza" 39–40:59–114

Jeannel R, Racovitza E (1929) Enumération des grottes visitées, 1918–1927 (Septième série) Biospeologica LIV. Arch Zool Exp Gén 68:293–608

Mátyási L (2015) Pe drumul apei—Naşterea calcarului şi a Peşterii Meziad. In: Meleg IN (ed) Cartea Peşterii Meziad. CAPDD Bihor, Oradea, pp 20–26

Meleg IN (2013) Population dynamics of subterranean Crustacea: ecology, genetics and conservation. Ed Presa Univ Clujeană, Cluj-Napoca, p 87

Moldovan O (2002) Coleopterele subterane. In: Racoviţă G, Moldovan O, Onac BP (eds) Monografia carstului din Munţii Pădurea Craiului. Presa Univ Clujeană, Cluj-Napoca, pp 165–171

Onac BP, Bengeanu M (1992) Phosphate minerals in few caves from Pădurea Craiului Mountains (Romania)—preliminary note. Nymphaea 22:23–28

Racoviţă G, Moldovan O, Onac BP (eds) (2002) Monografia carstului din Munţii Pădurea Craiului. Ed Presa Univ Clujeană, Cluj-Napoca, p 264

Robu M (2015) The palaeontology of the cave bear bone assemblage from Urşilor Cave of Chişcău—Osteometry, palaeoichnology, taphonomy, and stable isotopes. Unpublished PhD Thesis, "Emil Racoviţă" Institute of Speleology, Romanian Academy

Rusu T (1988) Carstul din Munţii Pădurea Craiului. Ed Dacia, Cluj-Napoca, p 253

Rusu T, Racoviţă G, Crăciun V (1974) La Grotte du Meziad. Aspects physico-géographiques, génèse et évolution de la cavité. Trav Inst Spéol "Émile Racovitza" 13:147–173

Rusu T, Racoviţă G, Crăciun V (1981) Peştera Meziad. Ed Sport-Turism, Bucureşti, p 107

Pădurea Craiului Mountains

Gheorghe M. L. Ponta

Abstract

Pădurea Craiului has the largest plateau-type karst in the Apuseni Mountains developed at about 500–600 m altitude with numerous karst features as karren, dolines, ponors, springs, and caves. In the southern part of the plateau, Ciur Izbuc–Ciur Ponor–Topliţa de Roşia karst system includes two distinct caves. Ciur Izbuc is a through cave traversed by a stream that disappears in the middle of the cave to resurface in Ciur Ponor–Topliţa de Roşia Cave. The highlights of the system are the Bears Gallery (Galeria Urşilor) and the Footprints Hall, which housed many skeletal remains of cave bear and about 400 footprints as old as 27–31 ka. Ciur Ponor–Topliţa de Roşia is the longest through cave in Romania, with a stream that runs from Ciur Ponor entrance to Topliţa de Roşia Spring passing through large chambers with their floor covered with alluvial sediments and breakdown blocks and large passages and sumps in between. In the northern part of the plateau, the underground stream traversing Bătrânului Cave resurfaces a few km downstream in Vadu Crişului Cave ending its journey with a spectacular waterfall in the south bank of the Crişul Repede River.

Keywords

Cave tracer studies • Dolines • Ponor • Spring Ciur-Izbuc • Ciur-Ponor

Geographic, Geologic, and Hydrologic Settings

The Pădurea Craiului Mountains with an area of 1150 km^2 of which 425 km^2 are limestones (Rusu 1988; Racoviţă et al. 2002) are located in the northwestern part of the Apuseni Mountains, between Crişul Repede River in the north and Crişul Negru River in the south (Fig. 1). The Pădurea Craiului Mountains present the largest peneplains (leveling or erosional surfaces) in the Apuseni Mountains: a central one dipping toward northwest and a marginal one dipping toward the surrounding depression areas (north and south) (Rusu 1973).

The peneplain/erosional surface developed between 400 and 600 m elevation is a continuation of the one from Codru-Moma Mountains in the south and Bihor Mountains in the west. These peneplains can be correlated with the Gornoviţa Platform of the Southern Carpathians (Bleahu et al. 1976). Due to differential weathering, the peneplains are dominated by isolated massifs made of Lower Jurassic sandstones, of which elevations decline from east to west. Also, the elevation of the watershed dividing Crişul Repede from Crişul Negru basins decreases in altitude from 700 m in the east to 300 m in the west. These orographic elements indicate the presence of two important leveling periods. An initial one (Paleogene), when the surface water system flow followed a SE-NW direction, and a secondary one (Neogene) when the hydrographic network was oriented towards the Vad-Aleşd sedimentary basin in the north and Beiuş to the south (Rusu 1973).

Tributaries of the Crişul Negru and Crişul Repede rivers cut into limestone deposits, generating 200–300 m deep gorges, with large plateaus in between. The most frequent karst forms/features in the limestone plateaus are dolines (sinkholes), which are developed in large number on the Jurassic/Cretaceous carbonaceous deposits. Occasionally, two or more dolines are interconnected forming uvalas or closed depressions. Some of the closed depressions are at the interface between the karst and non-karst rocks, crossed by short rivers, which in their lowest point are sinking underground, through either a cave or an impenetrable ponor.

In the Pădurea Craiului Mountains, more than 850 caves and pits are known, of which almost 50% are stream caves or caves which intercept a stream. Among them is Romania's longest cave, the Wind Cave (Peştera Vântului, >50 km long; Szilágyi Palkó et al. 2007). Several karst

G. M. L. Ponta (✉)
Geological Survey of Alabama, 420 Hackberry Lane, Tuscaloosa, AL 35401, USA
e-mail: gponta@yahoo.com

G. M. L. Ponta and B. P. Onac (eds.), *Cave and Karst Systems of Romania*,
Cave and Karst Systems of the World, https://doi.org/10.1007/978-3-319-90747-5_43

Fig. 1 Location of caves discussed in this chapter within the karst of Pădurea Craiului Mountains

systems, caves, or potholes are presented as separate chapters by different authors in the Pădurea Craiului Mountains section.

Within the Alpine cycle in the northern part of the Apuseni Mountains, there are two structural units: the Bihor Autochthonous (Bihor Unit) and the system of Codru Nappes. The sedimentary sequences belonging to the Autochthonous are monoclinal or widely folded and present a few horsts and grabens. The geologic formations of the Bihor Unit, which crop out in the area, range in age from Paleozoic to Neogene, whereas in the southwestern corner they belong to the post-tectonic cover formations (Bordea et al. 1986). The unconsolidated/impervious sediments and rocks are not described and discussed in detail herein; the geology of the underlying carbonates rocks is the focus of this paper.

Upper Anisian–Lower Carnian (Middle to Upper Triassic) white reefal (coral reef) limestones (Wetterstein) outcrop only on a small area in the northeastern corner of the study area and are transgressively and discordantly overlain by Hettangian–Sinemurian (Lower Jurassic) quartzitic sandstones with clay layers and breccia at the base (Dragastan et al. 1982).

The Lower to Middle Jurassic deposits shown in Fig. 2 as Upper Sinemurian-Domerian, Toarcian-Aalenian, Bajocian-Lower Callovian are formed by 3–4-m-thick encrinites limestones at the base, which are overlain by a 20-m-thick marl sequence interbedded with limestones, and 5–10-m-thick oolitic ferruginous limestones. The whole block rarely surpasses 50–60 m in thickness and outcrops in small and unevenly spread areas (Bordea et al. 1986).

The Upper Jurassic deposits (Oxfordian-Lower Tithonian) are separated as Vad (pelletal dark gray with chert nodules), Galăşeni (pelletal light gray with rare chert nodules), Farcu (white reefal), Cornet (reefal light gray), and Albioara (dark gray micrite) limestones, with a total thickness of 200 m, and are disposed in stratigraphic continuity (Bordea et al. 1986). Based on more recent lithofacies studies, Bucur and Cociuba (2001), Bucur et al. (2010) combined the Cornet and Farcu limestones into Cornet Limestones. At the end of Jurassic, the entire region was uplifted and a well-developed karst formed on top of the

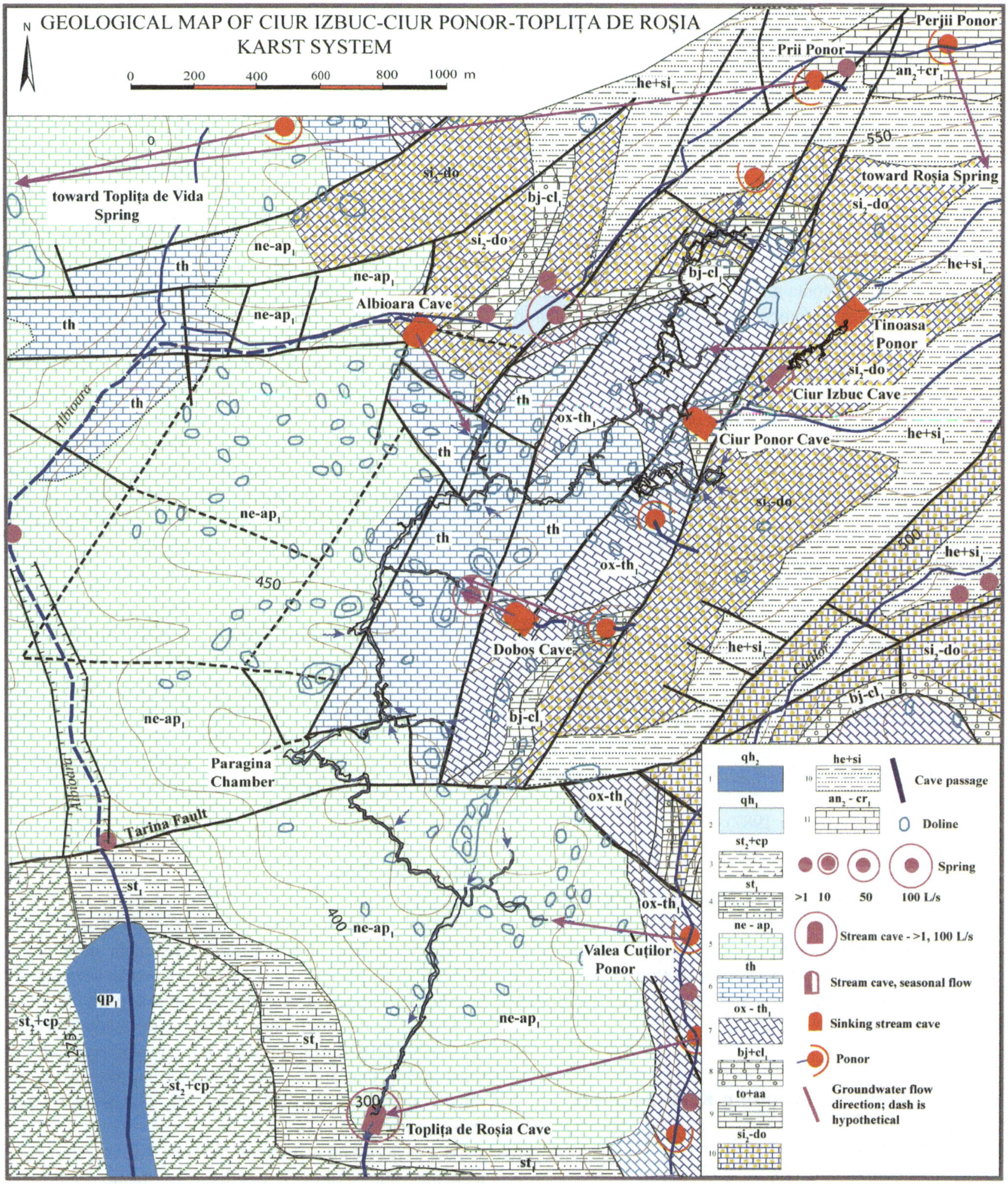

Fig. 2 Geological map of Ciur Izbuc–Ciur Ponor–Topliţa de Roşia karst system (geology modified after Bordea et al. 1986)

limestone sequence. During Neocomian (Lower Cretaceous), the paleokarst features (dolines, uvalas, potholes, etc.) were filled with aluminum-rich (±iron) sediments that formed bauxite ore deposits having lens-like shapes of variable sizes (max. 12 m wide, 130 m in diameter) (Papiu and Mînzatu 1969; Papiu 1970; Pop and Mârza 1977; Bordea et al. 1986; Onac and Popescu 1991).

The Neocomian-Lower Aptian (Lower Cretaceous) deposits make up the greatest portion of outcrops in the area of investigation and are the principal units containing surface and subsurface karst features. These limestones deposits are 35–500 m thick and contain Pachyodontes and Characeae at the base. The post-tectonic cover outcrop in the vicinity Roşia village, in the southern part of the area, and includes Upper Cretaceous (Lower Santonian) limestones with marl and sandstones layers and the Santonian and Lower Campanian in Codru facies. It is made up of limestones with Rudists and is overlain by breccia and sandstones. The alluvial deposits are found along stream valleys, in a narrow strip, or at the bottom of the large close depressions.

Karst Hydrogeological Maps

The availability of groundwater is controlled by the lithology of the subsurface material. The water-bearing characteristics of the aquifers in the study area are discussed based on the physical characteristics of the rocks or sediments in the area (porosity and permeability, granular or fractured, etc.), and the occurrence of karst features, springs, and sinking streams in carbonates rocks (Ponta 2003). This study area is part of the karst hydrogeological map scale 1:50,000, Zece Hotare Sheet (Ponta and Terteleac 1987).

Aquifers in Which Flow Is Mainly Intergranular

Small aquifers are located in 2–4-m-thick alluvial deposits along the Albioara Valley and at the bottom of large close depressions. A few dug wells and small springs were identified in this formation.

Fissured Aquifers, Including Karst Aquifers

Fissured Aquifers

Extensive and Highly Productive Aquifers

Sandstones and siltstones with some clay layers of Hettangian to Lower Sinemurian are present in the northeastern/central part of the study area. Several streams are formed on this formation (Albioara, Ciur Ponor, Tinoasa, Cuţilor), which at the interface with the limestones are disappearing underground (Fig. 3) recharging the mainstream of Ciur Ponor Cave.

Karst Aquifers

Highly Productive Karst Aquifers

Highly productive karst aquifers occur in limestones of Middle to Lower Triassic (Upper Anisian–Lower Carnian), limestone and dolomite deposits of Lower to Middle Jurassic (shown in Fig. 2 as Upper Sinemurian-Domerian, Toarcian-Aalenian, Bajocian-Lower Callovian), Upper Jurassic (Oxfordian-Lower Tithonian), and Lower Cretaceous (Neocomian-Lower Aptian) formations. These carbonates exhibit little intercrystalline porosity. Groundwater movement occurs along well-developed secondary porosity (fractures and joints). These interconnected features serve as conduits leading water from the sinking streams to the Topliţa de Roşia spring (Fig. 3). Carbonate rocks prevail for about 720 m in thickness. At the end of the Jurassic, bauxite lenses developed producing an impermeable horizon between the Tithonian and Neocomian-Aptian deposits, but due to the discontinuous nature of their deposition, water can flow between the two carbonaceous deposits.

Local or Discontinuous Productive Karst Aquifers

The limestones of the post-tectonic cover formed by Lower Santonian (part of the Upper Cretaceous) and the Upper Cretaceous/Senonian (Santonian and Lower Campanian) limestones located in the southern part of the area are included in this category. They host fewer surface and subsurface karst features and fewer springs.

Tracer Studies

The Topliţa de Roşia Spring is located in the Roşia/Albioara valleys watershed and constitutes the main outlet of the Runcuri karst plateau, which is developed in Jurassic–Cretaceous limestones. The recorded minimum discharge was 11 L/s, while the multi-annual average discharge is 105 L/s (Orăşeanu 2016). During long periods without rains, the main spring that appears through the entrance of the cave runs completely dry, and only the downstream spring remains active.

The cave stream is draining the waters from Albioara and Runcuri valleys through ponors, Ciur Izbuc, Doboş, and Tinoasa caves, and the losses of the Cuţilor Valley that is completely drained only during periods of low flow (Fig. 3). The catchment area of the Topliţa de Roşia Spring was defined by five dyes studies conducted by Rusu (1988) and tracer experiments conducted by Orăşeanu and Iurkiewicz (Orăşeanu 1991, 2010, 2016).

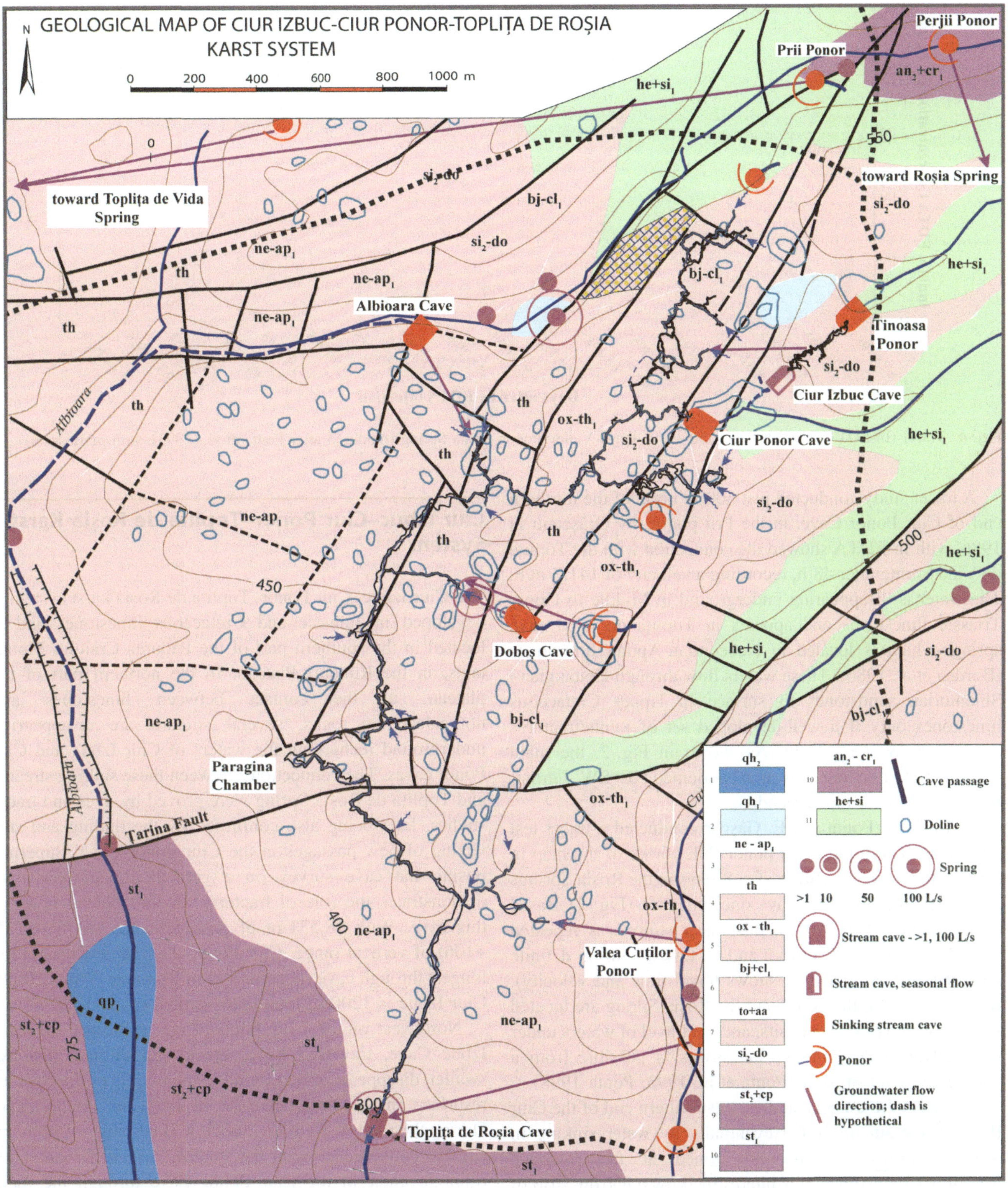

Fig. 3 Karst hydrogeologic map of Ciur Izbuc–Ciur Ponor–Topliţa de Roşia karst system, modified after Ponta et al. (1993), with permission (geology modified after Bordea et al. 1986). Key for the legend is available in the karst hydrogeology chapter

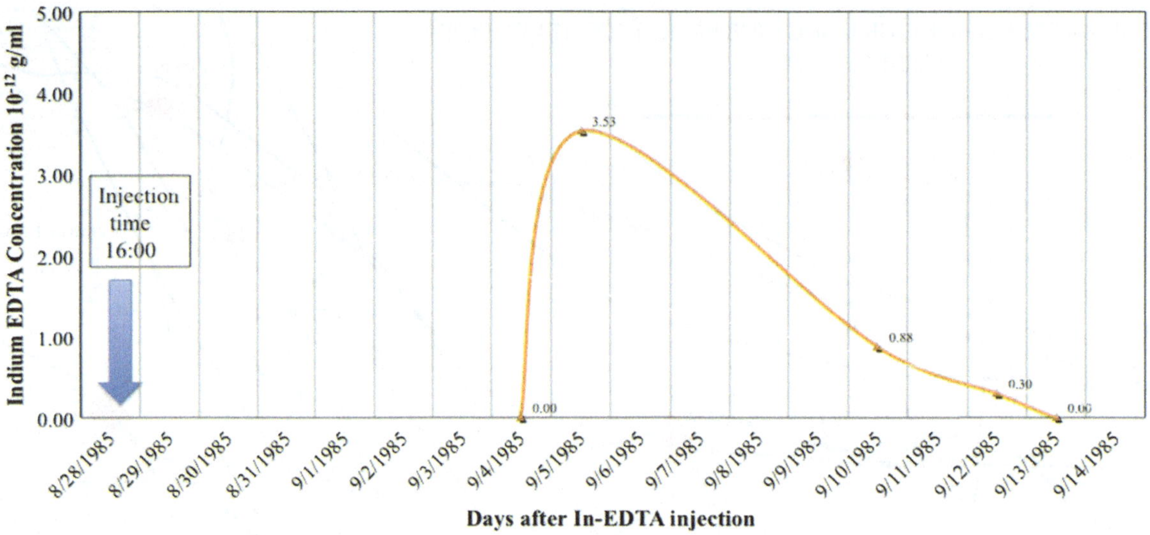

Fig. 4 Tracer (In-EDTA) breakthrough time curve Perjii Valley Ponor–Roşia Spring (modified after Ponta et al. (1993), with permission)

A tracer study conducted just 500 m north of the northern end of Ciur Ponor Cave, in the Prii ponor, by Orăşeanu in 1986, with In-EDTA showed the connection with the Topliţa de Vida Spring after 48 h, recording a velocity of 141.7 m/h. The water is disappearing underground in Middle to Upper Triassic limestones and appears in Topliţa de Vida cave spring, which is located in Neocomian-Aptian limestones (Bordea et al. 1986). These waters flow through Hettangian–Sinemurian sandstones to surface in Upper Cretaceous limestones only if a well-developed set of joints/fractures oriented E–W are present. As shown in Fig. 2, the main fracture pattern in the study area is oriented NE–SW. Further investigations are recommended.

In 1985, G. Ponta and E. Gaşpar conducted a tracer test with In-EDTA in the Perjii ponor (NE corner of the map in Fig. 3), the tracer appearing after 8 days in the Roşia Springs and lasting for almost 17 days since injection (Fig. 4). Based on Orăşeanu (1991, 2016), this tracer study was repeated with fluorescein by I. Povară and C. Lascu with no definite answer (the table does not shows travel time and velocity). In this case, both the ponor and the Roşia Spring are located in the Lower Triassic deposits, and the travel of waters under the Hettangian–Sinemurian sandstones is possible from a geological point of view (Ponta et al. 1993; Ponta 1998).

Based on these tracer studies, the northern end of the Ciur Ponor Cave stream is fed eventually by water sinking in some unknown yet ponors formed at the interface limestones/sandstones. It is interesting how a narrow strip of sandstones changed the flow direction toward the Roşia Spring instead Topliţa de Roşia Spring.

Ciur Izbuc–Ciur Ponor–Topliţa de Roşia Karst System

The Ciur Izbuc–Ciur Ponor–Topliţa de Roşia karst system is developed in Jurassic and Cretaceous limestones and is located in the southern part of the Pădurea Craiului Mountains, in the Runcuri Plateau. In the northern part of the plateau, at the contact between limestones and non-calcareous rocks, several streams are disappearing underground recharging the waters of Ciur Izbuc and Ciur Ponor caves. The connections between these sinking streams and Topliţa de Roşia Spring were proved by dyes and tracer studies, later being also confirmed by the finding and surveying of new passages in the Ciur Ponor Cave. Superimposing the cave survey on a geologic/topographic map demonstrates the role of fractures on the speleogenesis of this cave. With 18,531 m of passages and 309 m (−203, +106) of vertical range, Ciur Ponor–Topliţa de Roşia is the longest through cave (penetration) in Romania (Ponta 2009). Ciur Izbuc is 1200 m long with vertical range of 20 m.

Northwest of the Ciur Ponor Cave is located the Ciur Izbuc Cave, traversed by a stream fed through Tinoasa swallet, disappears underground in the middle of the cave, to resurface in the upstream part of the Ciur Ponor Cave. During the rainy season, the overflow is coming out from the cave to convergence with the stream crossing Groapa Ciurului, to enter in the Ciur Ponor Cave through the main entrance.

History of Exploration

The cave was first mentioned by P. A. Chappuis and R. Jeannel in 1951. In 1952, the cave was explored by T. Rusu and Ş. Avram on a length of 150 m, and then in 1965 was mapped by T. Rusu, I. Viehmann, G. Racoviţă, and V. Crăciun, team that discovered about 400 ancient human footprints (Fig. 5) (Rusu et al. 1969). Focul Viu Caving Club resurveyed the cave in the early eighties, and more recently, B. Tomuş and I. Neag produced a new map of the cave.

Ciur Ponor–Topliţa Roşia karst system is a through (penetration) cave resulted by linking the two caves (Ciur Ponor and Topliţa de Roşia), which until the summer of 1983 were unconnected. Scuba divers passed eight sumps, which led to the merger of the two caves in a single karst system (Halasi and Sârbu 1985).

In 1962, Ciur Ponor Cave was mapped by researchers of the "Emil Racoviţă" Institute of Speleology in Cluj-Napoca on a length of 220 m (Bleahu et al. 1976). Eighteen years later (1980) the cave is visited by H. Mitrofan, who free dove a short sump in a dry season, reaching the top of a 13 m waterfall. Subsequently, the exploration and mapping of the cave is continued with CSA Cluj-Napoca, and by the end of 1980, the cave was 2240 m long (Ponta 1984, 1989).

In April of 1981, H. Mitrofan and P. Brijan (Club Speodava Ştei) crossed a narrow passage downstream of the 13 m waterfall and explored the main gallery for another 1800 m. A month later, an expedition that includes four geologists from Bucharest Geological and Geophysical Prospecting Company (I. Orăşeanu A. Iurkiewicz, H. Mitrofan, and G. Ponta), accompanied by P. Brijan, passed the BH Terminus, going through Paragina Chamber and a large meandering borehole, reaching Sump 1.

In 1981, under the auspices of the Central Committee of Sportive Speleology (CCSS) within the Romanian Federation of Tourism-Climbing (FRTA), Focul Viu Caving Club, and CSER Bucharest began a systematic exploration and survey of the cave (Ponta 1994), which resulted with a detailed map of over 17 km of passages. In the summer of 1983, G. Halasi from Cristal Oradea Caving Club, A. Solomon, and G. Ponta from Focul Viu Caving Club, passed a sequence of three short Sumps (1–3), and surveyed 1561 m on the main gallery to a new sump (Sump 5 of Topliţa de Roşia) (Halasi and Sârbu 1985).

In Topliţa de Roşia Cave, divers from CSA Cluj-Napoca crossed the first three sumps in 1980. In the summer of 1983, G. Halasi dived the fourth sump, and in the fall, along with S. Sarbu (Underwater Speleological Diving Group Bucharest (GESS Bucureşti), crossed the fifth sump, making the connection with the Ciur Ponor Cave, which attained a length of 17,087 m and a vertical range of 228 m (−203 m, 25 m).

In 2007, T. Marin with a team of French cavers and divers crossed the sumps of Topliţa de Roşia and mapped the tributaries intercepted between the Sumps 3 and 4. Overall, the system between the Sumps 3 and 4 is 3005 m long with a vertical range of 106 m. Half of the surveyed length was explored for the first time during this expedition (Marin 2008). With these new discoveries, the cave reached its present length of 18,531 m.

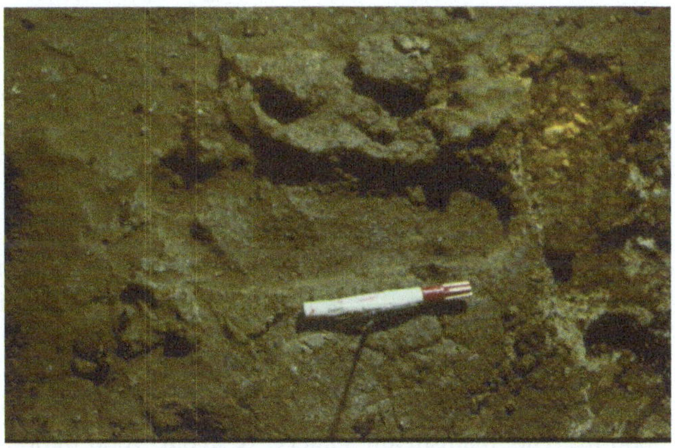

Fig. 5 Human footprint in Ciur Izbuc Cave (photograph by G. Ponta)

Description of the Ciur Izbuc Cave

Ciur Izbuc Cave is located in the Dermii Hill, between Tinoasa ponor and Ciur Ponor Cave. It is developed on two levels: a fossil upper one and a lower temporarily active (Fig. 6. The fossil level of phreatic origin has a total length of 605 m (Rusu 1988) and consists of the Footprints Hall (Sala Paşilor) and the Bears Gallery (Galeria Urşilor), which housed many skeletal remains of cave bear and about 400 human footprints. The age of these prints was originally estimated to 10–15 ka BP (Rişcuţia and Rişcuţia 1970) because at many locations they were intermixed with footprints and bones of cave bear (*Ursus spelaeus*). The anthropological site in Ciur Izbuc Cave was reinvestigated in 2012, and based on indirectly dating of footprints, these may be as old as 36.5 ka cal BP, a period when Neanderthals were being replaced by humans (Webb et al. 2014). More recently, Vremir (2008) identified what appears to be Paleolithic petroglyphs (impressions of palms, elbows, and knees).

The lower level has a total length of 425 m and corresponds to the current temporary-active gallery, traversed by the underground stream coming through Tinoasa Ponor, located at the upstream end of the cave. The junction between Tinoasa Ponor and Ciur Izbuc cave was done by the cavers from Focul Viu Caving Club in 1981. After emerging from a few narrow spaces, the stream flow through a

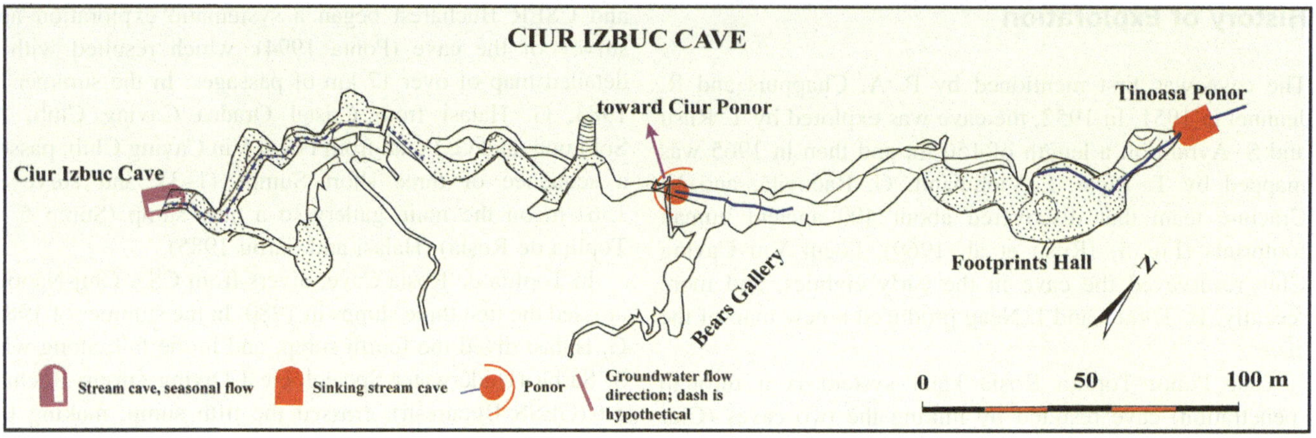

Fig. 6 Ciur Izbuc Cave (modified with permission from B-R Tomuş)

spacious gallery with horizontal ceilings, terraces, and erosion levels and disappears after 180 m, through an inaccessible ponor, heading to the Ciur Ponor Cave, as shown by the results of a dye study with fluorescein. During the rainy season, part of the stream cannot penetrate through this ponor (Rusu 1988), and flows along the main gallery, to converge with the stream of Ciurului Closed Depression, disappearing again underground through the main entrance of Ciur Ponor Cave, and resurfaces through the Topliţa de Roşia Spring (Fig. 3).

Description of the Ciur Ponor–Topliţa de Roşia Cave System

At 483 m above sea level (asl), elevation is located the 5 × 5 m entrance of the Ciur Ponor Cave, which leads to a 10 m diameter chamber, where the riverbed is covered by gravel with clay deposits on both sides. At its lower end, opens a narrow channel (only 0.8 high and 1.5 m wide) through which water flows rapidly, ending on the edge of a 13 m waterfall. At the base, there is a relatively round chamber, 5 m in diameter that divides the cave into two distinct sections, the downstream system and the upstream system (Fig. 7).

The downstream system begins with a narrow passage, which change its shape downstream of a free dive sump named the Fish's Sump (Sifonul lui Peşte), and reaching 2.5 × 2.5 m at the confluence with A_1 Affluent. The first left (East) side tributary is 1460 m long and ends at +30 m elevation from the junction point. The passage is disposed on two levels, a stream passage and a fossil/dry one. In the middle part of the gallery, the two levels merge, forming a labyrinth maze. Downstream the confluence with A_1 Affluent, the cave gallery has large dimensions (2 × 5 m). On the right/northern side is coming in the water of the third tributary, A_3 Affluent.

At 1300 m from the entrance, there are two chambers that are morphologically interconnected, known under the name of Heads of the Family Room (Sala Taţilor de Familie). Hundred meters downstream, the fourth left (east) side river appears (A_4 Affluent), which develops on several levels and totals 1 km in length. A new chamber crossed by a watercourse follows on the main gallery that has a clay beach on the right side.

From this point, the gallery has a sinuous aspect due to the presence of meanders with erosional levels. The gallery forks into a stream passage, where A_{5bis} Affluent is coming in, and the Whirlpool Gallery (Galeria Marmitelor) (a temporary stream passage/subfossil), merging again after 50 m in a single borehole. At the terminal point, named BH Terminus, the water is disappearing underground in a sump that can be overpassed through a short dry passage, which communicates with the river through a 4-m-deep shaft.

The main borehole suddenly enlarges in the biggest chamber of the cave named Paragina Chamber. The area has a chaotic aspect with fallen blocks pilling up to more than 25 m high, through which the river penetrates with difficulty to the Meanders Passage. This is a wide gallery (4–6 m) with erosional levels on walls and quite well decorated. Its terminal point is a lake, 18 m in diameter, which is the first sump, of the cave. Next, follows a succession of three short sumps spread over 100 m, beyond which a gallery with the same shape likes the Meanders Passage begins, having a length of 1540 m. In the central part of this passage, two left-side tributaries are coming in. At the junction point, a chamber with a large collapse of blocks is present. The section between Sumps 4–8 is 140 m long. The eighth sump is at the entrance of the Topliţa de Roşia Cave (6 × 5 m) and is located at 280 m elevation.

The upstream system has a total length of more than 3000 m. The mainstream gallery is about 1000 m long, smaller in size than the downstream section. Three streams

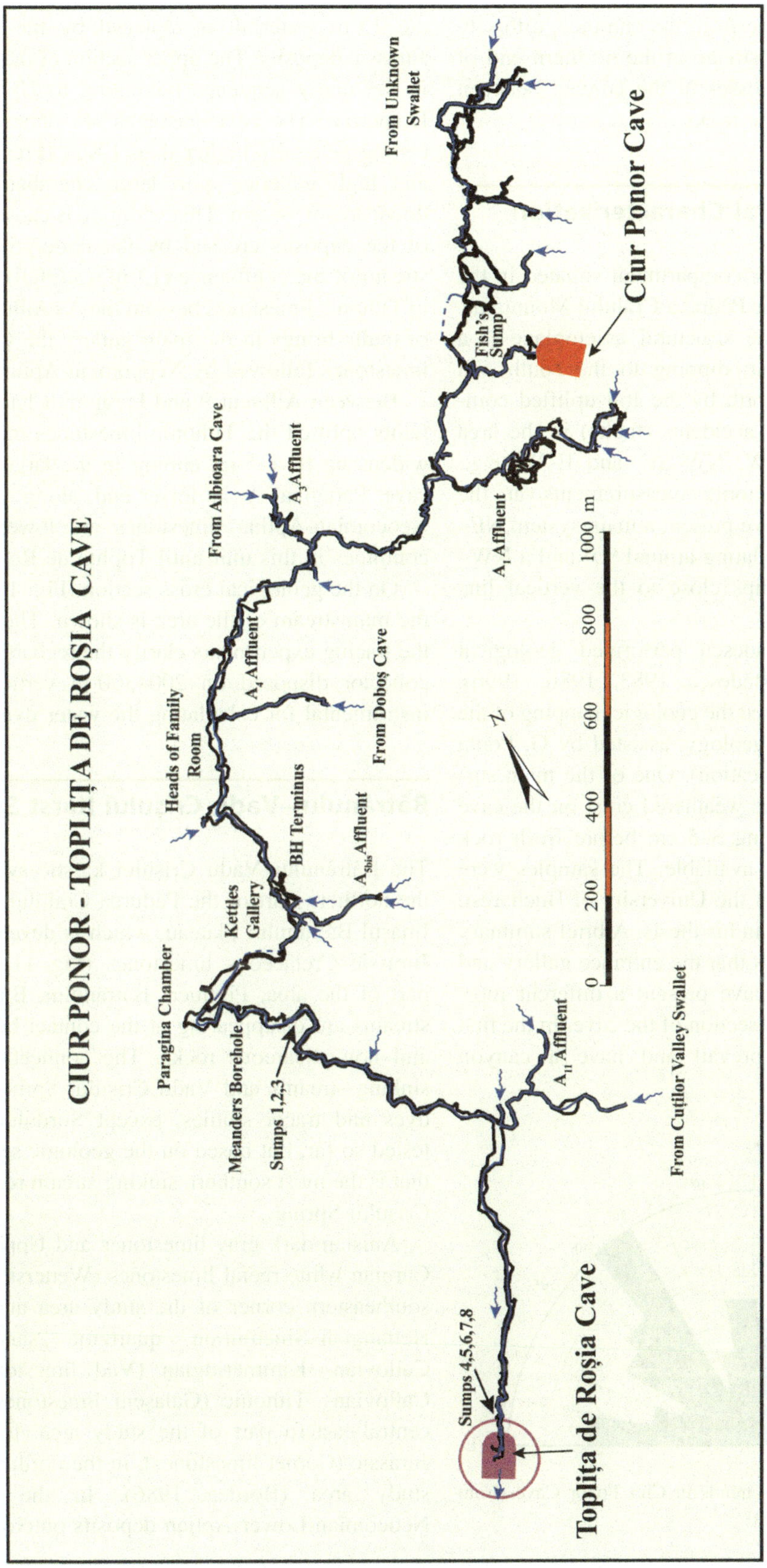

Fig. 7 Map of the Ciur Ponor–Toplița de Roșia Cave (surveyed by Focul Viu, CSER București, and Cristal Oradea Caving Clubs)

on the left side are coming in, one originating in Ciur Izbuc Cave. The main passage ends in a labyrinth maze, where the dry and stream passages alternate. The river appears in the cave at −0.5 m relative altitude from the entrance, diffusely through blocks of stone. The stream at the northern end of the cave is fed by diffused losses at the contact between limestone and the impermeable rocks.

Geological and Structural Characterization

The area is part of the tectonic compartment situated in the southern and eastern part of the Pădurea Craiului Mountains, named "antithetic faults." The structural assemblage is a generally orientated east–west, dipping to the south and southwest, delimited to the north by the Jofi uplifted compartment. The major tectonic accidents (faults) in the area are mainly orientated NE–SW, NW–SE, and E–W (e.g., Ţarina Fault). The microtectonic measurements at the entrance of the Ciur Ponor Cave present a main system NE–SW orientated, with dips oscillating around 90° and a NW–SE orientated system with dips close to the vertical line (Fig. 8) (Ponta et al. 1993).

In the early eighties, Bădescu performed geological observations in the cave (Bădescu 1983, 1986, 1996), whereas O. Enculescu conducted the geologic mapping of the cave for his master thesis in geology, assisted by G. Ponta (Enculescu personal communication). One of the main surprises was the thickness of the weathered crust on the cave wall, most of the time reaching 3–5 cm before fresh rock sample for thin sections was available. The samples were analyzed in the laboratories of the University of Bucharest, and the results being reported in his thesis. A brief summary of the geologic findings shows that the entrance gallery and the upstream system of the cave present a different morphology from the downstream section of the cave. In the first case, the narrow galleries prevail and have a canyon morphology, with erosional levels present in the walls (Figs. 8 and 9). The entrance gallery is developed in the Albioara limestone of Tithonic age, which in the vicinity of the 13 m waterfall is replaced by the Lower to Middle Jurassic deposits. The upper section of the cave is developed in the marly sequence belonging to Upper Sinemurian to Domerian. The cave passages are rather narrow, with the ceiling generally higher than 1.8 m (Figs. 9 and 10). Wide and high galleries with large chambers characterize the downstream system. This variation is caused by the lithology of the deposits crossed by the underground river. Downstream of the confluence (13 m waterfall) the cave develops in Tithonic limestones beyond the A_3 Affluent, where a series of faults brings in the main gallery the Oxfordian-Tithonic limestones followed by Neocomian-Aptian ones.

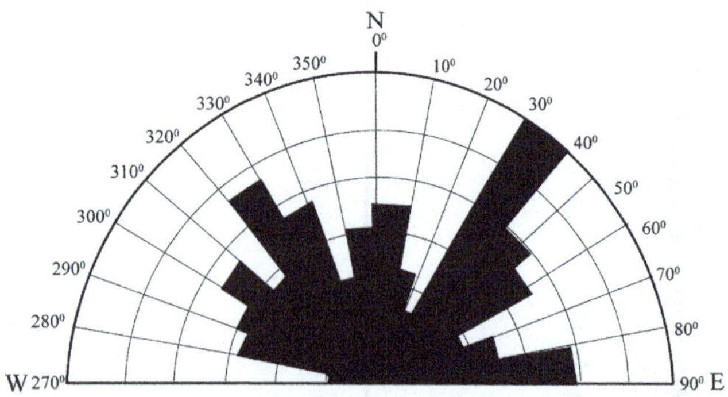

Fig. 8 Rose diagram on passages trends in Ciur Ponor Cave (from Ponta et al. (1993), with permission)

Between Affluent 9 and Paragina Chamber, a new set of faults uplifted the Tithonic limestones in which the gallery widens up to 4–5 m, ending in the largest chamber of the cave, Paragina. At its lower end, along a vertical fault, the Neocomian-Aptian limestones are lowered and the cave continues in this unit until Topliţa de Roşia entrance.

On the geological cross section (Fig. 10), the direction of the mainstream of the area is shown. The caves survey and the tracing experiments clarify the recharge area of the main collector disposed on 200–300 m vertical range and are instrumental for calculating the water dynamic reserve.

Bătrânului–Vadu Crişului Karst System

The Bătrânului–Vadu Crişului karstic system is located in the northern part of the Pădurea Craiului Mountains, in the Imaşul Bătrânului Plateau, which is developed on Triassic–Jurassic–Cretaceous limestones (Fig. 11). In the southern part of the area, Pintiuca, Bătrânului, Birtin, and Surdului streams are disappearing at the contact between limestones and non-calcareous rocks. The connection between these sinking streams and Vadu Crişului Spring were proved by dyes and tracer studies, except Surdului, which was not tested so far, but based on the geologic settings, we assume that is the most southern sinking stream recharging the Vadu Crişului Spring.

Anisian dark gray limestones and Upper Anisian–Lower Carnian white reefal limestones (Wetterstein) outcrop in the southeastern corner of the study area and are overlain by Hettangian–Sinemurian quartzitic sandstones. Middle Callovian—Kimmeridgian (Vad limestones) and Middle Callovian—Tithonic (Galaşeni limestones) outcrop in the central-eastern part of the study area, followed by Upper Jurassic (Cornet limestones), in the northwestern half of the study area (Bordea 1986). In the same area, the Neocomian-Lower Aptian deposits outcrop as two islands.

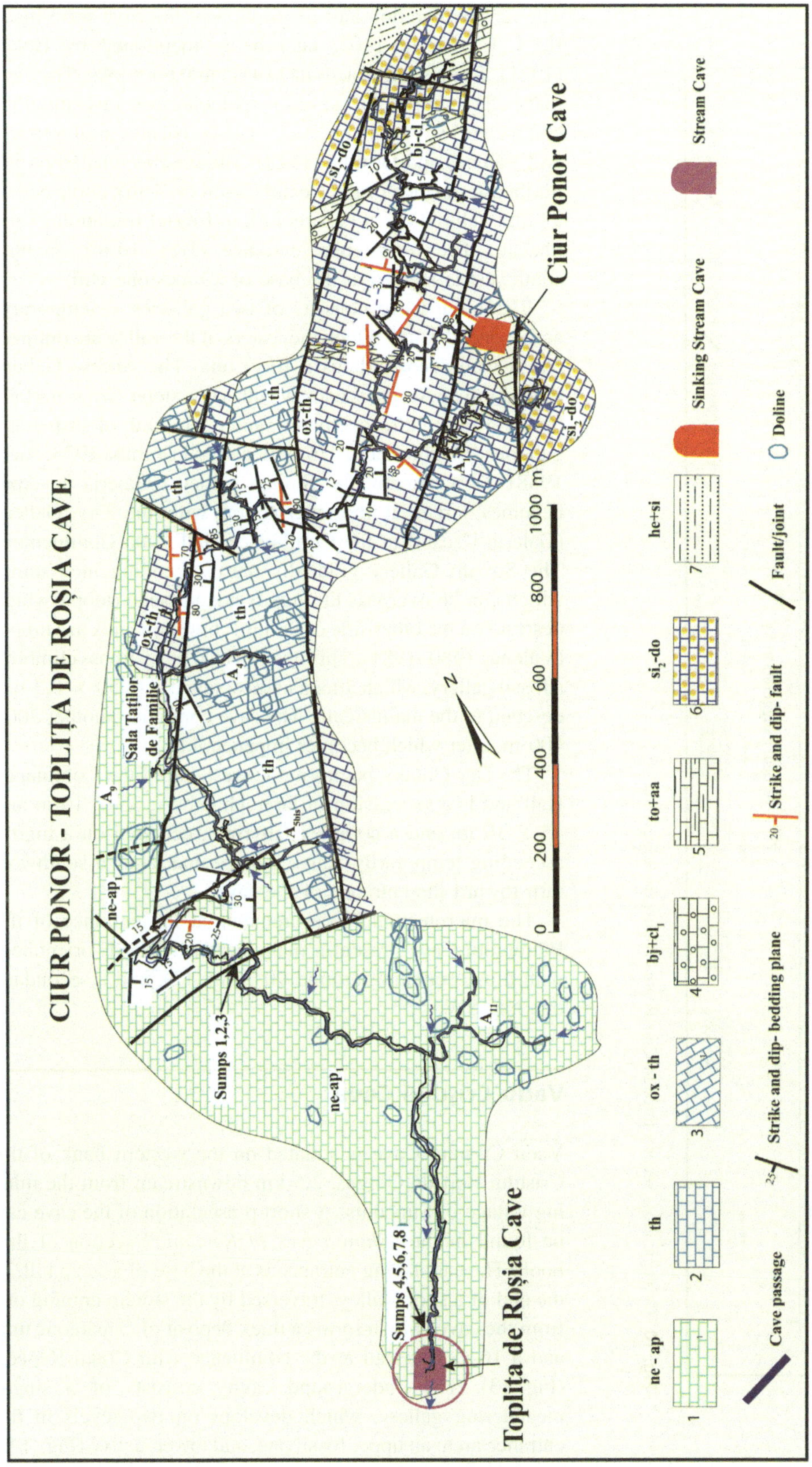

Fig. 9 Geologic pap of Ciur Ponor–Toplița de Roșia Cave

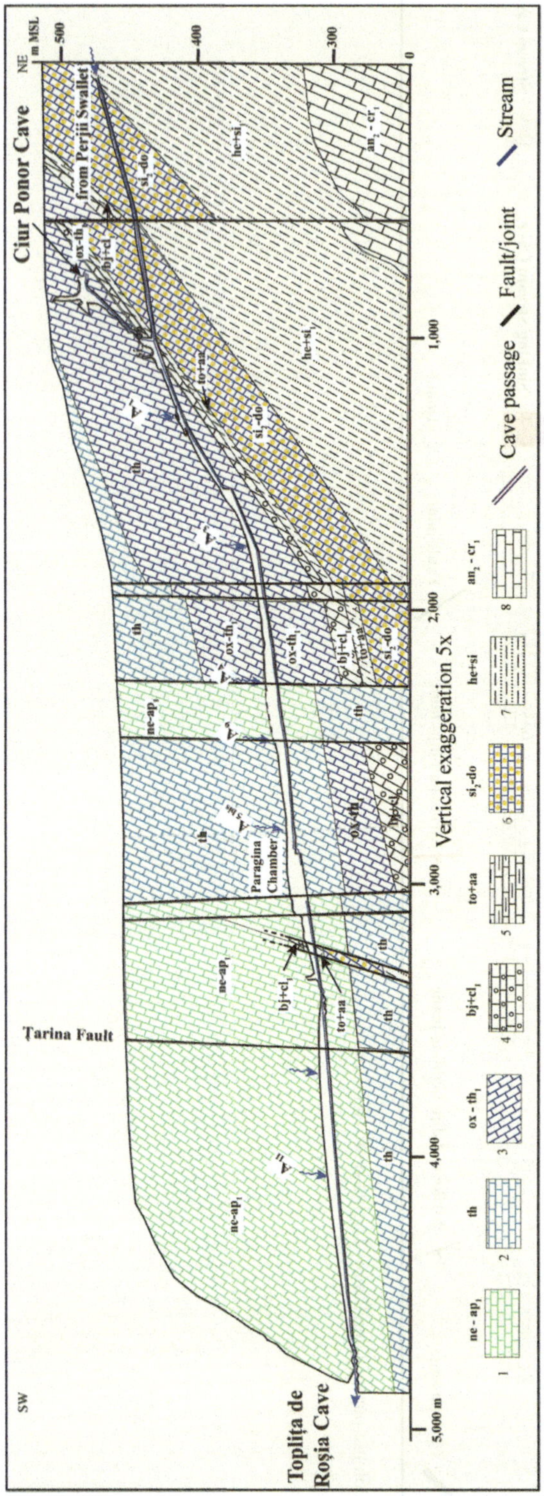

Fig. 10 Geologic cross section through Ciur Ponor–Toplița de Roșia Cave

Bătrânului Cave

The first exploration and description of the cave were made by J. Czaran in 1903, later being mentioned by Bokor (1921), Jeannel and Racoviță (1951) in their books. The cave made the subject of several speleological investigations conducted by Viehmann et al. (1964), Bleahu et al. (1976), and Vălenaş and Drîmba (1978). The cave is situated on the northeastern slope of the Dealul Crucii (724 m), at the end of a blind valley (Valea Peştireului), in Imaşul Bătrânului karst plateau. The rectangular entrance (18 × 10 m) oriented southwest is located at the base of a limestone cliff.

Bătrânului Cave consists of two galleries: a temporary-active one, through which the waters of the valley are drained, and a well-decorated fossil (dry) one. The Access Gallery (Coridorul de Acces) is narrow with a few steps, two waterfalls (6 and 5 m high, respectively), and a labyrinth of suspended galleries (Bleahu et al. 1976; Vălenaş and Drâmba 1978; Rusu 1988). Downstream, the Stream Gallery (Galeria cu Apă) becomes larger at the intersection with the Dry Gallery (Galeria Uscată) in the Confluence Hall (Sala Confluenței). The Stream Gallery gradually narrows ending in a sump, which can be overpass through a side passage (loop), which descend on the other side of Sump 1, but soon ends in Sump 2 (Vălenaş 1980–1981). This one can also be overpass climbing a fossil gallery, which after a narrow squeeze (0.50 × 0.3 m), descend to the mainstream. The cave continues another 300–400 m, after which becomes impenetrable.

The Dry Gallery begins at the west end of the Confluence Hall and is a succession of large chambers, 35 × 10 m and 45 × 19 m, and a complex labyrinth, centered on a highly ascending temporarily active gallery that takes a southwest turn toward the entrance of the cave.

The microtectonic measurements at the entrance of the Bătrânului Cave present a main system NW–SE orientated, with dips oscillating around 90° (Fig. 12), and a secondary one NE–SW.

Vadu Crişului Cave

Vadu Crişului Cave is situated on the western bank of the Crişului Repede Gorges, 2.5 km downstream from the mining village of Şuncuiuş. A short presentation of the cave can be found in the "*Show caves of Romania*" section of this book. The east-facing entrance is at the base of a steep hill, at the end of a short valley, traversed by the stream coming out from the cave, which forms a thick deposit of calcareous tufa and a 16-m waterfall at the confluence with Crişul Repede (Fig. 13). The underground cavity consists of a single meandering gallery, which develops on two levels in the entrance area: an upper fossil one, and lower, active (Fig. 14).

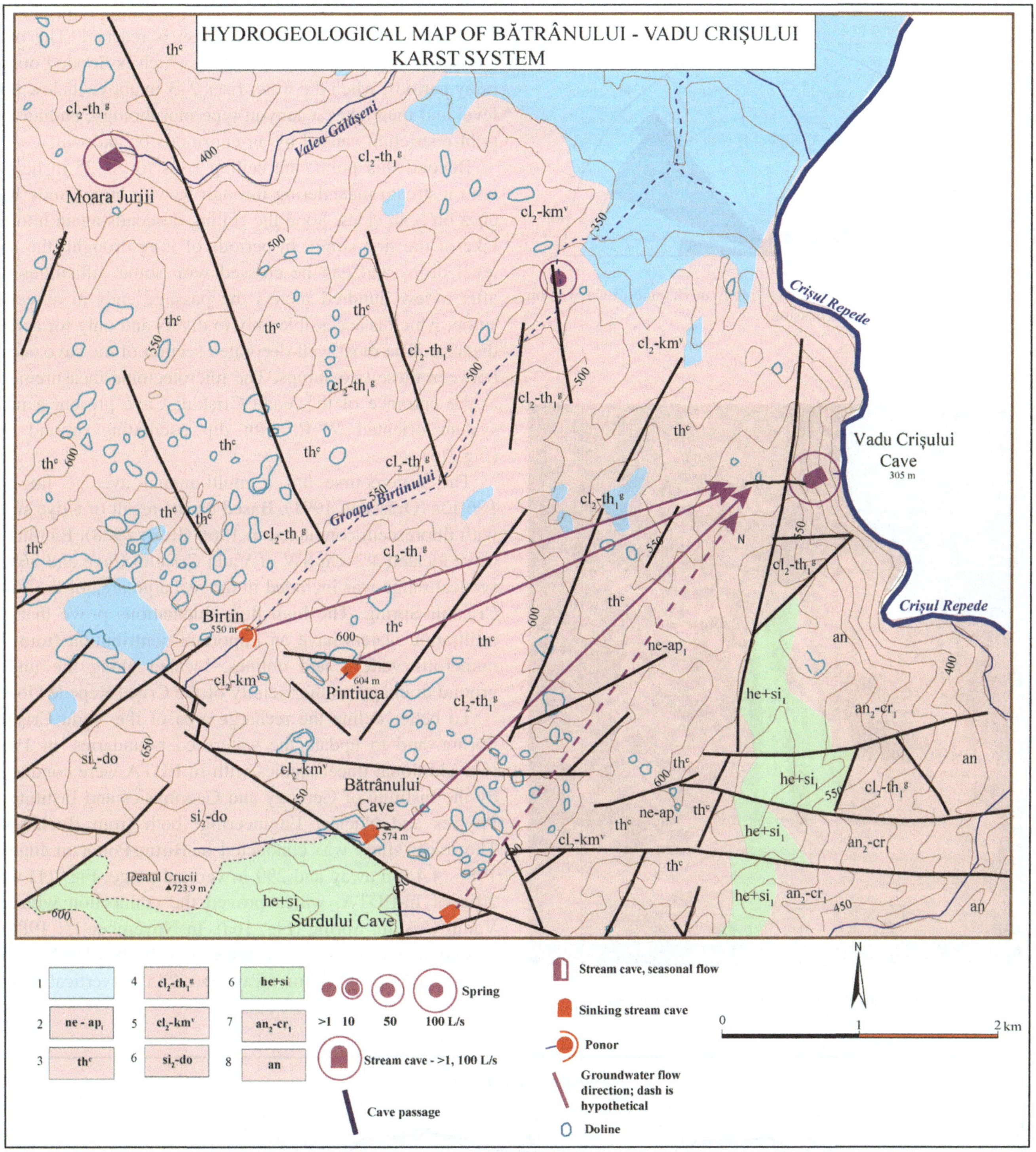

Fig. 11 Karst hydrogeologic map of Bătrânului–Vadu Crişului karst system, modified after Ponta et al. (1993), with permission (geology modified after Bordea et al. 1986). Key for the legend is available in the karst hydrogeology chapter

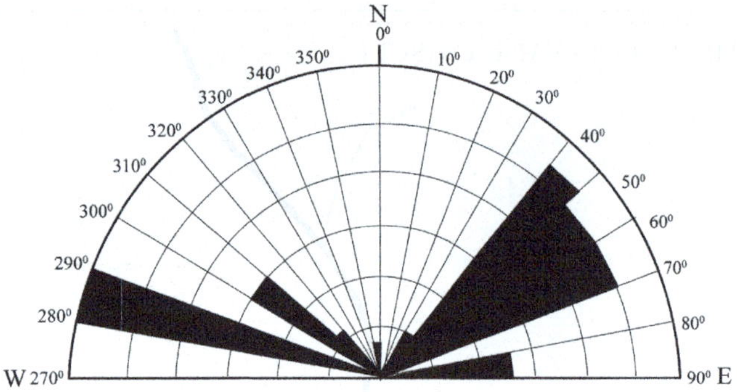

Fig. 12 Rose diagram on passages trends in Bătrânului Cave (from Ponta et al. (1993), with permission)

Fig. 13 Vadu Crişului Waterfall (photograph by G Ponta)

Descending from the upper floor, a bridge built along a very high (10–15 m) vadose canyon is reached. The main passage continues with a chamber, which is flooded during rainy periods, when the water rises 2–3 m above its low flow level and then takes a canyon-type morphology, ending in a tight restriction caused by an enormous breakdown.

Beyond this point, the ceiling rises to 20–25 m height, with a stream meandering through it, with large sandy beaches on both sides. Soon the ceiling descends again into the lake of the first sump. In periods of long drought, the lake level drops and can be crossed with some difficulties but after a few hundred meters the passage ends in a second sump, which is accessible only to divers and only for a short distance. The most well-decorated section of the cave occurs between these two sumps. The microtectonic measurements at the entrance of the Vadu Crişului Cave present a main system oriented W–E, with dip oscillating around 90° (Fig. 15).

The watercourse has a multi-annual average flow of 186 L/s (Orăşeanu 1991). Based on the result of a dye study with fluorescein conducted in 1966 (Rusu 1988), Bătrânului Cave located 4.2 km SW of Vadu Crişului Cave and 269 m vertical range was included in the recharge area of the Vadu Crişului Spring. The high flow fluctuations prove that the infiltration waters have an important contribution from the numerous valleys and dolines, located along the underground drainage, in the vicinity of the Crişul Repede Gorge.

To better define the recharge area of the Vadu Crişului Spring, and to update the watershed boundaries, in 1986, two additional tracer studies with In-EDTA were conducted by the Institute of Geology and Geophysics and Institute of Physics and Nuclear Engineering (both from Bucharest). The tracer study was conducted in Birtin Ponor in June 4, 1986 (4.4 km away and 299 m vertical range; Fig. 11) with 20 g of In-EDTA, which proved the connection with the Vadul Crişului Spring (Fig. 16a). In November 15, 1986, a second tracer study was conducted with In-EDTA in the Pintiuca Ponor (4.5 km away and 245 m vertical range;

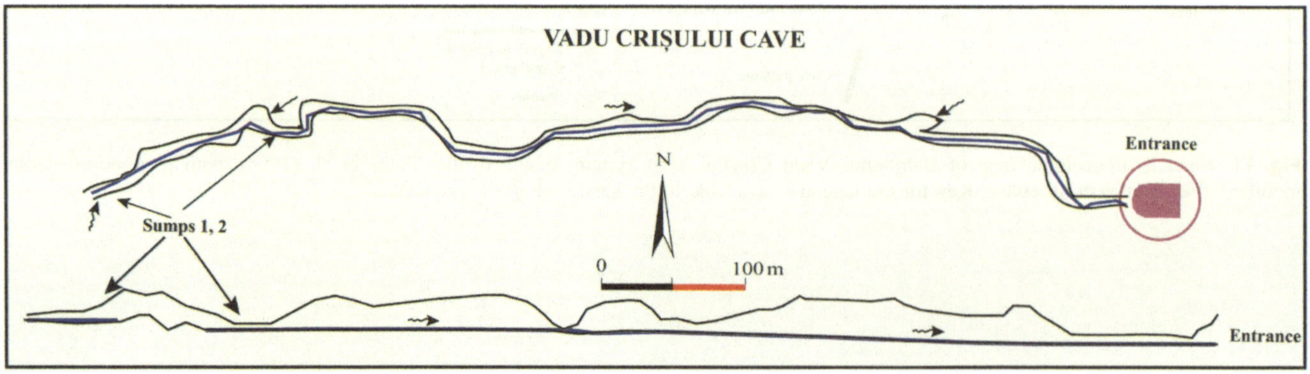

Fig. 14 Map of the Vadu Crişului Cave (modified with permission from Viehmann et al. 1964)

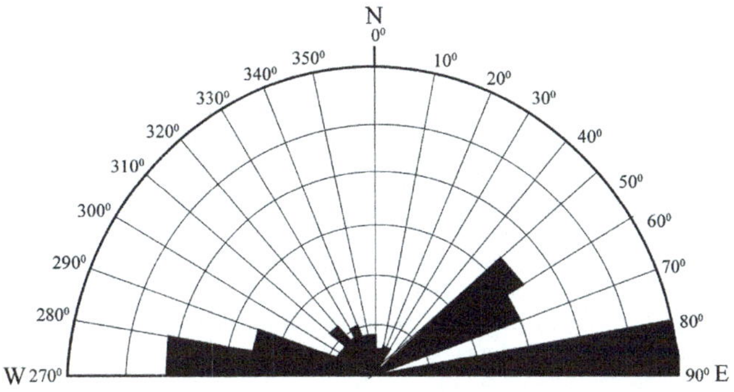

Fig. 15 Rose diagram on passages trends in Vadu Crişului Cave (from Ponta et al. (1993), with permission)

Fig. 11), showing the connection to Vadu Crişului Spring as well (Fig. 16b), and extending the recharge area of the spring toward north.

Two additional tracer studies were conducted in the area: one in Tomnatic Ponor with 40 g of La-EDTA and second one in cooperation with I. Orăşeanu in Pintiuca Ponor, with 5.5 kg of rhodamine. These tracers were not detected at any resurgence. More than likely, the clayey deposits encountered in the subsurface retained the dye.

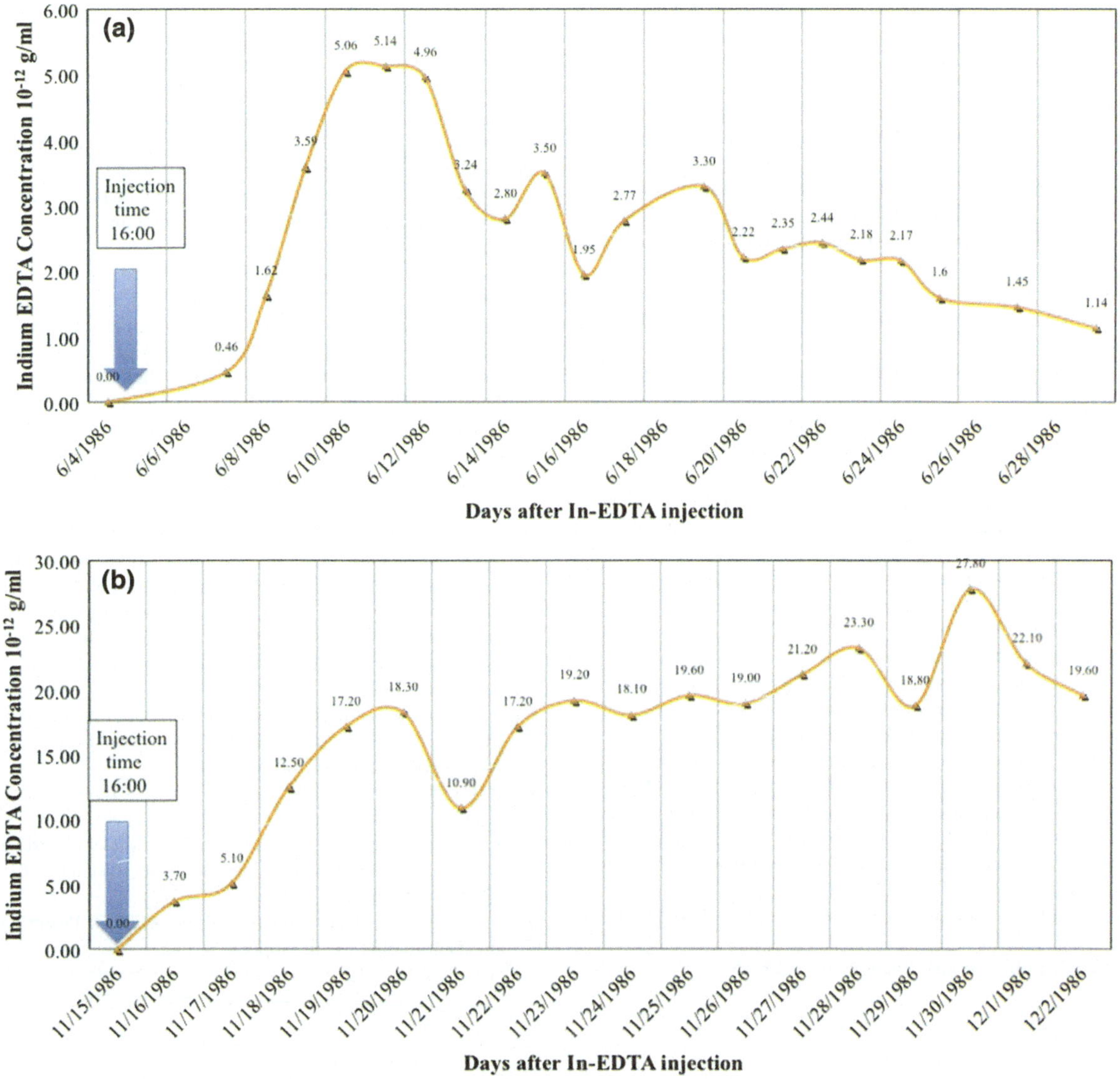

Fig. 16 Tracer (In-EDTA) breakthrough time curves for **a** Birtin Ponor–Vadu Crişului Spring (modified after Ponta et al. (1993), with permission) and **b** Pintiuca Ponor–Vadu Crişului Spring

Cave Conservancy

Ciur Ponor and Ciur Izbuc are class A protected caves (scientific reserves). Access is permitted for exploration/survey and scientific purposes, but requires special permissions from the Romanian Speleological Heritage Commission. Vadul Crişului is a touristic cave that offers guided tours.

Acknowledgements The author would like to acknowledge the cavers of Focul Viu and CSER Bucureşti Caving Clubs that made this work possible. In addition, I would like to acknowledge all the other Romanian and foreign cavers who provided support for exploring and mapping Ciur Ponor–Topliţa de Roşia karst system.

References

Bădescu D (1983) Studiu preliminar privind condiţionarea tectonică a peşterii Ciur Ponor (Munţii Pădurea Craiului). St Cerc Speol 3:28–36

Bădescu D (1986) Le karst des Monts Padurea Craiului et le reseau Tinoasa-Ciur-Topliţa de Roşia (Roumanie). Karstologia 7:41–48

Bădescu D (1996) Romanian karst: special zones and caves. Geol Balcanica 26(2):29–38

Bleahu M, Decu V, Negrea Ş, Pleşa C, Povară I, Viehmann I (1976) Peşteri din România. Ed Ştiinţifică şi Enciclopedică, Bucureşti, p 415

Bokor E (1921) Arthropoden der ungarischen Grotten. Barlang Kutatas Bd IX:1–4

Bordea S, Bordea J, Mantea G, Costea C (1986) Geological map of Romania, Zece Hotare Sheet. Scale 1:50.000, Inst Geol Geofiz, Bucureşti

Bucur II, Cociuba I (2001) Upper Jurassic-Lower Cretaceous deposits from Pădurea Craiului Mountains in algae and carbonate platforms in the western part of Romania. In: Bucur II, Filipescu S, Săsăran E (eds) 4th Regional meeting of IFAA, Cluj-Napoca (field trip guidebook), pp 40–49

Bucur II, Săsăran E, Cociuba I, Lazăr DF, Felner A (2010) Foraminiferal assemblage from the Upper Jurassic limestones of Pădurea Craiului Mountains (Northern Apuseni). Biostratigrafic, paleoecologic and paleoenvironmental significance. In: Seghedi A, Roban RD (eds) Geol Soc Rom GeoEcoMar, pp 29–32

Chappuis PA, Jeannel R (1951) Enumération des grottes visitées, 1927–1949 (huitième série). Arch Zool Exp Gén 88:81–230

Czaran J (1903) Kalauz Biharfuredi Kirandulasokra, Beiuş, p 262

Dragastan O, Diaconu M, Popa E, Damian R (1982) Biostratigraphy of the Triassic formations in the east of Pădurea Craiului Mountains. DS Inst Geol Geofiz LXVII(4):29–61

Halasi G, Sârbu S (1985) Joncţiunea: Izbucul Topliţei–Peştera Ciurului Ponor. Stix 2:48–53

Marin T (2008) Infinit de apă. Speomond 13:14–18

Onac BP, Popescu V (1991) Relations between the overall and local tectonic pattern of the Şuncuiuş deposit and the paleokarst at its bottom. Trav Inst Spéol "Émile Racovitza" 30:185–187

Orăşeanu I (1991) Hydrogeological map of the Pădurea Craiului Mountains (Romania). Theor Appl Karstol 4:97–127

Orăşeanu I (2010) Karst hydrogeology of Pădurea Craiului Mountains. In: Orăşeanu I, Iurkiewicz A (eds) Karst hydrogeology of Romania. Ed Belvedere, Oradea, pp 199–215

Orăşeanu I (2016) Hidrogeologia carstului din Munţii Apuseni. Ed Belvedere, Oradea, p 289

Papiu VC (1970) Litologia calcarelor mezozoice associate bauxitelor din masivul Pădurea Craiului (M. Apuseni). Rev Roum Géol Géophys Géogr 55(1):187–208

Papiu VC, Mînzatu S (1969) Ipoteză asupra genezei bauxitelor din masivul Pădurea Craiului (Munţii Apuseni). An Com Stat Geol XXXVII:37–69

Ponta G (1984) Ciur Izbuc—Ciur Ponor. Un compleso carsico rumeno che dischoperta na superato i 10 km. Speleologia 10:20–22

Ponta G (1989) Sistemul carstic Ciur Ponor—Topliţa de Roşia (M-ţii Pădurea Craiului). Bul Speol 12:61–75

Ponta G (1994) The Ciur Ponor—Topliţa de Roşia Cave system. J Westminster Speleol Group 9(7):56–61

Ponta G (1998) Karst of Romania. In: Yuan D, Liu Z (eds) Global karst correlation IGCP 299. Science Press, Beijing, pp 193–207

Ponta G (2003) Proposed modifications of the international standard legend for large scale hydrogeological maps in carbonate rocks. Western Kentucky University, Bowling Green, Kentucky (Int Conf Karst Hydrogeol Ecosys)

Ponta G (2009) Ciur Ponor. In: Dumitru R (ed) Focul Viu 40 de ani de speologie. Ed RAO, Bucureşti, pp 71–75

Ponta G, Terteleac N (1987) Hydrogeological map of Zece Hotare. Scale 1:50,000, Inst Geol Geofiz Bucureşti (unpublished)

Ponta G, Terteleac N, Gaspar E (1993) Three karstic systems (Roşia, Topliţa de Roşia and Vadul Crişului) in the Padurea Craiului Mountains, Romania. Theor Appl Karstol 4:129–143

Pop G, Mârza I (1977) La paléocarst éocrétace des Monts Pădurea Craiului et sa signification dans le mécanisme de génèse des bauxites. Bull Soc Géol Bull 30(1):51–58

Racoviţă G, Moldovan O, Onac BP (eds) (2002) Monografia carstului din Munţii Pădurea Craiului. Ed Presa Univ Clujeană, Cluj-Napoca, p 264 (in Romanian with extended English abstracts)

Rişcuţia C, Rişcuţia I (1970) Etude anthropologique des empreintes humanines de la grotte Ciurului Izbuc, Roumanie. Émile G. Racovitza, Livre de Centenaire 1868–1968. Ed Acad RSR, Bucureşti, pp 529–533

Rusu T (1973) La genèse du réseau hydrographique des monts Pădurea Craiului. Livre du cinquantenaire Inst Spéol "Émile Racovitza", Ed Acad RSR, Bucureşti, pp 575–589

Rusu T (1988) Carstul din Munţii Pădurea Craiului. Ed Dacia, Cluj, p 254

Rusu T, Viehmann I, Racoviţă G, Crăciun V (1969) Primele urme de paşi ale omului preistoric din peşterile României. Ocrotirea Nat 13:191–200

Szilágyi Palkó P, Deák Z, Dascăl T (eds) (2007) Peştera Vântului 50 ani. Delroti SRL, Cluj-Napoca, p 112

Vălenaş L (1980–1981) Noi cercetări de speologie fizică în Munţii Pădurea Craiului. Nymphaea VIII–IX:265–310

Vălenaş L, Drâmba G (1978) Cercetări de speologie fizică în Munţii Pădurea Craiului. Nymphaea VI:279–328

Viehmann I, Pleşa C, Rusu T (1964) Peştera de la Vadu-Crişului. Lucr Inst Speol "Emil Racoviţă" III:49–81

Vremir M (2008) Petroglife Paleolitice în Peştera Ciur—Izbuc? Speomond 13:50–51

Webb D, Robu M, Moldovan O, Constantin S, Tomuş B, Neag I (2014) Ancient human footprints in Ciur-Izbuc Cave, Romania. Am J Phys Anthropol 155:128–135

Further Reading

Bleahu M (1972) Karst of Rumania. In: Herak M, Springfield VT (eds) Important karst regions of the Northern Hemisphere. Elsevier, Amsterdam, pp 341–353

Cocean P (1995) Peşterile României. Potenţial turistic. Ed Dacia, Cluj-Napoca, p 264

Goran C (1982) Catalogul sistematic al peşterilor din Romania. Ed CNEFS, Bucuresti, p 496

Orghidan T, Negrea Ş, Racoviţă G, Lascu C (1984) Peşteri din România. Ghid Turistic. Ed Sport-Turism, Bucureşti, p 504

Useful Websites

www.speologie.org
www.karstgeology.com

Pădurea Craiului Mountains: Stanul Foncii Shaft (Avenul din Stanul Foncii)

Felix Papiu

Abstract

Stanul Foncii is the deepest shaft (−339 m) in the Pădurea Craiului karst with a total of 4106 m of passages. It is a typical vertical cave formed by vadose water flowing downward along large fractures or at intersecting fissures. The presence of smooth and sometimes rilled wall morphologies in all eight shafts is indicative of dissolution by calcite-undersaturated groundwater descending into the cave. Speleothems were encountered only in the Parallel System, where bones of *Ursus spelaeus* are also present.

Keywords

Vadose shaft • Stanul Foncii • Pădurea Craiului Mountains • Romania

Introduction

Stanul Foncii Shaft is situated in the northern extremity of the Farcu Plateau, on the left side of the Piatra Neagră Valley, in the upstream section of the Cuţilor Valley, at 660 m elevation (Fig. 1). To access the cave, one should exit the national road Cluj-Napoca-Oradea (E60) in Borod and drive south to Damiş via Bratca. Approximately 5 km after leaving Damiş toward Roşia on a very scenic route, the entrance to the shaft is situated at 60 m relative altitude from the Piatra Neagră Valley's thalweg (Rusu 1988).

Geologic and Hydrologic Settings

Pădurea Craiului Mountains are part of the Preapulian Craton, which includes the following tectonic units: Bihor Unit in an autochthonous position, as well as the Apusenides and the Transylvanides (Balintoni 2001). Relevant to the cave region is the Bihor Unit, characterized by a crystalline basement over which is disposed transgressive and discordant sedimentary deposits of Permian and Mesozoic age. The majority of the karst phenomena occur within the middle Triassic (Anisian to Ladinian) carbonates, represented by thick, black micritic limestones developed in the so-called facies of Gutenstein (Ianovici et al. 1976). The sedimentary cover of the Bihor Unit is slightly folded and crisscrossed by many vertical and subvertical faults that played a major role in the development of numerous caves and shafts (Iurkiewicz and Mitrofan 1984).

Hydrologically, the underground river intercepted at the bottom of the shaft has its origins in the ponor of Iezere Creek and losses in the Piatra Neagră Valley. Tracer tests performed by Rusu (1981) and Orăşeanu and Iurkiewicz (1987) indicated that Stanul Foncii belongs to a larger hydrogeological karst system that discharges its waters through the Roşia Spring, situated ~3.8 km SSW from the cave.

History of Exploration

In September 1979, Dan Soriţău, then a geology student and member of the "Emil Racoviţă" Speleological Club from Cluj-Napoca (CSER) discovered the main entrance of the shaft. The first exploration begun in May 1980 and the team composed of D. Soriţău, D. Milea, V. Jucan, and D. Nicoară descend the 65 m deep entrance shaft (P1 in Fig. 2) and begun to enlarge the passage by removing the sediments and gravels accumulated at the bottom (Soriţău et al. 1984). In June 1980, three other shafts (P2: −23 m; P3: −35 m; P4: −16 m, see Fig. 2) are explored down to −190 m where the team reached the entrance in the Crawl (Târâşul). The rainy weather stopped the exploration at this point, but later that month, I. Pop succeeded to pass the Crawl, but once again he had to stop at the edge of P5 (−14 m) because of high water

F. Papiu (✉)
"Omega"/CSER Speleological Club, N. Iorga 1A, Cluj-Napoca, Romania
e-mail: papiufelix@yahoo.com

G. M. L. Ponta and B. P. Onac (eds.), *Cave and Karst Systems of Romania*,
Cave and Karst Systems of the World, https://doi.org/10.1007/978-3-319-90747-5_44

Fig. 1 Location of Stanul Foncii Shaft in the central part of the Pădurea Craiului karst region

level. At this last attempt, members of the Speleological Amateur Caving Club from Cluj-Napoca (abbreviated CSA in Romanian) also participated.

The team composed of V. Jucan and D. Nicoară descended P5, P6 (−7 m), and P7 (−9 m) during a new expedition in August 1980. In order to continue the exploration, they started to dig through the sediment infilling at the bottom of P7 (~220 m below surface) in the point called "At window" (La Fereastră). After several digging trips, in September 1980 the same team descends P8 (−55 m), wherein a big chamber intercepted the underground stream passage, accessible both upstream and downstream. The Upstream Gallery (Galeria Activă Amonte) ends in a massive breakdown, whereas the Downstream Gallery (Galeria Activă Aval) terminates (at least at the time of that expedition) in a sump.

A new attempt to continue the exploration was led by D. Nicoară and V. Jucan, along with two members of the Vertikum Speleo Club from Budapest in May 1981. However, because of a thunderstorm, they only reached (with difficulty) the Crawl, which was impenetrable. The digging performed by D. Soriţău, D. Selişcan, and C. Popa at the surface near the main entrance, allowed them to discover the Parallel Entrance (Intrarea Paralelă). After descending 40 m, they explored and surveyed the Parallel System (Sistemul Paralel). In August 1981, the entire cave was resurveyed by I. Povară ("Emil Racoviţă" Institute of Speleology), D. Nicoară, I. Fetilă, and E. Silvestru. During this and another successive trip, a short digging in a passage above the final sump allowed to bypass it, but a new sump stopped the advance (Moldovan 1984). This last expedition was also attended by cavers from Vertikum and Ceska Speleogicki Spolecnost Suchy Zleb from Brno (Czech Republic) who helped CSER Cluj to set a phone cable between surface and the bottom of P8. The complete re-surveying of the Downstream Gallery was conducted by D. Moldovan and P. Deriaz (Troglolog Neuchâtel, Switzerland).

During an expedition with bivouac at −300 m (August 1983), T. Botezan, M. Someşan, I. Pop, and C. Popa successfully passed the final sump (temporarily open), but after several tens of meters and two other short sumps, the gallery stops in an impenetrable whirlpool, the cave reaching here

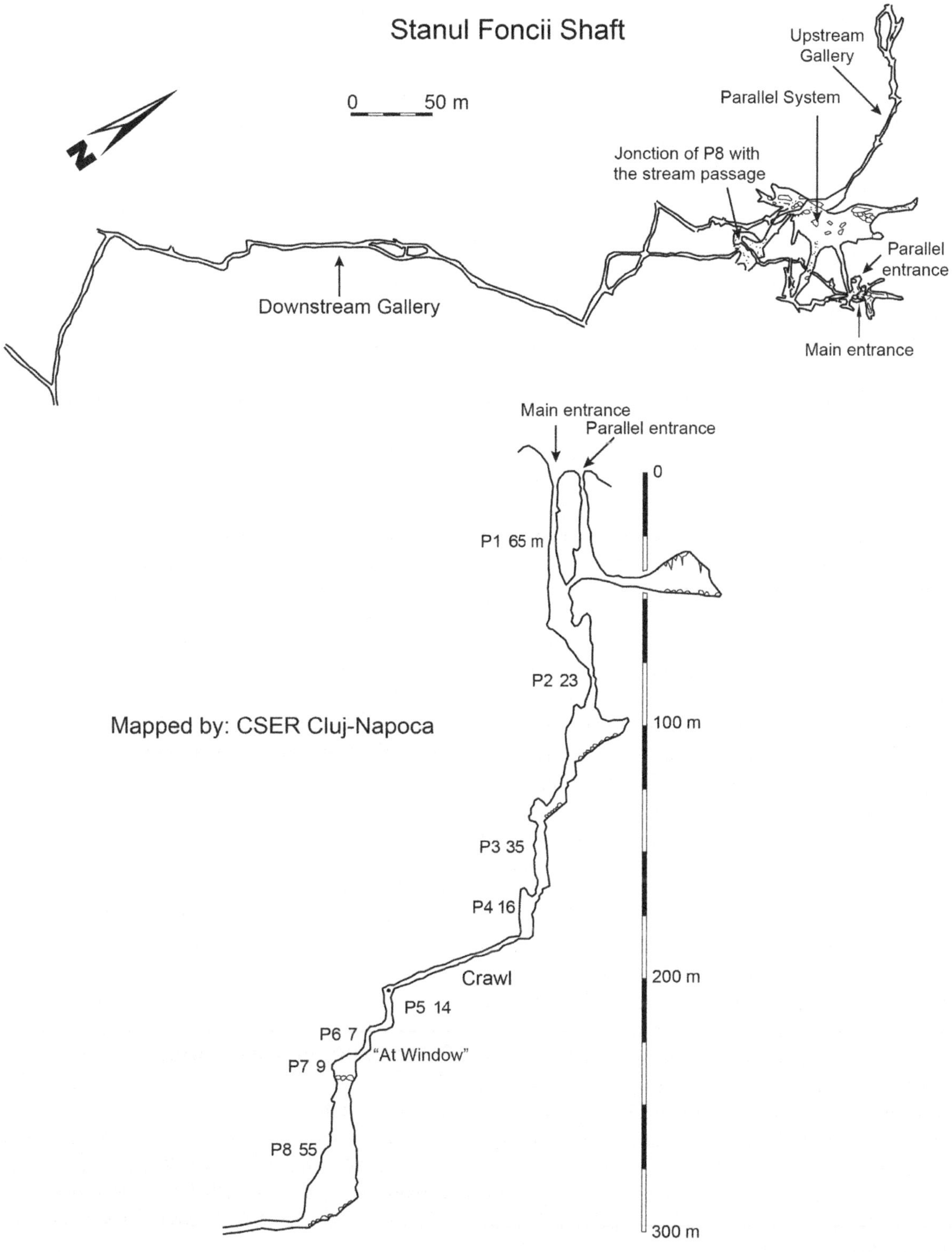

Fig. 2 Plan and profile of Stanul Foncii Shaft

its deepest point at −338.9 m (Popa 1984). Between 1984 and 1985, several exploration trips by members of the CSER Cluj (D. Moldovan, D. Coţop, F. Papiu, C. Patalita, T. Dan, and V. Breazu) led to the discovery of few minor side passages, with at least one connecting to cave's stream passage (Papiu 2001).

Cave Description

The entrance shaft (P1) leads to the middle part of a room with the floor covered by a large scree deposit and descending at 45°. At the lower end of the slope, follows a succession of three other pits of 23, 35, and 14 m, respectively, interconnected by subvertical passages. Beginning with P3, a perennial stream is present, but the discharge depends on the pluvial regime. In the southern extremity of the room from the base of P3 was intercepted another stream, which comes in from a perpendicular direction to the other watercourse. These streams are coming from a suspended gallery and ends after about 100 m into a sump.

Morphologically, the walls of all three pits display evidence of vertical vadose flow. The most "horizontal" section in what otherwise is a predominantly vertical developed cave is the so-called Crawl, which connects P4 with P5 via a narrow, descending gallery with a medium height of 0.5 m. The succession of verticals (P5, P6, P7, and P8) between the Crawl and the stream passage are textbook examples of vadose shafts. P5, P6, and P7 have perfect circular form, whereas P8 is in fact a bell-shaped large room, which is entered through an extremely narrow passage located at its very top. The room constitutes a former meander of the stream gallery, presenting abundant alluvial deposits. The Upstream Gallery (260 m) is slightly meandered, with passages having 2–10 m in width and heights of 10–40 m; its terminus is represented by a massive limestone breakdown. The Downstream Gallery is more sinuous, with widths of 1–2 m and heights of 10–20 m. A first sump can be bypassed through a short fossil passage. Until reaching the terminal sump, the underground river has four other insignificant tributaries.

A second entrance in the system is through the Parallel Entrance, a very narrow opening situated 7 m away from the main entrance. A 38 m deep shaft reaches a steeply descending slope from where the cave branches out in three directions: (1) a gallery that intercepts P1, (2) a short passage that ends after 13 m, and (3) a gallery that leads into the Stanul Foncii's largest room (60 × 20 × 20 m), which is nicely decorated with a variety of speleothems and also hosts *Ursus spelaeus* bones. Collectively, all these voids form the Parallel System.

Speleogenesis

The main role in the genesis of the Stanul Foncii Shaft was played by the local tectonic elements (fissures, fractures, and faults) along which surface water and groundwater descended vertically generating the system of eight shafts. Some of the shorter shafts (e.g., P3) might have been created by a small stream flowing along a more resistant limestone bed. The Crawl passage is a major break between the two vertical sections above and below it. The entrance in the Crawl is at ~185 m below surface, virtually at the bottom of the same fracture or fault on which P1 to P4 formed. Instead P5, it is vertically offset by 50 m indicating the underground stream continued to dissolve along a different fracture. A reasonable way to explain the existence of the Crawl is that the fracture extended only to this depth, and the groundwater encountered a more impure limestone bed along which it had to cut its way until reaching the fracture that initiated P5.

Based on the discussion above, Stanul Foncii Shaft could be classified as a drawdown vadose cave (Onac 2002), typical for low plateaus and hilly regions. Such karst features develop under hydrological settings where the water table is at the beginning close to the surface but then draw down until stabilizes at a lower elevation next to a spring (Ford and Williams 2007). Iurkiewicz and Mitrofan (1984) suggested that vertical extent of some caves and shafts in the S and SW part of the Pădurea Craiului Mountains was controlled by the local base level. Indeed, the deepest part of the Stanul Foncii (321 m absolute altitude) is just a few tens of meters above the elevation of Roșia Spring (290 m), through which the underground stream resurfaces.

Cave Climate Measurements

There is only one set of temperature measurements throughout the shaft collect by the Czech caver V. Kahle on August 17, 1981. At 3 pm, the temperature next to the entrance was 23 °C, whereas down the shaft, the following values were recorded: base of P1: 8.6 °C, base of P4: 9.2 °C, bottom of P8: 10.2 °C, downstream sump: 10.7 °C; the water temperature was 9.0 °C.

Paleontology/Cave Fauna

Bones belonging to *U. spelaeus* were found in the Parallel System. Among the taxa documented in Stanul Foncii Shaft, we mention *Duvalius redtenbacheri* and *D. paroecus* (Moldovan 2002) and *Pholeuon (Parapholeuon) gracile chappuisi* Jeannel, 1930 (Racoviță 2011).

Cave Conservancy

Stanul Foncii Shaft is an unprotected cavity; however, for any type of activities (including visits) you should obtain the permission(s) from the Romanian Speleological Heritage

Commission. Visiting the shaft requires very good skills of single rope technique. Presently, the shaft is entirely equipped with steel anchors, but these have neither hangers nor screws. It is highly recommended to avoid visiting the shaft during rainy events, but in case you are in and flooding is imminent, you must stay on the upper part of the Crawl or wait (without any problems) the end of the flood pulse in the big room at the bottom of P8. The time required for a visit varies depending on the purpose, but you may want to allocate at least 8 h.

References

Balintoni I (2001) Short outlook on the structure of the Apuseni Mountains. In: Bucur II, Filipescu S, Săsăran E (eds) Field trip guide, 4th Reg Meeting of IFAA, Ed Presa Univ Clujeană, Cluj-Napoca, pp 9–17

Ford DC, Williams P (2007) Karst hydrogeology and geomorphology. John Wiley and Sons, Chichester, p 562

Ianovici V, Borcoş M, Bleahu M, Patrulius D, Lupu D, Dimitrescu R, Savu H (1976) Geologia Munţilor Apuseni. Ed Acad RSR, Bucureşti, p 632

Iurkiewicz A, Mitrofan H (1984) On karstic cavities vertical distribution regularities in southern and south-western Pădurea Craiului Mountains. Theor Appl Karstol 1:77–82

Moldovan A (1984) Memoriile unei echipe de explorare. Peştera 1:113–117

Moldovan O (2002) *Duvalius* - the only carnivorous subterranean beetles from Pădurea Craiului Mountains. In: Racoviţă G, Moldovan O, Onac BP (eds) The karst of Pădurea Craiului Mountains. Monographic study. Ed Presa Univ Clujeană, Cluj-Napoca, pp 165–171

Onac BP (2002) Endokarst (hydrogeologic settings, speleogenetic models, cave deposits). In: Racoviţă G, Moldovan O, Onac BP (eds) The karst of Pădurea Craiului Mountains. Monographic study. Ed Presa Univ Clujeană, Cluj-Napoca, pp 35–92

Orăşeanu I, Iurkiewicz A (1987) Hydrogeological karst systems in Pădurea Craiului Mountains. Theor Appl Karstol 3:215–222

Papiu F (2001) Avenul din Stanu Foncii. Speomond 6:5–7

Popa C (1984) Stanul Foncii. Peştera 1:101–112

Racoviţă G (2011) Révision systématique des Leptodirinae souterrains des Monts Apuseni. VIII. Aperçu synthétique sur le genre *Pholeuon*. Trav Inst Spéol "Émile Racovitza" 50:37–59

Rusu T (1981) Les drainages souterraines de Monts Pădurea Craiului. Trav Inst Spéol "Émile Racovitza" XX:187–205

Rusu T (1988) Carstul din Munţii Pădurea Craiului. Ed Dacia, Cluj-Napoca, p 254

Soriţău D, Nicoară D, Silvestru E, Demeter I, Popa C, Viehmann I (1984) Avenul din Stanul Foncii. Peştera 1:79–88

Pădurea Craiului Mountains: Izbândis Cave System

Ferenc L. Forray

Abstract

Despite its modest appearance and relatively difficult to explore fossil section, the underwater part of the Izbândiş system makes this cave special, as it hosts the deepest sump in Romania (−105 m). Besides, archeological findings in this cave had helped understand the spread of the Igriţa-type culture and its distinctive objects. The cave is also shelter for five bat species (*Barbastella barbastellus*, *Myotis myotis*, *M. oxygnathus*, *Rhinolophus hipposideros*, and *Rh. ferrumequinum*) and habitat for eleven insect species.

Keywords

Cave • Bat • Archaeology • Speleothems
Climatology • Romania

Introduction

Pădurea Craiului Mountains with a surface area of ~1150 km^2 (Rusu 1988) and with more than 900 caves (www.speologie.org) is one of the most spectacular karst area of Romania. This mountain also hosts the longest cave in Romania, Vântului (Wind Cave), with more than 50-km-long passages. The Izbândiş Spring is one of the largest karst resurgence in the Pădurea Craiului Mountains (Fig. 1), besides which the dry cave located on the slope above was for a long time a karst cavity with modest dimensions. Due to the efforts of many cave explorers and scientist, our knowledge of the Izbândiş Cave has improved considerably, but there are many unanswered questions that keep the flame of exploration alive.

F. L. Forray (✉)
Department of Geology, Babeş-Bolyai University,
M. Kogălniceanu 1, 400084 Cluj-Napoca, Romania
e-mail: ferenc.forray@ubbcluj.ro

Geographic, Geologic, and Hydrologic Settings

The Izbândiş cave system (ICS), with a known total length of 2296 m, is located south of Şuncuiuş village, on the right side of the Măguran Valley (Fig. 2). As of 2017, upstream from the junction with the water coming from Izbândiş karst spring, the valley is dry throughout the year, except when there are heavy rains in the region or during springtime, when is fed by snowmelt water. From the lake formed at the base of a limestone cliff, which represents the karst spring with the underwater cavern system, the fossil part of the cave is located 30 m higher (Vălenaş and Drîmba 1978) in the limestone wall. This fossil entrance can be reached by climbing a ~10 m limestone wall, a natural obstacle that prevented humans to routinely access and inhabit the cave. Thus, very few archeological and paleontological artifacts were found inside.

The cave develops in Triassic limestones (Upper Campilian–Anisian age) and dolomites (Anisian age) (Patrulius et al. 1973; Bordea et al. 1986). There are several fault lines crossing this formation and influencing the underground water drainages in the area (Vălenaş and Iurkiewicz 1980–1981) (Fig. 2).

The hydrogeology of the ICS was investigated using fluorescein and indium EDTA (In-EDTA) tracers between 1964 and 1986 (Rusu 1988; Orăşeanu 1991, 2016). The tracer experiments indicated that the ICS is supplied with water from a number of ponors (see Fig. 2 for details). Rusu (1988) considered the infiltration water from Măguran Valley drainage basin to contribute water to the ICS.

The tracer studies revealed a shorter travel time (2.6–3.3 days) and a longer one (32 days) (Orăşeanu 2016). For the time being, there are no tracer studies for the water infiltrating in the Măguran Valley drainage basin. After heavy rainfall, the discharge peak occurs in 1–3 days and subsides to normal levels after 9–21 days (Rusu 1988). In 1982–1983, the maximum recorded discharge was 3980 L/s

G. M. L. Ponta and B. P. Onac (eds.), *Cave and Karst Systems of Romania*,
Cave and Karst Systems of the World, https://doi.org/10.1007/978-3-319-90747-5_45

Fig. 1 Location of Izbândiş Cave in the karst of northern Pădurea Craiului Mountains

and the minimum 49 L/s; the average discharge of the karst resurgence is 346 L/s (Orăşeanu 1991).

History of Exploration

Certainly, the fossil cave was known to the local villagers for a long time, but the earliest study was published by Bokor only in 1921. Over time, interest in the exploration and scientific research of this complex karst system has nevertheless gained a constant interest. The first systematic research of the fossil/dry part of the cave was carried out in 1942 by H. Kessler from the Geological Institute in Budapest, who made the first map of the cave. In 1973, J. Talmács, G. Drîmba, and L. Berecz resurvey the cave and made a new map (Vălenaş and Drîmba 1978). The Club of Amateur Speleologists (CSA) Caving Club from Cluj-Napoca explored the cave, dug out sediments in 70′ (L. Csoltkó), 80′ (A. Varga), and 90′ (Z. Deák), and mapped new sections (Deák 2001). The underwater cave system was explored by G. Halasi and G. Rajka in the early 80s (Vincze et al. 1987). At the time of writing this book, the deepest dive in the cave was at −105 m (diver L. Retegan).

Cave Description

The explored underwater section of the Izbândiş Cave is 437 m long (Fig. 3a, b), and its first part develops along a fracture zone that controls the morphology of the passages up to a depth of −40 m. Then, the passage morphology becomes an elliptic conduit, which continues to depths between −37 and −42 m. At about 170 m from the entrance and at a depth of −40 m, the passage splits in two: the Halasi Gábor Passage, which soon change into a very narrow conduit filled with thick fine sediments and a second passage developed along a fracture. The latter one, at the depth of −61 m turns into a phreatic conduit with a diameter between 4 and 6 m, descending sometimes vertically. At the depth of −105 m, the gallery is 1–2 m high and 2–3 m wide; beyond this point, it continues to descend, but for the time being the exploration ended here.

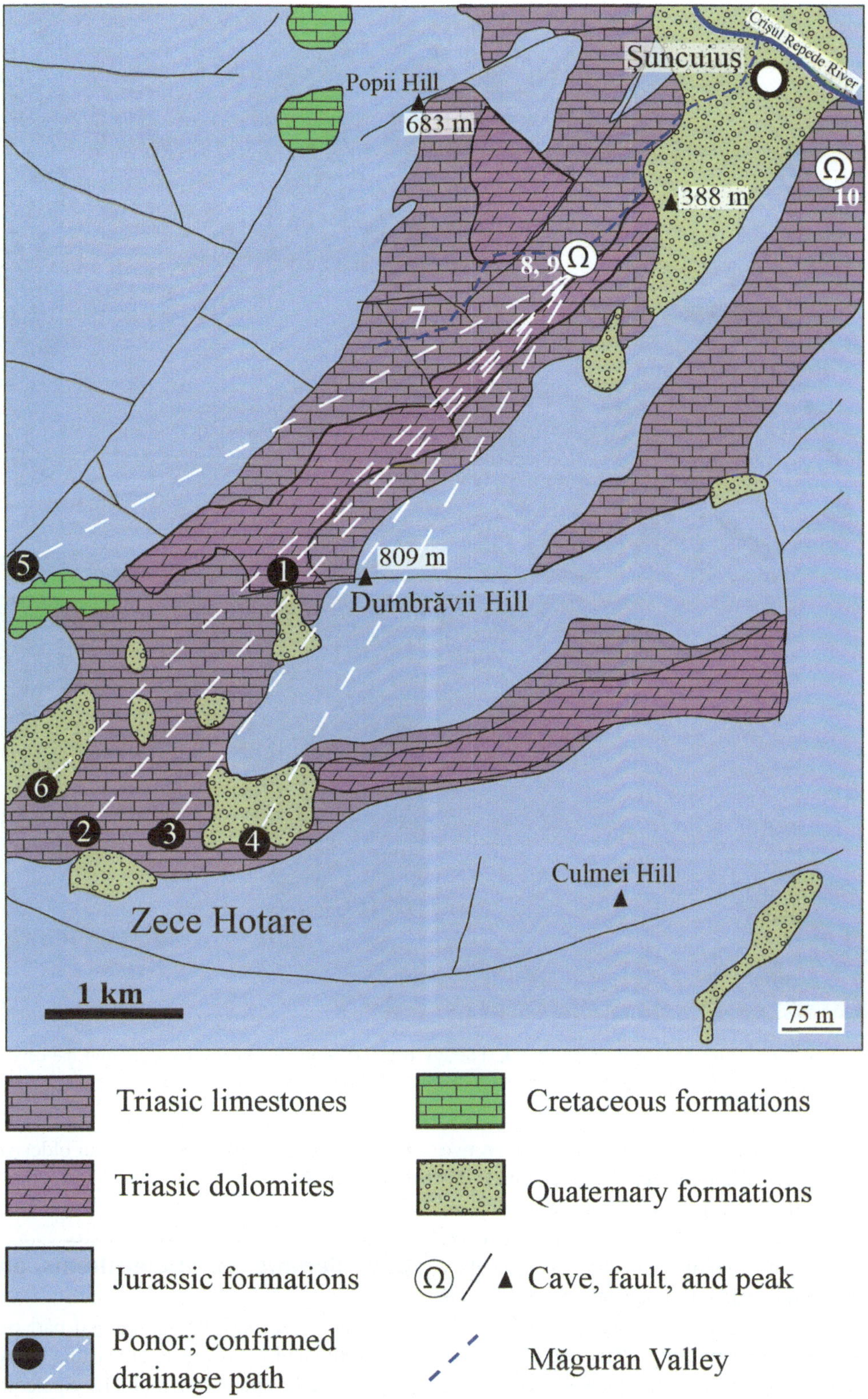

Fig. 2 Simplified geological map of the Izbândiş karst system and water recharge areas, modified after Patrulius et al. (1973) and Bordea et al. (1986). 1—Groapa Blidireşti ponor; 2—Pârâul Brezului ponor; 3—Pârâul Olfului ponor; 4—Pârâul Birăului ponor; 5—Pârâul Tomii ponor; 6—Iacoboaia ponor; 7—Măguran Valley drainage basin; 8/9—Izbândiş Cave/Spring; 10—Vântului Cave. The underground drainage paths are based on tracer tests performed by Orăşeanu (2016)

Fig. 3 Izbândiş Cave. **a** Karst resurgence (photograph by F.L. Forray); **b** Image from the underwater section of the cave (photograph by F.L. Forray); **c** Speleothems in the dry cave (photograph courtesy of P. Szilágyi–Palkó)

The fossil sections of the cave (1859 m long) have two main branches: the longer Main Gallery (Galeria Principală) and a shorter one called Divers Gallery (Galeria Scafandrilor). At the end of the Main Gallery, there is a maze spreading at various levels, ending with the Halasi Gallery from which a short underwater passage begins. Sediments (breakdown blocks, sand, and clay) are present almost everywhere in the cave. The majority of the passages have modest sizes, even the biggest chamber located close to the entrance is only 25 by 15 m and a height of 8 m (Deák 2001).

Unfortunately, the map of Izbandiş cave system was not provided by the CSA Caving Club; instead, the readers are encouraged to consult the Web site: www.speologie.org/Pestera-Izbucul-Izbandis, where an older version of the map is included.

Cave Sediments, Speleothems, and Minerals

Both the fossil and the submerged part of the cave system are dominated by deposits consisting of pebbles, sand, and clay, as well as breakdown blocks. Speleothems are found in only a couple of sectors within the fossil cave and are represented by gours (rimstone pools), calcite flowstones, pools with crystals, and helictites (Vălenaş and Drîmba 1978; Deák 2001; Rajka 2002), stalactites, columns, soda straws (some 50 cm long; Fig. 3c) (Deák 2001).

Speleogenesis

The Izbândiş Cave was created by the drainage that carried the water fed by several ponors (Orăşeanu and Iurkiewicz 1987; Rusu 1988). There was several evolution stages as the underground drainages that sculpted the caves system changed over time. The spring branch of the cave system is connected with the fossil part through several passages, some of which are too narrow to be explored. A connection with the Vântului Cave (located nearby) is not yet established, despite the fact that between the easternmost point of the Izbândiş Cave and the Vântului Cave's Youths Passage, the distance is just around 500–600 m.

Cave Climatology

There is no climate data on the fossil part of the Izbândiş Cave, but the close-by Vântului Cave, with which it may be genetically linked (Orăşeanu and Iurkiewicz 1987), has an average temperature of about 9 °C and the relative humidity around 98% (Onac and Racoviţă 1992). This is in accordance with the regions average annual temperature (9–10 °C). Due to the Izbândiş Cave descendent profile at the beginning, it is possible to have a slightly lower temperature closer to the cave entrance, caused by descendent air currents during winter time. Based on data recorded by an underwater temperature logger set up at ~120 m from the entrance at a depth of −18 m, the water temperature varies between 9.31 and 10.07 °C, with an average of 9.82 °C (data from September 2011 to March 2013).

Paleontology

Due to its difficult to reach entrance, it is not surprising that paleontological material is scarce. Skeletal remains of *Ursus spelaeus* and *Cervus elaphus* (Rusu 1988) were found in this cave. The bones of several bird species, *Lyrurus tetrix*, *Gallus gallus*, *Columba palumbus*, *Coloeus monedula*, *Cinclus cinclus*, *Turdus pilaris*, *T. viscivorus* (possibly Pleistocene–Holocene age) (Jurcsák and Kessler 1986–1987) were recovered from the cave sediments of the fossil section by I. Emödi between 1970 and 1980.

Cave Fauna

In a study of several caves fauna, Bokor (1921) mentioned only the presence of *Pholeuon* sp. from the order Coleoptera in the Izbândiş Cave. Later Racoviţă (1996) attributed this specie to *Pholeuon moczaryi*. Subsequently, other species described are *Cryptops croaticus* (Matic 1960), *Troglohyphantes racovitzai* (Dumitrescu and Georgescu 1970), *Drimeotus gracilis* (Moldovan 2000), *Deutonura conjuncta* and *Lepidocyrtus serbicus* (Gruia 2003), *Campodea spelaea* (Sendra et al. 2012), *Enchytraeus argenteus*, *Onychiurus armatus* (database of the "Emil Racoviţă" Institute of Speleology in Cluj-Napoca), *Duvalius redtenbacheri* collected and identified by O. Moldovan (pers. comm. 2017). In the underwater passage of the cave Lascu and Sârbu (1987) and later Sârbu (pers. comm.) reported the presence of *Niphargus* sp.

The cave is shelter for five bat species: *Barbastella barbastellus* (western barbastelle), *Myotis myotis* (greater mouse-eared bat), *M. oxygnathus* (lesser mouse-eared bat), *Rhinolophus hipposideros* (lesser horseshoe bat), and *Rh. ferrumequinum* (greater horseshoe bat) (Bücs et al. 2012).

Anthropology/Archeology

The Romanian National Archeological Repertory database (RAN code 31,529.04) identifies this site as a Neolithic settlement. Excavations were carried out by Emődi (1984, 1985, 1987–1988, 1992), who collected archeological and paleontological artifacts. He argued that the cave entrance is hard to reach, and the artifacts were sign that the cave was used for funeral or ritual purposes. Contrary, Boroneanţ (2000) considers the ceramic fragments, bone tools (shoulder blade sickle), stone tools, perforated bone beads, and different animal bones found in this cave as proof that the cave was inhabited by humans.

Ceramic fragments decorated with strings (Emődi 19871988, 1992), fragments from an amphora (Emődi 1984), needles with thick head (Chidioşan and Emődi 1983) indicate the Igriţa-type culture, specific to Roşia region (Chidioşan and Emődi 1982, 1983), which have similarities with the early Coţofeni Culture (Chalcolitic and Early Bronze age) from Transylvania (Emődi 1984).

Cave Conservancy

All bat species, excepting *Myotis oxygnatus*, reported from the cave are in the European Council Directive 92/43/EEC (EEC 1992) on the Conservation of Natural Habitats and of wild flora and fauna list. The bat species and their habitat need to be protected. To protect the cave, two special metallic gates have been installed on the cave's major passages. The cave system is located on the NATURA 2000 site ROSCI0062 (Defileul Crişului Repede–Pădurea Craiului) and have protected status. Research and visit permits need to be obtained from the Romanian Speleological Heritage

Commission and the "Centrul pentru Arii Protejate şi Dezvoltare Durabilă Bihor".

Acknowledgements We are thankful to past and present members of the CSA Caving Club Cluj-Napoca for their exploratory work and contribution to the understanding of this cave and karst system. The digging activities in the fossil part of the cave were carried out by L. Csoltkó, A. Varga, Á. Szilágyi, K. Kerekes, T. Jankovics, Z. Bogáti, M. Balogh, L. Vári, and Z. Deák, whereas the mapping was completed by A. Varga, Z. Deák, and G. Rajka. The underwater cave was explored and mapped by Z. Gyurka and G. Rajka from CSA Cluj-Napoca, K. Kéri, and L. Retegan. We are also thankful to G. Rajka for providing several rare publications and helpful information, all used to complete this chapter. Special thanks are due to Drs O. Moldovan, T. Brad, and D. Borda ("Emil Racoviţă" Institute of Speleology, Cluj-Napoca) for their helpful information provided on the cave fauna.

References

Bokor E (1921) A magyarhoni barlangok ízeltlábui. Barlangkutatás IX (1–4 füzet):1–22

Bordea S, Bordea J, Mantea G, Costea C (1986) Geological map of Romania, Zece Hotare Sheet (40b). Scale 1:50.000 (L-34-45-B), Inst Geol Geofiz, Bucureşti

Boroneanţ V (2000) Arheologia peşterilor şi minelor din România. Ed cIMeC, Bucureşti, p 355

Bücs S, Jére C, Csősz I, Barti L, Szodoray-Parádi F (2012) Distribution and conservation status of cave-dwelling bats in the Romanian Western Carpathians. Vespertilio 16:97–113

Chidioşan N, Emődi J (1982) Grupul cultural Igriţa de la sfârşitul epocii bronzului. Crisia XII:61–86

Chidioşan N, Emődi J (1983) Descoperirile arheologice din peştera Izbândiş (com. Şuncuiuş) aparţinând grupului cultural Igriţa. Crisia XIII:17–32

Deák Z (2001) Istoricul cercetărilor întreprinse de C.S.A. Cluj în Peştera Izbândiş. Meandre (Bul Speol C.S.A. Cluj) 1:8–10

Dumitrescu M, Georgescu M (1970) Révision des représentants du genre *Troglohyphantes* des grottes de Roumanie. Livre du centenaire Émile G. Racovitza 1868–1968. Romanian Academy, Bucharest, pp 313–331

Environment European Commission (1992) Council Directive 92/43/EEC of 21 May 1992 on the conservation of natural habitats and of wild fauna and flora, p 44

Emődi J (1984) Descoperiri ale culturilor Coţofeni şi Baden în peşterile Igriţa şi Izbândiş. Acta Mvsei Napocensis XXI:405–431

Emődi J (1985) Asupra începutului epocii bronzului în Bihor. Thraco-Dacica VI(1–2):123–144

Emődi J (1987–1988) Ceramica ornamentată cu şnur din peşterile Igriţa şi Izbândiş. Acta Mvsei Napocensis XXIV–XXV:485–506

Emődi J (1992) Die Schnurkeramik aus der Igriţa- und Izbîndiş-Höhle. In: Roman P, Dodd-Opriţescu AJP, Hauptmann H, Schrickel W (eds) Beiträge zur Problematik der schnurverzierten Keramik Südosteuropa von Zabern, Mainz am Rhein, pp 192–194

Gruia M (2003) Collembola from Romanian caves. Trav Mus Natl Hist Nat "Grigore Antipa" 35:139–158

Jurcsák T, Kessler E (1986–1987) Evoluţia avifaunei pe teritoriul României. Crisia 16:577–615

Kessler H (1942) Az északbihari forrásbarlang. Beszámoló a M. Kir. Intéz. Vitaül. Munkál., Budapest

Lascu C, Sârbu Ş (1987) Peşteri scufundate. Ed Acad R.S. România, Bucureşti, p 255

Matic Z (1960) Die Cryptopiden (Myriopoda, Chilopoda) der Sammlung des Speleologischen Institutes "E. Gh. Racovita" aus Cluj. Zoologischer Anzeiger 165:442–447

Moldovan OT (2000) Révision de *Drimeotus* s.s. Miller, 1856 (Coleoptera, Cholevidae, Leptodirinae) de Transylvanie (Roumanie) avec description de deux nouvelles espèces et clé de détermination des taxa. Zoosystema 22(1):139–152

Onac BP, Racoviţă G (1992) Recherches climatologiques dans la Peştera Vântului (Monts Pădurea Craiului). Theor Appl Karstol 5:167–192

Orăşeanu I (1991) Hydrogeological map of the Pădurea Craiului Mountains (Romania). Theor Appl Karstol 4:97–127

Orăşeanu I (2016) Hidrogeologia carstului din Munţii Apuseni. Ed Belvedere, Oradea, pp 17–61

Orăşeanu I, Iurkiewicz A (1987) Hydrological karst systems in Pădurea Craiului Mountains. Theor Appl Karstol 3:215–222

Patrulius D, Popa E, Câmpeanu S, Orăşeanu T (1973) Geological map of Romania, Remeţi Sheet (41a). Scale 1:50.000 (L-34-46-A). Inst Geol Geofiz, Bucureşti

Racoviţă G (1996) Révision systématique des Bathysciinae souterrains des Monts Apuseni. II. Le sous-genre *Parapholeuon* Ganglb. du bassin de Crişul Repede (Monts Pădurea Craiului). Trav Inst Spéol "Émile Racovitza" 35:69–105

Rajka G (2002) Drenaje sifonate. In: Racoviţă G, Moldovan OT, Onac BP (eds) The karst of Pădurea Craiului Mountains. Ed Presa Univ Clujeană, Cluj-Napoca, pp 39–48

Rusu T (1988) Carstul Munţilor Pădurea Craiului. Ed Dacia, Cluj-Napoca, p 254

Sendra A, Nitzu E, Sanjuan A (2012) Half a century after Ionescu's work on Romanian Diplura—A faunal contribution based on material collected from karst areas. Trav Inst Spéol "Émile Racovitza" 51:37–66

Vălenaş L, Drîmba G (1978) Cercetări de speologie fizică în Munţii Pădurea Craiului. Nymphaea VI:279–328

Vălenaş L, Iurkiewicz A (1980–1981) Studiu complex al carstului din zona Şuncuiuş-Mişid (Munţii Pădurea Craiului). Nymphaea 8–9:311–378

Vincze J, Vincze Z, Gyurka Z, Rajka G (1987) The Izbandis Cave system. VII Int Camp of Cave Diving. UIS Commission for Cave Diving, Gorizia 23–29 aug. 1987, pp 18–25

Useful links

https://www.speologie.org/Pestera-Izbucul-Izbandis

Pădurea Craiului Mountains: Vântului Cave (Wind Cave)

Bogdan P. Onac

Abstract

Peştera Vântului lies in a ~200-m-thick middle Triassic limestone and is the longest cave in Romania, with a surveyed length of over 50 km and vertical extent of 162 m. It is a multi-level cave, which consists of passages that are mainly sinuous tubes and canyons having their floor covered with alluvial sediments and breakdown blocks. Vadose flow is present in the lowest level of the cave, where significant amounts of black iron–manganese-rich deposits coat the gravels and the cave walls along the underground stream. Fourteen oxide and hydroxide minerals have been described in their composition, with braunite first documented in a cave environment worldwide. Gypsum speleothems are abundant particularly along the upper levels 1 and 2. Typical for Vântului Cave is its deep meandering canyons, which are best displayed in the Racoviță Meanders section of the second level. From a speleogenetic point of view, it is a base-level cave, having its all four major levels developed in the epiphreatic zone, later connected in between by vadose canyons and shafts.

Keywords

Cave minerals • Morphology • Meanders
Base-level cave

B. P. Onac (✉)
Karst Research Group, School of Geosciences, University of South Florida, 4202 E. Fowler Ave., NES 107, Tampa, FL 33620, USA
e-mail: bonac@usf.edu

B. P. Onac
"Emil Racoviță" Institute of Speleology, Romanian Academy, Clinicilor 5, 400006 Cluj-Napoca, Romania

Geomorphological, Geological, and Hydrogeological Settings

Peştera Vântului (Wind Cave) is located on the northern extremity of the Pădurea Craiului (King Forest) Mountains at their boundary with the Vad-Borod Neogene Depression (Fig. 1). From a geomorphological point of view, they represent a large fragmented Mesozoic platform comprising a series of summits and isolated massifs of which elevations decrease SE to NW from 1027 to 350 m (Rusu 1988). These relatively low elevations are partly compensated by areas with a rather high relief, which gives the landscape of the Pădurea Craiului a mountainous aspect. The great diversity of rocks making up the geological structure, later affected by tectonic processes, is expressed morphologically by a chaotic landscape that lacks any generally unique features. The summits are mainly made of sandstones, conglomerates, and igneous/metamorphic rocks, whereas at lower elevations, plateaus dotted with hundreds of dolines and catchment depressions dominate the karst landscape (Orăşeanu 1991; Onac 2002a).

The Apuseni Mountains, likewise other major geomorphic units of Romania are characterized by nappe structures, which were generated during the Alpine terranes amalgamation (*see* chapter by Balintoni). According to the provenance of the nappe sequence, Pădurea Craiului Mountains are part of the Apusenides (Balintoni 1997). The sedimentary rocks consist of shallow-marine carbonates and non-marine siliciclastic deposits belonging to the Bihor Unit (Săndulescu 1984), earlier known as the Bihor Autochthonous (Ianovici et al. 1976), which are overlain in the SW part by the Codru nappe system. Within the cave area, the sedimentary succession of the Bihor Unit includes:

(1) Non-karst rocks (conglomerates, sandstones, shales, and clastic carbonates) of upper Permian–lower Triassic age (Werfen facies; Popa 1981);

G. M. L. Ponta and B. P. Onac (eds.), *Cave and Karst Systems of Romania*,
Cave and Karst Systems of the World, https://doi.org/10.1007/978-3-319-90747-5_46

Fig. 1 Location of the cave in the karst area of northern Pădurea Craiului Mountains

(2) Medium- to thick-bedded intertidal dolomites and limestones (Anisian; middle Triassic), followed by a massive carbonate sequence (up to 250 m) composed of white-pinkish Wetterstein-type reefal limestone assigned to Ladinian and lower Carnian (Bleahu et al. 1970; Burchfiel and Bleahu 1976; Dragastan et al. 1982);
(3) Conglomerates and sandstones (rich in pyrite and marcasite nodules) of typical Gresten facies (lower Jurassic) overlie discordantly a paleokarst landscape developed during the upper Triassic when the entire region was uplifted and in which fireclay deposits accumulated (Papiu et al. 1988; Onac and Popescu 1991);
(4) Sandy and clayey limestones of middle Jurassic age rarely exceed 10 m in thickness (Ianovici et al. 1976);
(5) Thick (60–250 m) reef limestones (Stramberk facies) of upper Jurassic (Cociuba 2000).

The tectonostructural setting of the area south of Șuncuiuș played an important role in the development of the Vântului Cave. Two major fault lines (oriented NE–SW) are intersected by secondary fractures (W–E or NW–SE) causing the lithological units to partition into several blocks (Patrulius et al. 1973). The one in which the cave develops is bounded to the south by the Șesii Fault, on the east by the Mișid Valley, which incises along the axial plane of an anticline, whereas its west-distant boundary is represented by a fault, likely responsible for the development of Izbândiș Cave system (*see* chapter by Forray). Apart from the main set of faults and fractures, there is also a complex network of tension, shear, and strike joints developed in response to both folding and faulting. These are orientated oblique to the axial plane of the Recea Syncline and faults (NW–SE and SW–NE), and their presence is well reflected in the general configuration of the cave (Szilágyi Palkó et al. 2007).

With very few exceptions, similar to other karst massifs, the hydrological network of Pădurea Craiului Mountains is characterized by a lack of permanent rivers and a long evolution, during which stream piracy repeatedly changed flow direction in the surface valleys. Furthermore, the tectonic activity (folding and faulting) has turned the whole unit into several blocks, with karst/non-karst formations alternating and thus forcing karst drains along distinct alignments (Rusu 1988).

To establish the recharge area of the Vântului Cave, Orăşeanu and Gaşpar (1980–1981) conducted tests using radioactive tracers. The results indicated that the water disappearing in the Recea Ponor, as well as the diffuse losses along Şesii Valley (at the lower Jurassic siliciclastic and Ladinian limestone boundary) are responsible for the present-day origin of the underground watercourse. This one emerges in a group of springs next to the cave entrance, in the Poiana Frânturii (Orăşeanu 1991). Subsequent studies determined the boundaries of the hydrogeological karst system of the Vântului Cave, but could not identify any other major contributor to the cave stream (Orăşeanu and Iurkiewicz 1987; Orăşeanu 2010).

Vântului Cave develops in Ladinian to lower Carnian bioclastic limestones in the south side of the Crişul Repede River Gorge, 2.5 km upstream of Şuncuiuş, a former fireclay mining center. Its passages extend parallel to the Mişid Valley (a left-side tributary of the Crişul Repede) and follow the main features of the syncline west of the valley that is cut along the hinge line of the Mişid Anticline (Vălenaş and Iurkiewicz 1980–1981). The cave opens at 320 m above sea level (asl) and its over 50 km of galleries spread on four karstification levels, with an underground stream flowing along the lower one (Bagameri et al. 1961; Szilágy et al. 1979; Szilágyi Palkó et al. 2007).

History of Exploration

As with most large cave systems, the exploration and survey activities in the Vântului Cave were the result of a common effort to which many generations of very dedicated cavers contributed. For this reason, a complete account for the history of exploration is beyond the scope of this chapter, but readers are encouraged to examine the 50-year anniversary book dedicated to Vântului Cave, a bilingual (Hungarian/Romanian) publication that provides in details its exploration timeline (Szilágyi Palkó et al. 2007). The most significant milestones are listed below in chronological order.

The official cave discovery date is April 7, 1957, when B. Bagameri, a caver from Cluj first entered and explored ~500 m along the stream passage. For protection reasons, the cave was gated on April 29, 1957. A breakthrough discovery was made by G. Ferenc who found in 1958 the entrance in the first upper (fossil) level (Stairs/La Scări), which by the end of that year extended the length of the cave to 3.2 km. In 1959, a team led by E. Cristea spent 70 h underground advancing along the stream passage up to the C.C.A. Gallery. Adding all the newly explored passages on the first map produced by I. Viehmann, the cave reached 4.5 km. In 1960, B. Bagameri and I. Ghinculov climbed to the point now called The Anvil (La Nicovală) and found parts of the second fossil level, including the Hidden Room (Sala Ascunsă). Probably, the most spectacular morphological feature in the entire cave, the Racoviţă Meanders (Fig. 2), was discovered in 1962 and the total cave length passed 6.5 km. Two years later, July 2, 1964, Vântului Cave is declared a Natural Monument and in February 22, 1966, the Club of Amateur Speleologists (abbreviated CSA in Romanian) was founded, being one of the first of this kind in Romania. 1967 was another milestone in the cave history, as with its 13.7 km became the longest in Romania. CSA members continued to explore and map year after year, discovering new passages and rooms. Szilágy et al. (1979) published a description of the cave, including a color map of ~21.5 km of galleries. The cave passed 40 km in length in 1989 after several new findings in the second and third levels. Beginning with 1996, CSA focused its effort in resurveying the entire cave network using high precision instruments. By 2007, when the anniversary book was released, Vântului Cave was over 50 km in length, with a vertical relief of 162 m (−11; +151 m), and a total extension of only 2.8 km.

Cave Description

The following description concentrates only on some notable characteristics of the four main cave levels. Given the complexity of the cave, a detailed presentation would be

Fig. 2 "Torpedo" (photograph by BP Onac)

beyond the scope of this chapter. Three texts published over a 3-year interval provide good descriptions of the known parts of the cave at those respective times (Bleahu et al. 1976; Coman and Crăciun 1978; Szilágy et al. 1979). The most complete presentation of the cave, however, belongs to Szilágyi Palkó et al. (2007) and include all new discoveries between 1979 and 2007. Unfortunately, due to the ongoing resurvey of the cave, the CSA Cluj-Napoca has decided not to provide a recent cave map, which would have allowed readers to locate themselves and better understand many of the aspects discussed in this review. However, to overcome this unusual situation, any person interested in visualizing the map of Vântului Cave is directed to any of the following publications: Bleahu et al. (1976), Coman and Crăciun (1978), Szilágy et al. (1979), and Szilágyi Palkó et al. (2007).

Cave's main network of passages is oriented NE–SW being mainly developed along the axial plane of the Recea Syncline. The galleries in the upstream sector, beyond the Advanced Camp (Tabăra Înaintată), as well as the 7th of November and P galleries have a SSE–NNW direction. Vântului Cave consists of branchwork passages organized on four main levels, connected one to another by shafts and or vadose canyons. In the median section traversed by the Măguran Fault, up to nine distinct tiers were recognized (Szilágyi Palkó et al. 2007). These different levels not only control the vertical layout of the cave, but also provide a timescale for its evolution. Our presentation begins with the youngest part of the cave, which is the stream passage or Level 0.

Level 0 The underground river is encountered within 30 m after entering the cave and can be followed, upstream for slightly over 5 km. On the first 790 m, the stream passage is wide and spacious, and the morphology is characterized by both vadose (canyon, speleothems, sediment banks, and terraces) and phreatic (notches, cupolas, scallops) features. Something that certainly draws attention is the black sediment covering gravels and boulders, in and along the stream bed, as well as the black-stained cave walls to the top of the flood level. Remarkable is also the "Torpedo" (Torpila; Fig. 2), one of the largest speleothems in the cave, and a section of the gallery with a perfectly flat ceiling (Fig. 3). The genesis of the latter was interpreted as reflecting long-term stable hydrologic and tectonic conditions (Cocean 1979).

Before reaching Sump 1 (Sifonul 1), a system of chimneys gives access to the upper levels of the cave in two points: at Stairs and Stairwell (Casa Scărilor). The stream passage is blocked by Sump 1 at ~690 m from the entrance, but can be passed when water level is low. However, over the next 100 m, there are two more sumps, the last one impenetrable. After ~150 m of an unexplored passage, the stream can be reached again (through Level 1) and followed upstream all the way to the Speoterminus, where a massive breakdown stops the access forward. Along this section, the underground stream flows on a mainly vadose gallery that often changes its size. Narrow passages alternate with wide meandering sectors and areas with abundant collapses. The only two left-side (west) tributaries come along the Novi Sad and C.C.A., 1000 and 200 m long galleries, respectively. The third section of the underground river is accessible from the Advanced Camp and continues SSE (toward Mişid Valley) as a generally large vadose

Fig. 3 Section of the stream passage showing a splendid flat ceiling, wall notch, and alluvial sediments (photograph courtesy of AN Palmer)

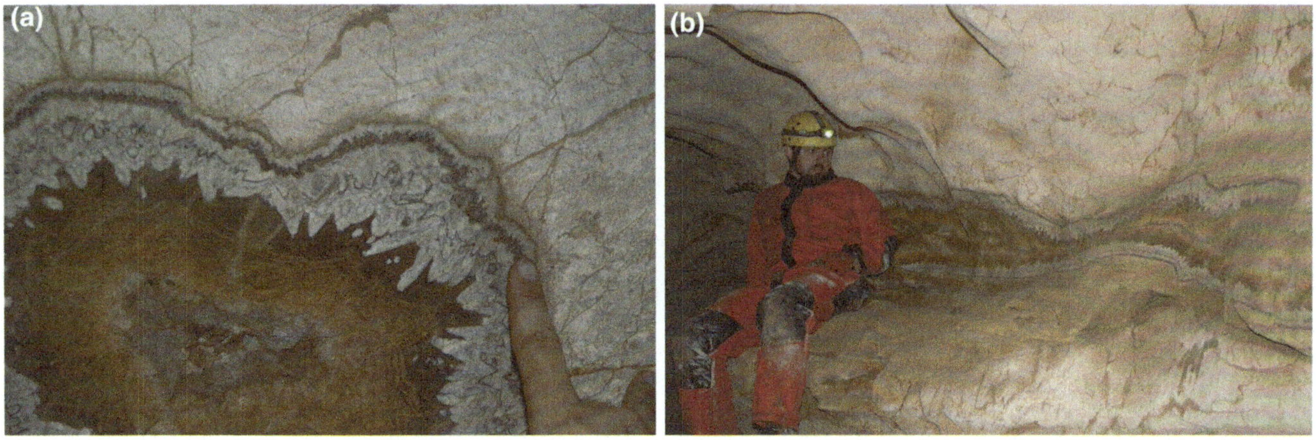

Fig. 4 Calcite-lined geode exposed in the walls of the Racoviţă Meanders (photograph by BP Onac)

gallery with frequent alluvial terraces, up to Terminus 68, the upstream end of the cave's lower level, first reached in 1968.

Level 1 The gallery of the first level is very sinuous and follows the stream passage almost over its entire length. In few points, however, sections of the first level were not yet discovered or are inaccessible. One of these is above the Big Room (Sala Mare); the collapse of its ceiling likely destroyed a segment of the gallery above, giving access to the underground river. Up to this point, the passage morphology is a typical vadose one with complex horizontal and vertical meandering (Şerban 1984) and big chambers with large breakdown blocks, such as: Titans (Sala Titanilor), Big (Sala Mare), or Black (Sala Neagră) rooms. Between the *Stairs* and the Big Room, several shafts of 25–30 m in depth (e.g., Bagameri) connect this upper level with the lower one. The mineralogical highlight of this gallery is the presence of gypsum as crusts, balloons, blisters, flowers, and crystals (Onac 1991). These are especially abundant in the Room (Sala Metalul) and Hippodrome (La Hipodrom). Beyond the Big Room, the fossil passage continues through the Amphitheater with the White (Galeria Albă) and Lakes (Galeria cu Lacuri) galleries. Along them, both phreatic and vadose morphologies are well preserved and some sections are richly decorated with helictites, coralloids, pool spar, and eccentrics made up of calcite and aragonite.

Level 2 The main gallery on the second level stretches between the Anvil and the end of the 7th of November Gallery. It superposes Level 1 up to the Small Oasis (Oaza Mică) from where it makes a SSE turn heading toward the Mişid Valley. Morphologically, the gallery is irregular and has an uneven floor with many breakdowns and sediments sequences up to 10 m in thickness. 1st of May Gallery is a short, but spectacular section of Level 2 accessible right after climbing the "Stairs" from the river passage into the Level 1. Its first part has a typical keyhole cross section, with a canyon that cut in the floor of perfectly rounded tube as the flow changed from phreatic to vadose. After this narrow section the gallery is large, well decorated with speleothems, abundant gypsum crusts, and red clay deposits.

Without doubts, the most spectacular part of Level 2 is represented by the Racoviţă Meanders, a complex system of horizontal and vertical (up to 30 m high) meander loops on which walls beautiful geodes lined internally with calcite crystals are exposed (Fig. 4). Other major galleries at this horizon are: the Red I, Red II, and Lakes II galleries.

Level 3 This level has a limited extension and a discontinuous development, with the longest passages being mapped in the median section of the cave. At several locations, the galleries formed right above the other levels, but also in between them (M and MP galleries) or in areas where the lower tiers were not yet discovered.

Cave Climate

A 2-year climatological research program was undertaken between 1988 and 1990 by Onac and Racoviţă (1992) showing that the mean annual temperature and relative humidity in the cave's stable meroclimate are 9.5 °C and 99%, respectively. The temperatures measured three times a month revealed a unidirectional thermocirculation, the cave entrance acting as the lower opening. Winter ventilation has an ascendant character; thus, the cold air entering the cave promotes the development of ice speleothems (stalagmites, stalactites, draperies, and hexagonal crystals) within 100 m from the entrance. Along the first 400 m of the cave, a seasonal disturbance meroclimate is established, following the winter time influences of the external climate. In summertime (air moving from the upper levels toward the entrance), the same passage is characterized by highly stable thermo-hygrometric conditions. Nevertheless, a cold air (~9.8 °C) is blown out, hence the Vântului Cave's name. The seasonal dynamics of evapo-condensation is directly related to the general ventilation regime. The maximum evaporation rate was measured in December 1989 (40 mL/dm^2/10 days), whereas condensation reached its highest value (1.5 mL/dm^2/10 days) in July of the same year (Onac and Racoviţă 1992).

In a recent study aiming at monitoring the radon (Rn) concentrations along the first 800 m of passages (including 450 m along the first fossil level), it was found that the values perfectly reflect the seasonal ventilation pattern (Cucoş et al. 2016). Thus, during winter when fresh, cold air enters the cave, Rn values increase progressively from 410 Bq/m^3 at the entrance to 866 Bq/m^3 toward the end of the monitored section of the cave. The situation changes dramatically during the warm season when the average concentration is 2063 Bq/m^3, regardless of the station. Because this value translates into fairly high effective dose to which cavers are exposed, the study recommends that in order to prevent any health issues those entering in the cave should not spend more than 7–9 h/month during summer.

Speleothems and Minerals

Apart from its length and majestic passages morphology, the cave is renown for a wealth of minerals mostly forming crusts and earthy masses (Onac 1992, 1996, 2003). In fact, Vântului Cave is home for as many as 11 oxides and hydroxides (Table 1), one of them having their cave-type locality here (Iosof et al. 1974; Onac and Forti 2011). Among carbonates, calcite, aragonite, hydromagnesite, and monohydrocalcite were identified so far, but only the first two form a wide range of speleothems such stalactites, stalagmites, columns, flowstones, helictites, eccentrics, and pools. In a study targeting the origin of aragonite eccentrics, Viehmann (1975) concluded that the following climatic conditions need to be met: (1) minimal air circulation, (2) annual variation of cave air temperature to be less than 1 °C, and (3) negligible evaporation.

Gypsum is the only representative of the sulfate group, forming granular and fibrous crusts, anthodites (gypsum flowers), helictites, as well as prismatic and tabular crystals (Onac 1991). Gypsum is considered to be genetically related to SO_4^{2-}-rich solutions produced by oxidation of pyrite and marcasite (iron sulfides) in the overlying detrital formations (Coman 1979; Onac and Viehmann 1987). In several places within the cave, "gypsum balloons" have also been noticed (Fig. 5a). These speleothems often burst because of gypsum crystals growing beneath them.

Samples of gypsum flowers and oulopholites (single curved or contorted crystal; Fig. 5b) collected from the Metal Hall were used for a detailed crystallographic study, aiming to explain the origin of these particular speleothems (Ghergari and Onac 1995). It was found that the curvature of the gypsum crystals is controlled by both concentration of the calcium sulfate and its flow rate, which highly depends on how the solution reaches the cave (capillary or larger pores). The type of bedrock (compacted vs. fractured limestone) and the ventilation regime are also important.

The vast majority of the mineralogical treasure of Vântului Cave is hidden in some nondescript black crusts, and especially in earthy masses that cover gravels, boulders, and the cave walls along the underground stream (Fig. 6). In the riverbed, the black deposits have a jelly appearance and can reach up to 3–5 cm in thickness. Black-stained crusts also cover the cave walls in the upper levels of the cave providing indications on the location of former stream beds. The investigation of earthy masses with a combination of X-ray diffraction, X-ray fluorescence, and scanning electron microscope analysis led to the identification of the following manganese (Mn) and iron (Fe) oxides and hydroxides: braunite, hausmannite, hematite, hollandite, goethite, lepidocrocite, lithiophorite, pyrolusite, romanèchite, and todorokite (Coman and Crăciun 1978; Onac 1996, 1998; Diaconu and Morar 1997; Onac et al. 1997a). Subordinately, reddish brown to black nodules and short stalactites composed of hematite, goethite, lepidocrocite, braunite, and hausmannite were described by Diaconu and Morar (1997).

The deposition of black precipitate was studied by Onac et al. (1997b) who placed ceramic tiles at different depths in the water column and monitored them over a year period (Fig. 7), while also collecting pH, Eh, water temperature, and flow rate data once a month. The observations were conducted near the place called the "Torpedo." The results suggest that the amount of material accumulated decreases from water surface toward the stream floor concomitant with the lowering of Eh from 0.83 to 0.11 volts, while pH (~5) remained constant. The reason why the black sediments cannot exceed a certain thickness is because during wet periods, the stream water flows faster and removes it. The source of iron in the black deposits from Vântului Cave is related to the sulfides disseminated in the sandstones, microconglomerates, fireclay, and bauxitic and red residual clays that overlie the limestones. It is brought into the cave via seeping water or/and acid mine drainage along the underground stream originating from nearby mining area. Regarding manganese's origin, it could come either from some Fe-rich clays with up to 0.5% Mn or from decaying of

Table 1 List of minerals identified in speleothems and earthy masses from Vântului Cave

Chemical class	Mineral	Chemical formula[a]
Carbonates	Aragonite	$CaCO_3$
	Calcite	$CaCO_3$
	Hydromagnesite	$Mg_5(CO_3)_4(OH)_2 \cdot 4H_2O$
	Monohydrocalcite	$CaCO_3 \cdot H_2O$
Sulfates	Gypsum	$CaSO_4 \cdot 2H_2O$
Phosphates	Brushite	$Ca(PO_3OH) \cdot 2H_2O$
Oxides & hydroxides	Birnessite	$(Na, Ca, K)_{0.6}(Mn^{4+}, Mn^{3+})_2O_4 \cdot 1.5H_2O$
	Braunite[b]	$Mn^{2+}Mn^{3+}_6O_8SiO_4$
	Diaspore	$AlO(OH)$
	Gibbsite	$Al(OH)_3$
	Goethite	$FeO(OH)$
	Hausmannite	$Mn^{2+}Mn^{3+}_2O_4$
	Hematite	Fe_2O_3
	Hollandite	$Ba(Mn^{4+}_6, Mn^{3+}_2)_8O_{16}$
	Ice	H_2O
	Lepidocrocite	$Fe^{3+}O(OH)$
	Lithiophorite	$(Al, Li)(Mn^{4+}, Mn^{3+})_2O_2(OH)_2$
	Pyrolusite	MnO_2
	Romanèchite	$(Ba, H_2O)_2(Mn^{4+}, Mn^{3+})_5O_{10}$
	Todorokite	$(Na, Ca, K, Ba, Sr)_{1-x}(Mn, Mg, Al)_6O_{12} \cdot 3\text{-}4H_2O$
Silicates	Allophone	$Al_2O_3(SiO_2)1.3\text{–}2.0 \cdot 2.5\text{-}3H_2O$
	Halloysite-10 Å	$Al_2Si_2O_5(OH)_4 \cdot 2H_2O$
	Kaolinite	$Al_2Si_2O_5(OH)_4$
	Montmorillonite	$(Na,Ca)_{0.3}(Al,Mg)_2Si_4O_{10}(OH)_2 \cdot nH_2O$
	Saponite	$(Ca,Na)_{0.3}(Mg,Fe)_3(Si,Al)_4O_{10}(OH)_2 \cdot 4H_2O$

[a]According with the International Mineralogical Association List
[b]First documented in a cave environment worldwide

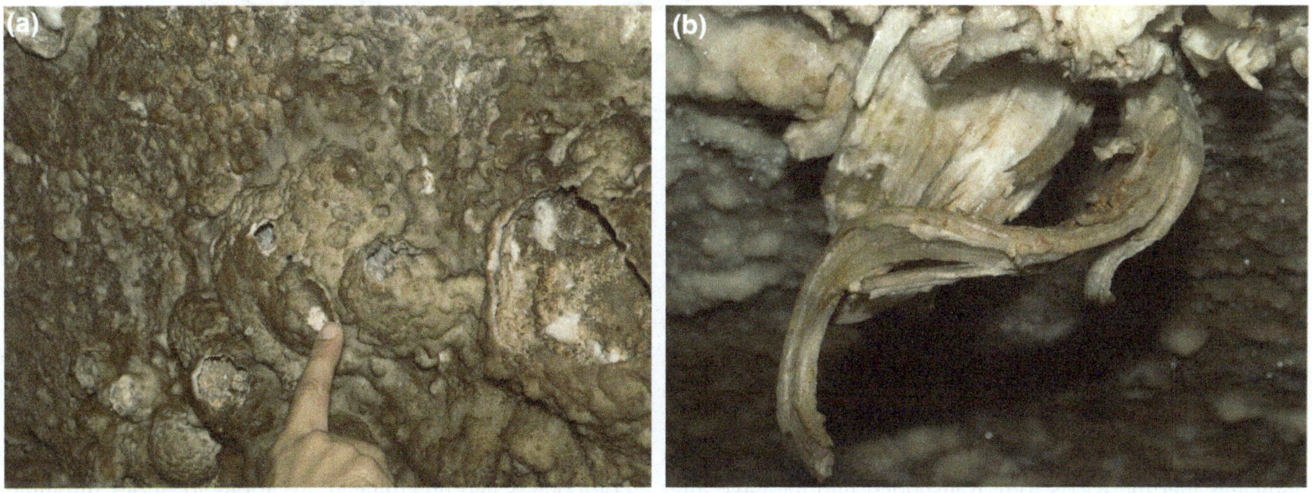

Fig. 5 A gypsum balloons and blisters and B oulopholite (photograph by BP Onac)

Fig. 6 Black ferromanganese deposits coating alluvial sediments and boulders (**a**) and cave wall (**b**) near the "Torpedo" (photograph by BP Onac)

Fig. 7 In-cave experimental setting for monitoring black ferromanganese deposition under different redox conditions (Eh, volts) in the water column (see text for more information)

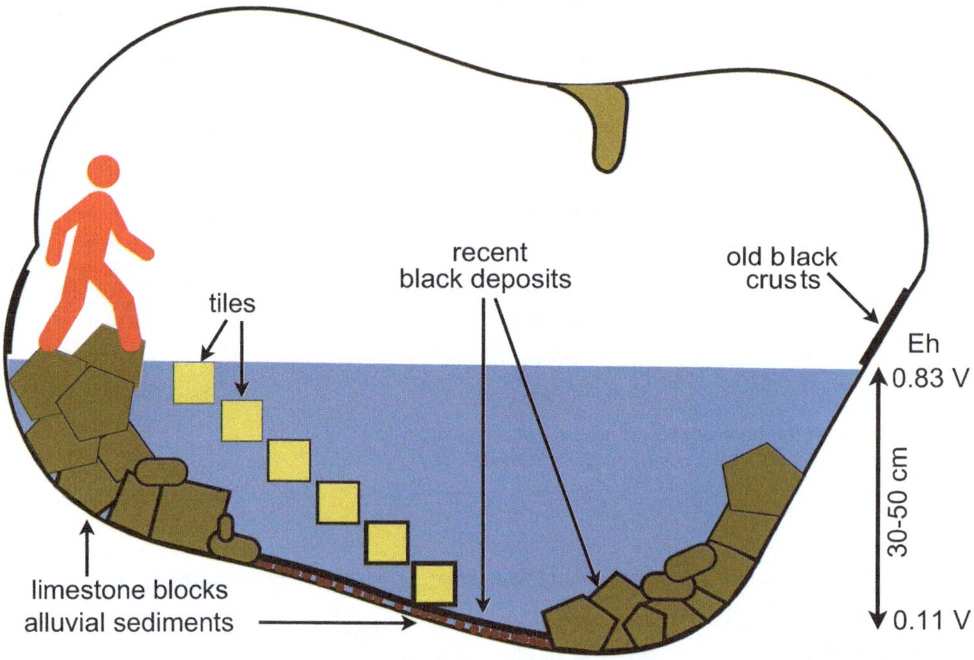

plants and other soluble/insoluble metallo-organic complexes. Microbiological activity was suggested to be responsible for the precipitation of at least few of the Fe–Mn-bearing minerals (Coman 1979).

Scanning electron microscope and secondary X-ray analyses performed on the black ferromanganese sediments show the material to have concentrated considerable amounts of rare earth elements (REEs) such as La, Ce, Sm, and Nd, in iron-rich spheres that build up botryoidal-like aggregates. The correlation of $^{143}Nd/^{144}Nd$ ratio for six different samples indicates that the REEs were transported into the cave after being leached from bauxitic and red residual clays from above the cave. Based on our observations, we conclude that an increase in pH resulted in adsorption of REEs onto the surface of ferromanganese minerals (Onac et al. 1997b).

Five silicate minerals were identified in the Vântului Cave (Table 1). Two of them, allophone and saponite, were described for the second time in the cave science literature (Iosof et al. 1974; Onac 1994). The presence of all silicates is intimately related to the fireclay deposits accumulated in the upper Triassic paleokarst (Onac and Popescu 1991). It is believed that K, Al, Na, Si, and Mg from clay minerals have been leached and transported by acidic waters in colloidal compounds and precipitated at various locations throughout the cave, when the solutions were buffered after interacting with limestone.

Geochronology

A short (46 cm tall) brownish stalagmite collected from the "1st of May" Gallery was dated by U-series disequilibrium method (Lauritzen and Onac 1995). Due to high contamination with ^{232}Th, the age obtained (24.6 kyr) showed large errors (±4 kyr). Nevertheless, the age was useful when collectively analyzing the results from several other caves in the Pădurea Craiului Mountains, an exercise that allowed reconstructing the active growing periods over the late Pleistocene (Onac 1996, 1998; Onac and Lauritzen 1996). Notably different is a new stalagmite sampled from the same cave location, which is much shorter (7.8 cm), but has well-developed flat growth layers. The preliminary U-series data indicate that it grew over three distinct intervals: 140–134 ka, 116–110 ka, and 2800 to present. A high-resolution oxygen and carbon stable isotope study is now under investigation (Moore et al. in prep).

A stalagmite from an undisclosed location in Vântului Cave was dated by means of thermoluminescence method (Labău et al. 1996), which measures the amount of natural dose accumulated by the calcite speleothem over its growth period (Aitken et al. 1968). The result indicated an age of 59,052 years, but the authors did not provide the error associated with this measurement.

Evidence of an early karstification stage has been recognized in the upper levels of Vântului Cave, where along sections of some galleries (e.g., "7th of November", Hippodrome) and in the Titans Hall, old karst voids and fractures are filled with limestone fragments cemented into a red clay matrix (Vremir and Damm 2000) (Fig. 8). The authors suggested an upper Triassic age based on the relationship between the paleokarst voids and the Ladinian limestone. Three clay samples collected from locations mentioned above were dated by K–Ar method at the Danish Lithosphere Center in Copenhagen (Denmark). The K–Ar ages range between 220 and 212 ± 4.5 Ma, confirming the upper Triassic, more exactly middle Norian age of the paleokarst infilling (Onac, in prep).

Fig. 8 Upper Triassic paleokarst infilling exposed along the Hippodrome (photograph by BP Onac)

Speleogenesis

Despite the fact Vântului Cave reached the top list of Romania's longest caves soon after its discovery, speleogenetic-dedicated studies are completely missing. In his karst morphology monograph, Bleahu (1974) included the cave among those having temporarily flooded passages, suggesting that it could be an epiphreatic system. Few years later, in an overview aiming to emphasize the biogeochemical complexity of karst, Coman (1979) proposed a sulfuric acid speleogenesis origin. Using examples from caves in North America, he argued that oxidation of pyrite produces sulfuric acid, which in turn dissolves limestone creating/enlarging underground voids (Coman 1984). To support his hypothesis, he invoked the presence of gypsum speleothems throughout the cave. Although this line of thinking is correct, it has been abundantly documented in recent years that pyrite oxidation (with very few exceptions) can only generate scattered porosity, not real large cave systems (Palmer 2007). Next, Cocean (1990) published a speleogenetic classification of caves and shafts in the Apuseni Mountains, in which Vântului Cave is included in the so-called lateral divergence type. Caves assigned to this group develop parallel to a slope, and their underground streams have tributaries resulting from laterally captured surface streams. The cave evolution occurs mainly under epiphreatic conditions and is controlled by the position of the water table. A similar interpretation, but which elaborates on pre- and post-Quaternary evolution of the cave, was put forward by Szilágyi Palkó et al. (2007).

Plugging all available information and morphological evidences into the most recent speleogenetic models reviewed by Palmer (2007) and Audra and Palmer (2013), it appears that the pattern of Vântului Cave passages when

Fig. 9 Spectacular meanders in the upper levels of Vântului Cave (photograph courtesy of AN Palmer)

seen in horizontal dimension is typical for a branchwork cave (Onac 2002b). The main and the most stable (in time) recharge point for the underground stream appears to be the Recea Ponor. However, throughout its long evolution, several right-side (east) tributaries fed by sinkholes or ponors (located at the karst/non-karst boundary) recharged and extend SE–NW mainly in the dip direction of strata. This setting was favored by the incision of Mişid Valley on the crest of an anticline of which left flank plunges toward WNW (Vălenaş and Iurkiewicz 1980–1981). The scallops on the cave walls of these side galleries are oriented in the direction of the mainstream supporting this assertion (Şerban and Domşa 1985). Apart from the recharge type and structural control (fractures and bedding planes), two other key elements played a major role in the genesis of the Vântului Cave. First, all sinking streams recharging the underground river collect highly aggressive water from the insoluble lower Jurassic rocks that cover the limestone. Second, the presence of the impervious siliciclastic caprock has allowed passages to increase in size (length and width) over time while protecting them from erosional destruction. Other evidences of an epiphreatic, periodically back-flooded development of Vântului Cave are: (1) multiple cave levels, (2) abundant stream-deposited sediments that reflect the geology of the overlaying non-karst rocks, (3) grain-size matches the local gallery gradient and scallop size, (4) no blind termination of passages, and (5) occurrence of speleothems is limited.

The most distinctive morphologic feature of the cave passages is the presence of meanders. These are well developed and displayed along most of the first level, but reaches a spectacular climax in the second level in the place known as the Racoviţă Meanders (Fig. 9). Şerban (1984) dedicated an entire study in an attempt to explain how both horizontal and vertical meanders formed along a section of the cave's first level. He emphasized an alternation of cross-sectional morphology of the primary conduit, from elliptical to canyon types; this suggests both vertical and horizontal undulations. The author considers this feature to be typical for karst drains in which pipe-full flow conditions may generate vertical meandering, analogous to the horizontal ones, which are characteristic to vadose flow conditions.

In contrast to this view, the author of this chapter believes that they are typical ingrowing meandering canyons as described by Ford and Williams (2007). The meanders were created by downward and downstream entrenchment, while often showing lateral wander if water encountered particularly penetrable bedding planes. In addition, it is also possible that during back-flooding events, the sediment accumulation may have forced upward dissolution of passages by paragenesis, overprinting any preexisting phreatic or vadose morphologies. The size and cross-sectional variability of the ingrowing meanders are likely controlled by different discharge regimes and by the position of the guiding elements (bedding plane or fracture).

Analyzing the longitudinal cross section through the cave between its entrance and Hippodrome, it becomes evident that the incision of Crişul Repede Valley controlled the underground drainage as the water table position changed, thus leading to the genesis of a typical base-level cave with four main levels. Each of these has a sinuous pattern and formed in the epiphreatic zone, on top of the water table; the connection between the various cave levels is via vadose canyons and shafts.

Microbiology

Without presenting microbiological evidences, Coman and Crăciun (1978) suggested that bacteria such as *Thiobacillus thiooxydans* and *T. ferrooxidans* may have played a role in the precipitation of black Fe–Mn deposits from Vântului Cave. This assertion was reiterated in another paper that discusses the karst as an biogeochemical environment (Coman 1984).

Using molecular methods to investigate the black sediments in the Vântului Cave, four species of bacteria and one of fungus have been identified (Manolache and Drăgan-Bularda 1994). Three of the bacterial species (*Hyphomicrobium* sp., *Pedomicrobium fusiforme, Pedomicrobium manganicum*) and *Cladosporium* sp. are known to mediate the oxidation and precipitation of manganese by enzymatic or nonenzymatic mechanisms in different environments. *Sphingomonas mali* could possibly be another bacterium that catalyzes manganese precipitation in the Vântului Cave (Manolache and Onac 2000). The fluctuation of the pH and/or Eh in the water column of the subterranean stream may control all these biologically mediated processes. A possible implication of these microorganisms in the retention of REEs within the black sediments is also possible and was previously suggested by Coman (1979) and Onac et al. (1997b).

Cave Fauna

The biospeleological investigations revealed interesting copepods of the Cyclopoida (*Acanthocyclops kieferi, Diacyclos bisetosus, Speocyclops troglodites, Acanthocyclops stygius deminutus*) and Harpacticoida: (*Paracamptus schmeili, Spelaeocamptus spelaeus*) collected from the underground stream interstitial and pools (Pleşa et al. 1964; Iepure and Oarga 2011), two species of Coleoptera (cave beetles): *Pholeuon (Parapholeuon) moczaryi* (Racoviţă 2011) and *Duvalius redtenbacheri almosi* (Moldovan 2002), Diplopoda: *Typhloiulus serbani unilineatus* (Nitzu et al. 2016), Nematoda: *Mylonchulus cavensis, Triply filicaudata, Tobrilus gracilis, Stenonchulus troglodytes*, Araneida: *Carpathonesticus biroi, Nesticus biroi*, Collembola: *Acherontides spelaea, Onychiurus granulosus minor, Lepidocyrtus serbicus*, Gastropoda: *Paladilhia transylvanica*, and Amphipoda: *Niphargus* sp. These species are rare and can be found only in certain parts of the cave.

Cave Conservancy

Vântului Cave is a protected area included in the "Defileul Crişul Repede—Pădurea Craiului" Natura 2000 site (ROSCI 0062). Except for the first 800 m of galleries (350 m on the stream passage and 450 m along the first fossil level), which have a B-class protection status, the rest of the cave is a scientific reserve (class A). Access to the cave requires authorization from the Speleological Heritage Commission and the Center for Protected Areas and Sustainable Development (Bihor), and it is permitted only in the presence of a guide from the CSA Cluj-Napoca, the custodian of the cave.

Acknowledgements CSA Cluj-Napoca is thanked for granting permission to enter the cave during my numerous research trips. Generations of cavers from this foremost Romanian speleological club shared with me countless hours in different parts of the cave in an effort to discover and understand the mineralogy and speleogenesis of this vast underground labyrinth. I am particularly grateful to my former students M. Vremir and T. Dascăl, as well as to many others who accompanied my 300+ professional trips in the Vântului Cave. I thank Dr. O.T. Moldovan for providing the list of cave fauna from the Biospeleological Database of the "Emil Racoviţă" Institute of Speleology in Cluj-Napoca.

References

Aitken MJ, Zimmerman DW, Fleming SJ (1968) Thermoluminescent dating of ancient pottery. Nature 219:442–445

Audra P, Palmer AN (2013) The vertical dimension of karst: controls of vertical cave patterns. In: Shroeder J (ed) Treatise of geomorphology. Academic Press, San Diego, pp 186–206

Bagameri B, Coman D, Toth K (1961) Szelek barlangja (Vântului Cave). Ed Tineretului, Bucureşti, p 85

Balintoni I (1997) Geotectonica terenurilor metamorfice din Romanian. Ed Carpatica, Cluj-Napoca

Bleahu MD (1974) Morfologia carstică. Ed Ştiinţifică, Bucureşti, p 590

Bleahu M, Decu V, Negrea S, Pleşa C. Povară I, Viehmann I (1976) Peşteri din Romania. Ed Ştiinţifică Enciclopedică, Bucureşti, p 415

Bleahu M, Patrulius D, Tomescu C, Bordea J, Panin S, Rădan S (1970) Date noi asupra depozitelor triasice din Munţii Apuseni. DS Inst Geol Geofiz LVI(4):29–41

Burchfiel BC, Bleahu M (1976) Geology of Romania. GSA Spec Paper 158:83

Cocean P (1979) Les voûtes planes-horizontales et le niveau de base karstique. Trav Inst Spéol "Émile Racovitza" 18:219–224

Cocean P (1990) Tripuri genetics de peşteri şi avene din carstul Munţilor Apuseni. Studia UBB Geogr 35(2):19–29

Cociuba I (2000) Upper Jurassic-Lower Cretaceous deposits in the south-western part of Pădurea Craiului. Formal lithostratigraphic units. Studia UBB Geologia XLV(2):33–61
Coman D (1979) Essai sur une interprétation écologique de l'origine des grottes. Trav Inst Spéol "Émile Racovitza" 18:191–199
Coman D (1984) Le karst—peysage biogéochimique. Theor Appl Karstol 1:23–28
Coman D, Crăciun V (1978) Peştera Vîntului. Ed Sport-Turism, Bucureşti, p 49
Cucoş (Dinu) A, Călugar MI, Burghele BD, Dumitru OA, Cosma C, Onac BP (2016) Radon levels in Romanian caves: an occupational survey survey. Environ Geochem Health 39(5):1085–1099. https://doi.org/10.1007/s10653-016-9878-1
Diaconu G, Morar M (1997) Nouveaux minéraux de fer et de manganèse dans la Grotte du Vent (Monts Padurea Craiului, Roumanie). Karstologia 29(1):51–53
Dragastan O, Diaconu M, Popa E, Damian R (1982) Biostratigraphy of the Triassic formations in the east of Pădurea Craiului Mountains. DS Inst Geol Geofiz LXVII(4):29–61
Ford DC, Williams P (2007) Karst hydrogeology and geomorphology. Wiley, Chichester, p 562
Ianovici V, Borcoş M, Bleahu M, Patrulius D, Lupu M, Dimitrescu R, Savu H (1976) Geologia Munţilor Apuseni. Ed Acad RSR, Bucureşti, p 631
Iepure S, Oarga A (2011) A new *Acanthocyclops* Kiefer, 1927 (Copepoda: Cyclopidae) from caves in Apuseni mountains (North–Western Romania). Ann Zool 61:427–438
Iosof V, Coman D, Ianc R (1974) Note sur la présence de l'allophane dans la grotte du Vent des Monts Apuseni. Trav Inst Spéol "Émile Racovitza" 13:183–189
Labău V, Gaspar E, Păunica T (1996) Speleothems dating using the thermoluminiscence method. Theor Appl Karstol 9:29–34
Lauritzen S-E, Onac BP (1995) Uranium series dating of some speleothems from Romania. Theor Appl Karstol 8:25–36
Manolache E, Drăgan-Bularda M (1994) Microbiological analysis of some clay samples from Vântului Cave. An Univ Oradea, Biologie 1:57–61
Manolache E, Onac BP (2000) Geomicrobiology of black sediments in Vântului Cave (Romania): preliminary results. Cave Karst Sci 27 (3):109–112
Moldovan O (2002) The beetles. In: Racoviţă G, Moldovan O, Onac BP (eds) The karst of Pădurea Craiului Mountains. Monographic study. Ed Presa Univ Clujeană, Cluj-Napoca, pp 165–171
Nitzu E, Giurginca A, Nae A, Popa I, Baba Ş, Meleg IN, Vlaicu M (2016) The catalogue of caves with endemic cavernicolous arthropod fauna of Romania. Trav Inst Spéol "Émile Racovitza" LV:3–62
Onac BP (1991) New data on some gypsum spelothems in the Vântului (Pădurea Craiului mountains) and Răstoci (Somesan plateau) caves. Trav Inst Spéol "Émile Racovitza" 30:189–193
Onac BP (1992) Mineralogy of the Apuseni Mountains caves. Theor Appl Karstol 5:193–201
Onac BP (1994) Peştera Vîntului (Vântului Cave). Theor Appl Karstol Program-Abstracts and Excursion, Băile Felix, Romania, pp 52–56
Onac BP (1996) Mineralogy of speleothems from caves in the Padurea Craiului Mountains (Romania), and their palaeoclimatic significance. Cave Karst Sci 23(3):109–124
Onac BP (1998) Speleothems from caves in Pădurea Craiului Mountains. Ed Acad Române, Bucuresti, p 176 (in Romanian with English abstract)
Onac BP (2003) Checklist of the Romanian cave minerals. Theor Appl Karstol 16:83–89
Onac BP (2002a) Exokarst (hydrogeologic settings, speleogenetic models, cave deposits). In: Racoviţă G, Moldovan O, Onac BP (eds) The karst of Pădurea Craiului Mountains. Monographic study. Ed Presa Univ Clujeană, Cluj-Napoca, pp 18–34
Onac BP (2002b) Endokarst (hydrogeologic settings, speleogenetic models, cave deposits). In: Racoviţă G, Moldovan O, Onac BP (eds) The karst of Pădurea Craiului Mountains. Monographic study. Ed Presa Univ Clujeană, Cluj-Napoca, pp 35–92
Onac BP, Forti P (2011) State of the art and challenges in cave minerals studies. Studia UBB Geologia 56(1):33–42
Onac BP, Lauritzen S-E (1996) The climate of the last 150,000 years recorded in speleothems: preliminary results from North–Western Romania. Theor Appl Karstol 9:9–21
Onac BP, Popescu V (1991) Relations between the overall and local tectonic pattern of the Suncuius deposit and the paleokarst at its bottom. Trav Inst Spéol "Émile Racovitza" 30:185–187
Onac BP, Racoviţă G (1992) Recherches climatologiques dans la Peştera Vântului (Monts Pădurea Craiului). Theor Appl Karstol 5:167–192
Onac BP, Viehmann I (1987) Origines et formes d'apparition du gypse dans Pestera Vintului (Monts Padurea Craiului). Theor Appl Karstol 3:243–245
Onac BP, Pedersen RB, Tysseland M (1997a) Presence of rare-earth elements in black ferromanganese coatings from Vântului Cave (Romania). J Caves Karst Stud 59(3):128–131
Onac BP, Tysseland M, Bengeanu M, Hofenpradli A (1997b) Deposition of black manganese and iron-rich sediments in Vântului Cave (Romania). In: Proceedings of the 12th International Congress of Speleology, La Chaux-de-Fonds 1:235–238
Orăşeanu I (1991) Hydrogeological map of the Pădurea Craiului Mountains (Romania). Theor Appl Karstol 4:97–127
Orăşeanu I (2010) Karst hydrogeology of Pădurea Craiului Mountains. In: Orăşeanu I, Iurkiewicz A (eds) Karst hydrogeology of Romania. Ed Belvedere, Oradea, pp 199–215
Orăşeanu I, Gaşpar E (1980–1981) Cercetări cu trasori radioactivi privind stabilirea zonei de alimentare a cursului subteran din Peştera Vîntului (Munţii Pădurea Craiului). Nymphaea 8–9:379–386
Orăşeanu I, Iurkiewicz A (1987) Hydrological karst systems in Pădurea Craiului Mountains. Theor Appl Karstol 3:215–222
Palmer AN (2007) Cave geology. Cave Books, Dayton, p 454
Papiu CV, Iosof V, Rădan S (1988) Constitution and genesis of the lower Jurassic fireclay formation of the Padurea Craiului massif. DS Inst Geol Geofiz 72–73:73–89
Patrulius D, Popa E, Cîmpeanu Ş, Orăşanu T (1973) Geologic map of Romania, Remeţi Sheet (41a). Scale 1:50.000, L-34-46-A. Inst Geol, Bucureşti
Pleşa C, Botea F, Racoviţă G (1964) Cercetări asupra faunei biotopurilor acvatice subterane din bazinul Crişului Repede. I. Valea Mişidului şi afluenţi. Lucr Inst Speol "Emil Racoviţă" 3:367–396
Popa E (1981) La biostratigraphie des formations mésozoiques de la partie orientale de Pădurea Craiului (Monts Apuseni). An Inst Geol Geofiz LVIII:203–282
Racoviţă G (2011) Révision systématique des Leptodirinae souterrains des Monts Apuseni. VIII. Aperçu synthétique sur le genre *Pholeuon*. Trav Inst Spéol "Émile Racovitza" L:37–59
Săndulescu M (1984) Geotectonica Romaniei. Ed Tehnică, Bucureşti, p 336
Szilágy A, Kömives E, Nagy I, Varga A, Kerekes K (1979) "Peştera Vîntului". Trav Inst Spéol "Émile Racovitza" 18:259–266
Szilágyi Palkó P, Deák Z, Dascăl T (eds) (2007) Peştera Vântului 50 ani. Delroti SRL, Cluj-Napoca, p 112
Şerban M (1984) Sur les méandres de la Peştera Vintului (Monts Padurea Craiului). Theor Appl Karstol 1:107–115
Şerban M, Domşa M (1985) Sur le microrelief de corrosion de la Peştera Vântului (Monts Padurea Craiului, Roumanie) et la

morphogénèse de la voute plane dans les conduites forcées. Theor Appl Karstol 2:97–121

Vălenaş L, Iurkiewicz A (1980–1981) Studiu complex al carstului din zona Şuncuiuş-Mişid (Munţii Pădurea Craiului). Nymphaea 8–9:311–378

Viehmann I (1975) Quelques considérations sur l'aragonite et les cristallisations excentriques des grottes. Trav Inst Spéol "Émile Racovitza" 14:141–147

Vremir M, Damm P (2000) Late Triassic palaeoendokarst network in Vantului Cave (Suncuius, Padurea Craiului Mountains, Romania). In: Onac BP, Tămaş T (eds) Karst Studies and Problems: 2000 and Beyond. Ed Presa Univ Clujeană, Cluj-Napoca, pp 156–157

http://www.pesteravantului.ro/index.html

Somes Plateau: Măgurici Cave (Pestera de la Măgurici de la Răstoci)

Bogdan P. Onac and Tudor Tămaş

Abstract

Măgurici Cave in the Someş Plateau of NW Romania hosts a large deposit of bat guano. Pollen and stable isotope investigations on this cave material provide meaningful information regarding the paleoclimate and paleoenvironmental conditions over the past 1200 years. Apart for some gypsum crystals, the cave is deprived of spectacular speleothems, but it contains nodules, crusts, and earthy masses in whose composition eight phosphate and four sulfate minerals were identified.

Keywords

Bat guano • Cave minerals • Speleothems
Paleoclimate • Someş Plateau • Romania

Geographic, Geologic, and Hydrologic Settings

The Someş Plateau in northwestern Romania is an extensive geomorphologic unit that hosts a variety of karst features, developed mainly on Paleogene (Eocene and Oligocene) to Miocene (Badenian and Sarmatian) carbonate rocks (Ghiurcă 1970; Giurgiu et al. 1983; Todoran and Onac 1987; Onac et al. 2010; Tămaş and Ungureanu 2010). Măgurici Cave (543 m in length), also known as "The Cave on the Meadow" (Peştera de pe Poiană), is located in the north-eastern part of this unit, more precisely in the southeastern corner of the Purcăreț–Boiu Mare karst plateau (Fig. 1), at ~2 km east of Răstoci Village (Baboş 1981; Giurgiu et al. 1983; Onac and Todoran 1987).

The geology of the area consists of a thick unfolded sedimentary sequence that forms a large monocline gently dipping (<20°) towards south and east (Rusu 1977). The main karst forming lithology is represented by a 40–60-m-thick reefal-bioclastic unit of upper Eocene–lower Oligocene age known as the Cozla Limestone (Bucur et al. 1989; Prică 2001). A biostratigraphic study conducted in Măgurici Cave and few other locations nearby established the Eocene/Oligocene boundary within the Cozla Limestone to lie immediately above a *Nummulites fabianii*-rich stratum (Todoran and Onac 1989). Since this horizon is easily distinguished inside caves, it allows for quickly assigning the age of the rock in which the cave formed. Based on this research, it was concluded that Măgurici Cave develops along the Eocene/Oligocene limit.

The Oligocene succession also includes two other carbonate sequences, namely the Cuciulat Formation comprising an alternation of medium-bedded limestone (2–3 m), marl, sandstone, and coal and Bizuşa Formation (calcareous marls), both hosting significant karst features (Onac et al. 1989). These two units are overlain by the Ileanda Formation (upper part of the lower Oligocene), a thick (50–60 m), yet finely laminated and fissile bituminous shale deposit. Efflorescences of native sulfur, gypsum, and jarosite develop between and along bedding planes (Ghergari et al. 1989; Tămaş and Ungureanu 2010), and their presence are critical for the deposition of gypsum in Măgurici Cave (Onac 1991).

Presently, the cave has no permanent stream. A small, temporary active stream originates in a spring situated just above the cave entrance, from where it reaches to the Entrance Passage via a 6-m waterfall. After a subaerial route

The authors contributed equally to this work.

B. P. Onac (✉)
Karst Research Group, School of Geosciences, University of South Florida, 4202 E. Fowler Ave., NES 107, Tampa, FL 33620, USA
e-mail: bonac@usf.edu

B. P. Onac
"Emil Racoviță" Institute of Speleology, Romanian Academy, Clinicilor 5, 400006 Cluj-Napoca, Romania

T. Tămaş
Department of Geology, Babeş-Bolyai University, Kogălniceanu 1, 400084 Cluj-Napoca, Romania
e-mail: tudor.tamas@ubbcluj.ro

G. M. L. Ponta and B. P. Onac (eds.), *Cave and Karst Systems of Romania*, Cave and Karst Systems of the World, https://doi.org/10.1007/978-3-319-90747-5_47

Fig. 1 Location of Măgurici Cave in NE part of the Someş Plateau

of about 20 m, it disappears under the scree deposited along the Entrance Passage and its route can be traced along the Lateral Passage until a small inlet at the base of its right, eastern wall. The only existing surface creek, which passes nearby the entrance was probably responsible for the formation of the cave in the past. Now it loses its water in a ponor just east of the cave entrance and resurfaces in the Izbucul de la Linie Spring, which reportedly has a permanent and constant discharge of 50 l/s (Baboş and Mureşan 1981).

History of Exploration and Research

Although the Măgurici Cave have been known to locals for a long time, it has only been explored and surveyed between 1978 and 1979 by the members of the "Emil Racoviţă" Speleological Club from Bucharest. Apparently, the cave was rediscovered and remapped in December 1979 by the Cepromin Caving Club in Cluj-Napoca (Baboş 1981; Giurgiu et al. 1983). Both teams reported similar lengths for the cave (507 m), but a slightly different vertical range. In 1985, one of the authors (BPO) initiated an extensive scientific campaign aiming to characterize the mineralogy of Măgurici Cave (see speleothem section below). In parallel, limestone samples for nannoplankton analysis were collected within the cave and used to define the Eocene/Oligocene boundary (Todoran and Onac 1989). Between 2002 and 2003, members of the "Emil Racoviţă" Speleological Institute in Cluj conducted a monthly cave climate monitoring and during the same period, the Montana Caving Club from Baia Mare remapped the cave to its current length (543 m). In the following years, a series of investigations targeting the large guano deposit in the Circular Room (Sala Circulară) begun and are currently still ongoing.

Cave Description

The entrance in Măgurici Cave opens at 319 m absolute altitude at the bottom of a ponor, through a portal of 2.5 × 4 m. The Entrance Passage (Galeria de Intrare) is a narrow (0.5–1 m) and high (6–7 m, then over 15 m in its downstream part) fissure oriented W-E on which the only morphologic feature is a 3 m drop, 20 m from the entrance

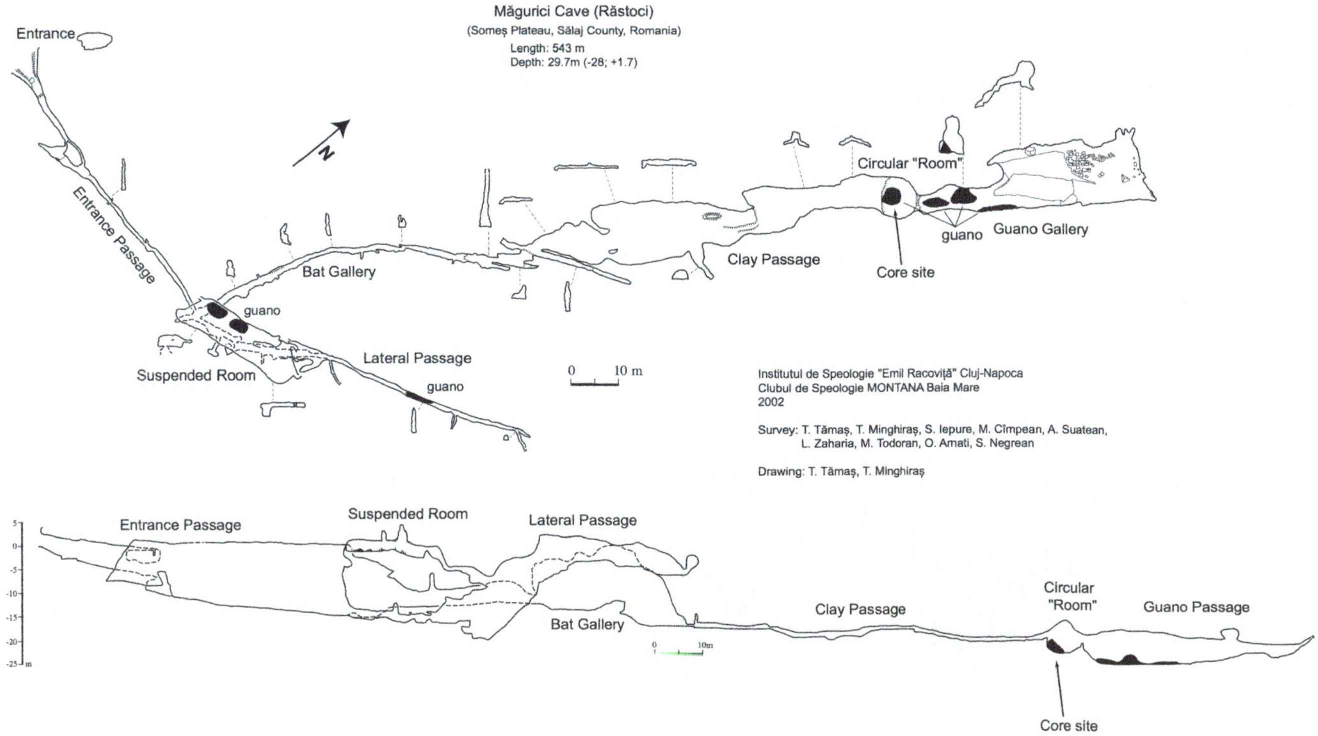

Fig. 2 Map of Măgurici Cave (used with permission from "Montana" Baia Mare Caving Club)

(Fig. 2). The end of this gallery opens in a larger space, the Bifurcation Room (Sala Bifurcației), from where a narrow/crawl passage leads into the Lateral Passage (Galeria Laterală), which continues in the same direction as the Entrance one, and may either continue along its floor or climb on a breakdown pile 15 m towards the Suspended Room (Sala Suspendată). The first half of the Lateral Passage includes a series of crawls under breakdown and is presently the route of the temporary stream flowing from the entrance. The Suspended Room, at the very top of the Lateral Passage, is an enlarged passage section along a bedding plane and is the part of the cave which is presently the closest to the surface. Past the Suspended Room, the Lateral Passage continues on for some 80 m. The temporary stream disappears under the right wall at about half the distance, into a tiny joint with air current to the east. The far end of the Lateral Passage, at about the same elevation as the Suspended Room, is reached after a 10-m-free climb on a slippery slope. Its last part of about 30 m is developed along a narrow canyon barely 0.5 m large and contains two guano accumulations.

The Bat Gallery (Galeria Liliecilor), where a bat hibernation colony is usually located during winter, is another high and narrow fissure type passage, 80 m long, that begins to the left of the Bifurcation Room and has a general N, then NE direction. There are two ways to advance along this gallery: 40 m from the Bifurcation Room continue straightforward through a 0.23 m wide tight section or climb 2–3 m above the cave floor and move forward along the fissure until reaching the other side of the constriction. At this point, a 3.5-m drop marks the entrance in the Clay Passage (Galeria cu Argilă). Between the Bifurcation Room and this point, the cave walls are covered with gypsum crusts and what once were some beautiful gypsum flowers.

Morphologically, the Clay Passage and the Guano Passage (together accounting for ~200 m) are significantly different from the rest of the cave in that the gallery develops along bedding planes, reaches up to 10 m in width and 7 m in height in its final part and hosts large accumulations of finely laminated, yellowish-brown to brown silt and clay deposits. The rich fossil association (corals, bivalves, gastropods, and foraminifera) of the Cozla Limestone is wonderfully exposed on walls and ceiling in this section of the cave. Crystals and aggregates of gypsum are abundant in the clay deposit that fills almost completely the gallery, requiring tight crawls. A particular dissolution feature is the sinuous ceiling channel (Fig. 3a) that can be observed all along the Clay Passage and into the Guano Gallery (Galeria cu Guano), where bats congregate along it (Fig. 3b). Its presence suggests an upward dissolution of the ceiling, a karst process known as paragenesis (not to be confused with mineral paragenesis) that happens when a cave passage is filled with sediments and water is forced to flow near the ceiling.

Fig. 3 Ceiling channel along the Clay Passage (**a**) and in the Guano Passage (**b**) where bats congregate along it (photograph by BP Onac)

The Clay Passage ends in the "Circular Room" (5 by 5 m), which hosts the largest and thickest (3 m) guano mound in the cave. Clay covers the floor of this hall and it appears to be interbedded with guano, at least at one level in the lower part of this organic deposit (Cleary et al. 2017). Based on the morphology of the Circular Room and how it connects to the Guano Gallery (via a +2-m step), we believe this was at one point the place where the underground river sunk. Further to the east, the cave continues with the Guano Gallery (Fig. 4) along which guano accumulates in three more locations and towards its end the floor is covered by large breakdown limestone blocks.

Speleogenesis

All morphological characteristics point to Măgurici Cave being dominantly formed by floodwater, which was periodically filling all opened passages (during snow melting and periods with high precipitation), while for the rest of the time vadose flow prevailed. The Entrance and the Lateral Passage, as well as the Bat Passage, have similar cross sections in the shapes of thin, high phreatic canyons, ending in blind pockets along joints, which in many places show obvious signs of sediment fill and enlargement by paragenesis. The few short lateral passages exhibit similar features. The sediment fillings are exclusively silts and clays, and they are commonly found suspended up to 5 m in some spots along the Bat Passage that were sheltered from subsequent flooding. At present, the fine sediments are also noticeable in the final part of the Lateral Passage, where they are several meters thick, whereas in the Suspended Room, a very interesting unsupported flowstone deposit attests for a former sediment fill period.

The Clay Passage was formed by floodwater circulating along the bedding plane. Subsequent breakdown along bedding planes, successive sediment fills, and ceiling enlargement through paragenesis contributed to the present shape of the passage. The ceiling channel in the Clay Passage as well as a sediment mound located just below a ceiling cupola next to it attests that this part of the cave was completely filled during periodic flooding events. This is also sustained by the sediment grain size, as only silt and clay are found in this part of the cave. At a later stage, an underground "sink" opened in the floor of the Guano Passage, and the sediments previously deposited there were eroded backward until the Circular Room, where they form a 1–1.5 m drop that is still visible today.

The Suspended Room is also formed along the bedding planes in the Eocene limestones, which were intersected by the joints controlling the orientation of the Entrance/Lateral Passages and the Bat Passage. Subsequent breakdown along the same bedding planes resulted in its present matchbox shape and several joints in the ceiling facilitated seepage water from the surface from which speleothems were deposited. As vadose flow conditions prevailed for some time along the Entrance Passage, all previous infilling was eroded down to the present floor, below the level of the Bat Passage, which at that stage probably only received water during flash floods. Most of the sediment previously filling the Lateral Passage was also eroded then, as the cave water followed the present path, towards the bottom of the Lateral Passage and into the joint eastern wall. Thus, another drop formed on sediments at the beginning of the Bat Passage,

Fig. 4 Guano accumulations along the Guano Gallery (photograph by T. Tămaş)

blocking cold winter air to proceed further. This allowed a stable climate zone to be establish between the first part of the Bat Passage and the Clay and Guano Passages, along the longer, northeastern branch of the cave.

Cave Climate

Temperature and relative humidity of the cave atmosphere were recorded monthly between February 2001 and March 2002 in 14 stations throughout the cave (Borda and Racoviţă 2000–2001; Borda et al. 2004). The first six locations spread along the Entrance Passage, three on the Bat Gallery, three on the Clay Passage, and one in the Circular Room and Guano Gallery, respectively. Data suggest that during the warm season, temperatures decrease between station 1 (10.94 °C) and 8 (7.76 °C) and increase thereafter to 11.75 °C. This trend reverses in winter when cold air enters the cave and progressively warms up over the first half of the cave (Station 1: 3.37 °C; Station 8: 5.22 °C), remaining constant after that. This implies that Măgurici Cave is characterized by two meroclimatic zones: the *perturbation* sector between the cave entrance and station 8 where a permanent bidirectional ventilation cell exist, and the *stable* zone (station 9–15) in which temperatures and relative humidity stay constant throughout the year (between 10 and 11.78 ± 0.25 °C and 94–98%; Borda and Racoviţă 2000–2001).

Speleothems and Minerals

Common calcite speleothems such as stalactites, stalagmites, and flowstones are very few in the cave. This is most likely due to the fact that the Eocene limestones consist of metric (submetric) banks separated by thin levels of detrital material; therefore, the seepage water does not reach the cave passages (Johnston et al. 2010). Calcite dripwater speleothems are restricted to two spots: the Bifurcation Room, where a massive calcite flowstone covers the southern wall, and the Suspended Room. The best display of calcite speleothems in the cave is in the Suspended Room where stalactites, stalagmites, and flowstones are abundant. Most, if not all of them are presently in a degrading state, as they are corroded by seepage water and/or by the acid aerosols related to the guano deposit located nearby.

Gypsum is the second most abundant speleothem-forming mineral in caves after calcite (Onac 2012). In Măgurici, gypsum is by far the most common presence occurring as crusts, crystals, gypsum flowers (anthodites; Fig. 5), and aggregates. Among these, very interesting are the so-called starburst crusts, which consist of gypsum crystals radiating away from a central point and growing parallel to the cave wall. It is the only location in Romania where this speleothem was encountered (Onac 1991). Large aggregates (up 15 cm in length) made of euhedral and subhedral gypsum crystals grow over or within the clay

Fig. 5 Cave wall covered by gypsum crust and flowers (photograph by T. Tămaş)

deposits accumulated along the Clay Passage. The X-ray diffraction analysis of some delicate fibrous crystals collected from the Ascending Gallery proved to be mirabilite, a mineral previously known only from Tăuşoare Cave in Romania. Later, it was also identified in association with ardealite in the Guano Passage. Two other rare sulfates (cesanite and bassanite) were also identified in association with phosphate minerals (Onac and Vereş 2003). Deposition of sulfate minerals in Măgurici Cave is related to the presence of sulfur and gypsum in the Ileanda Formation (Onac 1991). When found near bat guano deposits, the sulfates could be a by-product of the leaching of bat guano.

The chemical reactions taking place at the contact between guano and limestone or clay minerals generated an interesting phosphate mineral assemblage (Table 1). All these minerals occur as small crystals (<0.5 mm) on the surface of guano mounds or dispersed within them, coatings, and especially, unspectacular earthy masses surrounding the guano deposits. A thorough investigation conducted by Onac and Vereş (2003) showed that pH, relative humidity, and Ca/P ratio are the main parameters responsible for the deposition of this rich mineral paragenesis. The authors also reported the presence of phosphammite, Măgurici Cave being the second location worldwide in which this mineral was identified.

Cave Fauna and Air Microorganisms

Except for two beetles (*Catops* sp. and *Pholeuon* sp.) found along the Clay Gallery, no other biospeleological investigations were yet conducted in Măgurici Cave.

Table 1 List of phosphate minerals identified in Măgurici Cave

Mineral	Chemical formula*	Speleothem/abundance
Ardealite	$Ca_2(PO_3OH)(SO_4)\cdot 4H_2O$	Earthy masses/common
Brushite	$Ca(PO_3OH)\cdot 2H_2O$	Earthy masses/very common
Monetite	$Ca(PO_3OH)$	Earthy masses/rare
Hydroxylapatite	$Ca_5(PO_4)_3(OH)$	Crust/very common
Fluorapatite	$Ca_5(PO_4)_3F$	Crust/rare
Taranakite	$K_3Al_5(PO_3OH)_6(PO_4)_2\cdot 18H_2O$	Earthy masses/very common
Francoanellite	$K_3Al_5(PO_3OH)_6(PO_4)_2\cdot 12H_2O$	Earthy masses/rare
Phosphammite	$(NH_4)_2(PO_3OH)$	Crystals/rare

*According to the Commission on New Minerals and Mineral Names (http://nrmima.nrm.se)

The dominant bat species in Măgurici Cave are *Myotis myotis*, *Myotis blythii*, and *Miniopterus schreibersii.* Of these, the first two gather between April and end of August in large maternity colonies (hundreds to one thousand bats) located in the Suspended and Circular rooms (Borda et al. 2004). *M. myotis* forage ~5 km from its roosting site over deciduous woodlands, whereas *M. blythii* more commonly forage in grassland habitats. Apparently, *Miniopterus schreibersii* only enters the cave in autumn for mating. Few *Rhinolophus ferrumequinum* and *R. hipposideros* were also observed during winter along the Bat Gallery, where the major hibernation colony consists of *M. myotis* and *M. blythii.*

Seasonal measurements of airborne microorganisms were conducted in two stations near the nursery and mating bat colonies forming in the Suspended Room and Guano Gallery, respectively (Borda et al. 2004). At each location, five various culture media (i.e., agar, Levine, Chapmann, Holmes, and Sabouraud) pre-poured on Petri dishes were exposed for 15 min. Collectively, the results show that the highest values for all microorganisms (streptococci, staphylococci, fungi, etc.) occurred in the warm season nearby the nursery colony (Borda et al. 2004, 2014).

Guano and Paleoclimate

The most important scientific resource in Măgurici Cave is by far the large guano deposit accumulated in the Circular Room. Bats are feeding on insects whose dietary preferences reflect the vegetation in the foraging area, of which distribution is controlled by the local/regional climate. Given this association, Des Marais et al. (1980) first suggested guano might have paleoecological significance. The thickness of the guano heap in the Circular Room attracted the attention of scientists who applied different techniques to recover paleoenvironmental and paleo-hydroclimate information from it. One such study investigated a 2.7 m long guano core, which was dated using the radiocarbon method. It has been found that accumulation of guano began in the second part of the Medieval Warm Period (MWP), around AD 1195, ceased for almost 400 years during the cold interval known as the Little Ice Age (LIA), and resumed at increasing rates sometimes around AD 1700 (Johnston et al. 2010). A large peak in the radiocarbon activity found within the first 50 cm of the core was positively correlated with the nuclear weapons testing in the atmosphere during the 1950s and 60s. The study also attempted, for the first time in the world (in bat guano), to use ^{36}Cl as a possible solar irradiance proxy, but because of its post-depositional mobility, the method appears to have severe limitations.

Using the same core, Geantă et al. (2012) performed a detailed palynological investigation. Overall, the pollen assemblage is typical for the Subatlantic period (the last 2500 years of the Holocene), which in Romania is dominated by the presence of *Fagus sylvatica* (common beech tree). However, during the MWP, herbaceous taxa were well underrepresented compared to mesothermophilous taxa (*Quercus*, *Tilia*, and *Fraxinus*) suggesting dense forest nearby the cave. A marked decrease of tree pollen (especially the thermophilous species) concomitant with an increase of *F. sylvatica* was noted during the LIA. After AD 1810, the pollen assemblage resembles modern vegetation, but also contains indications of human impact, such as deforestation, agriculture, and grazing.

A more recent study conducted on a newer and slightly longer guano core (286 cm) suggests that the nitrogen isotope ratio (δ^{15}N) values likely reflect winter precipitation which are related to nitrogen mineralization prior to the growing season (Cleary et al. 2017). Using the relationship between the North Atlantic Oscillation (NAO) index and δ^{15}N values in the Măgurici guano for the instrumental period, the authors reconstructed wet or dry NAO-like phases back to AD 1650.

Geochronology and Past Human Activity in the Cave

Other than the guano, two more samples have provided chronological data. A broken stalagmite was recovered from the crawls at the bottom of the Lateral Passage, and due to its location was supposed to have fallen there from the Suspended Room. U–Th dating by alpha spectrometry provided the three ages, separated by growth hiatuses (Table 2). These place the onset of the stalagmite deposition in the marine isotope stage (MIS) 7, with a growth interruption during MIS 6. Calcite deposition resumed at about 122 ky BP, during MIS 5 and continued with at least one more interruption until 90 ky BP. The data obtained for the stalagmite base, 232 ky BP, set a minimum time for the breakdown which contributed to the actual shape of the Suspended Room before that period, most likely during MIS 8 (Table 2).

Incidentally, the reason the stalagmite was broken may be placed in time by another date: several pieces of wood caught underneath a boulder choke were noticed by cavers on a climb near the top of the Lateral Passage, next to the crawl leading into the Suspended Room. Unlike tree branches brought in the cave by natural causes, these pieces of wood were shaped as planks and therefore point to possible human activity in the past. Radiocarbon dating of one of these wood samples recovered from the choke gave an age of 378 ± 30 cal years BP. It is unclear why the planks were there and why were the locals visiting such a difficult spot (involving serious crawling and climbing), considering they

Table 2 U–Th ages and isotope ratios obtained on the stalagmite from Măgurici Cave

Sample	U (ppm) ± 1ss	$^{234}U/^{238}Ua$ ± 1ss	$^{234}U/^{238}Ui$ ± 1ss	$^{230}Th/^{234}U$ ± 1ss	$^{230}Th/^{232}Th$	Age (kyr BP)	+1ss (kyr)	−1ss (kyr)
Rm 0-a	14.57	1.0108 ± 0.011	1.021 ± 0.022	0.8844 ± 0.023	2347.6	**231.82**	23.46	19.3
Rm-2	6.225	1.0393 ± 0.008	1.055 ± 0.011	0.6794 ± 0.011	10 000	**122.09**	3.71	3.59
Rm-3	6.066	1.1415 ± 0.008	1.182 ± 0.011	0.5721 ± 0.01	607.2	**90.04**	2.61	2.55

could reach that section of the cave only from the present-day entrance. The fairest assumption would be that they mined guano for agriculture. So the stalagmite recovered from the Lateral Passage may have fallen either due to human activity or to the (more recent) breakdown that caught the planks underneath.

One other, later sign of human visits is an inscription carved on the wall of the Bat Passage, next to the 0.23 m crawl. It has a religious connotation and dates probably from World War 2.

Cave Conservancy

A detailed report, including much of the previous research mentioned in the text except for the more recent paleoclimatic data, was released by the members of the Institute of Speleology from Cluj-Napoca in order to document the scientific importance of the cave and the need to protect the bat population (Onac et al. 2004). At present, based on that documentation, the cave is a category B restricted site and visits are completely forbidden during bat roost period. The cave, nominated as Natura2000 site ROSCI0192, was declared a natural monument through the Government decision no. 2151/2004.

Acknowledgements Dr. O. Moldovan of the "Emil Racoviță" Institute of Speleology is thanked for kindly providing information on the cave fauna reported in this chapter.

References

Baboş R (1981) Sinteză a rezultatelor cercetărilor de speologie fizică în zona carstică Răstoci-Mesteacăn. Bul Speo 5:72–92

Baboş R, Mureşan M (1981) Zona carstică Mesteacăn-Răstoci. Stud Cerc Speol Rîmnicu Vîlcea 1:17–30

Borda D, Racoviță G (2000–2001) Données thermo-hygrométriques sur la Grotte de Ciungi et la Grotte de Măgurici (Plateau du Some, Transylvanie). Trav Inst Spéol "Émile Racovitza" XXXIX–XL:207–226

Borda D, Borda C, Tămaş T (2004) Bats, climate and air microorganisms in a Romanian Cave. Mammalia 68(4):337–343

Borda RD, Năstase-Bucur RM, Spînu M, Uricariu R, Mulec J (2014) Aerosolized microbes from organic rich materials: case study of bat guano from caves in Romania. J Cave Karst Stud 76(2):114–126. https://doi.org/10.4311/2013MB0116

Bucur II, Onac B, Todoran V (1989) Algues calcaires dans les dépôts oligocènes inférieurs de la région Purcăreț-Mesteacăn-Valea Chioarului (NW du Basin de Transylvanie). In: Petrescu I (ed) The Oligocene from the Transylvanian Basin, Cluj-Napoca, pp 141–148

Cleary DM, Wynn JG, Ionita M, Forray FL, Onac BP (2017) Evidence of long-term NAO influence on East-Central Europe winter precipitation from a guano-derived $\delta^{15}N$ record. Sci Rep 7:14095. https://doi.org/10.1038/s41598-017-14488-5

Des Marais DJ, Mitchell JM, Meinschein WG, Hayes JM (1980) The carbon isotope biogeochemistry of the individual hydrocarbons in bat guano and the ecology of the insectivorous bats in the region of Carlsbad, New Mexico. Geochim Cosmochim Acta 44:2075–2086. https://doi.org/10.1016/0016-7037(80)90205-7

Geantă A, Tanțău I, Tămaş T, Johnston VE (2012) Palaeoenvironmental information from the palynology of an 800 year old bat guano deposit from Măgurici Cave, NW Transylvania (Romania). Rev Palaeobot Palynol 174:57–66

Ghergari L, Petrescu I, Todoran V (1989) Mineralogical-petrographical studies on the Oligocene clay rocks of the Curtuiuş, Bizuşa and Ileanda Beds with a view to reconstitute the paleoenvironment. In: Petrescu I (ed) The Oligocene from the Transylvanian Basin, Cluj-Napoca, pp 459–468

Ghiurcă V (1970) Contribution géologique á l'étude du karst initial engendré par les formations miocènes du bassin de Baia Mare. In: Livre du centenaire Émile G. Racovitza 1868–1968, Ed Acad RSR, Bucharest, pp 669–673

Giurgiu I, Vădeanu T, Done A, Negru M, Sandeschi N, Silvăşanu G, Codescu M, Ciuculescu O, Solomon E, Sfâşie M, Cucu F (1983) Descoperiri şi explorări în podişul Someşan (bază pentru inventarul zonei). Bul Club Speol "Emil Racoviță" 8:11–82

Johnston VE, McDermott F, Tămaş T (2010) A radiocarbon dated bat guano deposit from N.W. Romania: implications for the timing of the Little Ice Age and Medieval Climate Anomaly. Palaeogeogr Palaeoclimatol Palaeoecol 291:217–227

Onac BP (1991) New data on some gypsum speleothems in the Vîntului (Pădurea Craiului Mountains) and Răstoci (Someşan Plateau) caves. Trav Inst Spéol "Émile Racovitza" XXX:189–193

Onac BP (2012) Minerals. In: Culver DC, White WB (eds) Encyclopedia of caves (2nd ed). Academic Press, New York, pp 499–508

Onac BP, Todoran V (1987) Contributions à la connaissance des formations de gypse de la Grote de Răstoci (NO de la Roumanie). In: Petrescu I (ed) The Eocene from the Transylvanian Basin, Cluj-Napoca, pp 301–306

Onac BP, Vereş D (2003) Sequence of secondary phosphates deposition in a karst environment: evidence from Măgurici Cave (Romania). Eur J Mineral 15:741–745

Onac BP, Todoran V, Onac L (1989) Review of the karst on the Oligocene formations from the Someş Plateau-Romania. In: Petrescu I (ed) The Oligocene from the Transylvanian Basin, Cluj-Napoca, pp 503–512

Onac BP, Borda D, Tămaş T (2004) Studiu de fundamentare ştiinţifică al peşterii Măgurici. Raport de cercetare, ISER Cluj-Napoca, p 9

Onac BP, Vereş D, Tămaş T, CS Montana Baia Mare, Todoran V (2010) The hydrogeology of Cuciulat- Purcăreţ-Boiu Mare karst area, northern part of the Someş plateau, Romania. In: Orăşeanu I, Iurkievicz A (eds) Karst hydrogeology of Romania. Ed Belvedere, Oradea, pp 329–341

Prică I (2001) Coralgal facies of the upper Eocene-lower Oligocene limestones in Letca-Răstoci area. Studia UBB Geologia XLVI (2):53–61

Rusu A (1977) La stratigraphie des depots oligocènes du nord-ouest de la Transylvanie (la region de Traznea-Hida-Poiana Blenchii). An Inst Geol Geogr LI:69–223

Tămaş T, Ungureanu R (2010) Mineralogy of speleothems from four caves in the Purcăreţ-Boiu Mare Plateau and the Baia Mare Depression. Studia UBB Geologia 55(2):43–49

Todoran V, Onac BP (1987) La morphologie de l'exo- et l'endokarst du Plateau Purcaret-Mesteacan (Plateau de Somes). Theor Appl Karstol 3:127–131

Todoran V, Onac BP (1989) Eocene/Oligocene boundary in the Purcaret-Mesteacan area (the Someş Plateau). In: Petrescu I (ed) The Oligocene from the Transylvanian Basin, Cluj-Napoca, pp 129–132

Dobrogea: Movile Cave

Serban M. Sarbu, Cristian Lascu, and Traian Brad

Abstract

Movile Cave (SE Romania) is the first known subterranean chemosynthesis-based ecosystem. Sulfur- and methane-oxidizing, as well as nitrifying microorganisms, form the base of the food web in this peculiar ecosystem. The bacteria use mainly oxygen from the cave's atmosphere as electron acceptor to oxidize H_2S, CH_4, and NH_4^+, which originate from the deep thermomineral aquifer. These microorganisms form microbial mats that cover the water surface and the cave walls adjacent to the water. For cave standards, this is an unusually abundant primary production that allows 51 invertebrates species to thrive here. Of these, 35 are endemic to this ecosystem and display advanced troglomorphy. Numerous aquatic and terrestrial species graze on the rich microbial mats present in the lower sections of the cave, while others predate on protozoa and smaller invertebrates. Speleogenesis in the Movile Cave area was initiated in the Late Miocene and continues today by the action of the sulfuric acid resulted from the oxidation of sulfide in the lower part of the cave. The cave was sealed off during the Quaternary by thick and impermeable layers of clays and loess.

Keywords

Cave • Chemosynthetic ecosystem • Hydrogen sulfide
Sulfuric acid speleogenesis • Fauna

S. M. Sarbu (✉) · C. Lascu
"Emil Racoviță" Institute of Speleology, Romanian Academy, Frumoasă 31, 010986 București, Romania
e-mail: serban.sarbu@yahoo.com

S. M. Sarbu
Department of Biological Sciences, California State University, Chico, USA

T. Brad
"Emil Racoviță" Institute of Speleology, Romanian Academy, Clinicilor 5, 400006 Cluj-Napoca, Romania

Introduction

Southern Dobrogea, Romania, is a limestone platform covered by Quaternary loess deposits. The upper part of the plateau consists of fossiliferous, oolitic limestones c. 12.5 Myr in age (Constantinescu 1989). The plateau is locally affected by solutional karstic features, such as dolines, dry valleys, and cliffs. Except for Limanu Cave (Fig. 1), a 4-km long maze network, Southern Dobrogea had never attracted cave explorers. The proposed construction of a power plant on the outskirts of the town of Mangalia in the mid-1980s required a thorough geological investigation of the area to avoid possible karst collapses. Several artificial shafts were dug in 1986 at two km from the shore of the Black Sea, and C. Lascu, a speleologist at the "Emil Racoviță" Institute of Speleology in Bucharest, performed the geological survey of these shafts. Eighteen meters beneath the surface, at the bottom of one of the shafts, he discovered a natural cave passage that had been isolated from the surface.

This underground void intercepted by the artificial shaft became known as Movile Cave ('Movile' means hillocks in Romanian). It is a 240 m long horizontal maze cave that has no natural entrance and consists of two levels (Fig. 2). The upper one is dry and lacks speleothems, and its galleries have diameters of 1–2 m with rounded and elliptical sections (Lascu et al. 1994). Red and yellowish fine clay is abundant, and the only secondary mineral deposits are calcite crusts that cover the cave floor and are associated with millimeter-sized aragonite needles and gypsum (Diaconu and Morar 1993; Sarbu and Lascu 1997). The 40 m long, partially flooded lower cave level has a classic morphology that indicates a more recent origin. It consists of a succession of submerged passages, 0.5–2 m wide and 1–3 m deep, alternating with air-bells. Very soft, gray sediment covers the floor of the submerged passages. Secondary gypsum crusts and needles cover the limestone cave walls near the lake and in the air-bells. Multidisciplinary investigations

G. M. L. Ponta and B. P. Onac (eds.), *Cave and Karst Systems of Romania*,
Cave and Karst Systems of the World, https://doi.org/10.1007/978-3-319-90747-5_48

Fig. 1 Location of Movile Cave in the karst area of south Dobrogea

performed in this region for more than three decades have shown that Movile Cave represents one of the few windows of access to a vast network of interconnected subterranean voids, developed in the Sarmatian limestones (Constantinescu 2002) and partially flooded by thermomineral waters, rich in reduced chemical compounds such as hydrogen sulfide (H_2S), methane (CH_4), and ammonia (NH_4^+).

The air temperature in Movile Cave is 21 °C, the relative humidity is approximately 100%, and no air movement can be detected (Boghean and Racoviță 1989). The atmosphere contains 20% oxygen and around 1% carbon dioxide (CO_2) in the upper level of the cave, while the air-bells become progressively depleted in oxygen (16–7%) and enriched in CO_2 (1.5–3.5%) due to the biological activity of the cave fauna and microbiota (Sarbu and Popa 1992; Sarbu 2000). Water droplets present on the cave walls tend to be acidic, displaying pH values that range between 3.5 and 4. The groundwater that floods the lower cave level contains 30 mg/L H_2S and its pH averages 7.2. A slight current (5 cm/s) is detected in the underwater cave passages. Atmospheric oxygen dissolves into the water at the air–water interface, but below the depth of 1 mm the cave water is completely anoxic (Riess et al. 1999). The thermomineral groundwater found in Movile Cave as well as in the entire area around the town of Mangalia originates from a deep (180–600 m) thermomineral artesian aquifer hosted in Mesozoic limestones (Lascu et al. 1994).

Energetics of the Movile Cave Ecosystem

Caves in general are aphotic environments devoid of photosynthetic primary production, and consequently, they lack abundant amounts of food (Culver and Pipan 2009). Life in caves is usually based on food that originates at the surface. Vegetal detritus and animal excreta, such as bat guano, and sometimes animal carcasses represent typical food resources for cave communities, and percolation water brings small amounts of organic matter along limestone fractures. In the case of Movile Cave, however, successive clay deposits alternating with the overlying limestone strata seal the rock fractures and pores and limit the infiltration of surface

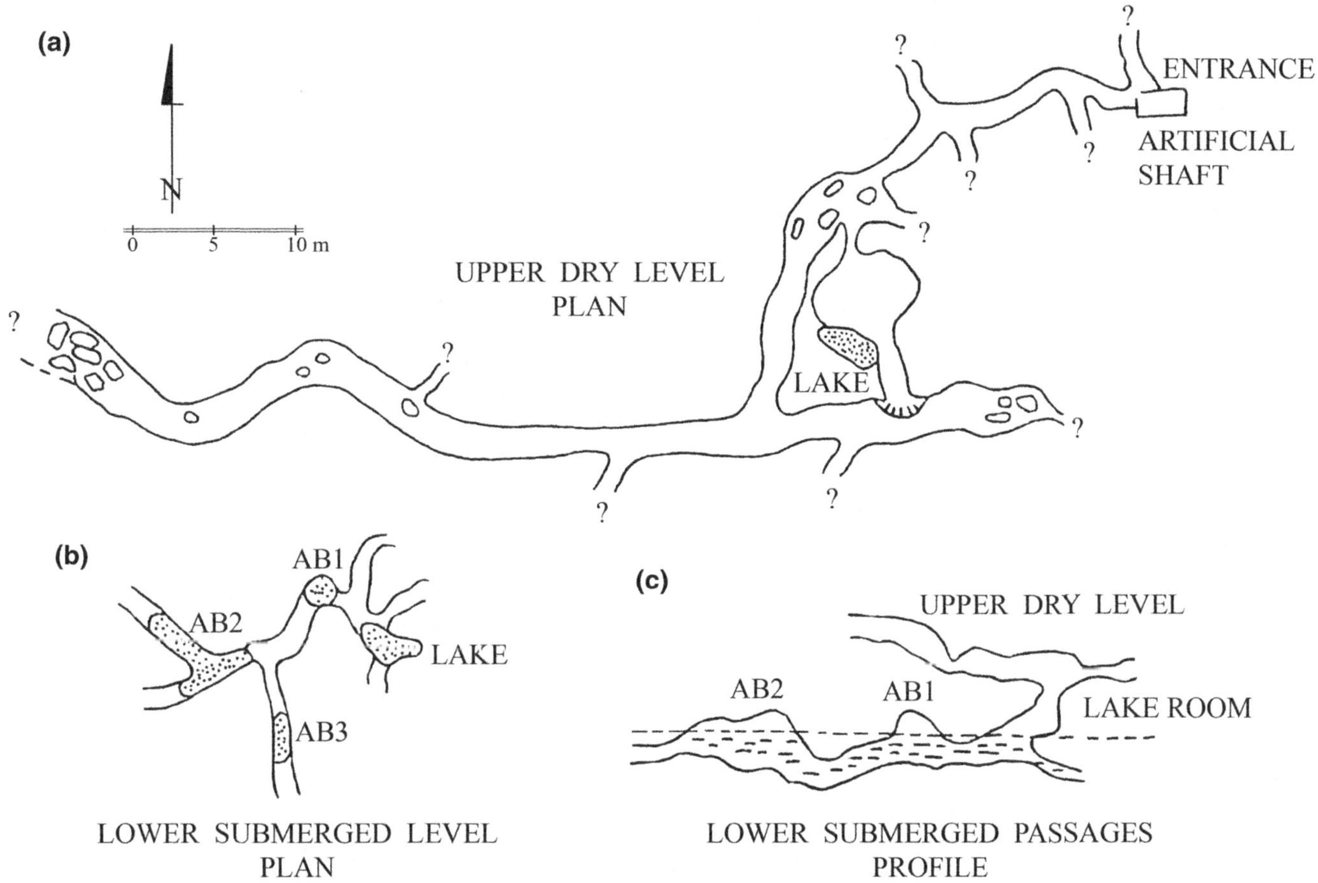

Fig. 2 Map of Movile Cave with depiction of the Lake Room and the three air-bells (AB 1 to 3) located in the lower submerged level. **a** Cave plan, **b** plan of the lower submerged level, **c** vertical profile for part of the cave

precipitation water into the limestones. Circumstantial evidence for the absence of input of surface materials into the groundwater aquifer is the absence of artificial radionuclides such as ^{137}Cs and ^{90}Sr in Movile Cave. Large amounts of such radionuclides resulted from the 1986 Chernobyl nuclear accident are abundant in the surface soil, plants, and sediments, as well as in other caves in the area (Lascu et al. 1994).

The presence of rich and abundant invertebrate communities thriving in the groundwater ecosystem associated with the thermomineral sulfidic aquifer at Mangalia suggests the existence of a large energy base for the subterranean ecosystem. High densities of invertebrate specimens and large numbers of predators with high metabolic activity require the presence of abundant sources of food. Stable isotope studies have demonstrated that the subterranean invertebrate community relies exclusively on in situ chemoautotrophic production by sulfur and methane oxidizers, as well as by nitrifying bacteria. Carbon and nitrogen stable isotopes ratios are significantly lighter in the cave organisms compared to the surface samples and to samples of cave fauna collected in Limanu Cave, which are of photosynthetic origin (Sarbu et al. 1996). Chemoautotrophic microorganisms use the isotopically light bicarbonate ($\delta^{13}C = -15‰$) present in the cave water as a carbon source and produce organic molecules that are very light isotopically ($\delta^{13}C = -44‰$). The process involves the RuBisCO enzyme which selects for the light isotope of carbon.

Microbiology

In the absence of allochthonous input of organic matter and energy from the surface, the food chain in Movile Cave ecosystem begins with the chemoautotrophic microorganisms living freely in water in the lower cave level, or gathered in complex biofilms floating at the water surface or attached to sediments and rock surfaces. The only energy sources available for these microorganisms are the reduced chemical compounds, such as H_2S, CH_4, and NH_4^+, originating from the thermomineral aquifer (Sarbu and Kane 1995; Sarbu 2000).

The microbial communities in the Movile Cave ecosystem (Kumaresan et al. 2014) consist mainly of sulfur-oxidizing bacteria, such as representatives of the genera *Thiobacillus*, *Thiothrix*, *Thioploca*, *Thiomonas*, and *Sulfurospirillum*. They use H_2S as energy source, and O_2 or NO_3^- as electron acceptors (Rohwerder et al. 2003; Chen et al. 2009; Flot et al. 2014). Dense populations of large microorganisms resembling *Thiovulum* sp. are present and thrive just below the water surface in Lake Room (Sala Lacului) and air-bells (unpublished data). Species of *Thiothrix* sp. live as epibionts on the body of the amphipod crustaceans *Niphargus dancaui* and *Pontoniphargus racovitzai* (Flot et al. 2014). Methane oxidation is another important primary production process within the Movile Cave ecosystem. Several microbial species of the genera *Methylomonas*, *Methylococcus* and *Methylocystis* (Hutchens et al. 2004), *Methanobacterium* (Schirmack et al. 2014) and *Methanosarcina* (Ganzert et al. 2014), together with other methylothrophs (Rohwerder et al. 2003; Chen et al. 2009), use CH_4 as their only source of carbon and energy via various biochemical pathways. Microorganisms involved in the nitrogen cycle have also been identified (Chen et al. 2009): ammonia (*Nitrosomonas* sp.)—and nitrite (*Nitrospira* sp.)—oxidizers, to denitrifying bacteria (*Denitratisoma* sp. and *Paracoccus denitrificans*). Archaea are represented in Movile Cave ecosystem by species of *Korarchaeota*, *Crenarchaeota*, and *Euryarchaeota* (Chen et al. 2009).

Several species of protozoan ciliates had been identified swimming at the water surface or through the complex microbial mats. These ciliates include large bacterivorous species of *Stentor coeruleus* (*Heterotricha*) and *Euplotes eurystomus* (*Hypotricha*), alongside other species, such as *Spirostomum teres*, *Blepharisma undulans*, *Oxytricha* sp., *Paramecium aurelia*, *Uronema* sp., *Cyclidium* sp., and a member of the *Trichopelmidae* (T. Fenchel, pers. comm.).

Fungi in Movile Cave ecosystem (Nováková et al. 2018) were encountered on various terrestrial substrates, such as cave soil, corroded limestone walls, or invertebrate feces and cadavers. Aquatic fungi were identified in the air-bells, in samples of sediments and floating microbial mats. Most of these fungal species were only identified in samples collected inside the cave and not in samples collected from the close-by surface habitats.

Cave Fauna

Movile Cave ecosystem hosts 51 invertebrate species, of which 34 species are endemic for this peculiar ecosystem (Table 1). As the entire primary production of organic molecules in Movile Cave takes place at the water surface in the lower level of the cave, the fauna is mainly present in and around the water in Lake Room and the neighboring cave's air-bells.

Aquatic Fauna The food web in Movile Cave ecosystem is very complex and far from being fully elucidated. Cave organisms in general cannot afford being specialists on one type of food, and they are rather omnivorous species, surviving on the limited organic matter that is usually introduced from the surface. In the Movile Cave ecosystem, however, the unusually rich chemoautotrophic microbiota generates copious amounts of food for both the aquatic and the terrestrial abundant and diverse invertebrate communities (Sarbu 2000). Nematodes and aquatic crustacean copepods, isopods, ostracods, and amphipods can be seen crawling on rock substrates along the shores of the lakes or swimming just below the water surface, grazing on sulfur-oxidizing microorganisms thriving freely, or congregated in thick biofilms floating on water surface, where H_2S oxidation occurs. Besides feeding on bacteria, cyclopoid copepods (*Eucyclops subterraneus scythicus*) predate actively on nematodes (*Panagrolaimus* sp. and *Poikilolaimus* sp.) (Muschiol et al. 2008). The predatory water scorpion *Nepa anophthalma* waits for its prey on the lake banks just under the water surface (Decu et al. 1994a, b), where a large population of snails (*Heleobia dobrogica*) is also present (Falniowski et al. 2008). Other top predators in this environment are the leech *Haemopis caeca* (Manoleli et al. 1998) and the flatworm *Dendrocoelum obstinatum* (Stocchino et al. 2017), which graze on bacteria from the water and biofilms, or predate on worms and crustaceans.

Terrestrial Fauna In the Movile Cave ecosystem, terrestrial fauna is better represented compared to the aquatic fauna. Terrestrial invertebrates roam especially in the Lake Room, but also in the adjacent cave galleries. Diplopods and terrestrial isopods graze on microbial mats and fungi that cover the cave walls in the lower cave level. They are very abundant in Movile Cave reaching densities of up to hundreds of specimens per square meter. Springtails (*Collembola*) can jump on the water surface and are common on the surface of the floating microbial mats where they graze on bacteria and fungi. The number of predatory species is present here, and their population densities are unusually high for cave standards. The list of predators includes spiders, pseudoscorpions, centipedes, mites, and carabid and staphylinid beetles. The top predator and largest invertebrate species in Movile Cave is the centipede *Cryptops anomalans*. It may grow up to 8-10 cm in length, and it predates especially on terrestrial isopods, such as *Trachelipus troglobius* and *Armadillidium tabacarui*, but also on smaller beetles, *Diplura* or spiders (for species names, see Table 1).

Table 1 Invertebrate species inhabiting the Movile Cave groundwater ecosystem

Species	Taxonomical affiliation	Reference
*Dendrocoelum obstinatum** Stocchino et al. 2017	Platyhelminthes, Tricladida, Dendrocoelidae	Stocchino et al. (2017)
Panagrolaimus c.f. *thienemani**	Aschelminthes, Nematoda, Panagrolaimidae	Muschiol et al. (2015)
*Chronogaster troglodytes** Poinar and Sarbu 1994	Aschelminthes, Nematoda, Chronogasteridae	Poinar and Sarbu (1994)
Udonchus tenuicaudatus Cobb. 1913	Aschelminthes, Nematoda, Rhabdolaimidae	Muschiol et al. (2015)
Poikilolaimus sp.	Aschelminthes, Nematoda, Rhabditidae	Muschiol et al. (2015)
Monhystrella sp.	Aschelminthes, Nematoda, Monhysteridae	Muschiol et al. (2015)
Habrotrocha rosa Donner 1949	Aschelminthes, Rotatoria, Habrotrochidae	C. Ricci pers. comm.
Habrotrocha bidens Gosse 1851	Aschelminthes, Rotatoria, Habrotrochidae	C. Ricci pers. comm.
*Haemopis caeca** Manoleli et al. 1998	Annelida, Hirudinea, Haemopidae	Manoleli et al. (1998)
Allolobophora oculata (Hoffmeister, 1845)	Annelida, Oligochaeta, Lumbricidae	E. Dumnicka pers. comm.
Aeolosoma hyalinum Bunke 1967	Annelida, Aphanoneura, Aelosomatidae	E. Dumnicka pers. comm.
Aeolosoma litorale Bunke 1967	Annelida, Aphanoneura, Aelosomatidae	E. Dumnicka pers. comm.
*Heleobia dobrogica** (Grossu and Negrea 1989)	Mollusca, Gastropoda, Moitessieriidae	Grossu and Negrea (1989)
Pseudocandona n.sp.*	Crustacea, Ostracoda, Cyprididae	D. Danielopol pers. comm.
*Eucyclops subterraneus scythicus** Pleşa 1989	Crustacea, Copepoda, Cyclopidae	Pleşa (1989)
Tropocyclops prasinus (Fischer 1860)	Crustacea, Copepoda, Cyclopidae	Pleşa (1989)
Parapseudoleptomesochra italica Pesce and Petkovski 1980	Crustacea, Copepoda, Harpacticoida	D.M.P. Galassi pers. comm
*Pontoniphargus racovitzai** Dancău 1970	Crustacea, Amphipoda, Niphargidae	Dancau (1970)
*Niphargus dancaui** Brad et al. 2015	Crustacea, Amphipoda, Niphargidae	Brad et al. (2015)
*Asellus aquaticus infernus** Turk-Prevorčnik and Blejec 1998	Crustacea, Isopoda, Asellidae	Turk-Prevorčnik and Blejec (1998)
Caucasonethes n.sp.*	Crustacea, Isopoda, Trichoniscidae	I. Tabacaru pers. comm.
*Haplophthalmus movilae** Gruia et Giurginca, 1998	Crustacea, Isopoda, Trichoniscidae	Gruia and Giurginca (1998)
*Trachelipus troglobius** Tabacaru and Boghean, 1989	Crustacea, Isopoda, Trachelipidae	Tabacaru and Boghean (1989)
*Armadillidium tabacarui** Gruia et al. 1994	Crustacea, Isopoda, Armadillidiidae	Gruia et al. (1994)
*Chthonius monicae** Boghean 1989	Arachnida, Pseudoscorpiones, Chthoniidae	Boghean (1989)
*Roncus dragobete** Ćurčić et al. 1993	Arachnida, Pseudoscorpiones, Neobisiidae	Ćurčić et al. (1993)
*Roncus ciobanmos** Ćurčić et al. 1993	Arachnida, Pseudoscorpiones, Neobisiidae	Ćurčić et al. (1993)
Carniella brignilii Thaler and Steinberger, 1988	Arachnida, Araneae, Theridiidae	Georgescu (1989)
*Leptyphantes constantinescui** Georgescu 1989	Arachnida, Araneae, Linyphiidae	Georgescu (1989)
*Agraecina cristiani** (Georgescu 1989)	Arachnida, Araneae, Clubionidae	Georgescu (1989)
*Kryptonesticus georgescuae**	Arachnida, Araneae, Nesticidae	Nae et al. (2018)
*Hahnia caeca** (Georgescu and Sarbu 1992)	Arachnida, Araneae, Hahniidae	Georgescu and Sarbu (1992)
Dysdera crocata Koch 1838	Arachnida, Araneae, Dysderidae	A. Nae pers. comm.
*Labidostoma motasi** Iavorschi 1992	Arachnida, Acarina, Nicoletiellidae	Iavorschi (1992)
Geophilus alpinus Meinert 1870	Chilopoda, Geophilomorpha, Geophilidae	M. Zapparoli pers. comm.
Clinopodes carynthiacus (Latzel 1880)	Chilopoda, Geophilomorpha, Geophilidae	M. Zapparoli pers. comm.
Cryptops anomalans Newport 1848	Chilopoda, Scolopendromorpha, Cryptopidae	Negrea (1993)
Symphylella sp.	Symphyla, Scolopendrellidae	U. Sheller pers. comm.
Archiboreoiulus n.sp.*	Diplopoda, Julida, Julidae	I. Tabacaru pers. comm.
Pygmarrhopalites pygmaeus (Wankel 1860)	Hexapoda, Collembola, Arrhopalitidae	Gruia (2003)
*Onychiurus movilae** Gruia 1989	Hexapoda, Collembola, Onychiuridae	Gruia (2003)
Heteromurus nitidus (Templeton 1835)	Hexapoda, Collembola, Entomobryidae	Gruia (2003)

(continued)

Table 1 (continued)

Species	Taxonomical affiliation	Reference
*Oncopodura vioreli** Gruia 1989	Hexapoda, Collembola, Oncopoduridae	Gruia (2003)
*Plusiocampa isterina** Conde 1993	Hexapoda, Diplura, Compodeidae	Conde (1993)
*Plusiocampa euxina** Conde 1996	Hexapoda, Diplura, Compodeidae	Conde (1996)
*Medon dobrogicus** Decu and Georgescu 1994	Hexapoda, Coleoptera, Staphylinidae	Decu and Georgescu (1994)
*Tychobythinus sulphydricus** Poggi and Sarbu 2013	Hexapoda, Coleoptera, Staphylinidae, Pselaphinae	Poggi and Sarbu (2013)
*Decumarellus sarbui** Poggi 1994	Hexapoda, Coleoptera, Staphylinidae, Pselaphinae	Poggi (1994)
*Bryaxis dolosus** Poggi and Sarbu 2013	Hexapoda, Coleoptera, Staphylinidae, Pselaphinae	Poggi and Sarbu (2013)
*Clivina subterranea** Decu et al. 1994	Hexapoda, Coleoptera, Carabidae	Decu et al. (1994a, b)
*Nepa anophthalma** Decu et al. 1994	Hexapoda, Hemiptera, Nepidae	Decu et al. (1994a, b)

Endemic species are marked with asterisk (*), and species with descriptions in progress are marked with 'n.sp.'

Speleogenesis of Movile Cave and of the South Dobrogean Karst

Movile Cave represents a recent, shallow, and limited segment of the extensive and deep karst system located in Southern Dobrogea, Romania. This platform unit consists of Mesozoic and Cenozoic limestones up to 800 m thick and is bordered by the Black Sea to the east and by the Danube to the west. It was affected by several karstification events related to oscillations in the level of the Black Sea, which have generated large karst networks. The analysis of 52 boreholes has proven the presence of several different karst levels.

The lowest level, located at an average depth of 200 m, was formed during the Upper Miocene when the Black Sea as well as the entire Mediterranean Basin experienced a dramatic hydrological and climatic event known as the Messinian Crisis (5.9–5.3 Ma). The northward movement of the African Plate closed the Strait of Gibraltar, separating the Mediterranean Sea from the Atlantic Ocean. Subsequently, the Mediterranean Sea lost nearly all of its water due to evaporation, leaving behind thick evaporite sequences (Govers 2009). The Black Sea level also experienced a drop of several hundred meters (Hsu 1978). The analysis of shallow water deposits in boreholes in the Danube Delta has shown that the Black Sea level had dropped approximately 200 m (Lascu et al. 1994). Water from the Dacian Lake, another basin located west of Dobrogea, drained toward the Black Sea through large karstic passages carved in the Dobrogean carbonate platform. Some of these karst conduits appear to be still active, draining large amounts of water from the Danube River toward the Black Sea (up to 10×10^6 L/s). Several hydrogeological boreholes that penetrate this level provide large amounts of water ranging between 110 and 225 L/s of sulfidic, thermal water (24–25 °C), at pressures averaging 1.7 atm.

More recent oscillations of the Black Sea level led to new karstification periods. A middle-Quaternary speleogenetic episode is believed to be responsible for the formation of the upper level of Movile Cave. Another karstification episode took place during the Würm II glacial, approximately 15,000 years ago, when the Black Sea level had dropped 50–60 m. This resulted in the initiation of collapse processes that generated large sinkholes in this area. One of these was the Movile sinkhole collapse that cuts the preexisting phreatic conduits, and water drainage was lowered to its present level.

Lastly, speleogenesis in modern Movile Cave has been influenced by microbial activity. Sulfuric acid resulting from microbial sulfur-oxidizing bacteria reacts with the bedrock accelerating the rate of carbonate dissolution. The effects of this process known as sulfuric acid speleogenesis, first proposed by Egemeier (1981), are evident primarily in the lower level of Movile Cave in the vicinity of the sulfidic lakes where the cave walls are covered by gypsum crusts and crystals (Sarbu 2000). The cave passages in the upper level of the cave are currently affected by condensation corrosion (Sarbu and Lascu 1997).

Cave Conservancy

Movile Cave is a strictly protected site (class A), accessible for scientific research only on basis of official authorizations. The area is part of Natura 2000 site (Code ROSCI0114—

Mlaştina Hergheliei—Obanul Mare şi Peştera Movile) and covers 232 ha.

Acknowledgements The authors of this chapter thank GESS Group for permanent support during the field trips that represented the beginning of every study included herein. The effort of the specialists that have identified the specimens of the various invertebrate species is highly appreciated. Their names appear in the species list included in this chapter, followed by 'personal communication' (pers. comm.) for each identified species.

References

Boghean V (1989) Sur un pseudoscorpion cavernicole nouveau, *Chthonius* (*C.*) *monicae* n.sp. (*Arachnida Pseudoscorpionida Chthonidae*). Misc Speol Rom 1:77–83

Boghean V, Racoviţă G (1989) Données préliminaires sur le topoclimat de la grotte "Peştera de la Movile" (Mangalia - Dobrogea du sud - Roumanie). Misc Speol Rom 1:19–31

Brad T, Fišer C, Flot J-F, Sarbu SM (2015) *Niphargus dancaui* sp. nov. (*Amphipoda, Niphargidae*)—a new species thriving in sulfidic groundwaters in southeastern Romania. Eur J Taxon 164:1–28

Chen Y, Wu L, Boden R, Hillebrand A, Kumaresan D, Moussard H, Baciu M, Lu Y, Murrell JC (2009) Life without light: microbial diversity and evidence of sulfur-and ammonium-based chemolithotrophy in Movile Cave. ISME J 3:1093–1104

Conde B (1993) Une lignée danubiene de genre *Plusiocampa* (*Diploures, Campodeides*). Rev Suisse Zool 100:735–745

Conde B (1996) Diploures Campodéides de la Peştera de la Movile (Movile Cave), Dobroudja méridionale (Roumanie). Rev Suisse Zool 103:101–114

Constantinescu T (1989) Considerations sur la zone karstique de "La Movile" (Mangalia, Dobrogea du Sud, Roumanie). Misc Speol Rom 1:7–12

Constantinescu T (2002) Le karst de la zone Mangalia et la Grotte de Monde (Dobrogea de Sud-Roumanie). Ann Valahia Univ Târgovişte Geogr Ser 2:147–156

Culver DC, Pipan T (2009) The biology of caves and other subterranean habitats. Oxford University Press, Oxford, p 273

Ćurčić BPM, Poinar GO Jr, Sarbu SM (1993) New and little known species of *Chthoniidae* and *Neobisiidae* (*Pseudoscorpiones, Arachnida*) from Movile Cave in southern Dobrogea, Romania. Bijdr Dierkd 63:221–241

Dancău D (1970) Sur un nouvel Amphipode souterrain de Roumanie, *Pontoniphargus racovitzai* n.g., n.sp. Livre du centenaire Emil G. Racovitza, Bucharest, pp 275–285

Decu V, Georgescu M (1994) Deux espèces nouvelles de *Medon* (*M. dobrogicus* et *M. paradobrogicus*) (*Coleoptera, Staphylinidae*) de la grotte "Peştera de la Movile", Dobrogea Meridionale, Roumanie. Mém Biospéol 21:47–51

Decu V, Gruia M, Keffer SL, Sarbu SM (1994a) Stygobiotic waterscorpion, *Nepa anophthalma*, n. sp. (*Heteroptera*: *Nepidae*), from a sulfurous cave in Romania. Ann Entomol Soc Am 87:755–761

Decu V, Nitu E, Juberthie C (1994b) *Clivina subterranea* (*Caraboidea, Scaritidae*) nouvelle espèce de la grotte "Pestera de la Movile", Dobrogea Meridionale, Roumanie. Mém Biospéo 21:41–45

Diaconu G, Morar M (1993) Analyses mineralogiques dans la grotte "Peştera de la Movile" (Mangalia - Dobrogea de Sud). Karstologia 22:15–20

Egemeier SJ (1981) Cavern development by thermal waters. Natl Speleol Soc Bull 43:31–51

Falniowski A, Szarowska M, Sirbu I, Hillebrand A, Baciu M (2008) *Heleobia dobrogica* (Grossu & Negrea, 1989) (*Gastropoda*: *Rissooidea*: *Cochliopidae*) and the estimated time of its isolation in a continental analogue of hydrothermal vents. Molluscan Res 28:165–170

Flot J-F, Bauermeister J, Brad T, Fišer C, Hillebrand A, Sarbu SM, Dattagupta S (2014) *Niphargus-Thiothrix* associations may be widespread in sulfidic groundwater ecosystems: evidence from southeastern Romania. Mol Ecol 23:1405–1417

Ganzert L, Schirmack J, Alawi M, Mangelsdorf K, Sand W, Hillebrand-Voiculescu A, Wagner D (2014) *Methanosarcina spelaei* sp. nov., a methanogenic archaeon isolated from a floating biofilm of a subsurface sulphurous lake. Int J Syst Evol Microbiol 64:3478–3484

Georgescu M (1989) Sur trois taxa nouveaux d'Araneides troglobies de Dobrogea (Roumanie). Misc Speol Rom 1:85–102

Georgescu M (1994) Sur la présence d'une espèce troglobionte de *Nesticus* (*Araneae, Nesticidae*) dans la grotte "Pestera de la Movile" (Dobrogea, Roumanie). Trav Inst Spéol "Émile Racovitza" 33: 85–87

Georgescu M, Sarbu SM (1992) Description d'un noveau taxon: *Iberina caeca* de la grotte: "Peştera de la Movile" (*Araneae - Hahniidae*). Mém Biospéol 19:319–141

Govers R (2009) Choking the Mediterranean to dehydration: the Messinian salinity crisis. Geology 37(2):167–170

Grossu AV, Negrea A (1989) *Paladilhia* (*Paladilhiopsis*) *dobrogica* - une nouvelle espèce de la famille *Moitessieriidae* (*Gastropoda, Prosobranchia*). Misc Speol Rom 1:33–37

Gruia M (1989) Nouvelles espèces troglobiontes des Collemboles de Roumanie. Misc Speol Rom 1:103–111

Gruia M (1998–1999) Collembola of the karstic system from Movile. Trav Inst Speol "Émile Racovitza" 37–38:167–174

Gruia M (2003) Collembola from Romanian caves. Trav Mus Nat Hist Nat "G. Antipa" 45:139–158

Gruia M, Giurginca A (1998) *Haplophthalmus movilae* (*Isopoda, Trichoniscidae*), a new troglobitic species from Movile Cave, Dobrogea, Romania. Mitt Hamb Zool Mus Inst 95:133–142

Gruia M, Iavorschi V, Sarbu SM (1994) *Armadillidium tabacarui* (*Isopoda, Oniscidea, Armadillidiidae*), a new troglobitic species from a sulfurous cave in Romania. Proc Biol Soc Wash 107: 699–706

Hsu KJ (1978) When the Black Sea was drained. Sci Am 238:52–63

Hutchens E, Radajewski S, Dumont MG, McDonald IR, Murrell JC (2004) Analysis of methanotrophic bacteria in Movile Cave by stable isotope probing. Environ Microbiol 6:111–120

Iavorschi V (1992) *Labidostoma motasi* n.sp. (*Nicoletiellidae*) a new species of the mite of Romania. Trav Inst Spéol "Émile Racovitza" 31:47–51

Kumaresan D, Wischer D, Stephenson J, Hillebrand-Voiculescu A, Murrell JC (2014) Microbiology of Movile Cave—a chemolithoautotrophic ecosystem. Geomicrobiol J 31:186–193

Lascu C, Popa R, Sarbu SM (1994) Le karst de Movile (Dobrogea de Sud). Rev Roum Geogr 38:85–94

Manoleli D, Klemm DJ, Sarbu SM (1998) *Haemopis caeca* sp.nov. (*Annelida, Hirudine, Haemopidae*) a troglobitic leech, from a chemoautotrophically based groundwater ecosystem in Romania. Proc Biol Soc Wash 111:222–229

Muschiol D, Marković M, Threis I, Traunspurger W (2008) Predatory copepods can control nematode populations: a functional-response experiment with *Eucyclops subterraneus* and bacterivorous nematodes. Fund Appl Limnol 172:317–324

Muschiol D, Giere O, Traunspurger W (2015) Population dynamics of a cavernicolous nematode community in a chemoautotrophic groundwater system. Limnol Oceanogr 60:127–135

Nae A, Sarbu SM, Weiss I (2018) *Kryptonesticus georgescuae* spec. nov. from Movile Cave, Romania (Araneae: Nesticidae). Arachnol Mitteil 55:22–24

Negrea Ş (1993) Sur une population troglobionte de *Cryptops anomalans* Newport, 1844 (*Chilopoda, Scolopendromorpha*) trouvee dans la grotte "Pestera de la Movile" (Dobrogea, Roumanie). Trav Inst Spéol "Émile Racovitza" 32:87–94

Nováková A, Hubka V, Valinová Š, Kolařík M, Hillebrand-Voiculescu AM (2018) Cultivable microscopic fungi from an underground chemosynthesis-based ecosystem: a preliminary study. Folia Microbiol 63(1):43–55. https://doi.org/10.1007/s12223-017-0527-6

Pleşa C (1989) Étude preliminaire des cyclopides (*Crustacea, Copepoda*) de la grotte "Peştera de la Movile", Mangalia (Roumanie). Misc Speol Rom 1:13–18

Poggi R (1994) Descrizione di un nuovo pselafide rumeno, primo rappresentante cavernicolo della tribù Tyrini. Boll Soc Ent Ital 125:221–228

Poggi R, Sarbu SM (2013) Two new *Pselaphine* beetles from Movile Cave (Romania) (*Coleoptera, Staphilinidae, Pselaphinae*). Ann Mus Civ Stor Nat "Giacomo Doria" 105:115–121

Poinar GO Jr, Sarbu SM (1994) *Chronogaster troglodytes* sp. n. (*Nematoda, Chronogasteridae*) from Movile Cave, with a review of cavernicolous nematodes. Fund Appl Nematol 17:231–237

Riess W, Giere O, Kohls O, Sarbu SM (1999) Anoxic thermomineral cave waters and bacterial mats as habitat for freshwater nematodes. Aquat Microb Ecol 18:157–164

Rohwerder T, Sand W, Lascu C (2003) Preliminary evidence for a sulfur cycle in Movile Cave, Romania. Acta Biotechnol 23:101–107

Sarbu SM (2000) Movile Cave: a chemoautotrophically based groundwater ecosystem. In: Wilken H, Culver DC, Humphreys WF (eds) Subterranean ecosystems. Elsevier, Amsterdam, pp 319–343

Sarbu SM, Kane TC (1995) A subterranean chemoautotrophically based ecosystem. Natl Speleol Soc Bull 57:91–98

Sarbu SM, Lascu C (1997) Condensation corrosion in Movile Cave, Romania. J Cave Karst Stud 59:99–102

Sarbu SM, Popa R (1992) A unique chemoautotrophically based cave ecosystem. In: Camacho AI (ed) The natural history of biospeleology. Mus Nat Hist Naturales, Madrid, pp 637–666

Sarbu SM, Kane TC, Kinkle BK (1996) A chemoautotrophically based cave ecosystem. Science 272:1953–1955

Schirmack J, Mangelsdorf K, Ganzert L, Sand W, Hillebrand-Voiculescu AM, Wagner D (2014) *Methanobacterium movilense* sp. nov., a hydrogenotrophic, secondary-alcohol-utilizing methanogen from the anoxic sediment of a subsurface lake. Int J Syst Evol Microbiol 64:522–527

Stocchino GA, Sluys R, Kawakatsu M, Sarbu SM, Manconi R (2017) A new species of freshwater flatworm (*Platyhelminthes, Tricladida, Dendrocoelidae*) inhabiting a chemoautotrophic groundwater ecosystem in Romania. Eur J Taxon 342:1–21

Tabacaru I, Boghean V (1989) Découverte en Dobrogea (Roumanie) d´une espèce troglobie du genre *Trachelipus* (*Isopoda Oniscoidea Trachelipidae*). Misc Speol Rom 1:53–75

Turk-Prevorčnik S, Blejec A (1998) *Asellus aquaticus infernus*, new subspecies (*Isopoda*: *Asellota*: *Asellidae*), from Romanian hypogean waters. J Crustac Biol 18:763–773

Useful Websites

http://www.bbc.com/earth/story/20150904-the-bizarre-beasts-living-in-romanias-poison-cave

https://en.wikipedia.org/wiki/Movile_Cave

http://www.cultureunplugged.com/documentary/watch-online/play/6889/Life-in-Hell-Survivors-of-Darkness

Dobrogea: Limanu Cave

Virgil Drăguşin, Daniela Dimofte, and Ionuţ Mirea

Abstract

Limanu Cave is a network maze situated in the southern part of the Dobrogea Plateau (SE Romania), close to the Black Sea coast. It is hosted in Sarmatian (Kersonian) lumashelic and oolithic limestones, interbedded with marls. The total length of its passages is close to 3.5 km, and it preserves evidences of important human impact on its wall morphology. Throughout the cave, human intervention can be identified in the form of carving of walls and ceiling, as well as the buildup of walls made of stacked limestone slabs. Mineralogical data reveal an important presence of magnesian calcite and dolomite. The cave is a Site of Community Importance (ROSCI0191) and hosts the last colony of *Rhinolophus mehelyi* in Romania, which gathers ~100 individuals, in a steep decline from ~5000 individuals reported from Dobrogea in the 1960s.

Keywords

Limanu Cave • Maze cave • *Rhinolophus mehelyi* Dobrogea • Romania

Geographic, Geologic, and Hydrologic Settings

Peştera Limanu (Limanu Cave), also known as Peştera de la Icoane, Peştera de la Baltă, or Peştera Caracicola, opens in the southern part of the Dobrogea Plateau, on the south side of Limanu Lake, near the village with the same name (Fig. 1). The coordinates of the main entrance are 43° 48′ 29″N and 28° 31′ 17″E. It is hosted in Sarmatian carbonates pertaining to the Kersonian substage (Mutihac 1990), of which a section composed of twelve layers was described from within the cave by Dumitrescu et al. (1965). These are mostly lumashelic and oolitic limestones, interbedded with greenish bentonite clay. The most notable layer from the section described by Dumitrescu et al. (1965) is #VI, defined by the presence of rounded limestone concretions with diameters of some tens of centimeters that can be observed throughout the cave (Fig. 2).

At present, the cave is dry (fossil), no flowing or stagnant water being present inside it. Small springs appear underneath the cave level, on the northern and western slopes of the cave hill, implying active karstification processes below the present-day cave. The average cave temperature is between 13 and 14 °C (Dumitrescu et al. 1965).

History of Exploration

The cave was known by locals since ancient times, as revealed by the discovery of Roman and medieval artifacts (Boroneanţ and Ciuceanu 1977). However, the first description dates from 1925, when C. N. Ionescu published a sketch of the passages from the entrance area, as well as a list of fauna. Later, Tafrali (1928) undertook an archeological survey within the cave, in connection with the study of the nearby ancient Greek city of Callatis (now Mangalia). Between 1958 and 1964, a group of researchers from the "Emil Racoviţă" Institute of Speleology led by M. Dumitrescu made a detailed study of the cave, including the topographical survey of the cave network. The authors reported a total length of the passages of ~3200 m. In 1975, I. Giurgiu, C. Roman, and E. Roman from the "Emil Racoviţă" Caving Club in Bucharest surveyed new passages, with a total length of 205 m (Giurgiu et al. 1976).

V. Drăguşin (✉) · D. Dimofte · I. Mirea
"Emil Racoviţă" Institute of Speleology, Romanian Academy, Frumoasă 31, 010986 Bucharest, Romania
e-mail: virgil.dragusin@iser.ro

D. Dimofte
Faculty of Geology and Geophysics, University of Bucharest, Nicolae Bălcescu 1, 010041 Bucharest, Romania

G. M. L. Ponta and B. P. Onac (eds.), *Cave and Karst Systems of Romania*, Cave and Karst Systems of the World, https://doi.org/10.1007/978-3-319-90747-5_49

Fig. 1 Location of Limanu Cave in the karst area of South Dobrogea

Fig. 2 Morphology of Passage 55. On the left wall, there could be seen the rounded concretions from layer VI, while in the background there is a pile of limestone slabs. The floor is covered by a ~20-cm thick layer of guano overlaying limestone rubble left over from human activities (photograph by I. Mirea)

Cave Description

The main cave entrance is located on the western side of the hill, toward Valea La Peşteră (the Cave Valley). Another entrance, of very small dimensions, is located on the opposite side (east) of the hill, toward Limanu Lake, and gives access to the second sector. The main entrance passage was altered through the removal of clay from the floor, in order to increase its height and facilitate access.

Although it is a maze, the cave has three distinct sectors linked via single passages (Fig. 3). The first sector is the most labyrinthine, with many loops and galleries intersecting at right angles. The crossing between the first and second sector could also be made via an unsurveyed small maze with very low ceiling south of Chamber 26. Throughout the cave, there are evidences of human intervention in the form of carving of walls and ceiling, as well as the buildup of unconsolidated walls made of stacked limestone slabs. Some of these walls appear to support a collapse-prone ceiling, whereas others were probably made to store unnecessary materials, as they are not built up to the ceiling.

The third sector preserves numerous evidences of human intervention, with carved walls and ceiling, as well as niches of different sizes. Some of the larger niches were interpreted as altars by O. Tafrali, as in the case of Altar C (Fig. 3).

Cave Sediments and Speleothems

The cave lacks any important speleothems, possibly due to reduced percolation and thin soil cover, as is also the case with most of the caves in Dobrogea. Since the cave is hosted by an almost completely isolated hill, presently there are no conditions for the development of a perched aquifer that could feed drip sites inside the cave. The most widespread speleothem type is represented by botryoidal calcite, which appears between the carbonate layers, a feature that is specific for the wider region around the cave. Its presence indicates that the Sarmatian deposits were at some point saturated with water, from which calcite precipitated. They are possibly older than the cave itself, and their dating could provide an estimate of the maximum age of the cave.

The floor is almost completely covered in clay or limestone boulders. It is difficult to assess to which extent these sediments were produced by human activity or by natural processes.

Cave Mineralogy

The only mineralogical study of the cave was published by Diaconu et al. (2008) and was focused on the mineralogy associated with guano deposits. By X-ray diffractometric analysis, the authors identified calcite, hydroxylapatite (accompanied by brushite), dolomite, gypsum (associated with guano deposits), quartz, and illite.

In order to bring more information on the mineralogy of the cave, we analyzed seven mineral samples, labeled L1–L7 (Fig. 3), using microscopy, X-ray diffraction (XRD) using a PANalytical diffractometer, and X-ray fluorescence (XRF) using a Horiba XGT 7000 analyzer. The first sample, L1, is a concretion from layer VI taken from the wall of Passage 10. Sample L2 was collected from Chamber 7 from a layer of bentonite clay referred to by Dumitrescu et al.

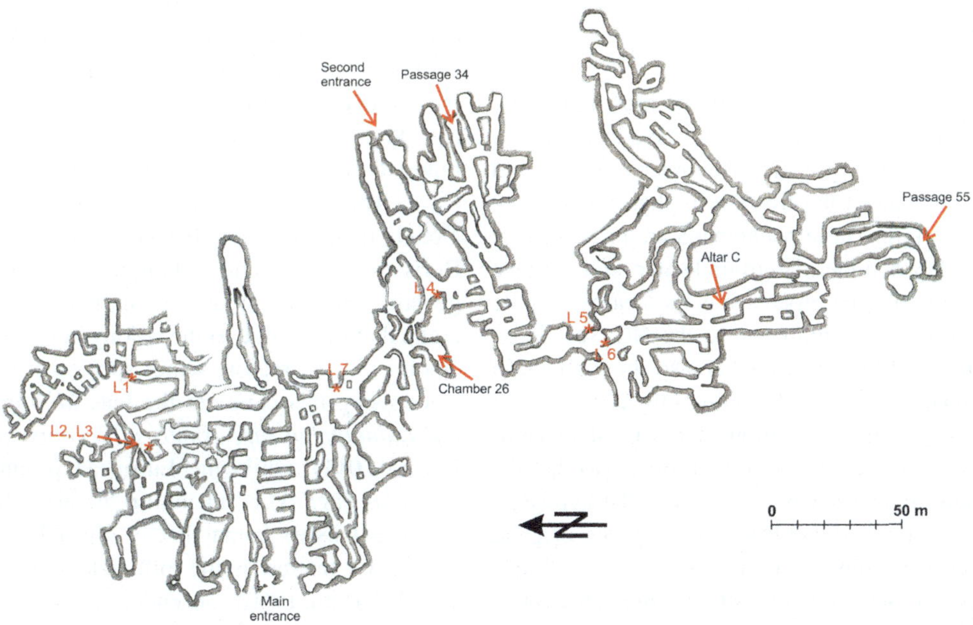

Fig. 3 Map of Limanu Cave (modified from Dumitrescu et al. 1965). Red arrows: chambers and passages cited in text; stars: locations of the mineralogical samples (Color figure online)

Table 1 Chemical composition of samples L1–L7 determined by XRF. Concentrations below 5% are not shown

Sample number	MgO (%)	Al_2O_3 (%)	SiO_2 (%)	CaO (%)
L1	–	–	–	99
L2	5	9	29	52
L3	13	–	11	71
L4	28	–	9	57
L5	–	–	5	90
L6	–	–	–	95
L7	14	–	–	83

(1965), whereas L3 is a fragment of limestone from the ceiling of the same room. Sample L4 is made of crystals deposited on the limestone wall at the western end of Passage 29, and L5 and L6 are botryoidal calcite samples from Room 41. L5 has a more yellowish hue due to the presence of iron oxyhydroxides, whereas L6 is white. Sample L7 is a thin, translucent crust formed on the eastern wall of Chamber 24.

Microscopic and XRD analyses of the L1 sample show that it is made of two types of limestone: a compact, holocrystalline, sparitic calcite (<20% porosity) that makes up the concretions and a more porous calcite (>30%) with occasional opaque, oxidic microconcretions incorporated into the calcite mass which surrounds the concretions.

XRF revealed a high content of Ca in sample L2 (Table 1), which together with Si and Al indicates that it is rather marl and not clay. Sample L3 includes iron oxide clasts, manganese microconcretions, and clay minerals. XRD analysis of sample L4–L7 indicates the presence of magnesian calcite and dolomite, whereas samples L5 and L6 are composed of magnesian calcite.

Cave Speleogenesis

The genesis of Limanu Cave is probably its most intriguing feature, but the least studied. Based on the cave location in a region with significant hypogene phenomena, Onac and Drăguşin (2017) advanced the hypothesis that it may, too, have a hypogene origin. These phenomena are promoted by the existence of H_2S-rich thermal groundwater, and the best example of active hypogene processes is the nearby Movile Cave, situated only a few kilometers to the NE (see Sarbu et al. 2018). Onac and Drăguşin (2017) also drew attention to the resemblance between the labyrinth character of Limanu passages and the disposition of the residual landscape near Movile Cave, which might prove to be the remnant of an unroofed maze cave. At present, the hypogene origin of Limanu Cave is difficult to assess, as the original wall morphology (including that of ceiling and of floor), which may have preserved diagnostic features, has been modified by human intervention. Also, mineralogical evidence like the presence of gypsum speleothems might have been removed too. Regarding the maze character of the cave, it is worth noting the striking resemblance between Limanu and Cserszegtomaji-kút Cave (Hungary), a hypogene cave developed under a non-carbonate cap rock.

A possible cave development scenario for Limanu Cave requires the existence of H_2S-rich thermal groundwater and a stable water table, giving the corrosion processes enough time to enlarge the most permeable paths (joints and bedding planes) creating a 2D network maze (Ford and Williams 1989; Gunn 2004). After the water-fed corrosion processes subside, the tabular, weak limestone is likely prone to collapse, leading to the trapezoidal section of the passages (Fig. 4a, b). While rectangular sections could also evolve under natural conditions, the human factor seems to have been significant at Limanu Cave in most of the passages (Fig. 2).

Cave Fauna

The occurrence here of a bat population of *Rhinolophus mehelyi* marks the northernmost occurrence of this circum-Mediterranean species and is known to be the only colony in Romania. This bat population declined abruptly in the last decades, from around 5000 reported from Dobrogea by Dumitrescu et al. (1963) to around 100 individuals reported by Nagy et al. (2006), all living within Limanu Cave. This puts more pressure on this species, which is already included on the IUCN Red List. In 2011, Dragu and Borissov used genetic techniques to show that there is low genetic variability among the individuals of this colony and argue that this is due to the colony being isolated from other colonies in the region. Other species of community importance using this cave are the bats *Miniopterus schreibersii* and *Rhinolophus hipposideros*, *Plecotus auritus* (Pocora and Pocora 2011), while the European ground squirrel (*Spermophilus citellus*) is inhabiting the area above the cave.

An extensive faunal list is given by Dumitrescu et al. (1965). Of interest is the millipede *Trachysphaera dobrogica* (Tabacaru 1960), an endemic species found only here and in the nearby Peştera Hoţilor (Thieves' Cave). Gruia

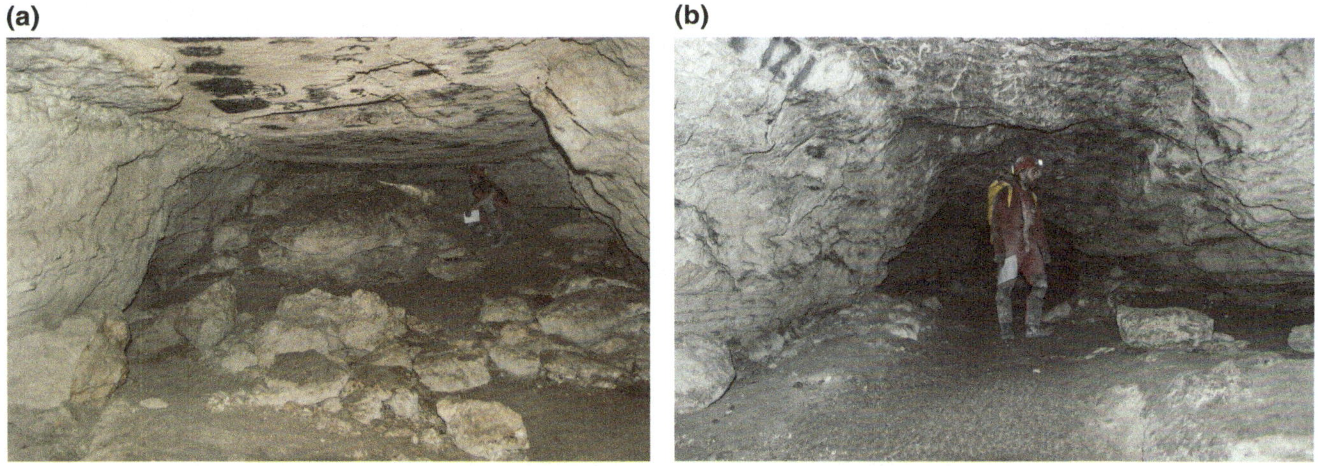

Fig. 4 **a** Trapezoidal section of Passage 34, a typical breakdown morphology in tabular stratification. The walls at this location are made of marl and not limestone. Several fissures in the limestone ceiling allowed for the detachment and collapse of the boulders seen on the floor. **b** Trapezoidal section of the entrance passage, a typical breakdown morphology in tabular stratification (photographs by I. Mirea)

(2003) signals the presence of six more species of Collembola.

Cave Conservancy

The cave is a Site of Community Importance (ROSCI0191) since 2008 and is currently in the custody of the Underwater and Cave Exploration Group (Bucharest), and a permit from the custodian is needed for cave visits.

Acknowledgements We thank the Underwater and Cave Exploration Group for permission to work inside Limanu Cave. This work was financially supported by the CARPATHEMS project (PCE-2016-0197 to S. Constantin).

References

Boroneanţ V, Ciuceanu R (1977) Cercetările arheologice din Peştera de la Limanu, Judeţul Constanţa. Rev Muz Monum - Monumente Ist Artă 46(2):49–57

Diaconu G, Marincea Ş, Dumitraş D (2008) Data on the Limanu Cave mineralogy, South Dobrogea. Trav Inst Spéol "Émile Racovitza" 47:65–77

Dragu A, Borissov I (2011) Low genetic variability of *Rhinolophus mehelyi* (Mehely's horseshoe bat) in Romania. Acta Theriol 56:383–387

Dumitrescu M, Tanasache J, Orghidan T (1962–1963) Răspândirea chiropterelor în Republica Populară Română. Lucr Inst Speol "Emil Racoviţă" 1–2:509–576

Dumitrescu M, Orghidan T, Tanasachi J, Georgescu M (1965) Contribuţii la studiul monograpic al Peşterii de la Limanu. Trav Inst Spéol "Émile Racovitza" 4:21–58

Ford DC, Williams PW (1989) Karst geomorphology and hydrology. Chapman and Hall, London, p 601

Giurgiu I, Ceară E, Roman C (1976) Noi descoperiri în Peştera de la Limanu. Bul Club Speol "Emil Racoviţă" Bucureşti 4:170–172

Gruia M (2003) Collembola from Romanian caves. Trav Mus Hist Nat "Grigore Antipa" 35:139–158

Gunn J (2004) Encyclopedia of caves and karst science. Taylor and Francis, New York, p 902

Ionesco CN (1925) La grotte de Caracicola (pres de Mangalia-Dobrogea). Ann Sci Univ Jassy 13(1–4):474–478

Mutihac V (1990) Structura geologică a teritoriului României. Ed Tehnică, Bucureşti

Nagy Z, Barti L, Dóczy A, Jére Cs, Postawa T, Szántó L, Szodoray-Parádi A, Szodoray-Parádi F (2006) Survey of Romania's underground bat habitats. Status and distribution of cave dwelling bats. Report for BP Conservation Programme, pp 1–44

Onac BP, Drăguşin V (2017) Hypogene caves of Romania. In: Klimchouk A (eds) Hypogene karst regions and caves of the world. Springer International Publishing, Cham, pp 257–265

Pocora I, Pocora V (2011) Seasonal distribution of cave-dwelling bats and conservation status of underground habitats in Moldova and Dobrogea (Romania). Stud Cerc, Univ "Vasile Alecsandri" Bacău 20(2):72–83

Sarbu S, Lascu C, Brad T (2018) Movile Cave. In: Ponta GML, Onac BP (eds) Karst and cave systems of Romania. Springer International Publishing, Cham, pp 429–436

Tabacaru I (1960) Neue cavernicolen *Gervaisia*– Arten: *G. racovitzai* n. sp., *G. spelaea* n. sp. und *G. dobrogica* n. sp. Zool Anz 165 (7/8):311–320

Tafrali O (1928) Les Tumuli de Callatis. Arta şi Arheol I(II):23–53

Evaporite Karst of Romania

Maria-Laura Tîrlă

Abstract

Evaporite karst in Romania is mostly related to Miocene occurrences in the intra-Carpathian Basins, Subcarpathian Nappe, northern Dacian Basin and, to a minor extent, the Eocene in the NW Transylvanian Basin. Evaporitic karst areas account for 259 km^2 (salt karst 48%, gypsum and anhydrite karst 52%). The most typical landforms are: dolines, karren, and caves. Until now, 53 salt and 20 gypsum and anhydrite caves, respectively, were recorded in Romania. The salt karst from Meledic Plateau (Mânzălești, Vrancea region) records the highest number of caves (47) and karst landscape complexity, followed by the Praid–Corund and Slănic–Prahova salt mountains. Romania's most notable evaporite endokarst feature is the 3234-m-long 6S Cave from Mânzălești. Karren morphology is sharp-edged, particularly dominated by rills, runnels, and straight wall channels. Mining and controlled flooding of abandoned mined chambers have significantly influenced the evolution and functioning of salt karst systems by formation of anthropogenic saline lakes and new active underground conduits (Slănic-Prahova, Ocna Mureș). Solution of gypsum and anhydrite has a more discrete imprint on the related karst landscape, outlined in the Meseș Mountains, Turzii Gorge, Tazlău Subcarpathians, Maramureș Depression, and Hunedoara Basin. Dolines and karren are the most prevalent exokarst landforms in gypsum, whereas caves are short (up to 30 m) and speleothems rather poor.

Keywords

Salt karst • Gypsum • Landforms • Meledic Plateau Subcarpathians • Romania

M.-L. Tîrlă (✉)
Faculty of Geography, University of Bucharest,
Nicolae Bălcescu 1, 010041 Bucharest 1, Romania
e-mail: tirla@geo.unibuc.ro

M.-L. Tîrlă
Research Institute of the University of Bucharest (ICUB),
Mihail Kogălniceanu 36-46, 050107 Bucharest 5, Romania

Introduction

Evaporite karst is commonly referred to as the *pseudokarst*, which consists, *sensu largo*, of karst landforms developed on other types of rocks than carbonates (Onac 2000). These rocks can be volcanic, metamorphic, or sedimentary, with variable contributions to cave-forming processes and related surface karst features. Although it involves mostly water dissolution and rapid precipitation, the evaporite karst resembles many features of typical carbonate karst compared to other pseudokarst landforms.

Evaporite karst in Romania is related to rock salt (halite), gypsum, and anhydrite deposits. These are generally overlain by nonkarst rocks (e.g., clay and marl, with their petrologic variations), and outcrop over small areas, sparsely distributed across the country. The total area of halite outcrops was estimated by Sencu et al. (1973) at 150 km^2, meaning 0.3% of the soluble rocks and 0.06% of the country area. Later, Ponta (1986) reported a 5% area covered by evaporite karst. This chapter aims to reassess evaporite outcrop distribution based on data integration from the geological maps of Romania.

Geographic, Geologic, and Hydrologic Settings

The four-evaporite sequences distinguished in Romania are significant stratigraphic markers of the Central and Eastern Paratethys Basins of Eocene and Miocene Age. However, karst outcrops are rather isolated, because evaporites are much more soluble than carbonates (Calaforra 1998; Klimchouk 2004).

Eocene (Priabonian) gypsum outcrops in the NW Transylvanian Basin. The Miocene deposits include the Lower Salt Formation (Aquitanian–Early Burdigalian), the Grey Schlier Formation (Late Burdigalian-Middle Badenian), and the Upper Salt Formation (Middle Badenian, corresponding

G. M. L. Ponta and B. P. Onac (eds.), *Cave and Karst Systems of Romania*,
Cave and Karst Systems of the World, https://doi.org/10.1007/978-3-319-90747-5_50

to Langhian-Serravallian Mediterranean stages). These were deposited in the Carpathian Foredeep and later thrusted within the Subcarpathian Nappe over the margin of the East European Platform during the late Miocene events. The most extensive evaporites are part of the Măghirești-Perchiu Digitation, the innermost tectonic unit of the Subcarpathian Nappe (Săndulescu 1984).

The depositional thickness of the Badenian rock salt layer in the Transylvanian Basin was estimated at $\sim$300 m (Krézsek and Bally 2006). Halite precipitated preferentially in the center part of deep lacustrine/marine basins (Balintoni and Petrescu 2002; Krézsek and Filipescu 2005), whereas gypsum and anhydrite on basin margins (Krézsek and Bally 2006), in a lagoon or sabkha-type environment (Ghergari et al. 1991). Superposition and lateral variations of the evaporites and related deposits allowed separation of distinct parasequences, developed under the control of sea-level oscillations. The sequence stratigraphy of Transylvanian Basin was recently performed and used to decipher its depositional evolution, including that of the evaporites (Maţenco et al. 2010).

The topmost evaporite deposits are commonly related to a major drought event followed by sea-level fall in the Paratethys domain during the Badenian stage. Large-scale tectonic movements and isolation of the Central and Eastern Paratethys Basins from the Indo-Pacific and Mediterranean seaways have triggered the onset of Badenian Salinity Crisis at 13.81 $\pm$ 0.08 Ma, which lasted between 200 and 600 kyr (Rögl et al. 1978; de Leeuw et al. 2010). Afterward, the salt tectonics has led to formation of diapirs and crypto-diapirs in the Transylvanian Basin, Carpathian Foredeep, and northern Dacian Basin. Salt karst developed on the exhumed diapirs sparsely distributed from Reghin to Sovata, Praid, Corund (Stoica 1938; Irimuş 1998), Odorhei, Ocna Sibiului (Alexandru 1960), Ocna Mureşului (Nagy 1980), Turda, Cojocna, Ocna Dejului, and other smaller areas (Balintoni and Petrescu 2002). Other crypto-diapirs are clustered into the basin center at various depths (Dumitrescu et al. 1976), but their role on karst development is less relevant. The Maramureş Basin hosts exhumed salt diapirs at Ocna Şugatag and Coştiui, and a crypto-diapir at Vlad (Drăgănescu 1997). The Aquitanian and Badenian rock salt formations of the Carpathian Foredeep outcrop in several notable saline areas: Solca-Cacica, Târgu Ocna-Oneşti, Vintileasca, Sările, Jghiab, Lopătari-Mânzăleşti (Meledic Plateau), Băicoi, Slănic-Prahova, etc. (Fig. 1). An isolated diapir surrounded by crypto-diapirs outcrops at Ocnele Mari-Ocniţa, in the northern Dacian Basin. Most salt massifs were historically mined, and anthropogenic imprints accelerated or expanded natural karst-forming processes. Anthropo-saline lakes resulted by collapsing of mined chambers are frequent.

Gypsum layers contain anhydrite, a diagenetic mineral, which forms by gypsum dehydration depending on thermal and pressure conditions and water activity during and after precipitation in the oversaturated brine (Hardie 1967). As a rule, in concentrated seawater in the evaporating basins of lagoons, primarily anhydrite already forms at 25 °C (Bögli 1980). Therefore, when referring to gypsum karst, its self-evident that anhydrite is also included.

Karst features on Priabonian gypsum are well exposed in the Meseş Mountains, nearby Stâna village (Viehmann and Mac 1966). Badenian gypsum layers (10–15 m) occur in the Şimleu Basin (NW Romania) around the Hăghişa horst and on the valleys incised into the western side of the Meseş Mountains: Ragului, Blidăreasa, Maşchii, etc. (Clichici 1973). The karst landforms on these last locations were investigated in 1989 by Onac (unpublished material). The Ca-sulfates of the Cheia Formation outcropping at Cheia (Turda) consist of a gypsiferous facies of 15–20 m in thickness with anhydrite nodules (Ghergari et al. 1991). The Eocene evaporite sequence at Stâna and Badenian layers at Cheia (Meseş and Turda, NW Romania) were described as fine-grained gypsum of high purity (more than 50% alabaster) and perfect cleavage. Differences in structure and habit occur: the Cheia alabaster displays an irregular texture with grains and lamellae in micro-folded structures, while the Stâna alabaster is made of plane-parallel lamellae (Viehmann and Mac 1966). A calcite content up to 20% increased rock hardness to at least 2 (Stoicovici and Moţiu 1957). Karst developed on Badenian gypsum was also reported from the Maramureş Depression, southwest of Botiza (Onac and Istvan 1994), and Hunedoara Basin, where gypsum and anhydrite layers vary between 2 and 40 m in thickness (Băicoană et al. 2000). Caves and karst from the Tazlău and Oituz Basins (Moldovei Subcarpathians) are related to Badenian Perchiu Unit that includes gypsum, marl, and sandstone of 0.1–0.5 m in thickness (Brânduş 1981; Grozavu 1990). Aquitanian gypsum stripes occur in the northern Dacian Basin at Nucşoara and Câmpulung-Suslăneşti (south of Făgăraş Mountains).

Cave development is strongly related to local groundwater levels, river activity, saline lakes water, and precipitation water that infiltrate along fractures and fill in the underground voids. The development of lakes on the collapsed bell-shaped chambers in the old salt mines has influenced groundwater levels, accelerated the dissolution process, and contributed to salt karst expansion.

Distribution of evaporite karst in Romania is shown on Fig. 1. The most important and well-developed salt karst landforms occur on the Meledic Plateau (3.47 km^2), located in the Bend Carpathian–Subcarpathian contact area at 500–650 m asl. However, the Lopătari-Mânzăleşti salt karst includes areas of neighboring basins—Meledic, Jghiab, and

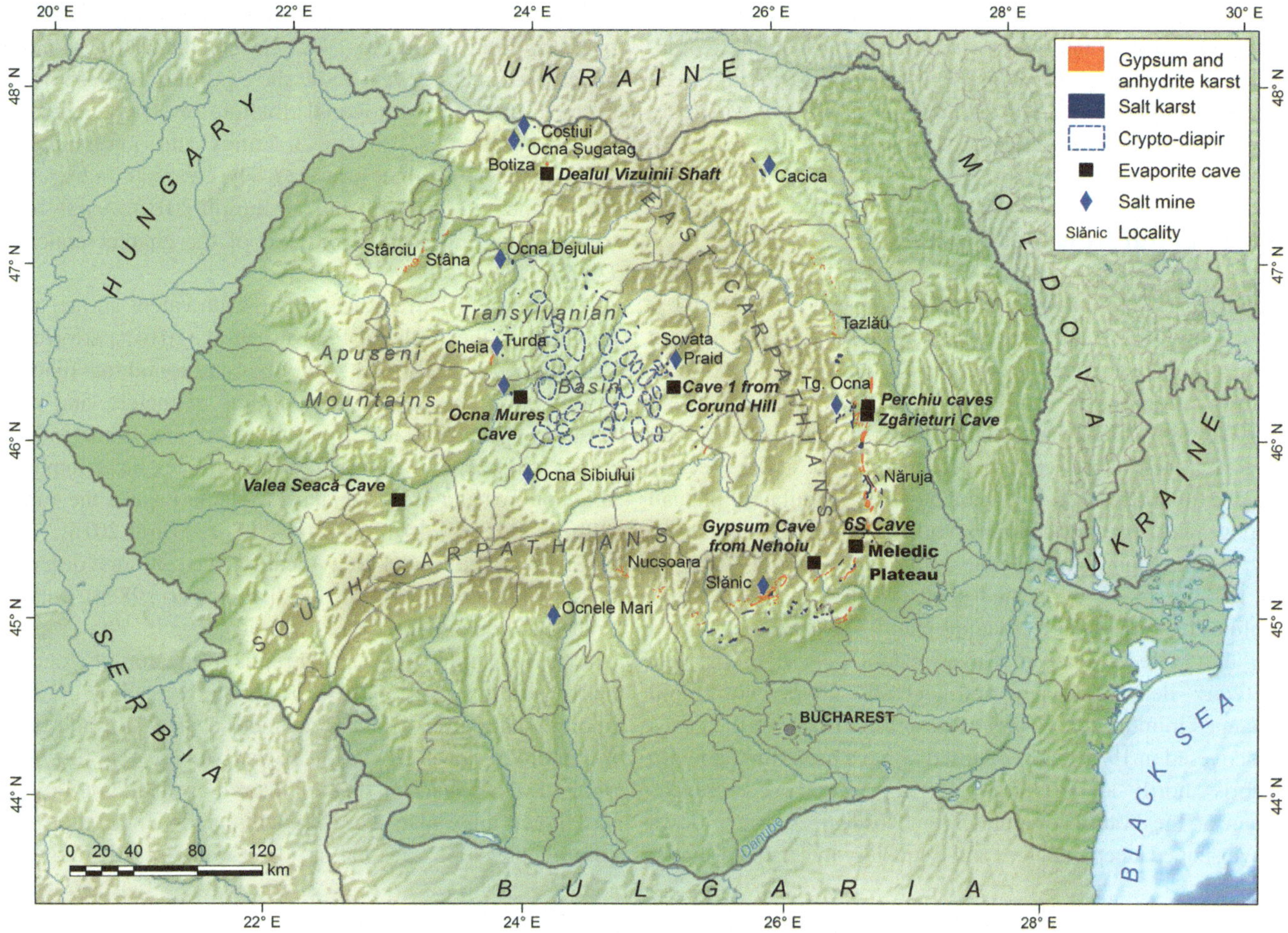

Fig. 1 Distribution of the evaporite karst in Romania, with important caves and salt mines. Relief map was used with permission under the CC 3.0 license

Sărăţel, totaling a surface of 6 km^2. The spatial extent of evaporite karst is related to evaporite outcrops distribution, synthesized from the geological maps of Romania at the scales of 1:200,000 (1966–1968) and 1:50,000 (1976–1998).

History of Exploration and Research

The caves of Lopătari-Mânzăleşti in the Meledic Plateau are the most numerous and record the highest density of karst conduits. These caves were discovered and explored in 1978–1980 by members of the "Emil Racoviţă" Caving Club in Bucharest. The 6S Cave was surveyed between 1980 and 1992 by I. Giurgiu, G. Silvăşanu, G. Miclăuş, E. Solomon, E. Roman, C. Radu, M. Vlădulescu, M. Popescu, A. Rădulescu, G. Chiriloi, L. Minoniu, R. Rădulescu, V. Oprea, L. Grad, and M. Vasilică (Giurgiu 2010). Morphological descriptions of salt karst landforms were made by Ielenicz (1975), Giurgiu et al. (1980), Ielenicz (1985), Giurgiu (1985b), etc. Chemical analyses on halite speleothems were performed by Todor et al. (1993) and Istvan (pers. comm.) on samples taken from 6S Cave.

Viehmann and Mac (1966) investigated the gypsum surface karst from Meseş Mountains (Stâna) and Cheile Turzii (Cheia), contributing with mineralogical insights on the gypsum varieties and their crystallographic properties. Brânduş (1981) described for the first time the gypsum karst in the Tazlău Subcarpathians (Perchiu Hill and Oituz Basin). Later, the members of the Montana-Oneşti Caving Club discovered the longest gypsum and anhydrite cave in Romania, Zgârieturi Cave, and six other shorter cavities (Puşcarciuc and Buzdugan 1987; Grozavu 1990). In a wider context on karst landforms developed on the Miocene rocks in the NW Romania, Onac and Istvan (1994) explored and discussed the gypsum caves and karst from Botiza (Maramureş Depression). The gypsum and anhydrite karst identified in the outskirts of Hunedoara City (Valea Seacă) was explored by Proteus Hunedoara Caving Club (Băicoană et al. 2000).

The current list includes 73 caves and shafts in evaporites, 53 of which developed in rock salt and 20 in gypsum and anhydrite. Except for nine of these cavities, all the others were announced and have a confirmed status in the Romanian caves catalog (Goran 1982; Vlaicu pers. comm.). The most important ones are listed in Table 1, some of which are presented later in this chapter.

Karst Landforms Description

Karst developed on rock salt (124.3 km^2) is very complex, especially where salt underwent recent exhumation from beneath the overlying shales and is subjected to a new karstification phase.

The salt karst at Mânzăleşti, Vrancea region, developed on the Meledic Plateau and adjacent units, is the most spectacular evaporite karst in Romania. The undulated plateau topography strongly contrasts with its margins, heavily cut by deep salt gorges (Meledic, Jghiab, and Sărăţel streams). A number of 47 caves and shafts (Giurgiu 2010) and 87 dolines hosting water or swamps (Strat 2016) were identified on the Meledic Plateau. Caves record here high passage network densities. They are generally short, with rapid changes in morphology and accessibility. Many of the caves discovered in 1980 have collapsed or have been filled up with debris shortly after (Giurgiu 1985a). Most cave entrances open at the bottom of dolines, and organized water flows have shaped steep passages and frequently meter-deep shafts.

The 6S Cave (3234 m) is considered the longest salt cave in Europe (e.g., Giurgiu 1995). Major dissolution conduits developed on 3–4 levels under the control of a NE–SW- and NW–SE-oriented orthogonal fracture system created by extensional forces. Pinnate fractures and joints allowed water to dissolve minor passages. The cave has an extremely high branching index of 11.26 (Giurgiu 2010). A link to the cave map is provided at the end of the chapter under the *Useful websites*.

Dolines and karren are the most prevalent surface karst features on the Meledic Plateau. Strat (2016) performed a morphometric assessment of dolines and found that diameter ranges between one and several tens of meters, and depth between 1 and 16 m. Most of them are funnel-shaped, round, ovate, or asymmetric. Some dolines have impermeable bottoms and preserve lakes; Meledic Lake is the largest, with an area of 1.14 ha and 6.9-m deep (Strat 2016).

Sharp-edged karren are typical for the salt karst described at Meledic (Giurgiu et al. 1980; Giurgiu 1985a), Slănic--Prahova, and Praid-Corund (Stoica 1938; Veress et al. 2011). Development of salt karren is a function of slope and halite purity, where pure salt is interbedded with thinly impure salt-and-clay layers, a micro-scaled cliff-and-bench morphology develops on the karren slopes (Fig. 2a). Morphological analysis of the salt karren from Slănic-Prahova has shown three main development stages, under the strong control of slope steepness: (1) formation of individual

Table 1 Morphometry of some relevant evaporite caves in Romania

Catalog no.	Name	Evaporite rock	Geographic unit	Development (m)	Depth/Height (m)	References
01-610/14	6S Cave	Salt	Meledic Plateau	3234	−40/+2	Giurgiu et al. (1980)
01-610/6	Three Entrance Cave from Săreni	Salt	Meledic Plateau	285	−45	
01-117/1	Cave 1 from Corund Hill	Salt	Corund Hill, Transylvanian Basin	217.5	+13.2	Giurgiu (1992)
01-117/2	Cave No. 2 from Praid	Salt	Transylvanian Basin	208.4	+21	Vlaicu (pers. comm.)
01-611/7	The Salt Cave from Miroiului Valley	Salt	Sării Hill, Buzău Subcarpathians	204	−22	Ristivan et al. (2003)
01-610/28	20S Cave	Salt	Meledic Plateau	180	+25.25	Giurgiu (1985a)
01-610/43	Izvorul Sărăţelului Cave	Salt	Meledic Plateau	153	−6.5	Giurgiu (1985a)
01-610/15	7S Cave	Salt	Meledic Plateau	152	−11/+3	Giurgiu (1985a)
01-/-	Ocna Mureş Cave	Salt	Ocna Mureş	50	−20	Nagy (1980)
02-601/8	Zgârieturi Cave	Gypsum Anhydrite	Oituz Basin, Tazlău Subcarpathians	30	−2.5/+1.5	Puşcarciuc and Buzdugan (1987)
02-/-	Dealul Vizuinii Shaft	Gypsum Anhydrite	Maramureş Depression	26	−10	Onac and Istvan (1994)
02-/-	Valea Seacă Cave	Gypsum Anhydrite	Hunedoara Basin	8.2	+0.65	Băicoană et al. (2000)

Fig. 2 Salt karst features in the Meledic Plateau: **a** Complex karren morphology; **b** cave passage section with abundant redeposited salt crusts; **c** anemolites; **d** stalactite formed by imbricated halite cubic crystals. Photographs courtesy of A. C. Mitrofan (**b–d** Three Entrance Cave from Săreni)

micro-pits (rain pits) by solution; (2) interconnection of rain pits by linear channels or rills; and (3) organization of a complicated dendritic network (Sencu 1967). Veress et al. (2011) have further detailed three generations of karren in the Sării Hill from Praid. Salt karren most often resembles the following types: rillenkarren (rills), rinnenkarren (runnels), and a sub-type of the latter named wandkarren (wallkarren) developed on the steepest slopes (65–90°). Extremely sharp edges and needle-like pinnacles are also common positive forms of the salt karren.

The salt karst from Băile Verzi-La Noroaie (Slănic-Prahova) is controlled by the activity of groundwater in a particular way. The presence of karst landforms and especially the activity of ponors were related to a permanent seepage and, moreover, a highly variable discharge of two water sources, validated by dye (fluorescein) studies (Povară et al. 1985).

Karst developed on gypsum and anhydrite (134.6 km^2) accounts for 52% of the total karst evaporite occurrences, but significant landmarks occur only in several places, as described below.

A comparative study of gypsum surface karst in the Meseş Mountains and similar carbonate karst landforms show the different behavior of sulfate versus carbonate minerals during karstification, especially the development of micro-karren and scallops. In gypsum, the latter were shaped by straightforward solution process rather than carbonic acid dissolution, which dominantly acts on limestones (Viehmann and Mac 1966).

The caves from Tazlău Subcarpathians developed in the Perchiu Unit along bedding planes dipping 30°. The karst system typically includes dolines, ponors, and karren (Brânduş 1981). Funnel-shaped dolines have a diameter/depth ratio of 4:1, and their bottoms generally end with ponors showing recent drainage activity (Grozavu 1990). Seven caves and shafts were discovered in this area (Perchiu caves on Fig. 1), Zgârieturi Cave being the longest of all, 30 m (Puşcarciuc and Buzdugan 1987).

In the north of Maramureş Depression, the gypsum layers near Botiza preserve typical karst landforms: karren, dolines, ponors, and springs between the Sas and Mireş valleys. The fusion of four dolines resulted in a 400-m-diameter depression. Three NE–SW-oriented short cavities (11–26 m in length) are strongly controlled by tectonics and structure: Dealul Vizuinii Shaft, Vizuina Bursucilor Cave, and Pădurea Cărbunar Cave (Onac and Istvan 1994).

The Valea Seacă Cave in the Cerna Valley, Hunedoara Basin, is a remnant of a formerly longer cavity. After the recent collapse of the main chamber, its length is only 8.2 m. The karst plateau topography is undulated by prominent dolines of 50–100 m in diameter (Băicoană et al. 2000).

The gypsum dolines and Învârtita Lake in Nucşoara, south of Făgăraş Mountains, used to be considered a landmark of the evaporite karst in Romania (Trufaş 1960). A large doline functioned as a drainage-controlling landform until a heavy rainfall triggered its complete filling and formation of the lake, at the middle of the nineteenth century (Trufaş 1963).

Speleothems and Cave Sediments

Speleothems are abundant especially in salt caves. Almost entirely, the cave passages are coated by a white crust made up of halite (Fig. 2b). Anemolites are the most frequent halite speleothems, best described and analyzed in the caves of the Meledic Plateau (Fig. 2c). Conic or tubular stalactites reach about 1.0–1.3 m in length and 0.15–0.2 m in width (Ielenicz 1975; Giurgiu 1985b). Spectacular stalactites made of imbricated cubic crystals indicate a very rapid growth rate (Fig. 2d). Small eccentrics such as helictites do not exceed 20 mm in length and 2 mm in width (Giurgiu 1985b). Comparatively, stalagmites are much rarer and smaller.

Fine-grained sediments in evaporite caves consist mainly of argillaceous material carried by streams. Coarser sediments originated from the rudites, arenites, and breccia often included in or overlying the salt formations.

Cave Speleogenesis

Despite its high solubility, halite forms impervious rock bodies, which lack primary porosity and prevent water from infiltrating and creating underground karst features. Sencu (1965) underlined the contribution of secondary porosities (especially of unloading fractures) to speleogenesis. Giurgiu (1985a) hypothesized on the genesis of the 6S Cave in the Meledic Plateau, as being formed initially under phreatic conditions and then continued in vadose regime. The surface drainage patterns were systematically transferred to the underground, where the conduits organized at the contact between the salt layer and overlying shales, then deepening into the salt body along cracks and fractures. The same model was also considered for the other caves in the area.

The impact of anthropogenic works on speleogenesis in salt massifs was confirmed by most studies. Mining activities have opened new fractures into the underground, disturbing pre-existent groundwater equilibrium and forcing it to achieve new drainage paths, e.g., the Ocna Mureş Cave (Nagy 1980).

Based on the assumption that water chemistry has a key role on the hydrodynamic processes responsible for salt dissolution and karst development, Povară et al. (1997) performed analysis of Cl concentrations in the groundwater of Unirea Mine (Slănic-Prahova) and obtained the highest values in the salt topography depressions and plateau areas, 122.8–161.8 g/l Cl^-. Results outlined the undersaturated status of groundwater and its high solution capacity at all sampling sites. The lower the NaCl content in water, the higher its dissolution capacity (Sencu 1965).

Evaporite Speleothem Mineralogy

Chemical analysis of 20 halite samples from the Meledic caves (speleothems and other precipitation forms) has shown the variable participation of halogenic and non-halogenic compounds: NaCl (53.9–99.9%), Fe_2O_3 (0.009–8.884%), MnO (0–0.045%), $CaSO_4$ (from less than 0.001% to 6.379%), $CaCO_3$ (0.012–40.56%), and $MgCO_3$ (0.01–0.694%) (Todor et al. 1993). This was the first mention of the rare mineral manganosite (MnO) in halite speleothems from Romania. Later, Istvan (pers. comm.) noted the constant presence of the halite–gypsum association in five analyzed speleothems and reported kieserite, $Mg(SO_4)\cdot H_2O$, traces in one of them. The evaporites from Valea Seacă (Hunedoara Basin) consist of interbedded anhydrite and gypsum layers displaying sugar-like, fibrous and lamellar textures, white to white-yellowish and yellowish to brown in color, depending on the weathering degree and argillaceous minerals content. The gypsum crystals that decorate the cave interior have similar color transitions and range from transparent to opaque (Băicoană et al. 2000).

6S Cave Climatology

Continuous, long-term datasets on cave air temperatures are lacking. Spontaneous surveys have shown that in the 6S Cave, air temperature varies between 8.5 and 10.2 °C (Giurgiu 2010), slightly lower than the outside mean annual temperature (10.7 °C). This offset is explained by the high surface temperature values, which are common in the external Subcarpathian Bend area due to the persistence of a heat island generated by the foehn wind effects.

Cave and Karst Conservancy

Meledic Plateau is a National Geoheritage area (Mărunţeanu and Ioane 2010), a Natura 2000 site (ROSCI0199 Meledic Plateau), and an esteemed geomorphosite (Irimia and Irimuş 2012). However, frequent collapses and breakdown are major risks for cavers, scientists, and potential cave visitors, reasons why no evaporite show caves exist in Romania. This statute serves, more or less, to their natural long-term conservation. The salt mountains in Slănic-Prahova and Praid-Corund are also geological and geomorphological reserves of national importance.

Conclusions

An updated map of the evaporite karst and caves of Romania was produced in this study. Accordingly, the gypsum and anhydrite karst landforms cover 134.3 km^2, and the saline karst 124.3 km^2. Salt caves and karst developed on diapirs and crypto-diapirs at shallow depths in the intra-Carpathian Basins and Subcarpathian Nappe. Gypsum karst usually occurs as stripes, parallel to the strike of the sedimentary layers. Dolines, karren, and caves are the most prevalent karst landforms. Salt caves reach considerable lengths and branching indices, whereas the gypsum caves do not exceed several tens of meters in length.

Meledic Plateau (Vrancea region) was designated a National Geoheritage Site and exhibits the most spectacular saline karst in Romania. It has the highest number (47) and density of evaporite caves (7.83 caves/km^2). The 6S Cave (3234 m) is considered the longest salt cave in Europe and one of the longest in the world (Giurgiu 2010). Other notable and complex karst areas are the salt mountains in Slănic--Prahova (Bend Subcarpathians) and Praid-Corund (Transylvanian Basin). The Ocna Mureş Cave represents a typical salt cave developed in a short time under the human-controlled flooding of mine chambers.

While the salt caves in Romania have been investigated to some extent, the knowledge on gypsum and anhydrite caves is still poor and lack consistency. Although not numerous and generally short, these caves may provide interesting information on mineralogy, speleogenesis, and karst geomorphology in local and regional context.

Acknowledgements We would like to thank A.C. Mitrofan for granting permission to use high-quality photographs of the Meledic salt karren and speleothems. We express our sincere gratitude to B.P. Onac for constructive suggestions and discussions, which helped improving the chapter quality. I. Povară, C. Goran, and N. Cruceru are acknowledged for providing support on materials and cave catalogue data search.

References

Alexandru M (1960) Fenomene carstice pe masivul de sare de la Ocna Sibiului. Natura 2:135–136

Balintoni I, Petrescu I (2002) A hypothesis on the Transylvanian halite genesis. Studia UBB Geol 1(Special Issue):51–61

Băicoană M, Breban RC, Bosdoc T, Tomuş B (2000) Pseudocarstul - Fenomene endo- şi exo-. Clubul Speologilor Proteus Hunedoara, Hunedoara, p 45

Bögli A (1980) Karst hydrology and physical speleology. Springer International Publishing, Cham, p 284

Brânduş C (1981) Subcarpaţii Tazlăului. Studiu geomorfologic. Ed Acad RSR, Bucureşti, p 141

Calaforra JM (1998) Karstología de yesos. Monografías Ciencia y Tecnología. University of Almeria, Almeria, p 392

Clichici O (1973) Stratigrafia Neogenului din estul Bazinului Şimleu. Ed Acad RSR, Bucureşti, p 178 de Leeuw A, Bukowsi K, Krijgsman W, Kuiper KF (2010) Age of the Badenian salinity crisis; impact of Miocene climate variability on the circum-Mediterranean region. Geology 38(8):715–718

Drăgănescu L (1997) Originea sării şi geneza masivelor de sare. Grafica Prahoveană, Ploieşti, p 226

Dumitrescu I, Săndulescu M, Popescu I (1976) Tectonica şi evoluţia paleogeografică. Plate II-2, Scale 1:1.500.000, Inst Geol Geofiz, Bucureşti

Ghergari L, Mészáros N, Hosu A, Filipescu S, Chira C (1991) The gypsiferous formation at Cheia (Cluj County). Studia UBB Geol XXXVI(1):13–28

Giurgiu I (1985a) Peşteri în sare din Subcarpaţii Vrancei. Bul Club Speol "Emil Racoviţă" 9:5–34

Giurgiu I (1985b) Grottes du sel de Roumanie. Spelunca 20:34–39

Giurgiu I (1992) Pseudocarst în România. Cerc Speol 1:85–88

Giurgiu I (1995) Peştera 6S de la Mânzăleşti (Subcarpaţii Vrancei, România), una dintre cele mai importante cavităţi în sare din lume. Cerc Speol 3:13–17

Giurgiu I (2010) Cea mai mare peşteră în sare din România. Natura României 21:123–215

Giurgiu I, Silvăşanu G, Vădeanu T (1980) Observaţii asupra carstului pe sare din zona Meledic (M. Buzău). Peştera 6S de la Mânzăleşti. Comparaţie între formele carstice dezvoltate în sare şi calcare. Bul Club Speol "Emil Racoviţă" 7:55–66

Goran C (1982) Catalogul sistematic al peşterilor din România. Ed CNEFS, Bucureşti, p 496

Grozavu A (1990) Considérations sur le relief karstique en gypse dans les Souscarpathes de Tazlău. Lucr Sem Geogr "Dimitrie Cantemir" 10:33–38

Hardie LA (1967) The gypsum-anhydrite equilibrium at one atmosphere pressure. Am Mineral 52:171–200

Ielenicz M (1975) Problèmes de morphologie karstique et pseudokarstique dans le nord des Subcarpates du bassin du Buzău. Univ Bucureşti, St Geogr, pp 179–193

Ielenicz M (1985) Karst et pseudokarst dans des montagnes et les collines subcarpatiques des bassins du Buzău et du Teleajen. Theor Appl Karstol 2:47–53

Irimia D-N, Irimuş I (2012) The touristic area Lopătari-Mânzăleşti. The role of tourism in territorial development. In: Vth International conference (October 10, 2012), Gheorgheni, Romania. Presa Univ Clujeană, Cluj-Napoca, pp 134–140

Irimuş A (1998) Relieful pe domuri şi cute diapire în Depresiunea Transilvaniei. Presa Univ Clujeană, Cluj-Napoca, p 311

Klimchouk A (2004) Evaporite karst. In: Gunn J (ed) Encyclopedia of caves and karst science. Fitzroy Dearborn, New York, pp 343–347

Krézsek C, Bally AW (2006) The Transylvanian Basin (Romania) and its relation to the Carpathian fold and thrust belt: insights in gravitational salt tectonics. Mar Petrol Geol 23(4):405–442

Krézsek C, Filipescu S (2005) Middle to late Miocene sequence stratigraphy of the Transylvanian Basin (Romania). Tectonophysics 410(1–4):437–463

Mărunţeanu C, Ioane D (2010) Salt karst in Mânzăleşti - Romania. In: Evelpidou N, DeFigueiredo T, Mauro F, Tecim V, Vassilopoulos A (eds) Natural heritage from east to west. Springer International Publishing, Cham, pp 105–109

Maţenco LC, Krézsek C, Merten S, Schmid S, Cloetingh S, Andriessen P (2010) Characteristics of collisional orogens with low topographic build-up: an example from the Carpathians. Terra Nova 22:155–165

Nagy I (1980) Carstul salin din zona zăcământului de la Ocna Mureş. Bul Club Speol "Emil Racoviţă" 7:87–90

Onac BP (2000) Geologia regiunilor carstice. Ed Didactică şi Pedagogică, Bucureşti, p 399

Onac BP, Istvan D (1994) Le karst développé sur des roches miocènes du nord-ouest de la Roumanie. Trav Inst Spéol "Émile Racovitza" XXXIII:173–178

Ponta G (1986) Evaporites karst from Romania. In: ATTI - Simposio Internazionale sul Carsismo nelle Evaporiti "Le Grotte d'Italia". Bologna, 2–26 ottobre 1985. Fed Speleol Regionale dell' Emilia-Romagna; Inst Ital Spelol, Bologna, pp 407–415

Povară I, Cosma R, Lascu C, Bulgăreanu VA (1985) Un cas particulier de karst dans les dépôts de sel (Slănic-Prahova – Roumanie). Trav Inst Spéol "Émile Racovitza" XXI:87–93

Povară I, Horoi V, Marin C, Nicolescu T, Baciu M (1997) Investigations on the covered saline karst in Slănic Prahova area (Romania). Theor Appl Karstol 10:167–175

Puşcarciuc R, Buzdugan C (1987) Fenomene carstice în bazinul mijlociu al văii Trotuşului. Bul Club Speol Montana Oneşti 1:3–23

Ristivan V, Stan M, Iordache A (2003) Observaţii asupra Peşterii în Sare din Valea Miroiului. Ecocarst 4:57–59

Rögl F, Steininger FF, Müller C (1978) Middle Miocene salinity crisis and paleogeography of the Paratethys (Middle and Eastern Europe). Initial Rep Deep-Sea Drill Proj 42(1):985–990

Săndulescu M (1984) Geotectonica României. Ed Tehnică, Bucureşti, p 335

Sencu V (1965) Carstul din sarea gemă în România. St Cerc Geogr XII (1):47–63

Sencu V (1967) Morphologie und entstehung des steinsalz-karstes bei Slănic-Prahova. Rev Roum Géol, Géophys Géogr 2(1):49–65

Sencu V, Rusu T, Ion ID, Bojoi I (1973) Carstul. In: Atlasul RSR (ed) Acad RSR, Bucureşti, Plate III-4

Stoica N (1938) Aspectul fizic al masivului de sare Praid. Bul Soc Reg Rom Geogr LVII:322–325

Stoicovici E, Moţiu A (1957) Date noi în legătură cu câteva zăcăminte de gips din regiunea Cluj. Bul Univ Cluj 1:303–311

Strat D (2016) Wetlands—as significant features of the salt karst landscape in the Meledic Plateau, Vrancea Subcarpathians, Romania. Proc Environ Sci 32:59–69

Todor D, Giurgiu I, Silvăşanu G (1993) L'analyse quantitative de quelques spéléothèmes et échantillons de la grotte en sel 6S de Mânzăleşti (Subcarpathes de Vrancea - Roumanie). Theor Appl Karstol 6:211–212

Trufaş V (1960) Lacul Învârtita de la Nucşoara. Natura 5:78–81

Trufaş V (1963) Dolinele de la Nucşoara - Argeş. Probl Geogr X:239–244

Veress M, Puskás J, Zentai Z, Benkó Z (2011) Development of karren formation on the Saltic Hill of Praid (Transylvanian Basin, Romania). Carpath J Earth Environ 6(2):183–194

Viehmann I, Mac I (1966) Observaţii asupra carstului din ghipsurile Munţilor Mezeş şi cheilor Turzii. Lucr Inst Speol "Emil Racoviţă" V:67–74

Useful Websites

https://sites.google.com/site/romanianatura9/carpatii-rasariteni/subcarpatii-vrancei/pestera-6s-de-la-manzalesti-cea-mai-mare-pestera-in-sare-din-lume

Cave map: https://docs.google.com/viewer?a=v&pid=sites&srcid=ZGVmYXVsdGRvbWFpbnxyb21hbmlhbmF0dXJhOXxneDoxMWU4YmYyYjQzYmE2NDY3

Eastern Subcarpathians Bend: Salt Karst: Meledic Plateau and Slănic Prahova

Gheorghe M. L. Ponta

Abstract

Romania hosts some of the largest salt deposits in Europe, most of them located in the Eastern Carpathians and the Transylvania Depression. These are thick Miocene age evaporates precipitated in a lagoon facies. Since the end of the eighteenth century, geologist extensively mapped the salt deposits for mining purposes, gaining worldwide recognition. In fact, the term diapir was introduced in the geological literature by the Romanian scientist Ludovic Mrazec in 1906. The karst formed on evaporite rocks represents about 5% of the Romania's exposed karst and is located mostly in the lowland hills of Eastern Subcarpathians Bend. The salt outcrops on ~150 km^2, just 0.06% of Romanian territory. A few details about Meledic Plateau, which is part of the Buzău Land Geopark, and the Slănic Prahova Salt Mine, are the subject of this brief chapter.

Keywords

Salt karst • Dolines • Caves • Meledic Plateau Romania

Meledic Plateau

In the Slănicul de Buzău Basin, salt outcrops in several places; the largest surface is Săreni-Trestioara, situated between Slănic, Sări, and Meledic valleys (Ponta 1986; Ponta and Ursu 2017), and is known as the Meledic Plateau (Fig. 1). Geomorphologically, the region is located at the contact between the Carpathians and Subcarpathians Bend (see Fig. 1 in the Chapter by Tîrlă). Massive salt deposits, interbedded with reddish, green, and gray clays of Aquitanian age (early Miocene), surrounded by Burdigalian deposit of gray marls and calcareous sandstones outcrop on the plateau (Melinte-Dobrinescu et al. 2017). The occurrence of the Aquitanian salt is linked to diapirism, as indicated by the presence of chaotic salt breccia deposits. Both nappes that occur in the area belong to the Outer Moldavides (Săndulescu 1984).

Dolines

The most widespread karst feature in Meledic Plateau is dolines, which occur in areas where salt is exposed at land surface or is covered by thin layers of soil. They frequently develop where joints, fractures, or other openings are present, allowing water to easily move into the subsurface. Two types of dolines that prevail are those of collapse and solution origin. Collapse dolines generally are circular or oval in shape with almost vertical walls. Solution dolines usually form bowl-shaped depressions. The largest dolines (50 m in diameter and 15 m deep) are exclusively developed in salt and function as temporary ponors during the rainy season (Fig. 2). Overall, the dolines and the sharp, sinuous ridges that separate them, create chaotic landscape morphology.

Surface runoff carries sand and clay particles into the dolines, which form a relatively impermeable seal at the bottom. Two of these dolines became lakes with fresh water, the Great Lake (Lacul Mare), also known as the Bottomless Lake with a maximum depth of 5.4 m (maximum 0.7 ha), and the Castle Lake, having a maximum depth of 3.9 m (Melinte-Dobrinescu et al. 2017). When the dolines are developed along fractures/faults, doline valleys appear. Due to rapid dissolution processes, they quickly become real valleys with a surface flow. Occasionally, the thalweg is covered by salt crystals forming a white carpet (Fig. 3) up to 10 m long and 1–3 m wide (Ponta 1998).

G. M. L. Ponta (✉)
Geological Survey of Alabama, 420 Hackberry Lane, Tuscaloosa, AL 35401, USA
e-mail: gponta@yahoo.com

G. M. L. Ponta and B. P. Onac (eds.), *Cave and Karst Systems of Romania*,
Cave and Karst Systems of the World, https://doi.org/10.1007/978-3-319-90747-5_51

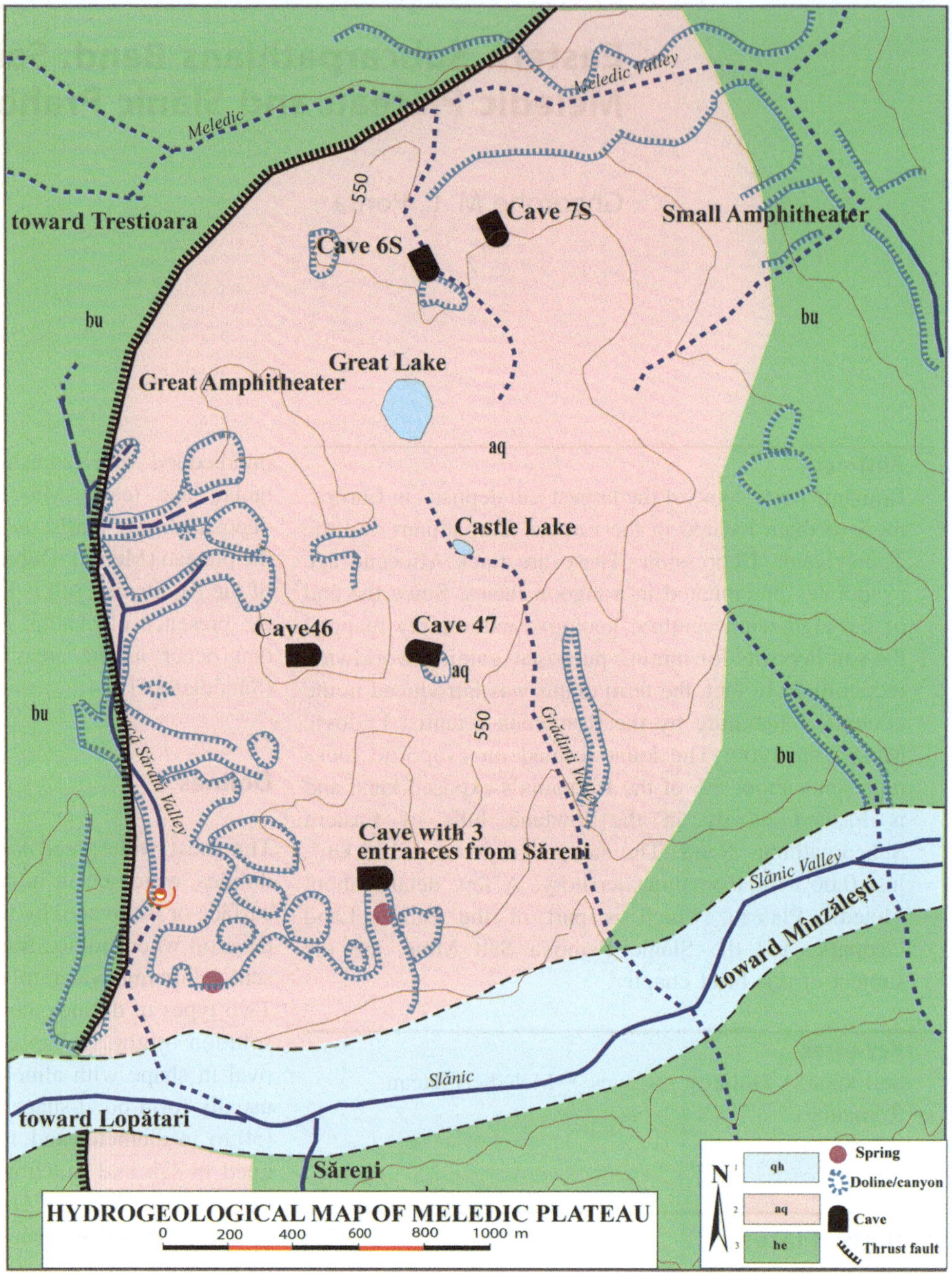

Fig. 1 Karst hydrogeologic map of the Meledic Plateau (geology modified after Murgeanu et al. 1968). Key for the legend is available in the karst hydrogeology chapter

Caves

Forty-seven salt caves are known in the Meledic Plateau, having a cumulated length of 4000 m and a total vertical development of 300 m. The longest one is the 6S Cave from Mânzălești, located in the northern part of the plateau with 3234 m of passages (Giurgiu 1995, 2010). Due to the high solubility of the salt rock, most of the surface and underground karst features are ephemeral. For example, the entrance of 6S Cave collapsed in October 1980, to be accessible again through a different entrance 5 years later. By June 14, 1986, the total length of the cave was 3118 m, and in June 1992, after being connected with the 4S Cave, it became 3234 m long and its vertical range reached −42 m. By the end of June 1992, the entrance collapsed again and was reopened after diggings in December 1992.

In places where the walls have shelters, corallites and salt stalactites are formed. The most common speleothems in salt caves are stalactites, tubular or conical in shape, with or without central supply channel. Their length is maximum

Fig. 2 Dolines in salt (photograph by G. Ponta)

Fig. 3 Streams thalweg covered with salt crusts (photograph by G. Ponta)

60 cm and 8 cm in diameter. There are also anemolites, 40 cm in length composed of interpenetrated salt crystal cubes that have their long diagonal coincident with the stalactite axis (see Fig. 2c, d in Chap. Evaporite Karst of Romania). The longest one discovered so far is 2 m and 20 cm in diameter (Giurgiu et al. 1980). Stalagmites are rare; they are a few tens of centimeters high and 10 cm in width. Generally, minute crystals cover the surface of most of the formations. A handful of flowstones, salt crusts, isolated eccentrics, and salt powder occur as well.

Slănic Prahova

Given halite's high solubility, a well-developed karst occurs at Slănic Prahova, where a large salt dome appears in the axis of the Slănic Syncline. Three centuries ago, the mining of salt began first on the left side of the Slănic Valley at Baia Verde and later on the right side at Baia Baciului (Bleahu et al. 1976). The mining resulted in a bell-shaped mine due to the work that progressed vertically, from the lower to the

(a)

(b)

Fig. 4 Bride's Cave in 1984 (a) and in 2013 (b) (photographs by G. Ponta)

upper part of the salt dome. The name of the mine was Bride's Cave (Grota Miresei; Fig. 4a), which along with a salt lake constituted a major touristic attraction in the region. By 2013, the combined action of dissolution, collapse, and landslide caused the Bride's Cave to vanish (Fig. 4b).

References

Bleahu M, Decu V, Negrea Ş, Pleşa C, Povară I, Viehmann I (1976) Peşteri din România. Ed Ştiinţifică şi Enciclopedică, Bucureşti, p 415

Giurgiu I (1995) Peştera 6S de la Mânzăleşti (Subcarpaţii Vrancei, România), una dintre cele mai importante cavităţi în sare din lume. Cerc Speol 3:13–17

Giurgiu I (2010) Cea mai mare peşteră în sare din România. Natura României 21:123–215

Giurgiu I, Silvăşanu G, Vădeanu T (1980) Observaţii asupra carstului pe sare din zona Meledic (M. Buzău). Peştera 6S de la Mânzăleşti. Comparaţie între formele carstice dezvoltate în sare şi calcare. Bul Club Speol "Emil Racoviţă" 7:55–66

Melinte-Dobrinescu MC, Brustur T, Jipa D, Macaleţ R, Ion G, Ion E, Popa A, Stănescu I, Briceag A (2017) The geological and palaeontological heritage of the Buzău Land Geopark (Carpathians, Romania). Geoheritage 9:225–236

Murgeanu G, Dumitrescu I, Săndulescu M, Bandrabur T, Săndulescu J (1968) Geological map of Romania, Covasna Sheet. Scale 1:200.000. Inst Geol Geofiz, Bucureşti

Ponta G (1986) The evaporite karst from Romania. Le Grotte d'Italia (4) XII:407–415

Ponta G (1998) Karst of Romania. In: Yuan D, Liu Z (eds) Global karst correlation IGCP 299. Science Press, Beijing, pp 193–207

Ponta G, Ursu V (2017) Highlights of the Romanian caves and karst. In: Moore K, White S (eds) Proceedings of the 17th International Congress of Speleology. Sydney July, 22–28, pp 78–81

Săndulescu M (1984) Geotectonica României. Ed Tehnică, Bucureşti, p 334

Ice Caves in Romania

Aurel Perşoiu and Bogdan P. Onac

Abstract
Ice caves represent a unique karst environment in which the morphological characteristics of cave passages influence the underground air circulation to such a degree that allows large perennial ice deposits to accumulate and be preserved for thousands of years. Scărişoara in the Bihor Mountains hosts the largest and the oldest ice deposit documented in a cave worldwide and is probably the world's best investigated ice cave.

Keywords
Ice caves • Climate • Stable isotopes • Ikaite Palaeoclimate

Introduction

Perennial snow and ice is a peculiar feature of a small number of caves in the Romanian Carpathians. It occurs either as firnified snow in high-altitude (2000 m above sea level) shafts of the Retezat Mountains, or as congelation ice in mid-altitude (~1200 m above sea level), single entrance, descendant caves in the Apuseni Mountains (Fig. 1). While the snow shafts in Retezat Mountains attracted little scientific attention due to their remote location and difficult access, the ice caves in the Apuseni Mountains have been the subject of intensive studies since the early 1920s; the results of these investigations setting a model for ice caves studies throughout the world. In this chapter, we present an overview of the results of these studies, by discussing the factors behind the genesis of ice in caves, the related processes, and the paleoclimatic significance of perennial ice accumulations in caves.

Genesis of Ice in Caves

The genesis and persistence of ice in caves is the results of a complex interplay between cave morphology and underground climatic conditions. These factors are discussed in detail below.

Morphologic Conditions

The accumulation of snow in caves of Retezat Mountains is favored by the existence of numerous vertical shafts located at altitudes above 1700 m, where mean annual temperatures are around 2 °C and most of precipitation falls as snow during the winter months. The descendent morphology of the caves allows for temperature inversion to persist, and the resulting cold environment conserves the snow (in a feedback loop, as the slowly melting snow further leads to subfreezing temperatures). Of the numerous shafts in Retezat Mountains that have semi-perennial snow at bottom, in two the snow has partly transformed into ice, through snow firnification and freezing of percolating water: Avenul Mare cu Zăpadă din Albele-Găuroane (AZA) (1890 m asl) and Avenul cu Gheaţă din Dâlma Brazii cei Vineţi (ABV) (1750 m asl). Both shafts open to the surface through a 20 × 25 m entrance, continued by vertical drops of 10–20 m. The bottom of the shafts is occupied by perennial snow. In AZA, a rimaye between the snow plug and the rock leads, after a ~10 m descent, to a room located under the snow accumulation, with the bottom occupied by a second snow deposit, ca. 40 m thick (Fig. 2a). ABV (Fig. 2b) has a similar morphology, the bottom of the 20 m deep entrance shaft being fully occupied by a snow plug, ~30 m thick.

A. Perşoiu (✉) · B. P. Onac
"Emil Racoviţă" Institute of Speleology, Romanian Academy, Clinicilor 5, Cluj-Napoca, 400006, Romania
e-mail: aurel.persoiu@gmail.com

B. P. Onac
e-mail: bonac@usf.edu

B. P. Onac
Karst Research Group, School of Geosciences, University of South Florida, 4202 E. Fowler Ave., NES 107, Tampa, FL 33620, USA

G. M. L. Ponta and B. P. Onac (eds.), *Cave and Karst Systems of Romania*, Cave and Karst Systems of the World, https://doi.org/10.1007/978-3-319-90747-5_52

Fig. 1 Distribution of ice caves in Romania. *A* Bihor Mountains (Scărişoara, Focul Viu, Barsa, Borţig, Vârtop), *B* Retezat Mountains (Avenul Mare cu Zăpadă din Albele-Găuroane and Avenul cu Gheaţă din Dâlma Brazii cei Vineţi). Map of Romania is available under CC BY-SA 3.0 license (https://creativecommons.org/licenses/by-sa/3.0/) from https://upload.wikimedia.org/wikipedia/commons/5/59/Relief_Map_of_Romania.png

The crevasses that allowed the exploration of the caves were sealed-off by snow in the mid-80s, and no further investigations have been carried out since. These two snow and ice deposits are estimated to be between 6000–10,000 m^3.

Contrary to the snow/firn/ice deposits of the Retezat Mountains, those in the Apuseni Mountains are formed by the freezing of water in winter. They are also larger in terms of volume and have been intensively investigated in the past 100 years. Perennial ice in caves in Apuseni Mts. has been described in six caves, as follows: Scărişoara Ice Cave (SIC), Focul Viu Ice Cave (FV), Borţig Ice Cave (BIC), Adevăratul Gheţar de la Vârtop, Gheţarul de la Barsa, and Peştera cu Gheaţă din Poiana Vârtop. Of these, recent surveys have shown that ice in Adevăratul Gheţar de la Vârtop ($\sim$1600 m^3, Moreh 1999) and Gheţarul de la Barsa has all but disappeared in recent years, persisting in Scărişoara Ice Cave (over 100,000 m^3, Hubbard 2017), Focul Viu Ice Cave ($\sim$30,000 m^3), Borţig Ice Cave ($\sim$25,000 m^3, Orghidan et al. 1984), and Peştera cu Gheaţă din Poiana Vârtop ($\sim$12,000 m^3, Feier et al. 2000–2001). All these four caves have a similar morphology, with an entrance shaft, varying in diameter and depth leading to large cavities, with the floor fully occupied by ice (Fig. 3).

The access in SIC is possible through a large (64 m in diameter, 46 m deep) shaft that connects through a 17 × 17 m entrance arch with the $\sim$3000 m^2 large Great Hall (Racoviţă and Onac 2000), occupied by the underground glacier (Fig. 3a). The sides of the ice block lead into the lower sections of the cave (Fig. 4), where positive temperatures preclude the formation and persistence of ice. FV Cave has a similarly descendant morphology, with a small entrances leading to a large (68 × 46 m) room, hosting a $\sim$20 m thick ice block (Fig. 3b). The ceiling of this room opens to the surface, allowing for snow and detritus to fall into the cave and accumulated into a pile of frozen material (Fig. 5). BIC has a simpler morphology, with a 44 m deep shaft ending in an ice block that almost completely occupies the floor (Fig. 3c, Kern et al. 2009). A series of narrow rimayes allow access to the lower, non-glaciated sections of the cave, located $\sim$15 m below

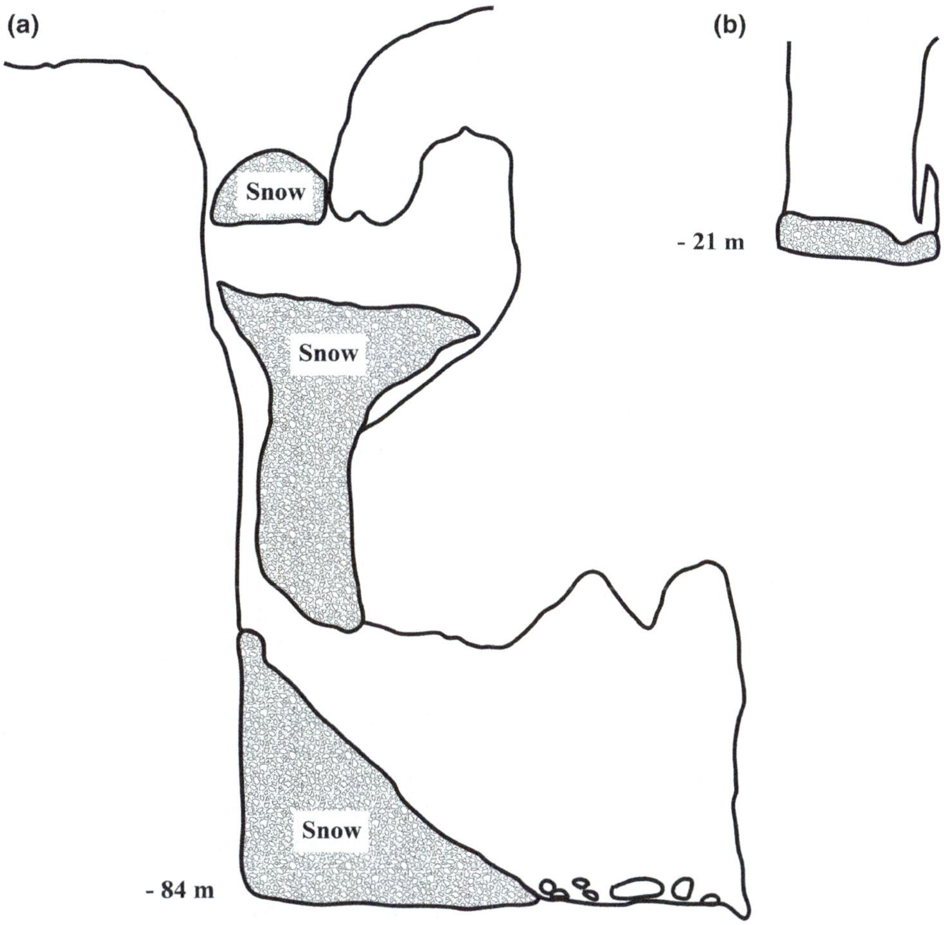

Fig. 2 Cross-section through **a** Avenul Mare cu Zăpadă din Albele-Găuroane and **b** Avenul cu Gheaţă din Dâlma Brazii cei Vineţi, Retezat Mountains, SW Romania (by permission from Ponta et al. 1984)

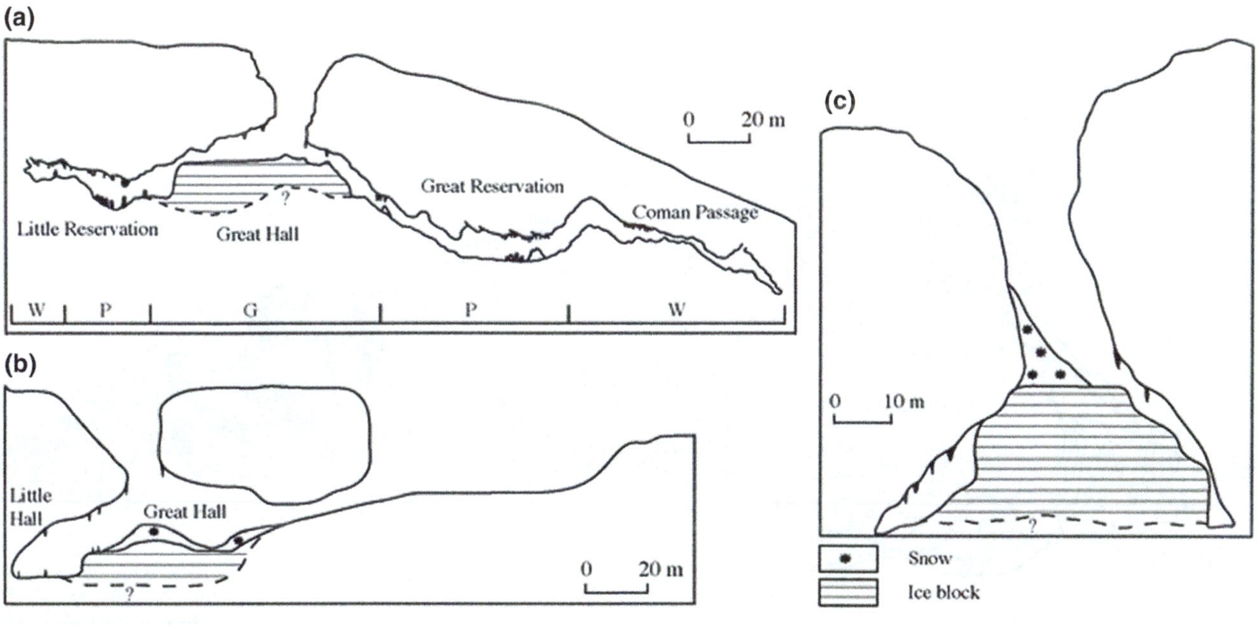

Fig. 3 Cross section through **a** Scărişoara Ice Cave, **b** Focul Viu Ice Cave, and **c** Borţig Ice Cave

the surface of the ice block. The morphologies of these caves determine peculiar climatic conditions, favorable for the accumulation and persistence of ice that have been studied in detail in SIC and FV caves.

Cave Climate

The peculiar climatic characteristics of Scărişoara Ice Cave result from the combination of external climatic conditions and the cave's morphology. The mean annual temperature outside the cave is ~5.2 °C, with the temperatures of the coldest month (January), and warmest (July) being around −4 and 15 °C, respectively (Perşoiu et al. 2011a). The mean annual precipitation amount is ~1200 mm (Perşoiu et al. 2007). The vertical development of the cave determines the circulation of air and subsequent climatic parameters, in a strong positive feedback loop. In winter, cold air avalanches, triggered by the higher density of the external cold air are tumbling the eastern side of entrance shaft, displacing the warmer air inside the cave, which rises along the western wall of the same shaft. Once reaching the Great Hall, cold air descends on the sides of the ice block toward the Little and Great reservations, cooling these sections of the cave and further pushing out the warm air along the ceiling. In summer, the cooling effect of the ice block makes air inside the cave denser than the external one, and thus, no circulation exists between the cave and the exterior. However, the cooling effect of the ice block does not extend to the deepest sections of the cave that have warmer (and thus, lighter) air that rises toward the Great Hall, where it cools and sinks to the floor and subsequently flows down the cave the same "route" it takes in winter. By this mechanism, two convective cells develop inside the cave in summer, that supply warm air to the Great Hall, enhancing the melting of ice induced by heat conduction through the rock and air in the entrance shaft.

The above-described mechanisms of air circulation are both the consequence and cause of the thermal characteristics of SIC. The air temperature has the greatest variations in the Great Hall (Figs. 4 and 6). During winter, the temperature follows closely the external ones, reaching the minimum January, while during summer, the underground temperature is controlled only by the thermal inertia of the ice block and of the overcooled walls of the cave, never increasing above +0 °C (Fig. 6, Perşoiu et al. 2011a). The rapid inflow of cold air in winter reaches the bottom of the Great Hall in less than 1 h and the Great Reservation in less than 2 h (Perşoiu et al. 2011a).

In FV, the presence of two entrances at different elevations allows for a stronger air circulation than in SIC, with the lower entrance acting as a cold air sink and the upper one as the outflow for warm air in winter. In summer, the heavier cold air inside the cave's Great Hall determine the cessation of the flow, and a slow warming due to conduction through the large opening in the ceiling (Fig. 7). Further, this natural skylight allows sunlight to reach the surface of the ice block warming the air column above the ice and thus preventing the creation of convective cells similar to those in SIC. In winter, negative air temperatures outside the cave determine rapid inflow of cold air through both the lower and upper entrances (visible in Fig. 5), leading to a rapid thermal response of the cave's air temperature (Fig. 7).

As a direct consequence of the peculiar climate inside SIC and FV percolating water freezes and layers of ice accumulate to build up large deposits of perennial ice. The main water sources that feed the ice in both caves (as well as in Borţig) are (1) rainwater infiltrating through fissures in the

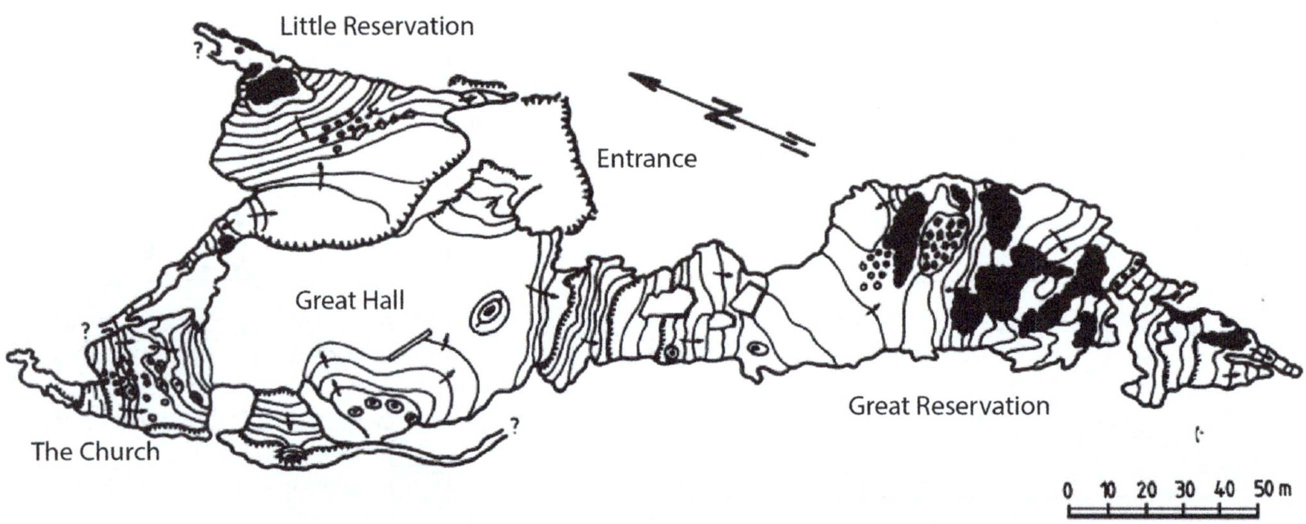

Fig. 4 Map of the Scărişoara Ice Cave (modified from Rusu et al. 1970)

Fig. 5 The Great Hall of Focul Viu Ice Cave. The lower entrance is visible to the left of the sunlight entering through the cave's skylight (upper entrance)

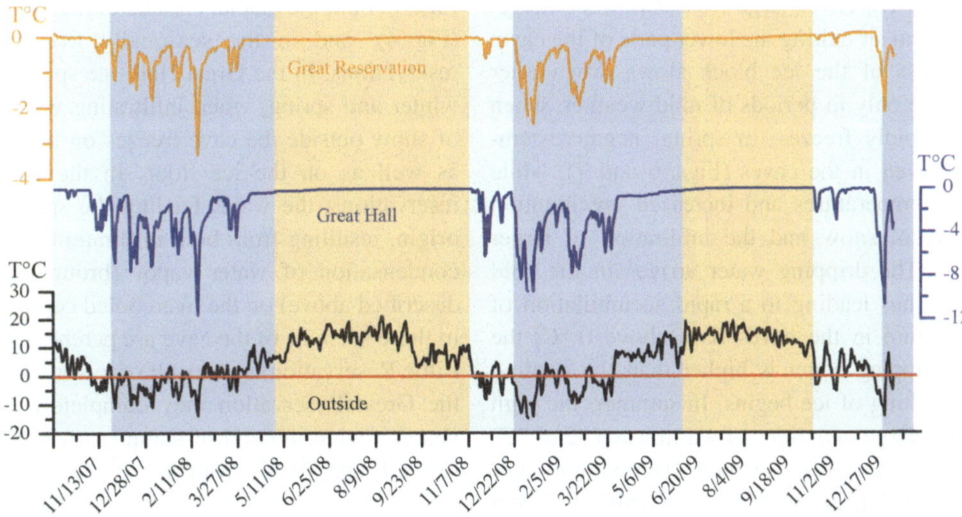

Fig. 6 Air temperature variations in the Great Hall and Great Reservation of Scărişoara Ice Cave (modified from Perşoiu et al. 2011a)

limestone or reaching the surface of the ice through the entrance shaft and (2) water derived from the partial thawing of the surface ice and snow accumulated below the entrance shafts of these caves (Figs. 3 and 5). The ice deposits consists of a sequence of laminated layers, each one containing a couplet of clear ice and sediments (organic matter, calcite, soil, and pollen). The formation of ice in the caves in Apuseni Mountains starts in autumn, when the main process is the rapid freezing of water accumulated in shallow lakes on top the ice blocks. As these lakes freeze, a layer of ice up to 15 cm thick is added to the ice blocks, showing a typical structure for a downward-freezing pond, with large hexagonal crystals (Perşoiu et al. 2011b). In winter, temperatures bellow 0 °C lead to the overcooling and freezing of the walls

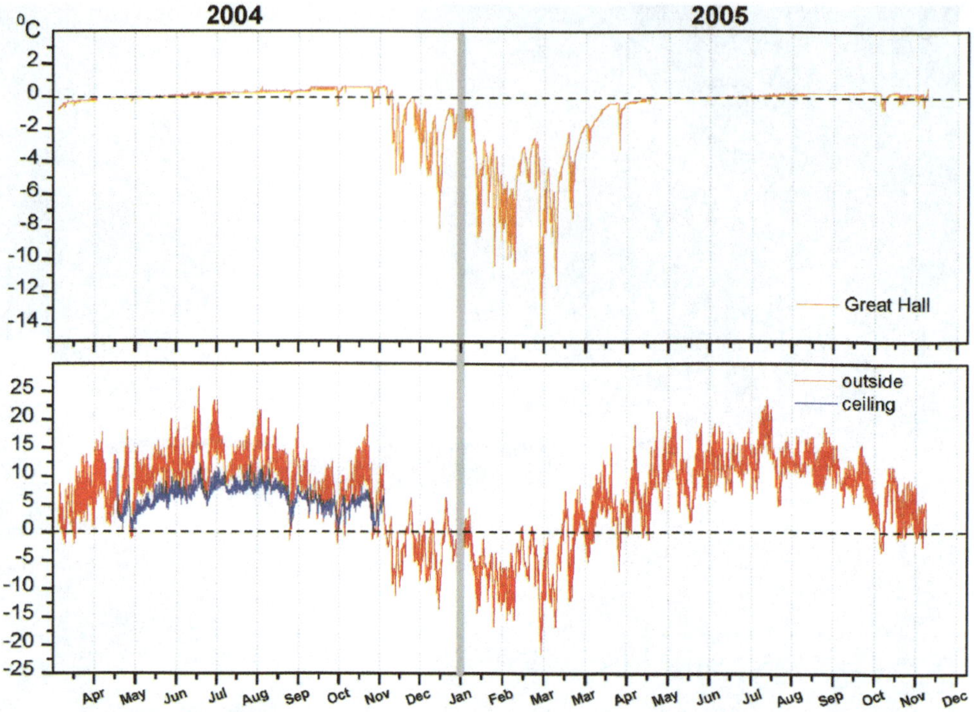

Fig. 7 Air temperature variations in Focul Viu Ice Cave

and thus the dripping of water stops. Inflow of cold and dry air leads to intensive ablation of ice, while on the overcooled walls of the SIC and FV caves large frost deposits emerge, generated by the warm air exiting the lower parts of the cave. The growing process of the ice block slows down after autumn, taking place only in periods of mild weather, when infiltrating water rapidly freezes. In spring, negative temperatures are preserved in the caves (Figs. 6 and 7), while outside, rising air temperatures and increased precipitation causes the melting of snow and the infiltration of larger amounts of water. The dripping water arrives in the cold cave environment, thus leading to a rapid accumulation of ice. When temperature in the caves rises above 0 °C, the heat induced by dripping water is higher than the cooling effect of ice and melting of ice begins. In summer, the high dripping rates bring large quantities of warm water (5–6 °C) into the caves, causing the ice to melt. Temporary lakes (10–15 cm deep) develop in all three caves (SIC, FV, and BIC) due to both infiltrating waters and melting of ice. The lake's water furthermore contributes to the melting of ice. Conductive transfer of geothermal heat (as well as, to a lesser extent, heat conduction through the air column in the entrance shaft) leads to an increase of temperature, which in turn induces a slow melting of the ice. In autumn, freezing of these lakes begins, reinitiating the cycle described above.

Apart from the perennial ice blocks, annual (or semi-perennial) ice speleothems form in all caves described above. While in FV and BIC they mostly occur as semi-perennial ice mounds (Fig. 5, foreground), in SIC a variety of shapes and sizes develop (Fig. 8). Ice speleothems form in both the glacial climate of the Great Hall and Church (Fig. 4), and in the seasonally frozen Great and Little reservations. In the Great Hall, ice speleothems form in late winter and spring, when infiltrating water from the melting of snow outside the cave freezes on the ceiling of the cave, as well as on the ice floor. In the Church and the two reservations, the water feeding the speleothems has a dual origin, resulting from both infiltration of external water and condensation of water vapor (brought by the circulation described above) on the overcooled ceiling. The speleothems in these sections of the cave are perennial (in the Church and Little Reservation) as a result of lower temperatures, while in the Great Reservation they completely melt (in wet/warm years) toward the end of summer, due to the heat delivered by infiltrating warm waters.

Processes in Ice Caves

Ice Dynamics

The snow and ice accumulations in the Retezat Mountains are rather static, with basal and lateral melting being the only processes affecting them. However, these processes are acting on a very long timescale, as proved by the closure of the two largest ice caves (see above) in the early to

Fig. 8 A field of ice stalagmite in the Great Reservation, Scărişoara Ice Cave

mid-1980s, without further opening. We hypothesize that continuous undercooling of the walls during winter prevents melting at the surface, and thus melting could occur only in the lower parts of the caves, which are shielded from the external source of cold air. The melting of snow from underneath leads to progressive opening of the crevasse, bottom to top, on multidecadal timescales.

Contrary, ice in caves in Apuseni Mts. has a very active dynamics, on timescales ranging from annual to centennial. In all caves, within one year, the level of ice reaches a maximum in late spring and a minimum in late summer/autumn. In SIC, ice level measurements have been initiated in 1947 and continued, with variable frequency, until present. Data on the long-term dynamics of the ice block in SIC indicate a clear tendency of melting since at least 1921 (Şerban et al. 1967; Racoviţă 1994; Perşoiu and Pazdur 2011). Thus, between 1921 and 1947, the ice level lowered by ca. 50 cm, leading to the opening of the entrance toward the Little Reservation (Şerban et al. 1948), unknown in the 1920s, when Emil Racoviţă visited the cave a couple of times. During his visits, he only noticed the entrance into the Great Reservation, and not the one toward the Little Reservation (Racoviţă 1927). This melting tendency accelerated after 1947 and up to ca. 1980; in the ~30 years encompassing, this interval the ice level lowering by ca. 140 cm. Between 1982 and 2007, the ice level was relatively stable (Perşoiu and Pazdur 2011), but over the past decade the melting tendency accelerated again, with the present-day ice level standing ca. 180 cm below the 1947 level. It is difficult to attribute the melting of the ice block to a single factor—as noted above, the yearly dynamic of the ice is controlled by a combination of winter and spring accumulation and summer melting, both of these being the result of temperature and precipitation amount variations. Thus, very cold winters, while inducing low temperatures in the cave, also lead to less water infiltration and lack of ice accumulation on top of the ice block (but induce vigorous air circulation and subsequent increase of condensation and enhanced formation of speleothems). Wet, but not necessarily warm summers, favor rapid melting of ice due to the increased infiltration of warm waters. A second factor affecting the long-term dynamics of the ice block is basal melting. Measurements performed in the past 40 years have shown that geothermal heat determines the melting of the sole of the ice block at a constant value of ~1.5 cm/year. However, this melting is not affecting evenly the sole of the glacier, being stronger at the contact with the lateral walls (where heat is delivered both from below and from the side) and thus allowing for the development of cavities that would promote flow of the ice and formation of folds (Fig. 9). Inflow of cold and dry air plays an important role in ablating the sides of the ice block, leading to ice sublimation and development of alveoli, which are very similar to the scallops developed by underwater corrosion.

Contrary to the ice block, the ice speleothems in SIC are more dynamic, following closely the temperature variations.

Fig. 9 Side view (in the Little Reservation) of the ice block in Scărişoara Ice Cave

Their height reaches a maximum in early spring, when dripping water encounters subfreezing temperatures in the cave, and subsequently recedes throughout the summer and autumn, reaching a minimum in early winter, before the onset of cooling. Ice stalagmites in the Little Reservation, and the Church are permanent, while those in the Great Reservation disappearing intermittently, especially after very wet and warm summers.

Generally, the glaciation of caves in the Apuseni Mountains has declined in the past decade, with permanent ice in several caves (Adevăratul Gheţar de la Vârtop, Gheţarul de la Barsa) disappearing altogether, and reaching a minimum in others (e.g., in Scărişoara Ice Cave). This situation is most dramatically seen in Adevăratul Gheţar de la Vârtop, where between 1999 and 2008, a 5 m thick ice block (~ 1600 m^3) vanished completely.

Speleothems and Minerals

Apart from stalagmites, stalactites, domes and crystals all made of ice; the Romanian caves hosting ephemeral and perennial ice deposits also contain a large variety of carbonate speleothems. The only cave in which these were thoroughly investigated is SIC. One of the earliest studies was conducted on a very particular type of cave pearls forming in the periglacial (temperatures fluctuates below and above freezing) part of the Great Reservation (Viehmann 1958). In a series of subsequent publications, Viehmann (1963, 1993) argued that freezing/melting processes rather than dripping water are responsible for moving the pearls. More recently, using isotopic analysis it has been documented that part of the calcite in these pearls is cryogenically precipitated (Žák et al. 2008, 2013). In fact, freezing of bicarbonate-rich solutions in ice caves triggers the deposition of the so-called cryogenic cave carbonates (CCC) that come in a variety of morphologies, sizes, and mineralogical composition (Žák et al. 2017). Two subtypes of CCC were described: fine-grained (below 1 mm) and coarse crystals (up to several cm), each characterizing different environmental conditions and falling in two distinct fields when their isotopic composition ($\delta^{18}O$, $\delta^{13}C$) is plotted. The first type is produced whenever freezing of water happens very rapidly and is accompanied by quick kinetic CO_2 degassing and water evaporation. Under these conditions, both $\delta^{18}O$ and $\delta^{13}C$ values are high, with the latter one reaching the most positive values known for carbonates (> +13‰). CCC_{coarse} precipitates during slow water freezing allowing calcite crystals to grow larger and imprinting it a completely different isotopic signature. The $\delta^{18}O$ values for this category can be as low as −30‰, whereas $\delta^{13}C$ values range between −10 and +10‰.

From a mineralogical point of view, the composition of CCCs is dominated by calcite. Two other minerals, namely monohydrocalcite and hydromagnesite, were described from the Small Reservation of SIC, but only the first one has a cryogenic origin (Onac 2000–2001). A few years later, white-light cream patchy accumulations of very small crystals (<0.5 mm) within certain ice layers and at the surface of ice stalagmites/domes (Fig. 10a) were investigated by means of X-ray diffraction and environmental scanning electron microscope analyses. The results indicated the presence of ikaite ($CaCO_3{\cdot}6H_2O$), a very rare hydrated calcium carbonate never mentioned before from a cave environment, which is only stable at temperature below 4 °C (Onac 2008; Onac et al. 2011). Another unusual discovery was made at

Fig. 10 **a** Cryogenic cave calcite in the Great Hall, **b** calcite pseudomorph after ikaite (glendonite) recovered from ice in the Great Reservation (photos B. P. Onac)

the entrance in the Great Reservation, where aggregates composed of calcite scalenohedrons clusters (up to 4.7 cm in size) were found protruding from the ice tongue (Fig. 10b). Based on their morphology and isotopic composition, the author suggested that these coarse CCC are glendonites, in other words, a pseudomorph of calcite after ikaite (Onac 2008). Considering ikaite's temperature-restricted field of formation, future isotopic work on this mineral may serve as basis to explore paleoclimatic and paleoenvironmental implications its presence in the perennial ice cave deposits might have.

Paleoclimatic Significance of Ice Accumulations in Caves

Perennial ice accumulations in caves host a vast array of proxies for both past climate and environmental changes, the most significant being the isotopic composition of ice and the pollen entrapped in it; the ice block in Scărişoara Cave was the first to be investigated as such. Thus, a study published in 1950 (Pop and Ciobanu 1950) examined the pollen trapped in the SIC ice block and attempted to reconstruct the vegetation history of the Apuseni Mountains. However, while the pollen assemblages and reconstructed biomes were correct, lack of age control (the study was done before the development of the radiocarbon dating method) gave an erroneously young age of about 3500 years for the base of the analyzed pollen sequence from the Little Reservation. Follow-up studies were published in 2011 when the authors have shown that ice has accumulated with varying rates during the past 1000 years, being lower during the Medieval Warm Period (MWP) and higher during the Little Ice Age (LIA) (Perşoiu and Pazdur 2011; Feurdean et al. 2011). Recent climate reconstruction from NW Romania (Feurdean et al. 2015) have shown that summers during the MWP were rather wet and during the LIA somewhat drier. Thus, during MWP summers, when the mass balance of ice in caves (see above) is influenced by precipitation amount (rather than temperature) inflow of warm waters led to ice ablation at a faster rate than during the subsequent cold and dry LIA. However, preservation of large amounts of macrofossils in the ice layers accumulated during the LIA (Feurdean et al. 2011) indicates possible inflow of large amounts of water, most likely during single summer storm events.

A recent study of stable isotope composition of cave ice and CCC (Bădăluţă et al. in prep) in Scărişoara and Focul Viu Ice Caves has further investigated the climate of the last 1000 years. In both caves, the oxygen and hydrogen stable isotope data indicate higher values (and thus warmer conditions) prior to AD 1200, and lower and highly variable ones (indicating colder conditions), between AD 1200 and 1800. Further, both summer and winter temperatures during the MWP show less variability than during the subsequent LIA. The coldest decades occurred in the sixteenth and seventeenth centuries. The d-excess parameter, interpreted as indicating the moisture source, has lower values during the MWP and higher during the LIA, suggesting a predominance of isotopically enriched moisture sources during the coldest periods of the past 1000 years. These data further indicate that during the warm MWP, SE Europe received more Atlantic precipitation than during the cold LIA, when it was receiving increased amounts of Mediterranean precipitation. A similar partition of moisture sources between the North Atlantic and the Mediterranean has been shown to extend across the entire Holocene (Perşoiu et al. 2017).

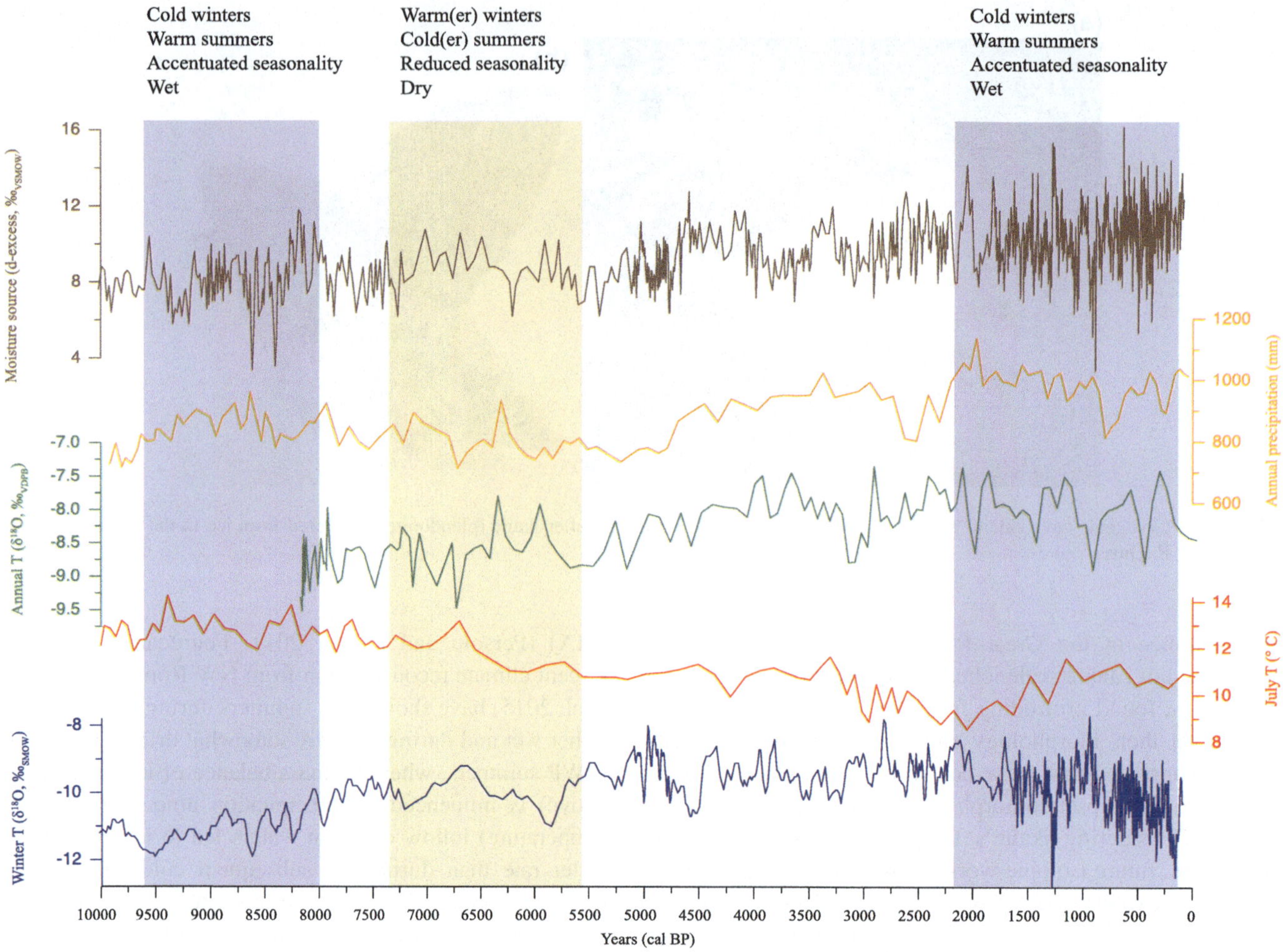

Fig. 11 Holocene climate variability in NW Romania. **a** d-excess in SIC (Perşoiu et al. 2017), **b** Pollen–based mean annual precipitation (Feurdean et al. 2008), **c** speleothem-based $\delta^{18}O$ annual temperature (Drăguşin et al. 2014), **d** Chironomid-based summer temperature (Tóth et al. 2015), **e** cave ice $\delta^{18}O$-derived winter temperature (Perşoiu et al. 2017)

These authors have analyzed $\delta^{18}O$ and $\delta^{2}H$ on a 22.5 m long, radiocarbon-dated ice core extracted from the Great Hall of SIC in 2003 (Holmlund et al. 2005) to reconstruct winter climate variability (temperature and moisture sources) during the past 10,000 years (Fig. 11). They have found that winter temperatures followed insolation throughout the Holocene, rising from an early Holocene minimum, peaking at ~5000 cal BP and decreasing toward the Neoglacial of the last 2000 years (accelerated melting in the second century led to the loss of the upper layers of ice; hence, the recent warming is not being recorded).

Contrary to the $\delta^{18}O$ record, the deuterium excess ($d = \delta^{2}H - 8.\ \delta^{18}O$; Dansgaard 1964) shows a differed trend, with low values in the early Holocene, and a sharp and sustained increase since 4700 cal BP. Similar to the findings for the last 1000 years, these indicate the predominance of mostly westerly circulation during the early Holocene and an increase of Mediterranean influence in the second half of the Holocene. The contrasting summer and winter patterns of temperature variability in Romania during the Holocene (Fig. 11) allow us the subdivision of the Holocene in three periods: (1) early Holocene (10,000–8000 cal BP), with warm summers, cold winters and high precipitation rates (with a strong Mediterranean component); (2) early to late-Holocene (8000–2000 cal BP), with decreasing summer and increasing winter temperatures and drying trend (with moisture originating mostly from the Atlantic Ocean); (3) late-Holocene (>2000 cal BP), with renewed amplifying seasonality and high precipitation rates with increasing source from the Eastern Mediterranean (Perşoiu et al. 2017).

Acknowledgements The authors wish to dedicate this chapter to the memory of Iosif "Pepi" Viehmann (1925–2016) and Gheorghe Racoviţă (1940–2015), who continued the early studies of Emil G. Racoviţă and Mihai Şerban in Scărişoara Ice Cave. The studies leading to the results reported in this chapter were partly provided by UEFISCDI Romania (PNII-RU-TE-2014-4-1993 and PN-III-P1-1.1-

TE2016-2210 grants to AP) and the EU ERA-NET LAC program (ELAC2014/DCC-0178 grant), as well as to BPO by the CNCSIS (16/1327) and NSF (AGS 0823253).

References

Dansgaard W (1964) Stable isotopes in precipitation. Tellus 16:436–468

Drăguşin V, Staubwasser M, Hoffmann DL, Ersek V, Onac BP, Vereş D (2014) Constraining Holocene hydrological changes in the Carpathian-Balkan region using speleothem $\delta^{18}O$ and pollen-based temperature reconstructions. Clim Past 10:1363–1380. https://doi.org/10.5194/cp-10-1363-2014

Feier N, Munteanu G, Onac BP (2000–2001) Mineralogy and speleogenesis of the ice-cave from Poiana Vârtop Bihor Mountains, Romania). Theor Appl Karstol 13–14:83–87

Feurdean A, Klotz S, Mosbrugger V, Wohlfarth B (2008) Pollen-based quantitative reconstructions of Holocene climate variability in NW Romania. Palaeogeogr Palaeoclimatol Palaeoecol 260:494–504

Feurdean A, Perşoiu A, Pazdur A, Onac BP (2011) Evaluating the palaeoecological potential of pollen recovered from ice in caves: a case study from Scărişoara Ice Cave Romania. Rev Palaeobot Palynol 165(1–2):1–10

Feurdean A, Galka M, Kuske E, Tanţău I, Lamentowicz M, Florescu G, Liaka J, Hutchinson SM, Mulch A, Hickler T (2015) Last millennium hydro-climate variability in Central-Eastern Europe (northern Carpathians, Romania). The Holocene 25:1179–1192

Holmlund P, Onac BP, Hansson M, Holmgren K, Mörth M, Nyman M, Perşoiu A (2005) Assessing the palaeoclimate potential of cave glaciers: the example of Scărişoara Ice Cave (Romania). Geogr Ann 87A(1):193–201

Hubbard JD (2017) 3D cave and ice block morphology from integrated geophysical methods: a case study at Scărişoara Ice Cave, Romania. Unpublished MS Thesis, Univ South Florida, Tampa, p 108

Kern Z, Molnár M, Svingor É, Perşoiu A, Nagy B (2009) High resolution, well preserved tritium record in the ice of Borţig Ice Cave (Bihor Mts., Romania). The Holocene 19(5):729–736

Moreh K (1999) Adevăratul Gheţar de la Vârtop. Speomond 4:25

Orghidan T, Negrea Ş, Racoviţă G, Lascu C (1984) Peşteri din Romania. Ghid turistic. Ed Sport-Turism, Bucureşti, pp 85–87

Onac BP (2000–2001) Mineralogical studies and Uranium-series dating of speleothems from Scărişoara Glacier Cave (Bihor Mountains, Romania). Theor Appl Karstol 13–14:33–38

Onac BP (2008) Ikaite in the Scărişoara ice deposit: precipitation and significance. In: Turri S (ed) 3rd International Workshop on Ice Caves (IWIC-III), Vol Abstr Perm State University, Perm, p 28

Onac BP, Wynn JG, Citterio M (2011) Ikaite in the Scărişoara Ice Cave (Romania): origin and significance. Geophys Res Abstr 13: EGU2011-5188

Perşoiu A, Pazdur A (2011) Ice genesis and its long-term mass balance and dynamics in Scărişoara Ice Cave, Romania. The Cryosphere 5 (1):45–53

Perşoiu A, Bojar A-V, Onac BP (2007) Stable isotopes in cave ice: What do they tell us? Stud UBB Geologia 52(1):59–62

Perşoiu A, Onac BP, Perşoiu I (2011a) The interplay between air temperature and ice dynamics in Scărişoara Ice Cave, Romania. Acta Carsologica 40(3):445–456

Perşoiu A, Onac BP, Wynn JG, Bojar A-V, Holmgren K (2011b) Stable isotopes behavior during cave ice formation by water freezing in Scărişoara Ice Cave. J Geophys Res-Atmos 116:D02111

Perşoiu A, Onac BP, Wynn JG, Blaauw M, Ionita M, Hansson M (2017) Holocene winter climate variability in Central and Eastern Europe. Sci Rep 7:1196. https://doi.org/10.1038/s41598-017-01397-w

Ponta G, Aldica G, Bădescu D, Panaiotu C, Solomon A (1984) Cercetări speologice în perimetrul Jiul de Vest–Cernişoara. Bul Speol 8:77–118

Pop E, Ciobanu I (1950) Analize de polen în ghiaţa dela Scărişoara: Anal Acad RPR, ser Geol, Geogr, Biol, Şt Tehn Agric III:23–50

Racovitza EG (1927) Observations sur la glacière naturelle dite "Gheţarul de la Scărişoara". Bull Soc Sci Cluj III:75–108

Racoviţă G (1994) Eléments fondamentaux dans la dynamique des spéléothèmes de glace de la grotte de Scărişoara, en relation avec la météorologie externe. Theor Appl Karstol 7:133–148

Racoviţă G, Onac BP (2000) Scărişoara Glacier Cave. Monographic study. Ed Carpatica, Cluj-Napoca, p 140. http://digital.lib.usf.edu/SFS0063515/00001/pdf

Rusu T, Racoviţă G, Coman D (1970) Contributions à l'étude du complexe karstique de Scărişoara. Ann Spéléol 25(2):383–408

Şerban M, Coman D, Givulescu R (1948) Découvertes récentes et observations sur la glacière naturelle dite "Gheţarul de la Scărişoara". Bull Soc Sci Cluj X:174–210

Şerban M, Blaga L, Blaga L, Chifu A, Ciobotaru T (1967) Contribuţii la stratigrafia depozitelor de gheaţă din Gheţarul de la Scărişoara. Lucr Inst Speol "Emil Racoviţă" VI:107–140

Tóth M, Magyari EK, Buczkó K, Braun M, Panagiotopoulos K, Heinri O (2015) Chironomid-inferred Holocene temperature changes in the South Carpathians (Romania). The Holocene 25:569–582. https://doi.org/10.1177/0959683614565953

Viehmann I (1958) Perlele de cavernă din Gheţarul de la Scărişoara. DS Şed Com Geol XLV:283–295

Viehmann I (1963) Un nouveau processus de genese des perles de caverne. Lucrările Institutului de Speologie "Emil Racoviţă" I-II:295–303

Viehmann I (1993) Note sur des perles de caverne particulieres. Soc Géogr Liège Bull 29:105–107

Žák K, Onac BP, Perşoiu A (2008) Cryogenic carbonates in cave environments: a review. Quat Int 187:84–96

Žák K, Orvošová M, Filippi M, Vlček L, Onac BP, Perşoiu A, Rohovec J, Světlík I (2013) Cryogenic cave pearls in the periglacial zone of ice caves. J Sed Res 83(2):207–220

Žák K, Onac BP, Kadebskaya OI, Filippi M, Dublyansky Y, Luetscher M (2017) Cryogenic mineral formation in caves. In: Perşoiu A, Lauritzen S-E (eds) Ice caves. Elsevier, Amsterdam, pp 123–162

Other Resources

http://digital.lib.usf.edu/SFS0063515/00001/pdf

Cave Minerals of Romania

Bogdan P. Onac

Abstract

The chapter presents an updated list of all minerals occurring under various settings in caves from Romania.

Keywords

Cave • Minerals • Classification • Database

The last published list tabulating 104 minerals from various caves of Romanian is 15 years old (Onac in Theor Appl Karstol 16:83–89, 2003). An up-to-date compilation is needed because: (i) New cave minerals (24) were updated or identified and described since 2003, (ii) previous species were re-examined and reconsidered, (iii) some minerals were unintentionally omitted, or errors occurred with respect to their location and/or author of first description, and (iv) the International Mineralogical Association (IMA) Commission on New Minerals, Nomenclature and Classification (CNMNC) has approved, discredited, redefined, and renamed minerals (Burke in Can Mineral 44:1557–1560, 2006, Mineral Rec 39:131–135, 2008; Mills et al. in Eur J Mineral 21:1073–1080, 2009; Pasero et al. in Eur J Mineral 22:163–179, 2010).

Since from a mineralogical point of view, the most significant caves are presented in this book (Cioclovina, Diana, Humpleu, Măgurici, Valea Rea, and Vântului); this chapter will only include the list of all minerals reported prior to March 2018. The format of mineral presentation follows the crystallochemical classification in classes (native elements, sulfides, oxides and hydroxides, halides, carbonates, nitrates, phosphates, silicates, sulfates, vanadates, chromates, and organic compounds) scheme presented in the *Cave minerals of the world* book by Hill and Forti (1997). Within each class, the minerals are alphabetically ordered, followed but their chemical formula (according to IMA-CNMNC), the cave where it was first identified, and the reference in which it was described. The crystal system was included only for polymorphs (e.g., calcite, aragonite, vaterite), and abbreviations are as follow: triclinic (tric.), monoclinic (mon.), rhombic (rhom.), tetragonal (tetr.), trigonal (trig.), and hexagonal (hex.). Names of mineral species that were omitted in the last checklist of Onac (2003) along with those reported since then are in **bold** face. Grandfathered species and species that have been discredited, redefined, or renamed by the CNMNC, as well as names used to designate a group of species (series), are in *italics*. Where necessary to understand the stoichiometry of mineral formulae, the charge for altervalent elements is provided.

The present list includes 118 minerals, representing almost half of the world's latest cave mineral inventory (319; Onac and Forti 2011). A quick surfing of the entries below reveals that as early as the eighteenth century, sulfur, alum, and calcite from caves in Romania were documented in the first mineralogical treatise compiled by Fridvaldszky (1767) and dedicated to Transylvania (Fig. 1). A more systematic investigation of cave minerals begun in the early 1970s and continued to grow ever since. At the beginning, the focus was on simply describing the mineralogy of some particular speleothems and moonmilk deposits. Then, the attention moved toward caves influenced by thermo-mineral waters, those located near or within mining areas, or cavities hosting significant guano accumulations. Consequently, a great number of minerals were described over the last two decades. While the instrument needs in cave mineralogical research will depend greatly upon the samples investigated, many of the minerals recently described would have been utterly impossible without the tremendous progress experienced by the analytical facilities. In addition, the likelihood to precisely diagnose minerals using just a few milligrams from a sample, especially when this is an unremarkable earthy mass, greatly increased the number of rare/uncommon cave minerals (Onac 2012).

B. P. Onac (✉)
Karst Research Group, School of Geosciences, University of South Florida, 4202 E. Fowler Ave., NES 107, Tampa, FL 33620, USA
e-mail: bonac@usf.edu

B. P. Onac
"Emil Racoviță" Institute of Speleology, Clinicilor 5, 4000006 Cluj-Napoca, Romania

G. M. L. Ponta and B. P. Onac (eds.), *Cave and Karst Systems of Romania*, Cave and Karst Systems of the World, https://doi.org/10.1007/978-3-319-90747-5_53

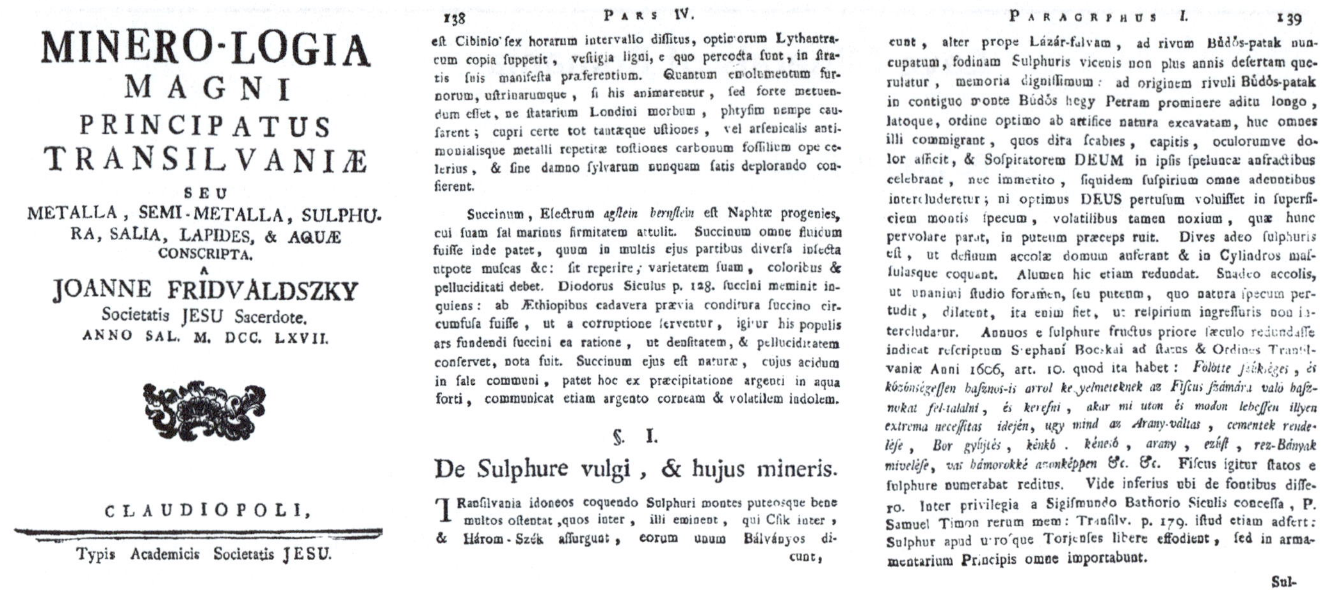
MINERO-LOGIA
MAGNI
PRINCIPATUS
TRANSILVANIÆ
SEU
METALLA, SEMI-METALLA, SULPHURA, SALIA, LAPIDES, & AQUÆ
CONSCRIPTA.
A
JOANNE FRIDVALDSZKY
Societatis JESU Sacerdote.
ANNO SAL. M. DCC. LXVII.

CLAUDIOPOLI,
Typis Academicis Societatis JESU.

138 PARS IV.

eſt Cibinio ſex horarum intervallo diſſitus, optimorum Lythantracum copia ſuppetit, veſtigia ligni, e quo percocta ſunt, in ſtratis ſuis manifeſta præferentium. Quantum emolumentum furnorum, uſtrinarumque, ſi his animarentur, ſed forte metuendum eſſet, ne ſtatarium Londini morbum, phtyſim nempe cauſarent; cupri certe tot tantæque uſtiones, vel arſenicalis antimonialisque metalli repetitæ toſtiones carbonum foſſilium ope celerius, & ſine damno ſylvarum nunquam ſatis deplorando conficerent.

Succinum, Electrum *agſtein bernſtein* eſt Naphtæ progenies, cui ſuam ſal marinus firmitatem attulit. Succinum omne fluidum fuiſſe inde patet, quum in multis ejus partibus diverſa inſecta utpote muſcas &c: ſit reperire; varietatem ſuam, coloribus & pelluciditati debet. Diodorus Siculus p. 128. ſuccini meminit inquiens: ab Æthiopibus cadavera prævia conditura ſuccino circumfuſa fuiſſe, ut a corruptione ſerventur, igitur his populis ars fundendi ſuccini ea ratione, ut denſitatem, & pelluciditatem conſervet, nota fuit. Succinum ejus eſt naturæ, cujus acidum in ſale communi, patet hoc ex præcipitatione argenti in aqua forti, communicat etiam argento corneam & volatilem indolem.

§. I.

De Sulphure vulgi, & hujus mineris.

TRanſilvania idoneos coquendo Sulphuri montes puteosque bene multos oſtentat, quos inter, illi eminent, qui Cſik inter, & Három-Szék aſſurgunt, eorum unum Bálványos dicunt,

PARAGRPHUS I. 139

cunt, alter prope Lázár-falvam, ad rivum Büdös-patak nuncupatum, fodinam Sulphuris vicenis non plus annis deſertam querulatur, memoria digniſſimum: ad originem rivuli Büdös-patak in contiguo monte Büdös hegy Petram prominere aditu longo, latoque, ordine optimo ab artifice natura excavatam, huc omnes illi commigrant, quos dira ſcabies, capitis, oculorumve dolor afficit, & Soſpitatorem DEUM in ipſis ſpeluncæ anfractibus celebrant, nec immerito, ſiquidem ſuſpirium omne adeuntibus intercluderetur; ni optimus DEUS pertuſum voluiſſet in ſuperficiem montis ſpecum, volatilibus tamen noxium, quæ hunc pervolare parat, in puteum præceps ruit. Dives adeo ſulphuris eſt, ut deſiuum accolæ domum auferant & in Cylindros maſſulasque coquant. Alumen hic etiam redundat. Suadeo accolis, ut unanimi ſtudio foramen, ſeu puteum, quo natura ſpecum pertudit, dilatent, ita enim fiet, ut reſpirium ingreſſuris non intercludatur. Annuos e ſulphure fructus priore ſæculo redundaſſe indicat reſcriptum Stephani Bocskai ad ſtatus & Ordines Tranſilvaniæ Anni 1606, art. 10. quod ita habet: *Fölötte ſzükséges, és közönségeſſen haſznos-is arrol kegyelmeteknek az Fiſcus ſzámára való haſznokat fel-talalni, és keresni, akar mi uton és modon lebeſſen illyen extrema neceſſitas idején, ugy mind az Arany-váltas, cementek rendeléſe, Bor gyüjtés, kénkő. kéneső, arany, ezüſt, rez-Bányak mivelése, vas bámorokké aſonképpen &c. &c.* Fiſcus igitur ſtatos e ſulphure numerabat reditus. Vide inferius ubi de fontibus diſſero. Inter privilegia a Sigiſmundo Bathorio Siculis conceſſa, P. Samuel Timon rerum mem: Tranſilv. p. 179. iſtud etiam adfert: Sulphur apud utroque Torjenſes libere effodient, ſed in armamentarium Principis omne importabunt.

Sul-

Fig. 1 Title cover and pages from Fridvaldszky's (1767) book where the presence of sulfur in caves from Büdös hegy (Puturosu Mountain) is mentioned

The distribution of cave minerals on various classes shows that the most numerous group belongs to sulfates (28), followed by phosphates (22), carbonates (21), oxides/hydroxides (16), silicates (16), sulfides (5), whereas the other classes include only one to two minerals (Fig. 2). A closer inspection of these classes provides some interesting observations. For example, half of the sulfate minerals are almost equally shared by two caves Tăuşoare (8) and Diana (7), whereas for phosphates, 13 out 22 minerals are hosted by the Cioclovina Cave. An even more remarkable situation is recorded in the oxides/hydroxides group, where 12 out of 16 species are described from Vântului Cave. Thus, comes as no surprise to learn that in fact Cioclovina and Vântului caves are the second and third richest in terms of minerals, respectively. At the top of this ranking is Valea Rea Cave, which is home for 33 minerals belonging to eight classes. From the entire inventory of Romania's 118 minerals, thirty-five have never been documented before from cave environments worldwide; these are underlined in the checklist below.

Native Elements (2)

Sulfur—S; caves in Puturosu Mountain/Büdös Hegy (Fridvaldszky 1767)
Gold—Au; Valea Rea Cave (Ghergari et al. 1997)

Sulfides (5)

Luzonite—Cu_3AsS_4; Crystal's Cave (Codreanu Mine, Băiţa Bihor) (Onac 2002)
Marcasite—FeS_2; caves in Băiţa (Metaliferi Mountains) (Nedopaca 1987a)
Pyrite—FeS_2; caves in Băiţa (Metaliferi Mountains) (Nedopaca 1987a)
Sphalerite—ZnS; Rodna Veche (Mârza and Silvestru 1988)
Wittichenite—Cu_3BiS_3; Big Cave (Bolfu III Mine, Băiţa Bihor) (Onac 2002)

Oxides and Hydroxides (16)

Birnessite—$(Na, Ca, K)_{0.6}(Mn^{4+}, Mn^{3+})_2O_4 \cdot 1.5H_2O$; Vântului Cave (Onac 1996)
Braunite—$Mn^{2+}Mn^{3+}_6O_8(SiO_4)$; Vântului Cave (Diaconu and Morar 1997)
Diaspore—AlO(OH); Vântului Cave (Coman 1979)
Gibbsite—$Al(OH)_3$; Vântului Cave (described as *hydrargillite* by Coman 1979)
Goethite—FeO(OH); caves in Trestia-Băiţa (Metaliferi Mountains) (Nedopaca 1987b)
Hausmannite—$Mn^{2+}Mn^{3+}_2O_4$; Vântului Cave (Diaconu and Morar 1997)

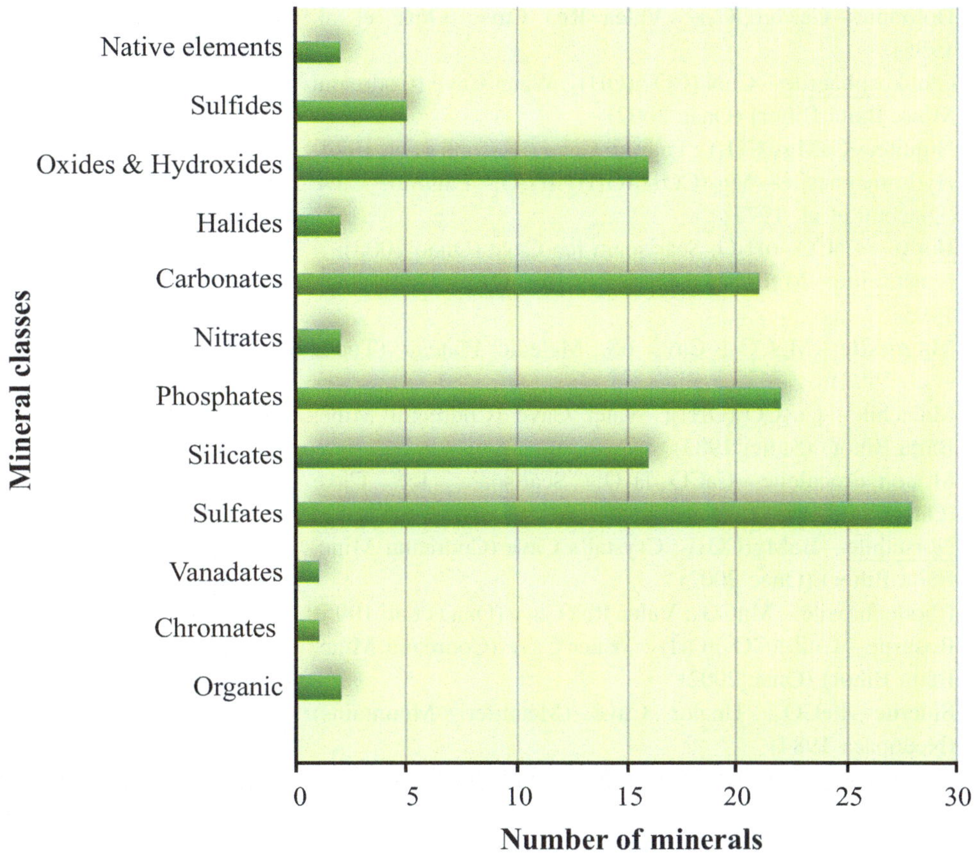

Fig. 2 Distribution of cave minerals on chemical classes

Hematite—Fe_2O_3; Vântului Cave (Diaconu and Morar 1997; Onac et al. 1997)

<u>Hollandite</u>—$Ba(Mn^{4+}_6, Mn^{3+}_2)O_{16}$; Vântului Cave (Onac 1996)

Ice—H_2O; Scărişoara Ice Cave (Racoviță 1927)

Lepidocrocite—$Fe^{3+}O(OH)$; Vântului/Valea Rea caves (Diaconu and Morar 1997; Ghergari et al. 1997)

*Limonite—**NOT** a mineral but generic term used for undifferentiated hydrated iron oxides; **NOT** approved by IMA-CNMNC*

Lithiophorite—$(Al, Li)(Mn^{4+}, Mn^{3+})_2O_2(OH)_2$; Vântului Cave (Onac 1994)

Magnetite—$Fe^{2+}Fe^{3+}_2O_4$; karst voids at Ocna de Fier (Mârza et al. 1995)

<u>Periclase</u>—MgO; Valea Rea Cave (Onac et al. 1995)

Pyrolusite—MnO_2; Vântului Cave (Onac 1992)

Romanèchite—$(Ba, H_2O)_2(Mn^{4+}, Mn^{3+})_5O_{10}$; Vântului Cave (Onac 1992)

Todorokite—$(Na, Ca, K, Ba, Sr)_{1-x}(Mn,Mg,Al)_6O_{12}\cdot 3\text{–}4H_2O$; Vântului Cave (Onac 1996)

*Wad—**NOT** a mineral but a generic term used for poorly crystallized, undifferentiated hydrated manganese oxides and hydroxides; **NOT** approved by IMA-CNMNC*

Halides (2)

Atacamite—$Cu_2Cl(OH)_3$; Cioclovina Cave (Onac et al. 2011)

Halite—NaCl; caves in the Meledic Plateau (Vrancei Subcarpathians) (Giurgiu et al. 1980)

Carbonates (21)

Ankerite—$Ca(Fe^{2+}, Mg)(CO_3)_2$; Runcului Cave (Onac 1992)

Aragonite—$CaCO_3$ (rhom.); Fagului, Micula, Baia lui Schneider caves

Aurichalcite—$(Zn, Cu)_5(CO_3)_2(OH)_6$; caves from Băița Bihor (Onac 2002)

Azurite—$Cu_3(CO_3)_2(OH)_2$; Water Cave (Codreanu Mine, Băița Bihor) (Stoici 1983)

<u>**Burbankite**</u>—$(Na, Ca)_3(Sr, Ba, Ce)_3(CO_3)_5$; Cioclovina Cave (Onac et al. 2009a)

Calcite—$CaCO_3$ (trig.); Dâmbovicioara Cave (Fridvaldszky 1767)

Dolomite—$CaMg(CO_3)_2$; Valea Rea Cave (Onac et al. 1995)

Glaukosphaerite—$CuNi(CO_3)(OH)_2$; Water Cave (Codreanu Mine, Băiţa Bihor) (Onac 2002)

Huntite—$CaMg_3(CO_3)_4$; Fagului Cave (Diaconu et al. 1977)

Hydromagnesite—$Mg_5(CO_3)_4(OH)_2{\cdot}4H_2O$; Fagului Cave (Diaconu et al. 1977)

Ikaite—$CaCO_3{\cdot}6H_2O$; Scărişoara Ice Cave (Onac 2008)

Lansfordite—$MgCO_3{\cdot}5H_2O$; Valea Rea Cave (Onac et al. 1995)

Magnesite—$MgCO_3$; Cave 6S, Meledic Plateau (Todor et al. 1993)

Malachite—$Cu_2CO_3(OH)_2$; Water Cave (Codreanu Mine, Băiţa Bihor) (Stoici 1983)

Monohydrocalcite—$CaCO_3{\cdot}H_2O$; Scărişoara Ice Cave (Onac 2000–2001)

Norsethite—$BaMg(CO_3)_2$; Crystal's Cave (Codreanu Mine, Băiţa Bihor) (Onac 2002)

Rhodochrosite—$MnCO_3$; Valea Rea Cave (Onac et al. 1995)

Rosasite—$CuZn(CO_3)(OH)_2$; Water Cave (Codreanu Mine, Băiţa Bihor) (Onac 2002)

Siderite—$FeCO_3$; Urşilor Cave (Metaliferi Mountains) (Nedopaca 1984)

Smithsonite—$ZnCO_3$; Cave #4, Runcului Hill (Zaharia et al. 2003)

Vaterite—$CaCO_3$ (hex.); Cloşani Cave (Diaconu 1990)

Nitrates (2)

Darapskite—$Na_3(SO_4)(NO_3){\cdot}H_2O$; Şălitrari Cave (Diaconu and Lascu 1998)

Nitratine (Soda niter)—$NaNO_3$; Sălitrari Cave (Diaconu and Lascu 1998)

Phosphates (22)

Ardealite—$Ca_2(PO_3OH)(SO_4){\cdot}4H_2O$; Cioclovina Cave (Schadler 1932)

Berlinite—$AlPO_4$; Cioclovina Cave (Onac and White 2003; Onac and Effenberger 2007)

Bobierrite—$Mg_3(PO_4)_3{\cdot}8H_2O$; Valea Rea Cave (Onac et al. 1995)

Brushite—$Ca(PO_3OH){\cdot}2H_2O$; Cioclovina Cave (Halla 1931)

Carbonate-fluorapatite—Discredited by IMA-CNMNC

Carbonate-hydroxylapatite (dahllite)—Discredited by IMA-CNMNC

Churchite-(Y)—$Y(PO_4){\cdot}2H_2O$; Cioclovina Cave (Onac et al. 2005)

Collinsite—$Ca_2Mg(PO_4)_2{\cdot}2H_2O$; Cioclovina Cave (Onac et al. 2002)

Crandallite—$CaAl_3(PO_4)_2(PO_3OH)(OH)_6$; Cioclovina Cave (Constantinescu et al. 1999)

Fluorapatite—$Ca_5(PO_4)_3F$; Cioclovina Cave (Onac et al. 2002)

Foggite—$CaAl(PO_4)(OH)_2{\cdot}H_2O$; Cioclovina Cave (Onac et al. 2002)

Francoanellite—$K_3Al_5(PO_3OH)(PO_4)_2{\cdot}12H_2O$; Măgurici Cave (Onac and Vereş 2003)

Hydroxylapatite—$Ca_5(PO_4)_3(OH)$; Cioclovina Cave (Schadler 1929)

Leucophosphite—$KFe_2^{3+}(PO_4)_2(OH)\cdot 2H_2O$; Cioclovina Cave (Onac et al. 2002)

Monetite—$Ca(PO_3OH)$; Cioclovina Cave (Onac et al. 2002)

Montgomeryite—$Ca_4MgAl_4(PO_4)_6(OH)_4{\cdot}12H_2O$; Humpleu Cave (Moldovan et al. 2015)

Phosphammite—$(NH_4)_2(PO_3OH)$; Măgurici Cave (Onac and Vereş 2003)

Strengite—$Fe^{3+}PO_4{\cdot}2H_2O$; Valea Rea Cave (Onac et al. 1995)

Taranakite—$K_3Al_5(PO_3OH)_6(PO_4)_2{\cdot}18H_2O$; Stracoş Cave (Onac and Bengeanu 1992)

Tinsleyite—$KAl_2(PO_4)_2(OH){\cdot}2H_2O$; Cioclovina Cave (Marincea et al. 2002)

Variscite—$AlPO_4{\cdot}2H_2O$; Cioclovina Cave (Onac et al. 2004

Vashegyite—$Al_{11}(PO_4)_9(OH)_6{\cdot}38H_2O$; Gaura cu Muscă Cave (Onac et al. 2006a)

Vivianite—$Fe_3^{2+}(PO_4)_2{\cdot}8H_2O$; Valea Rea Cave (Onac et al. 1995)

Wavellite—$Al_3(PO_4)_2(OH)_3{\cdot}5H_2O$; Valea Rea Cave (Onac et al. 1995)

Silicates (16)

Allophane—$Al_2O_3(SiO_2)_{1.3\text{-}2.0}{\cdot}2.5\text{-}3H_2O$; Vântului Cave (Iosof et al. 1974)

Benitoite—$BaTiSi_3O_9$; Iza Cave (Viehmann et al. 1981)

Clinochlore—$Mg_5Al(AlSi_3O_{10})(OH)_8$; Valea Rea Cave (Feier 2003)

Cristobalite—SiO_2 (tetr.); Alum Cave, Turia (Szakáll et al. 2006)

Dickite—$Al_2Si_2O_5(OH)_4$; Iza Cave (Viehmann et al. 1981)

Endellite—Discredited by IMA-CNMNC (use Halloysite-10Å)

Epidote—$Ca_2(Al_2Fe^{3+})[Si_2O_7][SiO_4]O(OH)$; Valea Rea Cave (Feier 2003)

Halloysite-7Å—$Al_2Si_2O_5(OH)_4$ (mon.); Vale Rea Cave (Onac unpubl.)

Halloysite-10Å—$Al_2Si_2O_5(OH)_4{\cdot}2H_2O$; Vântului Cave (Onac and Polyak unpubl.)

Hydroxylellestadite—$Ca_5(SiO_4)_{1.5}(SO_4)_{1.5}OH$; Cioclovina Cave (Onac et al. 2006b)

*Illite—**NOT** not a mineral; Name used to designate a group of species*

Kaolinite—$Al_2Si_2O_5(OH)_4$ (tric.); Iza Cave (Viehmann et al. 1979)

Montmorillonite—$(Na, Ca)_{0.3}(Al, Mg)_2Si_4O_{10}(OH)_2{\cdot}nH_2O$; Iza Cave (Viehmann et al. 1981)

Nacrite—$Al_2Si_2O_5(OH)_4$ (mon.); Valea Rea Cave (Onac et al. 1995)

Natrolite—$Na_2(Si_3Al_2)O_{10}{\cdot}2H_2O$; Big Cave (Bolfu III Mine, Băița Bihor) (Onac 2002)

Opal—$SiO_2{\cdot}nH_2O$; caves in Trestia-Băița (Metaliferi Mountains) (Nedopaca and Grama 1987)

Quartz – SiO_2 (trig.); caves in Trestia-Băița (Metaliferi Mountains) (Nedopaca and Grama 1987)

Saponite—$(Ca, Na)_{0.3}(Mg, Fe)_3(Si, Al)_4O_{10}(OH)_2{\cdot}4H_2O$; Vântului Cave (Iosof et al. 1974)

Sulfates (28)

Alum-(K)—$KAl(SO_4)_2{\cdot}12H_2O$; Alum Cave, Turia (Fridvaldszky 1767)

Aluminite—$Al_2SO_4(OH)_4{\cdot}7H_2O$; Valea Rea Cave (Feier 2003)

Alunite—$KAl_3(SO_4)_2(OH)_6$; Șălitrari Cave (Onac et al. 2009b)

Alunogen—$Al_2(SO_4)_3(H_2O)_{12}{\cdot}5H_2O$; Alum Cave, Turia (Szakáll et al. 2010)

Anhydrite—$CaSO_4$; Diana Cave (Diaconu 1974)

Apjohnite—$Mn^{2+}Al_2(SO_4)_4{\cdot}22H_2O$; Diana Cave (Onac et al. 2009b)

Arcanite—K_2SO_4; Tăușoare Cave (Domșa 1988)

Barite—$BaSO_4$; caves around Trestia-Băița (Metaliferi Mountains) (Nedopaca 1986)

Bassanite—$CaSO_4{\cdot}0.5H_2O$; Tăușoare Cave (Jude 1972)

Celestine—$SrSO_4$; Valea Rea Cave (Onac et al. 1995)

Cesanite—$Ca_2Na_3(SO_4)_3OH$; Măgurici Cave (Onac and Vereș 2003)

Chalcanthite—$CuSO_4{\cdot}5H_2O$; Water Cave (Codreanu Mine, Trestia-Băița) (Onac 2002)

Epsomite—$MgSO_4{\cdot}7H_2O$; Diana Cave (Onac et al. 2009b)

Gypsum—$CaSO_4{\cdot}2H_2O$; Valea Rea, Tăușoare, Ponoraș, Topolnița

Halotrichite—$Fe^{2+}Al_2(SO_4)_4{\cdot}22H_2O$; Diana Cave (Povară et al. 1972)

Jarosite—$KFe^{3+}_3(SO_4)_2(OH)_6$; Iza Cave (Tămaș et al. 2011)

Kieserite—$MgSO_4{\cdot}H_2O$; 6S Cave, Meledic Plateau (Dumitru pers. comm.)

Konyaite—$Na_2Mg(SO_4)_2{\cdot}5H_2O$; Tăușoare Cave (Onac et al. 2001)

Kröhnkite—$Na_2Cu(SO_4)_2{\cdot}2H_2O$; Cioclovina Cave (Onac et al. 2011)

Leonite—$K_2Mg(SO_4)_2{\cdot}4H_2O$; Tăușoare Cave (Onac et al. 2001)

Meta-aluminite—$Al_2SO_4(OH)_4{\cdot}5H_2O$; Valea Rea Cave (Feier 2003)

Mirabilite—$Na_2SO_4{\cdot}10H_2O$; Tăușoare Cave (Moțiu et al. 1977)

Pickeringite—$MgAl_2(SO_4)_4{\cdot}22H_2O$; Diana Cave (Diaconu and Medeșan 1973)

Rapidcreekite—$Ca_2(SO_4)(CO_3){\cdot}4H_2O$; Diana Cave (Onac et al. 2013)

Serpierite—$Ca(Cu, Zn)_4(SO_4)_2(OH)_6{\cdot}3H_2O$; Cave #4, Runcului Hill (Zaharia et al. 2003)

Syngenite—$K_2Ca(SO_4)_2{\cdot}H_2O$; Tăușoare Cave (Onac et al. 2001)

Tamarugite—$NaAl(SO_4)_2{\cdot}6H_2O$; Diana Cave (Pușcaș et al. 2013)

Thenardite—Na_2SO_4; Tăușoare Cave (Moțiu et al. 1977)

Vanadates (1)

Metatyuyamunite—$Ca(UO_2)_2(VO_4)_2{\cdot}3H_2O$; Valea Rea Cave (Onac et al. 2000)

Chromates (1)

Crocoite—$PbCrO_4$; Scărișoara Ice Cave (Onac 2000–2001)

Organic Compounds (2)

Guaine—$C_5H_3(NH_2)N_4O$; Gaura cu Muscă Cave (Onac et al. in prep)

Uricite—$C_5H_4N_4O_3$; Gaura Țuranului Cave (Tămaș et al. 2017)

Acknowledgements I am grateful to Drs. T. Tămaş, F.L. Forray, and M.-L. Tîrlă who provided relevant cave minerals information. This note is part of the Cave Mineral Database (CAMIDA) project, a collaborative effort of the University of South Florida Libraries, UIS's Cave Minerals Commission, the Karst Information Portal, "Emil Racoviţă" Institute of Speleology (Romania), and the Karst Research Group at University of South Florida (USA).

References

Burke EAJ (2006) A mass discreditation of GQN minerals. Can Mineral 44:1557–1560

Burke EAJ (2008) Tidying up mineral names: an IMA-CNMNC scheme for suffixies, hyphens and diacritical marks. Mineral Rec 39:131–135

Coman D (1979) Essai sur une interprétation écologique de l'origine des grottes. Trav Inst Spéol "Émile Racovitza" 18:191–199

Diaconu G (1974) Quelques considérations sur la présence de l'anhydrite dans la grotte "Pestera Diana" (Baile Herculane-Roumanie). Trav Inst Spéol "Émile Racovitza" 13:191–194

Diaconu G (1990) Cloşani Cave. Mineralogical and genetic study of carbonates and clays. Misc Speol Rom 2:190

Diaconu G, Lascu C (1998) Preliminary data about nitrates in "Peştera Mare de la Şăitrari", Cerna Mountains, Romania. Theor Appl Karstol 11–12:47–52

Diaconu G, Medeşan A (1973) Sur la présence du pickeringite dans la grotte de Diana (Băile Herculane, Roumanie). Trav Inst Spéol "Émile Racovitza" 12:303–309

Diaconu G, Medeşan A, Viehmann I (1977) Une nouvelle paragenèse minéralogique dans la grotte "Peştera Fagului", Départament de Bihor (huntite, hydromagnesite, aragonite, calcite). Trav Inst Spéol "Émile Racovitza" XVI:203–210

Diaconu G, Morar M (1997) Nouveaux minéraux de fer et de manganèse dans la Grotte du Vent (Monts Pădurea Craiului, Roumanie). Karstologia 29:51–53

Domşa M (1988) Geneza mirabilitului din Peştera Tăuşoare, Munţii Rodnei. Peştera 2:190–204

Feier N (2003) Date noi privind mineralogia Peşterii Valea Rea, Munţii Bihor. Ecocarst 4:22–24

Fridvaldszky J (1767) Mineralogia Magni Principatus Transylvaniae. Claudiopoli, Typis Academicis Societatis Jesu 206

Ghergari L, Tămaş T, Damm P, Forray F (1997) Hydrothermal paleokarst in Peştera din Valea Rea (Bihor Mountains, Romania). Theor Appl Karst 10:115–125

Giurgiu I, Silvăşanu G, Vădeanu T (1980) Observaţii asupra carstului pe sare din zona Meledic (M. Buzău). Peştera 6S de la Mînzăleşti. Comparaţie între formele carstice dezvoltate în sare şi calcare. Bul Speol CSER Bucureşti 7:55–66

Halla F (1931) Isomorphe Beziehungen und Doppelsalzbildung zwischen Gips und Brushit. Z Krist 80:349–352

Hill CA, Forti P (1997) Cave minerals of the world, 2nd ed. Natl Speleol Soc, Huntsville, p 464

Iosof V, Coman D, Ianc R (1974) Note sur la présence de l'allophane dans la grotte du Vent des Monts Apuseni. Trav Inst Spéol "Émile Racovitza" 13:183–189

Jude L (1972) Les anthodites des Monts de Rodna. Trav Inst Spéol "Émile Racovitza" 11:343–347

Marincea Ş, Dumitraş D, Gibert R (2002) Tinsleyite in the "dry" Cioclovina Cave (Şureanu Mountains, Romania): the second occurrence. Eur J Mineral 14:157–164

Mârza I, Silvestru E (1988) First mention of the hydrothermal karst phenomenon in Romania, associated to the neogene metasomatic sulphide ore deposits from Rodna Veche. Studia UBB Geologia 33 (1):77–81

Mârza I, Nicolescu Ş, Gruescu C (1995) La magnétite - draperie stalactitique (Formation de karst hydrothermal) du gisement de Ocna de Fier (Banat). Rom J Mineral 76:101–103

Mills SJ, Hatert F, Nickel EH, Ferraris G (2009) The standardisation of mineral group hierarchies: application to recent nomenclature proposals. Eur J Mineral 21:1073–1080

Moldovan C, Tămaş T, Haeuselmann P (2015) The mineralogy of some particular phosphate deposits from the Humpleu Cave System, Romania. Acta Miner-Petrogr, Abs Ser 9:43

Moţiu A, Viehmann I, Strusiewicz R (1977) Découverte des nouveaux mineraux dans la grotte Tăuşoare (Monts Rodna). Trav Inst Spéol "Émile Racovitza" 16:211–216

Nedopaca M (1984) Observations minéralogiques dans la Grotte des Ours-Bulzeşti. Theor Appl Karstol 1:47–49

Nedopaca M (1986) Carst mineralizat în Munţii Metaliferi. Bul Speol 10:66–72

Nedopaca M (1987a) Paragénèse pyrite-marcassite du karst des Monts Metaliferi. Theor Appl Karstol 3:235–238

Nedopaca M (1987b) Note sur la présence de quelques minéraux de manganèse et de fer dans l'endokarst des calcaires cristallins dans les Monts Métalifères. Theor Appl Karstol 3:239–242

Nedopaca M, Grama C (1987) Mineralizaţiile hidrotermale cu cuarţ din peşterile de la Trestia-Băiţa (jud. Hunedoara). Bul Speol 11:9–14

Onac BP (1992) Mineralogy of the Apuseni Mountains caves. Theor Appl Karstol 5:193–201

Onac BP (1994) Peştera Vîntului (Wind Cave). Theor Appl Karstol Program-Abstracts and Excursion, Băile Felix, Romania, 52–56

Onac BP (1996) Mineralogy of speleothems from caves in the Padurea Craiului Mountains (Romania), and their palaeoclimatic significance. Cave Karst Sci 23(3):109–124

Onac BP (2000–2001) Mineralogical studies and Uranium-series dating of speleothems from Scărişoara Glacier Cave (Bihor Mountains, Romania). Theor Appl Karstol 13–14:33–38

Onac BP (2002) Caves formed within Upper Cretaceous skarns at Băiţa (Bihor county): mineral deposition and speleogenesis. Can Mineral 40(6):1551–1561

Onac BP (2003) Checklist of the Romanian cave minerals. Theor Appl Karstol 16:83–89

Onac BP (2008) Ikaite in the Scărişoara ice deposit: precipitation and significance. In: Turri S (ed) 3rd International workshop on ice caves (IWIC-III), vol Abstr. Perm State University, Perm, p 28

Onac BP (2012) Minerals. In: White WB, Culver DC (eds) Encyclopedia of minerals. Academic Press, Chennai, pp 499–508

Onac BP, Bengeanu M (1992) Phosphate minerals in few caves from Pădurea Craiului Mountains (Romania)—Preliminary note. Nymphaea XXII:23–28

Onac BP, Effenberger HS (2007) Re-examination of berlinite ($AlPO_4$) from the Cioclovina Cave, Romania. Am Mineral 92:1998–2001

Onac BP, Forti P (2011) State of the art and challenges in cave minerals studies. Studia UBB Geologia 56(1):33–42

Onac BP, Vereş D (2003) Sequence of secondary phosphates deposition in a karst environment: evidence from Măgurici Cave (Romania). Eur J Mineral 15:741–745

Onac BP, White WB (2003) First reported sedimentary occurrence of berlinite ($AlPO_4$) in phosphate-bearing sediments from Cioclovina Cave, Romania. Am Mineral 88:1395–1397

Onac BP, Bengeanu M, Botez M, Zih J (1995) Preliminary report on the mineralogy of Peştera din Valea Rea (Bihor Mountains, Romania). Theor Appl Karstol 8:75–79

Onac BP, Pedersen RB, Tysseland M (1997) Presence of rare-earth elements in black ferromanganese coatings from Vântului Cave (Romania). J Cave Karst Stud 59(3):128–131

Onac BP, Kearns J, Damm P, White WB, Matyasi S (2000) The first reported occurrence of metatyuyamunite $Ca(UO_2)_2(VO_4)_2{\cdot}3\text{-}5H_2O$ in Romania. Rom J Mineral 80:5–10

Onac BP, White WB, Viehmann I (2001) Leonite $K_2Mg[SO_4]_2{\cdot}4H_2O$, konyaite $Na_2Mg[SO_4]_2{\cdot}5H_2O$, and syngenite $K_2Ca[SO_4]_2{\cdot}H_2O$ from Tăuşoare Cave (Rodnei Mts., Romania). Min Mag 65(1):1–7

Onac BP, Breban R, Kearns Tămaş T (2002) Unusual minerals related to phosphate deposits in Cioclovina Cave, Şureanu Mts. (Romania). Theor Appl Karstol 15:27–34

Onac BP, Kearns J, Breban R, Cîntă-Pânzaru S (2004) Variscite ($AlPO_4{\cdot}2H_2O$) from Cioclovina Cave (Şureanu Mountains, Romania): a tale of a missing phosphate. Studia UBB Geologia 49(1):3–14

Onac BP, Ettinger K, Kearns Balasz II (2005) A modern, guano-related occurrence of foggite, $CaAl(PO_4)(OH)_2{\cdot}H_2O$ and churchite-(Y), $YPO_4{\cdot}2H_2O$ in Cioclovina Cave, Romania. Mineral Petrol 85: 291–302

Onac BP, Effenberger HS, Ettinger K, Cîntă-Pânzaru S (2006b) Hydroxylellestadite from Cioclovina Cave (Romania): Microanalytical, structural, and vibrational spectroscopy data. Am Mineral 91:1927–1931

Onac BP, Zaharia L, Kearns J, Veres D (2006a) Vashegyite from Gaura cu Musca Cave (Locvei Mountains, Romania): a new and rare phosphate occurrence. Int J Speleol 35(2):67–73

Onac BP, Bernhardt H-J, Effenberger HS (2009a) Authigenic burbankite in the Cioclovina Cave sediments (Romania). Eur J Mineral 21:507–514

Onac BP, Sumrall JG, Tămaş T, Povară I, Kearns J, Dârmiceanu V, Vereş DŞ, Lascu C (2009b) The relationship between cave minerals and H_2S-rich thermal waters along the Cerna Valley (SW Romania). Acta Carsologica 38(1):27–39

Onac BP, Effenberger HS, Collins N, Kearns J, Breban R (2011) Revisiting three minerals from Cioclovina Cave (Romania). Int J Speleol 40:99–108

Onac BP, Effenberger HS, Wynn JG, Povară I (2013) Rapidcreekite in the sulfuric acid weathering environment of Diana Cave, Romania. Am Mineral 98:1302–1309

Pasero M, Kampf AR, Ferraris C, Pekov IV, Rakovan J, White TJ (2010) Nomenclature of the apatite supergroup minerals. Eur J Mineral 22:163–179

Povară I, Diaconu G, Goran C (1972) Observations préliminaires sur les grottes influencées par les eaux thermo-minérales de la zone Băile Herculane. Trav Inst Spéol "Émile Racovitza" 11:355–365

Puşcaş CM, Onac BP, Effenberger HS, Povară I (2013) Tamarugite-bearing paragenesis formed by sulphate acid alteration in Diana Cave, Romania. Eur J Mineral 25:479–486

Racoviţă EG (1927) Observations sur la Glacière naturelle dite "Gheţarul de la Scărişoara". Bul Soc Ştiinţe Cluj III:75–108

Schadler J (1929) Mineralogische-petrographische carakteristik der Phosphat ablagerung in der Cioclovina-höhle bei Pui. Publ Muz Hunedoara 5:642

Schadler J (1932) Ardealit, ein neues Mineral $CaHPO_4$ $CaSO_4$ $+4H_2O$. Zentral-bl Miner A:40–41

Stoici SD (1983) Districtul metalogenetic Băiţa, Bihorului edn. Acad RSR, Bucureşti

Szakáll S, Kristály F, Papucs A, Jánosi C (2006) Postvolcanic solfatara type alterations and sulphate efflorescences at Puturosu Hill, Torja (Ciomadu Group). In: 8th Geol Meeting in Szeklerland, Sapientia EMTE, pp 81–83

Szakáll S, Kristály F, Papucs A (2010) Minerals of Székelyland. In: Szakáll S, Kristály F (eds) Mineralogy of Székelyland, Eastern Transylvania, Romania. Csík County Nature and Conservation Society, Târgu Mureş, pp 171–307

Tămaş T, Kristály F, Barbu-Tudoran L (2011) Mineralogy of Iza Cave (Rodnei Mountains, N. Romania). Int J Speleol 40(2):171–179

Tămaş T, Giurgiu AM, Barbu-Tudoran L, Mureşan B (2017) Uricite from Gaura Ţuranului, NW Romania. Proceedings of the 17th Int Cong Speleol, Sydney, pp 262

Todor D, Giurgiu I, Silvăşanu G (1993) L'analyse quantitative de quelques spéléothèmes et échantillons de la grotte en sel 6S de Mânzalesti (Subcarpathes de Vrancea-Roumanie). Theor Appl Karstol 6:211–212

Viehmann I, Silvestru E, Fabian C (1979) La «grotte d'Iza» (Monts de Rodna, Roumanie). Trav Inst Spéol "Émile Racovitza" XVII: 201–207

Viehmann I, Demeter I, Lungu V, Sarkady P (1981) Note préliminaire sur l'argile blanche de la grotte d'Iza (Monts de Rodna, Roumanie). Trav Inst Spéol. "Émile Racovitza" XX:213–215

Zaharia L, Tămaş T, Suciu-Krausz E (2003) Mineralogy of Cave No. 4 from Runcului Hill (Metaliferi Mts., Romania). Theor Appl Karstol 16:41–46

Useful Websites

http://caveminerals.rc.usf.edu/node (Cave minerals of the world, database)

http://nrmima.nrm.se//imalist.htm (Official list of IMA-approved minerals)

http://rruff.info (Database of Raman, X-ray diffraction, and chemistry data for minerals)

Caves Discovered by Mining Activities and Mined Caves

Bogdan P. Onac

Abstract

Evidences show humans must have entered caves in Romania prior to 65,000 years ago. Their interest in mining activities came, however, much later, with the first documented signs predating the arrival of Romans in Dacia (present-day Romania), in the second century BC. Although writings about minerals in Romanian caves date back to the eighteenth and nineteenth centuries, the first scientific texts on minerals found in caves discovered during mining and quarrying activities only appeared after 1850s. From a mineralogical point of view, two distinct categories are recognizable: (1) caves displaying speleothems of ordinary carbonate mineralogy and (2) caves with unusual mineral paragenesis. The latter group could further be subdivided into: (i) cavities located near or within nonmetalliferous or polymetallic ore fields, (ii) skarn-hosted caves, and (iii) caves in which H_2S-rich thermo-mineral waters discharge. The study of these caves resulted in the discovery of minerals, either new for science (ardealite) or to the cave environment (anhydrite, burbankite, foggite, ikaite, konyaite, etc.). However, the scientific relevance of those caves discovered in mines and quarries, along with the mined caves, is not restricted to mineralogy but also encompasses anthropology, archeology, Quaternary geology, biospeleology, karst science (speleothems, speleogenesis, etc.), and tourism.

Keywords

Mine caves • Mined caves • Quarrying • Science Romania

In the memory of Paul Erik DAMM who dedicated part of his karst research to mine caves.

Introduction

Early human footprints are rare in the fossil record. Evidences indicate that *Homo neanderthalensis* have entered caves in Romania prior to 65,000 years ago (Onac et al. 2005a). Other archaeological and anthropological findings indicate that the early modern humans had a more constant presence in the Romanian caves (Trinkaus et al. 2003; Soficaru et al. 2007; Olariu et al. 2005; Clottes et al. 2012; Webb et al. 2014). However, their interest in mining activities came much later, with the first documented signs predating the arrival of the Romans in Dacia (present-day Romania), in the second century BC (Cauuet 2002).

Over the last 150 years, a significant number of cavities were accidentally intercepted during mining and quarrying activities, especially in the karst regions of central and northwestern Romania. Fewer, but nevertheless important mine caves are also known from the East and South Carpathians as well as from Dobrogea (Fig. 1). The main driving forces behind these discoveries were the exploration and exploitation of limestone, bauxite, and sulfur in quarries or mining for polymetallic ore deposits, bauxite, salt, guano-phosphates, or saltpeter. Prospects for water, road constructions, and geotechnical studies also contributed to some spectacular discoveries. The scientific relevance of these cavities is multifold. After a brief presentation of the Romanian karst, the main areas of scientific interest

The text in this chapter is an updated version of an earlier publication by Onac (2013). It is reproduced here by kind permission of the editors of the *Memorie Istituto Italiano di Speleologia*

B. P. Onac (✉)
Karst Research Group, School of Geosciences, University of South Florida, 4202 E. Fowler Ave., NES 107, Tampa, FL 33620, USA
e-mail: bonac@usf.edu

B. P. Onac
"Emil Racoviţă" Institute of Speleology, Romanian Academy, Clinicilor 5, 400006 Cluj-Napoca, Romania

G. M. L. Ponta and B. P. Onac (eds.), *Cave and Karst Systems of Romania*,
Cave and Karst Systems of the World, https://doi.org/10.1007/978-3-319-90747-5_54

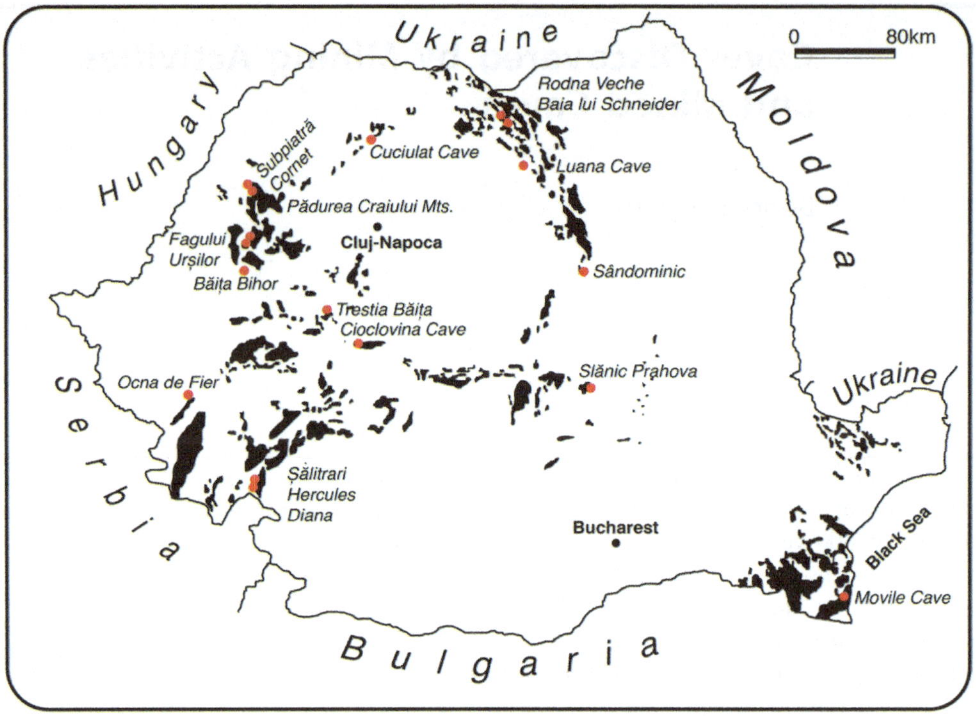

Fig. 1 Karst regions of Romania with the location of the most important mine/quarry caves

concerning mine/quarry and mined caves are assembled in several groups, then discussed, and exemplified.

Geographic and Geologic Settings

A concise presentation of the Romanian karst is given at the beginning of this book (see Chap. 3 by Onac and Goran); therefore, readers should refer to this text for information on karst rocks, landforms, and type of caves. The Romanian Cave Database administered by the "Emil Racoviță" Institute of Speleology (ERIS) includes over 13000 caves (Vlaicu pers. comm.); a very small fraction of them (less than 100) are caves discovered during mining operations either at surface or underground. These caves are further referred to as *mine or quarry caves*. For the purpose of this presentation, apart from the typical mine/quarry caves we also included common caves with known natural entrances but which were targeted by humans for mining out a variety of economically (more or less) valuable substances (ore deposits, phosphate or nitrate sediments, salt, or thermal water). In the process of excavation and extraction, some meaningful paleontological and anthropological sites were discovered. Furthermore, these activities provided scientists with good exposures of the sediment sequence or visual access to cave morphological features, thus allowing for detailed paleontological, mineralogical, or speleogenetic studies. This last group is discussed in a dedicated section titled *Mined caves*.

Caves with Mineralogical Relevance

This section discusses mine caves that contain unique speleothems from a genetic/morphologic, mineralogical, and crystallographic point of view. Also included are caves hosting sediments from which interesting mineral associations were described. Monographs and geological documents/reports confirm that in Romania it has been a continuous interest for mineralogical investigations in caves since the mid-eighteenth century (Fridvaldszky 1767; Pósepný 1874; Bieltz 1884; Koch 1886) up to present time (Schadler 1932; Nedopaca 1982; Diaconu 1985; Onac 2003). The first checklist of the Romanian cave minerals was published by Nedopaca (1982) and contained 29 species precipitated under various cave settings. Six other minerals (three of them from mine caves) were added by Diaconu (1985) who updated this cave minerals' list. In 2003, the number of minerals compiled by Onac increased to 104, with more than 25 described from mine or mined caves. Since this checklist was published, however, more than 15 new minerals were added, all collected from caves discovered during mining activities (see Chap. 53).

From a mineralogical point of view, two distinct categories are recognizable: (1) caves displaying speleothems of monotonous carbonate mineralogy (only calcite or aragonite, rarely, hydromagnesite and huntite) (Fagului, Urșilor, Lithophagus, Aurica, 1J2), and (2) caves with unusual mineral assemblages. The latter group is further subdivided into: (i) cavities located near or within nonmetalliferous or polymetallic ore fields, (ii) skarn-hosted caves, and (iii) caves in which H_2S-rich thermo-mineral waters discharge. Provided below is some supplemental information on each of these sub-categories.

(i) The exotic minerals associated with the polymetallic mine caves (e.g., Rodna Veche, Valea Vinului, Ocna de Fier, and Trestia-Băița regions) are all related to processes of oxidation and/or hydration of hydrothermal sulfides (Nedopaca 1986; Mârza and Silvestru 1988; Mârza et al. 1995; Iştvan and Tămaş 1996; Vişan 2011). As expected, the most common secondary cave minerals are sulfates, (gypsum, halotrichite group, barite, serpierite), some carbonates (malachite, azurite), and oxides (quartz, magnetite, goethite, hematite). Typically, these minerals form crusts, efflorescences, minute crystals, and aggregates.

(ii) In the Băița Bihor skarn district, many cavities were intercepted by adit and shafts (Stoici 1983; Onac and Damm 2002). The minerogenic processes and their associated products (recorded in brackets) described by Onac (2002) are: (a) *hydrothermal* (wittichenite, luzonite, quartz, and possibly some of the calcite), (b) *hydration or weathering* of hydrothermal minerals or primarily igneous rock constituents to form malachite, aurichalcite, rosasite, azurite, glaukosphaerite, chalcanthite, goethite, and natrolite, and (c) *precipitation* in low-temperature environment (calcite, aragonite, hydromagnesite, and norsethite). These minerals form crusts, coralloids, aggregates, and earthy masses (Onac 2002).

(iii) The caves along Cerna Valley (SW Romania) are the exponents for the third group, which is characterized by the presence of hot steam or/and thermal waters (<57 °C), which either pool or flow along cave passages reacting with the limestone/cave sediments to precipitate native sulfur, gypsum, anhydrite, or other sulfate minerals in the form of aggregates, crystals, rafts, and wall crusts (Povară et al. 1972; Diaconu 1974; Diaconu and Medeşan 1975; Onac et al. 2009a, b, 2013) (Fig. 2a). Also, efflorescences of tamarugite and halotrichite-group minerals are the main products of the bedrock weathering by acid sulfate condensate (Puşcaş et al. 2013).

A particular mineralogical case site is that of the Great Şălitrari Cave (Cerna Valley), where clastic sediments mixed or capped by abundant guano deposits (Fig. 2b) interacted with the H_2S-rich hot steam emerging in the cave to form a layered deposit with distinct phosphate/sulfate/nitrate horizons (Fig. 2b) (Diaconu and Lascu 1999; Onac et al. 2009a, b; Puşcaş et al. 2010). Twelve minerals (calcite, aluminite, alunite, gypsum, fluorapatite, hydroxylapatite, ardealite, brushite, taranakite, variscite) were identified in these sediments mined for saltpeter (darapskite and nitratine) between 1850 and the beginning of the twentieth century.

Cave with Significance in Quaternary and Deep-Time Paleobiology Studies

The first known written report on Romanian's cave paleontology belongs to P. Ranzanus (sixteenth century) who explained the massive accumulations of cave bear remains using Noah's biblical flood. Concentrated paleontological investigations were carried out during the nineteenth and early twentieth centuries, and especially in the second part of the twentieth century when scientists from the "Emil Racoviță" Institute of Speleology, "Țării Crişurilor" Museum, and the Department of Geology at Babeş-Bolyai University in Cluj-Napoca made important discoveries and studies in both ordinary and mine/quarry caves.

Two major sites were discovered in the Pădurea Craiului Mountains of NW Romania. One is related with the *Cornet mining district*, where in 1978 a gallery intercepted a lens of bauxite in which were concentrated huge amounts of well-preserved dinosaur, rare pterosaur, and bird fauna remains (Jurcsák and Kessler 1991; Posmoşanu 2003; Dyke et al. 2011). Based on the morphology of the bauxite lens and the condition of the bones surface (signs of abrasion, thus transport involved prior to deposition), it has been inferred that the fossil remains of Early Cretaceous age accumulated in large sinkholes formed on islands of the Tethys Ocean at the end of Jurassic. The insular environment was deduced based on adaptations recognized while studying the fossil remains (Benton et al. 1997; Posmoşanu and Cook 2000). The geographic and paleontological significance of this site lies in the fact that the dinosaur species discovered at Cornet share some common features with those from western Europe and Asia (Benton et al. 1997).

The second discovery took place in 1989 in a limestone quarry near the village of *Subpiatră* (Bihor County, Pădurea Craiului Mountains). At this location, the blasting activities exposed in the quarry's wall, a shaft completely filled with sediments and a very rich fossil assemblage that include large-size herbivores, omnivores, carnivores, and various species of amphibians, reptiles, insects, and rodents. Based on the fauna composition (mainly aquatic and semiaquatic), several lines of evidence indicate that the karst region must have had lakes, swamps, and widely forested terrains. The age of this fossil assemblage is Early Pleistocene (Venczel 1991).

Ever since its discovery, the *Urşilor (Bears) Cave* impressed by the huge amount of cave bears remains

Fig. 2 **a** Sulfate-rich mineral association in Diana Cave (Cerna Valley); **b** deposits of clastic sediments capped by phosphates and nitrate deposits in Great Şălitrari Cave (photographs by B. P. Onac)

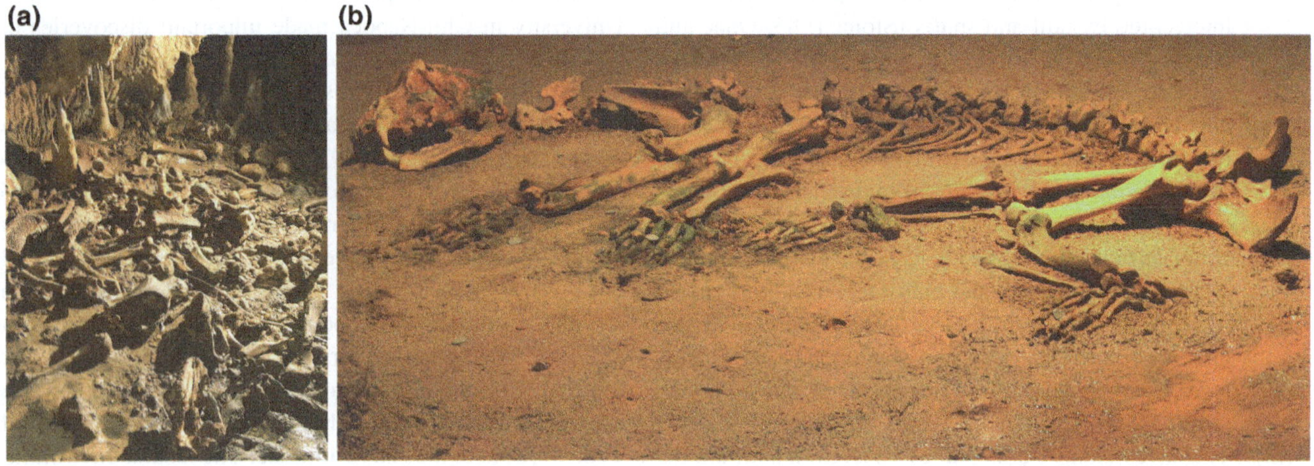

Fig. 3 Urşilor Cave, Bihor Mountains: **a** Bones and skulls along the touristic path; **b** cave bear skeleton in anatomic connection (photographs by B. P. Onac)

(Terzea 1978; Jurcsák et al. 1981). Two distinct areas are of interest to paleontologists: the Bone Gallery (through which the tourists access the cave) and the Scientific Reserve, in the lower level of the cave (Fig. 3). This section of the cave is gated, and access is restricted to scientific research only. It is this part that preserves not only traces of cave bear life (pes and footprints, scratch marks, stomach fur imprints, hibernation beds, etc.) but also three skeletons in anatomic connection and a rich assemblage of cave bears, lions, and hyenas (Diedrich 2011; Robu 2015).

Surprisingly, although the Urşilor Cave is such an impressive den, the paleontological investigations resumed only in 2007 and continued ever since with very detailed ichnological and ethological studies, supported by thorough sediment analyses, methodical excavations, and LIDAR mapping of the Scientific Reserve. The most thorough study so far belongs to Diedrich (2011) and Robu (2015) who abundantly documented all tracks and traces, illustrating the complete life cycle (before, during, and after hibernation) of the Upper Pleistocene cave bears and their dietary regimes.

Anthropology and Archeology

Romania has numerous caves, as well as mines and quarry caves from which significant archeological and anthropological vestiges were documented (Boroneanţ 2000). Two sites that fit the topic of this chapter deserve further attention, i.e., Cuciulat and Cioclovina caves.

Cuciulat Cave has been discovered in 1978 in a limestone quarry outside the village of Cuciulat (Sălaj County). A blast in the quarry face exposed a void that was explored and mapped (1707 m) by members of the "Emil Racoviță" Caving Club in Bucharest (Vădeanu and Done 1981; Done 1983). During the survey, cavers noticed some prehistoric paintings on the walls, which were later studied by Cârciumaru (1983). On a subsequent visit, a horse silhouette, the figures of a feline, and a bird (this one in a side passage), as well as other colored spots without clear outline, were also identified. Although the work in the surrounding of the Cuciulat Cave was stopped and the cave gated for conservation, the blasting in other parts of the quarry did eventually cause the collapse of the cave entrance; therefore now, the cave is sealed and certainly well preserved.

In the very early stage (prior to 1911) of guano-phosphate mining activities in *Cioclovina Cave*, seemingly a human skull was discovered within the mined sediment along with some cave bear remains and three lithic artifacts. The skull, however, was never found but definitely motivated archeologists to begin some very meticulous excavations, during which artifacts of Mousterian and Aurignacian age were recuperated (Roska 1923). Sometimes before 1942, another partial human skull was recovered from a silty clay sediment from a depth of ca. 2 m, where it was found along with *Ursus spelaeus* skulls and bones and some artifacts (Rainer and Simionescu 1942; Alexandrescu et al. 2010). The exact stratigraphic context is unclear, mainly because the discovery and all fossil materials were handled by the miners, which provided contradicting information. Nevertheless, the first direct AMS ^{14}C dating of the Cioclovina skull yielded an age of 29,000 ± 700 years (Olariu et al. 2005). Two years later, another attempt was made to radiocarbon-date the same skull using a newly developed sample preparation technique. The age obtained was very similar, but with a much lower error (28,510 ± 170 years; Soficaru et al. 2007). Both ages indicate the Cioclovina skull belongs to one of the earliest modern humans in Europe. Geometric morphometrics studies undertaken by Harvati et al. (2007) confirm that the skull is typical for a fully modern human with no indications that it may be a Neanderthal–early modern human hybrid as suggested by previous studies.

Caves with Relevance for Speleogenesis Studies

Among mine caves, probably the most interesting ones, i.e., from a speleogenetic point of view, are those discovered in the mining area of Rodnei, Bihor, Metaliferi, and Dognecea mountains (Onac and Drăgușin 2017). The size of cavities intercepted range from geodes of less than a meter in diameter to more than one-kilometer-long cave passages. The later ones went through a metasomatic/hydrothermal stage before phreatic and vadose processes shaped the caves to their present-day morphology.

The work by Mârza and Silvestru (1988) discusses in detail the assumption that the karst voids intercepted in the Rodna Veche mining area were generated in an early post-magmatic stage via corrosion exerted by acidic hydrothermal solutions rising along faults, joins, and other tectonic features. The karst cavity walls formed by hydrothermal-metasomatic weathering were afterward covered by micro-granular aggregates of pyrite, marcasite, marmatite, grossular, and quartz, as well as barite and calcite crusts. Based on the mineral precipitation sequence, the hydrothermal arguments supporting this speleogenetic mechanism are convincing. A few years later, Mârza et al. (1995) described another hydrothermal karst in the Ocna de Fier mining district. At this location, crusts and stalactites composed of magnetite cover the walls of a small cavity in the carbonate bedrock.

A particular three-stage cave development setting was documented in the Băița Bihor skarn mining district where the earliest part of the speleogenesis is considered to have been a metasomatic one (Onac 2002). The first stage is a deep-seated one, during which contact metamorphism of limestones causes decarbonation; the liberated CO_2 increases the acidity of the fluids escaping the system, and hence, the dissolution capacity of them is enhanced. At this stage, the circulation of hot metasomatic fluids may be responsible for some primitive dissolution-induced cavities within the skarn bodies or following bedding planes, faults, or geological boundaries (see Fig. 7a in Onac 2002). Subsequent phreatic and/or vadose processes likely obliterated the original morphology of these voids. In stage two (shallow setting), caves may have formed following two different processes: (a) upward flow of hydrothermal fluid, at which point the flow velocity was relatively high and dissolution outpaced deposition or replacement of the carbonate host-rocks, and (b) vigorous dissolution caused by mixing of ascending hydrothermal fluids with more oxygenated descending waters. During the last speleogenetic stage, the caves in the Băița Bihor skarn were modeled under phreatic and/or vadose conditions to their present appearance.

Mineralogical and isotope geochemistry studies on Șălitrari Cave in the upper Cerna Valley (SW Romania) strongly support the hypothesis that dissolution by sulfuric acid represented one of the stages in the cave evolution (Pușcaș et al. 2010). Two other caves (Diana and Hercules) located within the Herculane Spa (lower Cerna Valley) have typical sulfuric acid speleogenesis, and their thermal waters is directed towards various hotels where it is used for balneotherapy purposes.

Touristic Caves

Only two caves fall in this category. Urşilor Cave was discovered in 1975 after a blast in the Chişcău marble quarry (western Bihor Mountains). Immediately after the discovery, the work in the quarry ceased and a decision was taken by the Bihor County Council to study and prepare the cave for touristic exploitation. It took almost 5 years to complete the cave survey and plan the access path and electric lightning. In parallel, the ERIS and scientists from the "Ţării Crişurilor" Museum in Oradea conducted a detailed biospeleological, climatological, and paleontological study. The cave was included in the touristic circuit in the summer of 1980 (Rusu 1981). The major attractions for the visitors are the abundant fossil remains of *U. spelaeus* (skulls and various bones; Fig. 3a) exposed along the entrance gallery and the highly calcite-decorated cave passages (Fig. 4a). The candle-type stalagmites are of particular interest due to their size and high density/m^2. For paleontologists, the cave is a gold mine as its sediments conserve thousands of bones and hibernation beds (Diedrich 2011; Robu 2015), whereas for Quaternary geology scientists the candle stalagmites provide means to characterize the Holocene climate changes (Onac et al. 2002a).

The Cave with Crystals from Farcu Mine (Bihor County) was intercepted in 1987 during bauxite mining activities (Damm et al. 2003). It is a short (<300 m in length) and very nicely decorated geode-type cavity (Fig. 4b). Unfortunately, miners vandalized part of the speleothems before it was gated and protected. Even so, the cave remained attractive enough for tourists.

Biospeleology

In 1986, a shaft dug for geotechnical investigations in the proximity of Mangalia, south Dobrogea (Fig. 1), intersected a natural cave passage at a depth of 18 m (Constantinescu 1989). Movile Cave consists of a 200-m-long dry passage and a 40-m-long submerged level. The later one is flooded by thermo-mineral groundwater ascending from an aquifer located at a depth of ~200 m. The completely anoxic water has a temperature of 21 °C and contains high amounts of ammonia, methane, and hydrogen sulfide (Sarbu et al. 1996). The cave is famous for its rich ecosystem based on chemosynthesis, meaning the energy supporting the life in this unique environment comes from the oxidation of hydrogen sulfide dissolved in the thermo-mineral. For more information on this unique cave, readers are directed to *Movile Cave* chapter in this book.

Mined Caves

In contrast to the mine/quarry caves discussed above, Romania has a number of *mined caves*. The most significant ones are highlighted below.

Cioclovina Cave is renowned for its massive amount (>30,000 m^3) of phosphate-rich sediments (~30% P_2O_5) accumulated along its Main Gallery (Fig. 5a). Between World War I and II, Cioclovina Cave was subjected to extensive exploitation, aiming to mine out the guano-phosphate sediments as well as thousands of cave bear bones that were milled and used as fertilizers. This activity

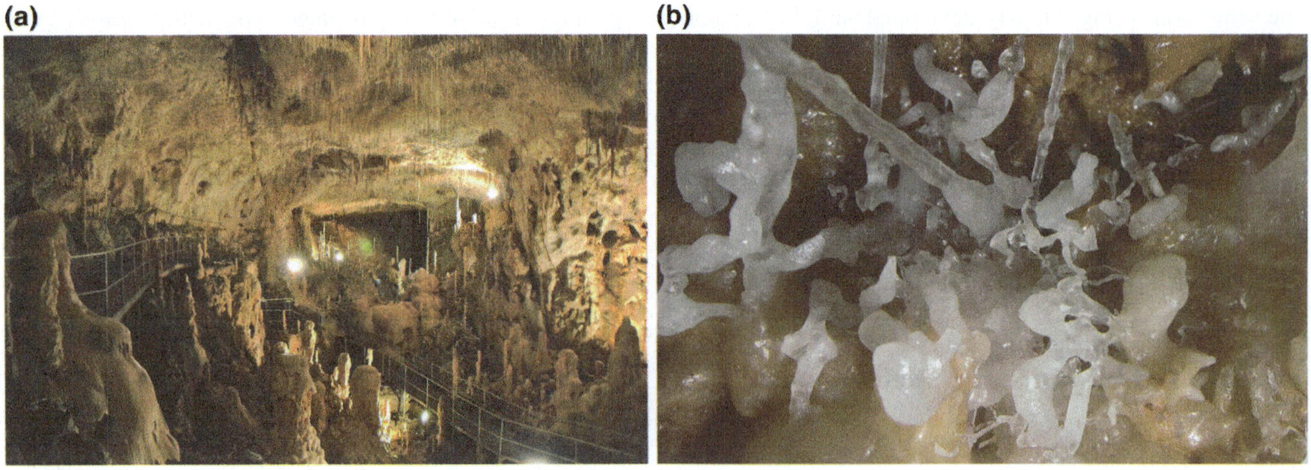

Fig. 4 **a** "Emil Racoviţă" Gallery in Urşilor Cave (photograph by B. P. Onac); **b** Calcite speleothems in the Cave with Crystals from Farcu Mine (photograph courtesy A. Posmoşanu)

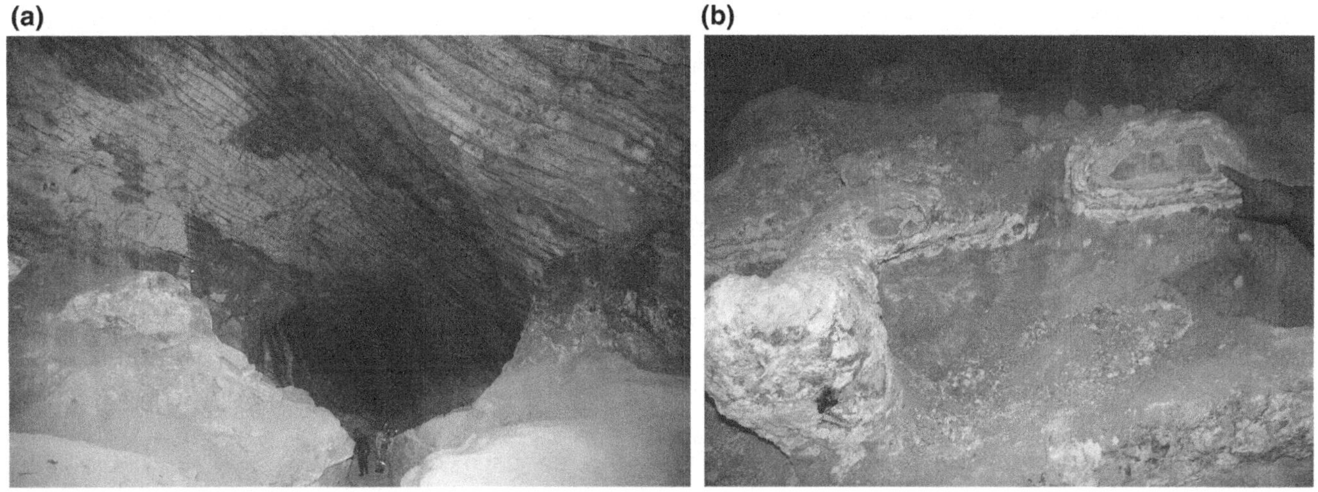

Fig. 5 a Passage in the Cioclovina Cave showing the guano-phosphate deposit; b close-up view of the highly phosphatized limestones and clastic sediments, host for a rich mineral assemblage (photographs by B. P. Onac)

led to the discovery of a new mineral, *ardealite* (Schadler 1932), and to some other common phosphate and sulfate species (brushite, gypsum, etc.) (Fig. 5b). Over the last decade, a significant number of studies identified and described a total of 29 species (see Table 2 in Chapter "Şureanu Mountains: Valea Stânii–Ponorici–Cioclovina cu Apă karst system"). This assemblage includes rare, high-temperature minerals (berlinite and hydroxylellestadite) formed during in situ guano combustion (Onac and White 2003; Onac et al. 2006; Onac and Effenberger 2007) as well as other common and exotic cave minerals, among which 14 phosphates (Marincea et al. 2002; Onac et al. 2002b, 2005b, 2009a, b, 2011; Dumitraş 2009).

Saltpeter had been exploited from Sălitrari Cave on Cerna Valley to manufacture gunpowder in the second part of the nineteenth century. The nitrates accumulated in the so-called Nitrate Passage (ca. 30 m in length) situated along the Main Gallery. Here, the upper part of a ~5-m-thick sediment deposit (Fig. 2b) was documented to have a high concentration of nitrates (darapskite and nitratine). This location and Saltpeter Cave from Olteţului Gorge are the only cave in Romania where nitrate minerals were identified (Diaconu and Lascu 1999; Ponta et al. 2018). Apparently, their presence is related to the mild sub-Mediterranean climate of this region and in particular to the topoclimate of the Nitrate Passage in which the relative humidity is below 75% year around and the temperature is rather high (~11.7 °C) compared to the rest of the cave (~7 °C), thus favoring a stable environment for nitrate minerals. These conditions prevailed at least over the last 125 years as wood fragments used by the Turkish while mining the saltpeter sediments are well preserved in this section of the cave.

An exceptional occurrence of goethite speleothems was described from Luana Cave (also known as Chocolate Palace) in Călimani Mountains, East Carpathians (Naum and Butnaru 1967; Hill and Forti 1997). Luana along with few other small cavities was discovered in a giant quarry extracting native sulfur. After the finding, Naum and Butnaru (1967) proposed that caves formed by dissolution of volcanoclastic deposits to be called *volcanokarst*. Unfortunately, the life of these cavities was short as the communist economy was in great need for sulfur in order to produce sulfuric acid for fertilizers, and therefore, the entire mountain (Negoiu Românesc) with its cavities was eventually completely destroyed. The caves formed in tuff pyroclastites by weathering of the volcanic constituents. An interesting contribution that examines the relationship between the presence of these cavities and the genesis of the sulfur deposit was published by Balintoni in 1968.

Acknowledgements I wish to thank Drs. I. Povară, T. Tămaş, and E. Posmoşan for sharing relevant references to the mine/quarry caves topic and also to many colleagues in the "Emil Racoviţă" Institute of Speleology, students, and cavers for their fieldwork support over the last 25 years.

References

Alexandrescu E, Olariu A, Skog G, Stenström K, Hellborg R (2010) Os fossiles humains des grottes Muierii et Cioclovina, Roumanie. L'Anthropologie 14:341–353

Balintoni I (1968) Relaţia dintre vulcano-carst şi epigeneză în zăcămîntul de sulf din Munţii Călimani. DS Şed Inst Geol LVI:5–11

Benton MJ, Cook E, Grigorescu D, Popa E, Tallódi E (1997) Dinosaurs and other tetrapods in an Early Cretaceous bauxite-filled fissure, northwestern Romania. Paloeogeogr Palaeoclimatol Palaeoecol 130:275–292

Bielz EA (1884) Beitrag zur Höhlenkunde Siebenbürgens. Jahrbuch des Siebenbürgischen Karpathen-Vereins IV:1–66

Boroneanţ V (2000) Arheologia peşterilor şi minelor din România. Ed cIMeC, Bucureşti, p 327
Cârciumaru M (1983) Peintures rupestres de la grotte Cuciulat (Roumanie). Bull Soc Préhist franç 80(3):827–834
Cauuet B (2002) Ancient gold mines of Dacia. Roşia Montană District (Apuseni Mountains, Romania). Report 2002. Available online at: http://www.rosiamontana.ro/img_upload/e3bca2bc61045730fbb1869240/Report_2002_English_Version.pdf
Clottes J, Besesek M, Gely B, Kenesz M, Lascu VT, Meyssonnier M, Philippe M, Plichon M, Prud'homme F, Radu VA, Rus T, Tociu RL (2012) Découverte d'une nouvelle grotte ornée paléolithique en Roumanie, dans le département du Bihor. In: Clottes J (ed) L'Art Pléistocéne dans le Monde. Bull Soc préhist l'Ariège 65–66:13–528
Constantinescu T (1989) Considération sur la zone karstique de "La Movile" (Mangalia, Dobrogea du sud, Roumanie). Misc Speol Rom 1:7–12
Damm P, Mărcuş O, Bruncsák T (2003) Carstul din zona Farcu. Bazinul văii Şteazelor, Munţii Pădurea Craiului. Speomond 7–8: 33–38
Diaconu G (1974) Quelques considérations sur la présence de l'anhydrite dans la grotte "Pestera Diana" (Baile Herculane-Roumanie). Trav Inst Spéol "Émile Racovitza" 13:191–194
Diaconu G (1985) Liste des minéreaux constitutifs des spéléothèmes. Theor Appl Karstol 2:11–24
Diaconu G, Lascu C (1999) Preliminary data about nitrates in "Peştera Mare de la Şălitrari", Cerna Mountains, Romania. Theor Appl Karstol 11–12:47–52
Diaconu G, Medeşan A (1975) Sur la présence du pickeringite dans la grotte de Diana (Băile Herculane, Roumanie). In: Proceedings of 6th International Congress of Speleology, Olomouc, pp 231–239
Diedrich CG (2011) An overview of the ichnological and ethological studies in the Cave Bear Den in Urşilor Cave (Western Carpathians, Romania). Ichnos 18:9–26
Done A (1983) Peştera din Cariera Cuciulat (4001/1). Bul Club Speol "Emil Racoviţă" Bucureşti 8:82–89
Dumitraş D-G (2009) Asociaţii mineralogice în sedimentele Peşterii Cioclovina Uscată. Unpublished Ph.D. Thesis, University of Bucharest
Dyke GJ, Benton MJ, Posmoşanu E, Naish D (2011) Early Cretaceous (Berriasian) birds and pterosaurs from the Cornet bauxite mine, Romania. Palaeontology 54:79–95
Fridvaldszky J (1767) Mineralogia Magni Principatus Transylvaniae. Claudiopoli Typis Acad Soc Jesu, p 206
Harvati K, Gunz P, Grigorescu D (2007) Cioclovina (Romania): affinities of an early modern European. J Hum Evol 53:732–746
Hill CA, Forti P (1997) Cave minerals of the world (2nd ed). Nat Speleol Soc, Huntsville, AL, p 464
Iştvan D, Tămaş I (1996) Date asupra carstului din Dealul Popii (Valea Vinului, Munţii Rodnei). Cerc Speol 4:5–9
Jurcsák T, Kessler E (1991) The Lower Cretaceous paleo-fauna from Cornet, Bihor County, Romania. Nymphaea 21:5–32
Jurcsák T, Polis R, Ignat D, Şerban M, Popa E (1981) Date privind fauna fosilă a Peşterii Urşilor (Munţii Bihor). Nymphaea 8–9:161–257
Koch A (1886) Mineralogical studies in Transylvania. Orvos-Termész Értes 11:211–226
Marincea Ş, Dumitraş D, Gibert R (2002) Tinsleyite in the "dry" Cioclovina Cave (Şureanu Mountains, Romania): the second occurrence. Eur J Min 14:157–164
Mârza I, Silvestru E (1988) First mention of the hydrothermal karst phenomenon in Romania, associated to the neogene metasomatic sulphide ore deposits from Rodna Veche. Studia UBB Geol 33(1): 77–81
Mârza I, Nicolescu Ş, Gruescu C (1995) La magnétite–draperie stalactitique (Formation de karst hydrothermal) du gisement de Ocna de Fier (Banat). Rom J Min 76:101–103
Naum T, Butnaru E (1967) Le volcano-karst des Calimani (Carpathes Roumaines). Ann Spéléol 22:727–755
Nedopaca M (1982) Contribuţii la cunoaşterea mineralogiei speleothemelor din România. Bul Speol 6:67–75
Nedopaca M (1986) Carst mineralizat în Munţii Metaliferi. Bul Speol 10:66–72
Olariu A, Skog G, Hellborg R, Stenström K, Faarinen M, Persson P, Alexandrescu E (2005) Dating of two Paleolithic human fossil bones from Romania by accelerator mass spectrometry. In: Olariu A, Stenström K, Hellborg R (eds) Applications of high-precision atomic and nuclear methods. Ed Acad Rom, Bucharest, pp 222–226
Onac BP (2002) Caves formed within Upper Cretaceous skarns at Băiţa (Bihor county): mineral deposition and speleogenesis. Can Min 40 (6):1551–1561
Onac BP (2003) Checklist of the Romanian cave minerals. Theor Appl Karstol 16:83–89
Onac BP (2013) The scientific relevance of mine, quarry, and mined caves of Romania: a review. Mem Ist Ital Speleol II 28:135–154
Onac BP, Damm P (2002) Preliminary data on the mineralogy of limestone and skarn-hosted caves from Băiţa (Bihor County, Romania). Studia UBB Geol 47(1):93–104
Onac BP, Drăguşin V (2017) Hypogene caves of Romania. In: Klimchouk A, Palmer AN, De Waele J, Auler A, Audra P (eds) Hypogene karst regions and caves of the world. Springer International Publishing, Cham, pp 257–265
Onac BP, Effenberger HS (2007) Re-examination of berlinite ($AlPO_4$) from the Cioclovina Cave, Romania. Am Min 92:1998–2001
Onac BP, White WB (2003) First reported sedimentary occurrence of berlinite ($AlPO_4$) in phosphate-bearing sediments from Cioclovina Cave, Romania. Am Min 88:1395–1397
Onac BP, Constantin S, Lundberg J, Lauritzen SE (2002a) Isotopic climate record in a Holocene stalagmite from Urşilor Cave (Romania). J Quat Sci 17(4):319–327
Onac BP, Breban R, Kearns JB, Tămaş T (2002b) Unusual minerals related to phosphate deposits in Cioclovina Cave, Şureanu Mts. (Romania). Theor Appl Karstol 15:27–34
Onac BP, Ettinger K, Kearns Balasz II (2005a) A modern, guano-related occurrence of foggite, $CaAl(PO_4)(OH)_2{\cdot}H_2O$ and churchite-(Y), $YPO_4{\cdot}2H_2O$ in Cioclovina Cave, Romania. Min Petrol 85:291–302
Onac BP, Viehmann I, Lundberg J, Lauritzen SE, Stringer C, Popiţă V (2005b) U-Th ages constraining the Neanderthal footprint at Vârtop Cave, Romania. Quat Sci Rev 24(10/11):1151–1157
Onac BP, Effenberger HS, Ettinger K, Cîntă-Pânzaru S (2006) Hydroxylellestadite from Cioclovina Cave (Romania): microanalytical, structural, and vibrational spectroscopy data. Am Min 91:1927–1931
Onac BP, Effenberger HS, Breban RC (2007) High-temperature and "exotic" minerals from the Cioclovina Cave, Romania: a review. Studia UBB Geologia 52:3–10
Onac BP, Bernhardt H-J, Effenberger HS (2009a) Authigenic burbankite in the Cioclovina Cave sediments (Romania). Eur J Min 21:507–514
Onac BP, Sumrall JG, Tămaş T, Povară I, Kearns J, Dârmiceanu V, Vereş DŞ, Lascu C (2009b) The relationship between cave minerals and H_2S-rich thermal waters along the Cerna Valley (SW Romania). Acta Carsologica 38(1):27–39
Onac BP, Effenberger HS, Collins N, Kearns Breban R (2011) Revisiting three minerals from Cioclovina Cave (Romania). Int J Speleol 40:99–108
Onac BP, Effenberger HS, Wynn JG, Povară I (2013) Rapidcreekite in the sulfuric acid weathering environment of Diana Cave, Romania. Am Min 98:1302–1309

Păunescu A (2001) Paleoliticul şi Mezoliticul din spaţiul Transilvan. Ed AGIR, Bucharest, p 573
Ponta MLP, Aldica G, Dumitru R (2018) Polovragi Cave-Olteţului Gorge karst area. In: Ponta GML, Onac BP (eds) Cave and Karst Systems of Romania. Springer International Publishing, Cham, pp 83–91
Pósepný F (1874) Geologisch-montanishe Studie der Erzlagerstattenlehre von Rézbánya. Budapest, Hungary
Posmoşanu E (2003) Iguanodontian dinosaurs from the Lower Cretaceous bauxite site from Romania. Acta Palaeontol Rom 4:431–439
Posmoşanu E, Cook E (2000) Vertebrate taphonomy and dinosaur paleopathology from a Lower Cretaceous bauxite lens, North West Romania. Oryctos 3:39–51
Povară I, Diaconu G, Goran C (1972) Observations préliminaires sur les grottes influencées par les eaux thermo-minérales de la zone Băile-Herculane. Trav Inst Spéol "Émile Racovitza" 11:355–365
Puscaş CM, Onac BP, Tămaş T (2010) The mineral assemblage of caves within Şălitrari Mountain (Cerna Valley, SW Romania): depositional environment and speleogenetic implications. Carb Evap 25(2):107–115
Puscaş CM, Onac BP, Effenberger HS, Povară I (2013) Tamarugite-bearing paragenesis formed by sulphate acid alteration in Diana Cave, Romania. Eur J Min 25:479–486
Rainer F, Simionescu I (1942) Sur le premier crâne d'homme paléolithique trouvé en Roumanie. Anal Acad Rom 3(17):489–507
Robu M (2015) The palaeontology of the cave bear bone assemblage from Urşilor Cave of Chişcău—osteometry, palaeoichnology, taphonomy, and stable isotopes. Ed Univ, Bucharest, p 248
Roska M (1923) Săpăturile din peştera de la Cioclovina. Publ Com Mon Ist 2:27–55
Rusu T (1981) La grotte des ours de Chişcău (Monts Apuseni). Rev roum géol, géophys géogr, Ser Géogr 25(2):193–204
Sarbu SM, Kane TC, Kinkle BK (1996) A chemoautotrophically based cave ecosystem. Science 272:1953–1955
Schadler J (1932) Ardealit, ein neues Mineral $CaHPO_4$ $CaSO_4$ $+4H_2O$. Zentral-bl Miner A:40–41
Soficaru A, Petrea C, Doboş A, Trinkaus E (2007) The human cranium from the Peştera Cioclovina Uscată, Romania: Context, age, taphonomy, morphology and paleopathology. Curr Anthropol 48:611–619
Stoici SD (1983) Districtul metalogenetic Baiţa Bihorului. Ed Acad RSR, Bucureşti
Terzea E (1978) Dépôts de remplissage et mammifères quaternaires de "Peştera Urşilor" de Chişcău, département de Bihor (note préliminaire). Trav Inst Spéol "Émile Racovitza" 17:139–144
Trinkaus E, Moldovan O, Milota S, Bâlgar A, Sarcină L, Athreya S, Bailey SE, Rodrigo R, Gherase M, Higham T, Ramsey CB, Van Der Plicht J (2003) An early modern human from the Pestera cu Oase, Romania. PNAS 100(20):11231–11236
Vădeanu T, Done A (1981) Peştera din cariera Cuciulat. Manifestări artistice preistorice. St Cerc Speol Rm Vâlcea 1:67–69
Venczel M (1991) New contributions to the fossil herpetofauna of Subpiatră (Bihor County, Romania). Nymphaea 21:81–88
Vişan I (2011) Mineralogia unor peşteri din sudul Munţilor Rodnei. Unpublished Master Thesis, Univ Babeş-Bolyai, Cluj-Napoca, p 44
Webb D, Robu M, Moldovan O, Constantin S, Tomuş B, Neag I (2014) Ancient human footprints in Ciur-Izbuc Cave, Romania. Am J Phys Anthropol 155:128–135

Cave Biology

Oana Teodora Moldovan and Ruxandra Năstase-Bucur

Abstract
An overview of Romanian cave fauna with the most important achievements in historical biogeography, diversity, ecology, phylogeny, and conservation is presented.

Keywords
Cave fauna • Biodiversity • Coleoptera • Crustacea Origin • Ecology • Phylogeny • Paleogeography

Introduction

Romania, with its geographic position in Europe, is rich in subterranean fauna. With almost 300 invertebrate species, the subterranean fauna (including cave fauna) is more diverse (Decu and Racovitza 1994) compared to other countries having considerably larger limestone coverage (i.e., Austria, Switzerland) (Fig. 1).

This unexpected high biodiversity can be explained by geographic and geologic features (Moldovan and Rajka 2007): (i) The geographic position of the country, with Atlantic and Mediterranean climatic influences; (ii) The high number of caves/km^2, with almost 12,000 caves discovered prior to 1989 (Goran 1989); (iii) The distribution of caves at low altitude, with 27% of karst rocks at altitudes below 500 m and 47% up to 1000 m.a.s.l. (Bleahu and Rusu 1965), the optimal altitude from the climatic point of view (temperature, precipitation) for species diversity; (iv) The patchy distribution of limestone areas; karst is distributed along the Romanian Carpathians and in Dobrogea, covering ~2.3% of the total country's surface (see also Chap. 4). From an ecological point of view, Romanian karst forms small continental islands scattered between non-karst areas, which act as natural barriers to migration of subterranean fauna, promoting evolution and speciation. The genetic differentiation between cave populations supports this assumption for both terrestrial and aquatic subterranean representatives (Bucur et al. 2003; Meleg et al. 2013). One example is the study on 11 populations of *Niphargus* amphipods analyzed by means of mitochondrial and nuclear molecular markers (Meleg et al. 2013). The haplotype network clearly separated two groups of populations, with a strong geographical pattern, identifying a split between the northern and the southern populations of the Apuseni Mountains (see also Fig. 2), with no indication of ongoing gene flow between them. The most diverse subterranean fauna can be found in the Apuseni Mountains and Southern Carpathians (Fig. 2).

Origin of Romanian Cave Fauna

Migration of Asian ancestors of cave fauna was not possible until the Upper Oligocene–Early Miocene (Fig. 3: 1) when a Dinaric–Pelagonia–Anatolian landmass was formed (Fig. 3: 2), making the connection with the rest of Europe by the recurring Slovenian corridor (Steininger and Rögl 1985). This was the first connection between Dinarides and Carpathians and lasted until Lower Badenian (16 Ma) (Fig. 3: 3–4), when the Central Paratethys was flooded. It provided the possibility of populating Southern Carpathians by Dinaric lineages and was also at the time when Carpathian system developed. During Miocene (Upper Burdigalian) (17–18 Ma), the first Alps-Bohemian Massif–Carpathians connection (Fig. 3: 2–3) provided conditions for the Apuseni Mountains colonization through the Bohemian Massif (Jeannel 1931). A connection between the Dinarides and the Carpathians was also established during the Messinian crisis (6–5.3 Ma) but an arid climate cannot explain large-scale processes of Southern Carpathian colonization. It

O. T. Moldovan (✉) · R. Năstase-Bucur
"Emil Racoviță" Institute of Speleology, Clinicilor 5,
400006 Cluj-Napoca, Romania
e-mail: oanamol35@gmail.com

R. Năstase-Bucur
e-mail: ruxandra.nastase.bucur@academia-cj.ro

G. M. L. Ponta and B. P. Onac (eds.), *Cave and Karst Systems of Romania*,
Cave and Karst Systems of the World, https://doi.org/10.1007/978-3-319-90747-5_55

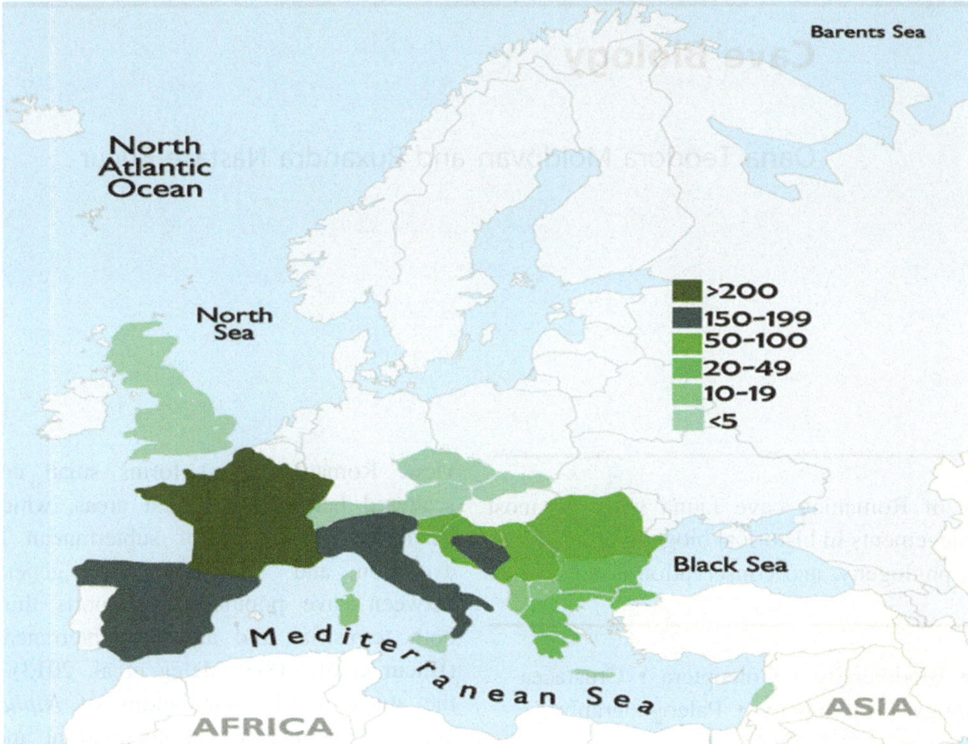

Fig. 1 Distribution of cave Leptodirini (Coleoptera) species and subspecies in Europe (data from Perreau 2000)

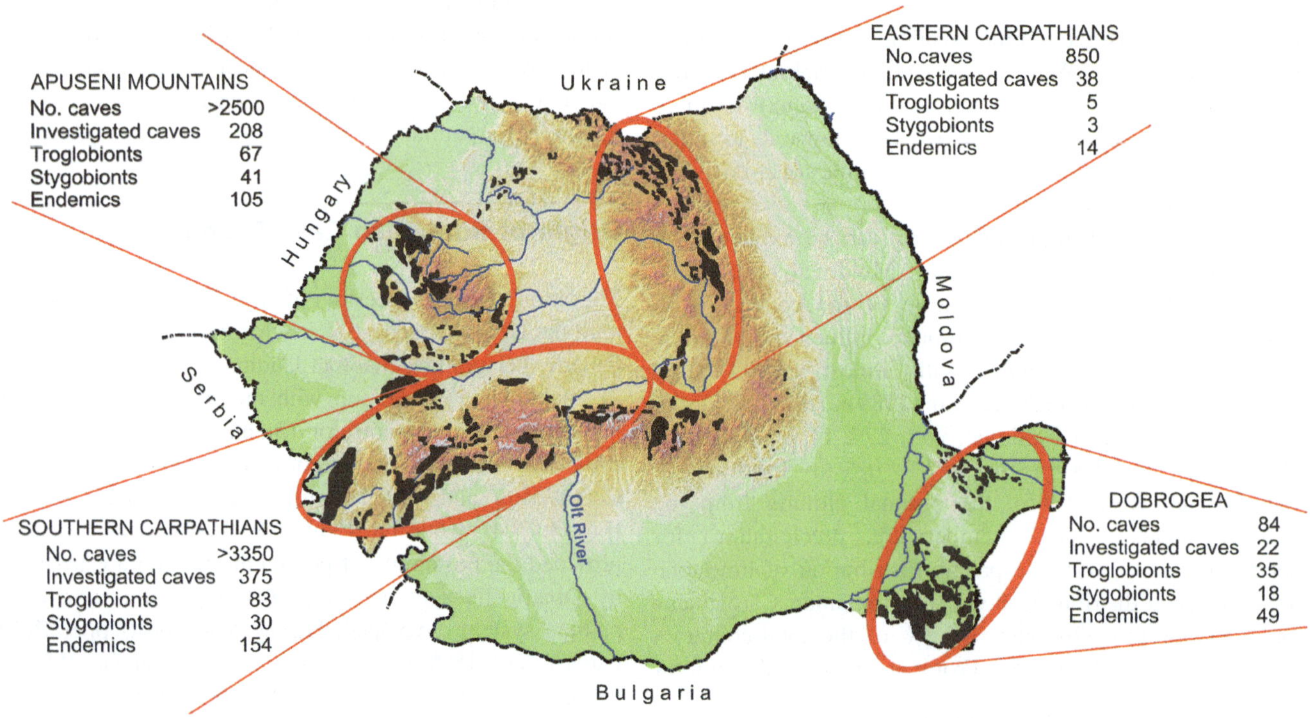

Fig. 2 Number of cave species (troglobionts and stygobionts) and endemics (also including troglo- and stygophiles) in the biologically investigated caves of the four main karst areas of Romania; limestone is represented in black, number of caves are from Goran (1989)

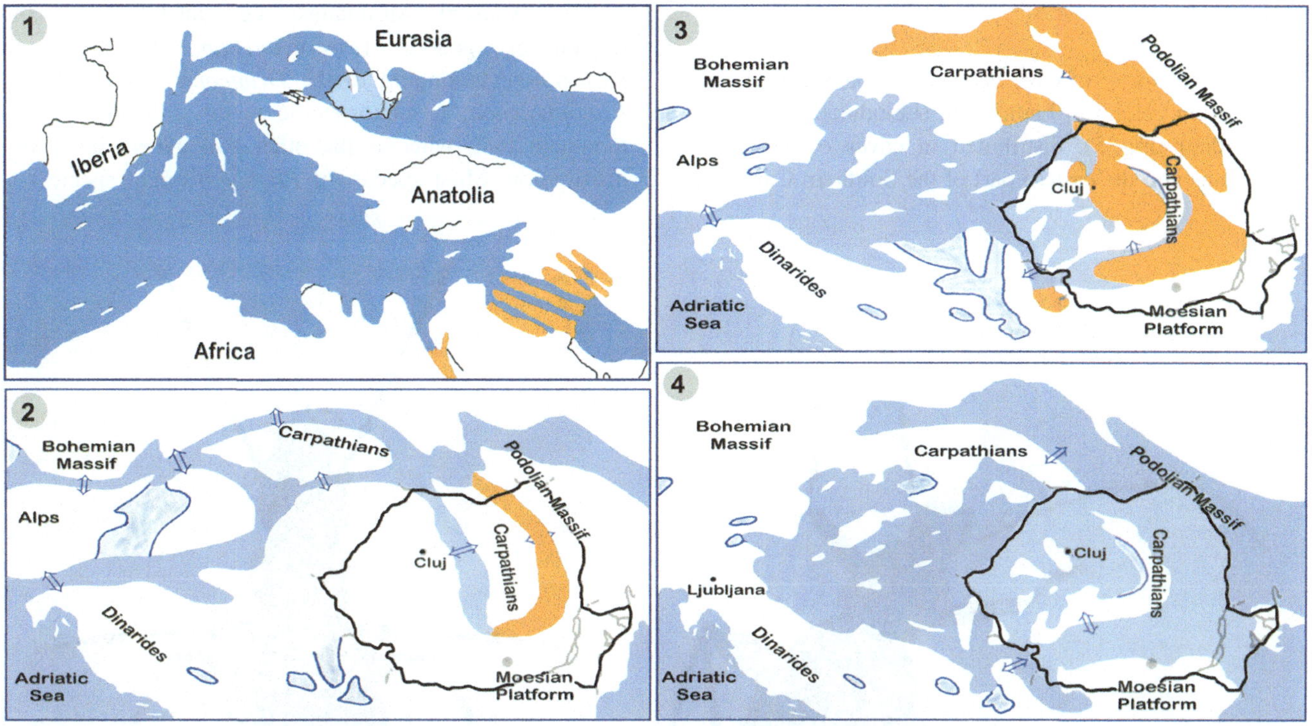

Fig. 3 Evolution of the Paratethys from Upper Oligocene to Early Miocene (1: 20–17 Ma), through Early (2) and Middle Miocene (3), and until Lower Badenian (4: 16 Ma) (modified after Steiniger and Rögl 1985): dark blue—marine realms, yellow—evaporitic basins, light blue—important areas with fluvio-terrestrial sedimentation and/or lignite formation, white—continental realms, ⇔—basins narrowed post-sedimentation by tectonic processes

can, however, explain processes of speciation due to appearance of natural barriers. Due to major geological and geographical landscape changes, extinction and vicariance alternated after the colonization of the new areas. A large and continuous distribution area of epigean and probably endogean ancestors of cave species was then fragmented, even before the colonization of the subterranean domain. The next stage of evolution was the colonization and speciation inside the subterranean domain. Climatic changes contributed significantly in shaping the distribution areas and breaking-off gene flow with surface relatives. The study concerning the origin of Western Mediterranean radiation of subterranean beetles (Coleoptera), without the inclusion of Dinaric taxa, indicated that Carpathian species are sister species with the monophyletic Pyrenean lineage (Ribera et al. 2010).

Subterranean Biodiversity in Romania

The most diverse groups of cave fauna in Romania are Coleoptera and Crustacea (Fig. 4).

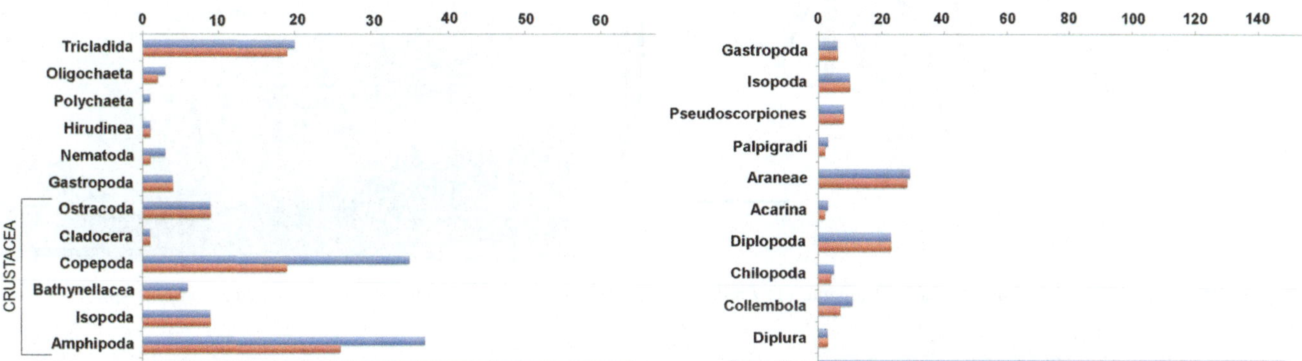

Fig. 4 Number of species (in blue) and of endemics (in red) for the terrestrial (left) and aquatic (right) groups inhabiting caves of Romania (data from Decu and Racovitza 1994 and the "Emil Racoviță" Institute of Speleology, Cluj-Napoca Department Database)

Coleoptera

The large majority of cave beetles troglobiont (Coleoptera) of Romania belong to Leptodirini (Leiodidae) and Trechini (Carabidae) (Fig. 5), distributed in caves of the Apuseni Mountains and the eastern part of the Southern Carpathians.

The Apuseni Mountains are inhabited by three Leptodirini genera (*Protopholeuon*, *Pholeuon*, and *Drimeotus*), belonging to the *Drimeotus* phyletic lineage. *Protopholeuon*, a monospecific genus, inhabits only one mountain unit, whereas the other two genera have larger distribution. Most species of *Pholeuon* (eight species) and

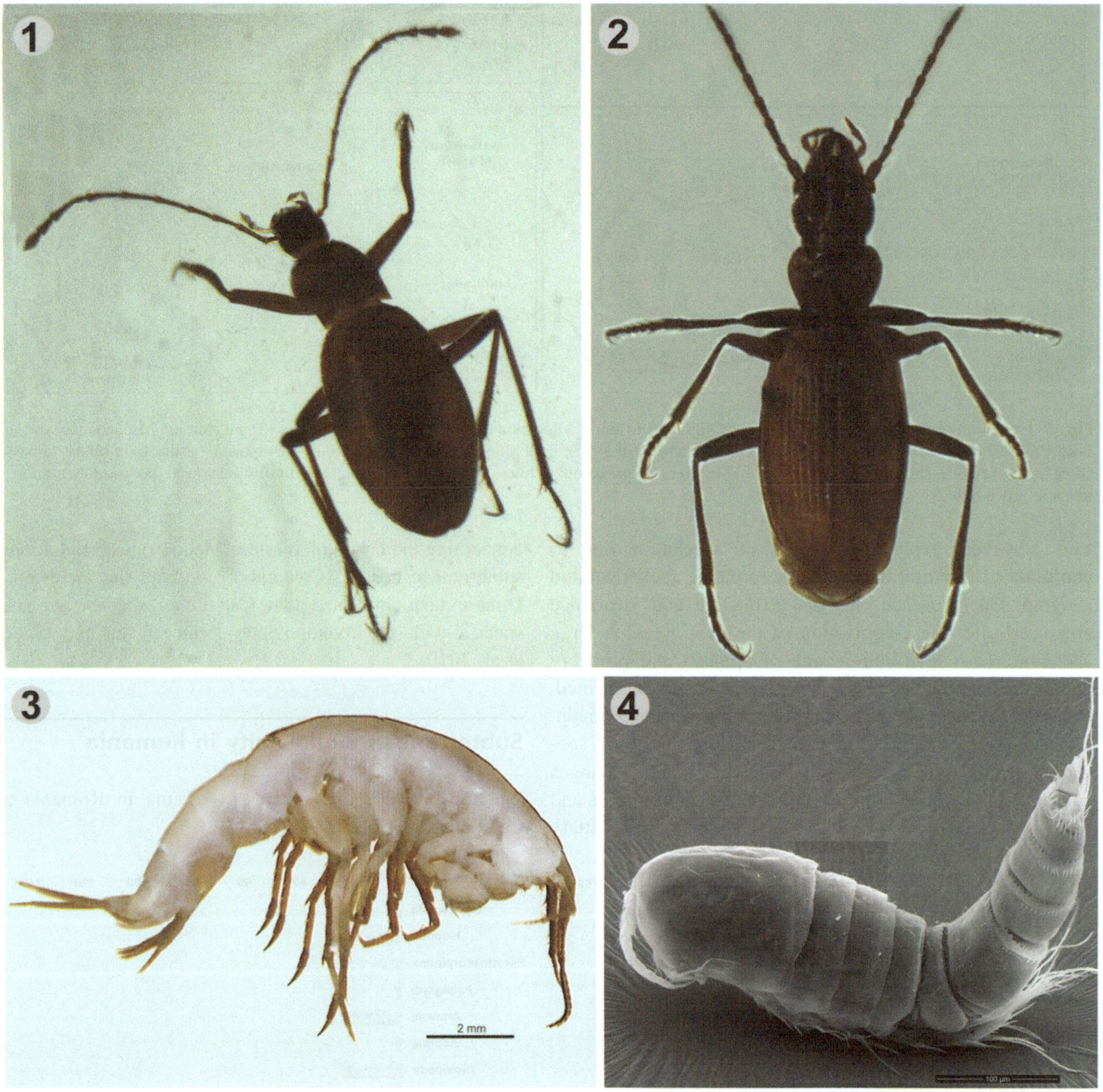

Fig. 5 Representatives of the cave fauna in Romania: (1) *Pholeuon* sp., (2) *Duvalius* sp., (3) *Niphargus* sp., (4) *Bryocamptus* sp.

Drimeotus (19 species) are distributed in Pădurea Craiului and Bihor Mountains. The Southern Carpathians, including Banat Mountains, are inhabited by four Leptodirini genera, *Sophrochaeta* with 16 species, *Closania* and *Tismanella*, each with two species, and *Banatiola vandeli*.

Duvalius, widely distributed in Europe, is the most speciose Trechini in Romania with 57 endemic species and subspecies. *Duvalius* (*Biharotrechus*) is the subgenus with most species (29) in entire Europe (Guéorguiev and Kostova 2011).

Crustacea

More than 85% of the known crustaceans living in Romanian groundwater are strictly adapted to this environment (stygobionts). The highest diversity, with more than 20% of all groundwater crustaceans, is that of harpacticoids, amphipods, and cyclopoids. The most elevated number of groundwater taxa is known from caves and phreatic habitats. Similar to cave beetles, crustaceans are best represented in the Apuseni Mountains and the Southern Carpathians. These two regions have also the highest degree of endemic taxa, with ca. 45% for the Apuseni Mountains and with ca. 34% for the Southern Carpathians. The endemic taxa belong mostly to amphipods, harpacticoids, and ostracods.

Ecology and Paleoecology

Ecological researches in Romania have a long tradition in understanding the impact of different caves and groundwater environmental features on the dynamics of terrestrial cave populations (Racoviţă 1971, 1973, 1978; Racovitza 1980; Decou and Herdlicka 1978; Ruşdea 1989; Fejér and Moldovan 2013) or on the groundwater communities (Pleşa et al. 1964; Pleşa and Racoviţă 1973; Iepure 1999; Moldovan and Rajka 2007; Moldovan et al. 2012), respectively. Racovitza (1983), in a study designed right after the discovery of the meso-shallow subterranean substratum (milieu souterrain superficiel, MSS) by Juberthie et al. (1980), proposed a model of seasonal migration from caves to MSS for cave beetles. He concluded that the effective of cave beetles population in the MSS is antagonistic to that encountered in the cave population, as a result of the physical conditions in the two habitats in different seasons (see also Fig. 6).

The importance of the forest cover on the survival of groundwater populations was emphasized by superimposing the distribution of the surface vegetation on the actual and probable distribution of groundwater copepods (Meleg et al. 2014), reinforcing the conclusions of Decu and Racovitza (1983). The first paleoecological researches in Romanian caves were undertaken on fossil invertebrates and microorganisms from sediments (Epure et al. 2014, 2015; Moldovan et al. 2016).

Molecular Phylogeny and Phylogeography

One of the main directions in the modern study of Romanian cave fauna implies the use of molecular methods for clarifying aspects concerning speciation and evolution in the subterranean environment. A comprehensive study of molecular phylogeny of cave beetles from Leptodirini family was undertaken by Bucur et al. (2003) on a total of 51 populations belonging to *Drimeotus* phyletic lineage. The results suggested a unique origin from a common ancestor of *Drimeotus* phyletic lineage. Still, the two genera of the series, *Pholeuon* and *Drimeotus*, proved not to be monophyletic, fact that can be explained by different cave colonization episodes. Based on molecular genetical techniques, one can infer the phylogeographic reconstruction; (Fig. 7: 1) shows the main genealogical phyletic lineages of *Drimeotus* (s.str.) subgenus superposed with the geographical location of taxa. The red haplotypes are hypothetical haplotypes that are either extinct or can be identified through further sampling. In the phylogenetic tree (Fig. 7: 3), two main groups of species have been identified, one composed of *Drimeotus chyzeri*, *Drimeotus entzi,* and *Drimeotus racovitai*, and the second of *Drimeotus horvathi*, *Drimeotus viehmanni* with the more genetically distant *Drimeotus puscariui*. *Drimeotus kovacsi* has the most basal position in the phylogenetic tree.

Conservation and Protection

More than 160 caves have at least three endemic species, representing more than a quarter of the biospeleologically prospected caves, and 53 caves have more than five cave-adapted species. The caves with the highest number of cave-adapted species are: Movile, Cloşani, Vadu Crişului, and Coliboaia. Romania has no Red List of vulnerable cave taxa, but interdiction for collection of any cave specimen

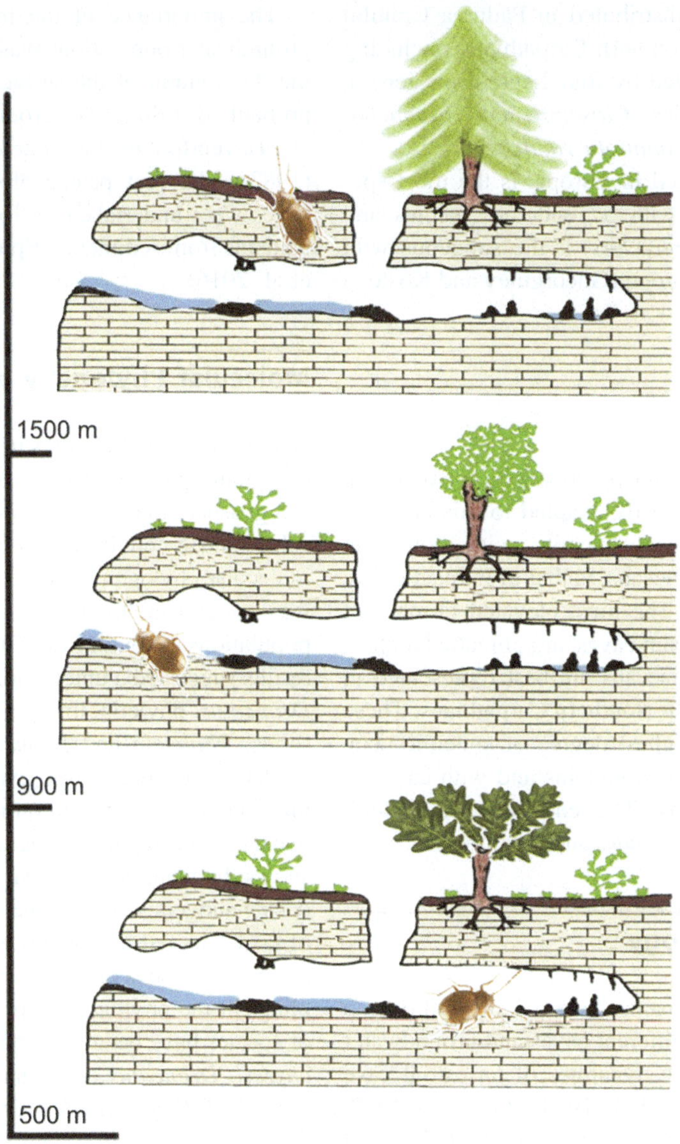

Fig. 6 Habitat change of *Drimeotus* representatives in Apuseni Mountains at different altitudes; cave species at lower altitudes have longer antennae and slender body shape and are restricted to deep cave habitats, while species at intermediary altitudes can be found also near the cave entrance and those at higher altitudes even in surface humid habitats

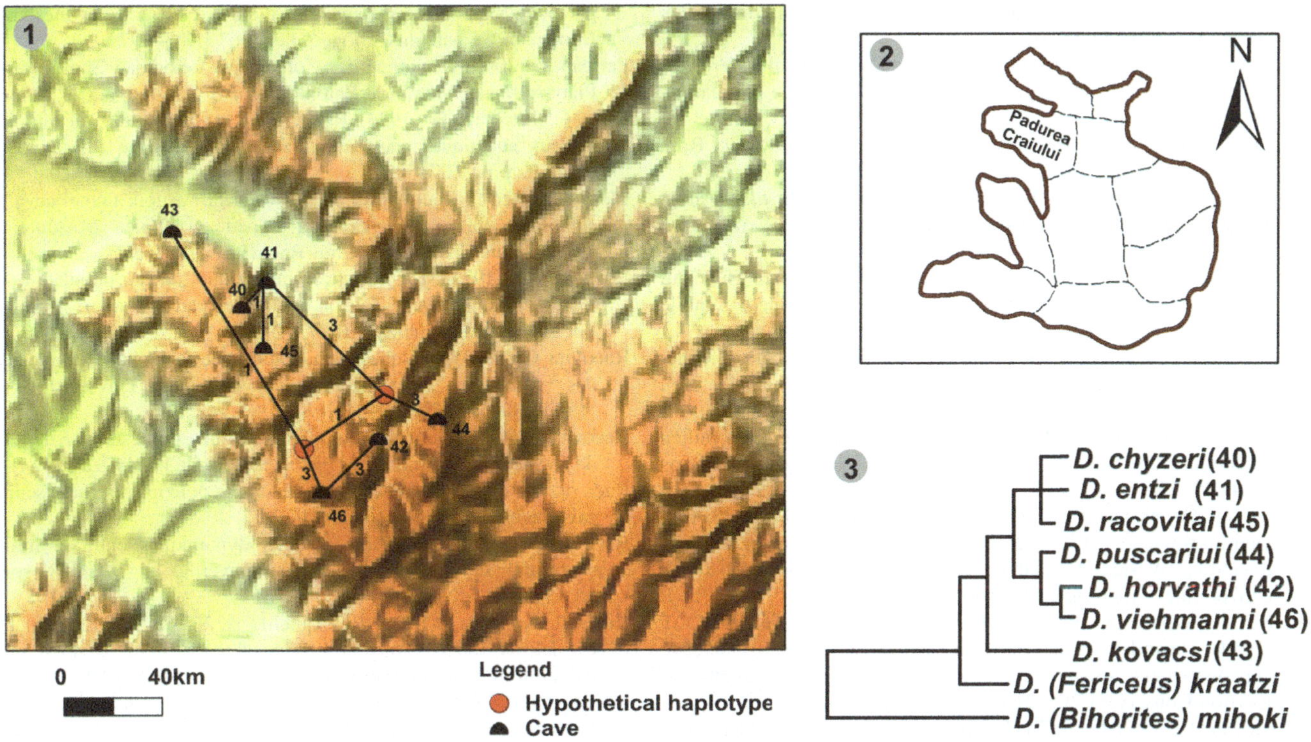

Fig. 7 Phylogeography of *Drimeotus* (s.str.) subgenus (1) in Pădurea Craiului Mountains, a unit of the Apuseni Mountains: 2). Caves are numbered as in the phylogenetic tree based on 1100 bp of mitochondrial DNA. *D. kraatzi* and *D. mihoki* were used as outgroup

without approval from The Commission of the Speleological Heritage (Comisia Patrimoniului Speologic) is stipulated in the Law 49/2011.

Conclusions

The significant biological diversity noted in the Romanian caves has a twofold explanation: (1) the highly fragmented nature of the karst areas (caves as islands) and (2) the steady efforts of several generations of biospeleologists. Romania is one of the best-investigated European countries from a biospeleological point of view. The beginning was in 1920 when E. G. Racovitza founded in Cluj the world's first Speleological Institute and along with R. Jeannel and P. A. Chappuis, started the first cave inventories and ecological and biogeographical studies. The second generation, represented by C. Motaş, T. Orghidan, and M. Dumintrescu, was followed by a third one, with G. Racoviţă, V. Decu, I. Tabacaru, C. Pleşa, Ş. Negrea, and many others, who greatly advanced the biospeleological researches in Romania. Taxonomical work was always completed by ecological and biogeographical studies. The use of molecular techniques brought over the past decades new approaches in cave biology.

Acknowledgements We thank S. Bercea, D. Borda, M. Cîmpean, S. Iepure, and late G. Racoviţă for the help in compiling the Romanian cave fauna database and to F. Fiers for the scanning microscope photograph of Fig. 5: 4. This chapter was completed thanks to CAVE-MONITOR Grant 17SEE/2014, funded through the 2009–2014 SEE mechanism.

References

Bleahu M, Rusu T (1965) Carstul din Romania. O scurtă privire de ansamblu. Lucr Inst Speol "Emil Racoviţă" 4:59–73

Bucur R, Kosuch J, Seitz A (2003) Molecular phylogenetic relationships of Romanian cave Leptodirinae (Coleoptera: Cholevidae). Atti Mus Civ Stor Nat Trieste 50:231–265

Decou V, Herdlicka W (1978) Recherches écologiques dans les grottes des Monts Mehedinţi (Carpates Méridionales). Trav Inst Spéol "Émile Racovitza" 17:95–116

Decu V, Racovitza G (1983) Sur la conservation des écosystèmes forestiers et souterrains des zones calcaires. Mem Biospeol 10: 177–178

Decu V, Racovitza G (1994) Roumanie. In: Juberthie C, Decu V (eds) Encyclopaedia Biospeologica, 1, Soc Biospéol Moulis-Bucarest, pp 779–802

Epure L, Meleg IN, Munteanu CM, Roban RD, Moldovan OT (2014) Bacterial and fungal diversity of quaternary cave sediment deposits. Geomicrobiol J 31:116–127

Epure L, Munteanu V, Constantin S, Moldovan OT (2015) Ecophysiological groups of bacteria from cave sediments as potential indicators of paleoclimate. Quat Int 432A:20–32. https://doi.org/10.1016/j.quaint.2015.04.016

Fejér A, Moldovan OT (2013) Population size and dispersal patterns for a *Drimeotus* (Coleoptera, Leiodidae, Leptodirini) cave population. Subterr Biol 11:31–44

Goran C (1989) La spéléogramme de la Roumanie. Miscell Speol Rom 1:335–346

Guéorguiev B, Kostova R (2011) A new species of *Duvalius*, subg. *Biharotrechus* (Coleoptera: Carabidae) from Bulgaria. Acta Zool Acad Sci Hung 57:305–313

Iepure S (1999) Recherches sur la stygofauna de la grotte Huda lui Papara (Monts Trascau, Roumanie). Trav Inst Spéol "Émile Racovitza" 38:133–141

Jeannel R (1931) Origine et évolution de la faune cavernicole du Bihar et des Carpathes du Banat. Arch Zool Ital (Atti XI Congr Int Zool Padova, 1930) 16:47–60

Juberthie C, Delay B, Bouillon M (1980) Sur l'existence d'un milieu souterrain superficiel en zone non calcaire. CR Acad Sci Paris 290D:49–52

Meleg IN, Zakšek V, Fišer C, Kelemen BS, Moldovan OT (2013) Can environment predict cryptic diversity? The case of *Niphargus* inhabiting Western Carpathian groundwater. PLOS ONE 8(10): e76760. https://doi.org/10.1371/journal.pone.0076760

Meleg IN, Năpăruş M, Fiers F, Meleg IH, Vlaicu M, Moldovan OT (2014) The relationships between land cover, climate and cave copepod spatial distribution and suitability along the Carpathians. Environ Conserv 41:206–216

Moldovan OT, Rajka G (2007) Historical biogeography of subterranean beetles—between "Plato's cave" and scientific evidence. Acta Carsologica 36:77–86

Moldovan OT, Meleg IN, Perşoiu A (2012) Habitat fragmentation and its effects on groundwater populations. Ecohydrology 5:445–452

Moldovan OT, Constantin S, Panaiotu C, Roban RD, Frenzel P, Miko L (2016) Fossil invertebrates records in cave sediments and paleoenvironmental assessments—a study of four cave sites from Romanian Carpathians. Biogeosciences 13:483–497

Perreau M (2000) Catalogue des Coléoptères Leiodidae Cholevinae et Platypsyllinae. Mém Soc Entomol France 4, p 460

Pleşa C, Racoviţă G (1973) Recherches sur les gours, en tant que milieux aquatiques souterrains. Note I. Livre Cinq Inst Spéol É. Racovitza, pp 487–503

Pleşa C, Botea F, Racoviţă G (1964) Cercetari asupra faunei acvatice subterane din bazinul Crisului Repede. I. Valea Misidului si afluenti. Lucr Inst Speol "Emil Racoviţă" 3:367–397

Racoviţă G (1971) La variation numérique de la population de *Pholeuon moczaryi* Cs. de la grotte de Vadu-Crişului. Trav Inst Spéol "Émile Racovitza" 10:273–279

Racoviţă G (1973) Quelques significations de la variation numérique des populations de Coléoptères cavernicoles. Trav Inst Spéol "Émile Racovitza" 12:9–31

Racoviţă G (1978) Données écologiques sur une population souterraine de *Drimeotus horvathi*. Trav Inst Spéol "Émile Racovitza" 17:81–94

Racovitza G (1980) Étude écologique sur les coléoptères Bathysciinae cavernicoles. Mem Biospeol 6:1–199

Racovitza G (1983) Sur les relations dynamiques entre le milieu souterrain superficiel et le milieu cavernicole. Mém Biospéol 10:85–89

Ribera I, Fresneda J, Bucur R, Izquierdo A, Vogler AP, Salgado JM, Cieslak A (2010) Ancient origin of a Western Mediterranean radiation of subterranean beetles. BMC Evol Biol 10:29

Ruşdea E (1989) Dynamique de la population de *Pholeuon (s. str.) knirschi* Breit (Coleoptera, Bathysciinae) de la grotte Ghetarul de la Virtop (Monts du Bihor, Roumanie). Mem Biospeol 16:201–213

Steininger FF, Rögl F (1985) Paleogeography and palinspastic reconstruction of the Neogene of the Mediterranean and Paratethys. In: Dixon JE, Robertson AHF (eds) The geological evolution of the Eastern Mediterranean, Special Publication of the Geological Society, No. 17, Blackwell Scientific Publications, Oxford, pp 659–668

Bat Fauna in Caves of Romania

Ioan Coroiu

Abstract

A review of the research and protection on bat fauna in the Romanian caves has been conducted. A few relevant arguments are given as to why the most abundant bat colonies in the European Union are located in Romania. Of the 31 bat species in Romania, 28 use caves as shelters, especially in the hibernating period. Over 100,000 (up to 170,000) individuals were counted in two of the 70 caves that are considered shelters of major importance. A group of five important caves and also several smaller others, along with the natural habitats in the area, form a bat hot spot in south-western Romania.

Keywords

Bat fauna • Nursery shelter • Hibernacula
Huda lui Papară Cave • Şura Mare Cave • Bat protection

Introduction

Contrary to appearances (the "handicap" of flight, the "hidden" life at night), bats are an evolutionary success because they represent a fifth of the extant mammal species, the second order after the rodents. Of the many types of shelters used by bats, caves represent a main category for the sub-order Microchiroptera given their sophisticated ultrasound orientation system (echolocation). Bats use caves only as shelter (trogloxenes), but choosing the cave depends on its characteristics (air temperature and humidity, length, height and air currents, favorable microhabitats, altitude, etc.) and on the types of habitats surrounding the cave (forest, open landscapes, agricultural areas, type of agriculture—traditional or intensive). Bats from temperate areas use caves as summer shelters (nursery, raising and nursing of their young, courtship and mating) and winter shelters (hibernacula during torpor).

I. Coroiu (✉)
Department of Taxonomy and Ecology, Babeş-Bolyai University, Clinicilor 5-7, 400006 Cluj-Napoca, Romania
e-mail: icoroiu@biolog.ubbcluj.ro

Facts

Certainly, Europe's largest bat communities are in Romania and this fact has multiple causes: (1) Romania has five biogeographical regions, the largest number in the European Union (EU); (2) it has the greatest biodiversity among the EU countries (along with Poland); (3) chemical pollution of agroecosystems in the context of intensive agriculture is significantly lower than in other EU countries due to the low performances in agriculture during the communist period and fragmentation of agricultural real estate in the post-communist period (especially in hill and mountain areas); (4) real estate heritage system which favoured the fragmentation of agricultural property from one generation to another and, implicitly, growth of biodiversity; (5) there is a large number of caves, which offer summer and winter shelters for bats; (6) the fragmented agricultural practice in the hilly and mountain areas offers the landscape a mosaic-like appearance (crop lands alternating with natural areas), with a beneficial effect upon biodiversity; (7) over 27% of Romania's surface is covered with forests; (8) the practice of speleo-tourism is reduced, especially in winter, and (9) adoption of the European protection measures for chiroptera (according to Habitats Directive).

Of the 31 bats species known in Romania, 28 (90%) of them use the cave as winter shelter (mandatory or optional) of which 10 species also use the caves during the active period (nursery, summer roosts, swarming) (Table 1). There is also a special category of bats (strictly cave dwelling), which always use caves as shelter during the summer and winter periods: all horseshoe bats and two species of vesper bats (Long-fingered bat and Geoffroy's bat, eventually Schreiber's bat) have this practice. Due to the fact that climatically Romania is situated in an intermediary position between northern and southern Europe, some of the horseshoe bat

G. M. L. Ponta and B. P. Onac (eds.), *Cave and Karst Systems of Romania*,
Cave and Karst Systems of the World, https://doi.org/10.1007/978-3-319-90747-5_56

Table 1 Bat species in Romania and their preference for caves as nursery shelters and hibernacula in winter period (after Schober and Grimmberger 1993)

Nr.	Specia	Summer (nurseries)	Winter
1	Blasius's horseshoe bat (*Rhinolophus blasii*)	*	*
2	Mediterranean horseshoe bat (*R. euryale*)	*	*
3	Greater horseshoe bat (*R. ferrumequinum*)	±	*
4	Lesser horseshoe bat (*R. hipposideros*)	±	*
5	Mehely's horseshoe bat (*R. mehelyi*)	*	*
6	Alcathoe whiskered bat (*Myotis alcathoe*)	–	+
7	Bechstein's bat (*M. bechsteinii*)	–	+
8	Brandt's bat (*M. brandtii*)	–	+
9	Long-fingered bat (*M. capaccinii*)	*	*
10	Pond bat (*M. dasycneme*)	–	+
11	Daubenton's bat (*M. daubentonii*)	–	+
12	Geoffroy's bat (*M. emarginatus*)	±	*
13	Greater mouse-eared bat (*M. myotis*)	+	+
14	Whiskered bat (*M. mystacinus*)	–	+
15	Natterer's bat (*M. nattereri*)	–	+
16	Lesser mouse-eared bat (*M. oxygnathus*)	+	+
17	Greater noctule (*N. lasiopterus*)	–	–
18	Leisler's bat (*N. leisleri*)	–	–
19	Noctule (*N. noctula*)	–	+
20	Serotine (*Eptesicus serotinus*)	–	+
21	Northern bat (*E. nilssonii*)	–	+
22	Parti-coloured bat (*Vespertilio murinus*)	–	+
23	Kuhl's pipistrelle (*Pipistrellus kuhlii*))	–	–
24	Nathusius' pipistrelle (*P. nathusii*)	–	+
25	Pipistrelle (*P. pipistrellus*)	–	+
26	Soprano pipistrelle (*P. pygmaeus*)	–	+
27	Savi's pipistrelle (*Hypsugo savii*)	–	+
28	Long-eared bat (*Plecotus auritus*)	–	+
29	Grey long-eared bat (*P. austriacus*)	–	+
30	Barbastelle (*Barbastella barbastellus*)	–	+
31	Schreiber's bat (*Miniopterus schreibersii*)	+	+

* Strictly cave dwelling bats; ± shelters in caves and/or other sites (depending on latitude); + shelters in caves and other sites (without altitudinal criteria); – no use of caves as shelters

species (Greater and Lesser horseshoe bats) and also Geoffroy's bat may use other types of shelters as nurseries. In northern Europe, these nurseries are not located in caves, whereas in southern Europe they are only in caves. These are used because they provide a structurally and climatically stable environment [temperature and relative humidity (RH)] and protection from predators (Baudinette et al. 2000).

The way a cave is used as shelter varies according to species and season. During the summer, individuals of certain species seek shelter alone, other species form small or large clusters depending on specific characteristics, caves and microclimate. Other species gather in colonies, sometimes quite large, which are loose during summer and very compact during hibernation. Winter colonies determine collective thermoregulation, which is a strategy for survival in cave hibernation time. Because of the high surface-to-volume ratio of their bodies and the large wing and tail membrane surfaces, the evaporative rate of body water is very high. This is the reason why many bat species choose active caves as shelters, especially during the wintertime when they rarely drink water. Furthermore, bats prefer caves for hibernation because these trap the warm temperature (Mitchell-Jones et al. 2007).

Bat Shelter Caves

The most important bat shelter caves of Romania are Huda lui Papară in the Trascău Mountains (Apuseni Mountains) and Şura Mare (Şureanu Mountains, part of the South Carpathians, also known as the Transylvanian Alps) (Fig. 1).

The Huda lui Papară Cave has an imposing entrance (56 m in height), an underground river and an ascendant gallery of over 1200 m in length. This is why it is used by bats as both hibernaculum and nursery. The winter data gathered during the period 1995–2015 show that 11 bat species use this cave, with numbers fluctuating between 32,000 and over 170,000 individuals (Coroiu and David 2008). The winter colonies are impressive in size, especially for Pipistrelle (Fig. 2a) (up to 84,000 individuals), Schreiber's bats (Fig. 2b) (up to 82,000), Greater mouse-eared bat (Fig. 2c) (10,000), Noctule (1500), and Greater horseshoe bat (Fig. 2d) (1360). The large opening of the cave allows sudden and extensive drops in temperature during the winter, affecting the great colonies inside. For example, a sudden drop in temperature during the 2011–2012 winter has led to the death of some thousand of Pipistrelle. Even though there are numerical fluctuations from one winter to the next, the general trend during the observation period is on rise. The cave is a nursery for Greater mouse-eared bat, Schreiber's bat and Mediterranean horseshoe bat. The last two species form common colonies. *The Huda lui Papară Cave is, obviously, the largest hibernacula in Europe.*

Forty out of over 5000 caves in the Apuseni Mountain karst have significant importance regarding winter and/or summer shelters for bat colonies, representing Romania's richest area in bat fauna (Bücs et al. 2012). Besides the presence of the karst, the abundance of species and individuals is also favoured by the fact that these low mountains are almost fully covered with forests. In this area, some important caves are as follows: Leşu Cave (hibernacula for Greater mouse-eared bat—Fig. 3a—and Greater horseshoe

Fig. 1 Location of bat shelter caves discussed in this chapter

Fig. 2 Outstanding examples of winter colonies in the Huda lui Papară Cave: **a** Pipistrelle and **b** Schreiber's bat. Characteristic clusters of Greater mouse-eared bat (**c**) and Greater horseshoe bat (**d**) in wintering colonies (Huda lui Papară Cave)

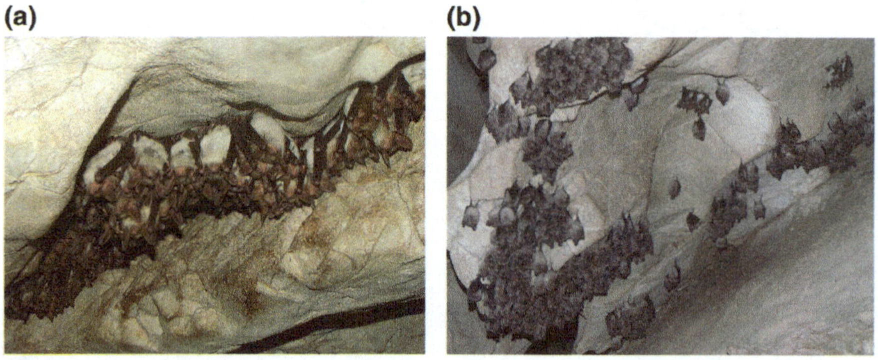

Fig. 3 Wintering colonies of Greater horseshoe bat (**a**) and Greater horseshoe bat (**b**) in Leşu Cave

bat—Fig. 3b), Meziad (nursery for Greater mouse-eared bat and Schreiber's bat, and hibernacula for Greater horseshoe bat and Schreiber's bat; Coroiu et al. 2007), Aştileu Cave (nursery for Greater mouse-eared bat and Schreiber's bat), Stracoş Cave (nursery for Mediterranean horseshoe bat), etc. (Nagy and Postawa 2010).

Fig. 4 A little part of a large colony of Pipistrelle in Şura Mare Cave (photograph courtesy of O. M Chachula)

Şura Mare Cave is similar in topography and climate with Huda lui Papară Cave, but has more stable climatic conditions during the winter. Inside the cave, 7 bat species were identified, two of which form colonies comparable in size with those of Huda lui Papară: Pipistrelle, up to 37,000 individuals (Fig. 4) and Schreiber's bat with over 67,000 individuals. Similar to the previous cave, Şura Mare is part of a network of caves (bats may change their shelter during the winter in case of sudden drops in temperature) and is located in an area with a complex natural habitat mosaic. It is remarkable that long-term estimations show a stability regarding bat colony sizes in this cave. In the 1950s, the total number of Pipistrelle and Schreiber's bats has been estimated to ~100,000 individuals (Dumitrescu et al. 1962–1963). Presently, their number appears to have remained the same (Murariu et al. 2007; Chachula unpublished; Coroiu unpublished).

There are five main caves used as bat shelters in the south-western part of Romania: Topolniţa Cave (on the Mehedinţi Plateau), Gura Ponivovei Cave (Almăjului Mts.), Gaura cu Muscă Cave (Locvei Mts.), Great Şălitrari Cave (Cerna Mts., Domogled-Valea Cernei National Park) and Comarnic Cave (Domanului Mts., Semenic—Cheile Caraşului National Park). Topolniţa Cave is a very important shelter for Mediterranean horseshoe bat, the maternity reaching about 1700 individuals (Fig. 5) (Coroiu unpublished).

The Bat's Gallery (an upper level of Ponicova Cave) shelters a Mediterranean horseshoe bat maternity of around 5000 bats (Coroiu unpublished) distributed in four compact colonies; this is probably the largest nursery shelter for this species in Europe. Mehely's horseshoe bat, a rare species in Romania, has also been found here (Decu et al. 2004–2005).

In the Gaura cu Muscă Cave, five bat species have been reported, of which three form significant maternities: Long-fingered bat, Schreiber's bat and Mediterranean horseshoe bat. The Long-fingered bat maternity (Fig. 6) reaches about 800 individuals and is the largest in Romania; the cave's location near the "Porţile de Fier" (Iron Gates) hydropower dam lake is ideal for this Mediterranean species, which is insectivorous, but also piscivorous (Aihartza et al. 2003).

Fig. 5 Nursery colony of Mediterranean horseshoe bat in Topolniţa Cave (photograph courtesy of F. Forray)

Fig. 6 Nursery colony of Long-fingered bat in Gaura cu Muscă Cave, the largest maternity of this species in Romania

Of the 11 bat species which hibernate in the Great Şălitrari Cave, some form spectacular colonies: Pipistrelle/Soprano pipistrelle/Schreiber's bat (33,000 individuals) and Mediterranean horseshoe bat (2500) (Fig. 7) (Chachula et al. 2016).

The Comarnic Cave is a hibernaculum for up to 500 Greater horseshoe bat and Greater mouse-eared bat (Pavel and Coroiu 2016). These five caves, along with several others with smaller colonies, make south-western Romania (i.e., Banatului Mountains) a true hot spot of European significance for biodiversity and bio-richness.

In the East Carpathians, the underground bat shelters are not so common due to the lack of extensive karst areas (most rocks are volcanic). This is why the few caves located here are densely populated with bats. For example, in the Vârghiş Gorge, there are over 20 caves in which 17 bat species have been found. The most important is Orbán Balázs Cave, with a maternity of up to 2500 Greater mouse-eared bat/Lesser mouse-eared bat individuals (Jére et al. 2006–2007). The Tăuşoare Cave in Rodnei Mountains is highly inadequate as a bat shelter (it is a steeply descendent cave), but because it is the only large cave available in the area,

Fig. 7 Wintering colony of Mediterranean horseshoe bat in Great Şălitrari Cave (photograph courtesy of R. Puşcaş)

it is used as a hibernaculum by the Greater mouse-eared bat/Lesser mouse-eared bat (8000–10,000 individuals) (Theodorescu pers. comm.). Another important shelter is Liliecilor Cave in Rarău Mountains, where 3300–3700 Greater/Lesser mouse-eared bats hibernate (Pocora et al. 2012).

Bat Protection

The bat conservation programs in Romania rely on protection for all bat species, protection of caves and other shelter types, habitats conservation of the feeding grounds and control of speleo-tourism. All of these are included in the Habitats Directive (Council Directive 92/43/EEC) of which Romania is part. Even though speleo-tourism is still at a moderate level, there are a few caves which have been profoundly affected by this type of activity: Polovragi and Muierii caves (speleo-tourism) and St. Grigorie Decapolitul (or Bistriţa Monastery) Cave (ecumenical tourism) in which maternities have been disturbed. One disturbing example is that of Gura Dobrogei Cave, which was one of the most important summer and winter bat shelters in the 1950s and maternity for Greater mouse-eared bat and Schreiber's bat (thousands of individuals), Greater horseshoe bat and Mehely's horseshoe bat. One of the most impressive hibernation colony was that of Mehely's horseshoe bat (around 5000 bats) (Dumitrescu et al. 1958). Červený (1982) estimates the total population during the summer of 1974 to about 8000 bats and in 1979 to only 800 individuals. Today, Mehely's horseshoe bat has disappeared from the cave, and the other species form colonies of only dozens of individuals, all because of farmers fair held continuously since the 1970s nearby this cave; and now, in addition, there is also a kind of ecumenical pilgrimage.

Romania has adopted the European legislation for bat protection (Council Directive 92/43/EEC, Bern Convention, etc.) and has a national network protection system for both bats and caves: e.g., Speleological Heritage Committee (Romanian Ministry of the Environment and Forests), Romanian Bat Protection Association, and "Emil Racoviţă" Institute of Speleology. All these organizations regulate the scientific activity, caving and speleo-tourism. A few large-scale projects contribute to the protection of bats by assuring ecological education and also limiting public.

References

Aihartza JR, Goiti U, Almenar D, Garin I (2003) Evidences of piscivory by *Myotis capaccinii* (Bonaparte, 1837) in Southern Iberian Peninsula. Acta Chiropterol 5(2):193–198

Baudinette RV, Churchill SK, Christian KA, Nelson JE, Hudson PJ (2000) Energy, water balance and the roost microenvironment in three Australian cave-dwelling bats (Microchiroptera). J Compar Physiol B 170:439–446

Bücs Z, Jére C, Csősz J, Barti L, Szodoray-Parádi F (2012) Distribution and conservation status of cave-dwelling bats in the Romanian Western Carpathians. Vespertilio 16:97–113

Červený J (1982) Notes on the bat fauna (Chiroptera) of Roumanian Dobrogea. Nyctalus (N.F.) 1(4–5):349–356

Chachula OM, Puşcaş R, Albuică A, Meşter LE (2016) Peştera Mare de la Şălitrari' (Cerna Valley)—an important bat shelter (Chiroptera, Mammalia) from SW Romania. Muzeul Olteniei, Craiova, Stud Com Şt Nat 32(2):121–126

Coroiu I, David A (2008) Long-term changes of hibernating bats in Huda lui Papară Cave (Apuseni Mountains, Romania). In: XIth European Bat Res Symp, Cluj-Napoca, 18–22 Aug 2008, p 39

Coroiu I, David A, Borda D (2007) Colonii de hibernare a chiropterelor in Peştera Meziad (Jud. Bihor). Misc Chiropterol 1:40–42

Decu V, Gheorghiu V, Murariu D (2004–2005) Contribution to the knowledge of the distribution of *Rhinolophus mehelyi* Matschie, 1901 species (Chiroptera: Rhinolophidae) in Romania. Trav Inst Spéol "Émile Racovitza", XLIII–XLIV:235–247

Dumitrescu M, Orghidan T, Tanasachi J (1958) Peştera Gura Dobrogei. An Com Geol 31:461–499

Dumitrescu M, Tanasachi J, Orghidan T (1962–1963) Răspândirea chiropterelor în R. P. Română. Lucr Inst Speol "Emil Racoviţă" 1–2:509–576

Jére C, Dóczy A, Barti L (2006–2007) Results on research on the bat fauna of the Vârghiş Gorge (Eastern Carpathians, Romania). Trav Inst Spéol "Émile Racovitza" XLV–XLVI:59–74

Mitchell-Jones T, Bihari Z, Masing M, Rodrigues L (2007) Protecting and managing underground sites for bats. EUROBATS Publ Series 2, UNEP/EUROBATS Secretariat, Bonn

Murariu D, Decu V, Gheorghiu V, Nistor V (2007) An impressive aggregation of bats hibernating in the Southern Carpathians (Şura Mare Cave—Romania). Stud Chiropterol 5:1–7

Nagy ZL, Postawa T (2010) Seasonal and geographical distribution of cave-dwelling bats in Romania: implications for conservation. Anim Conser 14(1):74–86

Pavel O, Coroiu I (2016) Seasonal dynamics of bat fauna in Comarnic Cave, Romania. Muzeul Olteniei, Craiova, Stud Com Şt Nat 32(1):110–114

Pocora I, Pocora V, Baltag ES (2012) Bat communities (Chiroptera: Vespertilionidae) from Liliecilor Cave of Rarău Mountains (Suceava County). An Şt Univ "A.I. Cuza" LVIII:137–150

Schober W, Grimmberger E (1993) Hamlyn guide to bats of Britain and Europe. Hamlyn, p 224

The Archaeology of Caves in Romania

Mircea Anghelinu and Adina Boroneanț

Abstract
The archaeological research of the substantial and diverse Romanian karst has a long history going back to nineteenth-century antiquarians. A more systematic interest emerged, however, in the interwar times and continues to the present day. The earliest proof for the human use of caves in the Romanian Carpathian area belongs to the Last Interglacial Neanderthals, but Last Glacial Mousterian presence was also reported. They all indicate short-lived and possibly recurrent excursions into generally low mountain environmental settings (<1000 m) in search of local game. Despite spectacular palaeoanthropological (including the earliest Anatomically Modern Humans in Europe) and parietal art finds, the intensity of cave use by Upper Palaeolithic hunter-gatherers (Aurignacian and Gravettian) was surprisingly low, with most consistent occupations dated only to the final stages of the Pleistocene (Epigravettian/Epipalaeolithic). The small Upper Palaeolithic inventories in caves indicate very short, exploratory stops that correlate to the documented focus of these communities on open-air settings at lower altitudes. From Mesolithic to Medieval times, caves were used with varying intensity, serving as temporary/seasonally residence, as well as for ritual or economic purposes. Thick Early Neolithic cultural sequences, occasionally spectacular Bronze and Iron Age depositions, much like the Roman and Palaeo-Christian finds are particularly telling for the important residential and ritual/religious role some caves played. Through time, apart from their topography and degree of accessibility, the importance granted to these natural shelters by the various communities depended on the continuous change of the sociocultural and economic contexts.

Keywords
Palaeolithic • Human fossils • Parietal art
Bronze Age • Iron Age • Depositions

Brief History of Research

The history of archaeological research in the Romanian caves (Fig. 1) generally followed the gradual separation of a local archaeological tradition from the antiquarian and natural sciences background, with an earlier start in nineteenth-century Transylvania—back then part of the Austrian, and later Austro-Hungarian Empire—followed by increasingly systematic research especially after World War II. Cave archaeology played an essential role in the professionalization of Romanian prehistoric and particularly Palaeolithic research. Archaeological research focusing on the karst features of present-day Romania have thus quite a long history that can be roughly subdivided into four major stages.

The first stage consists in early speleological and palaeontological explorations of Transylvanian caves in the late eighteenth and during the nineteenth century, long before the professionalization of the archaeological endeavour (Jungbert 1978, 1979, 1982). Work of naturalists, self-taught amateurs or enthusiasts connected to the Central and Western European intellectual milieu, these early studies were rarely, if ever, interested in the archaeological contents of caves. By the late nineteenth century, however, the works of A. Koch, C. Gooss, Z. Torma, G. Téglás (in Transylvania) or A. Odobescu (in Walachia) highlighted the caves' archaeological potential, particularly relevant for prehistoric research (for comprehensive reviews, see Boroneanț 2000;

M. Anghelinu (✉)
Faculty of Humanities, "Valahia" University of Târgoviște, Lt. Stancu Ion 34-36, Târgoviște, Romania
e-mail: mircea_anghelinu@yahoo.com

A. Boroneanț
"Vasile Pârvan" Institute of Archaeology of the Romanian Academy, Henri Coandă 11, Bucharest, Romania
e-mail: boro30@gmail.com

G. M. L. Ponta and B. P. Onac (eds.), *Cave and Karst Systems of Romania*,
Cave and Karst Systems of the World, https://doi.org/10.1007/978-3-319-90747-5_57

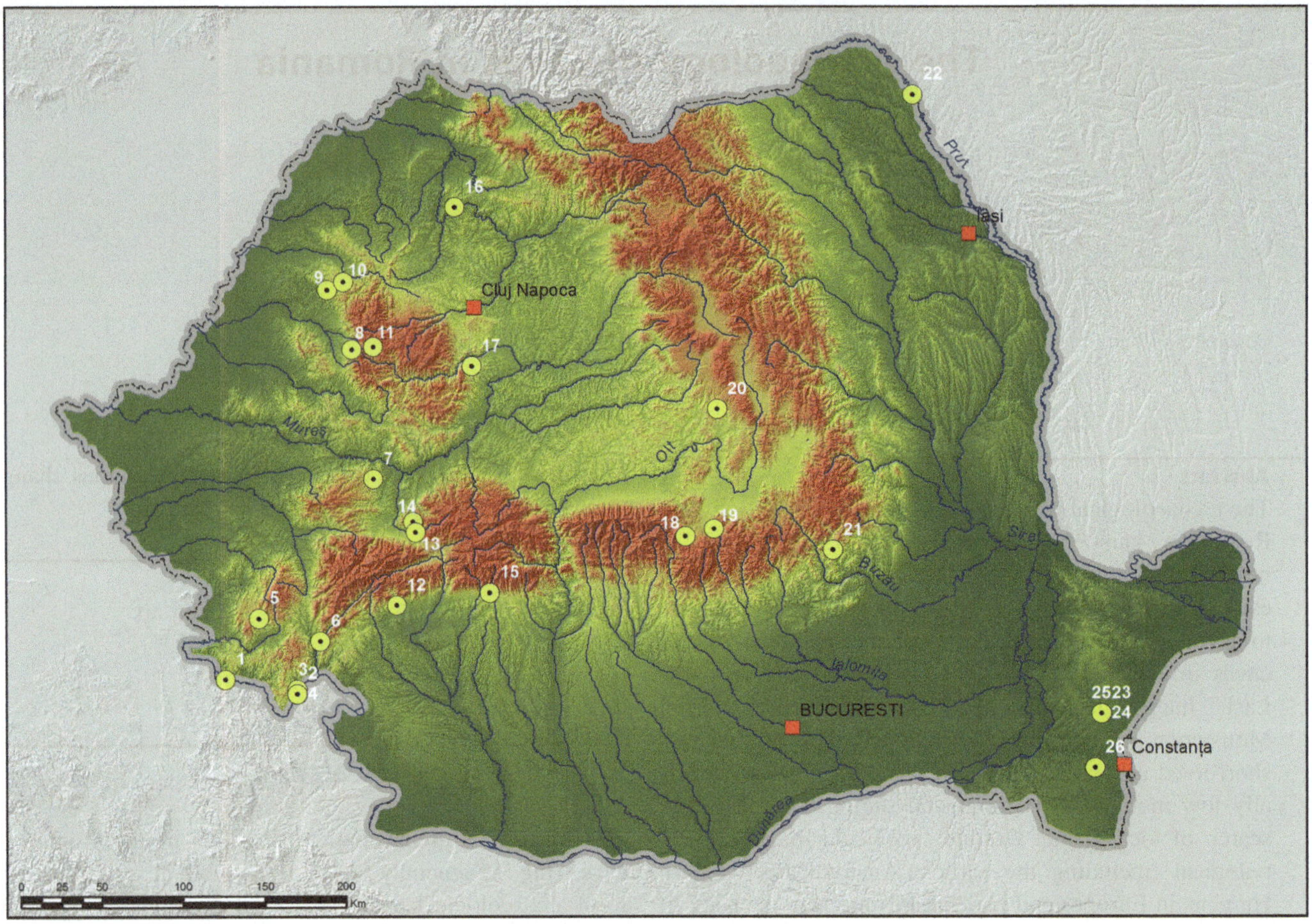

Fig. 1 Map of the key archaeological cave sites mentioned in the text: 1. Gaura cu Muscă; 2. Cuina Turcului; 3. Climente II; 4. Veterani; 5. Oase/Hoţu–Anina; 6. Hoţilor; 7. Curată/Spurcată (Nandru); 8. Coliboaia; 9. Ciur-Izbuc; 10. Mişidului; 11. Vârtop; 12. Cioarei–Boroşteni; 13. Ohaba-Ponor; 14. Cioclovina; 15. Muierii; 16. Cuciulat; 17. Piatra Secuiului–Râmetea; 18. Mare–Moieciu; 19. Gura Cheii–Râşnov; 20. Abri 122; 21. Nucu–Bozioru; 22. Stânca-Ripiceni; 23. Liliecilor; 24. La Adam; 25. Cheia (Dobrogea); 26. Basarabi (SRTM data generated from https://www2.jpl.nasa.gov/srtm/mou.html; map design G. Murătoreanu)

Păunescu 2001), a field slowly emerging from natural history all across Europe (Laming-Emperaire 1964; Trigger 1989).

A second stage of research encompasses the rich activity of archaeologist M. Roska and his co-workers in the interwar times of Transylvania. Apart from his many open-air excavations, Roska undertook systematic digging in several caves along Romanian Carpathians, some of which (e.g., Cioclovina, Bordul Mare–Ohaba-Ponor) were further explored in the following decades. According to the archaeological knowledge of the time, Roska proposed a wide-encompassing cultural-evolutionary framework for the cave finds in Romania, ranging from the Middle Palaeolithic to the Mesolithic (Table 1). Although many of his interpretations, much like the authenticity of some of his finds, failed the test of time, his work drew the map and secured the foundations for many subsequent investigations (cf. Păunescu 2001). The activity of his contemporary, geologist N. N. Moroşan, who excavated the currently destroyed Stânca-Ripiceni Cave, played a similar seminal role, certifying the preservation of Pleistocene archaeological remains in cave contexts distant from the Carpathian range (Moroşan 1938).

The third, and by far the most intensive stage of research, took place in the first decades after World War II and revolved around the activity of C. S. Nicolăescu-Plopşor and his co-workers. In contrast to previous efforts, Nicolăescu-Plopşor's research activity took a systematic, institutionally supported character (Nicolăescu-Plopşor 1957). His work benefited from a consistent sustain from the Communist authorities of the time, highly interested in providing archaeological support for their evolutionary tenets (Anghelinu 2003). Between 1951 and 1964, Nicolăescu-Plopşor conducted more or less extensive excavations in

Table 1 Main archaeological subdivisions

Period	Inner subdivisions	Chronology
Early Medieval		AD 400–1000
Roman Age		AD 106–400
Iron Age	La Téne Hallstatt	400 cal BC–AD 106 1200–400 cal BC
Bronze Age	Late Middle Early	1600/1500–1200 cal BC 2500–1600/1500 cal BC 3500–2500 cal BC
Neolithic/Chalcolithic	Late Chalcolithic Late Neolithic/Early Chalcolithic Middle Neolithic Early Neolithic	4500–3500 cal BC 5000–4500 cal BC 5500–5000 cal BC 6100–5500 cal BC
Epipalaeolithic/Mesolithic	Late Mesolithic Middle Mesolithic Epipalaeolithic/Early Mesolithic	7400–6100 cal BC 9700–7400 cal BC 12,700–9700 cal BC
Palaeolithic	Upper Palaeolithic Middle Palaeolithic Lower Palaeolithic	45–14 ka cal BP 250–45 ka cal BP 2.5 Ma–250 ka BP

several caves and rockshelters along the southern rim of the Carpathians (Muierii-Baia de Fier; Bordul Mare–Ohaba-Ponor; Hoţilor-Herculane; Nandru Spurcată and Curată; Mare, Mică, and Valea Coacăzii from Moieciu; Gura Cheii–Râşnov; Cioarei-Boroşteni; Veterani; Climente I and II; Cuina Turcului) and in Dobrogea (Liliecilor, Cheia), with archaeological remains ranging from Middle Palaeolithic to Roman and Medieval times. The investigations of many of these caves, regularly involving palaeontologists and geologists, were continued by Nicolăescu-Plopşor's students in the following decades (for reviews, see Boroneanţ 2000; Păunescu 1999, 2000, 2001).

The most recent stage corresponds to the last two decades of research, focusing mostly on the reassessments of previous archaeological and palaeoanthropological evidence (e.g., Onac et al. 2005; Soficaru et al. 2006, 2007; Doboş et al. 2010; Bonsall et al. 2012; Sîrbu and Matei 2012; Tuffreau et al. 2013; Webb et al. 2014), but also on some new and often spectacular cave finds (Trinkaus et al. 2003, 2007; Luca et al. 2005; Clottes et al. 2012; Cosac et al. 2018). Although less extensive in terms of scale and geographic coverage, the current stage of research witnessed an unprecedented involvement of interdisciplinary teams and allowed the rich archaeological heritage of Romanian caves to reach a wider scientific audience.

Given the long history of research into the Romanian caves, naturally involving old-fashioned research methodologies and publication standards, the amount and resolution of contextual information raise serious challenges for a synthetic approach. The present review, selective by necessity, aims at providing a broad diachronic picture of human cave use, and focuses only on several significant finds from an otherwise rich and diverse spectrum of cave archaeological contexts.

Middle and Upper Palaeolithic Cave Finds

Archaeology

Although open-air Lower Palaeolithic finds have been reported in the Romanian archaeological literature (Doboş 2008), no securely dated Lower or Middle Pleistocene archaeological contexts have been yet identified in cave settings across Romania. The oldest archaeological context in the Romanian karst is so far the lower Mousterian layer recently dated to Marine Isotope Stage (MIS) 5 at Abri 122 in the East Carpathians (Cosac et al. 2018; Veres et al. 2018). A huge chronological gap seems to separate this Last Interglacial occurrence from all other Middle Palaeolithic settlements located in South Carpathians, Apuseni Mountains, or in the Dobrogea karst (Cheia–Gura Dobrogei), all attributed to late stages of MIS 3 on the grounds of radiocarbon dates obtained in the 1980s. The methodological issues raised by these results, lying close to the limits of the radiocarbon method, have been, however, discussed on several occasions (e.g., Doboş et al. 2010; Anghelinu et al. 2012a; Doboş 2017). Unfortunately, only open-air Middle Palaeolithic occurrences in Romania have been reassessed chronologically in the last decades; they all point to a considerably older chronology for the Romanian Middle Palaeolithic, occasionally reaching MIS 6 (Doboş 2017). One may expect similar results for the Carpathian caves. While several Accelerator Mass Spectrometry

(AMS) radiocarbon ages from Muierii and Nandru-Curată caves, ranging around 45 ka cal BP, still give support to the idea of a Late Mousterian in Romania, their newly established chronology does not cross the conventional Upper Palaeolithic threshold as once thought (e.g., Mogoşanu 1978; Cârciumaru 1999).

Irrespective of their actual chronology, Middle Palaeolithic occupations of the Romanian caves have several major common features: (1) medium altitude settings (generally ranging between 300 and 950 m), with good accessibility from nearby river valleys or from the plateaus above; (2) shallow habitat traces, usually reduced to superficial hearths; and (3) small sized lithic assemblages, generally made in locally available raw materials (quartzite, diorite, cherts, opal, basalt, etc.), through expedient (multidirectional or discoid) knapping methods leading to flakes of various sizes and shapes, naturally backed knives and sub-triangular points; Levallois or bifacial technologies (Fig. 2) and worked bones have been occasionally reported (Păunescu 2000, 2001; Doboş 2017; Cosac et al. 2018). Only Muierii, Ohaba-Ponor, and Abri 122 provided larger lithic inventories (ca. 5000 items at Muierii; Riel-Salvatore et al. 2008), but their larger numbers expressing multiple occupations and distinct cultural layers (e.g., four at Ohaba-Ponor) do not alter the general pattern indicating short-lived camps.

Fig. 2 Mousterian leafpoints from Abri 122 (SE Transylvania) (photographs courtesy of M. Cosac)

The duration, seasonality, and function of these settlements have been only exceptionally discussed (e.g., Patou-Mathis 2001). Traditionally lumped under the Quartzite/Charentian Mousterian label (e.g., Mogoşanu 1978; Cârciumaru 1999), and comparable as such to several open-air settings (e.g., Zăbrani in Banat; Tuffreau et al. 2007), the Carpathian Mousterian seems to express less of a stylistic tradition and more of a redundant pattern in terms of shelter choice, mobility and lithic raw material use (Riel-Salvatore et al. 2008). Although the role of human agency in the formation of the associated faunal contexts (horse, bison, deer, ibex, etc.) has rarely been assessed (Patou-Mathis 2001; Cosac et al. 2018), these occupations point more to the opportunistic exploitation of the diverse topography/biota close to a mountain range than to a systematic appropriation of high altitudes. Only the tiny assemblages from the two caves at Moieciu (Mare and Valea Coacăzii—over 900 m in altitude) and the Vârtop footprints (Onac et al. 2005) indicate (brief?) explorations of the higher altitudes of the Carpathians by Neanderthals.

The first stages of the Upper Palaeolithic seemed to have brought a break in the human occupation of caves and presumably in the exploration/exploitation of the mountain environment in general, as the earliest dated archaeological contexts are no older than 34 ka cal BP (Păunescu 2000, 2001). Even after this date, the Late Aurignacian and Gravettian/Epigravettian traces in caves remain scanty, pointing to short-stops/ephemeral camps. Most assemblages are characterized by common and often culturally undiagnostic Upper Palaeolithic inventories: endscrapers and burins, (un)retouched blade and bladelets, flakes, and rarely cores, generally made in fine-grained and usually exotic flint, radiolarites, opal, and cherts. The largest (presumably Aurignacian) lithic assemblage at Peştera Mare–Moieciu amounts to only ca. 250 lithics; only a few bone items and several adornment objects (pierced carnivore canines, pendants) were found in these Upper Palaeolithic contexts (Păunescu 2000, 2001). Gravettian layers have been reported at Cioarei-Boroşteni (Cârciumaru 2000) and Gura Cheii–Râşnov caves (Păunescu 2001), but their taphonomical integrity remains unclear. A more consistent Epigravettian presence has been recorded only in several caves in south-western Romania (e.g., Hoţilor, Climente II) close to the Iron Gates (Mogoşanu 1978; Bonsall et al. 2016).

The light Upper Palaeolithic archaeological presence in caves is in stark contrast both to the relatively large number of palaeoanthropological finds (see below) and to the density of open-air settlements and isolated discoveries attributed to this time period (Păunescu 2000, 2001). While the Gravettian focus on steppe biomass and related open-air settings is largely acknowledged across East-Central Europe, Romania included (Anghelinu et al. 2012b), the scarcity of Aurignacian traces remains harder to explain, as consistent Aurignacian contexts, occasionally at high altitudes, have been recorded from Slovenia (Moreau et al. 2015) to the Swabian Jura (Conard and Bolus 2006). Moreover, the Epigravettian exploitation of mountain environment and the related cave use is well documented in many, especially Mediterranean, European areas (e.g., Mussi 2002). As both Middle Palaeolithic and post-Palaeolithic layers are regularly preserved in the same caves, preservation issues seem excluded. The resulting pattern suggests a general disinterest for cave use and the likely forested Carpathian environment (Feurdean et al. 2015) during the Upper Palaeolithic, in favour of low and middle altitude open landscapes and their related steppe-tundra biotopes.

The relatively recent (2009) identification of the Upper Palaeolithic parietal art at Coliboaia Cave (Clottes et al. 2012) provides, however, a cautionary note, suggesting that new field researcher might dramatically alter the image above. The 13 paintings at Coliboaia—among which a bison, a horse, a possible feline, bears, mammoth and two rhinoceros heads (Fig. 3)—made exclusively in black, evoke the Aurignacian and Gravettian parietal styles of the Franco-Cantabrian area. The direct and indirect dating of these representations (bracketed between 31–32 and 35–36 ka cal BP) strengthens their stylistic assessment. While in terms of antiquity, Coliboaia is so far unique in this part of Europe, the site is actually the second decorated Palaeolithic cave found in Romania: several red representations, including a horse of likely Epigravettian age were described at Cuciulat Cave (Cârciumaru and Bitiri 1980). However, none of these caves or their surrounding areas provided additional archaeological contexts.

To summarize, if one takes at face value the current state of knowledge, the Middle Palaeolithic humans were using caves as temporary shelters more often/consistently than their Upper Palaeolithic counterparts. However, the reported Palaeolithic presence in the Romanian karst is altogether meagre when measured against the actual potential, reaching to several thousand known fossil caves. Given the insufficient geographic coverage and the low resolution of many of the previous archaeological investigations, it is reasonable to expect a consistently changed picture in the years to come.

Fig. 3 Rhinoceros representation at Coliboaia Cave (photograph courtesy of A. Posmoşanu)

Palaeoanthropological Discoveries

In recent years, the re-examination and numerical dating of older palaeoanthropological finds in the Romanian caves reached some noticeable and occasionally spectacular results. Middle Palaeolithic (i.e., Neanderthal) human remains are rare. Only three phalanxes, found between 1923 and 1929 in a Mousterian layer at Bordul Mare–Ohaba-Ponor cave were attributed to Neanderthals (Păunescu 2001). Several charcoal samples recovered from the Mousterian layers here and dated by conventional radiocarbon method suggest ages close or beyond the limits of the method (>45 ka cal BP).

A spectacularly preserved proof on the presence of Neanderthal in caves at high altitudes is provided by the three footprints preserved in the Vârtop Cave (1170 m alt., Bihor Mountains). Although lacking an associated cultural context, the U–Th chronological constraints indicate ages in excess of 62 ka cal BP for the best-preserved footprint (Onac et al. 2005).

In contrast to the scanty Neanderthal osteological record, the Romanian Carpathians caves provided some of the earliest evidence for the presence of Anatomically Modern Humans (AMH) in Europe at the dawn of the Upper Palaeolithic. The first example is the partially preserved human skull (Fig. 4) found in the Cioclovina Uscată Cave (see Chapter "Şureanu Mountains: Valea Stânii–Ponorici–Cioclovina cu Apă Karst System"), during the mining of guano-phosphate sediments from the cave (1940–1941); it is worth mentioning that a now lost human skull had been found earlier in the twentieth century in the same cave (Soficaru et al. 2007). The actual context of both these finds, reportedly associated with *Ursus spelaeus* and carnivore bones, remains unclear even after the several archaeological surveys undertaken at the site since 1911. Scattered lithics attributed to both Middle Palaeolithic and Upper Palaeolithic were reported, with the latter (ca. 20 items) providing the most likely cultural proxy for the AMH findings (Păunescu 2001). Two different radiocarbon ages of ca. 33 ka cal BP of the Cioclovina skull secured its attribution to the Upper Palaeolithic (Soficaru et al. 2007), pointing to a Late Aurignacian/Early Gravettian time range. The taphonomic assessment indicates a secondary position for the human remains (Soficaru et al. 2007). The state of preservation of the fragile portions of the skull bone and the lack of carnivore interventions may suggest a burial context destroyed by the hydraulic transport of the sediment inside the cave.

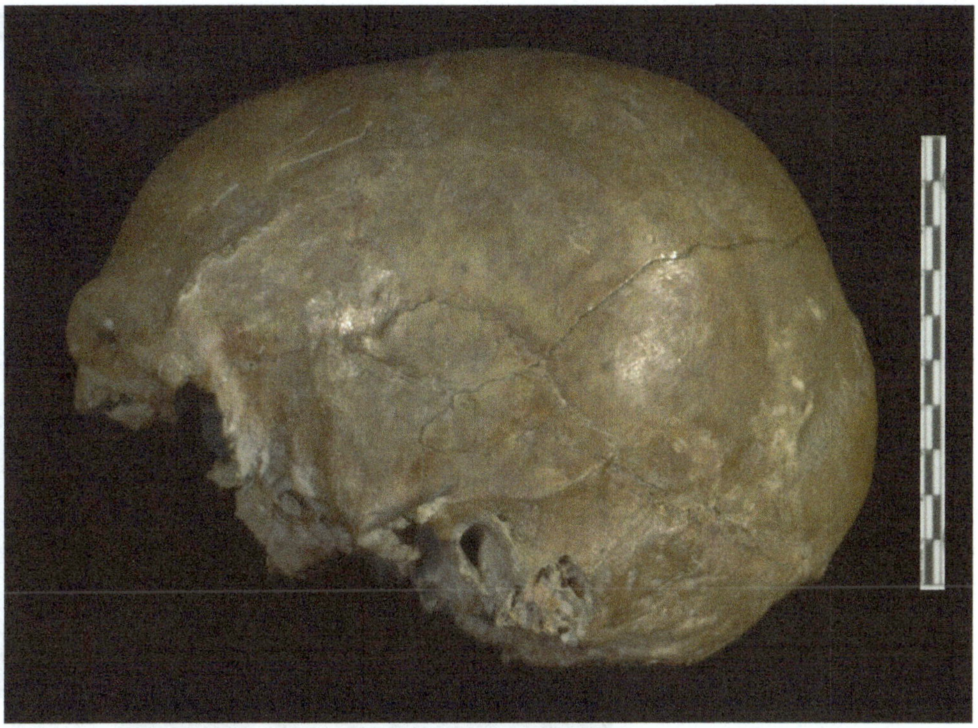

Fig. 4 Cioclovina AMH cranium (photograph courtesy of Erik Trinkaus)

Another key palaeoanthropological discovery took place in 1952 at the Muierii Cave, in the South Carpathians, where six skeletal elements belonging to three different individuals were found (Soficaru et al. 2006; Doboş et al. 2010). Four of these pieces (cranium, mandible, scapula, and tibia) belong to a female (Muierii 1) found in the so-called Mousterian Gallery (Fig. 5); a temporal bone and a fibular diaphysis originate from two other individuals (Muierii 2 and 3). While Muierii 3 failed in preserving enough collagen for dating, both Muierii 1 and Muierii 2 provided direct radiocarbon ages of 34–35 ka cal BP, suggesting a single/close in time depositional episode(s). Although originally considered a modern human found in a Mousterian setting (e.g., Nicolăescu-Plopşor 1968), despite convincing contextual arguments, the Muierii 1 and 2 were presumably correlated to the thin Upper Palaeolithic layer (Late Aurignacian?) in the Main Gallery (ca. 60 lithics and three bone items; Doboş et al. 2010).

The most recent palaeoanthropological discoveries took place at the Peştera cu Oase (Aninei Mountains). The remains of two AMH individuals (Oase 1 and Oase 2) were found during the speleological and archaeological exploration (2002–2004) of the complex karst system (Trinkaus et al. 2003, 2006). Although only the Oase 1 mandible was directly dated to ca. 40 ka cal BP, the many features in common between the two individuals suggest a comparable age for Oase 2 (Trinkaus et al. 2006; Rougier et al. 2007). Much like in the case of Cioclovina, although clearly displaced by hydraulic processes and found in a secondary position in a massive mammal bones accumulation, the human fossils at Oase are very well preserved, with minimum evidence of geological abrasion and no traces of carnivore gnawing (Fig. 6). Unfortunately, despite repeated attempts, including archaeological excavations in the karst features on top of the submerged Oase Cave, no relevant archaeological context has been yet found for these fossils (Băltean personal communication; Hauck pers. comm.).

The numerical chronology recommends the Oase fossils as the oldest indisputably directly dated AMH in Europe. Significantly, much like the AMH fossils in Muierii and Cioclovina caves, both individuals at Oase display a mosaic of anthropometric features suggesting Neanderthal/AMH admixture (Trinkaus 2007). Recent palaeogenetic studies confirmed both the partially Neanderthal ancestry of the Oase fossils and their belonging to a genetic lineage whose signature was subsequently lost among later Upper Palaeolithic moderns humans (Fu et al. 2015a, b). It remains unclear to which Early Upper Palaeolithic technocomplex

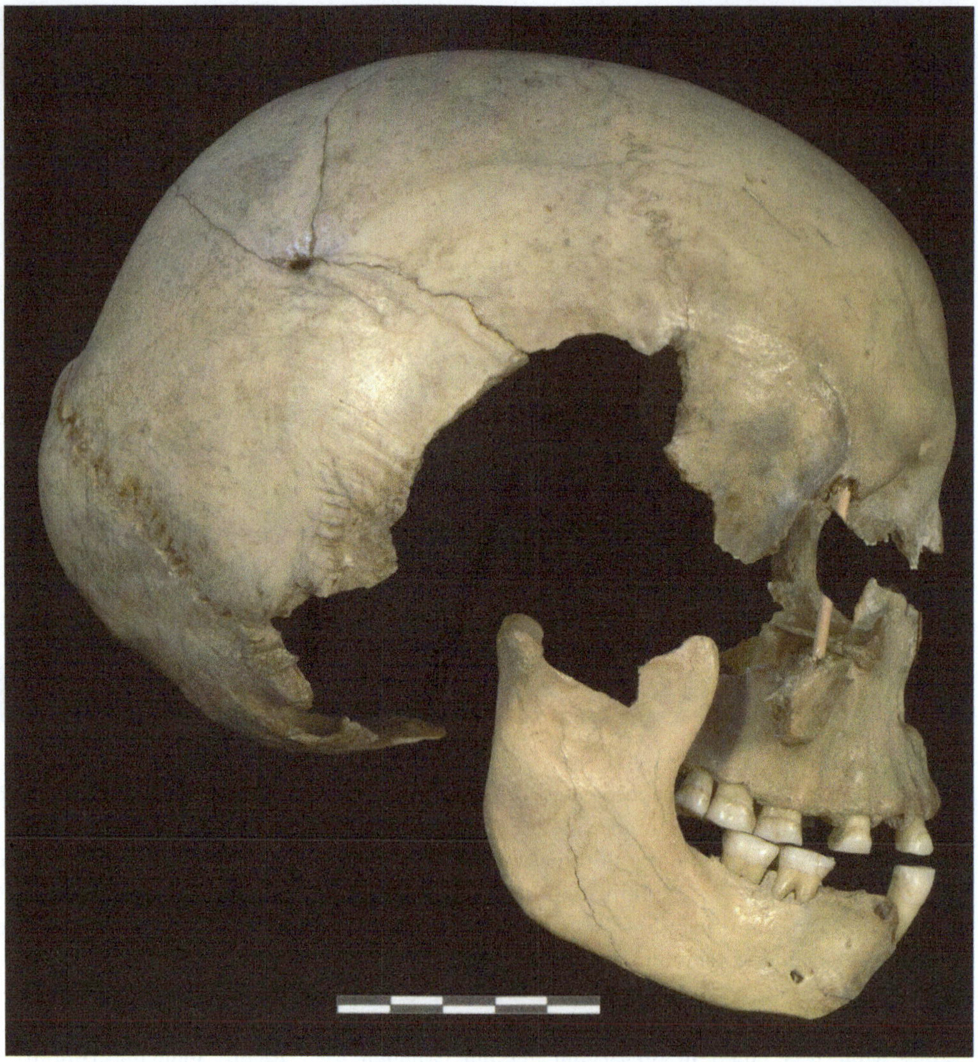

Fig. 5 Muierii 1 AMH cranium (photograph courtesy of Erik Trinkaus)

the Oase fossils should be correlated to, or if their population actually represented the first wave of AMH colonists in Europe. It should be noted, nevertheless, the recent chronological reassessment, at 40 ka cal BP, of the neighbouring Banat Early Aurignacian open-air settlements (Schmidt et al. 2013; Sitlivy et al. 2014), providing a parsimonious albeit hypothetical cultural correlate for the Oase fossils.

AMH seemed also to have been responsible for the ca. 400 footprints overlapping those of cave bear discovered in 1965 in the Ciur-Izbuc Cave (Pădurea Craiului Mountains; see Chapter "Pădurea Craiului Mountains"). A recent reassessment of the 51 still preserved footsteps identified traces of 7 children and adults, with a chronology bracketed between 36.5 and 28.7 ka cal BP (Webb et al. 2014). Much like in the case of Vârtop footprints, no supporting archaeological context has been found in this cave.

To conclude, although many of the early AMH finds in Romania do contribute to the better understanding of a crucial evolutionary stage notorious for its scarcity of human fossils, none of these discoveries can be properly interpreted in behavioural and palaeocultural terms, given the (doubtful) association with/absence of, diagnostic cultural contexts. For instance, despite their good state of preservation suggesting a protected environment during bodies decomposition, and the systematic presence of several individuals in each case, the original disposure of the corpses remains unknown (disturbed burials?).

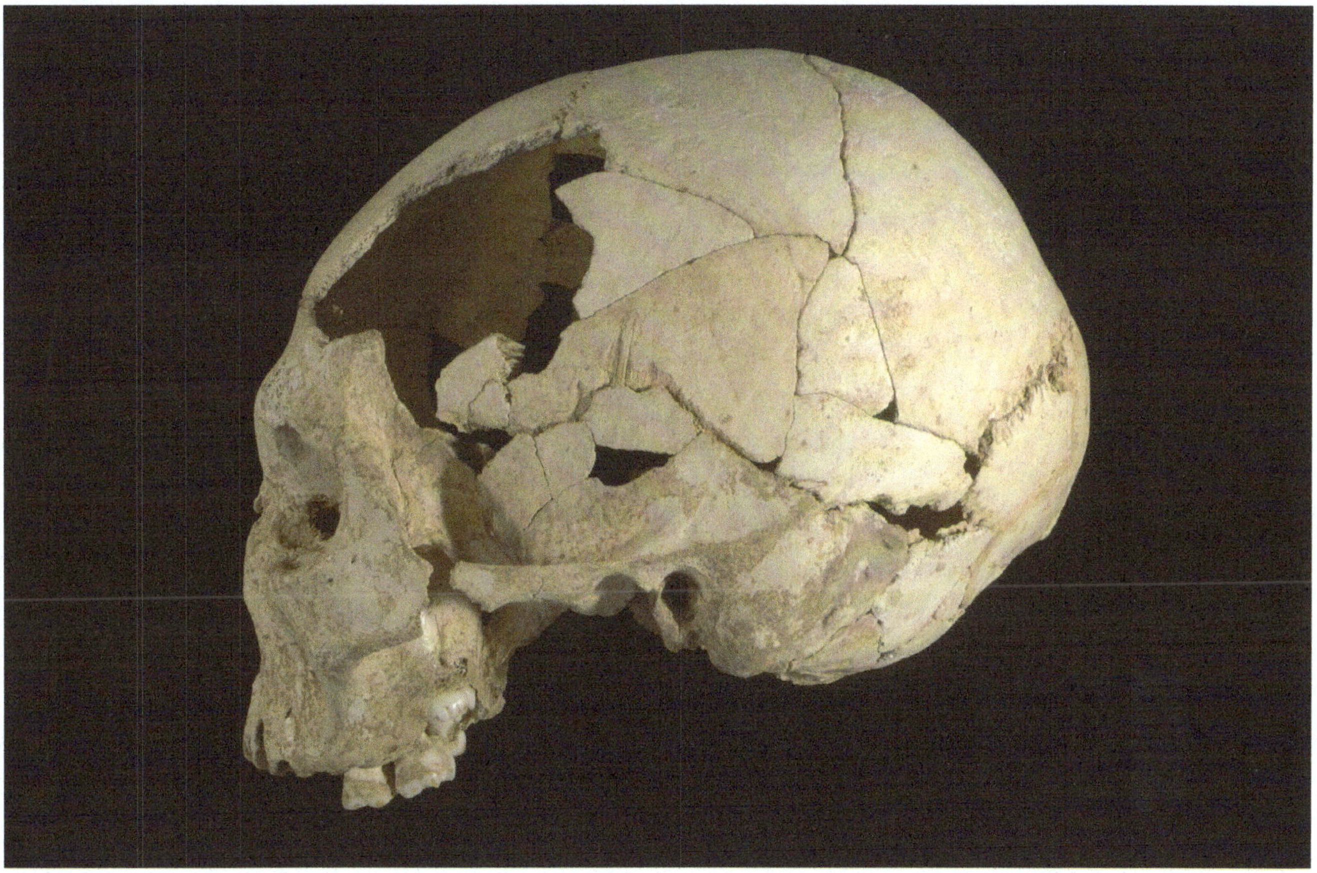

Fig. 6 Oase 2 AMH cranium (photograph courtesy of Erik Trinkaus)

Contemporary Aurignacian and Gravettian contexts, although reported as open-air occurrences in neighbouring areas (Păunescu 2000, 2001), are simply missing around these settlements—an absence to be hopefully remediated by further research.

Epipalaeolithic and Mesolithic

During the Epipalaeolithic and the Mesolithic, human presence in caves is only documented in Banat (south-western Romania), mainly in the Iron Gates region of the Danube. These cave sites had thick and complex stratigraphic sequences (Fig. 7), suggesting an intensive use during later periods such as: Early Neolithic, Chalcolithic, Bronze Age, Late Hallstatt period (Dacian Age), Roman, and Medieval times. But it is Climente II Cave and Cuina Turcului rockshelter in the Iron Gates that yielded important information regarding the Epipalaeolithic/Early Mesolithic cave use and burial practices.

At Climente II, the excavations uncovered the remains of a simple circular hearth, a substantial number of faunal remains and bone tools, ornaments (tooth pendants and a *Dentalium* shell bead), engraved bones, pebbles smeared with ochre and 600 chipped artifacts made of chert/flint, quartz/quartzite, radiolarite, and obsidian. The articulated skeleton of an adult male was uncovered toward the rear of the cave, although the cranium, scapulae, clavicles, and some other bones were missing. An infant skeleton that did not survive the lifting was also observed. Other fragmented human remains were recovered from various parts of the cave.

The body of the adult individual had been placed flexed on the left side, with the legs bent almost at right angles to

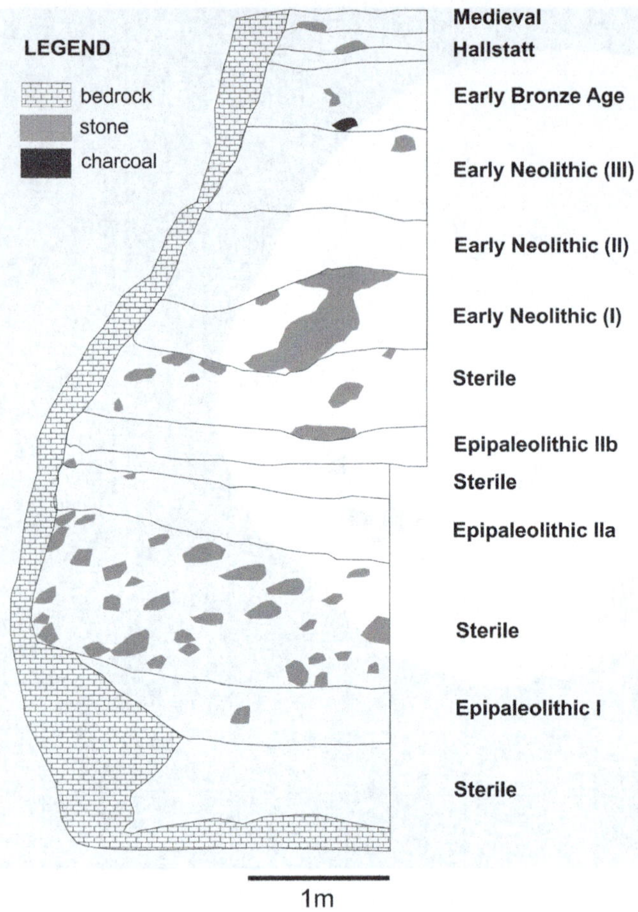

Fig. 7 Cultural sequence at Cuina Turcului rockshelter in the Iron Gates (drawing by A. Boroneanţ)

the torso, and the arms bent upwards with the hands in front of the face (Fig. 8). The grave was shallow and the base of the pit was lined with red ochre, also sprinkled over the corpse (Boroneanţ 1970). It is unclear whether the flint tools found in the soil around the skeleton were grave goods or just part of the infill of the pit. The freshwater reservoir effect corrected ^{14}C dates for the Climente II adult skeleton fall within the time range of the Bølling–Allerød warm period (ca 14.7–12.9 ka cal BP, Bonsall et al. 2016) making it the earliest formal human burial in Romania.

At Cuina Turcului, several humans remains recovered from the Epipalaeolithic and Early Neolithic layers might represent, according to more recent views, disturbed burials (Boroneanţ 2012). The thick Epipalaeolithic occupation layer also yielded the remains of numerous hearths associated with charcoal and burnt bone fragments, bone and antler tools, faunal remains, a large range of ornaments (various tooth and bone pendants, shell beads), engraved bones and ca. 39,000 chipped stone artifacts. The reservoir-corrected dates on some of the human bones from the Epipalaeolithic layer II here (Bonsall et al. 2016) fall at the beginning of the Holocene, and are similar to the dates on animal bones from the open-air Mesolithic sites of Padina, Lepenski Vir, and Vlasac on the Serbian right bank of the Iron Gates Gorge. It is interesting though that a conspicuous gap in the radiocarbon dates for the Iron Gates appears at the time of the Younger Dryas (Bonsall et al. 2016): occupation of caves seems to end with the Bølling–Allerød warm period and human presence resumes after the Younger Dryas cold period, with only Cuina Turcului (a rockshelter) and the open-air sites being occupied. But comparison of the earlier archaeological inventories resulted from caves, with those from later, open-air fisher-hunter-gatherer settlements in the Iron Gates suggests a continuity of mortuary ritual, lithic tradition, and subsistence practices from the Late Glacial into the Early Holocene (Bonsall et al. 2016).

Outside the Danube Gorges, at the recently excavated Hoţu Cave, a hearth and a few lithic artifacts were associated to the Final Mesolithic and provided a chronology ranging between 6.6 and 6.2 ka cal BC; an Epipalaeolithic presence was also suggested here based on the presence of two "atypical" flint pieces and some bone fragments affected by post-depositional processes and dated around 15–14.7 ka cal BC (Boroneanţ 2011).

Cave Use During the Second Part of the Holocene

In Banat, Transylvania, and in the Iron Gates mainly, caves continued to be used during the Early Neolithic (Cuina Turcului, Climente I and II, Veterani, Ponicova, Cauce, etc.). The Early Neolithic sequence at Cuina Turcului (1.15–2.05 m thick) consisted of three allegedly separate horizons, each interpreted as a period of occupation of the cave by Neolithic groups. The investigations uncovered 38 hearths, as well as a shallow pit-feature with vertical walls and flat base, interpreted as a dwelling (Boroneanţ 2012). Fragments of a human skull were found in the close proximity of one of the hearths suggesting a possible symbolic use of the cave,

Fig. 8 Burial M1 in Climente II Cave in the Iron Gates (photo from the archive of V. Boroneanț)

and several other human remains were recovered from various parts of the rockshelter. The archaeological material was incredibly rich, comprising complete and fragmented pots, numerous sherds, bone and antler tools, lithics, ornaments, and an impressive amount of faunal remains, all arguing for an intense occupation of the rockshelter.

During the later Neolithic and the Chalcolithic in Banat and Transylvania, the number of occupied caves decreases slightly, human presence leaving behind only scatters of pottery, a few lithic artifacts and occasional hearths. Some caves in southeastern Romania (Dobrogea) were also used during the later Neolithic and the Chalcolithic. The few potsherds and lithics at La Adam and Liliecilor caves (Harțuche 1976) suggest their use as temporary shelters, while the hearths and the rich archaeological material at Cheia and Baba caves (Voinea and Dobrinescu 2002–2003, Szmoniewski and Petcu 2008) indicate more permanent residential areas.

A major intensification of cave use is apparent in Banat and Transylvania during the Early Bronze Age (3500–2200 BC) and Middle Bronze Age (2200–1600/1500 BC), with most Neolithic caves continuing to be occupied during these later periods and the occupation extending to previously unused caves also. With a few exceptions, such caves were located at rather low altitudes (250–500 m) and at times, access to the visited areas was difficult. Other than the seasonal and temporary residential purpose, they served as places for human burials, deposition of various items, perhaps storage of goods/food, and as places with ritual and symbolic meaning. A few hypotheses might explain the shift in the occupation pattern of caves: (1) colder climatic episodes, such as the *Piora oscillation* (3200–2900 BC), the *4.2 ky event* (2200 BC) and the *Middle Bronze Age Cold Epoch* (2200–1800 BC) when weather in Europe was allegedly colder and wetter; (2) changes of social and cultural nature, which might be responsible for the human burials and the depositions; (3) economic changes that might explain the storage of goods.

Many of the cave sites have thick Bronze Age layers, comprising other than hearths and burials, rich archaeological material (pottery, bone tools and ornaments, occasional lithic implements, bronze items) as it was the case at Hoţilor Cave (Roman 1976). At Izbucul Topliţei de Vida (Bihor) access to the area occupied during the Middle Bronze Age was possible only after passing through the sinkholes at the entrance of the cave. In the main hall pottery, sherds and inhumation burials were observed on the rock bed and a hearth was identified in an adjacent gallery. Among the grave goods there were 15 complete pottery vessels, beads, chipped flint artifacts, two bronze axes (Boroneanţ 2000), and three gold hair-rings (Zsolt and Ghemiş 2003). The instance is not an isolated one—the area of the three Criş Rivers comprises a series of six caves and three stone quarries where "ritual burials" or ritual depositions were noted (Zsolt and Ghemiş 2003).

The cave at Nucu–Bozioru is a rare phenomenon for the Romanian Bronze Age, with weapon representations incised/engraved on two walls in a well-defined area at the cave entrance: over 140 daggers, spear and lance-points, and possibly halberds were identified so far and interpreted as *symbolic* depositions of weapons (Soroceanu and Sîrbu 2012).

Another interesting discovery was made at Mişidului Cave: at the end of a gallery with difficult access, on an area of ca 0.25 m^2 lying directly on the bed rock were a decorated bronze belt (found in a fragmentary state), 11 complete and one fragmented crescent-shaped bronze pendants, a fishing line made of thin bronze wire with a bronze hook at one end, a bronze pin and a bronze chain, two clay vessels, three clay spindles, two clay discs, five clay items of various sizes and shapes, 42 amber beads, one stone flake and two quartzite pebbles. Based on the bronze artifacts and the vessel types, the (allegedly ritual) deposition was attributed to the end of the Bronze Age/Early Iron Age (Chidioşan and Emödi 1981).

Caves continued to be occupied during the Early Iron Age throughout Romania. They were used for seasonal (as indicated by the hearths and pottery at Gaura Chindiei I and II in the Iron Gates region) or temporary activities (the few pottery finds at Ungurului Cave in Bihor, Climente I Cave in the Iron Gates region, Liliecilor Cave in Dobrogea), or perhaps holding ritual and symbolic roles. Deposition of various items (Cioclovina Uscată Cave, Cave with Stalactites at Pietroasa, Cave on Piatra Secuiului Mountain at Râmetea) could have had a symbolic, ritual or even practical (hiding or storage) meaning. The most substantial and impressive deposition was found in 1953 in Cioclovina cu Apă Cave: hidden among the boulders covering the floor of one of the galleries were discovered over 5000 artifacts made of bronze, glass, faience, deer antler, and amber. The bronze items comprised 68 large conical ampyx with four triangular openings each, 4 circular phalerae, 17 medium-size tutuli, 1528 small tutuli, 80 crescent-shaped pieces longitudinally perforated, 14 volutes from bronze wire, 251 saltaleoni in thin bronze wire, and 17 open rings made of bronze rods rhomboidal in section. There were also two cheek-pieces made of deer antler with incised decoration, ca. 1000 amber beads of various sizes and shapes, 500 faience beads, ca. 1500 small circular beads made of a greenish-blue glass (Comşa 1966). A small niche in one of the adjacent galleries (modified by closing one of the two openings with a wall made of boulders) had been presumably used as living area. Inside it, the few pots found were apparently characteristic of the Early Hallstatt and thus considered contemporary to the deposition.

In 1969, probably in the same gallery, a second similar discovery was made: one ampyx, a phalerae, 300 tutuli, 2 circular buttons, 10–12 crescent-shaped pieces, 4 complete saltaleoni and 2 in fragmentary state, one complete decorated cheekbone and another fragment from a second one, ca. 500 amber and glass beads and 100 faience beads (Petrescu-Dîmboviţa 1977). A complete publication of these two sets of finds (perhaps part of the same deposition) has never been attempted. The conditions of discovery remain unclear and the description of artifacts varies in different sources (compare Comşa 1966 to Petrescu-Dîmboviţa 1977).

Dacian (early La Tène period) occupation traces were known in Dobrogea (hearths, pottery, small items of domestic use, ornaments, weapons, and faunal remains) at Gura Dobrogei and Cheia caves (Harţuche 1976). Banat caves yielded poorer assemblages, comprising in most cases a few Dacian pottery sherds. Roman times added religious, military and *spa* dimensions to the use of caves: fragments of inscriptions attesting the practice of the cult of God Mithras were discovered at La Adam Cave in Dobrogea and Veterani Cave in the Iron Gates. A small garrison was probably located inside the latter, as fortification walls were built at the entrance and probably inside. At Germisara (Geoagiu Băi, Transylvania) a subterranean pool and baths were created in the local cave, taking advantage of the thermal hot springs (Boroneanţ 2000).

The archaeological record suggests a very diverse cave use during the Middle Ages. The residential and religious dimensions at times overlapped, with Christian hermits taking over the shallower caves and the rockshelters. New caves were carved in softer chalky hills to accommodate the new purpose, as it was the case of the Basarabi cave complex in Dobrogea. Dated to the tenth century, this cave complex consists of a series of man-made halls, galleries, and funerary chambers, all serving religious purposes. Occasionally, the walls display incised representations of religious and lay nature (Boroneanţ 2000). Similar representations were reported at Nucu–Bozioru, Fânaţe, and Chindiei caves (Fig. 9; Boroneanţ 2000; Soroceanu and Sîrbu 2012).

When geography was favourable, caves were fortified and used for defensive military purposes: Veterani (Fig. 10) and Gaura cu Muscă caves in the Iron Gates of the Danube were both turned into military strongholds during the Ottoman—Austrian conflicts at the end of the seventeenth century and served to survey the circulation along the Danube (Boroneanţ 1979).

From the end of the eighteenth century on, caves were the subject of the destructive actions of treasure hunters, visitors, and guano mining—as it was the case, for example, at Hoţilor Cave or Gaura cu Muscă in the Iron Gates. Cave walls were covered in graffiti (Fig. 11) or pierced by pits. Occasionally, local shepherds used the natural cavities as shelters for animals during summers. While this is part of the continuous use of caves by humans, it nevertheless affected and damaged the existing archaeological record.

Concluding Remarks

The human use of caves remained a constant feature in the archaeological record throughout Romania at least from the Upper Pleistocene on. In most cases, caves were used because they provided readymade and easy to use natural structures where a large range of activities could be performed. Some activities were economic, related to residence (longer or shorter), subsistence, acquisition of raw materials, storage, and waste disposal. A second aspect of maximum

Fig. 9 Medieval cave-wall painting at Chindiei Cave in the Iron Gates (photo from the archive of V. Boroneanţ)

Fig. 10 Austrian fortification wall built at the entrance of Veterani Cave in the Iron Gates (photograph by A. Boroneanţ)

interest is the ritual one: caves were used either as theaters for rituals (presence of cave art and/or votive deposits) or as burial places (Tolan-Smith and Bonsall 1997). Most interesting, many caves show records of both economic and ritual activities, whether synchronous (as it might have been the case during the Early prehistory), or with the ritual replacing the economic (e.g., Middle Ages). Location, shape and size of the caves influenced both aspects. Residential activities occur mostly toward the aired and lit part, and often in easily accessible caves. Ritual activities seemed to have required the most remote caves and/or the most difficult to access parts of the inhabited caves.

The occupation of Romanian caves clearly took place within larger environmental, socio-economic, and cultural contexts. Cave settlements have always acted as geographic, economic, or symbolic extensions of various cultural and adaptive systems essentially based on open-air settlement networks. This explains, for instance, the shift in intensity of cave use between Middle and Upper Palaeolithic, suggesting different mobility patterns in relation to the resources available in the mountain areas. A similar explanation holds for the unprecedented and consistent use of caves and rockshelters by the Epipalaeolithic/Mesolithic hunter-gatherers and Early Neolithic farmers (only) around the Iron Gates, the decreased Neolithic presence or the later, Bronze Age revival of cave use for both residential and ritual purposes etc. An accurate understanding of these diachronic changes requires, however, a continuous archaeological effort in both cave and open-air settings, involving high-resolution palaeoclimatic reconstructions, robust absolute chronologies and modern excavation techniques.

Fig. 11 Nineteenth-century graffiti inside Hoţilor Cave at Băile Herculane (photograph by A. Boroneanţ)

Acknowledgements Two anonymous reviewers helped in improving the content and style of our contribution. E. Trinkaus generously offered full access to his personal photo archive on Romanian palaeoanthropological finds. A. D. Popescu kindly offered support and advice on sourcing cave discoveries for the later Prehistory. We are grateful to all. All remaining errors or omission are of course our sole responsibility.

References

Anghelinu M (2003) Evoluţia gândirii teoretice în arheologia din România. Concepte şi modele aplicate în preistorie, Cetatea de Scaun, Târgovişte

Anghelinu M, Niţă L, Sitlivy V, Uthmeier T, Băltean I (2012a) Looking around Peştera cu Oase: the beginnings of Upper Palaeolithic in Romania. Quat Int 274:136–157

Anghelinu M, Niţă L, Steguweit L (2012b) Not that original after all: the chrono-cultural framework of the Upper Palaeolithic on the Bistriţa Valley (North-Eastern Romania). Arh Mold XXXV:7–46

Bonsall C, Boroneanţ A, Evatt A, Soficaru A, Nica C, Bartosiewicz L, Cook GT, Higham TFG, Pickard C (2016) The 'Clisurean' finds from Climente II cave, Iron Gates, Romania. Quat Int 423:303–314

Bonsall C, Boroneanţ A, Soficaru A, McSweeney K, Higham T, Miriţoiu N, Pickard C, Cook G (2012) Interrelationship of age and diet in Romania's oldest human burial. Naturwiss 99:321–325. https://doi.org/10.1007/s00114-012-0897-1

Boroneanţ A (2011) The Mesolithic of Banat. In: Drasovean F, Jovanovic B (eds) The Prehistory of Banat, I. The Palaeolithic and Mesolithic. Academiei, Bucharest, pp 103–142

Boroneanţ A (2012) Aspecte ale tranziţiei de la mezolitic la neoliticul timpuriu în zona Porţile de Fier. Ed Mega, Cluj-Napoca

Boroneanţ V (1970) Un mormânt din perioada de trecere de la paleoliticul superior la epipaleolitic. SCIVA 21(1):129–132

Boroneanţ V (1979) Descoperiri arheologice în unele peşteri din defileul Dunării. In: Orghidan T, Negrea Ş (eds) Speologia. Ed Acad RSR, Bucureşti, pp 140–185

Boroneanţ V (2000) Arheologia peşterilor şi minelor din România. Ed cIMEC, Bucharest

Cârciumaru M (1999) Le Paléolithique en Roumanie. J Millon, Grenoble

Cârciumaru M (2000) Boroşteni Peştera Cioarei. Paleomediul, cronologia şi activităţile umane în Paleolitic. Ed Macarie, Târgovişte

Cârciumaru M, Bitiri M (1980) Cele mai vechi picturi rupestre paleolitice din România. Rev Muz Monum 1:3–10

Chidioşan N, Emödi I (1981) O descoperire de la sfârşitul epocii bronzului şi începutul Hallstattului în peştera Mişidului, com. Şuncuiuş, jud. Bihor. Thraco-Dacica 2:161–167

Clottes J, Besesek M, Gély B, Ghemiş C, Kenesz M, Lascu VT, Meissonier M, Philippe M, Plichon V, Prud'homme F, Radu VA, Rus T, Tociu RL (2012) Découverte d'une nouvelle grotte ornée paléolithique en Roumanie, département du Bihor. In: Clottes J (ed), L'art pléistocène dans le monde, Actes du Congrès IFRAO, Tarascon-sur-Ariège, septembre 2010, N° spécial de Préhistoire, Art et Sociétés. Bull Soc Préhist Ariège-Pyrénées LXV–LXVI:511–528

Comşa E (1966) Le dépôt en bronze de Cioclovina (Carpates Méridionales). Acta Arch Carp 8(1–2):169–174

Conard NJ, Bolus M (2006) The Swabian Aurignacian and its place in European prehistory. In: Bar-Yosef O, Zilhão J (eds) Towards a definition of the Aurignacian, Trabalhos de Arqueologia 45. Instituto Portugues de Arqueologia/American School of Prehistoric Research, Lisbon, pp 211–239

Cosac M, Murătoreanu G, Veres D, Niţă L, Schmidt C, Hambach U, Radu A, Cuculici R, Buzea DL, Mărgărit M, Dumitraşcu V, Vasile Ş, Petculescu A, Dénes I (2018) Multi-proxy archaeological investigations of a Middle Palaeolithic occupation context in Eastern Transylvania, Romania. Quat Int 485:115–130. https://doi.org/10.1016/j.quaint.2017.09.014

Doboş A (2008) The Lower Palaeolithic of Romania: a critical review. PaleoAnthr 2008:218–233

Doboş, A (2017) The Middle Palaeolithic research in Romania. Past and Current Issues, MCA (SN) XIII:5–14

Doboş A, Soficaru A, Trinkaus E (2010) The prehistory and paleontology of the Peştera Muierii (Romania). ERAUL, Liège

Feurdean A, Perşoiu A, Tanţău I, Stevens T, Magyari EK, Onac BP, Marković S, Andrič M, Connor S, Fărcaş S, Gałka M, Gaudeny T, Hoek W, Kolaczek P, Kuneš P, Lamentowicz M, Marinova E, Michczyńska Perşoiu I, Płóciennik M, Słowiński M, Stancikaite M, Sümegi P, Svensson A, Tămaş T, Timar A, Tonkov S, Toth M, Veski Z, Willis KJ, Zernitskaya V (2015) Climate variability and associated vegetation response throughout Central and Eastern Europe (CEE) between 60 and 8 ka. Quat Sci Rev 106:206–222

Fu Q, Hajdinjak M, Moldovan OT, Constantin S, Mallick S, Skoglund P, Patterson N, Rohland N, Lazaridis I, Nickel B, Viola B, Prüfer K, Meyer M, Kelso J, Reich D, Pääbo S (2015a) An early modern human from Romania with a recent Neanderthal ancestor. Nature 524:216–219. https://doi.org/10.1038/nature14558

Fu Q, Posth C, Hajdinjak M, Petr M, Mallick S, Fernandes D, Fürtwangler A, Haak W, Meyer M, Mittnik A, Nickel B, Peltzer A, Rohland N, Slon V, Talamo et al (2015b) The genetic history of Ice Age Europe. Nature 534:200–205. https://doi.org/10.1038/nature17993

Harţuche N (1976) Unele probleme ale postpaleoliticului în lumina săpăturilor din peşterile Dobrogei. Pontica 9:13–21

Jungbert B (1978) Repertoriul localităţilor cu descoperiri paleolitice din Transilvania (I), AMN XV:1–19

Jungbert B (1979) Repertoriul localităţilor cu descoperiri paleolitice din Transilvania (II), AMN XVI:389–411

Jungbert B (1982) Repertoriul localităţilor cu descoperiri paleolitice din Transilvania (III), AMN XIX:543–557

Laming-Emperaire A (1964) Origines de l'archéologie préhistorique en France: Des superstitions médiévales à la découverte de l'homme fossile. Picard, Paris

Luca SA, Roman C, Diaconescu D (2005) Cercetări arheologice în peştera Cauce (I) (sat Cerişor, comuna Lelese, judeţul Hunedoara). Ed Economică, Bucureşti

Mogoşanu (1978) Paleoliticul din Banat. Ed Acad RSR, Bucureşti

Moreau L, Odar B, Higham T, Horvat A, Pirkmajer D, Turk P (2015) Reassessing the Aurignacian of Slovenia: techno-economic behavior and direct dating of osseous projectile points. J Hum Evol 78:158–180

Moroşan NN (1938) Le Pléistocene et le Paléolithique de la Roumanie de nord-est. An Inst Geo Rom XIX:8–52

Mussi M (2002) Earliest Italy. An overview of the Italian Palaeolithic and Mesolithic. Kluwer Academic, New York

Nicolăescu-Plopşor CS (1957) Le Paléolithique dans la République Populaire Roumaine à la lumiére des derniéres recherches, Dacia (N.S.) I:41–107

Nicolăescu-Plopşor D (1968) Les hommes fosiles découvertes en Roumanie. In: Actes de VIIème Congr Int des Sciences Anthropologiques et Ethnologiques, Moscou (III), pp 381–382

Onac B, Viehmann I, Lundberg J, Lauritze SE, Stringer C, Popiţă V (2005) U–Th ages constraining the Neanderthal footprint at Vârtop Cave, Romania. Quat Sci Rev 24:1151–1157

Patou-Mathis M (2001) Les grands mammiferes de la grotte Cioarei (Boroşteni, Roumanie): repaire de carnivores et halte de chasse. Prehist Eur 16–17:57–63

Păunescu A (1999) Paleoliticul şi mezoliticul de pe teritoriul Dobrogei. Satya Sai, Bucureşti

Păunescu A (2000) Paleoliticul şi mezoliticul de pe teritoriul cuprins între Carpaţi şi Dunăre. Satya Sai, Bucureşti

Păunescu A (2001) Paleoliticul şi mezoliticul din spaţiul transilvan. Ed Agir, Bucureşti

Petrescu-Dâmboviţa (1977) Depozitele de bronzuri din România. Ed Acad RSR, Bucureşti

Riel-Salvatore J, Popescu G, Barton M (2008) Standing at the gates of Europe: human behavior and biogeography in the Southern Carpathians during the Late Pleistocene. J Anthr Arch 27:399–417

Roman P (1976) Cultura Coţofeni. Ed Acad RSR, Bucureşti

Rougier H, Milotă S, Rodrigo R, Gherase M, Sarcină L, Moldovan O, Zilhão J, Constantin S, Franciscus RG, Zollikofer CPE, Ponce de Léon M, Trinkaus E (2007) Peştera cu Oase 2 and the cranial morphology of early modern Europeans. PNAS 104:1165–1170

Schmidt C, Sitlivy V, Anghelinu M, Chabai V, Uthmeier T, Hauck T, Băltean I, Hilgers A, Richter J, Radtke U (2013) First chronometric dates (TL and OSL) for the Aurignacian open-air site of Româneşti-Dumbrăviţa I, Romania. J Arch Sci 40:3740–3753

Sîrbu V, Matei S (2012) Un monument din Carpaţii Orientali cu reprezentări din preistorie şi evul mediu—Nucu—"Fundu Peşterii", judeţul Buzău. Istros, Brăila-Buzău

Sitlivy V, Chabai V, Anghelinu M, Uthmeier T, Kels H, Niţă L, Băltean I, Vesselsky A, Ţuţu C (2014) Preliminary reassessment of the Aurignacian in Banat (South-western Romania). Quat Int 351:193–212

Soficaru A, Doboş A, Trinkaus E (2006) Early modern humans from the Peştera Muierii, Baia de Fier, Romania. PNAS 103:17196–17201

Soficaru A, Petrea C, Doboş A, Trinkaus E (2007) The human cranium from the Peştera Cioclovina Uscată, Romania: context, age, taphonomy, morphology and paleopathology. Curr Anthr 48:611–619

Soroceanu T, Sîrbu V (2012) Peştera de la Nucu din neolitic până în epoca bronzului. In: Sîrbu V, Matei S (eds) Un monument din Carpaţii Orientali cu reprezentări din preistorie şi evul mediu—Nucu—"Fundu Peşterii", judeţul Buzău. Istros, Brăila-Buzău, pp 119–136

Szmonieswski BS, Petcu R (2008) Preliminary report from the new excavation in Baba cave, Grădina village, Constanţa county, Pontica 41:35–47

Tolan-Smith C, Bonsall C (1997) The human use of caves. In: Tolan-Smith C, Bonsall C (eds) The human use of caves. BAR-IS 667, Oxford, pp 316–319

Trigger BG (1989) A history or archaeological thought. Cambridge University Press, Cambridge

Trinkaus E (2007) European early modern humans and the fate of the Neandertals. PNAS 104:7367–7372. https://doi.org/10.1073/pnas.0702214104

Trinkaus E, Milotă S, Rodrigo R, Gherase M, Moldovan O (2003) Early modern human cranial remains from the Peştera cu Oase, Romania. J Hum Ev 45:245–253

Trinkaus E, Zilhão J, Rougier H, Rodrigo R, Milotă Ş, Gherase M, Sarcină L, Moldovan O, Băltean IC, Codrea V, Bailey SE, Franciscus RG, Ponce de Léon M, Zollikofer CPE (2006) The Peştera cu Oase and early modern humans in Southeastern Europe. In: Conard NJ (ed) When Neanderthals and modern humans met. Kerns, Tübingen, pp 145–164

Tuffreau A, Boroneanţ V, Goval E Lefevre B, Boroneanţ A, Doboş A, Popescu G (2007) Le gisement Paléolithique Moyen de Zăbrani (départment D'Arad), MCA (SN) III:5–18

Tuffreau A, Dobrescu R, Petculescu A, Ştiucă E, Balescu S, Lanoë F, Wismer M (2013) La grotte La Adam: un repaire de carnivores visité par les chasseurs du Paléolithique, In: Saint Martin JP (ed) Recherches croisées en Dobrogea. Ed Amanda, Bucarest, pp 75–86

Veres D, Cosac M, Schmidt C, Murătoreanu G, Hambach U, Hubay K, Wulf S, Karátson D (2018) New chronological constraints for Middle Palaeolithic (MIS 6/5–3) cave sequences in Eastern Transylvania, Romania. Quat Int 485:103–114. https://doi.org/10.1016/j.quaint.2017.07.015

Voinea V, Dobrinescu C (2002–2003) Aşezarea Hamangia III de la Cheia. Campaniile 2001–2002 (Raport preliminar), Pontica 35–36:9–22

Webb D, Robu M, Moldovan O, Constantin S, Tomus B, Neag I (2014) Ancient human footprints in Ciur-Izbuc Cave, Romania. Am J Phys Anthr 155:128–135

Zsolt M, Ghemiş C (2003) O descoperire aparţinând bronzului timpuriu din peştera Izbucu Topliţei (jud. Bihor). Consideraţii privind grupul Roşia. Eph Napoc 13:41–92

Show Caves of Romania

Ioana N. Meleg, Marius Robu, Daniela R. Borda, Călin Ghemiş, Ludovic Mátyási, and Viorel T. Lascu

Abstract

From prehistoric times, caves have been used as shelters and places for cultural, religious, and ceremonial manifestation. There is no ancient culture that does not mention in its history at least once "the afterlife," "the other world," or "the underground world." Starting with the eighteenth century, cave exploration and research began to reveal the subterranean beauty in all its elements: geology, biology, paleontology, archeology, and hydrology. In the last century, people started to use the caves for touristic purposes. Over the past decades, the concept of sustainable use of natural and cultural heritage has been introduced in show caves management to protect and preserve their heritage for future generations. In Romania, more than 12,500 caves have been discovered and explored. The first cave that opened its gates for tourism was Meziad Cave in 1903. Today, Romania has seventeen show caves, all included into a national protected area network. Apuseni Mountains and South Carpathians host most of the show caves of Romania, seven and eight, respectively. In these caves, the visitor has the opportunity to explore the ancient history of Europe and the today's beautiful subterranean landscapes. Some show caves (i.e., Muierii, Ungurului, Meziad) are emblematic for the distant past, with traces of human activities or skeletal remains ranging in age from the Paleolithic to the Neolithic. Scărişoara Ice Cave allows the visitors to discover one of the world's largest and oldest ice blocks, whereas Urşilor Cave is known for its important fossil assemblages, among which a cave bear skeleton in anatomic connection that is ca. 40,000 years old. All show caves of Romania host peculiar fauna and unique landscapes, and discovering their natural and cultural heritage while enjoying the beauty of the natural protected areas that host them is an experience worth taking.

Keywords

Show caves • Cultural trails • Heritage diversity
Natural protected areas • Network • Sustainable use

Introduction

In Romania, more than 12,500 caves are cataloged (Romanian Cave Catalog; see Chap. 5; Vlaicu pers. comm.) of which 132 represent natural protected areas of community interest (Natura 2000 network), and seventeen caves are developed for tourism (Fig. 1). These caves are located in the Romanian section of the Carpathians and, except for Şugău Cave (East Carpathians), which is formed in Paleozoic metamorphosed limestone, all the other ones develop in Mesozoic carbonate rocks.

The Romanian caves are well decorated and host unique minerals, rich fossil assemblages, artifacts, etc. They are also home for a diverse fauna that is a mixture of surface (epigeans) and cave-dwelling organisms (hypogeans). The

I. N. Meleg (✉) · M. Robu
"Emil Racoviţă" Institute of Speleology, Romanian Academy, Calea 13 Septembrie 13, 050711 Bucureşti 5, Romania
e-mail: ioana.meleg@iser.ro

M. Robu
e-mail: marius.robu@iser.ro

D. R. Borda
"Emil Racoviţă" Institute of Speleology, Romanian Academy, Clinicilor 5, 400006 Cluj-Napoca, Romania
e-mail: daniela.borda@academia-cj.ro

C. Ghemiş
Muzeul Ţării Crişurilor, Dacia Blvd. 1-3, 410464 Oradea, Romania
e-mail: ereshu@yahoo.com

L. Mátyási
Geo Prospect, 13 Septembrie 30/1, 415600 Ştei, Romania
e-mail: geoprospect@geoprospect.ro

V. T. Lascu
Centrul pentru Arii Protejate şi Dezvoltare Durabilă Bihor, Piaţa 1 Decembrie 4-6, Suite 8, 410068 Oradea, Romania
e-mail: violascu@spelemat.ro

G. M. L. Ponta and B. P. Onac (eds.), *Cave and Karst Systems of Romania*,
Cave and Karst Systems of the World, https://doi.org/10.1007/978-3-319-90747-5_58

Fig. 1 Location of Romanian's show caves; 1 Vadu Crişului, 2 Ungurului, 3 Peştera cu Cristale din Mina Farcu, 4 Meziad, 5 Urşilor, 6 Poarta lui Ionele, 7 Gheţarul de la Scărişoara, 8 Şugău, 9 Ialomiţei, 10 Valea Cetăţii 11 Dâmbovicioara, 12 Liliecilor de la Mănăstirea Bistriţa, 13 Polovragi, 14 Muierii, 15 Bolii, 16 Peştera de la Mănăstirea Tismana, 17 Veterani

epigeans species are either visiting the caves, using them as temporary shelter, or are trapped inside the caves. Most of the hypogean species are endemic to a cave, sometimes only to a specific site within a given cave. The bats are the most known visitors of caves, and all bat species inhabiting Romanian caves are protected by law.

Since the Paleolithic, a number of caves were used as shelter by humans and large mammals. Traces of humans and human activity (i.e., the Neanderthal footprint of Vârtop Cave dated at ~62,000 years; the remains and traces of *Homo sapiens* dated between ca. 35,000 and 30,000 years in Oase, Ciur Izbuc, Muierii, and Cioclovina caves; the rock art of Coliboaia Cave, one of the oldest in Europe dated at ~32,000 years, etc.) (Boroneanţ 2000; Trinkaus et al. 2003; Onac et al. 2005; Soficaru et al. 2007; Alexandrescu et al. 2010; Clottes et al. 2012; Webb et al. 2014) and bones of large mammals that went extinct around 25,000 years ago (Robu 2015, 2016) can be seen as duplications or authentic artifacts in some of the show caves or in adjacent museums.

In 2009, the Romanian Show Cave Association (http://www.pesteri-turistice.ro/) was established and became member of the International Show Caves Association. The goal of this organization is to promote cave tourism in Romania through a network of show caves, in connection with other touristic attractions within each area.

Apuseni Mountains

Pădurea Craiului Mountains

Vadu Crişului Cave

The cave is located on the left bank of the Crişul Repede River in the homonymous gorge, a natural reserve located between Oradea and Cluj-Napoca (Fig. 1; 1). The cave entrance opens at the base of a Cretaceous massive limestone cliff; between the cave entrance and Crişul Repede River the underground stream formed a large tufa deposit, including a spectacular, 15 m high waterfall. The cave can be reached by car up to Şuncuius or Vadu Crişului villages, from where only by foot thorough the gorge (~2.5 km from both directions). From Peştera railway station is a 5 min walk to the cave crossing an arched bridge over the river.

The cave was discovered in 1903, and by 1905, it has already been developed for tourism. Between 1963 and 1976, it was electrified with incandescent bulbs and represented the first cave developed for tourism in Romania based on conservation ethics. In 2008 cave's administration modernized and refurbished the show path, introducing "cold" bulbs and in 2012 LED lightning. The cave is administrated by the Ţării Crişurilor Museum (Oradea). Out of more than 3 km explored, only 630 m can be visited.

Vadu Crişului Cave (Fig. 2) is sub-horizontal and developed along a W–E-oriented fault. It has a total length of ca. 3000 m. The cave is hydrologically active, with alluvial sediments (pebbles, sand, and clay) present along the banks. It has some large flowstones of various morphologies, situated mainly in the first half of the cave [e.g., Canopies Corridor (Coridorul Baldachinelor), Big Chamber (Sala Mare), and Balcony Chamber (Sala Balconului)]. The main passage has three sumps, the last one located at the upstream end of the cave; this one has not been dived so far. Next to the last sump is a fossil (dry) passage with large fragments of collapsed limestone [Collapsed Blocks Corridor (Coridorul Blocurilor Prăbuşite)].

Two bat species, greater horseshoe (*Rhinolophus ferrumequinum*) and lesser horseshoe (*Rhinolophus hipposideros*), live in the cave all year long. Greater horseshoe species form a colony of ~100 bats during hibernation along the main cave corridor, in lateral galleries, or above the cave stream (visible from the touristic path).

Vadu Crişului Cave is home of 62 terrestrial and aquatic invertebrate species, belonging to 17 taxa, with 16 endemics (Fig. 3). Some of the hypogean invertebrate endemic species of Vadu Crişului Cave are: *Niphargus stygocharis* (Amphipoda), *Microcerberus plesai* (Isopoda), *Bathynella vaducrisensis* (Bathynellacea), *Kovalevskiella phreaticola* (Ostracoda), *Duvalius (Biharotrechus) redtenbacheri* (Coleoptera), *Paladilhiopsis transsylvanica* (Gastropoda) (Iepure and Moldovan 2002; Moldovan 2002). Bone remains belonging to the Upper Pleistocene cave bear (*Ursus spelaeus*) were found in this cave as revealed by studies of Czier and his collaborators (Czier and Gáspár 1993; Czier et al. 1994).

Archeological investigations carried out in Vadu Crişului Cave unveiled ceramic fragments at the entrance zone and a Copper Age female burial (beads necklace, copper armoring, and ceramic fragments) (Roska 1942). Vadu Crişului village is unique for its specific white pottery and pottery workshops. The white clay, modeled since prehistoric times, is dated as Vadu Crişului craft since 1693. It combines symbols from Early Bronze Age and Medieval decorative elements, all inspiring through their naturalness, finesse, and uniqueness.

Ungurului Cave

The large entrance of the Ungurului Cave opens in the Crişul Repede Gorge, ~2.5 km upstream Şuncuiuş village (Fig. 1; 2), where Crişul Repede is meandering toward south at its confluence with Mişid Valley. To access the cave, you

Fig. 2 Along Vadu Crişului Cave's show path (photograph courtesy of A. Posmoşanu)

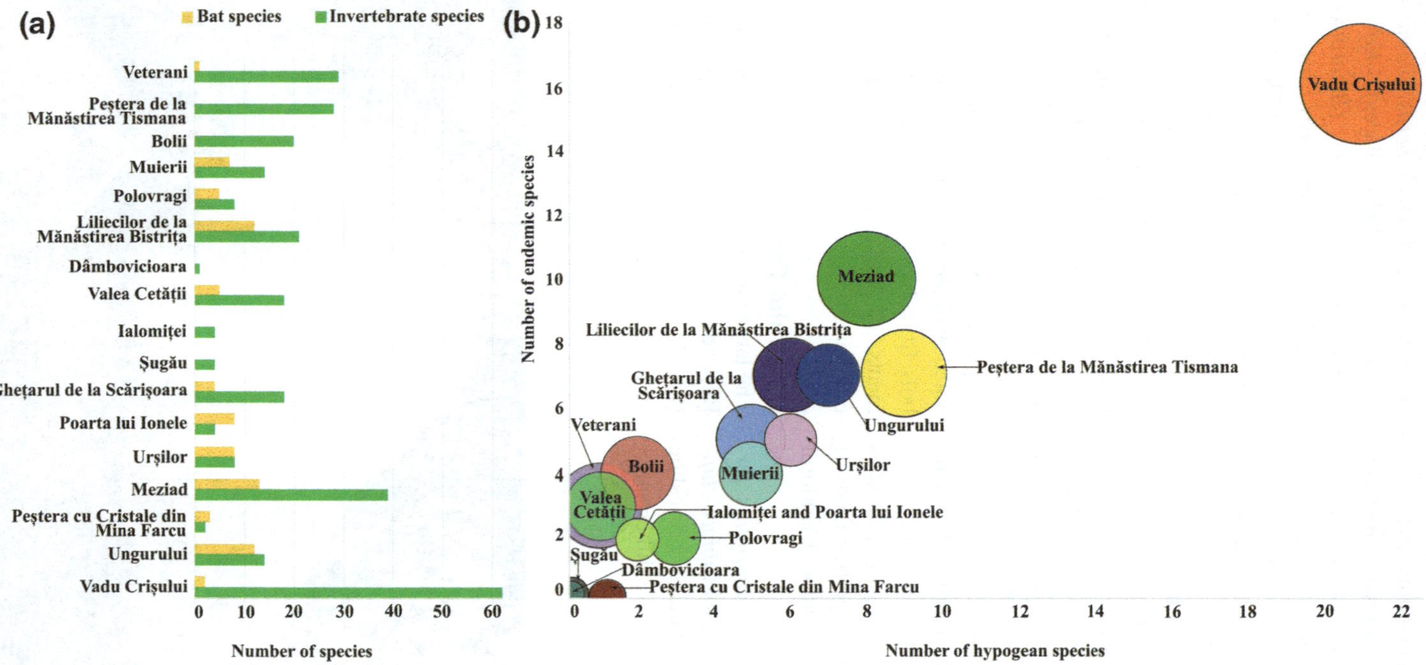

Fig. 3 Biodiversity of show caves of Romania. **a** Bat and invertebrate species richness in show caves. **b** Endemic and hypogean invertebrate species in show caves. The size of the circle corresponds to invertebrate species number indicated in plot (**a**)

Fig. 4 View of Ungurului Cave's entrance arch seen from the right bank of Crişul Repede River (photograph courtesy of A. Posmoşanu)

have to pass by a 200-year-old reconstructed and relocated wood house (built from a single oak) and to cross a suspended cable bridge that is 83 m long and sits 6 m above the river (Fig. 4). The cave is included in the Natural Reserve of the Crişul Repede Gorge.

Out of more than of ca. 1500 m of cvasi-horizontal explored passages that include a 195 m long temporary stream, 150 m have been developed for tourism in 2001. The show cave is administrated by the Şuncuiuş City Hall. The cave is attractive due to its impressive entrance (32 m high and 22 m width) and for the large main horizontal passage: 15–25 m in high and 10–33 m in width. In the rear part of the cave (upstream), the landscape is dominated by collapsed limestone fragments and steep slopes. The sediments are mainly alluvial. The bed of the underground stream along which the touristic path meanders hosts a milky yellowish white, sometimes light brown gelatinous deposit that is mainly composed of an amorphous mineral called *crisite* (Ghergari and Onac 1993). During heavy rainfalls, part of this mineral deposit is washed out of the cave.

The cave hosts twelve species of bats (Fig. 3): Greater horseshoe bat and Schreibers' bat (*Miniopterus schreibersii*) are present in both nurseries and hibernation colonies. The noctule (*Nyctalus noctula*), greater mouse-eared bat (*Myotis myotis*), lesser mouse-eared bat (*Myotis blythii*), barbastelle (*Barbastella barbastellus*), and long-eared bat (*Plecotus auritus*) use this cave mainly for hibernation. Lesser horseshoe bat, common pipistrelle (*Pipistrellus pipistrellus*), Geoffroy's bat (*Myotis emarginatus*), Daubenton's bat (*Myotis daubentonii*) and Bechstein's bat (*Myotis bechsteinii*) are occasional visitors of the cave (Barti 1998; Borda 2002). Excepting noctule and few solitaire horseshoe bats, the other bats gather in the non-touristic parts of the cave, choosing either to hang from the ceiling or to hide in fissures.

Invertebrate species have representatives from seven taxa, with seven endemic hypogean species (Fig. 3). The representative endemic hypogean species of Ungurului Cave are: *Troglohyphantes racovitzai* (Araneae), *Niphargus* sp. (Amphipoda), *Acanthocyclops transylvanicus,* and *Spelaeocamptus spelaeus* (Copepoda), *Duvalius (Biharotrechus) redtenbacheri* (Coleoptera) (Moldovan 2002; Meleg 2013).

Ungurului Cave shelters under its entrance arch, one of the largest prehistoric deposit recovered from a

Fig. 5 Under the translucent crystalictites of Crystal Cave (photograph courtesy of A. Posmoşanu)

Transylvanian cave. It includes artifacts belonging to Middle Neolithic (stone workshop, ceramic remains, lithic and bone tools, a fossil shell pendant, and a miniature boat) and Late Bronze (around 150 bronze pieces: weapons, jewelry, pottery, and an amber bead of Baltic origin) (Ghemiş 2003, 2009). The cave area is a favorite place for camping, photography, climbing, rafting, and hiking.

Peştera cu Cristale din Mina Farcu (Farcu's Mine Crystal Cave)

Farcu's Mine Crystal Cave, hereafter called Crystal Cave, one of the nine cavities intercepted by mining activities related to bauxite exploitation is located in the Farcu Hill, in the south-central part of Pădurea Craiului Mountains (Fig. 1; 3). This is part of a natural protected area situated in the upper basin of Roşia Valley (Bihor County). Farcu area is a karstic plateau, with rolling hills, large dolines, and karren fields, bordered by Şteazelor Valley to the north (which hosts the largest karstic spring in Pădurea Craiului Mountains, Roşia Spring) and Lazuri Valley to the south.

Discovered in 1987, the cave has been designated a natural monument in 1994 and has been developed as a show cave in 2012–2013, by its administrators the Centre for Protected Areas and Sustainable Development Bihor and the Village of Roşia. Access to the cave is through the bauxite mining gallery, where a museum can be visited, which preserves in situ the original mine with equipment, machines, galleries, pillars, and the former storage room for explosives. The mining gallery provides a unique visit into the geologic history of the region since the bauxite deposits are filling the paleokarst features (dolines, uvalas, etc.) formed on the Upper Jurassic limestones, which were then covered by Lower Cretaceous rocks.

The natural cavity parallels the main mine gallery oriented east–west and develops on an upper level and side passages interconnected through pits. Out of its total length of 251 m, about 100 m of galleries are open for tourists, and the rest of the cave remains under protection and preservation. Crystal Cave distinguishes itself through the diversity of calcite crusts and crystals, stalagmites covered by coralloids, large calcite monocrystals, alignments of flower-like crystalline aggregates, and up to 45 cm long white or translucent crystalictites (Fig. 5).

Three bat species hibernate in the cave: the greater horseshoe bat that use a social thermoregulation in colony, the lesser horseshoe bat hanging free, and the long-eared bat, usually hidden in rock cracks. In the mine galleries, one species of hypogean beetle belonging to *Duvalius* genus (Fig. 3) have been encountered.

Meziad Cave

Meziad Cave develops in the limestone block called Cave's Hillock (Gruiul Peşterii) (Meziad Village, Bihor County), an area included in the Natura 2000 site Crişului Repede Gorge–Pădurea Craiului (Fig. 1; 4). The cave can be accessed from the European road Oradea-Beiuş or Cluj-Napoca-Oradea through Remetea commune and Meziad village until the parking located 200 m from the cave entrance.

The cave is 6292 m long from which 1100 m are developed for tourism following an eight-shaped show path with ten ladders (365 steps) and three bridges. It has been opened for public for the first time in 1903 and a restoration of the trail took place in 2012. Since 2014, a visitor centre has been built for research and education purposes. Meziad Cave has large chambers traversed by natural suspended bridges, 15–20 m above the underground river from where the cave levels and arched walls can be observed. Wind-controlled speleothems (especially massive stalactites and stalagmites) are frequent (Rusu et al. 1981). They are often covered by newly precipitated white calcite crystals.

The cave is renown in Romania for its large accumulation of cave bear bone buried in a clay matrix, located in the upper level Bone's Passage (Galeria Oaselor). Along with the *Ursus spelaeus* remains (the main part of the paleontological deposit) originally discovered in the 1960s, fossil bones of carnivores, cervids, birds, and rodents have also been identified. The estimated age of the bones is considered to be Upper Pleistocene (Robu 2015).

Among all show caves, Meziad Cave shelters the highest bat diversity, with thirteen species (Fig. 3): Schreibers's bat, greater horseshoe bat, a medium horseshoe bat (*Rhinolophus euryale/blasii*), lesser horseshoe bat, greater mouse-eared bat, and lesser mouse-eared bat inhabit the cave all year long, part of them gathering in big colonies. Geoffroy's bat, whiskered bat (*Myotis mystacinus*), barbastelle, long-eared bat, noctule, common pipistrelle, and soprano pipistrelle (*Pipistrellus pygmaeus*) can be observed during the hibernation season (Dumitrescu et al. 1963; Borda 2002; Coroiu et al. 2007). The invertebrates are also represented by diverse communities. Among the 39 encountered species, the representative hypogean endemic species are: *Carpathonesticus biroi* (Araneae), *Niphargus bihorensis* (Amphipoda), *Bathynella chappuisi* (Bathynellacea), *Spelaeocamptus spelaeus* (Harpacticoida), *Onychiurus meziadicus* (Collembola), *Duvalius (Duvaliotes) redtenbacheri bihorensis* (Coleoptera) (Meleg 2015).

Archeological research revealed important discoveries about past human activity in the cave. These findings are made available to visitors through posters, and permanent or temporary exhibitions displayed under the cave porch or along the show path, and consist of pottery vessel of Copper Age–Coţofeni Culture, Early Bronze and Late Bronze Age, Dacian Period–La Tene, and Early Middle Age. The local authority and the NGO Center for the Protected Areas and Sustainable Development (CAPDD; Bihor County) jointly administer the cave.

Bihor Mountains

Peştera Urşilor (Bears Cave)

Bears Cave hereafter called Urşilor Cave is situated in the western part of the Bihor Mountains, within the Apuseni Mountains Natural Park, on the left side of the Crăiasa Valley, at 491 m asl (Fig. 1; 5). The cave was accidentally discovered in 1975 by blasting in a local marble quarry. The cave develops in slightly recrystallized Upper Jurassic (Tithonian) limestones of the Bihor Autochthonous unit (Vălenaş 1979). It has two distinct levels, separated by a 17 m drop and totals ~1500 m of large horizontal and sub-horizontal, extremely well-decorated passages. The upper level, hydrologically inactive, includes the Access Passage (Galeria de Acces), the Twisted Passage (Galeria Răsucită), the Passage of the Pits (Galeria Puţurilor), the Candlesticks Gallery (Galeria Lumânărilor), the "Emil Racoviţă" Gallery (Galeria "Emil Racoviţă"), and a large chamber (ca. 10 × 15 × 25 m) at the intersection of the last two galleries. The length of this upper level is ca. 1000 m, most of it being visited along a modern touristic path. The two artificial entrances correspond broadly with two former entrances that were naturally sealed before the discovery. In the farthest part of the "Emil Racoviţă" Passage, a smaller, ascending passage leads to a breakdown cone formed by blocks of gray wakes and flowstone fragments embedded in a thin clay matrix. This passage ends with a 17 m drop (the Shaft) that leads to the Scientific Reserve (Rezervaţia Ştiinţifică). This represents the lower level of the cave (ca. 500 m in length) and includes a relatively large horizontal gallery (10 × 15 m), with several small side passages and fossil meanders. A small stream (discharge of 1–10 L/s) flows through the lower level and sinks underneath the 17-m drop. Along the lower passage, the stream has deposited alluvial sediments consisting of fine sands, silts, and clays, on which often speleothems (flowstones, stalagmites, etc.) are formed (Constantin et al. 2014; Robu 2015).

The importance of the cave is due to its rich *Ursus spelaeus* bones deposits, which contains numerous and diverse cave bear fossils and ichnological traces (scratches, paw and claw marks, hair imprints, and cave bear beds; Diedrich 2011) (Fig. 6). Also, a rich Würmian–Holocene fauna found on the cave floor along the Access Gallery was described by Terzea (1978) and Jurcsák et al. (1980–1981). This contain: *Chiroptera, Talpa europaea, Glis glis, Apodemus sylvaticus, Clethrionomys glareolus, Arvicola terrestris, Microtus arvalis, M. agrestis, M. oeconomus, M. nivalis, Canis lupus, Vulpes vulpes, Panthera spelaea*, and *Ursus spelaeus.*

Bats were found only as fossils remains, because after cave's discovery the man-made gates have no bat friendly openings. Eight bat species were identified in Würmian alluvial deposits: greater and lesser horseshoe bats, greater mouse-eared bat, Bechstein's bat, Schreibers's bat, whiskered bat, long-eared bat, and common pipistrelle (Terzea 1978). Eight invertebrate species were recorded in the cave, with three terrestrial endemic hypogean species (Fig. 3): *Nesticus plesai* (Araneae) (Dumitrescu 1980), *Drimeotus bihorites mihoki* and *Pholeuon leptodirum* (Coleoptera) (Moldovan et al. 2007; Racoviţă 2010).

Artifacts include bone tools and a silex disk-shaped grating tool of the Mousterian culture (Jurcsák et al. 1980–1981), and three arrowheads assigned to the Upper Paleolithic Swiderian Culture (Ignat 2012).

Peştera Poarta lui Ionele (Gate of Ionele Cave)

Located within the Apuseni Natural Park, near one of the access roads to Scărişoara Ice Cave, Gate of Ionele Cave (hereafter Poarta lui Ionele Cave) has been developed for tourism in 1988 by cavers from Polaris Blaj Caving Club; now, it is administrated by the Gârda de Sus City Hall. The cave is easily accessible from Gârda de Sus (Fig. 1; 6), hiking 500 m upstream Gârda Seacă River until its confluence with Ordâncuşa, then following for another kilometer this tributary on the road that ultimately reaches the Scărişoara Ice Cave. Poarta lui Ionele Cave is one of the tourist attractions along the hiking trail that includes the Zgurăşti Cave with its famous 40 m diameter and 54 m deep entrance shaft, a natural bridge (across Zgurăşti), and Ordâncuşii Gorges.

Poarta lui Ionele Cave is a single passage cave, 390 m in length, traversed by a river on its first half. The cave is known for its large gothic-like entrance (22 m high) and for the travertine waterfalls situated between the cave entrance and Ordâncuşa River (Fig. 7). The first 150 m are developed for tourism. Several calcite formations (stalagmites and stalactites) occur in the rear part of the cave. From a hydrogeological point of view, Poarta lui Ionele is the outflow of Zgurăşti-Poarta lui Ionele cave system (5210 m; Damm et al. 1999; see also Chap. 34).

Eight bat species live in the cave: greater, lesser, and a medium horseshoe bats, barbastelle, greater and lesser mouse-eared bats, long-eared bat and Schreibers's bat (Borda et al. 2010). The nurseries and winter colonies are located in the upper level of the cave, which are not developed for tourism. Tourists can admire only a nursery located in the ceiling of the cave, over 20 m high, and few solitaire bats hanging on the walls. Four invertebrate species were reported here, with two hypogean endemics (Fig. 3): *Carpathonesticus spelaeus* (Araneae) and *Acanthocyclops stygius deminutus* (Cyclopoida).

Gheţarul de la Scărişoara (Scărişoara Ice Cave)

The Scărişoara Ice Cave is one of the most visited tourist attraction of the Apuseni Natural Park and Arieş Valley. It is

Fig. 6 *Ursus spelaeus* skeleton exhibited in Urşilor Cave (photograph courtesy of A. Posmoşanu)

Fig. 7 Spectacular entrance into the Gate of Ionele Cave (photograph courtesy of A. Posmoşanu)

easily accessible by road and hiking trails from the Padiş Plateau, Arieşeni, and Gârda de Sus. Gârda de Sus City Hall administrates the cave with adjacent welcome center and museum where traditional wooden objects can be purchased (alphorn, pottery, etc.).

Scărişoara Ice Cave has a total length of 700 m and is situated on the Scărişoara–Ocoale Plateau (Fig. 1; 7), an area that conserves archetypal elements of the archaic rural architecture of Geto-Dacian origin. It has a large vertical entrance (48 m diameter, 50 m depth), a chamber [Big Hall, (Sala Mare)], and two passages [Small Reserve (Rezervaţia Mică) and Big Reserve (Rezervaţia Mare)]. The show path consists of a wooden trail of 200 m, with a balcony that overlooks the ice speleothems in the so-called "Church" ("Biserica") (Fig. 8). The cave is famous for its >120,000 m^3 ice block, one of world's largest and oldest (Hubbard 2017; Perşoiu et al. 2017). A diversity of carbonate and ice speleothems enrich the underground landscape.

Four bat species often hibernate in glacier scientific reserves: greater and lesser mouse-eared bats, whiskered bat, and Brand's bat. Among 18 invertebrate species found in the cave (Fig. 3), 14 belong to Collembola group. The hypogean endemic species of the cave are: *Carpathonesticus racovitzai* and *Troglohyphantes racovitzai* (Araneae), *Niphargus laticaudatus* (Amphipoda), and the emblematic *Pholeuon proserpinae glaciale* (Coleoptera) (Racovitza 2000).

Fig. 8 Ice stalagmites in Scărişoara's "Church" (photograph courtesy of B. P. Onac)

East Carpathian Mountains (Carpaţii Orientali)

Şugău Cave, Giurgeului Mountains

The cave is located on the southern portion of the Şipoş Mountains that are part of Giurgeului Mountains (Fig. 1; 8), being accessible from the cities of Gheorghieni and Voşlobeni. The Gyilkosto Adventure Association administrates the cave. Every year it organizes concerts in the cave and provides guidance for ecotourism activities in the Şugău Natural Reserve where the cave is located.

Şugău Cave has almost 750 m of passages, divided into three sectors: Main Passage (Galeria Principală), Northern Branch (Ramificaţia Nordică), and Southern Branches (Ramificaţiile Sudice), of which 150 m are developed for tourism since 1974. The cave passages follow the direction of two main faults oriented E–W and NW–SE. The numerous calcite and aragonite speleothems (flowstones, stalagmites, and gours) are the most interesting underground features, which is a valuable asset in an area where few other caves exist.

In the Şugău Cave, three epigean species belonging to the Collembola group (Fig. 3) (Gruia and Ilie 2000–2001) were identified.

South Carpathian Mountains (Carpaţii Meridionali)

Ialomiţei Cave, Bucegi Mountains

Ialomiţei Cave is located in Ialomiţei Gorge of the Bucegi Natural Park (Fig. 1; 9). The cave is accessible by car from Sinaia or Târgovişte or by cable car from Buşteni or Sinaia resorts. An old monastery has been built in the sixteenth century under the cave's entrance arch by Prince Mihnea. In 2015, Dâmboviţa County Council, the administrator of the cave, undertook an extensive restoration of the ~500-m-long touristic path. The cave is one of the main tourist destinations in the area year around, because of its proximity to ski resorts and many natural attractions within the Bucegi Mountains (e.g., glacial lakes, waterfalls, gorges, the Sphinx and "Old Ladies" (Babele), well-known natural megaliths "sculpted" in conglomerates).

Ialomiţei Cave is carved in Jurassic fossiliferous limestone and has 1130 m of passages, mainly E–W oriented. The cave has a high entrance (>10 m) and several large halls. The Bear's Chamber (Sala Urşilor) is the largest measuring 60 m in length, 30 m in width, and 25 m in height. The calcite speleothems (soda straw, stalactites, stalagmites, moonmilk, and anemolites (Viehmann 1960) are widely represented in this room; unfortunately, many of them were vandalized by tourists.

The investigations carried out by the Romanian biospeleologist C. N. Ionescu in the early part of the twentieth century (1911) revealed a significant Upper Pleistocene bone deposit (mainly *Ursus spelaeus* remains). Since then, the paleontological heritage has been partially lost. Among four invertebrate species recorded in the cave (Fig. 3), two of them are hypogean endemics: *Niphargus carpathicus cavernicolus* (Amphipoda) and *Duvalius (Hungarotrechus) procerus* (Coleoptera) (Bleahu et al. 1976).

Peştera Valea Cetăţii (Citadel Valley Cave), Postăvaru Mountains

Citadel Valley Cave, hereafter Valea Cetăţii Cave, is located near the city of Râşnov (Fig. 1; 10), on the access road to Poiana Braşov, a famous Romanian ski resort located in the western part of the Postăvaru Mountains. The cave hosts weekly the quartet of Braşov Opera. Favored by its proximity to Râşnov Fortress, Prahova Valley, and Rucăr-Bran Couloir (Bran Castle, also known as Dracula's Castle), Valea Cetăţii is one of Romania's most visited cave.

Valea Cetăţii Cave is 958 m long and consists of eastern (entrance I) and western (entrance 2) passages, and a large chamber (40 × 30 × 20 m) traversed by a small stream. The show path that is ca. 300 m long and the lights reveal the white and translucent speleothems that decorates the walls and the gours in the Great Chamber (Sala Mare). In this room, the carbonate flowstones and alluvial clastic sediments are well represented.

Dumitrescu and Orghidan (1958) identified several cave bear bones on the cave floor. Recently, a number of cave bear hibernation nests were also discovered in the Great Chamber (Robu pers. comm.). Five bat species were observed during hibernation: greater and lesser horseshoe bats, greater mouse-eared bats, barbastelle, and long-eared bat (Dumitrescu et al. 1963; Barti 1998). Eighteen invertebrate species (Fig. 3), mainly Collembola, were described from this cave. Worth noting is the endemic gastropod *Vitrea transsylvanica* (Grosu and Negrea 1962–1963).

Dâmbovicioara Cave, Rucăr-Bran Couloir

Located in the Dâmbovicioara National Park, the homonymous cave is easily accessible from the road crossing the Rucăr-Bran Couloir (Fig. 1; 11). Set in a picturesque area that preserves traditional architecture specific to the mountain area, the cave is visited by tourists on their way to the wilderness of Piatra Craiului and Iezer-Păpuşa Mountains. The cave is administrated by the Podul Dâmboviţei City Hall.

Dâmbovicioara Cave has 555 m of passages of which 200 m are developed for tourism, distributed roughly on a single level (NNE–WSW oriented), with average heights of 4–5 m and widths of 3–4 m. Short lateral corridors (~5 m) are located on both sides of the main cave passage. The underground landscape is deprived by spectacular speleothems, but one can notice several areas with vermiculations and moonmilk on the southern wall of the main passage. Worth noting is that Fridvaldszky (1767) first mentioned the presence of calcite as a cave mineral from this cave.

Only one hypogean isopod has been reported here: *Mesoniscus graniger*, a species with wide distribution along the Carpathians, Dinaric Mountains, and Alps (Giurginca et al. 2015).

Peştera Liliecilor de la Mănăstirea Bistriţa (Bat's Cave of Bistriţa Monastery), Căpăţânei Mountains

Bat's Cave of Bistriţa Monastery hereafter called Liliecilor Cave opens in the west side of the Bistriţa Gorge at 640 m elevation. The cave is easily accessible on road from Râmnicu Vâlcea to Târgu Jiu, via Costeşti and Bistriţa villages, until Bistriţa Monastery. Monks inhabited the cave since 1635 when Macarius and Daniel built a church in the cave that has been restored in 1916. The cave is administrated by the Bistriţa Monastery, who organizes guided tours led by nuns. The show path preserves the religious and mystical spirit of the medieval period. The cave is located near Bistriţa, in the vicinity of Bistriţa and Arnota monasteries, Bistriţei Gorge, and Buila-Vânturariţa National Park (Fig. 1; 12).

The cave consists of a quasi-horizontal 400 m long main gallery (with some small lateral passages) disposed on two fossil (dry) levels, with many calcite speleothems, clastic sediments, and guano developed in upper Jurassic limestones; ~100 m of its passages are open for tourists. The access into the cave is through a small entrance (1.6 m height, 1.8 m width) opened in upper Jurassic limestones.

Twelve bat species were recorded in the cave; some stay all year long, while others only for hibernation. Greater and lesser mouse-eared bat, Geoffroy's bat, long fingered bat (*Myotis capaccinii*), pond bat (*Myotis dasycneme*), and Schreibers's bat form both nursery and hibernating colonies. The greater and lesser horseshoe bat, parti-colored bat (*Vespertilio murinus*), common pipistrelle, long-eared bat, and barbastelle were found only in winter (Dumitrescu et al. 1963; Decu and Gheorghiu 2003). Twenty-one invertebrate species have been reported in Liliecilor Cave (Fig. 3), six of which are hypogean endemic species: *Carpathonesticus hungaricus* and *Carpathonesticus simioni* (Araneae), *Deuteraphorura cloşanicus* (Collembola), *Sophrochaeta* sp. (Coleoptera), *Lithobius decapolitus* (Chilopoda), *Trachysphaera racovitzai* (Diplopoda) (Bleahu et al. 1976; Gruia and Ilie 2000–2001).

Polovragi Cave, Căpăţânei Mountains

Polovragi Cave is located on the left side of the Olteţului Gorge, in the southern part of the Căpăţânii Mountains (Fig. 1; 13) and is part of the Natura 2000 site North of Eastern Gorj (Nordul Gorjului de Est). The access to the cave is from Râmnicu Vâlcea-Târgu Jiu road, following the signs for Polovragi commune and Polovragi Monastery, from where a walk of about 20 min continues on the gravel road that crosses the gorge. Gorj County Museum is the cave's administrator.

Polovragi Cave is a complex karst system with more than 10 km of passages developed on four levels: Three are hydrologically inactive, and the lowest one is traversed by an underground stream. Only about 800 m of passages are open for tourists. The cavity has six entrances out of which three are overflows. The features that characterize the landscape of Polovragi Cave's passages are collapsed blocks, various carbonate speleothems, and alluvial sediments; large-sized halls mark the intersection of fissure systems, and the diversified morphology of passages indicates various stages of evolution.

The mineralogical analyses carried out in Polovragi Cave identified the presence of three mineral species: hydroxylapatite, brushite, and taranakite (Diaconu et al. 2008); however, calcite is by far the most common speleothem-building cave mineral.

Five bat species hibernate in cave: greater and lesser horseshoe bat, greater and lesser mouse-eared bat, and Geoffroy's bat. Solitaire horseshoe bats can be observed by tourists in summertime (Decu and Gheorghiu 2003). Out of the eight identified invertebrate species (Fig. 3), two are hypogean endemics: *Lithobius decapolitus* (Chilopoda) and *Trachysphaera spelaea* (Diplopoda) (Bleahu et al. 1976).

Although known since the nineteenth century, archeological investigations in Polovragi Cave were conducted only between 1965 and 1967; they revealed a variety of artifacts of Neolithic and Copper Age, as well as since Dacian Kingdom. Some of the archeological material is probably related to Polovragi Fortress that has been built around 150 BP (Boroneanţ 2000).

Peştera Muierii (Women's Cave), Parâng Mountains

Women's Cave, hereafter Muierii Cave, is located near Baia de Fier, on the western side of the Galbenului Gorge (Parâng Mountains; Fig. 1; 14) is a scientific reserve and archeological site belonging to the Natura 2000 site North of Eastern Gorj. Administrated by the Local Council, the cave was developed for tourism and electrified as early as 1963, being one of Romania's most visited caves. The cave is accessible from Râmnicu Vâlcea–Târgu Jiu highway, aiming for the Baia de Fier, from where road signs guide tourists all the way to the cave.

Muierii Cave develops in Upper Jurassic limestone and consists of a series of interconnected passages disposed on four levels with a total length of ~4.5 km. The main orientation of its cave passages is parallel to the Galbenu River. The longest passage, Galeria Electrificată (Illuminated Passage) (570 m in length), along with almost 1300 m of secondary lateral galleries (e.g., Mousterian and Secondary passages), situated at the same dry (fossil) level, represent less than half of the entire cave network. Tourists enter the cave through the upstream (north-facing) entrance, follow the main passage, and exit through a lower entrance oriented toward southeast. The lower level (mostly wet and ~2500 m in length) is represented by Bears' Passage (Galeria Urşilor), Electricians' Gallery (Galeria Electricienilor), Eccentrics Passage (Pasajul Excentritelor), The Gallery with Pools (Galeria cu Bazine), and Hades Passage (Pasajul lui Hades). Muierii Cave is decorated with speleothems and contains significant bone deposits, guano, and sediments.

The mineralogical analyses carried out in Muierii Cave revealed the presence of several mineral species: hydroxylapatite, brushite, calcite, and aragonite (Diaconu et al. 2008).

The archeo-paleontological excavations started in the 1920s and more systematically in the early 1950s. They yielded an abundance of Paleolithic and more recent archeological remains and a large quantities of Pleistocene faunal remains, such are *Canis lupus*, *Vulpes vulpes*, *Panthera spelaea*, *Crocuta crocuta spelaea*, and *Ursus spelaeus* (dominant species) (Nicolăescu-Plopşor 1938, 1956, 1957; Daicoviciu et al. 1953; Gheorghiu et al. 1954; Gheorghiu and Haas 1954; Nicolăescu-Plopşor et al. 1957).

Seven bat species were recorded in the cave: greater and lesser horseshoe bat, greater and lesser mouse-eared bat, Daubenton's bat, Schreibers's bat, and noctule. Old data and huge accumulation of guano are evidence of important nurseries in cave (Dumitrescu et al. 1963). Currently, Muierii Cave is important for bats only as hibernation site (Nagy et al. 2005). Fourteen invertebrate species have been reported (Fig. 3), with four hypogean endemics: *Duvalius (Biharotrechus) voitestii* and *Sophrochaeta chappuisi* (Coleoptera), *Lithobius decapolitus* (Chilopoda), *Trachysphaera spelaea* (Diplopoda) (Bleahu et al. 1976).

The archaeological research conducted in two campaigns, 1951–1953 and 1955, revealed a complex stratigraphy that contains Mousterian (Middle Paleolithic), Aurignacian layers (Upper Paleolithic), and a post-Paleolithic stratigraphy belonging to Coţofeni Culture (Păunescu 2000). A 30,000-year-old skeletal remains belonging to early modern humans, including a skull, were found mainly in the Mousterian Gallery (Galeria Musteriană) (Soficaru et al. 2006).

Bolii Cave, Sebeş Mountains

The cave is situated 6 km north of the town of Petroşani, not far from the Dinosaur Geopark in Haţeg and the Dacian fortresses area in the Orăştie Mountains within the Grădiştea Muncelului-Cioclovina Natural Park (Fig. 1; 15). Easily accessible by road, the cave area offers the possibility of visiting the karst of Şureanu Mountains and the Retezat National Park. The cave administrator, PetroAqua Association, organizes religious concerts around Christian holidays, classical music performances, and cultural events in the cave.

Bolii Cave is a 450 m long hydrologic penetration through the Dealul Bolii (Bolii Hill) and is traversed by Jupâneasa River (Fig. 9). It is a large (both in width and height), single meandering tunnel-type passage, carved in Jurassic limestones. The entire cave can be visited along a concrete pavement and wooden bridges, which were built using medieval mining engineering techniques. The general orientation of the main passage is ENE–VSV, and it appears to follow a local tectonic fault. Two large-sized entrances exist: the upstream one is 10 m high and 20 m wide, whereas the downstream one is slightly smaller (5 m high and 18 m wide). The speleothems are rare as a consequence of the strong air advection between the two entrances corroborated with a low secondary porosity of the limestone bedrock.

Out of 20 identified invertebrate species (Fig. 3), two are hypogean endemics: *Carpathonesticus puteorum* (Araneae) and *Duvalius (Biharotrechus) budae* (Coleoptera) (Bleahu et al. 1976).

Archeological studies revealed a layer containing a large deposit of artifacts belonging to the Coţofeni Culture (Copper Age) with remains of three households with wooden pillars and copper pieces (Andriţoiu and Mariş 1989; Popa 2011).

Peştera de la Mănăstirea Tismana (Cave of Tismana Monastery), Vâlcan Mountains

The Cave of Tismana Monastery (hereafter Tismana Cave) opens nearby the famous homonymous religious settlement, 6 km north of the town of Tismana (Fig. 1; 16). The cave is accessible by road from both Târgu Jiu or Baia de Aramă following the signs for Tismana Monastery. The cave is administrated by the National Bank of Romania, Târgu Jiu Branch, and is part of the Natura 2000 site North of Western Gorj (Nordul Gorjului de Vest). The cave became famous in 1944, when the National Bank of Romania's 212 tons gold reserve and three tons of gold of the Bank of Poland were hidden in the cave until 1947. This was part of the so-called Operation Neptune, a top secret action of the National Bank, Government, Military, the Railway Network of Romania, and the Orthodox Church of Oltenia. In 2015, the National Bank of Romania has finished a project started in 2007 that created a replica of those time gold deposits and also an adjacent museum. Near the cave entrance were found old monk shelters from different periods and traces of medieval inscriptions including those of Saint Nicodemus carved into the limestone. Apparently, Saint Nicodemus lived in the cave and was buried in the Tismana Monastery (Lecca 1937).

Tismana Cave is 930 m long and is developed in Lower Cretaceous massive limestones. The general orientation of the main passage is NE–SW. The cave is almost entirely traversed by an underground stream [excepting Collapsed Chamber (Sala Prăbuşită) and several other higher places] that ends upstream in a sump (Sump no. 1; currently under exploration). The transversal profile of the main passage is narrow and high suggesting a predominately tectonic control in the formation of the cave. The speleothems are rare, and alluvial sediments are present in terraces or along the underground streambed. Several fossil bones and bioglyphs (footprints and scratch marks) belonging to cave bears (*Ursus spelaeus*) were identified in the drier and higher places of the cave.

Twenty-eight invertebrate species are represented in the cave (Fig. 3), among which six hypogean endemics: *Haplophthalmus tismanicus* (Isopoda), *Parabathynella motaşi* (Bathynellacea), *Deuteraphorura cloşanicus* (Collembola), *Tismanella chappuisi chappuisi* (Coleoptera), *Trachysphaera jonescui tismanae* (Diplopoda), *Dendrocoelum tismanae* (Platyhelminthes) (Bleahu et al. 1976; Gruia and Ilie 2000–2001; Tabacaru and Giurginca 2013).

Fig. 9 Show path above the Jupâneasa River meandering through Bolii Cave (photograph courtesy of A. Posmoşanu)

Banatului Mountains

Veterani Cave, Almăjului Mountains

The cave develops in the Ciucaru Mare Hill, on the northern bank of the Danube River, in the Iron Gates Natural Park (Fig. 1; 17). The cave is managed by the Administration of the Iron Gates Natural Park. Cave access is only possible by boat, offering a unique opportunity to experience Danube Gorge (Cazanele Dunării). The boat trip also offers the chance to see Tabula Traiana and the old Roman road (~AD 100), the bas-relief of Dacian King Decebal (carved in a 55 m high limestone cliff facing the Danube), and the ruins of cave's medieval fortifications.

Veterani Cave is 87 m long, quasi-horizontal dry cave. There are two natural entrances: the first oriented toward southeast is the largest, and it is used for access; the other one is smaller in size and inconveniently situated on a steep slope. The cave has two small passages oriented west and north and a large-sized chamber (37 × 20 × 20 m), which represents the most impressive morphologic element of it (Fig. 10). Collapsed large blocks are present in the middle part of the room. The underground environment was highly disturbed by anthropogenic activities; thus, only few speleothems exist. The cave can be entirely visited by following a designated show path.

Only one bat species, the common pipistrelle, has been reported from the cave (Negrea and Negrea 1971). Most of the 29 identified invertebrate species (Fig. 3) are epigean. Only one hypogean endemic species that once inhabited the cave pools has been mentioned: *Parabathynella stygia* (Bathynellacea) (Bleahu et al. 1976).

During seventeenth–eighteenth centuries, Austrian troops had fortified the cave under the Marshal Federico Veterani, when the first and also one of the oldest descriptions of the cave have been made (Griselini 1984). The cave sheltered Romanello–Azilian, Starcevo Cris, and Coţofeni cultures. The terminal Upper Paleolithic and Neolithic (the Danube Valley Civilization) are represented by outstanding tools (Boroneanţ 2000).

Fig. 10 Light penetrating into the Veterani Cave (photograph courtesy of A. Posmoşanu)

Conclusions

Over 600,000 visitors enter annually the show caves of Romania. These caves belong to the natural heritage of Romania as part of the national protected area network. Show caves offer more than simply a cave tour; they are located on routes with touristic and spa destinations and thus provide a good opportunity to visit national and natural parks or natural reserves of Romania. Four show caves (Crystal, Meziad, Ialomiţei, and Valea Cetăţii) are developed in a modern and sustainable way. Others belong to a regional network. Visiting the show cave network of the Apuseni Mountains (Vadu Crişului Cave, Ungurului Cave, Crystal Cave, Meziad Cave, Urşilor Cave, Poarta lui Ionele Cave, Scărişoara Ice Cave), the visitor may come into contact with archaic tradition and unique traditional cultural landscape of natural park and reserves. The show cave network of South Carpathians and Banatului Mountains (Liliecilor, Polovragi, Muierii, Tismana, and Veterani) is part of a tourism network that includes medieval monastery, an ancient culture of ceramics, wild and unique landscapes of national parks, reserves or natural monuments. Ialomiţei, Valea Cetăţii and Dâmbovicioara caves are part of the Central South Carpathians network, and their visits may also include monasteries, fortresses, and castles.

Nevertheless, of ~ 100 caves with tourism potential in Romania, only 3–4% are under exploitation. A better consolidated network of show cave administrations could lead in the future to a higher number of caves developed for tourism based on sustainable concepts. In 2016, Romanian Government established the National Agency for Natural Protected Areas (ANANP). Among its objectives, one is targeting a better development of cave management integrated with the management of national protected area network.

Acknowledgements We are very grateful to A. Posmoşanu and B. P. Onac for providing photographs from some of the show caves. We are thankful to both editors and reviewers for critical comments and thoughtful suggestions that improved the manuscript quality and clarity.

References

Alexandrescu E, Olariu A, Skog G, Stentröm K, Hellborg R (2010) Os fossiles humains des grottes Muierii et Cioclovina, Roumanie. L'Anthropologie 114:341–353

Andriţoiu I, Mariş T (1989) Locuirea de tip Coţofeni din Peştera Bolii (com. Băniţa, jud.Hunedoara). Symp Thracol Tulcea 7:233–234

Barti L (1998) Liliecii. Speomond 3:32

Bleahu M, Decu V, Negrea Ş, Pleşa C, Povară I, Viehmann I (1976) Peşteri din Romînia. Ed Ştiinţifică şi Enciclopedică, Bucureşti, p 415

Borda D (2002) Chiropterele. In: Racoviţă G, Moldovan O, Onac BP (eds) Monografia carstului din Munţii Pădurea Craiului. Ed Presa Univ Clujeană, Cluj-Napoca, pp 172–182

Borda D, Racoviţă G, Năstase-Bucur R, Ciubotărescu C (2010) Ecological reconstruction of bat cave roost in western Carpathians. In: Bonacci O (ed) Sustainability of the karst environment, IHP-VII. Series on Groundwater 2:17–24

Boroneanţ V (2000) Arheologia peşterilor şi minelor din România. Ed cIMeC, Bucureşti, p 370

Clottes J, Besesek M, Gély B, Ghemis C, Kenesz M, Lascu VT, Meyssonnier M, Philippe M, Plichon V, Prud'homme F, Radu VA, Rus T, Ţociu RL (2012) Découverte d'une nouvelle grotte ornée paléolithique en Roumanie, dans le département du Bihor. In: Clottes J (ed) L'Art Pléistocène dans le Monde. Bull Soc Préhistor Ariège 65–66:513–528

Constantin S, Robu M, Munteanu CM, Petculescu A, Vlaicu M, Mirea I, Kenesz M, Drăguşin V, Hoffmann D, Anechitei V, Timar-Gabor A, Roban RD, Panaiotu CG (2014) Reconstructing the evolution of cave systems as a key to understanding the taphonomy of fossil accumulations: the case of Urşilor Cave (Western Carpathians, Romania). Quat Int 339–340:25–40

Coroiu I, David A, Borda D (2007) Colonii de hibernare a chiropterelor Peştera Meziad (Munţii Bihor). Misc Chiropter 2:40–42

Czier Z, Gáspár T (1993) Studiul morfologic şi paleontologic al Peşterii de la Cascadă (Peştera din Vadu Crişului) (Munţii Pădurea Craiului), Partea I. An Şt Soc Ardelene Speol 1:65–78

Czier Z, Gáspár T, Ungureanu M (1994) The fossiliferous deposit with bones of *Ursus spelaeus* from the cave "Peştera de la Cascadă" (Pădurea Craiului Mountains, Romania). Abst Symp Theor Appl Karstol 12:18–19

Daicoviciu C, Ferenczi Ş, Bodor A, Nicolăescu-Plopşor CS, Gostar N, Radu D, Detiu M, Duka P (1953) Şantierul Grădiştea Muncelului. Stud Cerc Ist Veche 4:153–219

Damm P, Lascu V, Ciubotărescu C (1999) Platoul Ocoale. Speomond 4:14–18

Decu V, Gheorghiu V (2003) Investigaţii asupra liliecilor din peşteri din sud-vestul Carpaţilor Meridionali: colonii de maternitate şi hibernare, habitate, statut, măsuri de protecţie. Studiu-Proiectul Life 00 NAT/RO/7187 "Conservarea Habitatelor Subterane ale Liliecilor din SV Carpaţilor". Inst Speol "Emil Racoviţă", p 144

Diaconu G, Dumitraş DG, Marincea Ş (2008) Data on the Limanu cave mineralogy, South Dobrogea. Trav Inst Spéol "Émile Racovitza" 47:89–105

Diedrich CG (2011) An overview of the iconological and ethological studies in the Cave Bear den in Urşilor Cave (Western Carpathians, Romania). Ichnos 18:9–26

Dumitrescu M (1980) Les Nesticus de Roumanie. Trav Inst Spéol "Émile Racovitza" 19:77–101

Dumitrescu M, Orghidan T (1958) Peştera din Valea Fundata (Rîşnov). An Com Geol 31:421–458

Dumitrescu M, Tanasachi J, Orghidan T (1963) Răspîndirea chiropterelor în R.P. Română. Lucr Inst Speol "Emil Racoviţă" 1-2:507–575

Fridvaldszky J (1767) Mineralogia Magni Principatus Transylvaniae. Claudiopoli: Typis Acad Soc Jesu, p 206

Gheorghiu A, Haas N (1954) Date privind omul primitiv de la Baia de Fier. Consideraţii paleoantropologice. Ses Sec Şt Med Acad RPR 41:641–663

Gheorghiu A, Nicolăescu-Plopşor CS, Haas N, Comşa E, Preda C, Bombiţă G, Enea G, Gheorghiu F, Iofcea S, Nicolăescu-Plopşor D, Neagoe A, Silveanu R, Surdu I (1954) Raport preliminar asupra cercetărilor de paleontologie umană de la Baia de Fier (Reg. Craiova) din 1951. Probl Antropol 1:73–86

Ghemiş C (2003) Piese inedite din primul depozit de la Şuncuiuş. Peştera Unguru Mare. Unpublished pieces from the first bronze Age deposit from Şuncuiuş—Unguru Mare cave. In: Cornea L, Ghemiş C, Moisa G (eds) In memoriam Nicolae Chidioşan. Ed Muzeului Ţării Crişurilor, Oradea, pp 101–113

Ghemiş C (2009) Amber discoveries from Transylvania. In: Palavestra A (ed) Amber in archaeology. Proc 5th Int Conf Amber in Archaeology, Belgrad, 5:124–131

Ghergari L, Onac BP (1993) Crisite—a new mineral species found in the Bolhac cave (Padurea Craiului mountains, Romania). Bull Soc Géogr Liege 29:97–104

Giurginca A, Munteanu CM, Vlaicu M, Tăbăcaru IG (2015) Cavernicolous Oniscidea of Romania. Ed Semne, Bucureşti, p 165

Griselini F (1984) Scrisoarea a II a. In: Feneşan C (ed) Încercare de istorie politică şi naturală a Banatului Timişoarei. Ed Facla, Timişoara, pp 237–245

Grosu AI, Negrea A (1962–1963) Contribuţii la cunoaşterea moluştelor din peşterile R.P. Române. Lucr Inst Speol "Emil Racoviţă" 1:503–508

Gruia M, Ilie V (2000–2001) Collembola from the karstic system of Romania. Trav Inst Spéol "Émile Racovitza" 39–40:59–114

Hubbard JD (2017) 3D cave and ice block morpohology from integrated geophysical methods: a case study at Scărişoara Ice Cave, Romania. Unpublished MS Thesis, School of Geosciences, University of South Florida, Tampa, p 97

Iepure S, Moldovan O (2002) Fauna subterană acvatică. In: Racoviţă G, Moldovan O, Onac BP (eds) Monografia carstului din Munţii Pădurea Craiului. Presa Univ Clujeană, Cluj-Napoca, pp 117–126

Ignat D (2012) Contribuţii la cunoaşterea Paleoliticului în nord-vestul României. In: Ştefănescu B, Goman I (eds) Interferenţe intelectuale. Studia in honorem Aurel Chiriac sexagenarii. Ed Muzeului Ţării Crişurilor, Oradea, pp 41–45

Jurcsák T, Poliş R, Ignat D, Şerban M, Popa E (1980–1981) Date privind fauna fosilă a Peşterii Urşilor. Nymphaea 8–9:161–257

Lecca OG (1937) Dicţionar istoric, arheologic şi geografic al României. Ed Universul, Bucureşti, p 630

Meleg IN (2013) Population dynamics of subterranean Crustacea: ecology, genetics and conservation. Ed Presa Univ Clujeană, Cluj-Napoca, p 87

Meleg IN (2015) The underground world-creatures of today. In: Meleg IN (ed) The story of Meziad Cave. CAPDD Bihor, Oradea, pp 28–41

Moldovan O (2002) Coleopterele subterane. In: Racoviţă G, Moldovan O, Onac BP (eds) Monografia carstului din Munţii Pădurea Craiului. Presa Univ Clujeană, Cluj-Napoca, pp 165–171

Moldovan OT, Viehmann I, Onac BP (2007) Peştera Urşilor de la Chişcău. Inst Speol "Emil Racoviţă" Cluj-Napoca, p 83

Nagy ZL, Larti L, Dóczy A, Jére C, Postawa T, Szántó L, Szodoray-Parádi A, Szodoray-Parádi F (2005) Survey of Romania's underground bat habitats. Status and distribution of cave dwelling bats. Report for BP conservation programme, p 44

Negrea A, Negrea Ş (1971) Sur la synusie des grottes du Banat (Roumanie). Trav Inst Spéol "Émile Racovitza" 10:81–122

Nicolăescu-Plopşor CS (1938) Le paleolitique en Roumanie. Dacia 5–6:41–107

Nicolăescu-Plopşor CS (1956) Rezultatele principale ale cercetarilor paleolitice in ultimii patru ani in R.P.R. Stud Cerc Ist Veche 7:7–39
Nicolăescu-Plopşor CS (1957) Şantierul arheologic Nandru. Mater Cerc Arheolog 3:281–290
Nicolăescu-Plopşor CS, Comsa E, Nicolăescu-Plopşor D, Bolomey A (1957) Şantierul arheologic Baia de Fier. Mater Cerc Arheolog 3:13–26
Onac BP, Viehmann I, Lundberg J, Lauritzen S-E, Stringer C, Popiţă V (2005) U–Th ages constraining the Neanderthal footprint at Vârtop Cave, Romania. Quat Sci Rev 24:1151–1157
Păunescu A (2000) Paleoliticul şi Mezoliticul cuprins între Carpaţi şi Dunăre. Ed Agir, Bucureşti, p 492
Perşoiu A, Onac BP, Wynn JG, Blaauw M, Ionita M, Hansson M (2017) Holocene winter climate variability in Central and Eastern Europe. Sci Rep 7:1196. https://doi.org/10.1038/s41598-017-01397-w
Popa CI (2011) Obiecte de metal dn locuirea Coţofeni de la Băniţa-Peştera Bolii (jud. Hunedoara). Crisia 41:33–45
Racovitza G (2000) Ice caves in temperate regions. In: Wilkens H, Culver DC, Humphreys B (eds) Ecosystems of the world, vol 30. Subterranean Ecosystems. Elsevier Sciences, Amsterdam, pp 567–574
Racoviţă G (2010) Révision systématique des leptodirinae souterrains des monts aspuseni. VII. Le sous-genre *Pholeuon* (s. str.) du bassin de Crişul Negru (Monts du Bihor) Trav Inst Spéol "Émile Racovitza" 49:3–27
Robu M (2015) The palaeontology of the cave bear bone assemblage from Urşilor Cave of Chişcău. Osteometry, palaeoichnology, taphonomy, and stable isotopes. Ed Univ Bucureşti, p 248
Robu M (2016) Age re-assessment of the cave bear assemblage from Urşilor Cave, northwestern Romania. Int J Speleol 45:123–133
Roska M (1942) Erdely Regeszeti Repertoriuma, nr. 31. Nagy Jeno es Fia, Kolozsvar, p 237
Rusu T, Racoviţă G, Crăciun V (1981) Peştera Meziad. Ed Sport-Turism, Bucureşti, p 107
Soficaru A, Doboş A, Trinkaus E (2006) Early modern humans from Peştera Muierii, Baia de Fier, Romania. PNAS 103:17196–17201
Soficaru A, Petrea C, Doboş A, Trinkaus E (2007) The human cranium from the Peştera Cioclovina Uscată, Romania. Curr Anthropol 48:611–619
Tabacaru I, Giurginca A (2013) Cavernicolous Oniscidea of Romania. Trav Inst Spéol "Émile Racovitza" 52:3–26
Terzea E (1978) Depot de remplisage et mammiferes quaternaires de "Peştera Urşilor" de Chişcău, departament de Bihor (Note preliminaire). Trav Inst Spéol "Émile Racovitza" 17:139–144
Trinkaus E, Moldovan O, Milota Ş, Bâlgăr A, Sarcină L, Athreya S, Bailey SE, Rodrigo R, Gherase M, Higham T, Ramsey CB, van der Plicht J (2003) An early modern human from the Peştera cu Oase, Romania. PNAS 100:11231–11236
Vălenaş L (1979) Studiu complex al zonei Valea Crăiasa-Valea Vârtoape, cu referire specială la Peştera Urşilor (Munţii Bihor). Nymphaea 7:139–176
Viehmann I (1960) Descrierea unor formaţiuni stalagmitice din Peştera Ialomiţei. Soc Şt Nat Geogr RPR, Com Geol Geogr 61–66
Webb D, Robu M, Moldovan O, Constantin S, Tomuş B, Neag I (2014) Ancient human footprints in Ciur-Izbuc Cave, Romania. Am J Phys Anthropol 155:128–135

Cave Protection in Romania

Oana Teodora Moldovan

Abstract

Presently, Romania experiences a substantial expansion of the so called speleological tourism in protected caves and opening of new caves with and without touristic infrastructure for which impact studies are of questionable quality or absent. These, along with the lack of microclimatic and biological monitoring, are the three main factors that jeopardize the protection of caves and their natural heritage represented by speleothems, sediments, rare minerals, human and animal fossils, endemic living fauna, etc. This chapter is the result of almost 100 years of experience gained in the study of caves by the researchers of the "Emil Racoviță" Institute of Speleology in Cluj and Bucharest and wishes to draw attention on the problems of Romanian caves protection and to point on some future directions to be followed in order to stop the actual destructive tendencies.

Keywords

Caves • Protection • Conservation • Tourism
Impacts • Legislation

Introduction

The cave environment is relatively static, is extremely fragile, and responds promptly to human presence, sometimes with permanent changes. Even if we trust cavers and speleologists have great skills and knowledge to avoid negative impacts, their simple presence means increasing the temperature and CO_2, reducing the relative humidity, producing air currents, introduction of allochthonous material, cave fauna, and sediment disturbances. How fast may the cave environment recover depends largely on multiple factors, such as the duration and size of impact, the microclimate and morphology/size of the passages. Ecological reconstruction (rebuilding the original conditions) is almost impossible in caves, not to mention that we have only vague ideas on the impact intensity and the original conditions. However, protection of caves cannot solely rely on the good will of cavers, but on a legislative framework that must be compiled under the supervision of specialists and, equally important, to be implemented and observed.

Human activities are placed in the top of the menaces on the subterranean domain with all its biotic and abiotic attributes. Anthropogenic degradation can be the result of direct actions, such as mining, quarrying, construction of dams, excessive visitation, excavations and opening of certain passages inside caves, or waste dumping. Other human actions are less direct, such as deforestation, soil pollution, water pumping, and road building. Therefore, caves protected separately, without considering their watershed, are extremely vulnerable to human impacts even if these are not acting right above the cave. Large areas of protection are beneficial not only because they encompass many subterranean habitats but also ensure the quality of the soil components and water that flows underground.

The idea of cave protection in Romania started in 1920 when Emil Racoviță founded world's first Institute of Speleology in Cluj that remains one of the very few worldwide even nowadays. The law of the Institute establishment stated the objectives that encompassed beside research also protection; "*any form of caves exploitation will not be done or continue, from the implementation of this law, without the approval of the Institute of Speleology*" This law is technically still in force, although it is completely ignored by those responsible for the protection and preservation of karst areas in Romania. Currently, although the multi-disciplinary Institute of Speleology has specialists that study caves in all their complexity, there is almost no consultation and no right of control granted to its staff over the cultural and natural use of caves in Romania.

O. T. Moldovan (✉)
"Emil Racoviță" Institute of Speleology, Romanian Academy, Clinicilor 5, 400006 Cluj-Napoca, Romania
e-mail: oanamol35@gmail.com

G. M. L. Ponta and B. P. Onac (eds.), *Cave and Karst Systems of Romania*,
Cave and Karst Systems of the World, https://doi.org/10.1007/978-3-319-90747-5_59

This chapter aims at drawing attention to the problems regarding Romanian caves protection, to emphasize the dangers of the so called-speleological tourism (minimal fitting of caves designed for adventure tourists and team-building activities) in protected caves or of the opening of new show caves in the absence of adequate impact studies and to propose future measures to stop the actual tendency in cave exploitation.

Romanian Legislation on Caves Protection

Before the political changes in Romania (1989), caves were protected at national, county, or local levels but there were no serious threats except for some caves rich in fossil remains (mostly cave bears) or which were heavily decorated with speleothems. The caving community was rather small and cavers were dedicated to protect the caves they discovered, with no interest in using them as a source of income. Accessing the very few touristic caves and especially those wild ones was challenging due to their remote location and difficulties posed by generally poor roads and transportation infrastructure.

In 2000, with the release of the National Landscaping Plan (Law no. 5/2000; Romanian: Legea Nr. 5/2000), a list of 83 caves/karst springs framed as natural heritage sites was published. The list also included other four archeological caves as cultural heritage sites. This was an incomplete list of caves to be protected, and furthermore, the buffer zone above each cave has been set up completely arbitrarily. From the very beginning, it was not clear who drafted those proposals, but most likely the local authorities in each of the 41 counties of the country. A Government Decision (HG 2151/2004; Romanian: Hotărârea nr. 2151/2004) published an additional list with ten caves/systems among other protected natural areas with the statute of scientific reserve, natural monuments, or natural reserve. A classification of some of the caves, grouped in four classes of protection (A through D; see Table 1 and the next section, *Classification of protected caves*) was provided later, through a Ministry Order 604/2005 (OM 604/2005).

Later, the Emergency Ordinance no. 57/2007 (OUG 57/2007; Romanian: Ordonanţa de urgenţă nr. 57/2007) included articles 42–45 that exclusively deal with caves. This was the first attempt to organize the speleological and research activities along with tourism in caves, under the scientific supervision of the Romanian Academy. Since the "Emil Racoviţă" Institute of Speleology is part of this academic organization, it was actively involved in providing decisions concerning all protected caves. Unfortunately, the political lobby of some cavers promoted the publication of the Law no. 49/2011 where incongruences of the OUG 57/2007 became even worse, especially those related with defining the protection classes and the activities in each of these classes. This law was published with changes affecting class A caves, which were not defined as scientific reserves any longer, and class B caves, for which permitted activities and access restrictions were not mentioned. Changes affected not only legal details concerning the protection of caves but also stipulated the founding of the Speleological Heritage Commission (Romanian: Comisia Patrimoniului Speologic—CPS) functioning under the coordination of the Ministry of the Environment. The organization and functioning of the CPS was detailed in the Ministry Order nr. 1044/2012.

The main objectives of the CPS are to: (a) determine the value of speleological assets based on scientific studies; (b) classify caves within protection classes; (c) authorize activities in caves; (d) protect speleological heritage by avoiding any negative impact; and (e) ask competent authorities to ensure the control and monitoring of the speleological heritage. For now, the Commission operates without budget and, except for authorizing activities in class A caves, no other tasks are currently undertaken. One of the most odd and dangerous legislative acts is the Government Decision no. 8/2012 (HG 8/2012) that established the value of protected areas, including caves. Since they were evaluated to extremely low prices (from 500 euros up to a few thousand euros), investors might find no reasons to protect them.

Table 1 Classes of cave protection and their main characteristics

Class of protection	Definition
A	Caves of exceptional value, which by their scientific or unique resources, are representative for the national heritage and international heritage (see also Table 2)
B	Caves of national importance, distinguished by size, scarcity of resources, and touristic potential (see also Table 2)
C	Caves of local importance, protected for their geological, landscaping, hydrological, historical, biodiversity significance, touristic potential or their dimensions
D	Small or medium caves without special value but important for the regional geology, biodiversity, and evolution that must be preserved and protected from pollution or destruction

Classification of Protected Caves

According to the law, Romanian caves are classified by their content values in classes of protection (Table 1; Fig. 1), where class A is the highest and D is the less important. Class A caves should be defined as scientific reserve as in Emergency Ordinance no. 57/2007, but this paragraph was repealed in the Law 49/2011. Caves grouped in class B are defined as natural monuments, previously as natural monuments or natural reserves, whereas all class C caves should be defined as natural reserves but there is no legal paragraph stipulating this.

For large underground networks, a cave can be further divided into sectors of different protection classes and the overall class of protection will be given by the sector with the highest level.

To date, only a limited number of caves are well-documented (Bleahu et al. 1976, Orghidan et al. 1984, and published papers of the "Emil Racoviță" Institute of Speleology) with respect to their scientific and esthetic importance; thus, their classification is justified. However, for other caves, the reasons they were included on this list with their respective class of protection remain obscure.

Most protected caves are concentrated in the northwestern part of Romania (Apuseni Mountains), where there is also the highest concentration of class A caves (Fig. 1; Table 2).

Gaps and Problems

The impact of tourism even in caves with no touristic infrastructure is tremendous and visible in many A caves which are known for a long time and monitored by the researchers of the "Emil Racoviță" Institute of Speleology. The duty for continuous monitoring of caves used for tourism on regular basis, with or without infrastructure, should be regulated and imposed by laws.

There is also a stringent need of re-classifying caves on scientific bases and to re-consider caves that were ignored but are important for the Romanian subterranean heritage.

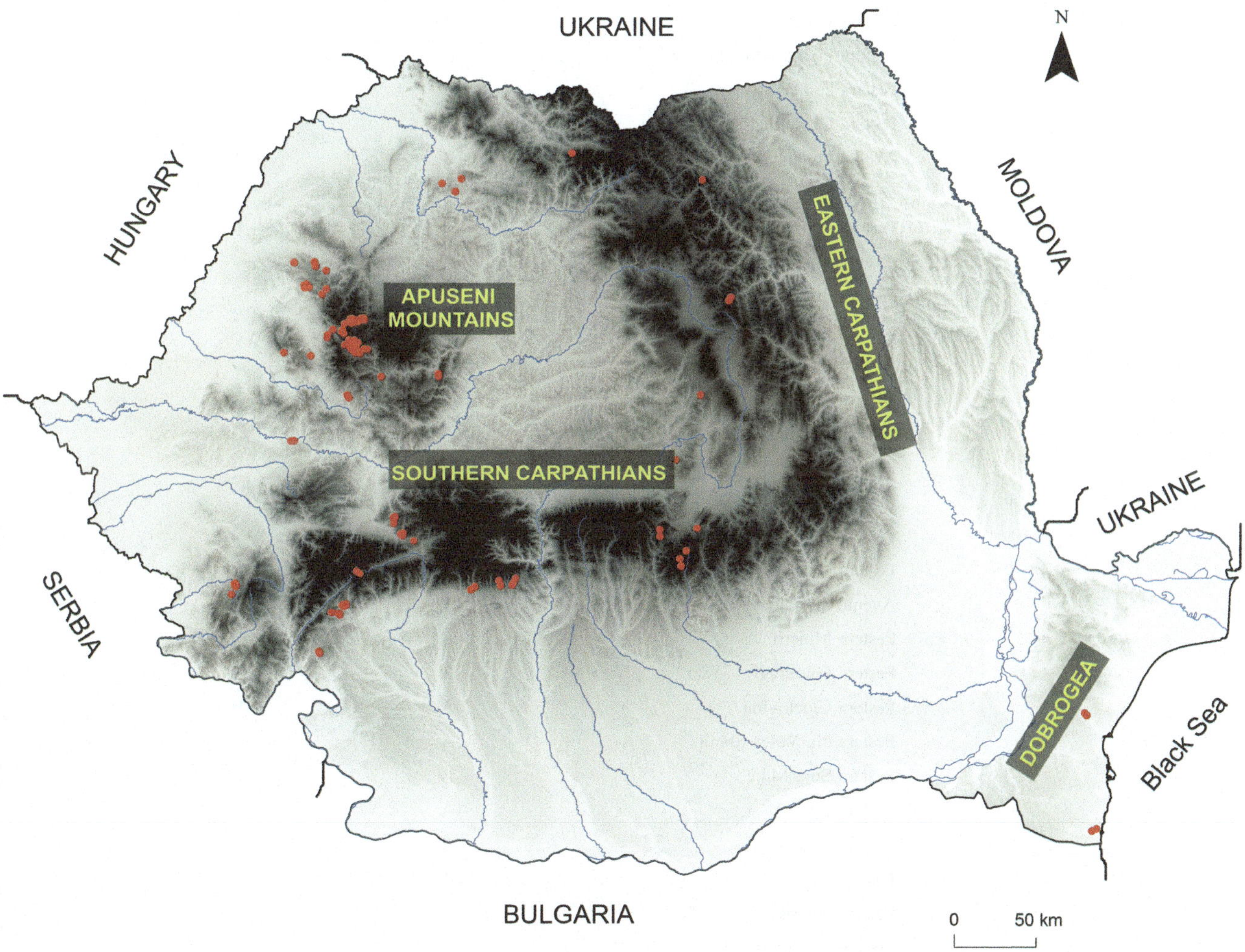

Fig. 1 Physical map of Romania with the distribution of the 132 protected caves (A, B, and C) (red dots) according to the OM (604/2005)

Table 2 Protected class A caves (or the equivalent scientific reserve) and caves with A sectors in Romania (OM 604/2005, HG 2151/2004)

Region	Cave	Class	With sector(s)
Apuseni Mountains	Huda lui Papară	A	B
	Peştera Gheţarul de la Scărişoara	A	B
	Hoanca Apei	A	
	Avenul de la Tău	A	
	Peştera Pojarul Poliţei	A	
	Peştera de sub Zgurăşti	A	B
	Peştera Dârninii	A	
	Peştera din Valea Morii	A	
	Gheţarul Focul Viu	A	
	Peştera Ciurului Ponor	A	
	Peştera Ciurului Izbuc	A	
	Peştera Urşilor-Chişcău	A	B
	Peştera Vântului	A	B
	Peştera lui Micula	A	
	Peştera Meziad	B	A
	Peştera Topliţa	A	
	Peştera Smeilor de la Onceasa	A	
	Peştera Altarului	A	
	Peştera Rece	B	A
	Peştera Poarta Alunului	B	A
	Peştera cu Oase	A	
	Peştera Cerbului	B	A
	Avenul cu Vacă	B	A
	Peştera Vârfuraşul	A	
	Peştera Mare (de pe Valea Firei)	A	B
	Avenul Poieniţa	A	
	Peştera din Piatra Ponorului	A	
	Peştera Cizmei	A	
Eastern Carpathians	Peştera de la Izvorul Tăusoarelor	A	
	Peştera Liliecilor	A	
Southern Carpathians	Peştera Barzoni (Peştera 40)	A	
	Peştera cu Apă din Valea Polevii	A	
	Peştera Răsuflătoarei	A	
	Peştera Comana	A	
	Peştera Cloşani	A	
	Avenul din Cioaca cu Brebenei	A	
	Peştera Muierii	A	B
	Peştera Polovragi	B	A
	Peştera Cioclovina	A	B
	Peştera din Valea Stânii	A	
	Peştera Şura Mare	A	B
	Peştera Epuran	A	
	Peştera Topolniţa	A	B
	Peştera Bulba	A	
	Peştera Izverna	A	B
	Peştera Pagodelor	A	
Dobrogea	Peştera Movile	A	

The classification and re-classification studies should be done by the researchers of the "Emil Racoviță" Institute of Speleology in association with cavers and caving clubs but independent of any organizations holding an interest in tourism or other cave-related business.

CPS was created to serve as a scientific and regulating organism for exploration, protection, and conservation of caves. In supporting the administration of the speleological heritage, it was stipulated to include specialists of the "Emil Racoviță" Institute of Speleology along with specialists (sic!) of the Romanian Federation of Speleology and other NGOs. This opened an unexpected and dangerous window for recreational cavers to control activities in protected caves. The cynical situation is that cavers have also the power to decide on the opportunity of research activities among other duties—a situation that must be corrected.

Owing to legislation gaps, presently, the only cave activities (highly restricted) are linked to research done almost exclusively by the specialists of the "Emil Racoviță" Institute of Speleology and their collaborators. In addition, only a handful of national and foreign specialists, including those conducting bat-monitoring activities, have a limited contribution to scientific research on Romanian caves.

Conclusions

Protection is a complex process that includes multi-disciplinary expertise and data. Conservation and protection of karst sites require proper management principles and technique, and specific measures. The protection area of a cave should extend well beyond the cave entrance through the whole catchment area or the karst massif.

Protection of caves in Romania lacks a simple and accurate legislation similar to that proposed on the foundation of the Institute of Speleology in 1920. Legislation has changed or adapted in agreement with the interests of some groups, interests which are linked to the opening of caves for tourism without control or in the absence of reliable impact studies and both climatic and fauna monitoring. Tourism organized by cavers in caves without touristic infrastructure is done almost exclusively in those classified as A and B, causing sizable damages on long term. Currently, Romania witnesses a generalized lack of interest for protection and conservation in general, and a complete lack of knowledge on the importance of caves for science and how fast they will degrade without urgent and professional decisions.

Since Romanian scientists that work in caves and cavers are both called "speleologists," misunderstandings arise and anybody may claim being a professional speleologist, even if they are recreational cavers, with either caving being a hobby or a source of income through NGO funding for various project and activities. Unfortunately, often such groups are targeting A and B protected caves to accomplish their goals.

However, this situation may change either way on short and medium term. The "Emil Racoviță" Institute of Speleology has conducted several multi-annual projects of show cave monitoring and is currently designing a best-practices manual for show cave operators. On the other hand, the so-called speleo-tourism may turn pristine caves in lucrative businesses with unforeseeable consequences. Speleology has long time been regarded as both a science and/or a non-lucrative hobby. Unfortunately, things look to change these days and it is sad that the lack of coherent registration makes this happened in the country that have given the world its first Institute of Speleology.

Acknowledgements I am grateful to S. Constantin for suggestions regarding the manuscript and to M. Kenesz for drawing the map in Fig. 1. This chapter was completed thanks to CAVEMONITOR Grant 17SEE/2014, funded through the 2009–2014 SEE mechanism.

References

Bleahu M, Decu V, Negrea Ș, Pleșa C, Povară I, Viehmann I (1976) Peșteri din România. Ed Științifică și Enciclopedică, p 415

Hotărârea nr. 2151/2004 privind instituirea regimului de arie naturală protejată pentru noi zone (Decision no. 2151/2004 regarding the creation of the protected area for new areas)

Hotărârea nr. 80 din 8 februarie 2012 privind modificarea anexei nr. 12 la Hotărârea Guvernului nr. 1.705/2006 pentru aprobarea inventarului centralizat al bunurilor din domeniul public al statului (Decision no. 80 of 8 February 2012 on the amendment on Annex no. 12 of the Government Decision no. 1.705/2006 approving the centralized inventory of goods from the public domain)

LEGE Nr. 5 din 6 martie 2000 privind aprobarea Planului de amenajare a teritoriului național - Secțiunea a III-a - zone protejate (LAW Nr. 5 of 6 March 2000 on the approval of the National Landscaping - Section III - Protected Areas

Legea nr. 49/2011 pentru aprobarea Ordonanței de urgență a Guvernului nr. 57/2007 privind regimul ariilor naturale protejate, conservarea habitatelor naturale, a florei și faunei sălbatice (Law no. 49/2011 for approving the Government Emergency Ordinance no. 57/2007 on the regime of protected natural areas, conservation of natural habitats, wild flora and fauna)

Ordinul nr. 604/2005 pentru aprobarea Clasificării peșterilor și a sectoarelor de peșteri - arii naturale protejate (Order no. 604/2005 for the approval of caves and cave sectors classification - protected areas)

Ordinul ministrului mediului și pădurilor nr. 1044/2012 privind aprobarea Regulamentului de organizare și funcționare a Comisiei Patrimoniului Speologic (Order of the Minister of Environment and Forests no. 1044/2012 on approval of the organization and functions of the Speleological Heritage)

Ordonanța de urgență nr. 57/2007 privind regimul ariilor naturale protejate, conservarea habitatelor naturale, a florei și faunei sălbatice (Emergency Ordinance no. 57/2007 on the regime of protected natural areas, conservation of natural habitats, wild flora and fauna)

Orghidan T, Negrea Ș, Racoviță G, Lascu C (1984) Peșteri din România. Ghid turistic. Ed Sport-Turism, p 504

Annex: Cave Survey

Gheorghe M. L. Ponta

Abstract

This document presents a brief description of methods, instruments and grading for a cave survey. It is intended to show to the present and future generation the way the caves were surveyed at the end of the XXth centuries, and what instruments were used the most. More then likely the new cave maps resurveyed with digital instruments will find discrepancies in the length or vertical development of a cave, but we did our best to do a good job with the technology available in that time.

Keywords

Cave survey • BCRA Grading • Map

A cave survey is a detailed map (mapping) of a cave or multiple cavities forming a karst system. Because a cave map is a two-dimensional representation of a three-dimensional object, cave map cannot ideally depict all angles and directions (Dasher 2011). Modern techniques using computer-aided design are increasingly important as they allow a more realistic representation of the three-dimensional cave pattern/network. Survey can be used to compare one cave to another by length, depth, and size. Furthermore, cave maps and longitudinal cross sections are crucial because they may reveal clues on speleogenesis and provide a spatial reference for other areas of scientific study.

The accuracy, or *grade*, of a cave survey is dependent on the methodology of measurement. A common survey classification, which uses a scale of seven grades, was proposed by the British Cave Research Association in the 1960s (Gunn 2003).

G. M. L. Ponta (✉)
Geological Survey of Alabama, 420 Hackberry Lane, Tuscaloosa, AL 35401, USA
e-mail: gponta@yahoo.com

Modified BCRA Gradings for a Cave Line Survey

Grade 1: Sketch of low accuracy where no measurements have been made (not at scale);
Grade 2: Sketch of intermediate in accuracy, some visual estimation of distance and angles (approximate scale);
Grade 3: Basic compass and clinometer. Horizontal and vertical angles measured to ±4.0°, distances measured with an error of ±10 cm for 10 m; station position error less than 25 cm;
Grade 4: Compass and clinometer. Horizontal and vertical angles measured to ±2.0°, distances measured to ±5 cm/10 m; station position error less than 10 cm;
Grade 5: A magnetic survey, clinometer. Horizontal and vertical angles measured to ±1°; distances should be observed and recorded to the nearest centimeter and station positions identified to less than 10 cm;
Grade 6: Mining compass magnetic survey, clinometer. Horizontal and vertical angles measured to ±0.5°; distances should be observed and recorded to the nearest centimeter and station positions identified to less than 1 cm;
Grade 7: A survey that is based primarily on the use of a theodolite or total station instead of a compass (Povară et al. 1990; Dasher 2011).

BCRA Grading for Recording Cave Passage Detail

Class A: All passage details based on memory;
Class B: Passage details estimated and recorded in the cave;
Class C: Measurements of detail made at survey stations only;
Class D: Measurements of detail made at survey stations and wherever else needed to show significant changes in passage dimensions.

G. M. L. Ponta and B. P. Onac (eds.), *Cave and Karst Systems of Romania*,
Cave and Karst Systems of the World, https://doi.org/10.1007/978-3-319-90747-5_60

The majority of the cave maps presented in this book were surveyed *BCRA Grade 4 D*. The most common compass used by cavers was Sport 4 (East Germany version of Silva) or occasionally geologic compass.

The main cave passages of Şura Mare, Ciur Ponor, Cioclovina were resurveyed at BCRA Grade 5D, using a mining compass, and the "Emil Racoviţă" Institute of Speleology mapped Topolniţa, Cloşani, Comarnic, Scărişoara, Meziad, etc., as BCRA Grade 6D, using theodolite. A few cavers with extensive survey skills like L. Kalinsky (Speotimiş) or cavers who were professional surveyors (V. Barbu, R. Perlic, and more recently B. Tomuş) remapped caves (section or complete) with theodolites as well as the distance between cave entrances (where needed) at land surface. They now also connect their underground data points to well-known GPS points outside the cave.

In recent years, the new digital technology is widespread rapidly. The cavers are using electronic distance finders, digital compasses, and the data processed by new and powerful software (Compass, Toporobot, Therion, AutoCad, etc.).

References

Dasher GR (2011) On station—a complete handbook for surveying and mapping caves. Natl Speleol Soc, Huntsville (AL), p 264. ISBN 1-879961-39-5

Gunn J (2003) Encyclopedia of caves and karst science. Routledge. ISBN 978-1579583996

Povară I, Goran C, Gutt FW (1990) Speologie. Ed Sport Turism, Bucureşti, p 237